Essential GENES
by
Benjamin Lewin

Thank you for purchasing Essential GENES. Each new copy of Essential GENES includes a 12-month, pre-paid subscription to an on-line version of the book.

Here is how to activate your subscription:

1. Go to **www.prenhall.com/lewin**

2. Click on **Essential GENES**.

3. Click on **Register**.

4. Enter your pre-assigned access code exactly as it appears below.

   PSLEG-JEHAD-BOTEL-SOARS-RUBBY-CAKES

5. Follow the on-screen instructions.

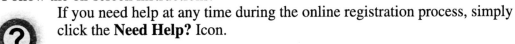 If you need help at any time during the online registration process, simply click the **Need Help?** Icon.

6. Complete the on-line registration form to create your own personal login name and password.

7. After you register, you can access Essential GENES anytime by entering your personal login name and password and then clicking **Login**.

Your access code can only be used once to establish your subscription, which is not transferable. Write down your personal login name and password. You will need them to access the site at each visit.

All new copies of the book include a 12-month subscription to the website. If you purchase a used copy of the textbook your access code many not be valid. You may purchase a subscription to the website by clicking **Buy Now** at **www.prenhall.com/lewin**.

If you are having difficulty logging in to this site, please contact technical support (**online.support@pearsoned.com**) and provide them with a detailed description of your computer system and the technical problem.

If you have questions about any of the content or need help while you are inside the site, please contact **support@ergito.com**.

Essential **GENES**

Essential **GENES**

Benjamin Lewin

Upper Saddle River, NJ 07458

Library of Congress Cataloging-in-Publication Data
Lewin, Benjamin.
    Essential genes / Benjamin Lewin.
        p. cm.
    Updated ed. of: Genes VIII.
    Includes index.
    ISBN 0-13-148988-7
        1. Genetics. 2. Genes. I. Lewin, Benjamin. Genes VIII. II. Title.

QH430.L4 2006
576.5—dc22

                                                    2004060004

Executive Editor: *Gary Carlson*
Editor in Chief: *John Challice*
Project Manager: *Crissy Dudonis*
Production Editor: *Caterina Melara/Preparé, Inc.*
Executive Managing Editor: *Kathleen Schiaparelli*
Assistant Managing Editor: *Beth Sweeten*
Managing Editor, Media: *Nicole M. Jackson*
Development Editor: *Elmarie Hutchinson*
Editor-in-Chief, Development: *Carol Trueheart*
Senior Media Editor: *Patrick Shriner*
Marketing Manager: *Andrew Gilfillan*
Manufacturing Buyer: *Alan Fischer*
Assistant Manufacturing Manager: *Michael Bell*
Director of Creative Services: *Paul Belfanti*

Art Director: *Jonathan Boylan*
Interior and Cover Design: *Kristine Carney*
Managing Editor, Audio and Visual Assets: *Patricia Burns*
AV Production Manager: *Ronda Whitson*
AV Production Editor: *Jessica Einsig*
Art Studio: *Artworks: Ryan Currier, Nathan Storck, Scott Wieber*
Freelance Artist: *Adam Steinberg/Jonathan Parrish*
Director, Image Resource Center: *Melinda Reo*
Manager, Rights and Permissions: *Zina Arabia*
Interior Image Specialist: *Beth Brenzel*
Cover Image Specialist: *Karen Sanatar*
Image Permission Coordinator: *Lashonda Morris*
Editorial Assistant: *Jennifer Hart*
Cover Image: *Argosy Publishing*

© 2006 by Benjamin Lewin
Published by Pearson Education, Inc.
Pearson Prentice Hall
Pearson Education, Inc.
Upper Saddle River, NJ 07458
This is a co-publication of the Publisher and Virtual Text.

Printed in the United States of America
10 9 8 7 6 5 4 3 2 1

ISBN 0-13-148988-7

Pearson Education LTD., *London*
Pearson Education Australia PTY, Limited, *Sydney*
Pearson Education *Singapore*
Pearson Education North Asia Ltd, *Hong Kong*
Pearson Education Canada, Ltd., *Toronto*
Pearson Educación de Mexico, S.A. de C.V.
Pearson Education—Japan, *Tokyo*
Pearson Education Malaysia, Pte. Ltd

Brief Contents

Contents

Part 2 Proteins

Protein Localization Requires Special Signals

10

Part 3 Gene Expression

Transcription

11

The Operon

12

Bacterial Replication Is Connected to the Cell Cycle

17

DNA Replication

18

Homologous and Site-Specific Recombination

19

Repair Systems Handle Damage to DNA

20

Transposons

21

Retroviruses and Retroposons

22

Recombination in the Immune System

23

Catalytic RNA
27

Part 6 The Nucleus

Chromosomes
28

Nucleosomes
29

Chromatin Structure Is a Focus for Regulation

30

Epigenetic Effects Are Inherited

31

Genetic Engineering

32

Glossary

Index

Preface

In the twenty years since the first edition of GENES, molecular biology has become a more mature discipline, and many of the important issues can now be explained in terms of simplifying principles. Understanding the fundamentals of molecular biology is prerequisite for many other fields in the life sciences, including cell biology, immunology, development, etc. The purpose of this edition is to give a mainstream account outlining the concepts that now govern thinking in the field.

This edition follows the general plan of the full edition, although it is less detailed and has some reorganization to enable readers to focus more sharply on individual topics. It takes a little further the move toward regarding genomic sequences as a starting point for analysis; and of course it has been generally updated where appropriate. To provide the background to this approach, a chapter has been added on genetic engineering. It also seems appropriate at this point to devote a separate chapter to epigenetic effects. The most obvious change in content is that this edition focuses more sharply on the molecular biology of the gene, taking its finishing point as being the production of protein.

Readers who wish to follow a particular topic in more detail will find that the Web site, available free with purchase of this book via *www.ergito.com*, has updated versions of both the full and essentials editions of *Genes*, with a concordance to make it easy to move between the corresponding sections in either edition.

The author would like to thank those who reviewed the text as well as those who served as accuracy checkers.

Reviewers:

Steve Ackerman	*University of Massachussetts, Boston*
Revi Allada	*Northwestern University*
Francis Choy	*University of Victoria*
Elliot Goldstein	*University of Arizona*
Robert Heath	*Kent State University*
David Herrin	*University of Texas*
Angel Islas	*Santa Clara University*
Steven Kilpatrick	*University of Pittsburgh, Johnstown*
Loren Knapp	*University of South Carolina*
Jocelyn Krebs	*University of Alaska, Anchorage*
Nandini Krishnamurthy	*University of California, Berkeley*
Thomas Leustek	*Rutgers University*
Reno Parker	*Montana State University, Northern*
Marilee Benore Parsons	*University of Michigan, Dearborn*

Kimmen Sjolander   *University of California,*
                               Berkeley

Ben Stark   *Illinois Institute of Technology*
Charles Toth   *Providence College*
Dennis Welker   *Utah State University*

Proofreaders:

Elliot Goldstein   *University of Arizona*
Jocelyn Krebs   *University of*
                               Alaska, Anchorage

Benjamin Lewin

For the Instructor:

Instructor Resource Center on CD: The instructor's resource center provides a variety of print and media resources to support your preparation of lecture and assessment materials. It includes the following items:

- Images in .jpeg format
- Images preloaded into PowerPoint™
- Lecture outlines
- Test Item File in Word™ format
- Key Concepts
- Key Terms

Website with E-Book (www.prenhall.com/lewin): This powerful Website contains an online version of the text supported by weekly updates to maintain currency on key topics. Links to original source material provide instant access to key research in molecular biology. The unique user interface allows you to view the site in three different formats: highlighting text, images, or a combination of both, to best support your teaching style.

Instructor's Manual: includes lecture outlines, additional resource suggestions, and suggestions on how to use the Companion Website in the classroom.

Test Item File: includes multiple-choice, short-answer, and essay questions that challenge students to apply their knowledge.

Transparency Package: includes a set of 400 four-color transparencies.

For the Student:

Student Handbook: This study tool provides students with the resources to review fundamental concepts from the text through practice questions and exercises. Also included are analytical and critical thinking questions to challenge students to apply their knowledge.

Website with E-Book (www.prenhall.com/lewin): This powerful Website, offered free with the purchase of this textbook, contains an online version of the text supported by weekly updates to maintain currency on key topics. Links to original source material provide instant access to key research in this field. The unique user interface allows you to view the site in three different formats, highlighting text, images or a combination of both, to best support your learning style.

Research Navigator (www.researchnavigator.com): Prentice Hall's Research Navigator™ gives your students access to the most current information available for a wide array of subjects via EBSCO's Center Select™ Academic Journal Database, The New York Times Search by Subject Archive, "Best of the Web" Link Library, and information on the latest news and current events. This valuable tool helps students find the most useful articles and journals, cite sources, and write effective papers for research assigments

ESSENTIAL GENES, is continuously updated on the ebook Web site which is accessible via www.prenhall.com/lewin, with updates posted weekly. One can also access Special Series components, "Great Experiments," "Techniques," and "Structures," as well as a glossary. The ebook on the Web site can be viewed as either sections from the book or as a slide show of figures. Some of the figures are animated, and there are additional references hyper-linked to original sources. You will find an access code for the web site on the card in the front of this book that is good for one year from the date of activation.

DNA Is the Hereditary Material

1.1 Introduction

Key Terms

- The **genome** is the complete set of sequences in the genetic material of an organism. It includes the sequence of each chromosome plus any DNA in organelles.
- **Nucleic acids** are molecules that encode genetic information. They consist of a series of nitrogenous bases connected to ribose molecules that are linked by phosphodiester bonds. DNA is deoxyribonucleic acid, and RNA is ribonucleic acid.
- A **chromosome** is a discrete unit of the genome carrying many genes. Each chromosome consists of a very long molecule of duplex DNA and an approximately equal mass of proteins. It is visible as a morphological entity only during cell division.
- A **gene** (cistron) is the segment of DNA specifying production of a polypeptide chain; it includes regions preceding and following the coding region (leader and trailer) as well as intervening sequences (introns) between individual coding segments (exons).

The hereditary nature of every living organism is defined by its **genome**, which consists of a long sequence of **nucleic acid** that provides the *information* needed to construct the organism. We use the term "information" because the genome does not itself perform any active role in building the organism; rather it is the *sequence* of the individual subunits (bases) of the nucleic acid that determines hereditary features. By a complex series of interactions, this sequence is used to produce all the proteins of the organism in the appropriate time and place.

A genome is divided physically into **chromosomes** and functionally into **genes**. Each chromosome is an independent physical unit that

1

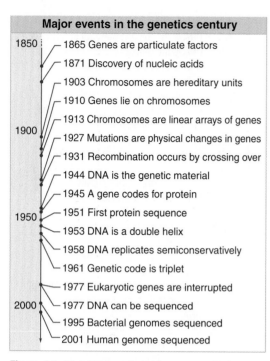

Major events in the genetics century

1850 — 1865 Genes are particulate factors
— 1871 Discovery of nucleic acids
— 1903 Chromosomes are hereditary units
— 1910 Genes lie on chromosomes
— 1913 Chromosomes are linear arrays of genes
1900 — 1927 Mutations are physical changes in genes
— 1931 Recombination occurs by crossing over
— 1944 DNA is the genetic material
— 1945 A gene codes for protein
1950 — 1951 First protein sequence
— 1953 DNA is a double helix
— 1958 DNA replicates semiconservatively
— 1961 Genetic code is triplet
— 1977 Eukaryotic genes are interrupted
2000 — 1977 DNA can be sequenced
— 1995 Bacterial genomes sequenced
— 2001 Human genome sequenced

Figure 1.1 A brief history of genetics.

carries a DNA sequence that contains many genes. The ultimate description of a genome specifies the DNA sequence of each chromosome.

During the last hundred years, we have progressed from Mendel's observation that the gene is a particulate structure, through the discovery that it consists of DNA, to Watson and Crick's model for the double helix, and most recently to the determination of the sequence of the human genome. **Figure 1.1** summarizes the stages in the transition from the historical concept of the gene to the modern definition of the genome.

The first definition of the gene as a functional unit followed from the discovery that individual genes are responsible for the production of specific proteins. The difference in chemical nature between the DNA of the gene and its polypeptide product led to the concept that a gene *codes* for a protein. This in turn led to the discovery of the complex apparatus that allows the DNA sequence of a gene to generate the amino acid sequence of a protein.

Understanding the process by which a gene is expressed allows us to define "gene" more rigorously. **Figure 1.2** shows the basic theme of this book. A gene is a sequence of DNA that produces another nucleic acid, RNA. The DNA has two strands of nucleic acid, and the RNA has only one strand. The sequence of the RNA is determined by the sequence of the DNA (in fact, it is identical to one of the DNA strands). In many—but not all—cases, the RNA is in turn used to direct production of a protein. *So a gene is a sequence of DNA that codes for an RNA; in protein-coding genes, the RNA in turn codes for a protein.*

1.2 DNA Is the Genetic Material of Bacteria

Key Terms

- **Transformation** of bacteria is the acquisition of new genetic material by incorporation of added DNA.

- The **transforming principle** is DNA that is taken up by a bacterium and whose expression then changes the properties of the recipient cell.

Key Concept

- Bacterial transformation provided the first proof that DNA is the genetic material by showing that DNA extracted from one bacterial strain can change the genetic properties of a second bacterial strain.

The idea that genetic material is nucleic acid had its roots in the discovery of **transformation** in 1928. The bacterium *Pneumococcus* kills mice by causing pneumonia. The virulence of the bacterium is determined by its *capsular polysaccharide*. This is a component of the surface that allows the bacterium to escape destruction by the host. Several types (I, II, III) of *Pneumococcus* have different capsular polysaccharides. They have a smooth (S) appearance.

Each of the smooth *Pneumococcal* types can give rise to variants that fail to produce the capsular polysaccharide. These bacteria have a rough (R) surface (consisting of the material that was beneath the capsular polysaccharide). They are *avirulent*. They do not kill the mice, because the absence of the polysaccharide allows the animals to destroy the bacteria.

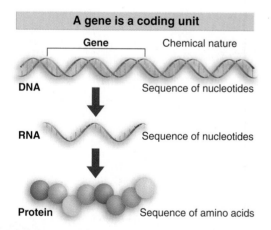

A gene is a coding unit

Gene   Chemical nature

DNA   Sequence of nucleotides

RNA   Sequence of nucleotides

Protein   Sequence of amino acids

Figure 1.2 A gene codes for an RNA, which may code for protein.

When smooth bacteria are killed by heat treatment, they lose their ability to harm mice. But inactive heat-killed S bacteria and the ineffectual variant R bacteria together have a quite different effect from either bacterium by itself. **Figure 1.3** shows that when they are jointly injected, the mouse dies as the result of a *Pneumococcal* infection. Virulent S bacteria can be recovered from the mouse postmortem.

This means that some property of the dead type S bacteria can *transform* the live R bacteria so that they become virulent. This property has in effect changed the genetic information in the bacteria, so that they can heritably produce the capsular polysaccharide that they need to cause a lethal infection. In other words, the component of the dead bacteria responsible for transformation is the genetic material.

The active component was called the **transforming principle**. **Figure 1.4** shows how it was identified. Extracts of the dead S bacteria were mixed with live R bacteria and the mixture was injected into a mouse, which died. Recovery of living S bacteria from the dead mouse and purification of the transforming principle in 1944 showed that it is *deoxyribonucleic acid (DNA)*.

1.3 DNA Is the Genetic Material of Viruses

Key Concept

- Phage infection proved that DNA is the genetic material of viruses. When the DNA and protein components of bacteriophages (bacterial viruses) are labeled with different radioactive isotopes, only the DNA is transmitted to the progeny phages produced within infected bacteria.

After the demonstration that DNA is the genetic material of bacteria, the next step was to demonstrate that DNA provides the genetic material in a quite different system. Phage T2 is a virus that infects the bacterium *E. coli*. When phages are added to bacteria, they adsorb to the outside surface, some material enters the bacterium, and then ~20 minutes later each bacterium bursts open (lyses) to release a large number of progeny phages.

Figure 1.5 illustrates the results of an experiment in 1952 in which bacteria were infected with T2 phages that had been radioactively labeled *either* in their DNA component (with ^{32}P) *or* in their protein component (with ^{35}S). The infected bacteria were agitated in a blender, and two fractions were separated by centrifugation. One contained the empty phage coats that were released from the surface of the bacteria. The other fraction consisted of the infected bacteria themselves.

Most of the ^{32}P label was present in the infected bacteria. The progeny phages produced by the infection contained ~30% of the original ^{32}P label. The progeny received very little—less than 1%—of the protein contained in the original phage population. This experiment therefore showed directly that only the DNA of the parent phages enters the bacteria and then becomes part of the progeny phages, exactly the pattern of inheritance expected of genetic material.

A phage reproduces by commandeering the machinery of an infected host cell to manufacture more copies of itself. The phage possesses genetic material whose behavior is analogous to that of cellular genomes: its traits are faithfully reproduced, and they are subject to the same rules that govern inheritance. The case of T2 reinforces the general conclusion that the genetic material is DNA, whether part of the genome of a cell or virus.

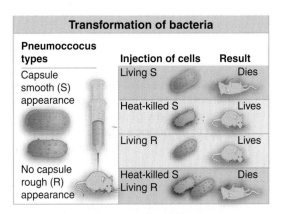

Transformation of bacteria

Pneumoccocus types	Injection of cells	Result
Capsule smooth (S) appearance	Living S	Dies
	Heat-killed S	Lives
No capsule rough (R) appearance	Living R	Lives
	Heat-killed S Living R	Dies

Figure 1.3 Neither heat-killed S-type nor live R-type bacteria can kill mice, but simultaneous injection of both can kill mice just as effectively as the live S-type.

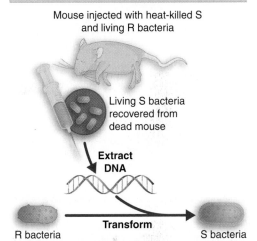

The transforming principle is DNA

Mouse injected with heat-killed S and living R bacteria

Living S bacteria recovered from dead mouse

Extract DNA

R bacteria   Transform   S bacteria

Figure 1.4 The DNA of S-type bacteria can transform R-type bacteria into the same S-type.

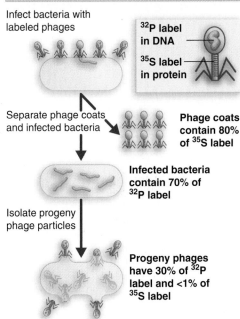

Only DNA is inherited by progeny phages

Infect bacteria with labeled phages

^{32}P label in DNA

^{35}S label in protein

Separate phage coats and infected bacteria

Phage coats contain 80% of ^{35}S label

Infected bacteria contain 70% of ^{32}P label

Isolate progeny phage particles

Progeny phages have 30% of ^{32}P label and <1% of ^{35}S label

Figure 1.5 The genetic material of phage T2 is DNA.

1.4 DNA Is the Genetic Material of Animal Cells

Key Concept

• DNA can be used to introduce new genetic features into animal cells or whole animals.

When DNA is added to populations of single eukaryotic cells growing in culture, the nucleic acid enters the cells, and in some of them results in the production of new protein. **Figure 1.6** depicts one of the standard systems, in which addition of a gene for thymidine kinase to mutant cells that do not have the enzyme results in the production of the corresponding protein.

Although for historical reasons these experiments are described as **transfection** when performed with eukaryotic cells, they are a direct counterpart to bacterial transformation. The DNA that is introduced into the recipient cell becomes part of its genetic material, and is inherited in the same way as any other part. Its expression confers a new trait upon the cells (synthesis of thymidine kinase in the example in Figure 1.6). At first, these experiments were successful only with individual cells adapted to grow in a culture medium. Since then, however, DNA has been introduced into mouse eggs by microinjection; introduced DNA may become a stable part of the genetic material of the mouse (see *32.5 Genes Can Be Injected into Animal Eggs*).

Such experiments show directly not only that DNA is the genetic material in eukaryotes, but also that *it can be transferred between different species and yet remain functional.*

Transfection introduces new DNA into cells

Cells that lack *TK* gene cannot produce thymidine kinase and die in absence of thymidine

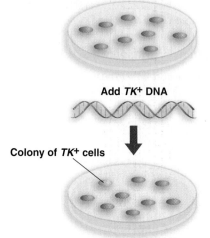

Add *TK*⁺ DNA

Colony of *TK*⁺ cells

Some cells take up *TK* gene; descendants of transfected cell pile up into a colony

Figure 1.6 Eukaryotic cells can acquire a new phenotype as the result of transfection by added DNA.

1.5 Polynucleotide Chains Have Nitrogenous Bases Linked to a Sugar-Phosphate Backbone

Key Concepts

• A nucleoside consists of a purine or pyrimidine nitrogenous base linked to position 1 of a pentose sugar.

• Positions on the ribose ring are described with a prime (') to distinguish them from positions on the base.

• The difference between DNA and RNA is in the group at the 2' position of the sugar. DNA has a deoxyribose sugar (2'—H); RNA has a ribose sugar (2'—OH).

• A nucleotide consists of a nucleoside linked to a phosphate group on either the 5' or 3' position of the (deoxy)ribose.

• Successive (deoxy)ribose residues of a polynucleotide chain are joined by a phosphate group between the 3' position of one sugar and the 5' position of the next sugar.

• One end of the chain (conventionally the left) has a free 5' end and the other end has a free 3' end.

• DNA contains the four bases adenine, guanine, cytosine, and thymine; RNA has uracil instead of thymine.

The basic building block of nucleic acids is the nucleotide. This has three components:

- a nitrogenous base;
- a sugar;
- and a phosphate.

The nitrogenous base is a purine or pyrimidine ring. The base is linked to position 1 on a pentose sugar by a glycosidic bond from N_1 of pyrimidines or N_9 of purines. To avoid ambiguity between the numbering systems of the heterocyclic rings and the sugar, positions on the pentose are given a prime (′).

Nucleic acids are named for the type of sugar; DNA has 2′–deoxyribose, whereas RNA has ribose. The difference is that the sugar in RNA has an OH group at the 2′ position of the pentose ring. The sugar can be linked by its 5′ or 3′ position to a phosphate group.

A nucleic acid consists of a long chain of nucleotides. **Figure 1.7** shows that the backbone of the polynucleotide chain consists of an alternating series of pentose (sugar) and phosphate residues. This is constructed by linking the 5′ position of one pentose ring to the 3′ position of the next pentose ring via a phosphate group. So the sugar-phosphate backbone is said to consist of 5′–3′ phosphodiester linkages. The nitrogenous bases "stick out" from the backbone.

Each nucleic acid contains four types of base. The same two purines, adenine and guanine, are present in both DNA and RNA. The two pyrimidines in DNA are cytosine and thymine; in RNA uracil is found instead of thymine. The only difference between uracil and thymine is that thymine has a methyl group at position C_5 and uracil has none. The bases are usually referred to by their initial letters. DNA contains A, G, C, T, while RNA contains A, G, C, U.

The terminal nucleotide at one end of the chain has a free 5′ group; the terminal nucleotide at the other end has a free 3′ group. It is conventional to write nucleic acid sequences in the 5′ → 3′ direction—that is, from the 5′ terminus at the left to the 3′ terminus at the right.

A polynucleotide has a repeating structure

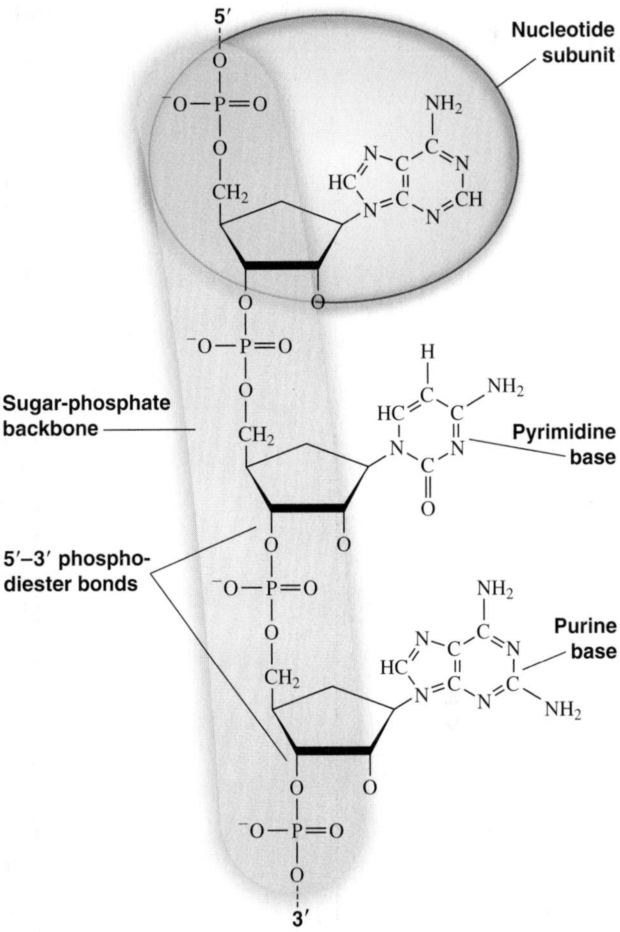

Figure 1.7 A polynucleotide chain consists of a series of 5′–3′ sugar-phosphate links that form a backbone from which the bases protrude.

1.6 DNA Is a Double Helix

Key Terms

- **Deoxyribonucleic acid (DNA)** is a nucleic acid molecule consisting of long chains of polymerized (deoxyribo)nucleotides. In double-stranded DNA the two strands are held together by hydrogen bonds between complementary nucleotide base pairs.
- **Base pairing** describes the specific (complementary) interactions of adenine with thymine or of guanine with cytosine in a DNA double helix (thymine is replaced by uracil in double helical RNA).
- **Antiparallel** strands of the double helix are organized in opposite orientation, so that the 5′ end of one strand is aligned with the 3′ end of the other strand.
- The **minor groove** of DNA is 12 Å across.
- The **major groove** of DNA is 22 Å across.
- A helix is said to be **right-handed** if the turns run clockwise along the helical axis.

- A stretch of **overwound** DNA has more base pairs per turn than the usual average (10 bp = 1 turn). This means that the two strands of DNA are more tightly wound around each other, creating tension.

- A stretch of **underwound** DNA has fewer base pairs per turn than the usual average (10 bp = 1 turn). This means that the two strands of DNA are less tightly wound around each other; ultimately this can lead to strand separation.

Key Concepts

- The B-form of DNA is a double helix consisting of two polynucleotide chains that run antiparallel.
- The nitrogenous bases of each chain are flat purine or pyrimidine rings that face inwards and pair with one another by hydrogen bonding to form A·T or G·C pairs only.
- The diameter of the double helix is 20 Å, and there is a complete turn every 34 Å, with 10 base pairs per turn.
- The double helix forms a major (wide) groove and a minor (narrow) groove.

Three notions converged in the construction of the double helix model for **DNA** by Watson and Crick in 1953:

- X-ray diffraction data showed that DNA has the form of a regular helix, making a complete turn every 34 Å (3.4 nm), with a diameter of ~20 Å (2 nm). Since the distance between adjacent nucleotides is 3.4 Å, there must be 10 nucleotides per turn.

- The density of DNA suggests that the helix must contain two polynucleotide chains. The constant diameter of the helix can be explained if the bases in each chain face inward and are restricted so that a purine is always opposite a pyrimidine, avoiding partnerships of purine-purine (too wide) or pyrimidine-pyrimidine (too narrow).

- Irrespective of the absolute amounts of each base, the proportion of G is always the same as the proportion of C in DNA, and the proportion of A is always the same as that of T.

Watson and Crick proposed that the two polynucleotide chains in the double helix associate by *hydrogen bonding between the nitrogenous bases*. G can hydrogen bond specifically only with C, while A can bond specifically only with T. These reactions are described as **base pairing**, and the paired bases (G with C, or A with T) are said to be *complementary*.

The Watson–Crick model proposed that the two polynucleotide chains run in opposite directions (**antiparallel**), as illustrated in **Figure 1.8**. One strand runs in the $5' \rightarrow 3'$ direction, while its partner runs $3' \rightarrow 5'$.

The sugar-phosphate backbone is on the outside and carries negative charges on the phosphate groups. When DNA is in solution *in vitro*, the charges are neutralized by the binding of metal ions, typically by Na^+. In the cell, positively charged proteins provide some of the neutralizing force. These proteins play an important role in determining the organization of DNA in the cell.

The bases are flat structures, lying in pairs perpendicular to the axis of the helix, inside the backbone. The double helix is often likened to a

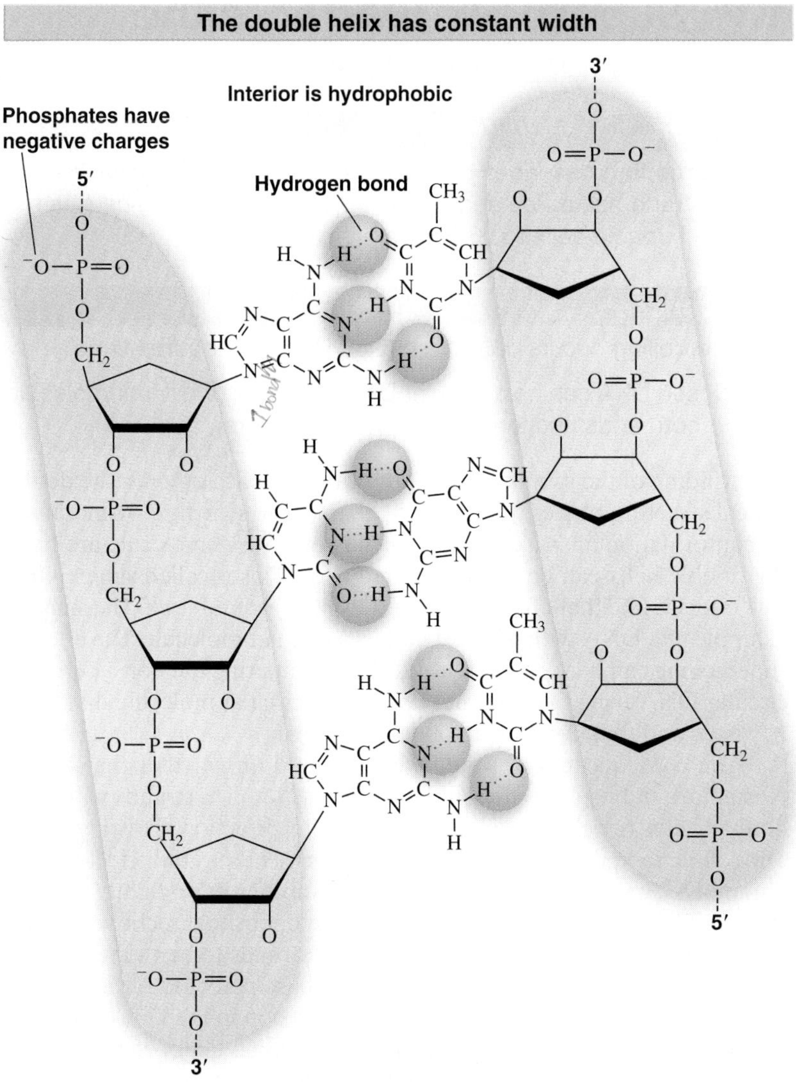

The double helix has constant width

Figure 1.8 The double helix maintains a constant width because purines always face pyrimidines in the complementary A · T and G · C base pairs. The sequence in the figure is T · A, C · G, A · T, G · C.

Flat base pairs connect the DNA strands

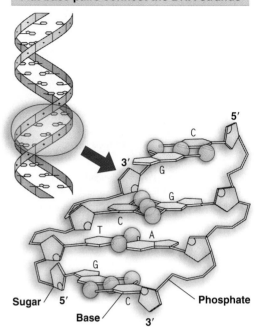

Figure 1.9 Flat base pairs lie perpendicular to the sugar-phosphate backbone.

spiral staircase: the base pairs form the treads, as illustrated schematically in **Figure 1.9**. Proceeding along the helix, bases are stacked above one another like a pile of plates.

Each base pair is rotated ~36° around the axis of the helix relative to the next base pair. So ~10 base pairs make a complete turn of 360°. The twisting of the two strands around one another forms a double helix with **minor groove** (~12 Å across) and a **major groove** (~22 Å across), as can be seen from the scale model of **Figure 1.10**. The double helix is **right-handed**; the turns run clockwise as viewed along the helical axis. These features represent the accepted model for what is known as the *B-form* of DNA.

It is important to realize that the B-form represents an *average*, not a precisely specified *structure*. DNA structure can change locally. If a double helix has more base pairs per turn it is said to be **overwound**; if it has fewer base pairs per turn it is **underwound**. Local winding can be affected by the overall conformation of the DNA double helix in space or by the binding of proteins to specific sites.

The DNA double helix has two grooves

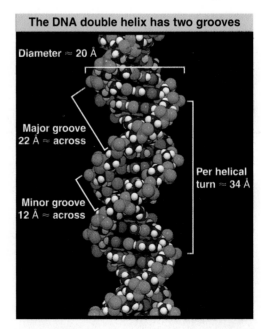

Figure 1.10 The two strands of DNA form a double helix.

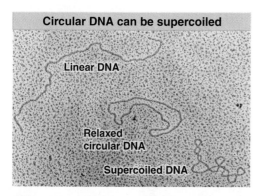

Circular DNA can be supercoiled

Linear DNA

Relaxed circular DNA

Supercoiled DNA

Figure 1.11 Linear DNA is extended, a circular DNA remains extended if it is relaxed (nonsupercoiled), but a supercoiled DNA has a twisted and condensed form.

Strand separation requires changes in topology

Rotation about a free end

Rotation at fixed ends

Strand separation compensated by positive supercoiling

Nicking, rotation, and ligation

Nick

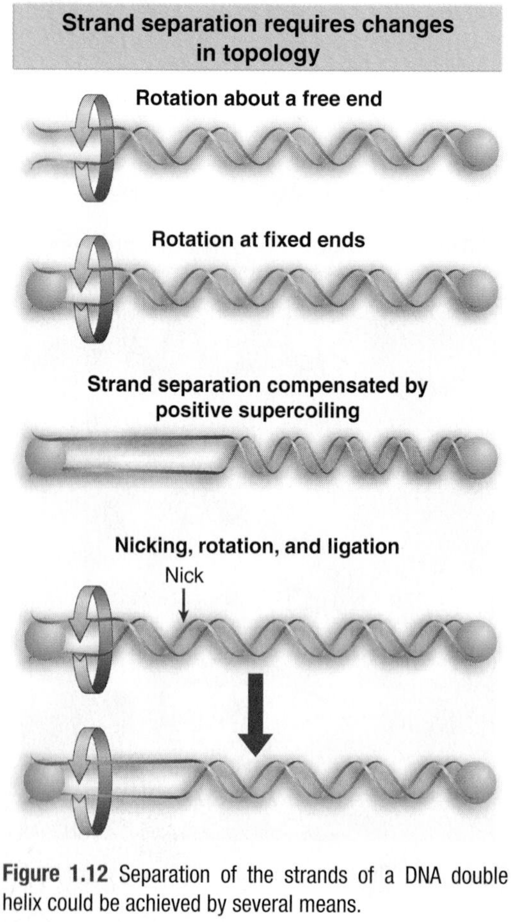

Figure 1.12 Separation of the strands of a DNA double helix could be achieved by several means.

1.7 Supercoiling Affects the Structure of DNA

Key Term

- **Supercoiling** describes the coiling of a double helix in which neither strand has a free end or any breaks; supercoiling causes the closed double helix to cross over its own axis.

Key Concepts

- Supercoiling occurs only in a closed DNA with no free ends.

- A closed DNA can be a circular DNA molecule or a linear molecule with both ends anchored in a protein structure.

The winding of the two strands of DNA around each other in the double helical structure makes it possible to change the structure by modifying its conformation in space. If the two ends of a DNA molecule are fixed, the double helix can be wound around itself. This is called **supercoiling**. The effect is like that of a rubber band twisted around itself. The simplest example of a DNA with no fixed ends is a circular molecule. The effect of supercoiling can be seen in **Figure 1.11** by comparing the nonsupercoiled circular DNA lying flat with the supercoiled circular molecule that forms a twisted and therefore more condensed shape.

The consequences of supercoiling depend on whether the DNA is twisted around itself in the same direction as the two strands within the double helix (clockwise) or in the opposite direction. Twisting in the same direction produces *positive supercoiling*. This has the effect of causing the DNA strands to wind around one another more tightly, so there are more base pairs per turn. Twisting in the opposite direction produces *negative supercoiling*. This causes the DNA strands to be twisted around one another less tightly, so there are fewer base pairs per turn. Negative supercoiling can be thought of as creating tension in the DNA that is relieved by unwinding the double helix. The ultimate effect of negative supercoiling is to generate a region in which the two strands of DNA have separated—that has, in other words, zero base pairs per turn.

Topological manipulation of DNA is part of all of its functions—recombination, replication, and transcription—and of its higher-order structure. DNA synthesis requires the double strands to separate. However, because the strands are intertwined, they must rotate about each other to separate. Some possibilities for the unwinding reaction are illustrated in **Figure 1.12**.

We might envisage the structure of DNA in terms of a free end that would allow the strands to rotate about the axis of the double helix for unwinding. Given the length of the double helix, however, this would involve the separating strands in a considerable amount of flailing about, which seems unlikely in the confines of the cell.

A similar result is achieved by placing an apparatus to control the rotation at the free end. However, the effect must be transmitted over a considerable distance, again involving the rotation of an unreasonable length of material.

Consider the effects of separating the two strands in a molecule whose ends are not free to rotate. When two intertwined strands are pulled apart from one end, the result is to increase their winding about each other farther along the molecule. The problem can be overcome by introducing a transient nick in one strand. An internal free end allows the nicked strand to rotate about the intact strand, after which the nick can be sealed. Each repetition of the nicking and sealing reaction releases one superhelical turn.

Supercoiling can be measured as the density of supercoils per unit length of DNA, which is called σ. We will see that the supercoiling density has important effects *in vivo*, and that specific enzymes are necessary to change it when the structure of DNA requires manipulation.

1.8 The Structure of DNA Allows Replication and Transcription

Key Terms

- A **parental strand** or duplex of DNA is the DNA that is replicated.
- The **template strand** (antisense strand) of DNA is complementary to the sense strand, and is the one that acts as the template for synthesis of mRNA.
- A **daughter strand** or duplex of DNA is the newly synthesized DNA.
- A **DNA polymerase** is an enzyme that synthesizes daughter strands of DNA (under direction from a DNA template). Any particular enzyme may be involved in repair or replication (or both).
- **RNA polymerases** are enzymes that synthesize RNA using a DNA template (formally described as DNA-dependent RNA polymerases).
- **Reverse transcriptase** is an enzyme that uses a template of single-stranded RNA to generate a double-stranded DNA copy.

Key Concepts

- A template strand of a nucleic acid directs synthesis of a complementary product strand by base pairing.
- Nucleic acids are synthesized by adding nucleotides one by one to the 3′—OH end of a polynucleotide chain.
- DNA polymerases use these reactions to synthesize DNA, and RNA polymerases use them to synthesize RNA.

It is crucial that the genetic material is reproduced accurately. Because the two polynucleotide strands are joined only by hydrogen bonds, they are able to separate without requiring breakage of covalent bonds. The specificity of base pairing suggests that each of the separated **parental strands** could act as a **template strand** for the synthesis of a complementary **daughter strand**. **Figure 1.13** shows the principle that a new daughter strand is assembled on each parental strand. The sequence of the daughter strand is dictated by the parental strand; an A in the parental strand causes a T to be placed in the daughter strand, a parental G directs incorporation of a daughter C, and so on.

The top part of the figure shows a parental (unreplicated) duplex that consists of the original two parental strands. The lower part shows the two daughter duplexes that are being produced by complementary base pairing. Each of the daughter duplexes is identical in sequence with the original parent, and contains one parental strand and one newly synthesized strand. *The structure of DNA carries the information needed to perpetuate its sequence.*

The same principle is used in all nucleic acid synthesis: a specific enzyme recognizes the template and undertakes the task of catalyzing the addition of subunits to a polynucleotide chain that is being synthesized. The enzymes are named according to the type of chain that is synthesized. **Figure 1.14** summarizes the types of enzymes and their substrates and products:

- **DNA polymerase** is responsible for the replication of double-stranded DNA. Each of the strands of the parental duplex acts as a

Base pairing accounts for specificity of replication

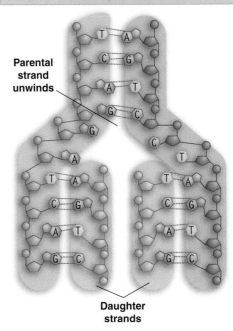

Figure 1.13 Base pairing provides the mechanism for replicating DNA.

Polymerases synthesize nucleic acids

DNA polymerase replicates dsDNA

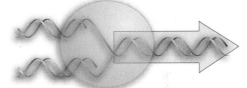

Daughter duplexes are identical to parental duplex

RNA polymerase transcribes dsDNA into ssRNA

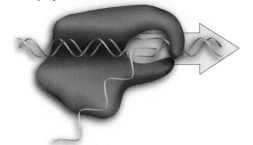

RNA is identical to parental green strand

Reverse transcriptase synthesizes dsDNA on ssRNA

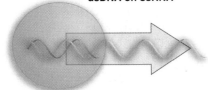

Figure 1.14 All nucleic acids are synthesized by polymerases that use base pairing to ensure that each new chain is complementary to its template chain.

template to synthesize a complementary daughter strand. The result is to produce two duplex DNAs, each of which is identical to the original duplex.

- **RNA polymerase** (or more properly, DNA-dependent RNA polymerase) copies a template strand of DNA into a complementary RNA. The RNA is identical in sequence to the other (nontemplate) strand of DNA. The process is called transcription. DNA structure is disrupted only transiently, and the RNA product is released as a single-stranded molecule.

- **Reverse transcriptase** can copy a template of single-stranded RNA into a complementary DNA strand. The reaction produces a double-stranded nucleic acid with one DNA strand and one RNA strand. A similar reaction is catalyzed by RNA replicases (more properly RNA-dependent RNA polymerases) that are responsible for replicating single-stranded RNA viruses.

The chemical reaction that synthesizes a nucleic acid is the same in all cases. Nucleotides are added one by one to the 3′ end of the chain. **Figure 1.15** shows that an incoming nucleotide has a triphosphate

Figure 1.15 A nucleic acid chain extends from a 5′ end to a 3′ —OH end. A new nucleotide has a 5′ triphosphate. When it is added to the chain, it breaks the bond to its terminal two phosphates, and the remaining phosphate is linked to the oxygen of the 3′ —OH terminus of the chain.

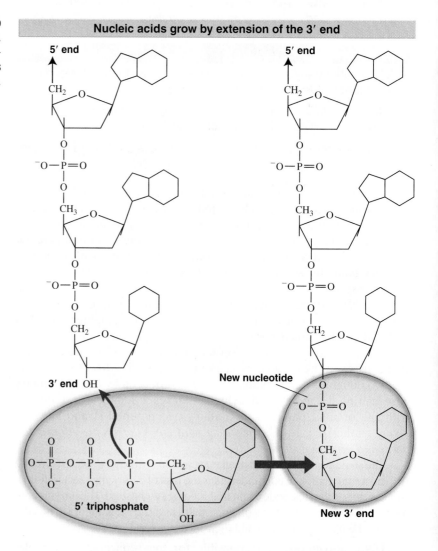

Nucleic acids grow by extension of the 3′ end

group at its 5′ position. It reacts with the 3′ — OH hydroxyl at the end of the polynucleotide chain. The two terminal phosphates are released, and a new bond is formed between the remaining phosphate and the oxygen atom of the 3′ terminal hydroxyl group. The polynucleotide chain is now one subunit longer.

1.9 DNA Replication Is Semiconservative

Key Term

- **Semiconservative replication** is accomplished by separation of the strands of a parental duplex, each then acting as a template for synthesis of a complementary strand.

Key Concepts

- The Meselson-Stahl experiment used density labeling to prove that the single polynucleotide strand is the unit of DNA that is conserved during replication.
- Each strand of a DNA duplex acts as a template to synthesize a daughter strand.
- The sequences of the daughter strands are determined by complementary base pairing with the separated parental strands.

A parental duplex of DNA replicates to form two daughter duplexes, each of which consists of one parental strand and one (newly synthesized) daughter strand. *The units that are conserved from one generation to the next are the two individual strands of the parental duplex.* This behavior is called **semiconservative replication**.

Experimental support for the model of semiconservative replication is illustrated in **Figure 1.16**. If a parental DNA carries a "heavy" density label because the organism has been grown in medium containing a suitable isotope (such as ^{15}N), its strands can be distinguished from those that are synthesized when the organism is transferred to a medium containing normal "light" isotopes.

The parental DNA consists of a duplex of two heavy strands (red). After one generation of growth in light medium, the duplex DNA is "hybrid" in density—it consists of one heavy parental strand (red) and one light daughter strand (blue). After a second generation, the two strands of each hybrid duplex have separated; each gains a light partner, so that now half of the duplex DNA remains hybrid while half is entirely light (both strands are blue).

This experimental confirmation demonstrating that the *individual strands of these duplexes are entirely heavy or entirely light* is the Meselson-Stahl experiment of 1958, which followed the semiconservative replication of DNA through three generations of growth of *E. coli.* When DNA was extracted from bacteria and its density measured by centrifugation, the DNA formed bands corresponding to its density—heavy for parental, hybrid for the first generation, and half hybrid and half light in the second generation.

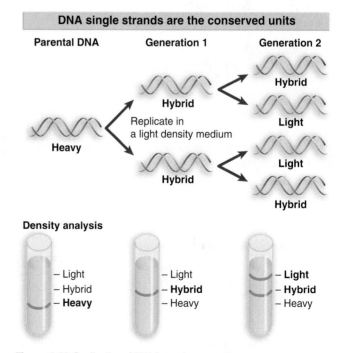

Figure 1.16 Replication of DNA is semiconservative.

1.10 DNA Strands Separate at the Replication Fork

Key Terms

- A **replication fork** (growing point) is the point at which strands of parental duplex DNA separate so that replication can proceed. A complex of proteins including DNA polymerase interact at the fork.
- A **deoxyribonuclease (DNAase)** is an enzyme that specifically attacks bonds in DNA. It may cut only one strand or both.
- **Ribonucleases (RNAase)** are enzymes that cleave RNA. They may be specific for single-stranded or for double-stranded RNA, and may be either endonucleases or exonucleases.
- **Exonucleases** cleave nucleotides one at a time from the end of a polynucleotide chain; they may be specific for either the 5′ or 3′ end of DNA or RNA.
- **Endonucleases** cleave bonds within a nucleic acid chain; they may be specific for RNA or for single-stranded or double-stranded DNA.

Key Concept

- Replication of DNA is accomplished by a complex of enzymes that separate the parental strands at the replication fork and synthesize the daughter strands.

Replication requires the two strands of the parental duplex to separate. This disruption of structure is only transient, however, and is reversed as the daughter duplex forms. Only a small stretch of the duplex DNA is separated into single strands at any moment.

The helical structure of a molecule of DNA engaged in replication is illustrated in **Figure 1.17**. The nonreplicated region consists of the parental duplex, opening into the replicated region where the two daughter duplexes have formed. The double helical structure is disrupted at the junction between the two regions, which is called the **replication fork**. As the DNA replicates, the replication fork moves along the parental DNA, so there is a continuous unwinding of the parental strands and rewinding into daughter duplexes.

Degradation of nucleic acids also requires specific enzymes: **deoxyribonucleases (DNAases)** degrade DNA, and **ribonucleases (RNAases)** degrade RNA. The nucleases fall into the general classes of **exonucleases** and **endonucleases**:

- Endonucleases cut individual bonds *within* RNA or DNA molecules, generating discrete fragments. Some DNAases cleave both strands of a duplex DNA at the target site, while others cleave only one of the two strands. Endonucleases catalyze cutting reactions, as shown in **Figure 1.18**.
- Exonucleases remove residues one at a time from the end of a molecule, generating mononucleotides. They always function on a single nucleic acid strand, and each exonuclease proceeds in a specific direction, that is, starting at either a 5′ or a 3′ end and proceeding toward the other end. They catalyze trimming reactions, as shown in **Figure 1.19**.

A replication fork moves along DNA

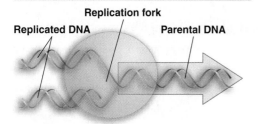

Figure 1.17 The replication fork is the region of DNA in which there is a transition from the unwound parental duplex to the newly replicated daughter duplexes.

Endonucleases attack internal bonds

Broken bond

Figure 1.18 An endonuclease cleaves a bond within a nucleic acid. This example shows an enzyme that attacks one strand of a DNA duplex.

1.11 Genetic Information Can Be Provided by DNA or RNA

Key Term

- The **central dogma** describes the basic nature of genetic information: sequences of nucleic acid can be perpetuated and interconverted by replication, transcription, and reverse transcription, but translation from nucleic acid to protein is unidirectional, because nucleic acid sequences cannot be retrieved from protein sequences.

Key Concepts

- Cellular genes are DNA, but viruses and viroids may have genes of RNA.
- DNA is converted into RNA by transcription, and RNA may be converted into DNA by reverse transcription.
- The translation of RNA into protein is unidirectional.

The **central dogma** defines the paradigm of molecular biology. Genes are perpetuated as sequences of nucleic acid, but function by being expressed in the form of proteins. Replication of DNA is responsible for the inheritance of genetic information. Transcription of DNA into RNA and translation of RNA into protein are responsible for the conversion from one form of genetic information to another.

Figure 1.20 illustrates the roles of replication, transcription, and translation, viewed from the perspective of the central dogma:

- *The perpetuation of nucleic acid may use either DNA or RNA as the genetic material.* Cells use only DNA. Some viruses use RNA, and replication of viral RNA occurs in an infected cell.
- *The expression of cellular genetic information usually is unidirectional.* Transcription of DNA generates RNA molecules that can be used further *only* to generate protein sequences; generally they cannot be retrieved for use as genetic information. Translation of RNA into protein is always irreversible.

Replication, transcription, and translation maintain and express the cellular genetic information of prokaryotes or eukaryotes and the information carried by viruses. The genomes of all living organisms consist of duplex DNA. Viruses have genomes that consist of DNA or RNA, and there are examples of each type that are double-stranded (ds) or single-stranded (ss). The general principle of the nature of the genetic material, then, is that it is always nucleic acid; in fact, it is DNA except in the RNA viruses. Details of the mechanism used to replicate the nucleic acid vary among the viral systems, but the principle of replication via synthesis of complementary strands remains the same, as illustrated in **Figure 1.21**.

Cellular genomes reproduce DNA by the mechanism of semiconservative replication. Double-stranded viral genomes, whether DNA or RNA, also replicate by using the individual strands of the duplex as templates to synthesize partner strands.

Viruses with single-stranded genomes use the single strand as a template to synthesize a complementary strand; and this complementary strand in turn is used to synthesize its complement, which is, of course, identical with the original starting strand. Replication may involve the formation of stable double-stranded intermediates or use double-stranded nucleic acid only as a transient stage.

The usual direction for information transfer is from DNA to RNA. However, this is reversed in the *retroviruses*, whose genomes consist of single-stranded RNA molecules. During the infective cycle, the RNA is converted by the process of *reverse transcription* into a single-stranded DNA, which in turn is converted into a double-stranded DNA. This duplex DNA becomes part of the genome of the cell, and is inherited like any other gene. *So reverse transcription allows a sequence of RNA to be retrieved and used as genetic information.*

The existence of RNA replication and reverse transcription establishes the general principle that *information in the form of either type of nucleic acid sequence can be converted into the other type.* In the usual course of events, however, the cell relies on the processes of DNA replication, transcription, and translation. But on rare occasions (possibly mediated by an RNA virus), information from a cellular RNA is converted into DNA and inserted into the genome. Although reverse transcription plays no role in the regular operations of the cell, it becomes a mechanism of potential importance when we consider the evolution of the genome.

Exonucleases nibble from the ends

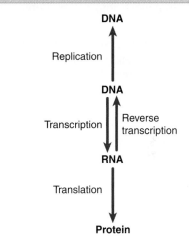

Figure 1.19 An exonuclease removes bases one at a time by cleaving the last bond in a polynucleotide chain.

The central dogma describes information flow

DNA

Replication

DNA

Transcription | Reverse transcription

RNA

Translation

Protein

Figure 1.20 The central dogma states that information in nucleic acid can be perpetuated or transferred, but the transfer of information into protein is irreversible.

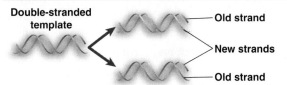

Nucleic acids replicate via complementary strands

Double-stranded template — Old strand / New strands / Old strand

Replication generates two daughter duplexes each containing one parental strand and one newly synthesized strand

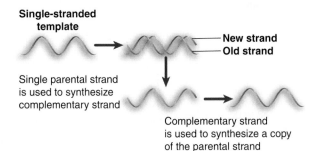

Single-stranded template — New strand / Old strand

Single parental strand is used to synthesize complementary strand

Complementary strand is used to synthesize a copy of the parental strand

Figure 1.21 Double-stranded and single-stranded nucleic acids both replicate by synthesis of complementary strands governed by the rules of base pairing.

1.12 Nucleic Acids Hybridize by Base Pairing

Key Terms

- **Denaturation** of DNA or RNA describes its conversion from the double-stranded to the single-stranded state; usually strand separation is caused by heating.
- **Renaturation** describes the reassociation of denatured complementary single strands of a DNA double helix.
- **Annealing** of DNA describes the renaturation of a duplex structure from single strands that were obtained by denaturing duplex DNA.
- **Hybridization** describes the pairing of complementary RNA and DNA strands to give an RNA-DNA hybrid.

Key Concepts

- Heating causes the two strands of a DNA duplex to separate.
- The T_m is the midpoint of the temperature range for denaturation.
- Complementary single strands can renature when the temperature is reduced.
- Denaturation and renaturation/hybridization can occur with DNA-DNA, DNA-RNA, or RNA-RNA combinations, and can be intermolecular or intramolecular.
- The ability of two single-stranded nucleic acid preparations to hybridize is a measure of their complementarity.

The concept of base pairing is central to all processes involving nucleic acids. Disruption of the base pairs is a crucial aspect of the function of a double-stranded molecule, while the ability to form base pairs is essential for the activity of a single-stranded nucleic acid.

A crucial property of the double helix is the ability to separate the two strands without disrupting covalent bonds. This makes it possible for the strands to separate and reform under physiological conditions at the (very rapid) rates needed to sustain genetic functions. The specificity of reformation is determined by complementary base pairing.

Formation of duplex regions from single-stranded nucleic acids is most important for RNA. **Figure 1.22** shows that base pairing enables complementary single-stranded nucleic acids to form a duplex structure. This can be either intramolecular or intermolecular:

- An intramolecular duplex region can form by base pairing between two complementary sequences at different positions along a single-stranded molecule.
- A single-stranded molecule may base pair with an independent, complementary single-stranded molecule to form an intermolecular duplex.

The lack of covalent links between complementary strands makes it possible to manipulate DNA *in vitro*. The noncovalent forces that stabilize the double helix are disrupted by heating or by exposure to low salt concentration. The two strands of a double helix separate entirely when all the hydrogen bonds between them are broken.

The process of separating the strands of DNA is called **denaturation** or (more colloquially) *melting*. ("Denaturation" is not restricted to describing changes in DNA, but is also used more generally to describe loss of authentic structure in any situation where the natural conformation of a macromolecule has been converted to some other form.)

DNA denatures over a narrow temperature range, bringing about striking changes in many of its physical properties. The midpoint of the temperature range over which the strands of DNA separate is called the *melting temperature* (T_m). It depends on the proportion of $G \cdot C$ base pairs. Because each $G \cdot C$ base pair has three hydrogen bonds, it is more stable than an $A \cdot T$ base pair, which has only two hydrogen bonds. The more $G \cdot C$ base pairs are contained in a DNA, the greater the energy that is needed to separate the two strands. In solution under physiological conditions, a DNA that is 40% $G \cdot C$—a value typical of mammalian genomes—denatures with a T_m of about 87°C. So duplex DNA is stable at the temperature prevailing in the cell.

The denaturation of DNA is reversible under appropriate conditions. The ability of the two separated complementary strands to reform into a

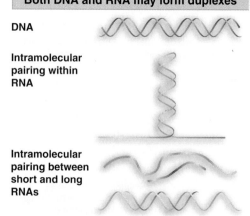

Both DNA and RNA may form duplexes

DNA

Intramolecular pairing within RNA

Intramolecular pairing between short and long RNAs

Figure 1.22 Base pairing occurs in duplex DNA and also in intra- and intermolecular interactions in single-stranded RNA (or DNA).

double helix is called **renaturation**. Renaturation depends on specific base pairing between the complementary strands. **Figure 1.23** shows that the reaction takes place in two stages. First, single strands of DNA in the solution encounter one another by chance; if their sequences are complementary, the two strands base pair to generate a short double-helical region. Then the region of base pairing extends along the molecule by a zipper-like effect to form a lengthy duplex molecule. Renaturation of the double helix restores the original properties that were lost when the DNA was denatured.

Renaturation describes the reaction between two complementary sequences that were separated by denaturation. However, the technique can be extended to allow any two complementary nucleic acid sequences to react with each other to form a duplex structure. This is sometimes called **annealing**, but the reaction is more generally described as **hybridization** whenever nucleic acids of different sources are involved, as in the case when one preparation consists of DNA and the other consists of RNA.

The principle of the hybridization reaction is to expose two single-stranded nucleic acid preparations to each other and then to measure the amount of double-stranded material that forms. **Figure 1.24** illustrates a procedure in which a DNA preparation is denatured and the single strands are adsorbed to a filter. Then the filter is immersed in a second denatured DNA (or RNA) preparation. The filter has been treated so that the second preparation can adsorb to it only if it is able to base pair with the DNA that was originally adsorbed. Usually the second preparation is radioactively labeled, so that the reaction can be measured as the amount of radioactive label retained by the filter.

The extent of hybridization between two single-stranded nucleic acids constitutes a precise measure of their complementarity. Two sequences need not be *perfectly* complementary to hybridize. If they are closely related but not identical, an imperfect duplex is formed in which base pairing is interrupted at positions where the two single strands do not correspond.

1.13 Mutations Change the Sequence of DNA

Key Terms

- A **mutation** is any change in the sequence of genomic DNA.
- **Spontaneous mutations** occur in the absence of any added reagent to increase the mutation rate, as the result of errors in replication (or other events involved in the reproduction of DNA) or by environmental damage.
- The **background level** of mutation describes the rate at which sequence changes accumulate in the genome of an organism. It reflects the balance between the occurrence of spontaneous mutations and their removal by repair systems, and is characteristic for any species.
- **Mutagens** increase the rate of mutation by inducing changes in DNA sequence, directly or indirectly.
- **Induced mutations** result from the action of a mutagen. The mutagen may act directly on the bases in DNA or it may act indirectly to trigger a pathway that leads to a change in DNA sequence.

Key Concepts

- Mutations provide the basis for evolution.
- Mutations may occur spontaneously or may be induced by mutagens.
- The rate of mutation in any organism depends on the balance between the occurrence of mutations and their removal by cellular systems.

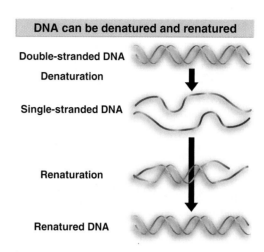

DNA can be denatured and renatured

Double-stranded DNA

Denaturation

Single-stranded DNA

Renaturation

Renatured DNA

Figure 1.23 Denatured single strands of DNA can renature to give the duplex form.

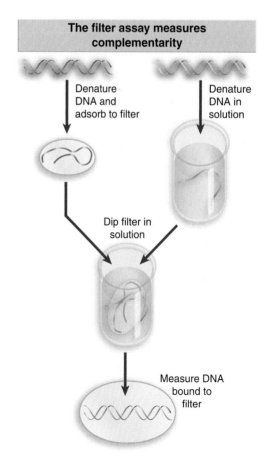

The filter assay measures complementarity

Denature DNA and adsorb to filter

Denature DNA in solution

Dip filter in solution

Measure DNA bound to filter

Figure 1.24 Filter hybridization establishes whether a solution of denatured DNA (or RNA) contains sequences complementary to the strands immobilized on the filter.

Mutation rates increase with target size

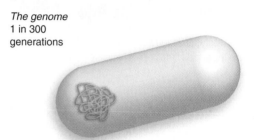

Mutation rate

Any base pair
1 in $10^9 - 10^{10}$

Any gene
1 in $10^5 - 10^6$
generations

...AGCTGTCATGGGTACATTA...
...TCGACAGTACCCATGTAAT...

The genome
1 in 300
generations

Figure 1.25 On the average, a base pair mutates at a rate of $10^{-9} - 10^{-10}$ per generation, a gene of 1000 bp mutates at $\sim 10^{-6}$ per generation, and a bacterial genome mutates at 3×10^{-3} per generation.

The sequence of DNA determines the nature of any organism. Differences between the genetic properties of the individual members of a species are the consequence of differences in their DNA sequences. Any change that is made in the sequence of a genome from one generation to the next is called a **mutation**.

Not all mutations affect the phenotype of an organism, but any mutation with phenotypic consequences means that individuals with the mutation differ in some way from individuals who do not have it. This provides the basis for evolution to select among individuals.

Most mutations that influence the phenotype do so by changing the properties or the amount of some protein. When a change in the sequence of DNA causes an alteration in the sequence of a protein, we may conclude that the DNA codes for that protein. The existence of many mutations in a gene may allow many variant forms of a protein to be compared, and a detailed analysis can be used to identify regions of the protein responsible for individual enzymatic or other functions.

All organisms suffer a certain number of mutations as the result of normal cellular operations or random interactions with the environment. These are called **spontaneous mutations**; the rate at which they occur is characteristic for any particular organism and is sometimes called the **background level**. Mutations are rare events, and those that damage a favorable phenotype are selected against during evolution. It is therefore difficult to obtain large numbers of spontaneous mutants to study from natural populations.

The occurrence of mutations can be increased by treatment with certain compounds. These are called **mutagens**, and the changes they cause are referred to as **induced mutations**. Most mutagens act directly either by modifying a particular base of DNA or by becoming incorporated into the nucleic acid. The effectiveness of a mutagen is judged by how much it increases the rate of mutation above background. By using mutagens, it becomes possible to induce many changes in any gene.

When a mutation occurs in a somatic cell, it can affect only the individual carrying that cell and other cells descended from it. When a mutation occurs in the germline, it may be inherited by the next generation. However, all organisms have cellular systems that counteract mutations by attempting to correct changes in DNA. The rate at which mutations accumulate depends on the balance between the rate at which they occur and the efficiency of the cellular system in removing them.

Spontaneous mutations that inactivate gene function accumulate in bacteriophages and bacteria at a relatively constant rate of $3-4 \times 10^{-3}$ per genome per generation. **Figure 1.25** shows that in bacteria the mutation rate corresponds to $\sim 10^{-6}$ events per gene per generation or to an average rate of change per base pair of $10^{-9} - 10^{-10}$ per generation. The mutation rate of individual base pairs varies very widely, however, over a 10,000 fold range. We have no accurate measurement of the rate of mutation in eukaryotes, although usually it is thought to be somewhat similar to that of bacteria on a per-locus per-generation basis.

1.14 Mutations May Affect Single Base Pairs or Longer Sequences

Key Terms

- A **point mutation** is a change in the sequence of DNA involving a single base pair.
- A **transition** is a mutation in which one pyrimidine is replaced by the other or in which one purine is replaced by the other.

- A **transversion** is a mutation in which a purine is replaced by a pyrimidine or vice versa.
- **Base mispairing** is a coupling between two bases that does not conform to the Watson-Crick rule, e.g., adenine with cytosine, thymine with guanine.
- An **insertion** is the addition of a stretch of base pairs in DNA. Duplications are a special class of insertions.
- A **transposon** (transposable element) is a DNA sequence able to insert itself (or a copy of itself) at a new location in the genome, without having any sequence relationship with the target locus.
- A **deletion** is the removal of a sequence of DNA, the regions on either side being joined together except in the case of a terminal deletion at the end of a chromosome.

Key Concepts

- Point mutations can be caused by the chemical conversion of one base into another or by mistakes that occur during replication.
- Insertions are the most common type of mutation, and result from the movement of transposable elements.

Mutations can take various forms, ranging from insertions or deletions of large amounts of DNA to changes in single base pairs.

Any base pair of DNA can be mutated. A **point mutation** changes only a single base pair, and can be caused by either of two types of event:

- Chemical modification of DNA directly changes one base into a different base.
- A malfunction during the replication of DNA causes the wrong base to be inserted into a polynucleotide chain during DNA synthesis.

Point mutations can be divided into two classes, depending on the type of change by which one base is substituted for another:

- The more common class is the **transition**, comprising the substitution of one pyrimidine by the other, or of one purine by the other. This replaces a G·C pair with an A·T pair or vice versa.
- The less common class is the **transversion**, in which a purine is replaced by a pyrimidine or vice versa, so that an A·T pair becomes a T·A or C·G pair.

The effects of nitrous acid provide a classic example of a transition caused by the chemical conversion of one base into another. **Figure 1.26** shows that nitrous acid performs an oxidative deamination that converts cytosine into uracil. In the replication cycle following the transition, the U pairs with an A, instead of with the G with which the original C would have paired. So the C·G pair is replaced by a T·A pair when the A pairs with the T in the next replication cycle. (Nitrous acid also deaminates adenine, causing the reverse transition from A·T to G·C.)

Transitions are also caused by **base mispairing**, when unusual partners pair in defiance of the usual restriction to Watson-Crick pairs. Base mispairing usually results from the incorporation into DNA of an abnormal base that has ambiguous pairing properties. **Figure 1.27** shows the example of bromouracil (BrdU), an analog of thymine that contains a bromine atom in place of the methyl group of thymine. BrdU is incorporated into DNA in place of thymine. But it has ambiguous pairing properties because the presence of the bromine atom allows a shift in which the base changes structure from a keto (=O) form to an enol (—OH) form. The enol form can base pair with guanine, which leads to substitution of the original A·T pair by a G·C pair.

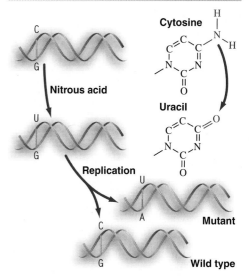

Nitrous acid deaminates cytosine to uracil

Figure 1.26 Mutations can be induced by chemical modification of a base.

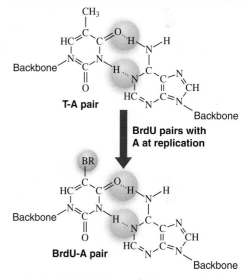

BrdU causes A-T to be replaced by G-C

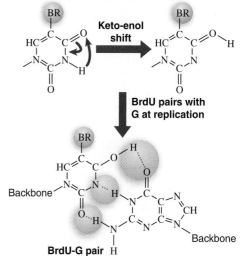

Figure 1.27 Mutations can be induced by the incorporation of base analogs into DNA.

The mispairing can occur either during the original incorporation of the base or in a subsequent replication cycle. The transition is induced with a certain probability in each replication cycle, so the incorporation of BrdU has continuing effects on the sequence of DNA.

Point mutations were thought for a long time to be the principal means of change in individual genes. However, we now know that **insertions** of stretches of additional material are quite frequent. The source of the inserted material lies with **transposons**, sequences of DNA with the ability to move from one site to another (see *Chapter 21 Transposons* and *Chapter 22 Retroviruses and Retroposons*). An insertion usually abolishes the activity of a gene. Where such insertions have occurred, **deletions** of part or all of the inserted material, and sometimes of the adjacent regions, may subsequently occur.

A significant difference between point mutations and insertions/deletions is that the frequency of point mutation can be increased by mutagens, whereas the occurrence of changes caused by transposable elements is not affected. However, insertions and deletions can also occur by other mechanisms—for example, involving mistakes made during replication or recombination—although probably these are less common.

1.15 The Effects of Mutations Can Be Reversed

Key Terms

- **Revertants** are derived by reversion of a mutant cell or organism to the wild-type phenotype.
- **Forward mutations** inactivate a wild-type gene.
- A **back mutation** reverses the effect of a mutation that had inactivated a gene; thus, it restores wild type.
- A **true reversion** is a mutation that restores the original sequence of the DNA.
- **Second-site reversion** is a second mutation that suppresses the effect of a first mutation.
- **Suppression** is a change that eliminates the effect of a mutation without restoring the original sequence of DNA.
- A **suppressor** is a second mutation that compensates for or alters the effects of a primary mutation.

Key Concepts

- Point mutations and insertions can be distinguished from deletions because they can be reverted.
- Suppression, when one mutation counteracts the effects of another, can be intragenic or intergenic.

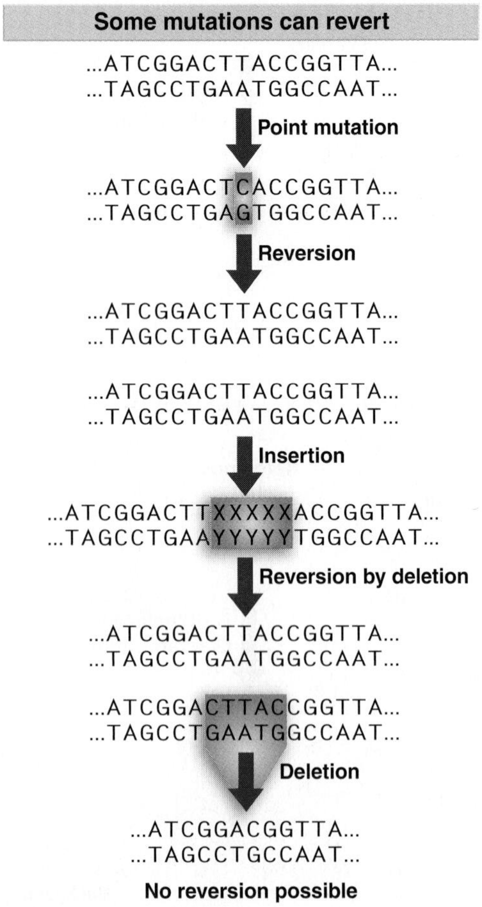

Figure 1.28 shows that from the observation of **revertants**, point mutations and insertions may be distinguished from deletions:

- A point mutation can revert by restoration of the original sequence or by acquisition of a compensatory mutation elsewhere in the gene.
- An insertion of additional material can revert by deletion of the inserted material.
- A deletion of part of a gene cannot revert.

Mutations that inactivate a gene are called **forward mutations**. Their effects are reversed by **back mutations**, which are of two types.

Some mutations can revert

```
...ATCGGACTTACCGGTTA...
...TAGCCTGAATGGCCAAT...
```
↓ **Point mutation**
```
...ATCGGACTCACCGGTTA...
...TAGCCTGAGTGGCCAAT...
```
↓ **Reversion**
```
...ATCGGACTTACCGGTTA...
...TAGCCTGAATGGCCAAT...
```

```
...ATCGGACTTACCGGTTA...
...TAGCCTGAATGGCCAAT...
```
↓ **Insertion**
```
...ATCGGACTTXXXXXACCGGTTA...
...TAGCCTGAAYYYYYTGGCCAAT...
```
↓ **Reversion by deletion**
```
...ATCGGACTTACCGGTTA...
...TAGCCTGAATGGCCAAT...
```

```
...ATCGGACTTACCGGTTA...
...TAGCCTGAATGGCCAAT...
```
↓ **Deletion**
```
...ATCGGACGGTTA...
...TAGCCTGCCAAT...
```
No reversion possible

Figure 1.28 Point mutations and insertions can revert, but deletions cannot revert.

An exact reversal of the original mutation is called **true reversion**. So if an A · T pair has been replaced by a G · C pair, another mutation that restores the A · T pair exactly regenerates the wild-type sequence.

Alternatively, another site elsewhere in the gene may mutate, and the effects of this mutation may compensate for the first mutation. This is called **second-site reversion**. For example, one amino acid change in a protein may abolish gene function, but a second alteration may compensate for the first and restore protein activity.

A forward mutation results from any change that inactivates a gene, whereas a back mutation must restore function to a protein damaged by a particular forward mutation. So the demands for back mutation are much more specific than those for forward mutation. The rate of back mutation is correspondingly lower than that of forward mutation, typically by a factor of ~10.

A mutation in another gene may circumvent the effect of a mutation in the original gene. This effect is called **suppression**. A locus in which a mutation suppresses the effect of a mutation in another locus is called a **suppressor**.

1.16 Mutations Are Concentrated at Hotspots

Key Term

- A **hotspot** is a site in the genome at which the frequency of mutation (or recombination) is very much increased, usually by at least an order of magnitude relative to neighboring sites.

Key Concept

- The frequency of mutation at any particular base pair is determined by statistical fluctuation, except for hotspots.

Most genes within a species show more or less similar rates of mutation. This suggests that the gene can be regarded as a target for mutation, and that damage to any part of it can abolish its function. As a result, susceptibility to mutation is roughly proportional to the size of the gene. But consider the sites of mutation within the sequence of DNA; are all base pairs in a gene equally susceptible or are some more likely to be mutated than others?

What happens when we isolate a large number of independent mutations in the same gene? Many mutants are obtained. Each is the result of an individual mutational event. Then the site of each mutation is determined. Most mutations lie at different sites, but some lie at the same site. Two independently isolated mutations at the same site may constitute exactly the same change in DNA (in which case the same mutational event has happened on more than one occasion), or they may constitute different changes (three different point mutations are possible at each base pair).

The histogram of **Figure 1.29** shows the frequency with which mutations are found at each base pair in the *lacI* gene of *E. coli*. The statistical probability that more than one mutation occurs at a particular site is given by random-hit kinetics (as seen in the Poisson distribution). So some sites will gain one, two, or three mutations, while others will not gain any. But some sites gain far more than the number of mutations expected from a random distribution; they may have 10× or even 100× more mutations than predicted by random hits. These sites are called **hotspots**. Spontaneous mutations may occur at hotspots; and different mutagens may have different hotspots.

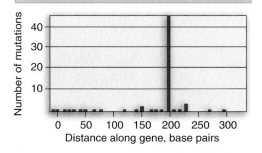

A hotspot has >10x increase in mutations

Number of mutations (y-axis): 10, 20, 30, 40

Distance along gene, base pairs (x-axis): 0, 50, 100, 150, 200, 250, 300

Figure 1.29 Spontaneous mutations occur throughout the *lacI* gene of *E. coli*, but are concentrated at a hotspot.

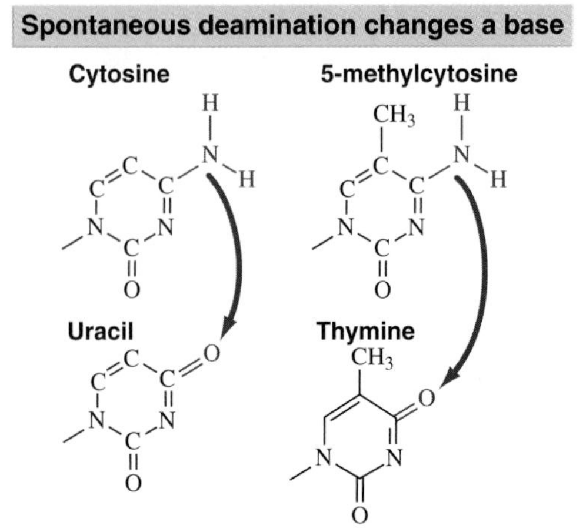

Spontaneous deamination changes a base

Cytosine   5-methylcytosine

Uracil   **Thymine**

Figure 1.30 Deamination of cytosine produces uracil, whereas deamination of 5-methylcytosine produces thymine.

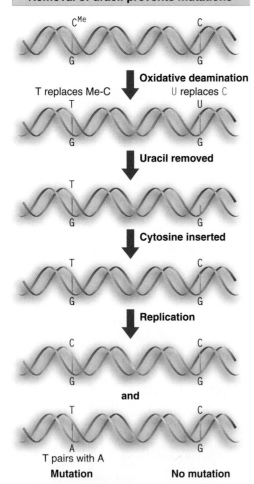

Removal of uracil prevents mutations

C^{Me}   C

G   G

Oxidative deamination

T replaces Me-C   U replaces C

T   U

G   G

Uracil removed

T

G   G

Cytosine inserted

T   C

G   G

Replication

C   C

G   G

and

T   C

A   G

T pairs with A

Mutation   **No mutation**

Figure 1.31 The deamination of 5-methylcytosine produces thymine (causing C · G to T · A transitions), while the deamination of cytosine produces uracil, which usually is removed and then replaced by cytosine.

1.17 Many Hotspots Result from Modified Bases

Key Terms

- **Modified bases** are all those except the usual four from which DNA (T, C, A, G) or RNA (U, C, A, G) are synthesized; they result from postsynthetic changes in the nucleic acid.

- A **mismatch** describes a site in DNA where the pair of bases does not conform to the usual G · C or A · T pairs. It may be caused by incorporation of the wrong base during replication or by mutation of a base.

Key Concept

- A common cause of hotspots is the modified base 5-methylcytosine, which is spontaneously deaminated to thymine.

A major cause of spontaneous mutation results from the presence of an unusual base in the DNA. In addition to the four bases that are inserted into DNA when it is synthesized, **modified bases** are sometimes found. The name reflects their origin; they are produced by chemically modifying one of the four bases already present in DNA. The most common modified base is 5-methylcytosine, generated by a methylase enzyme that adds a methyl group to cytosine residues at specific sites in the DNA.

Sites containing 5-methylcytosine provide hotspots for spontaneous point mutation in *E. coli*. The mutation takes the form of a G · C to A · T transition. The hotspots are not found in strains of *E. coli* that cannot methylate cytosine.

The reason for the existence of the hotspots is that cytosine bases suffer spontaneous deamination at an appreciable frequency. In this reaction, the amino group is replaced by a keto group. Recall that deamination of cytosine generates uracil (see Figure 1.26). **Figure 1.30** compares this reaction with the deamination of 5-methylcytosine, which generates thymine. The effect in DNA is to generate the base pairs G · U and G · T, respectively, in which there is a **mismatch** between the partners.

All organisms have repair systems that correct mismatched base pairs by removing and replacing one of the bases. The operation of these systems determines whether mismatched pairs such as G · U and G · T result in mutations.

Figure 1.31 shows that the consequences of deamination are different for 5-methylcytosine and cytosine. Deaminating the (rare) 5-methylcytosine causes a mutation, whereas deamination of the more common cytosine does not have this effect. This happens because the repair systems are much more effective in recognizing G · U than G · T, and they are efficient at converting a G · U pair to a G · C pair.

1.18 Genomes Vary Greatly in Size

Key Terms

- A **viroid** is a small infectious nucleic acid that does not have a protein coat.

- **Virion** is the physical virus particle (irrespective of its ability to infect cells and reproduce).

- A **subviral pathogen** is an infectious agent that is smaller than a virus, such as a virusoid.

- A **prion** is a proteinaceous infectious agent, which behaves as a heritable trait, although it contains no nucleic acid. Examples are PrP^{sc}, the agent of scrapie in sheep and of bovine spongiform encephalopathy, and Psi, which confers an inherited state in yeast.

Key Concepts

- Cellular genomes consist of double-stranded DNA.
- Some viruses have DNA genomes, but others have RNA genomes.
- Some viral genomes are double-stranded nucleic acid, others are single-stranded.
- Subviral agents consist of infectious RNAs.
- Some very small hereditary agents do not code for protein but consist of RNA or of protein that has hereditary properties.

The same principles are followed to perpetuate genetic information from the massive genomes of plants or amphibians to the tiny genomes of mycoplasma and the yet smaller genetic information of DNA or RNA viruses. **Figure 1.32** summarizes some examples that illustrate the range of genome types and sizes.

Throughout the range of organisms, with genomes varying in total content over a 100,000 fold range, a common principle prevails: *The DNA codes for all the proteins that the cell(s) of the organism must synthesize; and the proteins in turn (directly or indirectly) provide the functions needed for survival.* A similar principle describes the function of the genetic information of viruses, whether DNA or RNA: *The nucleic acid codes for the protein(s) needed to package the genome and also for any functions additional to those provided by the host cell that are needed to reproduce the virus during its infective cycle.*

Viroids are infectious agents that cause diseases in higher plants. They are very small circular molecules of RNA. Unlike viruses, in which the infectious agent consists of a **virion**, a genome encapsulated in a protein coat, *the viroid RNA is itself the infectious agent.* The viroid consists solely of the RNA, which is extensively but imperfectly base paired, forming a characteristic rod like the example shown in **Figure 1.33**. Mutations that interfere with the structure of the rod reduce infectivity.

A viroid RNA consists of a single molecular species that replicates autonomously in infected cells. Its sequence is faithfully perpetuated in its descendants. Viroids fall into several groups. A given viroid is identified with a group by its similarity of sequence with other members of

Genomes vary greatly in size		
Genome	**Gene Number**	**Base Pairs**
Organisms		
Plants	<50,000	<10^{11}
Mammals	30,000	~3 x 10^9
Worms	14,000	~10^8
Flies	12,000	1.6 x 10^8
Fungi	6,000	1.3 x 10^7
Bacteria	2-4,000	<10^7
Mycoplasma	500	<10^6
ds DNA Viruses		
Vaccinia	<300	187,000
Papova (SV$_{40}$)	~6	5,226
Phage T4	~200	165,000
ss DNA Viruses		
Parvovirus	5	5,000
Phage fX174	11	5,387
ds RNA Viruses		
Reovirus	22	23,000
ss RNA Viruses		
Coronavirus	7	20,000
Influenza	12	13,500
TMV	4	6,400
Phage MS2	4	3,569
STNV	1	1,300
Viroids		
PSTV RNA	0	359

Figure 1.32 The amount of nucleic acid in genomes varies over an enormous range.

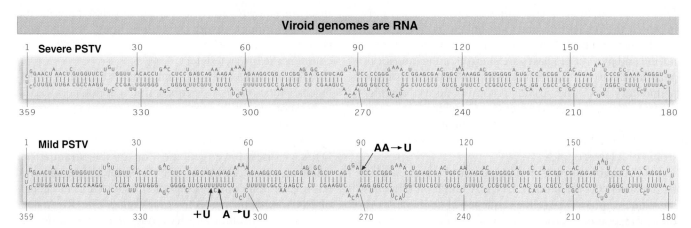

Figure 1.33 PSTV RNA is a circular molecule that forms an extensive double-stranded structure, interrupted by many interior loops. The severe and mild forms differ at three sites.

the group. For example, four viroids related to PSTV (potato spindle tuber viroid) have 70–83% similarity of sequence with it. Different isolates of a particular viroid strain vary from one another, and the differences may have an effect on infected cells. For example, there are *mild* and *severe* strains of PSTV differing by three nucleotide substitutions.

Viroids resemble viruses in having heritable nucleic acid genomes that transmit their genetic information. Yet viroids differ from viruses in both structure and function. They are sometimes called **subviral pathogens**. Viroid RNA does not appear to be translated into protein. So it cannot itself code for the functions needed for its survival. This situation poses two questions. How does viroid RNA replicate? And how does it affect the phenotype of the infected plant cell?

Viroids are presumably pathogenic because they interfere with normal cellular processes. They might do this in a relatively random way, for example, by sequestering an essential enzyme for their own replication or by interfering with the production of necessary cellular RNAs. Alternatively, they might behave as abnormal regulatory molecules, with particular effects upon the expression of individual genes.

An even more unusual agent is *scrapie*, the cause of a degenerative neurological disease of sheep and goats. The disease is related to the human diseases of kuru and Creutzfeldt-Jakob syndrome, which affect brain function.

The infectious agent of scrapie does not contain nucleic acid. This extraordinary agent is called a **prion** (proteinaceous infectious agent). It is a 28 kD hydrophobic glycoprotein, called PrP, which is coded by a cellular gene (conserved among the mammals) that is expressed in normal brain. The protein exists in two forms. The product found in normal brain is called PrP^c. It is entirely degraded by proteases. The protein found in infected brains is called PrP^{sc}. It is extremely resistant to degradation by proteases. PrP^c is converted to PrP^{sc} by a modification or conformational change that confers protease resistance, and which has yet to be fully defined.

As the infectious agent of scrapie, PrP^{sc} must in some way modify the synthesis of its normal cellular counterpart so that it becomes infectious instead of harmless (see *31.8 Prions Cause Diseases in Mammals*). Mice that lack a PrP gene cannot be infected to develop scrapie, which demonstrates that PrP is essential for development of the disease.

1.19 SUMMARY

DNA is a double helix consisting of antiparallel strands in which the nucleotide units are linked by 5′–3′ phosphodiester bonds. The backbone provides the exterior; purine and pyrimidine bases are stacked in the interior in pairs in which A is complementary to T while G is complementary to C. The strands separate and use complementary base pairing to assemble daughter strands in semiconservative replication. Complementary base pairing is also used to transcribe an RNA representing one strand of a DNA duplex.

A mutation consists of a change in the sequence of A · T and G · C base pairs in DNA. A mutation in a coding sequence may change the sequence of amino acids in the corresponding protein. A point mutation changes only a single amino acid in the protein. Point mutations may be reverted by back mutation of the original mutation.

Insertions may revert by loss of the inserted material, but deletions cannot revert. Mutations may also be suppressed indirectly when a mutation in a different gene counters the original defect.

The natural incidence of mutations is increased by mutagens. Mutations may be concentrated at hotspots. A type of hotspot responsible for some point mutations is caused by deamination of the modified base 5-methylcytosine.

Although all genetic information in cells is carried by DNA, viruses have genomes of double-stranded or single-stranded DNA or RNA. Viroids are subviral pathogens that consist solely of small circular molecules of RNA, with no protective packaging. The RNA does not code for protein and its mode of perpetuation and pathogenesis is unknown. Scrapie consists of a proteinaceous infectious agent.

Genes Code for Proteins

2.1 Introduction

Key Terms

- An **allele** is one of several alternative forms of a gene occupying a given locus on a chromosome.
- A **locus** is the position on a chromosome at which the gene for a particular trait resides; a locus may be occupied by any one of the alleles for the gene.
- **Linkage** describes the tendency of genes to be inherited together as a result of their location on the same chromosome; measured by percent recombination between loci.

The gene is the functional unit of heredity. Each gene is a sequence within the genome that functions by giving rise to a discrete product (which may be a protein or an RNA). Genomes for living organisms may contain as few as <500 genes (for a mycoplasma, a type of bacterium) to as many as 30,000 for man.

The basic behavior of the gene was defined by Mendel more than a century ago. Summarized in his two laws, the gene was recognized as a "particulate factor" that passes unchanged from parent to progeny. A gene may exist in alternative forms. These forms are called **alleles**.

In diploid organisms, which have two sets of chromosomes, one copy of each chromosome is inherited from each parent. This is the same behavior that is displayed by genes. One of the two copies of each gene is the paternal allele (inherited from the father); the other is the maternal allele (inherited from the mother). The equivalence led to the discovery that chromosomes in fact carry the genes.

Each chromosome consists of a linear array of genes. Each gene resides at a particular location on the chromosome. The location is more

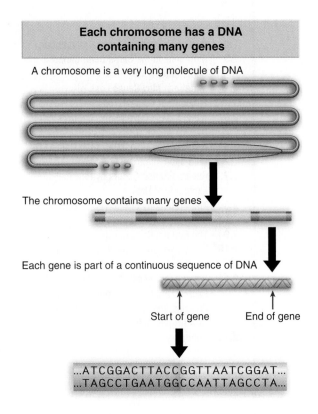

Each chromosome has a DNA containing many genes

A chromosome is a very long molecule of DNA

The chromosome contains many genes

Each gene is part of a continuous sequence of DNA

Start of gene   End of gene

...ATCGGACTTACCGGTTAATCGGAT...
...TAGCCTGAATGGCCAATTAGCCTA...

Figure 2.1 A genome is divided into chromosomes. Each chromosome has a single long molecule of DNA within which are the sequences of individual genes.

formally called a genetic **locus**. The alleles of a gene are the different forms that are found at its locus.

The key to understanding the organization of genes into chromosomes was the discovery of genetic **linkage**—the tendency for genes on the same chromosome to remain together in the progeny instead of assorting independently as predicted by Mendel's laws. Once the unit of recombination (reassortment) was introduced as the measure of linkage, the construction of genetic maps became possible.

The resolution of the recombination map of a higher eukaryote is restricted by the small number of progeny that can be obtained from each mating. Recombination occurs so infrequently between nearby points that it is rarely observed between different mutations in the same gene. As a result, classical linkage maps of eukaryotes can place the genes in order, but cannot determine relationships within a gene. By moving to a microbial system in which a very large number of progeny can be obtained from each genetic cross, researchers could demonstrate that recombination occurs within genes. It follows the same rules that were previously deduced for recombination between genes.

Mutations within a gene can be arranged into a linear order, showing that the gene itself has the same linear construction as the array of genes on a chromosome. So the genetic map is linear within as well as between loci: it consists of an unbroken sequence within which the genes reside. This conclusion leads naturally into the modern view summarized in **Figure 2.1** that the genetic material of a chromosome consists of an uninterrupted length of DNA representing many genes.

2.2 A Gene Codes for a Single Polypeptide

Key Concepts

- The one gene: one enzyme hypothesis summarizes the basis of modern genetics: that a gene is a stretch of DNA coding for a single polypeptide chain.
- Most mutations damage gene function.

The first systematic attempt to associate genes with enzymes showed that each stage in a metabolic pathway is catalyzed by a single enzyme and can be blocked by mutation in a different gene. This led to the *one gene: one enzyme hypothesis*. Each metabolic step is catalyzed by a particular enzyme, whose production is the responsibility of a single gene. A mutation in the gene alters the activity of the protein for which it is responsible.

A modification in the hypothesis is needed to accommodate proteins that consist of more than one subunit. If the subunits are all the same, the protein is a *homomultimer*, represented by a single gene. If the subunits are different, the protein is a *heteromultimer*. Stated as a more general rule applicable to any heteromultimeric protein, the one gene: one enzyme hypothesis becomes more precisely expressed as **one gene: one polypeptide chain**.

Identifying which protein represents a particular gene can be a protracted task. The mutation responsible for creating Mendel's wrinkled-pea mutant was identified only in 1990 as an alteration that inactivates the gene for a starch branching enzyme!

It is important to remember that a gene does not directly generate a protein. As shown previously in Figure 1.2, a gene codes for an RNA, which may in turn code for a protein. Most genes code for proteins, but some genes code for RNAs that do not give rise to proteins. These RNAs may be structural components of the apparatus responsible for synthesizing proteins or may have roles in regulating gene expression. The basic principle is that the gene is a sequence of DNA that specifies the sequence of an independent product. The process of gene expression may terminate in a product that is either RNA or protein.

A mutation is a random event with regard to the structure of the gene, so the greatest probability is that it will damage or even abolish gene function. Most mutations that affect gene function are recessive: *they represent an absence of function, because the mutant gene has been prevented from producing its usual protein.* **Figure 2.2** illustrates the relationship between recessive and wild-type alleles. When a heterozygote has one wild-type allele and one mutant allele, the wild-type allele is able to direct production of the enzyme. The wild-type allele is therefore dominant. (This assumes that an adequate *amount* of protein is made by the single wild-type allele. When this is not true, the smaller amount made by one allele as compared to two alleles results in the intermediate phenotype of a partially dominant allele in a heterozygote.)

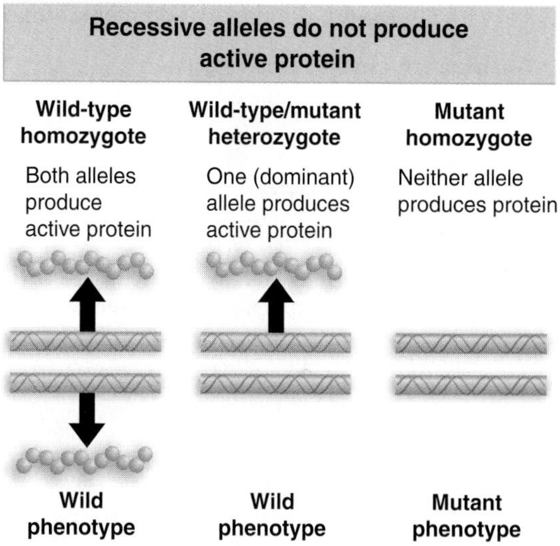

Figure 2.2 Genes code for proteins; dominance is explained by the properties of mutant proteins. A recessive allele does not contribute to the phenotype because it produces no protein (or protein that is nonfunctional).

2.3 Mutations in the Same Gene Cannot Complement

Key Concepts

- A mutation in a gene affects only the protein coded by the mutant copy of the gene and does not affect the protein coded by any other allele.
- Failure of two mutations to complement (produce wild-type phenotype) when they are present in *trans* configuration in a heterozygote means that they are part of the same gene.

How do we determine whether two mutations that cause a similar phenotype lie in the same gene? If they map close together, they may be alleles. However, they could also represent mutations in two *different* genes whose proteins are involved in the same function. The **complementation test** is used to determine whether two mutations lie in the same gene or in different genes. The test consists of making a heterozygote for the two mutations (by mating parents homozygous for each mutation).

If the mutations lie in the same gene, the parental genotypes can be represented as:

$$\frac{m_1}{m_1} \text{ and } \frac{m_2}{m_2}$$

The first parent provides an m_1 mutant allele and the second parent provides an m_2 allele, so that the heterozygote has the constitution:

$$\frac{m_1}{m_2}$$

No wild-type gene is present, so the heterozygote has mutant phenotype.

If the mutations lie in different genes, the parental genotypes can be represented as:

$$\frac{m_1 +}{m_1 +} \text{ and } \frac{m_2+}{m_2+}$$

Key Terms

- A **complementation test** determines whether two mutations are alleles of the same gene. It is accomplished by crossing two different recessive mutations that have the same phenotype and determining whether the wild-type phenotype can be produced. If so, the mutations are said to complement each other and are probably not mutations in the same gene.

- A **complementation group** is a series of mutations unable to complement when tested in pairwise combinations in *trans*; defines a genetic unit (the cistron).

- A **cistron** is the genetic unit defined by the complementation test; it is equivalent to the gene.

- A **gene (cistron)** is the segment of DNA necessary for producing a polypeptide chain; it includes regions preceding and following the coding region (leader and trailer) as well as intervening sequences (introns) between individual coding segments (exons).

Genes are defined by complementation

Two mutations are in the same gene

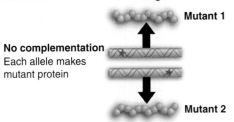

No complementation
Each allele makes
mutant protein

Mutant 1

Mutant 2

Two mutations are in different genes

Mutant 1   Wild type

Complementation
One copy of each
gene is wild type;
gives wild phenotype

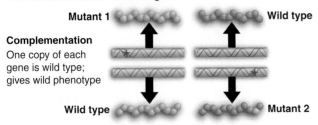

Wild type   Mutant 2

Figure 2.3 The cistron is defined by the results of a complementation test in *trans* configuration. Genes are represented by bars; red stars identify sites of mutation.

Mutations vary from silent to null function

Wild-type gene codes for protein

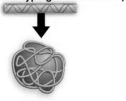

Silent mutation   Point mutation
does not affect protein   may damage function

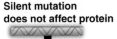

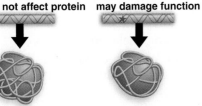

Null mutation   Point mutation
makes no protein   may create new function

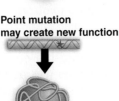

Figure 2.4 Mutations that do not affect protein sequence or function are silent. Mutations that abolish all protein activity are null. Point mutations that cause loss-of-function are recessive; those that cause gain-of-function are dominant.

Each chromosome has a wild-type copy of one gene (represented by the plus sign) and a mutant copy of the other. The heterozygote has the constitution:

$$\frac{m_1\ +}{+\ m_2}$$

in which the two parents between them have provided a wild-type copy of each gene. The heterozygote has wild phenotype; the two genes are said to *complement*.

The complementation test is shown in more detail in **Figure 2.3**. If two mutations lie in the same gene, the *trans* configuration is mutant, because each allele has a (different) mutation. If the two mutations lie in different genes, we see a wild phenotype, because there is one wild-type and one mutant allele of each gene.

Failure to complement means that two mutations are part of the *same* genetic unit. Mutations that do not complement one another are said to be part of the same **complementation group**. Another term that is used to describe the unit defined by the complementation test is the **cistron**. This is the same as the **gene**. Basically these three terms all describe a stretch of DNA that functions as a unit to give rise to an RNA or protein product. The properties of the gene with regards to complementation are explained by the fact that this product is a single molecule that behaves as a functional unit.

2.4 Mutations May Cause Loss-of-Function or Gain-of-Function

Key Terms

- A **null mutation** completely eliminates the function of a gene.
- **Leaky mutations** leave some residual function, for instance, when the mutant protein is partially active (in the case of a missense mutation), or when read-through produces a small amount of wild-type protein (in the case of a nonsense mutation).
- A **loss-of-function mutation** eliminates or reduces the activity of a gene. It is often, but not always, recessive.
- A **gain-of-function mutation** usually refers to a mutation that causes an increase in the normal gene activity. It sometimes represents acquisition of certain abnormal properties. It is often, but not always, dominant.
- **Silent mutations** do not change the sequence of a protein because they produce synonymous codons.
- **Neutral substitutions** in a protein cause changes in amino acids that do not affect activity.

Key Concept

- Testing whether a gene is essential requires a null mutation (one that completely eliminates its function).

The various possible effects of mutation in a gene are summarized in **Figure 2.4**.

When a gene has been identified, insight into its function can be gained by generating a mutant organism that entirely lacks the gene.

A mutation that completely eliminates gene function, usually because the gene has been deleted, is called a **null mutation**. If a gene is essential, a null mutation is lethal.

To determine what effect a gene has upon the phenotype, it is essential to characterize a null mutant. When a mutation fails to affect the phenotype, it is always possible that this is because it is a **leaky mutation**—enough active product is made to fulfill its function, even though the activity is quantitatively reduced or qualitatively different from the wild type. But if a null mutant fails to affect a phenotype, we may safely conclude that the gene function is not necessary.

Mutations that impede gene function (but do not necessarily abolish it entirely) are called **loss-of-function mutations**. A loss-of-function mutation is recessive (as in the example of Figure 2.2). Sometimes a mutation has the opposite effect and causes a protein to acquire a new function; such a change is called a **gain-of-function mutation**. A gain-of-function mutation is usually dominant.

Not all mutations in DNA lead to a detectable change in the phenotype. Mutations without apparent effect are called **silent mutations**. They fall into two types. Some are base changes in DNA that do not cause any change in the amino acid present in the corresponding protein. Others change the amino acid, but the replacement in the protein does not affect its activity; these are called **neutral substitutions**.

2.5 A Locus May Have Many Different Mutant Alleles

Key Concept

- The existence of multiple alleles allows the production of heterozygotes with any pairwise combination of alleles.

Key Term

- A locus is said to have **multiple alleles** when more than two allelic forms have been found. Each allele may cause a different phenotype.

If a recessive mutation is produced by every change in a gene that prevents the production of an active protein, there may be a large number of such mutations in any one gene. Many amino acid replacements may change the structure of the protein sufficiently to impede its function.

Different variants of the same gene are called **multiple alleles**, and their existence makes it possible to create a heterozygote with two different mutant alleles. The relationship between such multiple alleles takes various forms.

In the simplest case, a wild-type gene codes for a protein product that is functional. Mutant allele(s) code for proteins that are nonfunctional.

But there are often cases in which a series of mutant alleles have different phenotypes. For example, wild-type function of the *white* locus of *Drosophila melanogaster* is required for development of the normal red color of the eye. The locus is named for the effect of extreme (null) mutations, which cause homozygous mutant flies to have white eyes. (The gene codes for a protein involved in transporting pigment into the eye.)

To specify wild-type and mutant alleles, wild genotype is indicated by a plus superscript after the name of the locus (w^+ is the wild-type allele for [red] eye color in *D. melanogaster*). Sometimes + is used by itself to specify the wild-type allele, and only the mutant alleles are indicated by the name of the locus.

An entirely defective form of the gene (or absence of phenotype) may be indicated by a minus superscript. To distinguish among a variety of mutant alleles with different effects, other superscripts may be introduced, such as w^i or w^a.

In a heterozygote, the w^+ allele is dominant over any other allele. There are many different mutant alleles, including more than 250 that

Each allele has a different phenotype

Allele	Phenotype of homozygote
w^+	red eye (wild type)
w^{bl}	blood
w^{ch}	cherry
w^{bf}	buff
w^h	honey
w^a	apricot
w^e	eosin
w^l	ivory
w^z	zeste (lemon-yellow)
w^{sp}	mottled, color varies
w^1	white (no color)

Figure 2.5 The w locus has an extensive series of alleles, whose phenotypes extend from wild-type (red) color to complete lack of pigment.

change eye color. **Figure 2.5** shows a (small) sample. Although some alleles make eyes colorless, many alleles produce some color. Each of these mutant alleles must therefore represent a different mutation of the gene, which does not eliminate its function entirely, but leaves a residual activity that produces a characteristic phenotype. These alleles are named for the color of the eye in a homozygote. (Most w alleles affect the quantity of pigment in the eye, and the examples in the figure are arranged in [roughly] declining amount of color, but others, such as w^{sp}, affect the pattern in which pigment is deposited.)

When multiple alleles exist, an animal may be a heterozygote that carries two different mutant alleles. The phenotype of such a heterozygote depends on the residual activity of each allele. The relationship between two mutant alleles is no different from that between wild-type and mutant alleles: one allele may be dominant, there may be partial dominance, or there may be codominance.

2.6 A Locus May Have More than One Wild-Type Allele

Key Term

- **Polymorphism** (or, more fully, "genetic polymorphism") refers to the simultaneous occurrence in a population of genomes of alleles with variations at a given position. The original definition applied to alleles producing different phenotypes. Now it is also used to describe changes in DNA affecting the restriction pattern or even the sequence. For practical purposes, to be considered as being part of a polymorphism, an allele must be at a frequency >1% in the population.

Key Concept

- A locus may have a polymorphic distribution of alleles, with no individual allele that can be considered to be the sole wild type.

There is not necessarily a unique wild-type allele at any particular locus. Control of the human blood group system provides an example. Lack of function is represented by the null alleles, the O alleles. But the functional alleles A and B provide activities that are codominant with one another and dominant over O alleles.

The ABO blood-group precursor molecule is generated in all individuals and consists of a particular carbohydrate group that is added to proteins. **Figure 2.6** shows that the ABO locus codes for a galactosyltransferase enzyme that adds a sugar group to the precursor molecule. The specificity of this enzyme determines the blood group. The A allele produces an enzyme that uses the cofactor UDP-N-acetylgalactose, creating the A antigen. The B allele produces an enzyme that uses the cofactor UDP-galactose, creating the B antigen. The A and B versions of the transferase protein differ in four amino acids that presumably affect its recognition of the type of cofactor. The O allele has a mutation (a small deletion) that eliminates the enzymatic activity needed for production of antigen.

This explains why A and B alleles are dominant in the AO and BO heterozygotes: the corresponding transferase activity creates the A or B antigen. The A and B

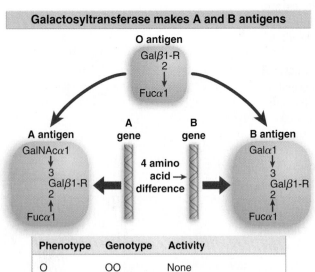

Figure 2.6 The ABO blood group locus codes for a galactosyltransferase whose specificity determines the blood group.

alleles are codominant in *AB* heterozygotes, because both transferase activities are expressed. The *OO* homozygote is a null that has neither activity, and therefore lacks both antigens.

Neither *A* nor *B* can be regarded as uniquely wild type, since they represent alternative activities rather than loss or gain of function. A situation such as this, in which there are multiple functional alleles in a population, is described as a **polymorphism** (see *4.3 Individual Genomes Show Extensive Variation*).

2.7 Recombination Occurs by Physical Exchange of DNA

Key Terms

- **Crossing-over** is a reciprocal exchange of material between chromosomes during prophase I of meiosis and is responsible for genetic recombination.

- A **bivalent** is the structure containing all four chromatids (two representing each homologue) at the start of meiosis.

- **Chromatids** are the copies of a chromosome produced by replication. The name is usually used to describe the copies in the period before they separate at the subsequent cell division.

- A **chiasma** (*pl.* chiasmata) is a site at which two homologous chromosomes appear to have exchanged material during meiosis.

- **Breakage and reunion** describes the mode of genetic recombination, in which two DNA duplex molecules are broken at corresponding points and then rejoined crosswise (involving formation of a length of heteroduplex DNA around the site of joining).

Key Concept

- Recombination is the result of crossing-over at chiasmata between two of the four chromatids.

Genetic recombination generates new combinations of alleles each generation in diploid organisms. The two copies of each chromosome may have different alleles at some loci. By exchanging corresponding parts between the chromosomes, recombinant chromosomes are generated that differ from the parental chromosomes.

Recombination results from a physical exchange of chromosomal material. This is visible as **crossing-over** during meiosis (the specialized division that produces haploid germ cells). Meiosis starts with a cell that has duplicated its chromosomes, so that it has four copies of each chromosome. Early in meiosis, all four copies are closely associated (synapsed) in a structure called a **bivalent**. Each individual chromosomal unit is called a **chromatid** at this stage. Pairs of the chromatids exchange material.

The visible result of a crossing-over event is called a **chiasma**, and is illustrated diagrammatically in **Figure 2.7**. A chiasma represents a site at which two of the chromatids in a bivalent have been broken at corresponding points. The broken ends have been rejoined crosswise, generating new chromatids. Each new chromatid consists of material derived from one chromatid on one side of the junction point, with material from the other chromatid on the opposite side. The two recombinant chromatids have reciprocal structures. The event is described as a **breakage and reunion**. When a crossover occurs, it involves only two of the four associated chromatids. This explains why a single recombination event can produce only 50% recombinants.

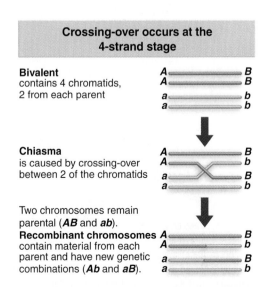

Figure 2.7 Chiasma formation represents the generation of recombinants.

2.8 The Probability of Recombination Depends on Distance Apart

Key Concepts

- Linkage between genes is determined by their distance apart.
- Genes that are close together are tightly linked because recombination between them is rare.
- Genes that are far apart may not be directly linked, because recombination is so frequent as to produce the same result as for genes on different chromosomes.

Over chromosomal distances, recombination events occur more or less at random, with a characteristic frequency. The probability that a crossing-over will occur within any specific region of the chromosome is more or less proportional to the length of the region, up to a saturation point. For example, a large human chromosome usually has three or four crossing-over events in a meiosis, but a small chromosome has only one on average.

Figure 2.8 compares three situations: two genes on different chromosomes, two genes that are far apart on the same chromosome, and two genes that are close together on the same chromosome. Genes on different chromosomes segregate independently according to Mendel's laws, resulting in the production of 50% parental types and 50% recombinant types during meiosis. When genes are sufficiently far apart on the same

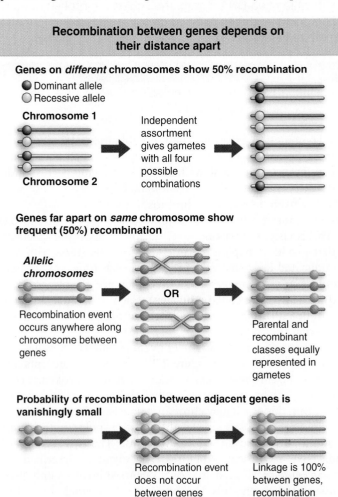

Figure 2.8 Genes on different chromosomes segregate independently so that all possible combinations of alleles are produced in equal proportions. Recombination occurs so frequently between genes that are far apart on the same chromosome that they effectively segregate independently. But recombination is reduced when genes are closer together, and for adjacent genes may hardly ever occur.

Recombination between genes depends on their distance apart

Genes on *different* chromosomes show 50% recombination

● Dominant allele
○ Recessive allele

Chromosome 1

Chromosome 2

Independent assortment gives gametes with all four possible combinations

Genes far apart on *same* chromosome show frequent (50%) recombination

Allelic chromosomes

Recombination event occurs anywhere along chromosome between genes

OR

Parental and recombinant classes equally represented in gametes

Probability of recombination between adjacent genes is vanishingly small

Recombination event does not occur between genes

Linkage is 100% between genes, recombination occurs elsewhere chromosome, the probability of one or more recombination events in the region between them becomes so high that they behave in the same way as those on different chromosomes and show 50% recombination.

But when genes are close together, the probability of a recombination event between them is reduced, and occurs only in some proportion of meioses. For example, if it occurs in one-quarter of the meioses, the overall rate of recombination is 12.5% (because a single recombination event produces 50% recombination, and this occurs in 25% of meioses). When genes are very close together, as shown in the bottom panel of Figure 2.8, recombination between them may never be observed in phenotypes of higher eukaryotes.

This leads us to the view that a chromosome contains an array of many genes. Each gene is an independent unit of expression, and is represented in a protein chain. The properties of a gene can be changed by mutation. The allelic combinations present on a chromosome can be changed by recombination. We can now ask what is the relationship between the sequence of a gene and the sequence of the protein chain it represents.

2.9 The Genetic Code Is Triplet

Key Concepts

- The genetic code is read in triplet nucleotides called codons.
- The triplets are nonoverlapping and are read from a fixed starting point.
- Mutations that insert or delete individual bases shift the triplet sets after the site of mutation.
- Combinations of mutations that together insert or delete three bases (or multiples of three) insert or delete amino acids but do not change the reading of the triplets beyond the last site of mutation.

Key Terms

- The **genetic code** is the correspondence between triplets in DNA (or RNA) and amino acids in protein.
- A **codon** is a triplet of nucleotides that represents an amino acid or a termination signal.
- **Frameshifts** are mutations that arise by deletions or insertions that are not a multiple of three base pairs and that change the frame in which triplets are translated into protein.
- **Acridines** are mutagens that act on DNA to cause the insertion or deletion of a single base pair. They were useful in defining the triplet nature of the genetic code.
- A **suppressor** is a second mutation that compensates for or alters the effects of a primary mutation.

Each gene represents a particular protein chain, and each protein consists of a particular series of amino acids. The discovery that a gene consists of DNA faces us with the issue of how a sequence of nucleotides in DNA represents a sequence of amino acids in protein. The relationship between a sequence of DNA and the sequence of the corresponding protein is called the **genetic code**.

The structure and enzymatic activity of a protein follow from its primary sequence of amino acids. By determining the sequence of amino acids in a protein, the gene is able to carry all the information needed to specify an active polypeptide chain. In this way, a single type of structure—the gene—is able to direct construction of innumerable polypeptide forms.

In any given region of DNA, only one of the two strands codes for protein, so we write the genetic code as a sequence of bases (rather than base pairs). The genetic code is read in groups of three nucleotides, each group representing one amino acid. Each trinucleotide sequence is called a **codon**. A gene includes a series of codons that is read sequentially from a starting point at one end to a termination point at the other end. Written in the conventional $5' \rightarrow 3'$ direction, the nucleotide sequence of the DNA strand that codes for protein corresponds to the amino acid sequence of the protein written in the direction from N-terminus to C-terminus.

The genetic code is read in *nonoverlapping triplets from a fixed starting point*:

- Nonoverlapping implies that each codon consists of three nucleotides and that successive codons are represented by successive trinucleotides.
- The use of a *fixed starting point* means that assembly of a protein must start at one end and work to the other, so that different parts of the coding sequence cannot be read independently.

The nature of the code predicts that two types of mutations will have different effects. If a particular sequence is read sequentially, such as:

UUU AAA GGG CCC (codons)

aa1   aa2   aa3   aa4 (amino acids)

then a point mutation will affect only one amino acid. For example, the substitution of an A by some other base (X) causes aa2 to be replaced by aa5:

UUU AAX GGG CCC

aa1   aa5   aa3   aa4

because only the second codon has been changed.

But a mutation that inserts or deletes a single base will change the triplet sets for the entire subsequent sequence. A mutation of this sort is called a **frameshift**. An insertion might take the form:

UUU AAX AGG GCC C

aa1   aa5   aa6   aa7

Because the new sequence of triplets following the insertion is completely different from the old one, the entire amino acid sequence of the protein is altered beyond the site of mutation. So the function of the protein is likely to be lost completely.

Frameshifts are induced by the **acridines**, compounds that bind to DNA and distort the structure of the double helix, causing additional bases to be incorporated or omitted during replication. Each mutagenic event induced by an acridine adds or removes a single base pair.

If an acridine mutant is produced by, say, addition of a nucleotide, it should revert to wild type by deletion of the nucleotide. But reversion can also be caused by deletion of a different base, at a site close to the first. Combinations of such mutations provided revealing evidence about the nature of the genetic code.

Figure 2.9 illustrates the properties of frameshifts. An insertion or a deletion changes the entire protein sequence following the site of mutation. But the combination of an insertion and a deletion causes the code to be read incorrectly only between the two sites of mutation; correct reading resumes after the second site.

Genetic analysis of acridine mutations in the *rII* region of the phage T6 in 1961 showed that all the mutations could be classified into one of two sets, described as (+) and (−). Either type of mutation by itself causes a frameshift, the (+) type by virtue of a base addition, the (−) type by virtue of a base deletion. Double mutant combinations of the types (++) and (−−) are mutant. But combinations of the types (+−) or (−+) suppress one another; in such cases one mutation is said to be a **suppressor** of the other.

These results show that the genetic code must be read as a sequence that is fixed by the starting point, so an addition and a deletion compensate for each other, whereas double additions and double deletions are mutant. But this does not reveal how many nucleotides make up each codon.

When triple mutants are constructed, only (+++) and (−−−) combinations show the wild phenotype, while other combinations remain mutant. If we take three additions or three deletions to correspond respectively to the addition or omission overall of a single amino acid, this implies that the code is read in triplets. An incorrect amino acid sequence is found between the two outside sites of mutation, and the sequence on either side remains wild type, as indicated in Figure 2.9.

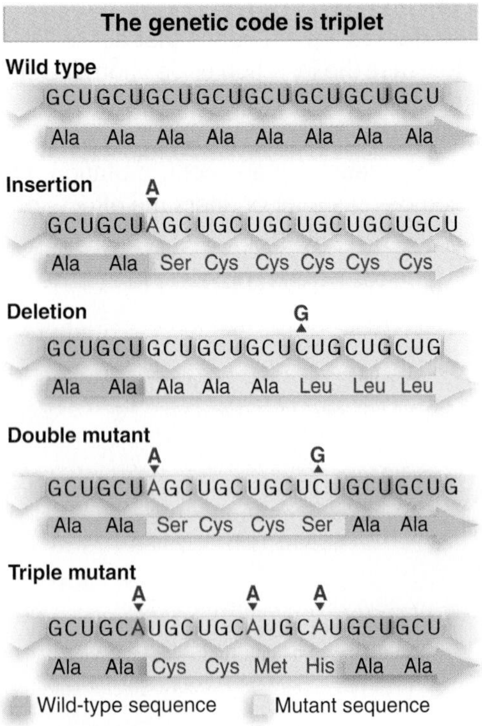

The genetic code is triplet

Wild type

GCUGCUGCUGCUGCUGCUGCUGCU

Ala Ala Ala Ala Ala Ala Ala Ala

Insertion   A

GCUGCUAGCUGCUGCUGCUGCUGCU

Ala Ala Ser Cys Cys Cys Cys Cys

Deletion   G

GCUGCUGCUGCUGCUCUGCUGCUG

Ala Ala Ala Ala Ala Leu Leu Leu

Double mutant   A   G

GCUGCUAGCUGCUGCUCUGCUGCUG

Ala Ala Ser Cys Cys Ser Ala Ala

Triple mutant   A   A   A

GCUGCAUGCUGCAUGCAUGCUGCU

Ala Ala Cys Cys Met His Ala Ala

Wild-type sequence   Mutant sequence

Figure 2.9 Frameshift mutations show that the genetic code is read in triplets from a fixed starting point.

2.10 Every Sequence Has Three Possible Reading Frames

Key Terms

- A **reading frame** is one of the three possible ways of reading a nucleotide sequence. Each reading frame divides the sequence into a series of successive triplets. There are three possible reading frames in any sequence, depending on the starting point. If the first frame starts at position 1, the second frame starts at position 2, and the third frame starts at position 3.
- An **open reading frame (ORF)** is a sequence of DNA consisting of triplets that can be translated into amino acids; an ORF starts with an initiation codon and ends with a termination codon.
- The **initiation codon** is a special codon (usually AUG) used to start synthesis of a protein.
- A **termination codon (stop codon)** is one of three triplets (UAG, UAA, UGA) that causes protein synthesis to terminate. They are also known historically as *nonsense codons*. The UAA codon is called ochre, and the UAG codon is called amber, after the names of the nonsense mutations by which they were originally identified.
- A **blocked reading frame** cannot be translated into protein because of the presence of termination codons.

Key Concept

- Usually only one reading frame is translated and the other two are blocked by frequent termination signals.

If the genetic code is read in nonoverlapping triplets, there are three possible ways of translating any nucleotide sequence into protein, depending on the starting point. These are called **reading frames**. For the sequence

ACGACGACGACGACGACG

the three possible reading frames are

ACG ACG ACG ACG ACG ACG ACG
CGA CGA CGA CGA CGA CGA CGA
GAC GAC GAC GAC GAC GAC GAC

A reading frame that consists exclusively of triplets representing amino acids is called an **open reading frame**, or **ORF**. A sequence that is translated into protein has a reading frame that starts with a special **initiation codon (AUG)** and that extends through a series of triplets representing amino acids until it ends at one of three types of **termination codon** (UAA, UGA, or UGA) (see *Chapter 7 Messenger RNA*).

A reading frame that cannot be read into protein because it contains termination codons is called a **blocked reading frame**. If a sequence is blocked in all three reading frames, it cannot have the function of coding for protein.

When the sequence of a DNA region of unknown function is obtained, each possible reading frame is analyzed to determine whether it is open or blocked. Usually no more than one of the three possible frames of reading is open in any single stretch of DNA. **Figure 2.10** shows an example of a sequence that can be read in only one reading frame, because the alternative reading frames are blocked by frequent termination codons. A long open reading frame is unlikely to exist by chance; if it were not translated into protein, there would have been no selective

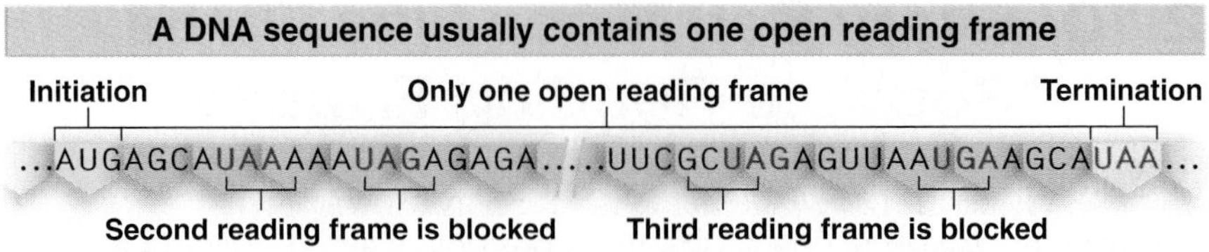

Figure 2.10 An open reading frame starts with AUG and continues in triplets to a termination codon. Blocked reading frames may be interrupted frequently by termination codons.

A DNA sequence usually contains one open reading frame

Initiation   Only one open reading frame   Termination

...AUGAGCAUAAAAAUAGAGAGA.....UUCGCUAGAGUUAAUGAAGCAUAA...

Second reading frame is blocked   Third reading frame is blocked

pressure to prevent the accumulation of termination codons. So the identification of a lengthy open reading frame is taken to be *prima facie* evidence that the sequence is translated into protein in that frame. An open reading frame (ORF) for which no protein product has been identified is sometimes called an unidentified reading frame (URF).

2.11 Several Processes Are Required to Express the Protein Product of a Gene

Key Terms

- **Messenger RNA (mRNA)** is the intermediate that represents one strand of a gene coding for protein. Its coding region is related to the protein sequence by the triplet genetic code.
- **Transcription** describes synthesis of RNA on a DNA template.
- **Translation** is synthesis of protein on an mRNA template.
- A **coding region** is the part of a gene that represents a protein sequence.
- The **leader (5′ UTR)** of an mRNA is the nontranslated sequence at the 5′ end that precedes the initiation codon.
- A **trailer (3′ UTR)** is a nontranslated sequence at the 3′ end of an mRNA following the termination codon.

Key Concepts

- A prokaryotic gene is expressed by transcription into mRNA and then by translation of the mRNA into protein.
- In eukaryotes, a gene is transcribed in the nucleus, but the mRNA must be transported to the cytoplasm to be translated.
- An mRNA consists of a nontranslated 5′ leader, one or more coding regions, and a nontranslated 3′ trailer.

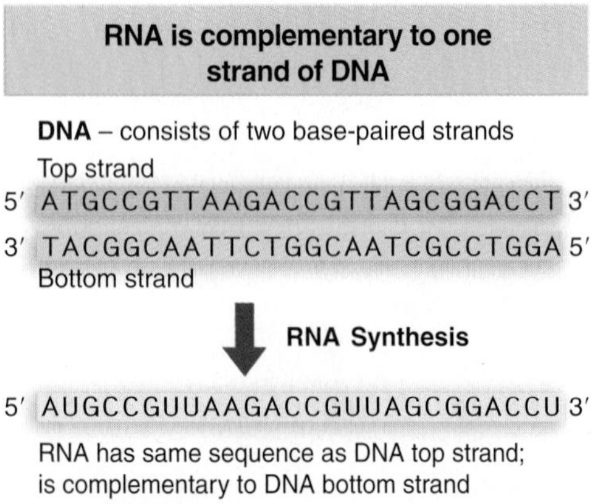

RNA is complementary to one strand of DNA

DNA – consists of two base-paired strands
Top strand
5′ ATGCCGTTAAGACCGTTAGCGGACCT 3′
3′ TACGGCAATTCTGGCAATCGCCTGGA 5′
Bottom strand

 RNA Synthesis

5′ AUGCCGUUAAGACCGUUAGCGGACCU 3′
RNA has same sequence as DNA top strand; is complementary to DNA bottom strand

Figure 2.11 RNA is synthesized by using one strand of DNA as a template for complementary base pairing.

In comparing gene and protein, we are restricted to dealing with the sequence of DNA stretching between the points corresponding to the ends of the protein. However, a gene is not directly translated into protein, but is expressed via the production of a **messenger RNA** (abbreviated to **mRNA**), a nucleic acid intermediate actually used to direct synthesis of a protein (as we see in detail in *Chapter 7 Messenger RNA*).

Messenger RNA is synthesized by the same process of complementary base pairing used to replicate DNA, with the important difference that only one strand of the DNA double helix participates. **Figure 2.11** shows that the sequence of messenger RNA is complementary with the sequence of one strand of DNA and is identical (apart from the replacement of T with U) with the other strand of DNA. The convention for writing DNA sequences is that the top strand runs $5' \rightarrow 3'$, with the sequence that is the same as RNA.

The process by which a gene gives rise to a protein is called *gene expression*. In bacteria, it consists of two stages. The first stage is **transcription**, when an mRNA copy of one strand of the DNA is produced. The second stage is **translation** of the mRNA into protein. This is the process by which the sequence of an mRNA is read in triplets to give the series of amino acids that make the corresponding protein.

A messenger RNA includes a sequence of nucleotides that corresponds with the sequence of amino acids in the protein. This part of the nucleic acid is called the **coding region**. But the messenger RNA includes additional sequences on either end; these sequences do not directly represent protein. The 5′ nontranslated region is called the **leader**, and the 3′ nontranslated region is called the **trailer**.

The *gene* includes the entire sequence represented in messenger RNA. Sometimes mutations impeding gene function are found in the additional, noncoding regions, confirming the view that these regions are a legitimate part of the genetic unit.

Figure 2.12 illustrates this situation, in which the gene is considered to comprise a continuous stretch of DNA, needed to produce a particular protein. It includes the sequence coding for that protein, but also includes sequences on either side of the coding region.

A bacterium consists of only a single compartment, so transcription and translation occur in the same place, as illustrated in **Figure 2.13**.

In eukaryotes transcription occurs in the nucleus, but the RNA product must be *transported* to the cytoplasm in order to be translated. **Figure 2.14** shows that this results in a spatial separation between transcription (in the nucleus) and translation (in the cytoplasm).

2.12 Proteins Are *trans*-Acting but Sites on DNA Are *cis*-Acting

Key Terms

- *cis* configuration describes two sites on the same molecule of DNA.
- *trans* configuration of two sites refers to their presence on two different molecules of DNA (chromosomes).
- A *cis*-acting site affects the activity only of sequences on its own molecule of DNA (or RNA); this property usually implies that the site does not code for protein.

Key Concepts

- All gene products (RNA or proteins) are *trans*-acting. They can act on any copy of a gene in the cell.
- *cis*-acting mutations identify sequences of DNA that are targets for recognition by *trans*-acting products. They are not expressed as RNA or protein and affect only the contiguous stretch of DNA.

A crucial step in the definition of the gene was the realization that all its parts must be present on one contiguous stretch of DNA. In genetic terminology, sites that are located on the same DNA are said to be in *cis*. Sites that are located on two different molecules of DNA are described as being in *trans*. So two mutations may be in *cis* (on the same

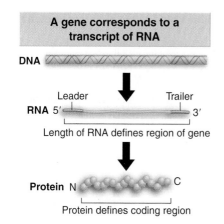

Figure 2.12 The gene may be longer than the sequence coding for protein.

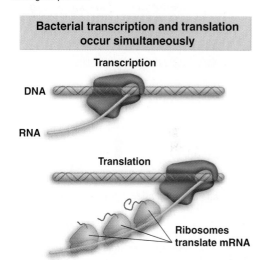

Figure 2.13 Transcription and translation take place in the same compartment in bacteria.

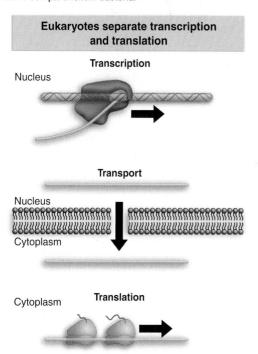

Figure 2.14 In eukaryotes, transcription occurs in the nucleus, and translation occurs in the cytoplasm.

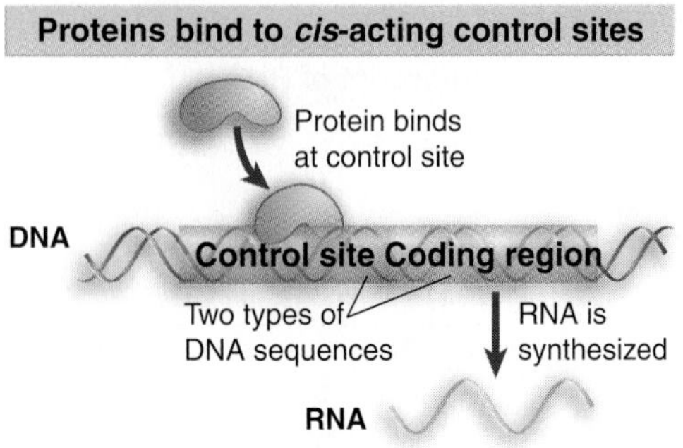

Proteins bind to *cis*-acting control sites

Protein binds at control site

DNA

Control site Coding region

Two types of DNA sequences

RNA is synthesized

RNA

Figure 2.15 Control sites in DNA provide binding sites for proteins; coding regions are expressed via the synthesis of RNA.

Mutations in control sites are *cis*-acting

Both alleles synthesize RNA in wild type

Control site mutation affects only contiguous DNA

Mutation

No RNA synthesis from Allele 1

RNA synthesis continues from Allele 2

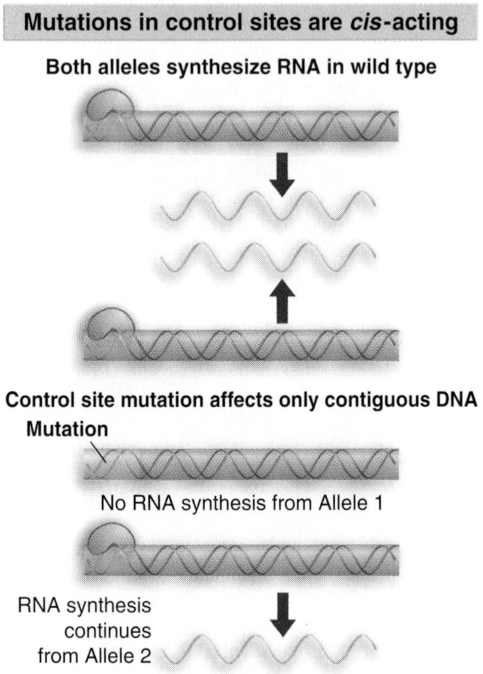

Figure 2.16 A *cis*-acting site controls the adjacent DNA but does not influence the other allele.

DNA) or in *trans* (on different DNAs). The complementation test uses this concept to determine whether two mutations are in the same gene (see *2.3 Mutations in the Same Gene Cannot Complement*). We may now extend the concept of the difference between *cis* and *trans* effects from defining the coding region of a gene to describing the interaction between regulatory elements and a gene.

Suppose that the ability of a gene to be expressed is controlled by a protein that binds to the DNA close to the coding region. In the example depicted in **Figure 2.15**, messenger RNA can be synthesized only when the protein is bound to the DNA. Now suppose that the DNA sequence to which this protein binds mutates, so that the protein can no longer recognize the DNA. The result is that the DNA can no longer be expressed.

So a gene can be inactivated either by a mutation in a control site or by a mutation in a coding region. The mutations cannot be distinguished genetically, because both have the property of acting only on the DNA sequence of the single allele in which they occur. They have identical properties in the complementation test, and a mutation in a control region is therefore defined as being in the gene in the same way as a mutation in the coding region.

Figure 2.16 shows that a deficiency in the control site *affects only the coding region to which it is connected; it does not affect the ability of the other allele to be expressed.* A mutation that acts solely by affecting the properties of the contiguous sequence of DNA is called ***cis*-acting**.

We may contrast the effect of the *cis*-acting mutation shown in Figure 2.16 with that of a mutation in the gene coding for the regulator protein. **Figure 2.17** shows that the absence of regulator protein would prevent *both* alleles from being expressed. A mutation of this sort is said to be *trans*-acting.

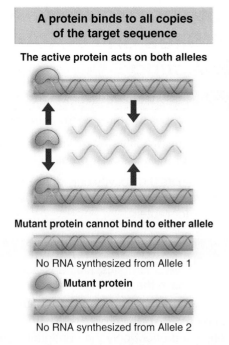

A protein binds to all copies of the target sequence

The active protein acts on both alleles

Mutant protein cannot bind to either allele

No RNA synthesized from Allele 1

Mutant protein

No RNA synthesized from Allele 2

Figure 2.17 A *trans*-acting mutation in a protein affects both alleles of a gene that it controls.

Reversing the argument, if a mutation is *trans*-acting, we know that its effects must be exerted through some diffusible product (typically a protein) that acts on multiple targets within a cell. But if a mutation is *cis*-acting, it must function via affecting directly the properties of the contiguous DNA, which means that it is *not expressed in the form of RNA or protein.*

2.13 SUMMARY

A chromosome consists of an uninterrupted length of duplex DNA that contains many genes. Each gene (or cistron) is transcribed into an RNA product, which in turn is translated into a polypeptide sequence if the gene codes for protein. An RNA or protein product of a gene is said to be *trans*-acting. A gene is defined as a unit on a single stretch of DNA by the complementation test. A site on DNA that regulates the activity of an adjacent gene is said to be *cis*-acting.

When a gene codes for protein, the relationship between the sequence of DNA and the sequence of the protein is given by the genetic code. Only one of the two strands of DNA codes for protein. A codon consists of three nucleotides that represent a single amino acid. A coding sequence of DNA consists of a series of codons, read from a fixed starting point. Usually only one of the three possible reading frames can be translated into protein.

A gene may have multiple alleles. Recessive alleles are caused by loss-of-function mutations that interfere with the function of the protein. A null allele has total loss-of-function. Dominant alleles are caused by gain-of-function mutations that create a new property in the protein.

Genes May Be Interrupted

3.1 Introduction

Key Terms

- A **colinear** relationship describes the 1:1 representation of a sequence of triplet nucleotides in a sequence of amino acids.
- An **exon** is a segment of an interrupted gene that is represented in the mature RNA product.
- An **intron (intervening sequence)** is a segment of DNA that is transcribed but later removed from within the transcript by splicing together the sequences (exons) on either side of it.
- A **transcript** is the RNA product produced by copying one strand of DNA. It may require processing to generate a mature RNA.
- **RNA splicing** is the process of excising introns from RNA and connecting the exons into a continuous mRNA.

Key Concepts

- Eukaryotic genomes contain interrupted genes in which exons (represented in the final RNA product) alternate with introns (removed from the initial transcript).
- The exon sequences are in the same order in the gene and in the RNA, but an interrupted gene is longer than its final RNA product because of the presence of the introns.

The simplest form of a gene is a length of DNA that is **colinear** with its protein product. Bacterial genes are almost always of this type, in which a continuous coding sequence of 3N base pairs represents a protein of N amino acids. But in eukaryotes, a gene may include additional sequences that lie within the coding region, interrupting the sequence that represents the protein. These sequences are removed from the RNA product during gene expression, generating an mRNA that includes a nucleotide sequence exactly corresponding with the protein product according to the rules of the genetic code.

The sequences of DNA composing an **interrupted gene** are divided into the two categories depicted in **Figure 3.1**:

- The **exons** are the sequences represented in the mature RNA. By definition, a gene starts and ends with exons, corresponding to the 5′ and 3′ ends of the RNA.
- The **introns** are the intervening sequences that are removed when the primary transcript is processed to give the mature RNA.

The expression of interrupted genes requires an additional step not needed for uninterrupted genes. The DNA of an interrupted gene gives rise to an RNA copy (a **transcript**) that exactly represents the genome sequence. But this RNA is only a precursor; it cannot be used for producing protein. First the introns must be removed from the RNA to give a messenger RNA that consists only of the series of exons. This process is called **RNA splicing**. It involves a precise deletion of an intron from the primary transcript and joining the ends of the RNA on either side to form a covalently intact molecule (see *Chapter 26 RNA Splicing and Processing*).

The gene comprises the region in the genome between points corresponding to the 5′ and 3′ terminal bases of mature mRNA. We know that transcription starts at the 5′ end of the mRNA, but usually it extends beyond the 3′ end, which is generated by cleavage of the RNA (see *26.14 The 3′ Ends of mRNAs Are Generated by Cleavage and Polyadenylation*). The gene is considered to include the regulatory regions on both sides of the gene that are required for initiating and (sometimes) terminating gene expression.

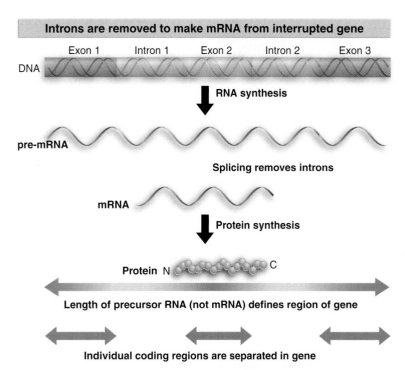

Figure 3.1 Interrupted genes are expressed via a precursor RNA. Introns are removed when the exons are spliced together. The mRNA has only the sequences of the exons.

3.2 Interrupted Genes Were First Detected by Comparing mRNA and DNA

Key Concepts

- Introns can be detected by the presence of additional regions when genes are compared with their RNA products by restriction mapping or electron microscopy, but the ultimate definition is based on comparison of sequences.
- Electron microscopy mapping of mRNA-genome hybrids shows loops of extra material in the genome DNA.
- Comparisons of the restriction maps of a cDNA made from mRNA with the genomic DNA show extra sites in the genome DNA.
- Introns do not usually code for proteins.

Key Terms

- **Restriction endonucleases** recognize specific short sequences of DNA and cleave the duplex (sometimes at target site, sometimes elsewhere, depending on type).
- A **restriction map** is a linear array of sites on DNA cleaved by various restriction enzymes.

The characterization of eukaryotic genes was first made possible by the development of techniques for physically mapping DNA. When an

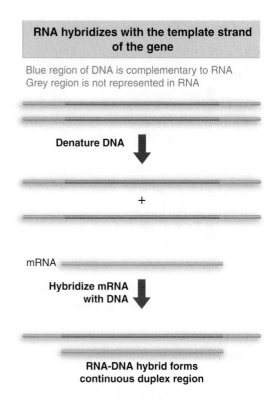

Figure 3.2 Hybridizing an mRNA from an uninterrupted gene with the DNA of the gene generates a duplex region corresponding to the gene.

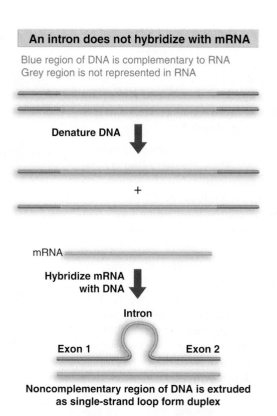

Figure 3.3 RNA hybridizing with the DNA made from an interrupted gene produces a duplex corresponding to the exons, with an intron excluded as a single-stranded loop between the exons.

mRNA is compared with the genomic sequence from which it was derived, the genomic sequence turns out to have extra regions that are not represented in the mRNA.

One technique for comparing mRNA with genomic DNA is to hybridize the mRNA with the complementary strand of the DNA. If the two sequences are colinear, a duplex is formed. **Figure 3.2** shows a typical result when an RNA made from an uninterrupted gene is hybridized with a DNA that includes the gene. The sequences on either side of the gene are not represented in the RNA, but the sequence of the gene hybridizes with the RNA to form a continuous duplex region.

Suppose we now perform the same experiment with RNA made from an interrupted gene. The difference is that the sequences represented in mRNA lie on either side of a sequence that is not in the mRNA. **Figure 3.3** shows that the RNA-DNA hybrid forms a duplex, but the nonreacting sequence in the middle remains single-stranded, forming a loop that extrudes from the duplex. The hybridizing regions correspond to the exons, and the extruded loop corresponds to the intron.

The structure of the mRNA-DNA hybrid can be visualized by electron microscopy. One of the very first examples of the visualization of an interrupted gene is shown in **Figure 3.4**. Tracing the structure in the lower part of the figure shows that three introns are located close to the very beginning of the gene.

Another way to compare the sequences of mRNA and DNA is to use restriction mapping, which enables a physical map of any DNA molecule to be obtained by breaking it at defined points whose distance apart can be accurately determined. The breaks are defined because **restriction endonucleases** recognize specific sequences of double-stranded DNA as targets for cleavage. The techniques can be extended to (single-stranded) RNA by making a (double-stranded) DNA copy of the RNA. By determining the lengths and order of the fragments generated by restriction cleavage, we can make a **restriction map** that represents a linear sequence of the cleavage sites.

When a gene is uninterrupted, the restriction map of its DNA corresponds exactly with the map of its mRNA.

When a gene possesses an intron, the map at each end of the gene corresponds with the map at each end of the message sequence. But within the gene, the maps diverge, because additional regions are found in the gene, but are not represented in the message. Each such region corresponds to an intron. The example of **Figure 3.5** compares the restriction maps of a β-globin gene and mRNA. (The way the mRNA is mapped is actually to make a copy of it, called cDNA, that is treated with restriction enzymes.) There are two introns. Each intron contains a series of restriction sites that are absent from the cDNA. The pattern of restriction sites in the exons is the same in both the cDNA and the gene.

Ultimately a comparison of the nucleotide sequences of the genomic and mRNA sequences precisely defines the introns. As indicated in **Figure 3.6**, an intron usually has no open reading frame and would block translation of the unspliced RNA. An intact reading frame is created in the mRNA sequence by the removal of the introns.

3.3 Interrupted Genes Are Much Longer than the Corresponding mRNAs

Key Concepts

- Introns are removed by RNA splicing, which occurs only in *cis* on an individual RNA molecule.
- Only mutations in exons can affect protein sequence, but mutations in introns can affect processing of the RNA and therefore prevent production of protein.

How does the existence of introns change our view of the gene? Following splicing, the exons are always joined together in the same order in which they lie in DNA. So the colinearity of gene and protein is maintained between the individual exons and the corresponding parts

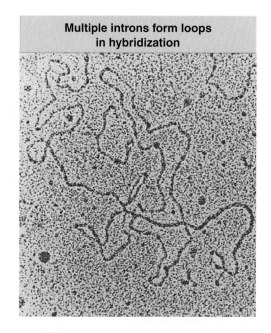

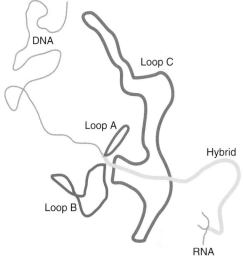

Figure 3.4 Hybridization between an adenovirus mRNA and its DNA identifies three loops corresponding to introns that are located at the beginning of the gene. Photograph kindly provided by Philip Sharp, Center for Cancer Research, Massachusetts Institute of Technology.

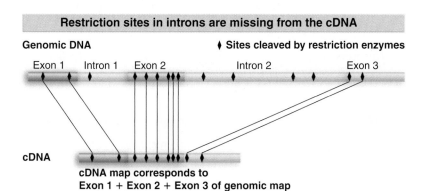

Figure 3.5 Comparison of the restriction maps of cDNA and genomic DNA for mouse β-globin shows that the gene has two introns that are not present in the cDNA. The exons can be aligned exactly between cDNA and gene.

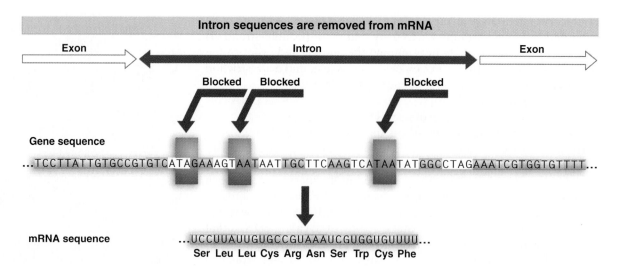

Ser Leu Leu Cys Arg Asn Ser Trp Cys Phe

Figure 3.6 An intron is a sequence present in the gene but absent from the mRNA (here shown in terms of the cDNA sequence). All three possible reading frames are blocked by termination codons in the intron.

of the protein chain. **Figure 3.7** shows that the *order* of sites in the gene remains the same as the order of corresponding points in the protein. But the *distances* in the gene do not correspond with the distances in the protein. Genetic distances, as seen on a recombination map, have no relationship to the distances between the corresponding points in the protein. The length of the gene is defined by the length of the initial (precursor) RNA instead of by the length of the messenger RNA.

All the exons are represented on the same molecule of RNA, and their splicing together occurs only as an *intra*molecular reaction. There is usually no joining of exons carried by *different* RNA molecules, so the mechanism excludes any splicing together of sequences representing different alleles.

Mutations that directly affect the sequence of a protein must lie in exons. What are the effects of mutations in the introns? Because the introns are not part of the messenger RNA, mutations in them cannot directly affect protein structure. However, they can prevent the production of the messenger RNA—for example, by inhibiting the splicing together of exons.

Mutations that affect splicing are usually deleterious. The majority are single base substitutions at the junctions between introns and exons. They may cause an exon to be left out of the product, cause an intron to be included, or make splicing occur at an aberrant site. The most common result is to introduce a termination codon that results in truncation of the protein sequence. About 15% of the point mutations that cause human diseases are caused by disruption of splicing.

Figure 3.7 Exons remain in the same order in mRNA as in DNA, but distances along the gene do not correspond to distances along the mRNA or protein products. The distance between A and B in the gene is smaller than the distance between B and C; but the distance between A and B in the mRNA (and protein) is greater than the distance between B and C.

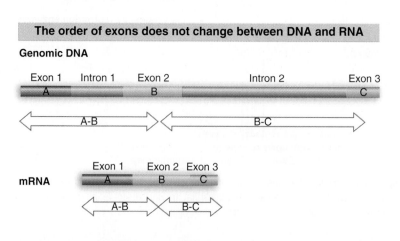

Interrupted genes are found in all classes of organisms. In higher eukaryotes, most nuclear genes are interrupted; the introns are usually much longer than exons, creating genes that are very much longer than their coding regions. Interrupted genes are found in lower eukaryotes (such as yeast), although they are a much smaller proportion of the total gene number. Interruptions also are found in mitochondrial genes in lower eukaryotes, and in chloroplast genes in plants. Interrupted genes have been found in bacteria and bacteriophages, although they are extremely rare in prokaryotic genomes.

3.4 Organization of Interrupted Genes Is Often Conserved

Key Concept

- The positions of introns may be conserved, as is shown by comparisons of homologous genes in different species.

The arrangement of introns and exons is often conserved in related genes. This means that the introns are found at the same locations relative to the exons, although they can vary widely in sequence or length.

The globin genes provide an extensively studied example (see *3.10 The Members of a Gene Family Have a Common Organization*). The two general types of globin gene, α and β, share a common type of structure. The consistency of the organization of mammalian globin genes is evident from the structure of the "generic" globin gene summarized in **Figure 3.8**.

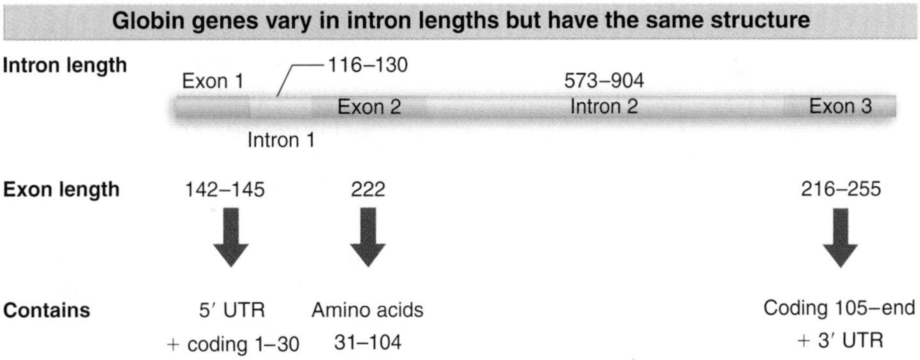

Interruptions occur at homologous positions (relative to the coding sequence) in all known active globin genes, including those of mammals, birds, and frogs. The first intron is always fairly short, and the second usually is longer, but the actual lengths can vary. Most of the variation in overall lengths between different globin genes results from the variation in the second intron. In the mouse, the second intron in the α-globin gene is only 150 bp long, so the overall length of the gene is 850 bp, compared with the major β-globin gene where the intron length of 585 bp gives the gene a total length of 1382 bp. The variation in length of the genes is much greater than the range of lengths of the mRNAs (α-globin mRNA = 585 bases, β-globin mRNA = 620 bases).

The example of DHFR, a somewhat longer gene, is shown in **Figure 3.9**. The mammalian DHFR (dihydrofolate reductase) gene is organized into six exons that correspond to the 2000 base mRNA. But the genes extend over a much greater length of DNA because the introns are very long. In three mammals the exons remain essentially the same, and the relative positions of the introns are unaltered, but the lengths of individual introns vary extensively, resulting in a variation in the length of the gene from 25–31 kb.

The globin and DHFR genes present examples of a general phenomenon: *Genes that are related by evolution have related organizations, with conservation of the positions of (at least some) of the introns. Variations in the lengths of the genes are primarily determined by the lengths of the introns.*

Globin genes vary in intron lengths but have the same structure

Intron length	Exon 1 — 116–130	573–904
	Exon 2	Intron 2	Exon 3
	Intron 1		
Exon length	142–145	222	216–255
Contains	5′ UTR + coding 1–30	Amino acids 31–104	Coding 105–end + 3′ UTR

Figure 3.8 All functional globin genes have an interrupted structure with three exons. The lengths indicated in the figure are those of the mammalian β-globin genes.

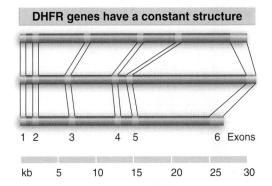

DHFR genes have a constant structure

1 2   3   4 5   6 Exons

kb   5   10   15   20   25   30

Figure 3.9 Mammalian genes for DHFR have the same relative organization of rather short exons and very long introns, but vary extensively in intron lengths.

3.5 Exon Sequences Are Conserved but Introns Vary

Key Concepts

- Comparisons of related genes in different species show that the sequences of corresponding exons are usually conserved, but the sequences of introns are not.
- Introns evolve much more rapidly than exons because there is no selective pressure operating on the introns to produce a protein with a useful sequence.

When two genes are related, their exons are usually much more similar than their introns. In an extreme case, the exons of two genes may code for the same protein sequence, but the intron sequences may differ. This implies that the two genes originated by a duplication of some common ancestral gene. Then differences from mutations accumulated between the copies, but mutations arising in the exons were mostly eliminated by the selective pressure against disfunctional proteins.

As we see later, when we consider the evolution of the gene, exons can be considered as basic building blocks that are assembled in various combinations. A gene may have some exons that are related to exons of another gene and other exons that are unrelated. Such genes may arise by dupliction and translocation of individual exons. Usually the introns are not related at all in such genes.

The relationship between two genes can be plotted in the form of the dot matrix comparison of **Figure 3.10**. A dot is placed to indicate each position at which the same sequence is found in each gene. The dots form a line at an angle of 45° if two sequences are identical. The line is absent from regions that lack similarity, and it is displaced laterally or vertically by deletions or insertions present in one sequence and absent from the other.

In the comparison of the two β-globin genes of the mouse in Figure 3.10, the plotted line extends through the three exons and through the small intron. The line peters out in the flanking regions and in the long intron. This is a typical pattern, in which coding sequences are well related, with the relationship extending beyond the boundaries of the exons but lost from longer introns and from the regions on either side of the gene.

The divergence between two exons is related to the differences between the proteins. It is caused mostly by base substitutions. In the translated regions, the exons are under the constraint of needing to code for amino acid sequences, so they are limited in their potential to change sequence. Changes occur more freely in nontranslated regions (corresponding to the 5′ leader and 3′ trailer of the mRNA).

In corresponding introns, the pattern of divergence involves both changes in length (due to deletions and insertions) and base substitutions. Introns evolve much more rapidly than exons. When a gene is compared in different species, sometimes the exons are homologous, while the introns have diverged so much that corresponding sequences cannot be recognized.

Mutations occur at the same rate in both exons and introns, but are removed more effectively from the exons by selection. However, in the absence of the constraints imposed by a coding function, an intron is able quite freely to accumulate point substitutions and other changes. These accumulations imply that the intron does not have a sequence-specific function.

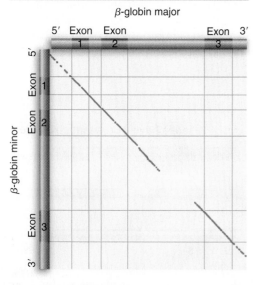

Figure 3.10 The sequences of the mouse β^{maj} and β^{min} globin genes are closely related in coding regions, but differ in the flanking regions and long intron. Data kindly provided by Philip Leder.

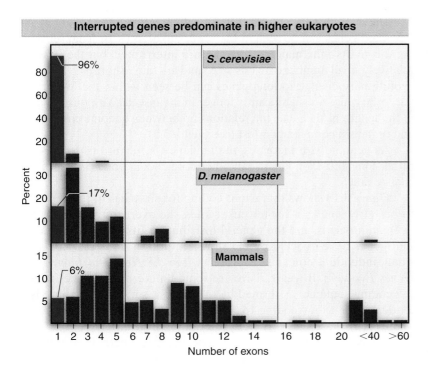

Interrupted genes predominate in higher eukaryotes

Figure 3.11 Most genes are uninterrupted in yeast, but most genes are interrupted in flies and mammals. (Uninterrupted genes have only one exon, and are totaled in the leftmost column.)

3.6 Genes Show a Wide Distribution of Lengths

Key Concepts

- Most genes in yeasts are uninterrupted, but most genes in higher eukaryotes are interrupted.
- Exons are usually short, typically coding for <100 amino acids.
- Introns are short in lower eukaryotes, but range up to several 10s of kb in length in higher eukaryotes.
- The overall length of a gene is determined largely by its introns.

Figure 3.11 summarizes the overall organization of genes in yeasts, insects, and mammals. In the yeast *Saccharomyces cerevisiae*, the great majority of genes (>96%) are not interrupted, and those that have exons have only a few. There are virtually no *S. cerevisiae* genes with more than four exons.

In insects (for example, the fly *Drosophila melanogaster*) and mammals, the situation is reversed. Only a few genes have uninterrupted coding sequences (6% in mammals). Insect genes tend to have a fairly small number of exons, typically fewer than 10. Mammalian genes are split into more pieces, and some have several tens of exons. About 50% of mammalian genes have >10 introns.

Examining the consequences of uninterrupted versus interrupted organization for the overall length of the gene, we see in **Figure 3.12** that there is a striking difference between yeast and the higher eukaryotes. The average yeast gene is 1.4 kb long, and very few are longer than 5 kb. The predominance of interrupted genes in higher eukaryotes, however, means that the gene can be much longer than the unit that codes for protein. Relatively few genes in flies or mammals are shorter than 2 kb, and many have lengths between 5 kb and 100 kb. The average human gene is 27 kb long (see Figure 5.10).

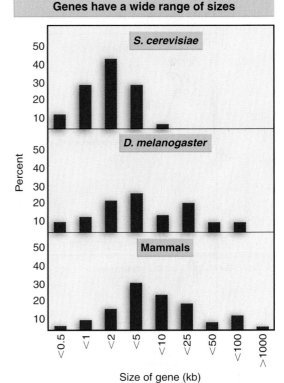

Genes have a wide range of sizes

Figure 3.12 Yeast genes are short, but genes in flies and mammals have a dispersed distribution extending from short to very long.

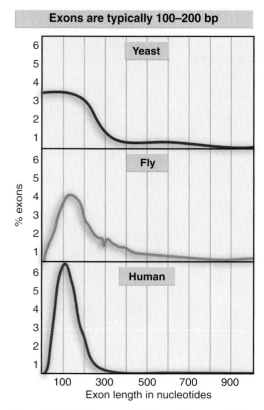

Figure 3.13 Most exons coding for proteins are short.

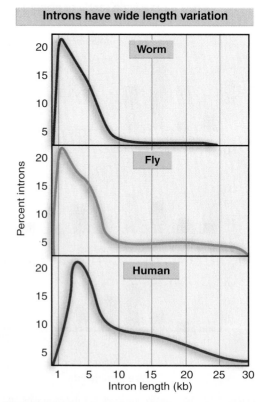

Figure 3.14 Introns range from very short to very long.

The evolutionary change from largely uninterrupted to largely interrupted genes can be seen in the lower eukaryotes. In fungi (excepting the yeasts), the majority of genes are interrupted, but they have a relatively small number of exons (<6) and are fairly short (<5 kb). The evolutionary change to long genes can be seen within the higher eukaryotes: genes are significantly longer in the insects. With this increase in the length of the gene, the relationship between genome complexity and organism complexity is lost (see Figure 5.5).

As genome size increases, the tendency is for introns to become rather long, but exon length (and therefore coding potential) does not change much.

Figure 3.13 shows that exons coding for stretches of protein tend to be fairly small. In higher eukaryotes, the average exon codes for ~50 amino acids, and the general distribution fits well with the idea that genes have evolved by the slow addition of units that code for small, individual domains of proteins (see *3.8 How Did Interrupted Genes Evolve?*). In yeast, there are some uninterrupted genes where the coding sequence is formed by one longer, single exon. There is a tendency for exons coding for untranslated 5′ and 3′ regions to be longer than those that code for proteins.

Figure 3.14 shows that introns vary widely in length. In worms and flies, the average intron is not much longer than the exons. There are no very long introns in worms, but flies contain a significant proportion. In vertebrates, the length distribution is much wider, extending from approximately the same length as the exons (<200 bp) to lengths measured in 10s of kbs, with the longest being 50–60 kb.

Very long genes are the result of very long introns, not the result of coding for longer products. There is no correlation between gene length and mRNA length in higher eukaryotes; nor is there a good correlation between gene length and number of exons. The length of a gene therefore depends primarily on the lengths of its individual introns. In mammals, insects, and birds, the "average" gene is ~5× the length of its mRNA.

3.7 Some DNA Sequences Code for More than One Protein

Key Concepts

- The use of alternative initiation or termination codons allows two proteins to be generated where one is equivalent to a fragment of the other.
- Nonhomologous protein sequences can be produced from the same sequence of DNA when it is read in different reading frames by two (overlapping) genes.
- Homologous proteins that differ by the presence or absence of certain regions can be generated by differential (alternative) splicing when certain exons are included or excluded. This may take the form of including or excluding individual exons or of choosing between alternative exons.

Most genes consist of a sequence of DNA devoted solely to the purpose of coding for one protein (although the gene may include noncoding regions at either end and introns within the coding region). However, there are some cases in which a single sequence of DNA codes for more than one protein.

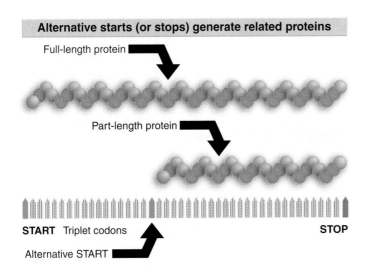

Alternative starts (or stops) generate related proteins

Full-length protein

Part-length protein

START Triplet codons

Alternative START

STOP

Figure 3.15 Two proteins can be generated from a single gene by starting (or terminating) expression at different points.

Overlapping genes exist in the relatively simple situation in which one gene is part of the other. The first half (or second half) of a gene is used independently to specify a protein that represents the first (or second) half of the protein specified by the full gene. This relationship is illustrated in **Figure 3.15**. The end result is much the same as that of a partial cleavage of a protein product that generates part-length protein products.

Two genes overlap in a more subtle manner when the same sequence of DNA is shared between two *nonhomologous* proteins. This situation arises when the same sequence of DNA is translated in more than one reading frame. In cellular genes, a DNA sequence usually is read in only one of the three potential reading frames, but in some viral and mitochondrial genes, there is an overlap between two adjacent genes that are read in different reading frames. This situation is illustrated in **Figure 3.16**. The distance of overlap is usually relatively short, so that most of the sequence coding for protein has a unique coding function.

In some genes, *alternative* patterns of gene expression create switches in the pathway for connecting the exons. A single gene may generate a variety of mRNA products that differ in their content of exons. The difference may be that certain exons are optional—they may be included or spliced out. Or there may be exons that are treated as mutually exclusive—one or the other is included, but not both. The alternative forms produce proteins in which one part is common while the other part is different.

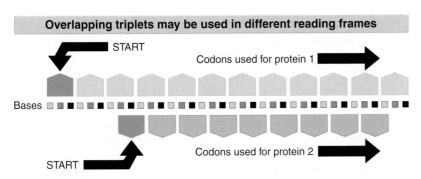

Overlapping triplets may be used in different reading frames

START

Codons used for protein 1

Bases

Codons used for protein 2

START

Figure 3.16 Two genes may share the same sequence by reading the DNA in different frames.

Alternative splicing can substitute exons

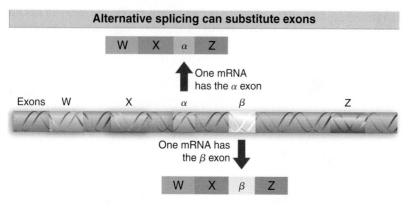

Figure 3.17 Alternative splicing generates the α and β variants of troponin T.

Figure 3.17 shows the example of the 3′ half of the troponin T gene of rat muscle, which contains 5 exons. Only 4 of the exons are used to construct an individual mRNA. Three exons, *WXZ*, are the same in both expression patterns. However, in one pattern the α exon is spliced between *X* and *Z*; in the other pattern, the β exon is used. The α and β forms of troponin T therefore differ in the sequence of the amino acids present between sequences *W* and *Z*, depending on which of the alternative exons, α or β, is used. Either one of the α and β exons can be used to form an individual mRNA, but both cannot be used in the same mRNA.

So alternative (or differential) splicing can generate proteins with overlapping sequences from a single stretch of DNA. Alternative splicing expands the number of proteins relative to the number of genes by ~15% in flies and worms, but has much bigger effects in humans, where ~60% of genes may have alternative modes of expression (see *5.4 The Human Genome Has Fewer Genes than Expected*). About 80% of the alternative splicing events result in a change in the protein sequence.

3.8 How Did Interrupted Genes Evolve?

Key Concepts

- The major evolutionary question is whether genes originated as sequences interrupted by introns or whether they were originally uninterrupted.
- Most protein-coding genes probably originated in an interrupted form, but interrupted genes for RNAs that do not code for proteins may have originally been uninterrupted.

The highly interrupted structure of eukaryotic genes suggests a picture of the eukaryotic genome as a sea of introns (mostly but not exclusively unique in sequence), in which islands of exons (sometimes very short) are strung out in individual archipelagoes that constitute genes.

What was the original form of genes that today are interrupted?

- The "introns early" model supposes that introns have always been an integral part of the gene. Genes originated as interrupted structures, and those without introns have lost them in the course of evolution.
- The "introns late" model supposes that the ancestral protein-coding units consisted of uninterrupted sequences of DNA. Introns were subsequently inserted into them.

The introns early model suggests that the mosaic structure of genes is a remnant of an ancient approach to the reconstruction of genes to make novel proteins. Suppose that an early cell had a number of separate protein-coding sequences. One aspect of its evolution is likely to have been the reorganization and juxtaposition of different polypeptide units to build up new proteins. This could happen by recombination of DNA to place an exon from one gene into another gene.

Figure 3.18 illustrates the outcome when a random sequence that includes an exon is translocated to a new position in the genome. Exons are very small relative to introns, so it is likely that the exon will find itself within an intron. Because only the sequences at the exon-intron junctions are required for splicing, the exon is likely to be flanked by functional 3′ and 5′ splice junctions, respectively (see *26.2 Nuclear Splice Junctions Are Short Sequences*).

Splicing junctions are recognized in pairs, so that the 5′ splicing junction of the original intron will interact with the 3′ splicing junction introduced by the new exon, instead of with its original partner. Similarly, the 5′ splicing junction of the new exon will interact with the 3′ splicing junction of the original intron. The result is to insert the new exon into the RNA product between the original two exons. So long as the new exon is in the same coding frame as the original exons, a new protein sequence will be produced.

This type of event could have been responsible for generating new combinations of exons during evolution. Evidence to support this idea is provided by the existence of related exons in different genes, as though the genes had been assembled by mixing and matching exons, as discussed in the next section.

Alternative forms of genes for rRNA and tRNA are sometimes found, with and without introns. In the case of the tRNAs, where all the molecules conform to the same general structure, it seems unlikely that evolution brought together the two regions of the gene. After all, the

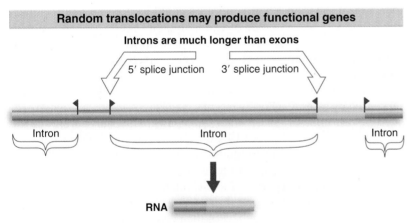

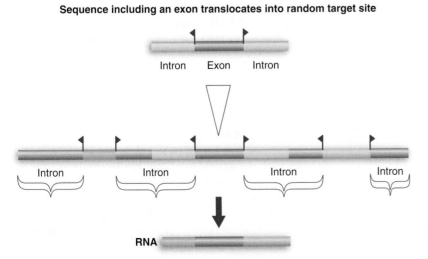

Figure 3.18 An exon surrounded by flanking sequences that is translocated into an intron may be spliced into the RNA product.

different regions are involved in the base pairing that gives significance to the structure. So here it must be that the introns were inserted into continuous genes.

3.9 Some Exons Can Be Equated with Protein Functions

> **Key Concepts**
>
> - Facts suggesting that exons were the building blocks of evolution are:
>   - Gene structure is conserved between genes in very distant species.
>   - Many exons can be equated with protein-coding sequences that have particular functions.
>   - Related exons are found in different genes.

It is relatively common for an exon to code for a distinct protein domain that is able to fold into its mature conformation independently. For example, in secreted proteins, such as insulin, the first exon, coding for the N-terminal region of the polypeptide, often specifies the signal sequence involved in membrane secretion. This sequence acts independently of the rest of the protein to sponsor its association with a membrane.

Exons tend to be fairly small (see Figure 3.13), around the size of the smallest polypeptide that can assume a stable folded structure, ~20–40 residues. Perhaps proteins were originally assembled from rather small modules. Each module need not necessarily correspond to a current function; several modules could have combined to generate a function. The number of exons in genes tends to increase with the length of their proteins, which is consistent with the view that proteins acquire multiple functions by successively adding appropriate modules.

The view that exons are the functional building blocks of genes is supported by cases in which two genes may have some exons that are related to one another, while other exons are found only in one of the genes. **Figure 3.19** summarizes the relationship between the receptor for human LDL (plasma low density lipoprotein) and other proteins. In the center of the LDL receptor gene is a series of exons related to the exons of the gene for the precursor for EGF (epidermal growth factor). In the N-terminal part of the protein, a series of exons codes for a sequence related to the blood protein complement factor C9. So the LDL receptor gene was created by assembling *modules* for its various functions. These modules are also used in different combinations in other proteins.

This situation favors the view that proteins generally evolve by the addition of exons as suggested in Figure 3.18. This means that introns were present in early evolution, have survived in the eukaryotes with various adjustments in individual proteins, but have been lost from the prokaryotes.

Can we use the features of exons to distinguish between models for gene evolution? For example, if introns had been inserted at random into a gene (introns late model), a relationship between the exons and the domains of the protein product would not be expected. However, if exons are related to the features of the protein, this suggests that they are individual units that preexisted together with the surrounding introns (introns early model).

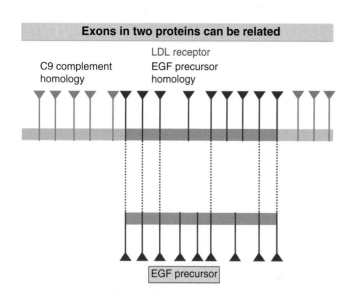

Exons in two proteins can be related

C9 complement homology

LDL receptor
EGF precursor homology

EGF precursor

Figure 3.19 The LDL receptor gene consists of 18 exons, some of which are related to EGF precursor exons and some to the C9 blood complement gene. Triangles mark the positions of introns. In the region of LDL and EGF homology, only some of the introns in the two genes are identical in position.

3.10 The Members of a Gene Family Have a Common Organization

Key Concepts

- A common feature in a set of genes is assumed to identify a property that preceded their evolutionary separation.
- All globin genes have a common form of organization with three exons and two introns, suggesting that they are descended from a single ancestral gene.

Many genes in a higher eukaryotic genome are related to others in the same genome. A **gene family** can be defined as a group of genes that code for related or identical proteins. A family originates when a gene is duplicated. Initially the two copies are identical, but then they diverge as mutations accumulate in them. Further duplications and divergence extend the family further. The globin genes are an example of a family that can be divided into two subfamilies (α-globin and β-globin), but all its members have the same basic structure and function. The concept can be extended further when we find genes that are more distantly related, but still can be recognized as having common ancestry; in this case, a group of gene families can be considered to make up a **superfamily**.

A fascinating case of evolutionary conservation is presented by the α- and β-globins and two other proteins related to them. Myoglobin is a monomeric oxygen-binding protein of animals, whose amino acid sequence suggests a common (though ancient) origin with the globin subunits. Leghemoglobins are oxygen-binding proteins present in the legume family of plants; like myoglobin, they are monomeric. They too share a common origin with the other oxygen-binding proteins. Together, the globins, myoglobin, and leghemoglobin constitute the globin superfamily, a set of gene families all descended from some (distant) common ancestor.

Both α- and β-globin genes have three exons (see Figure 3.8). The two introns are located at constant positions relative to the coding sequence. The central exon represents the oxygen-binding domain of the globin chain.

Myoglobin is represented by a single gene in the human genome, whose structure is essentially the same as that of the globin genes. The three-exon structure therefore predates the evolution of separate myoglobin and globin functions.

Leghemoglobin genes contain three introns, the first and last of which occur at points in the coding sequence that are homologous to the locations of the two introns in the globin genes. This remarkable similarity suggests an exceedingly ancient origin for the oxygen-binding proteins. As **Figure 3.20** shows, the central intron of leghemoglobin separates two exons that together code for the sequence corresponding to the single central exon in globin. Could the central exon of the globin gene have been derived by a fusion of two central exons in the ancestral gene? Or is the single central exon the ancestral form?

Cases in which homologous genes differ in structure may provide information about their evolution. An example is insulin. Mammals and birds have only one locus for insulin, except for the rodents, which have two loci. **Figure 3.21** illustrates the structures of these insulin genes.

The principle we use in comparing the organization of related genes in different species is that *a common feature identifies a structure that predated the evolutionary separation of the two species*. In chicken, the single insulin gene has two introns; one of the two rat genes has the same structure. The common structure implies that the ancestral insulin gene had two introns. However, the second rat gene has only one

Key Terms

- A **gene family** consists of a set of genes within a genome that code for related or identical proteins. The members were derived by duplication of an ancestral gene followed by accumulation of changes in sequence between the copies. Most often the members of a family are related but not identical.
- A **superfamily** is a set of genes all related by presumed descent from a common ancestor, but now showing considerable variation.

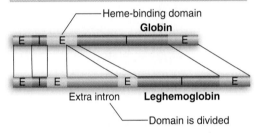

Figure 3.20 The exon structure of globin genes corresponds with protein function, but leghemoglobin has an extra intron in the central domain.

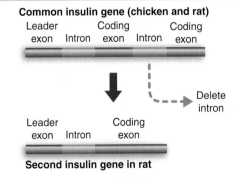

Figure 3.21 The rat insulin gene with one intron evolved by the loss of an intron from an ancestral sequence with two introns.

intron. It must have evolved by a gene duplication in rodents that was followed by the precise removal of one intron from one of the copies.

The relationship between exons and protein domains varies. In some cases there is a clear 1:1 relationship; in others no pattern is to be discerned. One possibility is that removal of introns has fused the adjacent exons. This means that the intron must have been precisely removed, without changing the integrity of the coding region. An alternative is that some introns arose by insertion into a coherent domain. Together with the variations that we see in exon placement when we compare related genes, this argues that intron positions can be adjusted in the course of evolution, possibly by changing the positions of the ends.

3.11 Pseudogenes Are Dead Ends of Evolution

Key Term

- A **processed pseudogene** is an inactive gene copy that lacks introns, contrasted with the interrupted structure of the active gene. Such genes originate by reverse transcription of mRNA and insertion of a duplex copy into the genome.

Key Concept

- Pseudogenes have no coding function, but they can be recognized by sequence similarities with existing functional genes. They arise by the accumulation of mutations in (formerly) functional genes.

Pseudogenes (ψ) are defined by their possession of sequences that are related to those of the functional genes, but that cannot be translated into a functional protein.

Some pseudogenes have the same general structure as functional genes, with sequences corresponding to exons and introns in the usual locations. They may have been rendered inactive by mutations that prevent any or all of the stages of gene expression. The changes can take the form of abolishing the signals for initiating transcription, preventing splicing at the exon-intron junctions, or prematurely terminating translation.

Usually a pseudogene has several deleterious mutations. Presumably once it ceased to be active, there was no impediment to the accumulation of further mutations. Pseudogenes that represent inactive versions of currently active genes have been found in many systems, including globin, immunoglobulins, and histocompatibility antigens, where they are located in the vicinity of the active genes.

A typical example is the rabbit pseudogene, ψβ2, which has the usual organization of exons and introns, and is related most closely to the functional globin gene β1. But it is not functional. **Figure 3.22** summarizes the many changes that have occurred in this pseudogene. The deletion of a base pair at codon 20 of ψβ2 has caused a frameshift that would lead to termination shortly after. Several point mutations have changed later codons representing amino acids that are highly conserved in the β globins. Neither of the two introns any longer possesses recognizable boundaries with the exons, so probably the introns could not be spliced out even if the gene were transcribed. However, there are no transcripts corresponding to the gene, possibly because there have been changes in the 5′ flanking region.

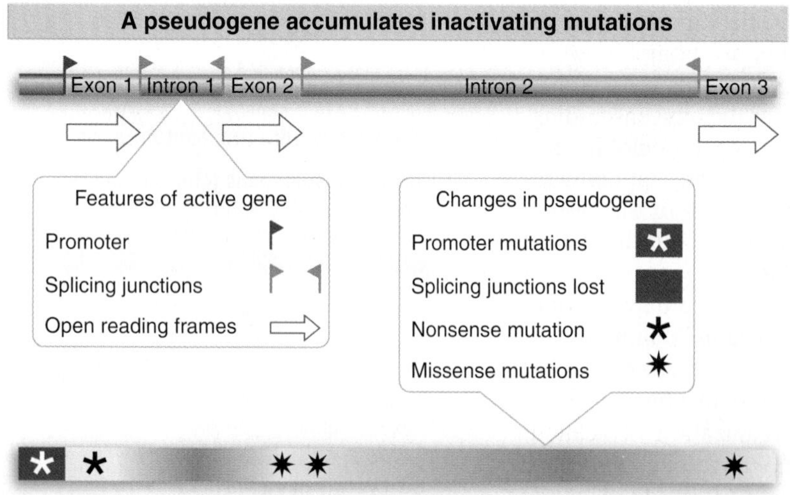

Figure 3.22 Many changes have occurred in a β globin gene since it became a pseudogene.

Because this list of defects includes mutations potentially preventing each stage of gene expression, we have no means of telling which event originally inactivated this gene. However, from the divergence between the pseudogene and the functional gene, we can estimate when the pseudogene originated and when its mutations started to accumulate. It seems that ψβ2 started out as an active gene when it diverged from the β1 gene, but was later inactivated to become a pseudogene.

In some cases, pseudogenes appear always to have been inactive. One example is the mouse ψα3 globin gene, which has an interesting property: it precisely lacks both introns. Its sequence can be aligned (allowing for accumulated mutations) with the α-globin mRNA. The apparent time of inactivation coincides with the original duplication, which suggests that the loss of introns was a part of the original inactivating event.

Inactive genomic sequences that resemble the RNA transcript are called **processed pseudogenes**. They originate by insertion at some random site of a product derived from the RNA (see *Chapter 22 Retroviruses and Retroposons*). Their characteristic features are summarized in Figure 22.18.

If pseudogenes are evolutionary dead ends, simply an unwanted accompaniment to the rearrangement of functional genes, why are they still present in the genome? Do they fulfill any function or are they entirely without purpose, in which case there should be no selective pressure for their retention?

We should remember that we see only those genes that have survived in present populations. In past times, any number of other pseudogenes may have been eliminated. This elimination could occur by deletion of the sequence as a sudden event or by the accretion of mutations to the point where the pseudogene can no longer be recognized as a member of its original sequence family (probably the ultimate fate of any pseudogene that is not suddenly eliminated).

3.12 SUMMARY

All types of eukaryotic genomes contain interrupted genes. The proportion of interrupted genes is low in yeasts and increases in other fungi; few genes are uninterrupted in higher eukaryotes.

Introns are found in all classes of eukaryotic genes. An interrupted gene contains exons that are joined together in RNA in the same order as in the DNA, and the introns usually have no coding function. Introns are removed from RNA by splicing. Some genes are expressed by alternative splicing patterns, in which a particular sequence is removed as an intron in some situations, but retained as an exon in others.

Positions of introns are often seen to be conserved when the organization of homologous genes is compared between species. Intron sequences vary between species, and may even be unrelated, although exon sequences remain well related. The conservation of exons can be used to isolate related genes in different species.

The length of a gene is determined primarily by the lengths of its introns. Early in the evolution of higher eukaryotes, introns became longer and gene lengths therefore increased as well. The range of gene lengths in mammals is generally from 1 to 100 kb, but some genes are even longer; the longest known is the dystrophin gene at 2000 kb.

Some genes share only some of their exons with other genes, suggesting that they have been assembled by addition of exons representing individual modules of the protein. Such modules may have been incorporated into a variety of different proteins. The idea that genes have been assembled by accretion of exons implies that introns were present in genes of primitive organisms. Some of the differences between homologous genes can be explained by loss of introns from the primordial genes, with different introns being lost in different lines of descent.

The Content of the Genome

4.1 Introduction

The key question about the genome is how many genes it contains. We can think about the total number of genes at four levels, corresponding to successive stages in gene expression:

- The *genome* is the complete set of genes of an organism. Ultimately, it is defined by the complete DNA sequence, although as a practical matter it may not be possible to identify every gene unequivocally solely on the basis of sequence.

- The *transcriptome* is the set of expressed genes. It is defined in terms of the RNA molecules that are encoded, and can refer to a single cell type or to any more complex assembly of cells up to the complete organism. Because some genes generate multiple mRNAs, the transcriptome is likely to be larger than the number of genes defined directly in the genome. The transcriptome includes noncoding RNAs as well as mRNAs.

- The *proteome* is the complete set of proteins. It should correspond to the mRNAs in the transcriptome, although there can be differences of detail reflecting changes in the relative abundance or stabilities of mRNAs and proteins. It can be used to refer to the set of proteins coded by the whole genome or produced in any particular cell or tissue.

- Proteins may function independently or as part of multiprotein assemblies. If we could identify all protein-protein interactions, we could define the total number of independent assemblies of proteins.

The number of genes in the genome can be identified directly by defining open reading frames. Large scale mapping of this nature is complicated by the fact that interrupted genes may consist of many separated open reading frames. Because we do not necessarily have information about the functions of the protein products, or indeed proof that the open reading frames are expressed at all, this approach is restricted to defining the *potential* of the genome. However, a strong presumption exists that any conserved open reading frame is likely to be expressed.

Another approach is to define the number of genes in terms of the transcriptome (by directly identifying all the mRNAs) or proteome (by directly identifying all the proteins). This allows us to ask how many genes are expressed in a particular tissue or cell type, what variation exists in the relative levels of expression, and how many of the genes expressed in one particular cell type are unique to that cell type and how many are also expressed elsewhere.

And we must remember, of course, that not all eukaryotic genes are in the nucleus. A small amount of additional information is carried in the DNA of the mitochondrion, and in plant cells a chloroplast carries somewhat more information.

Concerning the types of genes, we may ask whether a particular gene is essential: what happens to a null mutant? If a null mutation is lethal, or the organism has a visible defect, we may conclude that the gene is essential or at least conveys a selective advantage. But some genes can be deleted without apparent effect on the phenotype. Are these genes really dispensable, or does a selective disadvantage result from the absence of the gene, perhaps in other circumstances, or over longer periods of time?

4.2 Genomes Can Be Mapped by Linkage, Restriction Cleavage, or DNA Sequence

Key Concepts

- The key in mapping a genome physically is to use overlapping fragments of sequences to permit unequivocal joining of adjacent segments.

- To relate the physical map to the genetic linkage map requires placing mutations on the physical map.

Defining the contents of a genome essentially means making a map. We can think about mapping genes and genomes at several levels of resolution.

A genetic (or linkage) map identifies the distance between mutations in terms of recombination frequencies. It is limited by its reliance on mutations that affect the phenotype. Because recombination frequencies can be distorted relative to the physical distance between sites, a genetic map does not accurately represent physical distances along the genetic material. **Figure 4.1** summarizes the resolution of genetic mapping in *Drosophila*, where 270 map units, or centimorgans (cM), correspond to ~10^8 bp. In effect, genetic linkage can be established between mutations within ~25 cM, or half a chromosome arm apart; a mutation can be localized within ~0.5 cM, or <200 kb of DNA.

A linkage map can also be constructed by measuring recombination between sites in genomic DNA. These sites have sequence variations that generate differences in the susceptibility to cleavage by certain (restriction) enzymes. Because such variations are common, such a map can be prepared for any organism without using mutations. It has the same disadvantage as any linkage map that the relative distances are based on recombination. In practical terms, its resolution is

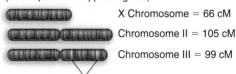

Genetic map resolution in *Drosophila* is 0.5–25 cM

Total genome = 1.4 × 10^8 bp
(~10^8 bp in the mapped regions)

X Chromosome = 66 cM

Chromosome II = 105 cM

Chromosome III = 99 cM

~25% recombination occurs
between sites separated
by half a chromosome arm   1 cM = 370,000 bp

Figure 4.1 *Drosophila* genes within ~25 cM of each other can be mapped by mutational crosses with ~0.5 cM accuracy.

better because the density of sites is greater than the density of observable mutations, but such a map is still limited by the fact that recombination between closely located sites is very rare.

A restriction map is constructed by cleaving DNA into fragments with restriction enzymes and measuring the distances between the sites of cleavage. This represents distances in terms of the length of DNA, so it provides a physical map of the genetic material. We might view a restriction map as beginning where the genetic map leaves off—a region of (say) 1 cM could be represented in a local restriction map.

The ultimate map is the sequence of the DNA. From the sequence, we can identify genes and the distances between them. By analyzing the protein-coding potential of a sequence of the DNA, we can deduce whether it represents a protein. The basic assumption here is that natural selection prevents the accumulation of damaging mutations in sequences that code for proteins. Reversing the argument, we may assume that an intact coding sequence is likely to be used to generate a protein.

A restriction map depends *only* on DNA sequence, without regard to the function of the sequence. For it to be related to the genetic map, mutations have to be characterized in terms of their effects upon the restriction sites. Large changes in the genome can be recognized because they affect the sizes or numbers of restriction fragments. Point mutations are more difficult to detect. Ultimately, by comparing the sequence of a wild-type DNA with that of a mutant allele, we can determine the nature of a mutation and its exact site. This defines the relationship between the genetic map (based on mutation sites) and the physical map (based on the sequence of DNA).

Similar techniques are used to identify and sequence genes and to map the genome, although there is of course a difference of scale. In each case, the principle is to obtain a series of overlapping fragments of DNA, which can be connected into a continuous map. The crucial feature is that each segment is related to the next segment on the map by characterizing the overlap between them, so that we can be sure no segments are missing. This principle is applied both at the level of ordering large fragments into a map, and in connecting the sequences that make up the fragments.

4.3 Individual Genomes Show Extensive Variation

Key Terms

- **Polymorphism** (more fully, genetic polymorphism) refers to the simultaneous occurrence in the population of genomes showing variations at a given position. The original definition applied to alleles producing different phenotypes. Now it is also used to describe changes in DNA affecting the restriction pattern or even the sequence. For practical purposes, to be considered as an example of a polymorphism, an allele should be found at a frequency >1% in the population.

- **Single nucleotide polymorphism (SNP)** is a polymorphism (variation in sequence between individuals) caused by a change in a single nucleotide. This is responsible for most of the genetic variation between individuals.

- **Restriction fragment length polymorphism (RFLP)** refers to inherited differences in cleavage sites for restriction enzymes (caused, for example, by base changes in the target site) that result in differences in the lengths of the fragments produced by cleavage with the relevant restriction enzyme. RFLPs are used for genetic mapping to link the genome directly to a conventional genetic marker.

Key Concepts

- Polymorphism may be detected at the phenotypic level when a sequence affects gene function, at the restriction fragment level when it affects a restriction enzyme target site, and at the sequence level by direct analysis of DNA.

- The alleles of a gene show extensive polymorphism at the sequence level, but many sequence changes do not affect function.

The coexistence of multiple alleles at a locus is called genetic **polymorphism**. Any site at which multiple alleles exist as stable components of the population is by definition polymorphic. An allele is usually defined as polymorphic if it is present at a frequency >1% in the population.

What is the basis for polymorphism among mutant alleles? They possess different mutations that alter the protein function, thus producing changes in phenotype. If we compare the restriction maps or the DNA sequences of these alleles, they too will be polymorphic in the sense that each map or sequence differs from the others.

When we consider the genome at the level of restriction sites or sequence, polymorphism is extensive. The wild-type population may include many differences in sequence that do not affect function, and which therefore do not produce phenotypic variants. Many different sequence variants may exist at a given locus; some of them are evident because they affect the phenotype, but others are hidden because they have no visible effect. There may be a continuum of changes at a locus, including those that change DNA sequence but do not change protein sequence, those that change protein sequence without changing function, those that create proteins with different activities, and those that create mutant proteins that are nonfunctional.

A change in a single nucleotide when alleles are compared is called a **single nucleotide polymorphism (SNP)**. One occurs every ~1330 bases in the human genome. Defined by their SNPs, every human being is unique.

One aim of genetic mapping is to obtain a catalog of common variants. The observed frequency of SNPs per genome predicts that, over the human population as a whole (taking the sum of all human genomes of all living individuals), there should be >10 million SNPs at a frequency >1%. Already >1 million have been identified.

Some polymorphisms in the genome can be detected by comparing the restriction maps of different individuals. The criterion is a change in the pattern of fragments produced by cleavage with a restriction enzyme. **Figure 4.2** shows that when a target site is present in the genome of one individual and absent from another, the extra cleavage in the first genome generates two fragments corresponding to a single fragment in the second genome.

Because the restriction map is independent of gene function, a polymorphism at this level can be detected *irrespective of whether the sequence change affects the phenotype*. Probably very few of the restriction site polymorphisms in a genome actually affect the phenotype. Most are sequence changes that have no effect on the production of proteins (for example, because they lie between genes).

A difference in restriction maps between two individuals is called a **restriction fragment length polymorphism (RFLP)**. Basically, an RFLP is an SNP that is located in the target site for a restriction enzyme. It can be used as a genetic marker in exactly the same way as any other marker. Instead of examining some feature of the phenotype, we directly assess the genotype, as revealed by the restriction map. **Figure 4.3** shows a pedigree of a restriction polymorphism followed through three generations. It displays Mendelian segregation at the level of DNA marker fragments.

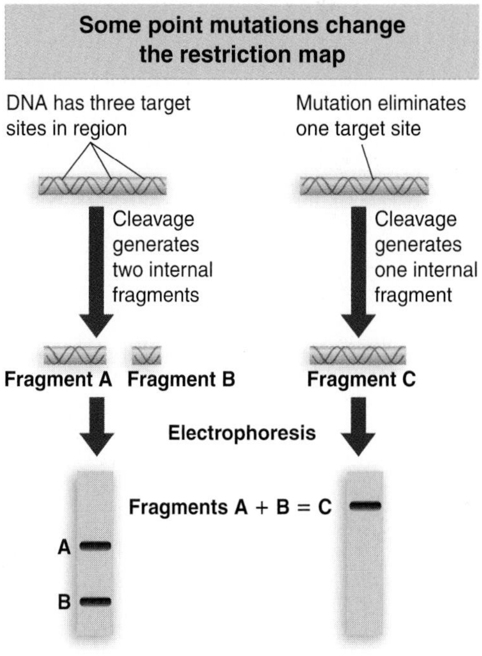

Figure 4.2 A point mutation that affects a restriction site is detected by a difference in restriction fragments.

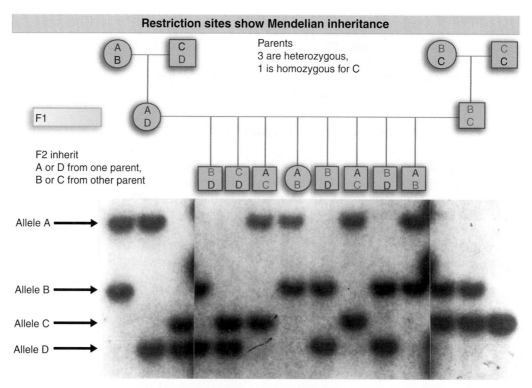

Figure 4.3 Restriction site polymorphisms are inherited according to Mendelian rules. Four alleles for a restriction marker are found in all possible pairwise combinations and segregate independently at each generation. Photograph kindly provided by Raymond L. White/Eccles Institute of Human Genetics.

4.4 RFLPs and SNPs Can Be Used for Genetic Mapping

Key Terms

- The **haplotype** is the particular combination of alleles in a defined region of some chromosome, in effect the genotype in miniature. Originally used to described combinations of MHC alleles, it now may be used to describe particular combinations of RFLPs, SNPs, or other markers.
- **DNA fingerprinting** analyzes the differences between individuals of the fragments generated by using restriction enzymes to cleave regions that contain short repeated sequences. Because these are unique to every individual, the presence of a particular subset in any two individuals can be used to define their common inheritance (for example, a parent–progeny relationship).

Key Concept

- RFLPs and SNPs can be the basis for linkage maps and are useful for establishing parent–progeny relationships.

Because restriction markers are not restricted to those genome changes that affect the phenotype, they provide the basis for an extremely powerful technique for identifying genetic loci at the molecular level. A typical problem concerns a mutation with known effects on the phenotype, where the relevant genetic locus can be placed on a genetic map, but for which we have no knowledge about the corresponding gene or protein. Many damaging or fatal human diseases fall into this category. For example, cystic fibrosis shows Mendelian inheritance, but the molecular nature of the mutant function was unknown until it could be identified as a result of characterizing the gene.

Recombination frequency can be measured between a restriction marker and a visible phenotypic marker, as illustrated in **Figure 4.4**. So a genetic map can include both genotypic and phenotypic markers.

If restriction polymorphisms are distributed at random in the genome, there should be some near any particular target gene. We can identify such restriction markers by virtue of their tight linkage to the mutant phenotype. If we compare the restriction map of DNA from patients suffering from a disease with the DNA of unaffected people, we may find that a particular restriction site is always present (or always absent) from the patients, and we conclude that the site is near the gene responsible for the disease phenotype.

A hypothetical example is shown in **Figure 4.5**. In the situation depicted, there is 100% linkage between the restriction marker and the phenotype, which implies that the restriction marker lies so close to the mutant gene that it is never separated from it by recombination.

Linkage of genetic markers to RFLPs or SNPs allows unknown genes to be located on the map. RFLP mapping in both man and mouse established maps for both genomes that have been used for this purpose. With a great enough density of RFLPs, any unknown marker can be rapidly located by linkage to the closest RFLPs. And now knowledge of the high frequency of SNPs in the human genome makes them even more effective than RFLPs for genetic mapping. From the 1.4×10^6 SNPs that have already been identified, there is on average an SNP every 1–2 kb. This should allow rapid localization of new disease genes by locating them between the nearest SNPs.

The identification of an RFLP that is linked to a disease has two important consequences:

- It may offer a diagnostic procedure for detecting the disease. Some of the human diseases that are genetically well characterized but ill defined in molecular terms cannot be easily diagnosed. If a restriction marker is reliably linked to the phenotype, then its presence can be used to diagnose the disease.
- It may lead to isolation of the gene. The restriction marker must lie relatively near the gene on the genetic map if the two loci rarely or never recombine. Although "relatively near" in genetic terms can be a substantial distance in base pairs of DNA, nonetheless it provides a starting point from which we can proceed along the DNA to the gene itself.

The frequency of polymorphism means that every individual has a unique constellation of SNPs or RFLPs. The particular combination of sites found in a specific region is called a **haplotype**, a genotype in miniature. It describes the particular combination of alleles or restriction sites (or any other genetic markers) present in some defined area of the genome.

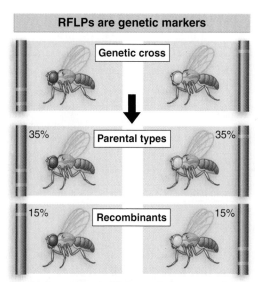

RFLPs are genetic markers

Genetic cross

Parental types   35%   35%

Recombinants   15%   15%

Restriction marker is 30 map units from eye color marker

Figure 4.4 A restriction polymorphism can be used as a genetic marker to measure recombination distance from a phenotypic marker (such as eye color). The figure simplifies the situation by showing only the DNA bands corresponding to the allele of one genome in a diploid.

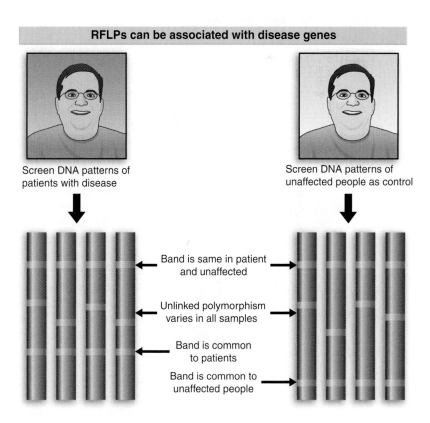

RFLPs can be associated with disease genes

Screen DNA patterns of patients with disease

Screen DNA patterns of unaffected people as control

Band is same in patient and unaffected

Unlinked polymorphism varies in all samples

Band is common to patients

Band is common to unaffected people

Figure 4.5 The mutation changing the band that is common in unaffected people into the band that is common in patients must be very closely linked to the disease gene.

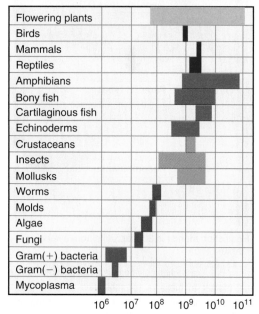

Figure 4.6 DNA content of the haploid genome increases with morphological complexity of lower eukaryotes, but varies extensively within some groups of higher eukaryotes.

The existence of RFLPs provides the basis for a technique to identify parent–progeny relationships unequivocally. In cases where parentage is in doubt, a comparison of the RFLP map in a suitable chromosome region between potential parents and child allows absolute assignment of relationship. The use of DNA restriction analysis to identify individuals has been called **DNA fingerprinting**. Analysis of especially variable "minisatellite" sequences is used for defining human relationships. (see *6.12 Minisatellites Are Useful for Genetic Mapping*).

4.5 Why Are Genomes So Large?

Key Terms

- The **C-value** is the total amount of DNA in the genome (per haploid set of chromosomes).
- The **C-value paradox** describes the lack of relationship between the DNA content (C-value) of an organism and its coding potential.

Key Concepts

- Minimum genome size increases with increasing complexity of organisms.
- There are wide variations in the genome sizes of organisms within many phylogenic groups, and little correlation between genome size and genetic complexity.

The total amount of DNA in the (haploid) genome is a characteristic of each living species known as its **C-value**. There is enormous variation in the range of C-values, from $<10^6$ bp for a mycoplasma to $>10^{11}$ bp for some plants and amphibians. **Figure 4.6** summarizes the range of C-values found in different evolutionary groups.

Plotting the *minimum* amount of DNA required for a member of each group of organisms suggests in **Figure 4.7** that increased genome size is associated with increased complexity.

Mycoplasma are the smallest prokaryotes, and have genomes only ~3× the size of a large bacteriophage. Bacteria start at ~2×10^6 bp. Unicellular eukaryotes (whose lifestyles may resemble those of prokaryotes) get by with genomes that are also small, although larger than those of the bacteria. Being eukaryotic *per se* does not imply a vast increase in genome size; a yeast may have a genome size of ~1.3×10^7 bp, only about twice the size of an average bacterial genome.

A further twofold increase in genome size is adequate to support the slime mold *Dictyostelium discoideum*, which is able to live in either unicellular or multicellular modes. Another increase in genome size characterizes the simplest fully multicellular organisms; the nematode worm *Caenorhabditus elegans* has a DNA content of 8×10^7 bp.

As we saw in Figure 4.6, genome sizes of insects, birds, amphibians, and mammals are greater than those of lower eukaryotes. Among higher eukaryotes there is no good relationship between genome size and morphological complexity of the organism, and within some of these groups genome size varies by an order of magnitude or more. **Figure 4.8** lists the genome sizes of some species of the most commonly analyzed organisms.

We know that genes are much larger than the sequences needed to code for proteins, because exons (coding regions) may occupy only a small part of the total length of a gene. And there may also be significant

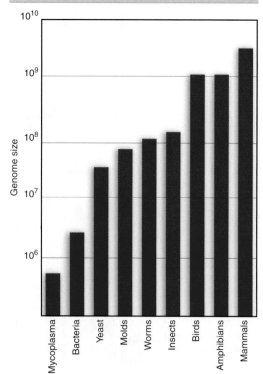
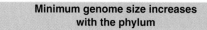

Figure 4.7 The minimum genome size found in each group increases from prokaryotes to mammals.

lengths of DNA between genes. So it is not possible to deduce anything about the number of genes from the overall size of the genome.

The **C-value paradox** refers to the lack of direct correlation between genome size and genetic complexity. There are some extremely curious variations in relative genome size. Toads of the species *Xenopus laevis* and humans have genomes of essentially the same size. But we assume that humans are more complex in terms of genetic development! And in some groups there are extremely large variations in DNA content between organisms that do not vary much in complexity (see Figure 4.6). For example, in amphibians, the smallest genomes are $<10^9$ bp, while the largest are $\sim 10^{11}$ bp, but there is unlikely to be a large difference in the number of genes needed to code for the development of the organisms. We do not know whether there are evolutionary consequences of this variation.

Useful genome sizes		
Phylum	Species	Genome (bp)
Algae	*Pyrenomas salina*	6.6×10^5
Mycoplasma	*M. pneumoniae*	1.0×10^6
Bacterium	*E. coli*	4.2×10^6
Yeast	*S. cerevisiae*	1.3×10^7
Slime mold	*D. discoideum*	5.4×10^7
Nematode	*C. elegans*	8.0×10^7
Insect	*D. melanogaster*	1.8×10^8
Bird	*G. domesticus*	1.2×10^9
Amphibian	*X. laevis*	3.1×10^9
Mammal	*H. sapiens*	3.3×10^9

Figure 4.8 The genome sizes of some common experimental organisms.

4.6 Eukaryotic Genomes Contain Both Nonrepetitive and Repetitive DNA Sequences

Key Terms

- **Nonrepetitive DNA** shows reassociation kinetics expected of unique sequences.
- **Moderately repetitive DNA sequences** are repeated in the haploid genome, usually in related rather than identical copies.
- **Highly repetitive DNA (simple sequence DNA)** consists of short, tandemly repeated, sequences.
- A **transposon (transposable element)** is a DNA sequence able to insert itself (or a copy of itself) at a new location in the genome, without having any sequence relationship with the target locus.
- **Selfish DNA** describes sequences that have no effect on the genotype of the organism but have self-perpetuation within the genome as their sole function.

Key Concepts

- The eukaryotic genome contains nonrepetitive, moderately repetitive, and highly repetitive DNA sequences.
- Genes are generally coded by sequences in nonrepetitive DNA.
- Larger genomes within a taxonomic group do not contain more genes, but have large amounts of repetitive DNA.
- A large part of repetitive DNA may be made up of transposons.

We can divide the eukaryotic genome into three general types of sequences:

- **Nonrepetitive sequences** are unique; the haploid genome contains only one copy of such a sequence.
- **Moderately repetitive sequences** are found in multiple copies. The copies may be identical to one another or, more typically, may be seen to be related although not identical.
- **Highly repetitive sequences** are very short and are present in large numbers of copies, often organized as tandem repeats. These include very large blocks of material organized as satellite DNA and smaller blocks that are called minisatellites or microsatellites, depending on their length.

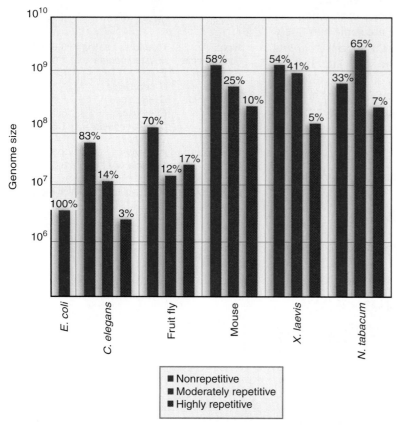

Nonrepetitive DNA is only part of the genome

- Nonrepetitive
- Moderately repetitive
- Highly repetitive

Figure 4.9 The proportions of different sequence types vary among eukaryotic genomes. The absolute content of nonrepetitive DNA increases with genome size, but reaches a plateau at ~2 × 10⁹ bp.

The proportion of the genome occupied by nonrepetitive DNA varies widely. **Figure 4.9** summarizes the genome organization of some representative organisms. Prokaryotes contain only nonrepetitive DNA. In lower eukaryotes, most of the DNA is nonrepetitive; <20% is repetitive. In many animal cells, up to half of the DNA is moderately or highly repetitive. In plants and amphibians, the moderately and highly repetitive DNA may account for up to 80% of the genome, so that the nonrepetitive DNA is a minority component.

A significant part of the moderately repetitive DNA consists of **transposons**, short sequences of DNA (~1 kb) that have the ability to move to new locations in the genome and/or to make additional copies of themselves. In some higher eukaryotic genomes, transposons occupy more than half of the genome (see *Chapter 21 Transposons* and *Chapter 22 Retroviruses and Retroposons*). Transposons are sometimes viewed as fitting the concept of **selfish DNA**, defined as sequences that propagate themselves within a genome without contributing to the development of the organism. Transposons may sponsor genome rearrangements, and these could confer selective advantages, but it is fair to say that we do not really understand why selective forces do not act against transposons becoming such a large proportion of the genome.

The length of the nonrepetitive DNA component tends to increase with overall genome size, as we proceed up to a total genome size ~3 × 10⁹ (characteristic of mammals). Further increase in genome size, however, generally reflects an increase in the amount and proportion of the repetitive components, so that it is rare for an organism to have a nonrepetitive DNA component >2 × 10⁹. The nonrepetitive DNA content of genomes therefore accords better with our sense of the relative complexity of the organism. *E. coli* has 4.2 × 10⁶ bp, *C. elegans* increases an order of magnitude to 6.6 × 10⁷ bp, *D. melanogaster* increases further to ~10⁸ bp, and mammals increase another order of magnitude to ~2 × 10⁹ bp.

4.7 Genes Can Be Isolated by the Conservation of Exons

Key Term

- A **zoo blot** describes the use of Southern blotting to test the ability of a DNA probe from one species to hybridize with DNA from other species.

Key Concept

- Conservation of exons can be used as the basis for identifying coding regions by identifying fragments whose sequences are present in multiple organisms.

The advent of large-scale sequencing has sharpened the question of how we can identify a gene solely on the basis of its sequence, that is, without information about its expressed products. Bacterial genes are relatively simple to identify. The transcription of mRNA is initiated at a

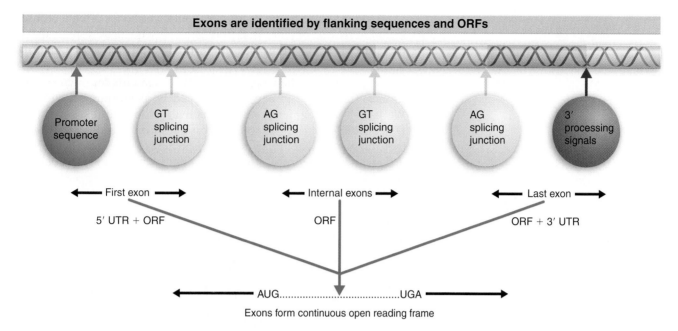

Figure 4.10 Exons of protein-coding genes are identified as coding sequences flanked by appropriate signals (with untranslated regions at both ends). The series of exons must generate an open reading frame with appropriate initiation and termination codons.

site (a promoter) that conforms to certain sequence requirements. Transcription terminates at a specific site that again fits specific sequence requirements. Within the mRNA sequence there are one or more open reading frames, each starting with an initiation signal (the nucleotide triplet AUG) and ending with a termination triplet (UAA, UAG, or UGA). Each open reading frame is preceded by a ribosome-binding sequence. Finding any one of these features is not sufficient, but their combination in the right order almost certainly identifies a functional gene. All of these elements of a bacterial gene are found within a relatively short distance, typically of the order of 1000 bp.

Eukaryotic genes that are interrupted have additional features. **Figure 4.10** summarizes the features we look for in order to identify such a gene. There is variability in the sequence of the promoter at the start of the gene, but in principle potential promoters can be identified by their sequence. Although there is usually no specific sequence for terminating transcription in mRNA-coding genes, they have specific sequences that are responsible for generating the 3′ end of the transcript. The promoter and terminator sequences may be very far apart, however, and between them may be a large number of individual exons. The sequences of the exons can be connected to give a continuous open reading frame. In the simplest cases, the first and last exons contain the start and end of the coding region, respectively (as well as the 5′ and 3′ untranslated regions), but in more complex cases the first or last exons may have only untranslated regions, and may therefore be more difficult to identify. The coding sequence starts with an initiation codon, and ends with a termination codon, just as in a bacterial gene. The introns between the exons start and end with very short specific sequences called splicing junctions.

Because an interrupted gene can be very long, the key to its identification is picking out the exons. Some major approaches to identifying genes are based on the contrast between the conservation of exons and the variation of introns. In a region containing a gene whose function has been conserved among a range of species, the sequence representing the protein should have two distinctive properties:

- It must have an open reading frame.
- It is likely to have a related sequence in more than one species.

The best guide to the validity of a putative exon is often its conservation in related organisms. Before genome sequences were available, the hunt for genes of medical importance often started by identifying genome fragments that cross-hybridize with the genomes of other species, and then examining these fragments for open reading frames. Conservation can be examined by performing a **zoo blot**. We use short fragments as (radioactive) probes to test for related DNA from a variety of species by Southern blotting. If we find hybridizing fragments in several species related to that of the probe—the probe is usually human—the probe is likely to contain a conserved exon.

4.8 Genes Involved in Diseases Can Be Identified by Comparing Patient DNA with Normal DNA

Key Concept

- Human disease genes are identified by mapping and sequencing DNA of patients to find differences from normal DNA that are genetically linked to the disease.

When a human disease is caused by a mutation in a known protein, the gene that is responsible can be identified because it codes for the protein, and its responsibility for the disease can be confirmed by showing that it has inactivating mutations in the DNA of patients but not in normal DNA. However, in many cases we do not know the cause of a disease at the molecular level, and it is necessary to identify the gene without any information about its protein product.

The basic criterion for identifying a gene involved in a human disease is to show that it has an inactivating mutation in every patient with the disease that is not present in normal DNA. However, the extensive polymorphism between individual genomes means that we may find many changes when we compare patient DNA with normal DNA. Before the sequencing of the human genome, genetic linkage could be used to identify a region containing a disease gene, but the region could contain many candidate genes. For a very large gene, with introns spread over a long distance of the genome, it was difficult to identify the critical mutations in patients. The availability of high-resolution SNP maps and of the genome sequence now make it much easier to pinpoint a smaller region containing the gene in which sequences of normal and patient DNA can be directly compared.

An example of the process by which a disease gene can be tracked down is provided by Duchenne muscular dystrophy (DMD), a degenerative disorder of muscle, which is X-linked and affects 1 in 3500 of human male births. The steps in identifying the gene are summarized in **Figure 4.11**.

In the first step, linkage analysis localized the DMD locus to chromosomal band Xp21. Patients with the disease often have chromosomal rearrangements involving this band. By comparing the ability of X-linked DNA probes to hybridize with DNA from patients and with normal DNA, cloned fragments were obtained that correspond to the region that was rearranged or deleted in patients' DNA.

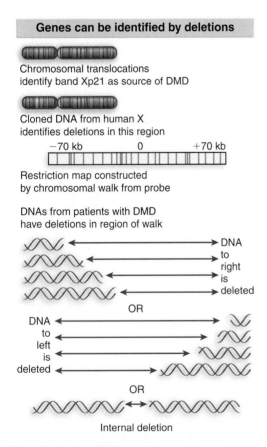

Genes can be identified by deletions

Chromosomal translocations
identify band Xp21 as source of DMD

Cloned DNA from human X
identifies deletions in this region

−70 kb   0   +70 kb

Restriction map constructed
by chromosomal walk from probe

DNAs from patients with DMD
have deletions in region of walk

DNA
to
right
is
deleted

OR

DNA
to
left
is
deleted

OR

Internal deletion

Figure 4.11 The gene involved in Duchenne muscular dystrophy was tracked down by chromosome mapping and walking to a region in which deletions can be identified with the occurrence of the disease.

Once some DNA in the general vicinity of a target gene has been obtained, it is possible to "walk" along the chromosome until the gene is reached. In the second step of identifying the DMD gene, a chromosomal walk was used to construct a restriction map of the region on either side of the probe, covering a region of >100 kb. Finally, analysis of the DNA from a series of patients identified large deletions in this region, most of which extended beyond an end of the mapped region. The most telling deletion is one contained entirely within the region, since this delineates a segment that must be important in gene function and indicates that the gene, or at least part of it, lies in this region.

Having come into the region of a target gene, we need to identify its exons and introns. A DMD zoo blot identified fragments that cross-hybridize with the mouse X chromosome and with other mammalian DNAs. As summarized in **Figure 4.12**, these were scrutinized for open reading frames and the sequences typical of exon-intron junctions. Fragments that met these criteria were used as probes to identify homologous sequences in a cDNA library prepared from muscle mRNA.

The cDNA corresponding to the gene identifies an unusually large mRNA, ~14 kb. Hybridization back to the genome shows that the mRNA is represented in >60 exons, which are spread over ~2000 kb of DNA. This makes DMD the longest gene identified.

The gene codes for a protein of ~500 kD, called dystrophin, which is a component of muscle, present in rather low amounts. All patients with the disease have deletions at this locus and lack (or have defective) dystrophin.

Muscle also has the distinction of having the largest known protein, titin, with almost 27,000 amino acids. Its gene has the largest number of exons (178) and the longest single exon in the human genome (17,000 bp).

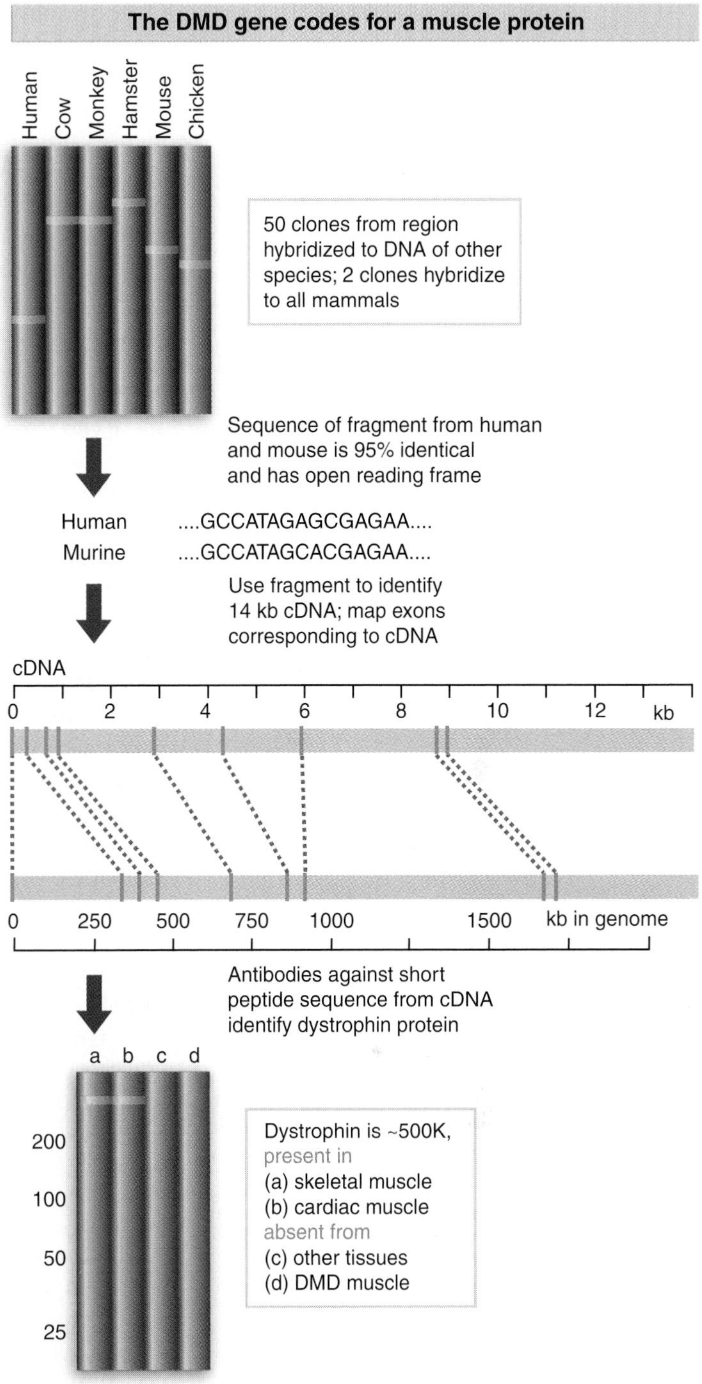

Figure 4.12 The Duchenne muscular dystrophy gene was characterized by zoo blotting, cDNA hybridization, genomic hybridization, and identification of the protein.

4.9 The Conservation of Genome Organization Helps to Identify Genes

Key Concepts

- Algorithms for identifying genes are not perfect and many corrections must be made to the initial data set.
- Pseudogenes must be distinguished from active genes.
- Syntenic relationships are extensive between mouse and human genomes, and most active genes are in a syntenic region.

Key Term

- **Synteny** describes a relationship between chromosomal regions of different species in which homologous genes occur in the same order.

Once we have assembled the sequence of a genome, we still have to identify the genes within it. Coding sequences represent a very small fraction. Exons can be identified as uninterrupted open reading frames flanked by appropriate sequences. What criteria need to be satisfied to identify an active gene from a series of exons?

The algorithms that are used to connect exons are not completely effective when the genome is very large and the exons may be separated by very large distances. For example, the initial analysis of the human genome mapped 170,000 exons into 32,000 genes. This is unlikely to be correct, because it gives an average of 5.3 exons per gene, whereas the average of individual genes that have been fully characterized is 10.2. Either we have missed many exons, or they should be connected differently into a smaller number of genes in the whole genome sequence.

Even when the organization of a gene is correctly identified, there is the problem of distinguishing active genes from pseudogenes. Many pseudogenes can be recognized by obvious defects in the form of multiple mutations that create an inactive coding sequence. However, pseudogenes that have arisen more recently, and which have not accumulated so many mutations, may be more difficult to recognize. In an extreme example, the mouse has only one active *Gapdh* gene (coding for glyceraldehyde phosphate dehydrogenase), but has ~400 pseudogenes. However, >100 of these pseudogenes initially appeared to be active in the mouse genome sequence. Individual examination was necessary to exclude them from the list of active genes.

Confidence that a gene is active can be increased by comparing regions of the genomes of different species. There has been extensive overall reorganization of sequences between the mouse and human genomes, as seen in the simple fact that there are 23 chromosomes in the human haploid genome and 20 chromosomes in the mouse haploid genome. However, at the local level, the order of genes is generally the same: when pairs of human and mouse homologues are compared, the genes located on either side also tend to be homologues. This relationship is called **synteny**.

Figure 4.13 shows the relationship between mouse chromosome 1 and the human chromosomal set. We can recognize 21 segments in this mouse chromosome that have syntenic counterparts in human chromosomes. The extent of reshuffling that has occurred between the genomes is shown by the fact that the segments are spread among six different human chromosomes. The same types of relationships are found in all mouse chromosomes, except for the X chromosome, which is syntenic only with the human X chromosome. This is explained by the fact that the X is a special case, subject to dosage compensation to adjust for the difference between males (one copy) and females (two copies) (see *31.4 X Chromosomes Undergo Global Changes*). This may apply selective pressure against the translocation of genes to and from the X chromosome.

Comparison of the mouse and human genome sequences shows that >90% of each genome lies in syntenic blocks that range widely in size (from 300 kb to 65 Mb). There is a total of 342 syntenic segments, with an average length of 7 Mb (0.3% of the genome). 99% of mouse genes have a homologue in the human genome; and for 96% that homologue is in a syntenic region.

A validation of the importance of syntenic blocks comes from pairwise comparisons of the genes within them. Looking for likely pseudogenes on the basis of sequence comparisons, a gene that is not in a syntenic location (that is, its context is different in the two species) is twice as likely to be a pseudogene. Put another way, translocation away

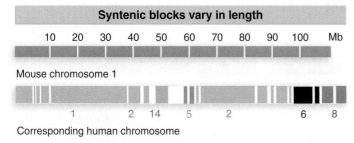

Figure 4.13 Mouse chromosome 1 has 21 segments of 1–25 Mb that are syntenic with regions corresponding to parts of 6 human chromosomes.

from the original locus tends to be associated with the creation of pseudogenes. The lack of a related gene in a syntenic position is therefore grounds for suspecting that an apparent gene may really be a pseudogene. Overall, >10% of the genes that are initially identified by analysis of the genome are likely to turn out to be pseudogenes.

As a general rule, comparisons between genomes add significantly to the effectiveness of gene prediction. When sequence features indicating active genes are conserved, for example, between man and mouse, there is an increased probability that they identify active homologues.

4.10 Organelles Have DNA

Key Concepts

- Mitochondria and chloroplasts have genomes that show non-Mendelian inheritance. Typically they are maternally inherited.
- Organelle genomes may undergo somatic segregation in plants.
- Comparisons of mitochondrial DNA suggest that humans are descended from a single female who lived 200,000 years ago in Africa.

Key Terms

- **Maternal inheritance** describes the preferential survival in the progeny of genetic markers provided by one parent.
- **Extranuclear genes** reside outside the nucleus in organelles such as mitochondria and chloroplasts.

The first evidence for the presence of genes outside the nucleus was provided by non-Mendelian inheritance in plants (observed in the early years of this century, just after the rediscovery of Mendelian inheritance). The extreme form of non-Mendelian inheritance is uniparental inheritance, when the genotype of only one parent is inherited and that of the other parent is permanently lost. In less extreme examples, the progeny of one parental genotype exceed those of the other genotype. Usually it is the mother whose genotype is preferentially (or solely) inherited. This effect is sometimes described as **maternal inheritance**. The important point is that the genotype contributed by the parent of one particular sex predominates, as seen in abnormal segregation ratios when a cross is made between mutant and wild type. This contrasts with the behavior of Mendelian genetics in which reciprocal crosses show the contributions of both parents to be equally inherited.

Non-Mendelian inheritance results from the presence in mitochondria and chloroplasts of DNA genomes that are inherited independently of nuclear genes. In effect, the organelle genome comprises a length of DNA that has been physically sequestered in a defined part of the cell, and is subject to its own form of expression and regulation. An organelle genome typically codes for some or all of the RNAs needed to perpetuate the organelle, but codes for only some of the needed proteins. The other proteins are coded in the nucleus, expressed via the cytoplasmic protein synthetic apparatus, and imported into the organelle.

Genes not residing within the nucleus are generally described as **extranuclear genes**; they are transcribed and translated in the *same* organelle compartment (mitochondrion or chloroplast) in which they reside. By contrast, *nuclear* genes are expressed by means of *cytoplasmic* protein synthesis. (The term **cytoplasmic inheritance** is sometimes used to describe the behavior of genes in organelles. However, we shall not use this description, since it is important to be able to distinguish between events in the general cytosol and those in specific organelles.)

Extranuclear DNA molecules do not attach to the meiotic or mitotic spindle, but are segregated stochastically to daughter cells. **Figure 4.14** shows an example in which this happens when the mitochondria inherited from the male and female parents have different alleles, and by chance a daughter cell receives an unbalanced distribution of mitochondria that represents only one parent (see *17.11 How Do Mitochondria Replicate and Segregate?*).

Uneven segregation causes somatic variation

Cell has mitochondria from both parents

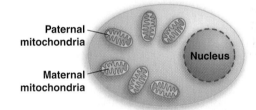

Possible outcomes of stochastic segregation

Cells usually have both types of mitochondria

Uneven distribution gives cells with only one type

Figure 4.14 When paternal and maternal mitochondrial alleles differ, a cell has two sets of mitochondrial DNAs. Mitosis usually generates daughter cells with both sets. Somatic variation may result if unequal segregation generates daughter cells with only one set.

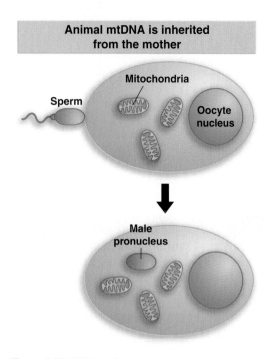

Animal mtDNA is inherited from the mother

Sperm

Mitochondria

Oocyte nucleus

Male pronucleus

Figure 4.15 DNA from the sperm enters the oocyte to form the male pronucleus in the fertilized egg, but all the mitochondria are provided by the oocyte.

Higher animals show maternal inheritance, which can be explained if the mitochondria are contributed entirely by the ovum and not at all by the sperm. **Figure 4.15** shows that the sperm contributes only a copy of the nuclear DNA. So the mitochondrial genes are derived exclusively from the mother; the mitochondrial genes in males are discarded each generation.

Non-Mendelian inheritance is sometimes associated with the phenomenon of somatic segregation, especially as seen in plants when there may be variation in somatic tissues that results from genetic differences that arise during development. Somatic segregation results from uneven distribution of genes during mitosis, comparable to the uneven distribution at meiosis that causes non-Mendelian inheritance.

Conditions in the organelle are different from those in the nucleus, and organelle DNA therefore evolves at its own distinct rate. If inheritance is uniparental, there can be no recombination between parental genomes; and usually recombination does not occur in those cases where organelle genomes are inherited from both parents. Because organelle DNA has a different replication system from that of the nucleus, the error rate during replication may be different. Mitochondrial DNA accumulates mutations more rapidly than nuclear DNA in mammals, but in plants the accumulation in the mitochondrion is slower than in the nucleus (the chloroplast is intermediate).

One consequence of maternal inheritance is that the sequence of mitochondrial DNA is more sensitive than nuclear DNA to reductions in the size of the breeding population. Comparisons of mitochondrial DNA sequences in a range of human populations allow an evolutionary tree to be constructed. The divergence among human mitochondrial DNAs spans 0.57%. A tree can be constructed in which the mitochondrial variants diverged from a common (African) ancestor. The rate at which mammalian mitochondrial DNA accumulates mutations is 2–4% per million years, >10× faster than the rate for globin. Such a rate would generate the observed divergence over an evolutionary period of 140,000–280,000 years. This implies that the human race is descended from a single female, who lived in Africa ~200,000 years ago.

4.11 Mitochondrial Genomes Are Circular DNAs that Code for Organelle Proteins

Key Terms

- **Mitochondrial DNA (mtDNA)** is an independent genome, usually circular, in a mitochondrion.
- **Chloroplast DNA (ctDNA)** is an independent genome, usually circular, in a plant chloroplast.

Key Concepts

- Organelle genomes are usually—but not always—circular molecules of DNA.
- Organelle genomes code for some but not all of the proteins found in organelles.
- Animal cell mitochondrial DNA is extremely compact and typically codes for 13 proteins, 2 rRNAs, and 22 tRNAs.
- Yeast mitochondrial DNA is 5× longer than animal cell mtDNA because of the presence of long introns.

Most organelle genomes are single circular molecules of DNA of unique sequence (denoted **mtDNA** in the mitochondrion and **ctDNA** in the chloroplast). In a few exceptions, mostly in lower eukaryotes, mitochondrial DNA is a linear molecule.

Usually there are several copies of the genome in an individual organelle. Because there are multiple organelles per cell, there are many organelle genomes per cell. Although the organelle genome itself is unique, it constitutes a repetitive sequence relative to any nonrepetitive nuclear sequence.

Mitochondrial genomes vary in total size by more than an order of magnitude. Animal cells have small mitochondrial genomes, ~16.5 kb in mammals. There are several hundred mitochondria per animal cell. Each mitochondrion has multiple copies of the DNA. The total amount of mitochondrial DNA relative to nuclear DNA is small, <1%.

In yeast, the mitochondrial genome is much larger. In *S. cerevisiae*, the exact size varies among different strains, but is ~80 kb. There are ~22 mitochondria per cell, which corresponds to ~4 genomes per organelle. In growing yeast cells, the proportion of mitochondrial DNA can be as high as 18%.

Plants show an extremely wide range of variation in mitochondrial DNA size, with a minimum of ~100 kb. The mitochondrial genome is usually a single sequence, organized as a circle. Within this circle, there are short homologous sequences. Recombination between these elements generates smaller, subgenomic circular molecules that coexist with the complete, "master" genome, explaining the apparent complexity of plant mitochondrial DNAs.

With mitochondrial genomes sequenced from many organisms, we can now see some general patterns in the representation of functions in mitochondrial DNA. **Figure 4.16** summarizes the distribution of genes in mitochondrial genomes. The total number of protein-coding genes is rather small, and does not correlate with the size of the genome. Mammalian mitochondria use their 16 kb genomes to code for 13 proteins, whereas yeast mitochondria use their 60–80 kb genomes to code for as few as 8 proteins. Plants, with much larger mitochondrial genomes, code for more proteins. Introns are found in most mitochondrial genomes, although not in the very small mammalian genomes.

The major part of the protein-coding activity is devoted to the components of the multisubunit assemblies of respiration complexes I–IV. The two major rRNAs are always coded by the mitochondrial genome. The number of tRNAs coded by the mitochondrial genome varies from none to the full complement (25–26 in mitochondria). Many ribosomal proteins are coded in protist and plant mitochondrial genomes, but there are few or none in fungi and animal genomes. There are genes coding for proteins involved in import in many protist mitochondrial genomes.

Animal mitochondrial DNA is extremely compact. There are extensive differences in the detailed gene organization found in different animal taxa, but the general principle is maintained of a small genome coding for a restricted number of functions. There are no introns, some genes actually overlap, and almost every single base pair can be assigned to a gene. With the exception of the D loop, a region concerned with the initiation of DNA replication, no more than 87 of the 16,569 bp of the human mitochondrial genome can be regarded as lying in intercistronic regions. The map of the human mitochondrial genome, summarized in **Figure 4.17**, shows 13 protein-coding regions, all representing components of the respiration apparatus.

The fivefold discrepancy in size between the *S. cerevisiae* (84 kb) and mammalian (16 kb) mitochondrial genomes alone alerts us to the fact that there must be a great difference in their genetic organization in spite of their common function. The map shown in **Figure 4.18** accounts for the major RNA and protein products of the yeast mitochondrion. The most notable feature is the dispersion of loci on the map. The two most prominent loci are the interrupted genes *box* (coding for cytochrome *b*) and *oxi3* (coding for subunit 1 of cytochrome oxidase). Together these two genes are almost as long as the entire mitochondrial genome in mammals! Many of the long introns in these

Mitochondria code for RNAs and proteins			
Species	Size (kb)	Protein-coding genes	RNA-coding genes
Fungi	19–100	8–14	10–28
Protists	6–100	3–62	2–29
Plants	186–366	27–34	21–30
Animals	16–17	13	4–24

Figure 4.16 Mitochondrial genomes have genes coding for (mostly complex I–IV) proteins, rRNAs, and tRNAs.

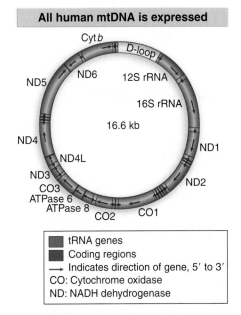

All human mtDNA is expressed

- tRNA genes
- Coding regions
- → Indicates direction of gene, 5′ to 3′
- CO: Cytochrome oxidase
- ND: NADH dehydrogenase

Figure 4.17 Human mitochondrial DNA has 22 tRNA genes, 2 rRNA genes, and 13 protein-coding regions. 14 of the 15 protein-coding or rRNA-coding regions are transcribed in the same direction. 14 of the tRNA genes are expressed in the clockwise direction and 8 are read counterclockwise.

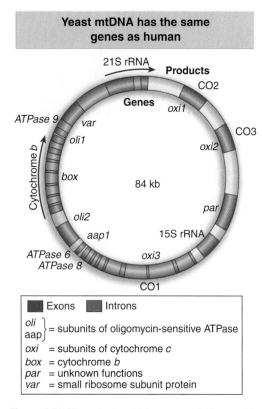

Figure 4.18 The mitochondrial genome of *S. cerevisiae* contains both interrupted and uninterrupted protein-coding genes, rRNA genes, and tRNA genes (positions not indicated). Arrows indicate direction of transcription.

Chloroplasts have >100 genes	
Genes	Types
RNA-coding	
16S rRNA	1
23S rRNA	1
4.5S rRNA	1
5S rRNA	1
tRNA	30–32
Gene expression	
r-proteins	20–21
RNA polymerase	3
Others	2
Chloroplast functions	
Rubisco and thylakoids	31–32
NADH dehydrogenase	11
Total	**105–113**

Figure 4.19 The chloroplast genome in plants codes for 4 rRNAs, 30 tRNAs, and ~60 proteins.

genes have open reading frames in register with the preceding exon (see *27.5 Some Group I Introns Code for Endonucleases That Sponsor Mobility*).

4.12 The Chloroplast Genome Codes for Many Proteins and RNAs

Key Concept

- Chloroplast genomes vary in size, but are large enough to code for 50–100 proteins as well as the rRNAs and tRNAs.

Chloroplast genomes are relatively large, usually ~140 kb in plants, and <200 kb in algae. This is comparable to the size of a large bacteriophage, for example, T4 at ~165 kb. There are multiple copies of the genome per organelle, typically 20–40 in a plant, and multiple copies of the organelle per cell, typically 20–40.

What genes are carried by chloroplasts? The sequenced chloroplast genomes (>30 in total) have a range of 87–183 genes. **Figure 4.19** summarizes the functions coded by the chloroplast genome in plants. There is more variation in the chloroplast genomes of algae.

The situation is generally similar to that of mitochondria, except that more genes are involved. The chloroplast genome codes for all the rRNA and tRNA species needed for protein synthesis. The ribosome includes two small rRNAs in addition to the major species. The tRNA set may include all of the necessary genes. The chloroplast genome codes for ~50–100 proteins, including RNA polymerase and ribosomal proteins. Again the rule is that organelle genes are transcribed and translated by the apparatus of the organelle. About half of the chloroplast genes code for proteins involved in protein synthesis.

4.13 Organelles Evolved by Endosymbiosis

Key Concepts

- Mitochondrial genomes are more closely related to bacterial genomes than to eukaryotic nuclear genomes.
- Mitochondria probably originated when a eukaryotic cell "captured" a bacterium.
- Integration of the mitochondrion has involved transfer of genetic information in both directions between it and the nucleus.

How did a situation evolve in which an organelle contains genetic information for some of its functions, while others are coded in the nucleus? **Figure 4.20** shows the endosymbiosis model for mitochondrial evolution, in which primitive cells captured bacteria that provided the functions that evolved into membrane-bounded mitochondria and chloroplasts. At this point, the proto-organelle must have contained all genes needed to specify its functions.

Sequence homologies suggest that mitochondria and chloroplasts evolved separately, from lineages that are common with eubacteria.

Mitochondria share an origin with α-purple bacteria. The closest known relative of mitochondria among the bacteria is *Rickettsia* (the causative agent of typhus), an obligate intracellular parasite that is probably descended from free-living bacteria. This reinforces the idea that mitochondria originated in an endosymbiotic event involving an ancestor that is also common to *Rickettsia*.

The endosymbiotic origin of the chloroplast is emphasized by the relationships between these genes and their counterparts in bacteria. The organization of the rRNA genes in particular is closely related to that of a cyanobacterium, which pins down more precisely the last common ancestor between chloroplasts and bacteria.

There must have been both loss and transfer of genes as the bacterium became integrated into the recipient cell and evolved into the mitochondrion (or chloroplast). The organelles have far fewer genes than an independent bacterium, and have lost many of the functions that are necessary for independent life (such as metabolic pathways). And since the majority of genes coding for organelle functions are in fact now located in the nucleus, these genes must have been transferred there from the organelle.

Transfer of a gene from an organelle to the nucleus requires physical movement of the DNA, of course, but successful expression also requires changes in the coding sequence. Organelle proteins that are coded by nuclear genes have special sequences that allow them to be imported into the organelle after they have been synthesized in the cytoplasm (see *10.6 Posttranslational Membrane Insertion Depends on Leader Sequences*). These sequences are not required by proteins that are synthesized within the organelle. Perhaps the gene transfer occurred at a period when compartments were less rigidly defined, so that it was easier both for the DNA to be relocated and for the proteins to be incorporated into the organelle irrespective of the site of synthesis.

Phylogenetic maps show that gene transfers have occurred independently in many different lineages. It appears that mitochondrial genes were transferred to the nucleus early in animal cell evolution, but the process may still be continuing in plant cells. The number of transfers can be large; there are >800 nuclear genes in the plant *Arabidopsis* whose sequences are related to genes in the chloroplasts of other plants. These genes are candidates for evolution from genes that originated in the chloroplast.

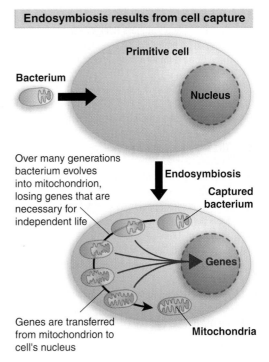

Figure 4.20 Mitochondria originated by a endosymbiotic event when a bacterium was captured by a eukaryotic cell.

4.14 SUMMARY

The DNA sequences composing a eukaryotic genome can be classified in three groups:

- nonrepetitive sequences are unique;
- moderately repetitive sequences are dispersed and repeated a small number of times as related but not identical copies;
- and highly repetitive sequences are short and usually repeated as tandem arrays.

The proportions of the types of sequence are characteristic for each genome, although larger genomes tend to have a smaller proportion of nonrepetitive DNA. Almost 50% of the human genome consists of repetitive sequences, the vast majority corresponding to transposon sequences. Most structural genes are located in nonrepetitive DNA. The amount of nonrepetitive DNA is a better reflection of the complexity of the organism than the total genome size; the greatest amount of nonrepetitive DNA in genomes is ~2×10^9 bp.

Non-Mendelian inheritance is explained by the presence of DNA in organelles in the cytoplasm. Mitochondria and chloroplasts are membrane-bounded systems in which some proteins are synthesized within the organelle, while others are imported. The organelle genome is usually a circular DNA that codes for all the RNAs and some of the proteins required by the organelle.

Mitochondrial genomes vary greatly in size from the 16 kb minimalist mammalian genome to the 570 kb genome of higher plants. The larger genomes may code for additional functions. Chloroplast genomes range from 120–200 kb. Those that have been sequenced have similar organization and coding functions. In both mitochondria and chloroplasts, many of the major proteins contain some subunits synthesized in the organelle and some subunits imported from the cytosol.

Rearrangements occur in mitochondrial DNA rather frequently in yeast, and recombination between mitochondrial or between chloroplast genomes has been found. Transfers of DNA have occurred between chloroplasts or mitochondria and nuclear genomes.

Genome Sequences and Gene Numbers

5.1 Introduction

Since the first genomes were sequenced in 1995, both the speed and range of sequencing have improved greatly. The first genomes to be sequenced were small bacterial genomes, <2 Mb. By 2002, the human genome of 3000 Mb had been sequenced. Genomes have now been sequenced from a wide range of organisms, including bacteria, archaea, yeasts, lower eukaryotes, plants, and animals including worms, flies, rodents, and mammals.

Figure 5.1 plots the number of bacterial sequences reported each year, and shows the years when some important eukaryotic genomes were sequenced. About 150 bacterial genomes have been sequenced, and more are being added at the rate of about one every week. More than 20 archaeal genomes have been sequenced. More than 10 important eukaryotic genomes have been sequenced. In addition, >1000 mitochondrial DNA sequences and >32 chloroplast genomes have been sequenced.

Perhaps the most important single piece of information provided by a genome sequence is the number of genes. Information from a range of sequenced genomes is summarized in **Figure 5.2**. *Mycoplasma genitalium*, an obligate intracellular bacterium, has the smallest known genome of any organism, with only ~470 genes. Free-living bacteria range from 1700 genes to 7500 genes. Archaea are in a similar range. Single-celled eukaryotes start with about 5300 genes. Worms and flies have roughly 18,500 and 13,500 genes, respectively, but the number rises only to ~30,000 for mouse and man.

Figure 5.3 summarizes the minimum number of genes found in six groups of organisms. It takes ~500 genes to make a cell, ~1500 to make a free-living cell, >5000 to make a cell with a nucleus, >10,000 to make a multicellular organism, and >13,000 to make an organism with a nervous system. Of course, many species may have more than the

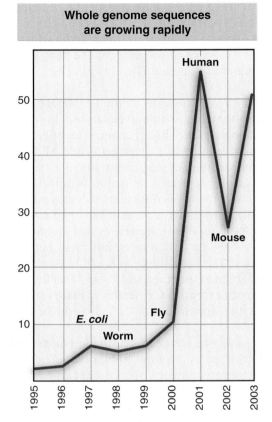

Whole genome sequences are growing rapidly

Figure 5.1 In less than a decade, more than 150 genomes have been fully sequenced.

Sequenced genomes vary from 470–30,000 genes			
Species	Genome (Mb)	Genes	Lethal loci
Mycoplasma genitalium	0.58	470	~300
Rickettsia prowazekii	1.11	834	
Haemophilus influenzae	1.83	1,743	
Methanococcus jannaschi	1.66	1,738	
B. subtilis	4.2	4,100	
E. coli	4.6	4,288	1,800
S. cerevisiae	13.5	6,034	1,090
S. pombe	12.5	4,929	
A. thaliana	119	25,498	
O. sativa (rice)	466	~30,000	
D. melanogaster	165	13,601	3,100
C. elegans	97	18,424	
H. sapiens	3,300	~30,000	

Figure 5.2 Genome sizes and gene numbers are based on complete sequences. Lethal loci are estimated from genetic data.

minimum number required for their type, so the number of genes can vary widely even among closely related species.

Within bacteria and the lower eukaryotes, most genes are unique. Within higher eukaryotic genomes, however, genes can be divided into families of related members. Of course, some genes are unique (formally the family has only one member), but many belong to families with 10 or more members. The number of different families may be better related to the overall complexity of the organism than the number of genes.

5.2 Bacterial Gene Numbers Range Over an Order of Magnitude

Key Concept

- Genome sequences show that there are 500–1200 genes in parasitic bacteria, 1500–7500 genes in free-living bacteria, and 1500–2700 genes in archaea.

**Minimum gene numbers
range from 500 to 30,000**

500 genes
Extracellular (parasitic)
bacterium

1,500 genes
Free-living bacterium

5,000 genes
Unicellular eukaryote

13,000 genes
Multicellular eukaryote

25,000 genes
Higher plants

30,000 genes
Mammals

Figure 5.3 The minimum gene number required for any type of organism increases with its complexity. Photograph of mycoplasma kindly provided by A. Albay, K. Frantz, and K. Bott. Photograph of bacterium kindly provided by Jonathan King.

The sequences of the genomes of bacteria and archaea show that almost all of the DNA (typically 85–90%) codes for RNA or protein. **Figure 5.4** shows that bacterial and archaeal genome sizes range over about an order of magnitude, and that genome size is proportional to number of genes. The typical gene is about 1000 bp in length.

All bacteria with genome sizes below 1.5 Mb are obligate intracellular parasites—they live within a eukaryotic host that provides them with small molecules. Their genomes identify the minimum number of functions required to construct a cell. All classes of their genes are reduced in number compared with bacteria with larger genomes, but the most significant reduction is in loci coding for enzymes concerned with metabolic functions (which are largely provided by the host cell) and with regulation of gene expression.

The archaea have biological properties that are intermediate between the prokaryotes and eukaryotes, but their genome sizes and gene numbers fall in the same range as bacteria. Their genome sizes vary from 1.5–3 Mb, corresponding to 1500–2700 genes. The archaeal apparatus for gene expression resembles eukaryotes more than prokaryotes, but the apparatus for cell division better resembles prokaryotes.

The archaea and the smallest free-living bacteria identify the minimum number of genes required to make a cell able to function independently in the environment. The smallest archaeal genome has ~1500 genes. The free-living bacterium with the smallest known genome is the thermophile *Aquifex aeolicus*, with 1.5 Mb and 1512 genes. A "typical" gram-negative bacterium, *Haemophilus influenzae*, has 1743 genes, each ~900 bp long. So we can conclude that ~1500 genes are required to make a free-living organism.

Bacterial genome sizes extend over almost an order of magnitude to ~9 Mb. The bacteria with the largest genomes, *S. meliloti* and *M. loti*, are nitrogen-fixing bacteria that live on plant roots. Their genome sizes and total gene numbers (>7500) are similar to those of yeasts.

The size of the genome of *Escherichia coli* is in the middle of the range. The common laboratory strain has 4288 genes, with an average length ~950 bp, and an average separation between genes of 118 bp. But there are differences between strains. The known extremes of *E. coli* are from the smallest strain that has 4.6 Mb with 4249 genes to the largest strain that has 5.5 Mb with 5361 genes.

We do not know the functions of all bacterial genes. In most of these genomes, ~60% of the genes can be identified on the basis of homology with known genes in other species. (Some archaea and bacteria live under extreme conditions, and in these cases it is more difficult to identify genes by comparison with other organisms.) The known genes fall about equally into classes whose products are concerned with metabolism, cell structure or transport of components, and gene expression and its regulation. In virtually every bacterial genome, >25% of the genes cannot be ascribed any function. However, many of these genes can be found in related organisms, which implies that they have a conserved function.

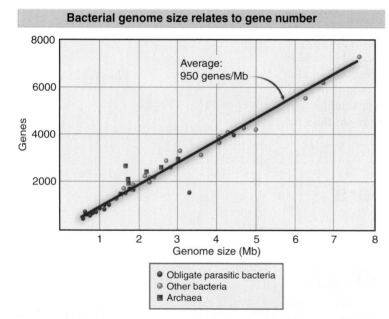

Figure 5.4 The sizes and number of genes increase with genome size in bacterial and archaeal genomes.

5.3 Total Gene Number Is Known for Several Eukaryotes

Key Concept

- There are 6000 genes in yeast, 18,500 in worm, 13,600 in fly, 25,000 in the small plant *Arabidopsis*, and probably 30,000 in mouse and man.

As soon as we look at eukaryotic genomes, we see that the relationship between genome size and gene number is lost. The genomes of unicellular eukaryotes fall in the same size range as the largest bacterial genomes. Higher eukaryotes have more genes, but the number does not correlate with genome size, as can be seen from **Figure 5.5**.

The most extensive data for lower eukaryotes are available from the sequences of the genomes of the yeasts *Saccharomyces cerevisiae* and *Schizosaccharomyces pombe*. **Figure 5.6** summarizes the most important features. The yeast genomes of 12.5 Mb and 13.5 Mb have ~6000 and ~5000 genes, respectively. The average open reading frame is ~1.4 kb, so that ~70% of the genome is occupied by coding regions. The major difference between them is that only 5% of *S. cerevisiae* genes have introns, compared to 43% in *S. pombe*. The density of genes is high; organization is generally similar, although the spaces between genes are a bit shorter in *S. cerevisiae*. About half of the genes identified by sequence were either known previously or related to known genes. The remainder are new, which gives some indication of the number of new types of genes that may be discovered.

The genome of *Caenorhabditis elegans* DNA varies between regions rich in genes and regions in which genes are sparse. The total sequence contains ~18,500 genes. Only ~42% of the genes have putative counterparts outside the Nematoda.

Although the fly genome is larger than the worm genome, there are fewer genes (13,600) in *Drosophila melanogaster*. The number of different transcripts is slightly larger (14,100) as the result of alternative splicing. We do not understand why the fly—a much more complex organism—has only 70% of the number of genes in the worm. This emphasizes forcefully the lack of an exact relationship between gene number and complexity of the organism.

The plant *Arabidopsis thaliana* has a genome size intermediate between the worm and the fly, but has a larger gene number (25,000) than either. This again shows the lack of a clear relationship, and also emphasizes the special quality of plants, which may have more genes (due to ancestral duplications) than animal cells. A majority of the *Arabidopsis*

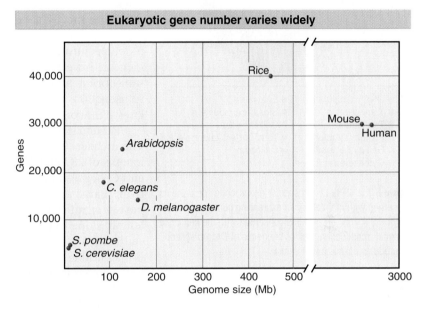

Figure 5.5 The number of genes in a eukaryote varies from 6000 to 40,000 but does not correlate with the genome size or the complexity of the organism.

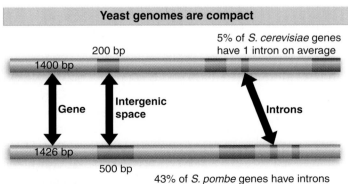

Figure 5.6 The *S. cerevisiae* genome of 13.5 Mb has 6000 genes, almost all uninterrupted. The *S. pombe* genome of 12.5 Mb has 5000 genes, almost half having introns. Gene sizes and spacing are fairly similar.

Functions are known for only half the fly genes

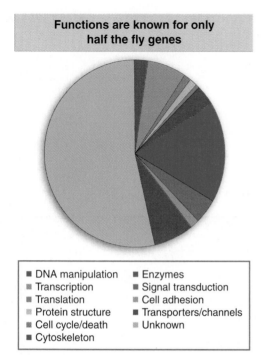

- ■ DNA manipulation
- ■ Transcription
- ■ Translation
- ■ Protein structure
- ■ Cell cycle/death
- ■ Cytoskeleton
- ■ Enzymes
- ■ Signal transduction
- ■ Cell adhesion
- ■ Transporters/channels
- ■ Unknown

Figure 5.7 ~20% of *Drosophila* genes code for proteins concerned with maintaining or expressing genes, ~20% for enzymes, <10% for proteins concerned with the cell cycle or signal transduction. Half of the genes of *Drosophila* code for products of unknown function.

genome is found in duplicated segments, suggesting that there was an ancient doubling of the genome (to give a tetraploid). Only 35% of *Arabidopsis* genes are present as single copies.

The genome of rice (*Oryza sativa*) is ~4× larger than that of *Arabidopsis*, but the number of genes is only ~50% larger, probably ~40,000. Repetitive DNA occupies 42–45% of the rice genome. More than 80% of the genes found in *Arabidopsis* are represented in rice. Of these common genes, ~8000 are found in *Arabidopsis* and rice but not in any of the bacterial or animal genomes that have been sequenced. These are probably the set of genes that code for plant-specific functions, such as photosynthesis.

From the fly genome, we can form an impression of how many genes are devoted to each type of function. **Figure 5.7** breaks down the functions into different categories. Among the gene products that are identified, we find ~2500 enzymes, ~750 transcription factors, ~700 transporters and ion channels, and ~700 proteins involved with signal transduction. But just over the half genes code for products of unknown function. ~20% of the proteins reside in membranes.

Protein size increases from prokaryotes and archaea to eukaryotes. The archaea *M. jannaschi* and bacterium *E. coli* have average protein lengths of 287 and 317 amino acids, respectively; whereas *S. cerevisiae* and *C. elegans* have average lengths of 484 and 442 amino acids, respectively. Large proteins (>500 amino acids) are rare in bacteria, but compose a significant component (~1/3) in eukaryotes. The increase in length is due to the addition of extra domains, with each domain typically constituting 100–300 amino acids. But the increase in protein size is responsible for only a very small part of the increase in genome size.

Another insight into gene number is obtained by counting the number of expressed genes. If we rely upon the estimates of the number of different mRNA species that can be counted in a cell, we would conclude that the average vertebrate cell expresses ~10,000–15,000 genes. The existence of significant overlaps between the messenger populations in different cell types would suggest that the total expressed gene number for the organism should be within a fewfold of this. The estimate for the total human genome of ~30,000 genes would imply that a significant proportion of the total gene number is actually expressed in any given cell.

In addition to functional genes, there are also copies of genes that have become nonfunctional (identified as such by interruptions in their protein-coding sequences). These are called pseudogenes (see *3.11 Pseudogenes Are Dead Ends of Evolution*). The number of pseudogenes can be large. In the mouse and human genomes, the number of pseudogenes is ~10% of the number of (potentially) active genes.

5.4 The Human Genome Has Fewer Genes than Expected

Key Concepts

- Only 1% of the human genome consists of coding regions.
- The exons compose ~5% of each gene, and genes (exons plus introns) compose ~25% of the genome.
- Current estimates suggest the human genome has ~30,000 genes.
- ~60% of human genes are alternatively spliced.
- Up to 80% of the alternative splices change protein sequence, so there may be ~50,000–60,000 different protein products.

The human genome was the first vertebrate genome to be sequenced. This massive task has revealed a wealth of information about the genetic makeup of our species, and about the evolution of the genome in general. Our understanding is deepened further by the ability to compare the human genome sequence with the more recently sequenced mouse genome.

Mammal and rodent genomes generally fall into a narrow size range, ~3 × 10⁹ bp (see *4.5 Why Are Genomes So Large?*). The genomes contain similar gene families and genes, with most genes having a corresponding gene in the other genome, but with differences in the number of members of a family, especially in those cases where the functions are specific to the species. The estimate of 30,000 genes for the mouse genome is about the same as current estimates for the human gene number. **Figure 5.8** plots the distribution of the mouse genes. The 30,000 protein-coding genes are accompanied by ~4000 pseudogenes, as well as the ~1600 RNA-coding genes; the other RNA genes are generally small. Almost half of these genes code for tRNAs, for which a large number of pseudogenes also have been identified.

The human (haploid) genome contains 22 autosomes plus the X or Y. The chromosomes range in size from 45 to 279 Mb of DNA, making a total genome content of 3286 Mb (~3.3 × 10⁹ bp). **Figure 5.9** shows that a tiny proportion (~1%) of the human genome is accounted for by the exons that actually code for proteins. The introns that constitute the remaining sequences in the genes bring the total of DNA concerned with producing proteins to ~25%. As shown in **Figure 5.10**, the average human gene is 27 kb long, with 9 exons that include a total coding sequence of 1340 bp. The average coding sequence is therefore only 5% of the length of the gene.

Based on comparisons with other species and with known protein-coding genes, there are ~24,000 clearly identifiable human genes. Sequence analysis identifies ~12,000 more potential genes. This number is much less than we had expected—most previous estimates had been ~100,000. It shows a relatively small increase over flies and worms (13,600 and 18,500, respectively), not to mention the plant *Arabidopsis* (25,000; see Figure 5.2). However, we should not be particularly surprised by the notion that it does not take a great number of additional genes to make a more complex organism. The difference in DNA sequences between man and chimpanzee is extremely small (there is >99% similarity), so it is clear that the functions and interactions between a similar set of genes can produce very different results.

The number of human genes is less than the number of potential proteins because of alternative splicing. The extent of alternative splicing is greater in man than in fly or worm; it may affect as many as 60% of the genes, so the increase in size of the human proteome relative to those of other eukaryotes may be larger than the increase in the number of genes. The proportion of the alternative splices that actually result in changes in the protein sequence may be as high as 80%. This could increase the size of the proteome to 50,000–60,000 members.

In number of gene families, however, the discrepancy between man and the other eukaryotes may not be so great. Many of the human genes belong to families. An analysis of ~25,000 genes identified 3500

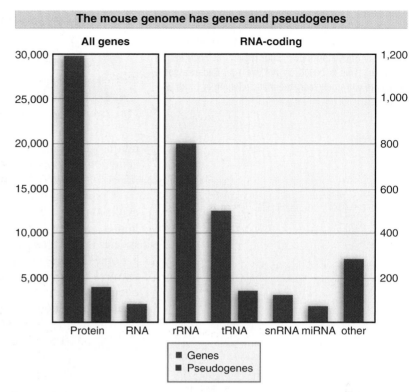

Figure 5.8 The mouse genome has ~30,000 protein-coding genes, and ~4000 pseudogenes. There are ~1600 RNA-coding genes. The data for RNA-coding genes are replotted on the right, at an expanded scale to show that there are ~800 rRNA genes, ~350 tRNA genes and 150 pseudogenes, and ~450 other noncoding RNA genes, including snRNAs and miRNAs.

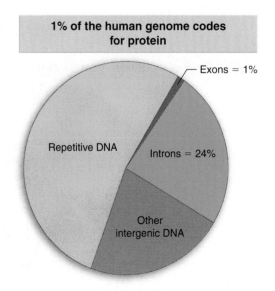

Figure 5.9 Genes occupy 25% of the human genome, but protein-coding sequences are only a tiny part of this fraction.

Figure 5.10 The average human gene is 27 kb long and has 9 exons, usually comprising two longer exons at each end and 7 internal exons. The UTRs in the terminal exons are the untranslated (noncoding) regions at each end of the gene. (This is based on the average. Because some genes are extremely long, the median length is 14 kb with 7 exons.)

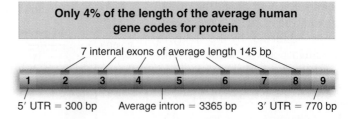

unique genes and 10,300 gene pairs. This extrapolates to a number of gene families only slightly larger than that of worm or fly (*see 5.7 How Many Different Types of Genes Are There?*).

Comparing genomes provides interesting information about the evolution of species. The number of gene families in the mouse and human genomes is the same, and a major difference between the species is the differential expansion of particular families in one of the genomes. This is especially noticeable in genes that affect phenotypic features that are unique to the species. Of 25 families whose size has been expanded in mouse, 14 contain genes specifically involved in rodent reproduction, and 5 contain genes specific to the immune system.

5.5 How Are Genes and Other Sequences Distributed in the Genome?

Key Concepts

- Repeated sequences (present in more than one copy) account for >50% of the human genome.
- The great bulk of repeated sequences consist of copies of nonfunctional transposons.
- There are many duplications of large chromosome regions.

Are genes uniformly distributed in the genome? Some chromosomes are relatively poor in genes, and have >25% of their sequences as "deserts"—regions longer than 500 kb where there are no genes. Even the most gene-rich chromosomes have >10% of their sequences as deserts. So overall ~20% of the human genome consists of deserts that have no genes.

Repetitive sequences account for >50% of the human genome, as seen in **Figure 5.11**. The repetitive sequences fall into five classes:

- Transposons (either active or inactive) are the largest class (45% of the genome). All transposons are found in multiple copies.
- Processed pseudogenes (~3000 in all) account for ~0.1% of total DNA. These are sequences that arise by insertion of a copy of an mRNA sequence into the genome (see *3.11 Pseudogenes Are Dead Ends of Evolution*).
- Simple sequence repeats (highly repetitive DNA such as $(CA)_n$) account for ~3%.
- Segmental duplications (blocks of 10–300 kb that have been duplicated into a new region) account for ~5%. A few of these duplications are found on the same chromosome; most are on different chromosomes.
- Tandem repeats form blocks of one type of sequence (especially found at centromeres and telomeres).

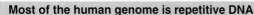

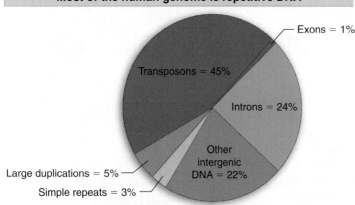

Figure 5.11 The largest component of the human genome consists of transposons. Other repetitive sequences include large duplications and simple repeats.

The sequence of the human genome emphasizes the importance of transposons. (Transposons have the capacity to replicate themselves and insert into new locations. They may function exclusively as DNA elements [see *Chapter 21 Transposons*] or may have an active form that is RNA [see *Chapter 22 Retroviruses and Retroposons*]. Their distribution in the human genome is summarized later in the chapter in Figure 5.17.) Most of the transposons in the human genome are nonfunctional; very few are currently active. However, the high proportion of the genome occupied by these elements indicates that they have played an active role in shaping the genome. One interesting feature is that some present genes originated as transposons, and evolved into their present condition after losing the ability to transpose. Almost 50 genes appear to have originated in this way.

Segmental duplication at its simplest is the tandem duplication of some region within a chromosome (typically because of an aberrant recombination event at meiosis; see *6.6 Unequal Crossing-Over Rearranges Gene Clusters*). In many cases, however, the duplicated regions are on different chromosomes, implying that either there was originally a tandem duplication followed by a translocation of one copy to a new site, or that the duplication arose by some different mechanism altogether. The extreme case of a segmental duplication is when a whole genome is duplicated, in which case the diploid genome initially becomes tetraploid. As the duplicated copies develop differences from one another, the genome may gradually become effectively a diploid again, although homologies between the diverged copies leave evidence of the event. This is especially common in plant genomes.

5.6 The Y Chromosome Has Several Male-Specific Genes

Key Concepts

- The Y chromosome has ~60 genes that are expressed specifically in testis.
- The male-specific genes are present in multiple copies in repeated chromosomal segments.
- Gene conversion between multiple copies allows the active genes to be maintained during evolution.

The sequence of the human genome has significantly extended our understanding of the role of the sex chromosomes. It is generally thought that the X and Y chromosomes have descended from a common (very ancient) autosome. In the course of their development, the X chromosome has retained most of the original genes, while the Y chromosome has lost most of them.

The X chromosome behaves like an autosome insofar as females have two copies and recombination can take place between them. The density of genes on the X chromosome is comparable to the density of genes on other chromosomes.

The Y chromosome is much smaller than the X chromosome and has many fewer genes. Its unique role results from the fact that only males have the Y chromosome, and there is only one copy, so Y-linked loci are effectively haploid, instead of diploid like all other human genes.

For many years, the Y chromosome was thought to carry almost no genes except for one (or more) sex-determining genes that determine maleness. Most of the length of the Y chromosome (>95% of its sequence)

Figure 5.12 The Y chromosome consists of X-transposed regions, X-degenerate regions, and amplicons. The X-transposed and X-degenerate regions have 2 and 14 single-copy genes, respectively. The amplicons have eight large palindromes (P1–P8), which contain nine gene families. Each family contains at least two copies.

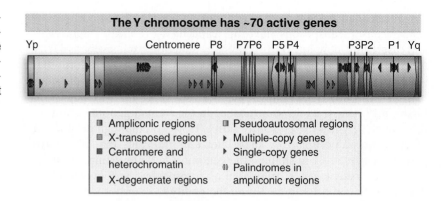

does not undergo crossing-over with the X chromosome, which led to the view that it could not contain active genes because there would be no means to prevent the accumulation of deleterious mutations. This region is flanked by short pseudoautosomal regions that exchange frequently with the X chromosome during male meiosis. The long region was originally called the nonrecombining region, but now has been renamed the *male-specific region*.

Detailed sequencing of the Y chromosome shows that the male-specific region contains three types of regions, as illustrated in **Figure 5.12**:

- The *X-transposed sequences* consist of a total of 3.4 Mb comprising some large blocks resulting from a transposition from band q21 in the X chromosome about 3–4 million years ago. This is specific to the human lineage. These sequences do not recombine with the X chromosome and have become largely inactive. They now contain only two active genes.

- The *X-degenerate segments* of the Y are sequences that have a common origin with the X chromosome (going back to the common autosome from which both X and Y have descended); the X-degenerate segments contain genes or pseudogenes related to X-linked genes. There are 14 active genes and 13 pseudogenes. The active genes have defied the trend for genes to be eliminated from chromosomal regions that cannot recombine at meiosis.

- The *ampliconic segments* have a total length of 10.2 Mb and are internally repeated on the Y chromosome. There are eight large palindromic blocks. They include nine protein-coding gene families, with copy numbers per family in the 2–35 range. The name "amplicon" reflects the fact that the sequences have been internally amplified on the Y chromosome.

From the total number of genes in these three regions, we see that the Y chromosome contains many more genes than had been expected. There are 156 transcription units, of which half represent protein-coding genes, and half represent pseudogenes.

The presence of the active genes is explained by the fact that the existence of closely related gene copies in the ampliconic segments allows gene conversion between multiple copies of a gene to be used to regenerate active copies. The most common needs for multiple copies of a gene are quantitative (to provide more protein product) or qualitative (to code for proteins with slightly different properties or that are expressed in different times or places). In this case, the existence of multiple copies allows recombination within the Y chromosome itself to substitute for the evolutionary diversity that is usually provided by recombination between allelic chromosomes.

5.7 How Many Different Types of Genes Are There?

Key Terms

- A **gene family** consists of a set of genes within a genome that code for related or identical proteins. The members were derived by duplication of an ancestral gene followed by accumulation of changes in sequence between the copies. Most often the members of a family are related but not identical.
- The **proteome** is the complete set of proteins that is expressed by the entire genome. Because some genes code for multiple proteins, the size of the proteome is greater than the number of genes. Sometimes the term is used to describe the complement of proteins expressed by a cell at any one time.
- **Orthologs** are corresponding proteins in two species as defined by sequence homologies.

Key Concepts

- Only some genes are unique; others belong to families with other members that are related but not usually identical.
- The proportion of unique genes declines with increasing genome size, and the proportion of genes in families increases.
- The minimum number of gene families coding for a bacterium is >1000, a yeast is >4000, and a higher eukaryote 11,000–14,000.

Because some genes are present in more than one copy or are related to one another, the number of different types of genes is less than the total number of genes. We can group some of the genes into **gene families** by comparing their exons. A family of related genes arises by duplication of an ancestral gene followed by accumulation of changes in sequence between the copies. Most often the members of a gene family are related but not identical. The number of types of genes is obtained by adding the number of unique genes (for which there are no related genes in the genome) to the number of gene families. **Figure 5.13** compares the total number of genes with the number of gene families in each of six genomes. In bacteria, most genes are unique, so the number of distinct families is close to the total gene number. The situation is different even in the lower eukaryote *S. cerevisiae*, in which there is a significant proportion of repeated genes. The most striking effect is that the number of genes increases quite sharply in the higher eukaryotes, but the number of gene families does not change much.

Figure 5.14 shows that the proportion of unique genes drops sharply with genome size. When genes are present in families, the

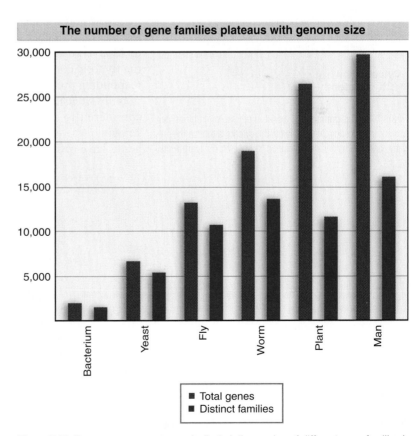

The number of gene families plateaus with genome size

(Bacterium, Yeast, Fly, Worm, Plant, Man)

- ■ Total genes
- ■ Distinct families

Figure 5.13 Because many genes are duplicated, the number of different gene families is much less than the total number of genes. The histogram compares the total number of genes with the number of distinct gene families.

Family size increases with genome size

	Unique genes	Families with 2–4 members	Families with >4 members
H. influenzae	89%	10%	1%
S. cerevisiae	72%	19%	9%
D. melanogaster	72%	14%	14%
C. elegans	55%	20%	26%
A. thaliana	35%	24%	41%

Figure 5.14 The proportion of genes that are present in multiple copies increases with genome size in higher eukaryotes.

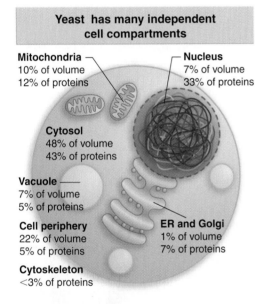

Yeast has many independent cell compartments

Mitochondria
10% of volume
12% of proteins

Nucleus
7% of volume
33% of proteins

Cytosol
48% of volume
43% of proteins

Vacuole
7% of volume
5% of proteins

Cell periphery
22% of volume
5% of proteins

ER and Golgi
1% of volume
7% of proteins

Cytoskeleton
<3% of proteins

Figure 5.15 Assignments of yeast proteins to different cell compartments show almost half in the cytosol, a third in the nucleus, and the remainder assigned to specific compartments.

number of members in a family is small in bacteria and lower eukaryotes, but is large in higher eukaryotes. Much of the extra genome size of *Arabidopsis* is accounted for by families with >4 members.

If every gene is expressed, the total number of genes should equal the total number of proteins required to make the organism (the **proteome**). However, two effects mean that the proteome is different from the total gene number. Because genes are duplicated, some of them code for the same protein (although it may be expressed in a different time or place) and others may code for related proteins that again play the same role in different times or places. And because some genes can produce more than one protein by means of alternative splicing, the proteome can be larger than the number of genes.

What is the core proteome—the basic number of the different types of proteins in the organism? A minimum estimate is given by the number of gene families, ranging from 1400 in the bacterium, >4000 in the yeast, and a range of 11,000–14,000 for the fly and worm.

What is the distribution of the proteome among types of proteins? The 6000 proteins of the yeast proteome include 5000 soluble proteins and 1000 transmembrane proteins. **Figure 5.15** summarizes the distribution of individual yeast proteins compared with the volumes of their compartments. The cytosol occupies about half of the cell and has a proportionate number of proteins. The nucleus is only ~7% of the cell's volume, but has a third of the protein diversity, probably representing many regulatory proteins that are present in low abundance. The number of proteins found in most organelles is more or less related to their volume, although the cell periphery occupies a large volume but has a simple structure with relatively few individual proteins.

How many genes are common to all organisms (or to groups such as bacteria or higher eukaryotes) and how many are specific for the individual type of organism? **Figure 5.16** summarizes estimates made for the fly from comparisons with yeast and worm. Genes that code for corresponding proteins in different organisms are called **orthologs**. Operationally, we usually reckon that two genes in different organisms can be considered to provide corresponding functions if their sequences are similar over >80% of the length. By this criterion, ~20% of the fly genes have orthologs in yeast. These genes are probably required by all

Figure 5.16 The fly genome can be divided into genes that are (probably) present in all eukaryotes, additional genes that are (probably) present in all multicellular eukaryotes, and genes that are more specific to subgroups of species that include flies.

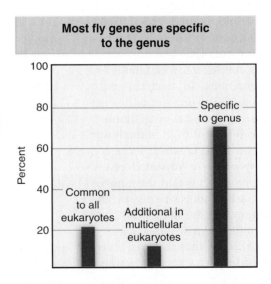

Most fly genes are specific to the genus

Common to all eukaryotes

Additional in multicellular eukaryotes

Specific to genus

eukaryotes. The proportion is 30% in comparisons of fly and worm, with the increase probably representing the addition of gene functions that are common to multicellular eukaryotes. This still leaves a major proportion of genes as coding for proteins that are required specifically by either flies or worms, respectively.

Once we know the total number of proteins, we can ask how they interact. By definition, proteins in structural multiprotein assemblies must form stable interactions with one another. Proteins in signaling pathways interact with one another transiently. In both cases, such interactions can be detected in test systems in which a readout system magnifies the effect of the interaction.

As a practical matter, assays of pairwise interactions can give us an indication of the minimum number of independent structures or pathways. An analysis of the ability of all 6000 (predicted) yeast proteins to interact in pairwise combinations shows that ~1000 proteins can bind to at least one other protein. Direct analyses of complex formation have identified 1440 different proteins in 232 multiprotein complexes. This is the beginning of an analysis that will lead to definition of the number of functional assemblies or pathways.

5.8 More Complex Species Evolve by Adding New Gene Functions

Key Concepts

- Comparisons of different genomes show a steady increase in gene number as additional genes are added to make eukaryotes, make multicellular organisms, make animals, and make vertebrates.

- Most of the genes that are unique to vertebrates are concerned with the immune or nervous systems.

Comparison of the human genome sequence with sequences of other species reveals how evolution has proceeded. **Figure 5.17** analyzes human genes according to the breadth of their distribution in living organisms. Starting with the most generally distributed (top right corner of the figure), 21% of genes are common to eukaryotes and prokaryotes. These tend to code for proteins that are essential for all living forms—proteins needed for basic metabolism, replication, transcription, and translation. Moving clockwise, another 32% of genes are added in eukaryotes in general—for example, they may be found in yeast. These code for proteins needed in eukaryotic cells but not in bacterial cells—for example, they may be concerned with specifying organelles or cytoskeletal components. Another 24% of genes are needed to specify animals. These include genes necessary for multicellularity and for development of different tissue types. And 22% of genes are unique to vertebrates. These mostly code for proteins of the immune and nervous systems; they code for very few enzymes, consistent with the

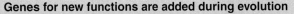
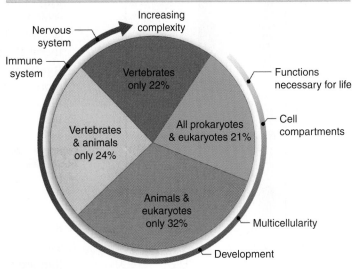

Figure 5.17 Human genes can be classified according to how widely their homologs are distributed in other species.

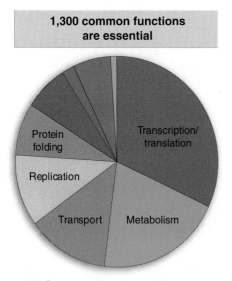

Figure 5.18 Common eukaryotic proteins are concerned with essential cellular functions.

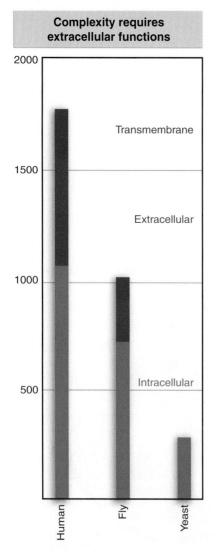

Figure 5.19 Increasing complexity in eukaryotes is accompanied by addition of new proteins for transmembrane and extracellular functions.

idea that enzymes have ancient origins, and that metabolic pathways originated early in evolution. We see, therefore, that the progression from bacteria to vertebrates requires addition of groups of genes representing the necessary new functions at each stage.

One way to define commonly needed proteins is to identify the proteins present in all proteomes. Comparing the human proteome in more detail with the proteomes of other organisms, 46% of the yeast proteome, 43% of the worm proteome, and 61% of the fly proteome is represented in the human proteome. A key group of ~1300 proteins is present in all four proteomes. The common proteins are basic housekeeping proteins required for essential functions, falling into the types summarized in **Figure 5.18**. The main functions are concerned with transcription and translation (35%), metabolism (22%), transport (12%), DNA replication and modification (10%), protein folding and degradation (8%), and cellular processes (6%).

One of the striking features of the human proteome, compared with those of other eukaryotes, is that it has many new proteins but relatively few new protein domains. Most protein domains appear to be common to the animal kingdom. However, there are many new protein architectures, defined as new combinations of domains. **Figure 5.19** shows that the greatest increase is in transmembrane and extracellular proteins. In yeast, the vast majority of architectures are concerned with intracellular proteins. About twice as many intracellular architectures are found in fly (or worm), and there is a very striking increase in transmembrane and extracellular proteins, as might be expected from the addition of functions required for the interactions between the cells of a multicellular organism. The increase in intracellular architectures required to make a vertebrate (man) is relatively small, but there is a large increase in transmembrane and extracellular architectures.

5.9 How Many Genes Are Essential?

Key Terms

- **Redundancy** describes the concept that two or more genes may fulfill the same function, so that no single one of them is essential.
- A **synthetic lethal** is created by combining two mutations that by themselves are viable, but which are lethal together.
- **Synthetic genetic array analysis (SGA)** is an automated technique in budding yeast whereby a mutant is crossed to an array of approximately 5000 deletion mutants to determine if the mutations interact to cause a synthetic lethal phenotype.

Key Concepts

- Not all genes are essential. In yeast and fly, deletions of <50% of the genes have detectable effects.
- When two or more genes are redundant, a mutation in one of them may not have detectable effects.
- We do not fully understand the survival in the genome of genes that are apparently dispensable.

Natural selection is the force that ensures that useful genes are retained in the genome. Mutations occur at random, and their most common effect in an open reading frame is to damage the protein product. An organism with a damaging mutation is at a disadvantage in evolution, and ultimately the mutation will be eliminated by the competitive failure of organisms carrying it. The frequency of a disadvantageous allele in the population is balanced between the generation of new mutations and the elimination of old mutations. Reversing this argument, whenever we see an intact open reading frame in the genome, we assume that its product plays a useful role in the organism. Natural selection must have prevented mutations from accumulating in the gene. The ultimate fate of a gene that ceases to be useful is to accumulate mutations until it is no longer recognizable.

The maintenance of a gene implies that it confers a selective advantage on the organism. But in the course of evolution, even a small relative advantage may be the subject of natural selection, and a phenotypic defect may not necessarily be immediately detectable as the result of a mutation. However, we should like to know how many genes are actually *essential*. This means that their absence is lethal to the organism. In the case of diploid organisms, it means of course that the homozygous null mutation is lethal.

One approach to the issue of gene number is to determine the number of essential genes by mutational analysis. If we saturate some specified region of the chromosome with mutations that are lethal, the mutations should map into a number of complementation groups that corresponds to the number of lethal loci in that region. By extrapolating to the genome as a whole, we may calculate the total essential gene number.

In the organism with the smallest known genome (*Mycoplasma genitalium*), random insertions have detectable effects only in about two-thirds of the genes. Similarly, fewer than half of the genes of *E. coli* appear to be essential. The proportion is even lower in the yeast *S. cerevisiae*. A systematic survey based on completely deleting each of 5916 genes (>96% of the identified genes) shows that only 18.7% are essential for growth on a rich medium (that is, when nutrients are fully provided). **Figure 5.20** shows that these include genes in all categories. The only notable concentration of defects is in genes coding for products involved in protein synthesis, where ~50% are essential. Of course, this approach underestimates the number of genes that are essential for the yeast to live in the wild, when it is not so well provided with nutrients.

Figure 5.21 summarizes the results of a systematic analysis of the effects of loss of gene function in the worm *C. elegans*. Detectable effects on the phenotype were only observed for 10% of genes whose function was prevented, suggesting that most genes do not play essential roles. There is a greater proportion of essential genes (21%) among those worm genes that have counterparts in other eukaryotes, suggesting that widely conserved genes tend to serve more basic functions.

How do we explain the survival of genes whose deletion appears to have no effect? The most likely explanation is that the organism has alternative ways of fulfilling the same function. The simplest possibility is that there is **redundancy**, and that some genes are present in multiple copies. This is certainly true in some cases, in which multiple (related) genes must be knocked out in order to produce an effect. In a slightly more complex scenario, an organism might have two separate pathways capable of providing some activity. Inactivation of either pathway by itself would not be damaging, but the simultaneous mutation of genes in both pathways would be deleterious.

Such situations can be tested by combining mutations. Deletions in two genes, neither of which is lethal by itself, are introduced into the same strain. If the double mutant dies, the strain is called a **synthetic lethal**. This technique has been used to great effect with yeast, where the isolation of double mutants can be automated. The procedure is called **synthetic**

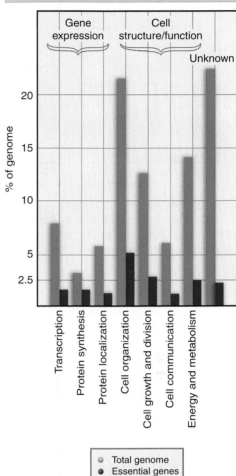

Figure 5.20 Essential yeast genes are found in all classes. Green bars show total proportion of each class of genes, red bars show those that are essential.

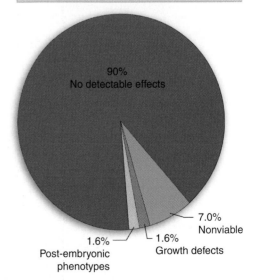

Figure 5.21 A systematic analysis of loss of function for 86% of worm genes shows that only 10% have detectable effects on the phenotype.

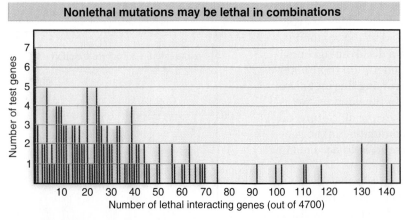

Nonlethal mutations may be lethal in combinations

Number of test genes (y-axis: 1–7)
Number of lethal interacting genes (out of 4700) (x-axis: 10–140)

Figure 5.22 Every one of 132 mutant test genes has some combinations that are lethal when it is combined with each of 4700 nonlethal mutations. The chart shows how many lethal interacting genes there are for each test gene.

genetic array analysis (SGA). **Figure 5.22** summarizes the results of an analysis in which an SGA screen was made for each of 132 viable deletions, by testing whether it could survive in combination with each of 4700 viable deletions. Every one of the test genes had at least one partner with which the combination was lethal, and most of the test genes had many such partners; the median is ~25 partners, and the greatest number is shown by one test gene that had 146 lethal partners. A small proportion (~10%) of the interacting mutant pairs codes for proteins that interact physically.

This result goes some way toward explaining the apparent lack of effect of so many deletions. Natural selection will act against these deletions when they find themselves in lethal pairwise combinations. To some degree, the organism has protected itself against the damaging effects of mutations by building in redundancy. However, it pays a price in the form of accumulating the "genetic load" of mutations that are not deleterious in themselves, but that may cause serious problems when combined with other such mutations in future generations. The theory of natural selection would suggest that the loss of the individual genes in such circumstances produces a sufficient disadvantage to maintain the active gene during the course of evolution.

5.10 About 10,000 Genes Are Expressed at Widely Different Levels in a Eukaryotic Tissue

Key Terms

- The **transcriptome** is the complete set of RNAs present in a cell, tissue, or organism. Its complexity is due mostly to mRNAs, but it also includes noncoding RNAs.

- The **abundance** of an mRNA is the average number of molecules per cell.

- **Abundant mRNAs** consist of a small number of individual species, each present in a large number of copies per cell.

- **Scarce mRNA (complex mRNA)** consists of a large number of individual mRNA species, each present in very few copies per cell. This accounts for most of the sequence complexity in RNA.

- **Housekeeping genes (constitutive genes)** are those (theoretically) expressed in all cells because they provide basic functions needed for sustenance of all cell types.

- **Luxury genes** are those coding for specialized functions synthesized (usually) in large amounts in particular cell types.

Key Concepts

- In any given cell, most genes are expressed at a low level.

- Only a small number of genes, whose products are specialized for the cell type, are highly expressed.

- mRNAs expressed at low levels overlap extensively when different cell types are compared.
- The abundantly expressed mRNAs are usually specific for the cell type.
- ~10,000 expressed genes may be common to most cell types of a higher eukaryote.

For a lower eukaryote such as yeast, the total number of expressed genes is ~4000. For somatic tissues of higher eukaryotes, the number usually is 10,000–15,000. The value is similar for plants and for vertebrates. How much overlap is there between the genes expressed in different tissues? For example, the expressed gene number of chick liver is ~11,000–17,000, compared with the value for oviduct of ~13,000–15,000. How many of these two sets of genes are identical? How many are specific for each tissue? These questions are usually addressed by analyzing the **transcriptome**—the set of sequences represented in RNA.

The transcriptome measures the number of genes that are expressed. Within this number, genes are expressed at widely differing levels. The average number of molecules of each particular mRNA per cell is called its **abundance**. We can divide the mRNA population into two general classes, according to their abundance:

- The **abundant mRNA** component typically consists of <100 different mRNAs present in 1000–10,000 copies per cell. It often corresponds to a major part of the mass, approaching 50% of the total mRNA.
- About half the mass of the mRNA consists of a large number of sequences, on the order of 10,000, each represented by only a small number of copies in the mRNA—say, <10. This is the **scarce mRNA** or **complex mRNA** class.

We see immediately that there are likely to be substantial differences among the genes expressed in the abundant class. Ovalbumin, for example, is synthesized only in the oviduct, not at all in the liver. It accounts for 50% of the mass of mRNA in the oviduct.

But the abundant mRNAs represent only a small proportion of the number of expressed genes. In terms of the total number of genes of the organism, and of the number of changes in transcription that must be made between different cell types, we need to know the extent of overlap between the genes represented in the scarce mRNA classes of different cell phenotypes.

Comparisons between different tissues show that, for example, ~75% of the sequences expressed in liver and oviduct are the same. In other words, ~12,000 genes are expressed in both liver and oviduct, ~5000 additional genes are expressed only in liver, and ~3000 additional genes are expressed only in oviduct.

The scarce mRNAs overlap extensively. Between mouse liver and kidney, ~90% of the scarce mRNAs are identical, leaving a difference between the tissues of only 1000–2000 in terms of the number of expressed genes. The general result obtained in several comparisons of this sort is that only ~10% of the mRNA sequences of a cell are unique to it. The majority of sequences are common to many, perhaps even all, cell types.

This suggests that the common set of expressed gene functions, numbering perhaps ~10,000 in a mammal, comprises functions that are needed in all cell types. Sometimes a gene encoding this type of function is referred to as a **housekeeping gene** or **constitutive gene**. Such functions contrast with specialized functions (such as those of ovalbumin or globin) needed only for particular cell phenotypes. Genes for specialized functions are sometimes called **luxury genes**.

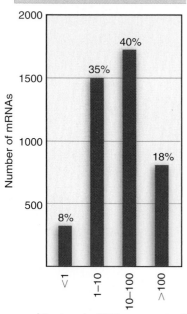

Figure 5.23 The abundances of yeast mRNAs vary from <1 per cell (meaning that not every cell has a copy of the mRNA) to >100 per cell (coding for the more abundant proteins).

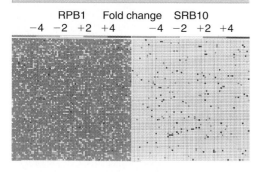

Figure 5.24 HDA analysis allows change in expression of each gene to be measured. Each square represents one yeast gene (in each panel, the left part is first gene on chromosome I, the right part is last gene on chromosome XVI). Change in expression relative to wild type is indicated by red (reduction), white (no change), or blue (increase). Photograph kindly provided by Rick Young.

5.11 Expressed Gene Number Can Be Measured *en masse*

Key Concepts

- "Chip" technology allows a snapshot to be taken of the expression of the entire genome in a yeast cell.
- ~75% (~4500 genes) of the yeast genome is expressed under normal growth conditions.
- Chip technology allows detailed comparisons of related animal cells to determine (for example) the differences in expression between a normal cell and a cancer cell.

Recent technology allows more systematic and accurate estimates of the number of expressed genes. One approach (SAGE, serial analysis of gene expression) allows a unique sequence tag to be used to identify each mRNA. The technology then allows the abundance of each tag to be measured. This approach identifies 4665 expressed genes in *S. cerevisiae* growing under normal conditions, with abundances varying from 0.3 to >200 transcripts/cell. **Figure 5.23** summarizes the proportions of mRNA at each abundance level. About 75% of the total gene number (~6000) can be detected in the expressed fraction.

The most powerful new technology uses chips that contain high-density oligonucleotide arrays (HDAs). Their construction is made possible by knowledge of the sequence of the entire genome. In the case of *S. cerevisiae*, each of 6181 ORFs is represented on the HDA by 20 25-mer oligonucleotides that perfectly match the sequence of the message, and 20 mismatch oligonucleotides that differ at one base position. The expression level of any gene is calculated by subtracting the average signal of a mismatch from its perfect match partner. The entire yeast genome can be represented on four chips. This technology is sensitive enough to detect transcripts of 5460 genes (~90% of the genome), and shows that many genes are are expressed at low levels, with abundances of 0.1–2 transcripts/cell. An abundance of <1 transcript/cell means that not all cells have a copy of the transcript at any given moment.

The technology allows not only measurement of levels of gene expression, but also detection of differences in expression in mutant cells compared with wild type, cells growing under different growth conditions, and so on. The results of comparing two states are expressed in a grid, in which each square represents a particular gene, and the relative change in expression is indicated by color. The upper part of **Figure 5.24** shows the effect of a mutation in RNA polymerase II, the enzyme that produces mRNA, which as might be expected causes the expression of most genes to be heavily reduced. By contrast, the lower part shows that a mutation in an ancillary component of the transcription apparatus (*SRB10*) has much more restricted effects, causing increases in expression of some genes.

The extension of this technology to animal cells will allow the general descriptions based on RNA hybridization analysis to be replaced by exact descriptions of the genes that are expressed, and the abundances of their products, in any given cell type.

5.12 SUMMARY

Genomes that have been sequenced include many bacteria and archaea, yeasts, and a worm, fly, mouse, and man. The minimum number of genes required to make a living cell (an obligatory intracellular parasite) is ~470. The minimum number required to make a free-living cell is ~1700. A typical gram-negative bacterium has ~1500 genes. Strains of *E. coli* vary from 4300 to 5400 genes. The average bacterial gene is ~1000 bp long and is separated from the next gene by a space of ~100 bp. The yeasts *S. pombe* and *S. cerevisiae* have 5000 and 6000 genes, respectively.

Although the fly *D. melanogaster* is a more complex organism and has a larger genome than the worm *C. elegans*, the fly has fewer genes (13,600) than the worm (14,100). The plant *Arabidopsis* has 25,000 genes, and the lack of a clear relationship between genome size and gene number is shown by the fact that the rice genome is 4× larger, but contains only a 50% increase in gene number, to ~40,000. Mouse and man have ~30,000 genes, which is much less than had been expected. The complexity of development of an organism may depend on the nature of the interactions between genes as well as their total number.

About 6000 genes are common to prokaryotes and eukaryotes and are likely to be involved in basic functions. A further 10,000 genes are found in multicellular organisms. Another 8000 genes are added to make an animal, and a further 6000 (largely involved with the immune and nervous systems) are found in vertebrates. In each organism that has been sequenced, only ~50% of the genes have defined functions. Analysis of lethal genes suggests that only a minority of genes are essential in each organism.

Genes are expressed at widely varying levels. There may be 10^5 copies of the mRNA for an abundant gene whose protein is the principal product of the cell, 10^3 copies of each mRNA for <10 moderately abundant messages, and <10 copies of each mRNA for >10,000 scarcely expressed genes. Overlaps between the mRNA populations of cells of different phenotypes are extensive; the majority of mRNAs are present in most cells.

Clusters and Repeats

6.1 Introduction

Key Terms

- A **gene family** consists of a set of genes whose exons are related; the members were derived by duplication and variation from some ancestral gene.

- A **translocation** is a rearrangement in which part of a chromosome is detached by breakage or aberrant recombination and then becomes attached to some other chromosome.

- A **gene cluster** is a group of adjacent genes that are identical or related.

- **Unequal crossing-over (nonreciprocal recombination)** results from an error in pairing and crossing-over in which nonequivalent sites participate in a recombination event. It produces one recombinant with a deletion of material and one with a duplication.

The initial event that creates related exons or genes is a duplication. A set of genes descended by duplication and variation from some ancestral gene is called a **gene family**. Its members may be clustered together or dispersed on different chromosomes (or a combination of both). Genome analysis shows that many genes belong to families; the 30,000 genes identified in the human genome fall into ~15,000 families, so the average gene has a couple of relatives in the genome (see Figure 6.13). Gene families vary enormously in the degree of relatedness between members, from those consisting of multiple identical members to those in which the relationship is quite distant. Genes are usually related only by their exons, with introns having diverged (see *3.5 Exon Sequences Are Conserved but Introns Vary*).

Tandem duplications are created when the duplicates remain together. They may arise through errors in replication or recombination.

Duplicates separate when a **translocation** transfers material from one chromosome to another. A duplicate at a new location may also be produced directly by a transposition event that is associated with copying a region of DNA from the vicinity of a transposon. Duplications may apply either to intact genes or to collections of exons or even individual exons. Immediately after duplication of a whole gene, there are two copies whose activities are indistinguishable, but then usually the copies diverge as each accumulates different mutations.

The members of a well-related structural gene family usually have related or even identical functions, although they may be expressed at different times or in different cell types. So different globin proteins are expressed in embryonic and adult red blood cells, and different actins are utilized in muscle and nonmuscle cells. When genes have diverged significantly, or when only some exons are related, the proteins may have different functions.

Some gene families consist of identical members. Clustering is a prerequisite for maintaining identity between genes, although clustered genes are not necessarily identical. **Gene clusters** range from extremes where a duplication has generated two adjacent related genes to cases where hundreds of identical genes lie in a tandem array. There may be extensive tandem repetition of a gene whose product is needed in unusually large amounts. Examples are the genes for rRNA or histone proteins. This creates a special situation with regards to the maintenance of identity and the effects of selective pressure.

Gene clusters offer us an opportunity to examine the evolution of the genome over larger regions than single genes. Duplicated sequences, especially those that remain in the same vicinity, provide the substrate for further evolution by recombination. A normal recombination event is an exact exchange between two chromosomes, generating recombinant chromosomes that have the same organization as the parental chromosome. However, the existence of repeated copies of a sequence makes **unequal crossing-overs** possible as the result of a recombination event between two sites that are not homologous.

Figure 6.1 shows that unequal crossing-over happens when one copy of a repeat in one chromosome misaligns for recombination with a different copy of the repeat in the homologous chromosome, instead of with the corresponding copy. Recombination then increases the number of repeats in one chromosome and decreases it in the other. In effect, one recombinant chromosome has a deletion and the other has an insertion. This mechanism is responsible for the evolution of clusters of related sequences. A repeat can be as substantial as a whole gene or as short as a simple oligonucleotide sequence. The principle is the same, that unequal crossing-over can expand or contract the size of an array in both gene clusters and regions of highly repeated DNA.

The highly repetitive fraction of the genome consists of multiple tandem copies of very short repeating units. These often have unusual properties. One is that they may be identified as a separate peak on a density gradient analysis of DNA, which gave rise to the name *satellite DNA*. Multiple tandem copies are often associated with inert regions of the chromosomes, and in particular with centromeres (which contain the points of attachment for segregation on a mitotic or meiotic spindle). Because of their repetitive organization, they show some of the same behavior with regard to evolution as the tandem gene clusters. In addition to the satellite sequences, there are shorter stretches of DNA that show similar behavior, called *minisatellites*. They are useful in showing a high degree of divergence between individual genomes that can be used for mapping purposes.

All of these events that change the constitution of the genome are rare, but they are significant over the course of evolution.

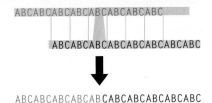

Unequal crossing-over changes the repeat number

Figure 6.1 Unequal crossing-over results from pairing between nonequivalent repeats in regions of DNA consisting of repeating units. Here the repeating unit is the sequence ABC, and the third repeat of the red chromosome has aligned with the first repeat of the black chromosome. Throughout the region of pairing, ABC units of one chromosome are aligned with ABC units of the other chromosome. Crossing-over generates chromosomes with 10 and 6 repeats each, instead of the 8 repeats of each parent.

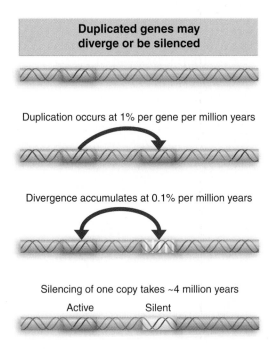

Duplicated genes may diverge or be silenced

Duplication occurs at 1% per gene per million years

Divergence accumulates at 0.1% per million years

Silencing of one copy takes ~4 million years

Active   Silent

Figure 6.2 After a gene has been duplicated, differences may accumulate between the copies. The genes may acquire different functions or one of the copies may become inactive.

6.2 Gene Duplication Is a Major Force in Evolution

Key Concept

- Duplicated genes may diverge, generating different genes, or one copy may become inactive.

When a gene is duplicated, mutations can accumulate in one copy without attracting the adverse attention of natural selection. This copy may then evolve to a new function; it may become expressed in a different time or place from the first copy, or it may acquire different activities.

Figure 6.2 summarizes our present view of the rates at which these processes occur. There is ~1% probability that a given gene will be included in a duplication in a period of 1 million years. After the gene has duplicated, differences develop as the result of the occurrence of different mutations in each copy. These accumulate at a rate of ~0.1% per million years (see *6.4 Sequence Divergence Is the Basis for the Evolutionary clock*).

The organism is not likely to need to retain two identical copies of the gene. As differences develop between the duplicate genes, one of two types of event is likely to occur

- Both of the genes become necessary. This can happen either because the differences between them generate proteins with different functions, or because they are expressed specifically in different times or places.
- If this does not happen, one of the genes is likely to be silenced because it will by chance gain a deleterious mutation. Typically this takes ~4 million years. In such a situation, it is purely a matter of chance which of the two copies becomes inactive. (This can contribute to incompatibility between different individuals, and ultimately to speciation, if different copies become inactive in different populations.) In the initial period after the gene has been silenced, it can be recognized as a pseudogene. Ultimately, it may diverge so far from the active gene that the relatedness is no longer recognizable.

6.3 Globin Clusters Are Formed by Duplication and Divergence

Key Concepts

- All globin genes are descended by duplication and mutation from an ancestral gene that had three exons.
- The ancestral gene gave rise to myoglobin, leghemoglobin, and α- and β-globins.
- The α- and β-globin genes separated in the period of early vertebrate evolution, after which duplications generated the individual clusters of separate α-like and β-like genes.
- Once a gene has been inactivated by mutation, it may accumulate further mutations and become a pseudogene, which is homologous to the active gene(s) but has no functional role.

Key Term

- **Nonallelic genes** are two (or more) copies of the same gene that are present at *different* locations in the genome (contrasted with alleles, which are copies of the same gene derived from different parents and present at the same location on homologous chromosomes).

The most common type of duplication generates a second copy of a gene located close to the first copy. In some cases, the copies remain associated, and further duplication may generate a cluster of related genes. The best characterized example of a gene cluster is presented by the globin genes, which constitute an ancient gene family, concerned with a function that is central to the animal kingdom: the transport of oxygen through the bloodstream.

The major constituent of the red blood cell is the globin tetramer, associated with its heme (iron-binding) group in the form of hemoglobin. Functional globin genes in all species have the same general structure, divided into three exons as shown previously in Figure 3.8. We conclude that all globin genes are derived from a single ancestral gene; so by tracing the development of individual globin genes within and between species, we may learn about the mechanisms by which gene families evolve.

In adult cells, the globin tetramer consists of two identical α chains and two identical β chains. Embryonic blood cells contain hemoglobin tetramers that are different from the adult form. Each embryonic tetramer contains two identical α-like chains and two identical β-like chains, each of which is related to the adult polypeptide and is later replaced by it. This is an example of developmental control, in which different genes are successively switched on and off to provide alternative products that fulfill the same function at different times.

The division of globin chains into α-like and β-like reflects the organization of the genes. Each type of globin is coded by genes organized into a single cluster. The structures of the two clusters in the higher primate genome are illustrated in **Figure 6.3**.

Stretching over 50 kb, the β cluster contains five functional genes (ε, two γ, δ, and β) and one pseudogene ($\psi\beta$). The two γ genes differ in their coding sequence in only one amino acid; the G variant has glycine at position 136, where the A variant has alanine.

The more compact α cluster extends over 28 kb and includes one active ζ gene, one ζ pseudogene, two α genes, two α pseudogenes, and the θ gene of unknown function. The two α genes code for the same protein. Two (or more) identical genes present on the same chromosome are called **nonallelic genes**.

The details of the relationship between embryonic and adult hemoglobins vary with the organism. The human pathway has three stages: embryonic, fetal, and adult. The distinction between embryonic and adult is common to mammals, but the number of preadult stages varies. In man, zeta and alpha are the two α-like chains. Epsilon, gamma, delta, and beta are the β-like chains. **Figure 6.4** shows how the chains are expressed at different stages of development.

In the human pathway, ζ is the first α-like chain to be expressed, but is soon replaced by α. In the β-pathway, ε and γ are expressed first, with δ and β replacing them later. In adults, the $\alpha_2\beta_2$ form provides 97% of the hemoglobin, $\alpha_2\delta_2$ is ~2%, and ~1% is provided by persistence of the fetal form $\alpha_2\gamma_2$.

What is the significance of the differences between embryonic and adult globins? The embryonic and fetal forms have a higher affinity for oxygen. This is necessary for obtaining oxygen from the mother's blood. This explains why there is no equivalent in (for example) chicken, where the embryonic stages occur outside the maternal body (that is, within the egg).

A similar general organization is found in other vertebrate globin gene clusters, but details of the types, numbers, and order of genes all vary, as illustrated in **Figure 6.5**. Each cluster contains both embryonic and adult genes. The total lengths of the clusters vary widely. The longest is found in the goat, where a basic cluster of four genes has been duplicated twice. The distribution of active genes and pseudogenes differs in each species, illustrating the randomness of the conversion of copies of a duplicated gene into the inactive state.

The characterization of these gene clusters makes an important general point: *there may be more members of a gene family, both functional and nonfunctional, than we would suspect on the basis of protein analysis.* The extra functional genes may represent duplicates that code for identical polypeptides; or they may be related to known proteins, although different

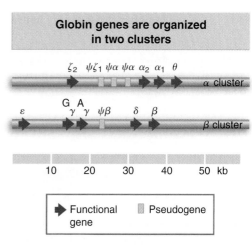

Globin genes are organized in two clusters

$\zeta_2 \quad \psi\zeta_1 \quad \psi\alpha \quad \psi\alpha \quad \alpha_2 \quad \alpha_1 \quad \theta$

α cluster

$\varepsilon \qquad \begin{matrix} G & A \\ \gamma & \gamma \end{matrix} \quad \psi\beta \qquad \delta \quad \beta$

β cluster

10   20   30   40   50 kb

Functional gene   Pseudogene

Figure 6.3 Each of the α-like and β-like globin gene families is organized into a single cluster that includes functional genes and pseudogenes (ψ).

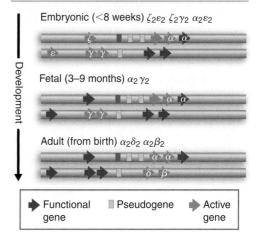

Hemoglobin expression changes during development

Embryonic (<8 weeks) $\zeta_2\varepsilon_2 \quad \zeta_2\gamma_2 \quad \alpha_2\varepsilon_2$

$\zeta \qquad\qquad \alpha \cdot \alpha$

$\varepsilon \qquad \gamma \cdot \gamma$

Fetal (3–9 months) $\alpha_2\gamma_2$

$\alpha \cdot \alpha$

$\gamma \cdot \gamma$

Adult (from birth) $\alpha_2\delta_2 \quad \alpha_2\beta_2$

$\alpha \cdot \alpha$

$\delta \cdot \beta$

Development

Functional gene   Pseudogene   Active gene

Figure 6.4 Different hemoglobin genes are expressed during embryonic, fetal, and adult periods of human development.

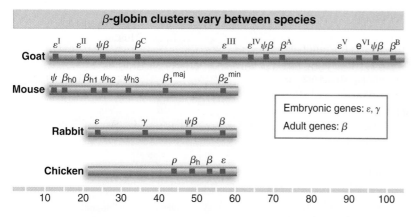

Figure 6.5 Clusters of β-globin genes and pseudogenes are found in vertebrates. Seven mouse genes include 2 early embryonic, 1 late embryonic, 2 adult genes, and 2 pseudogenes. Rabbit and chick each have four genes.

Globin genes have duplicated and diverged

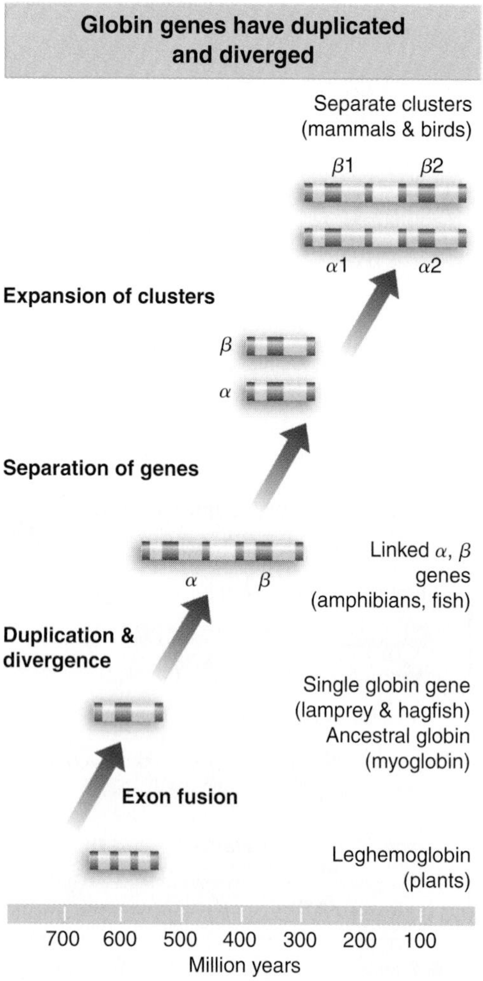

Figure 6.6 All globin genes have evolved by a series of duplications, transpositions, and mutations from a single ancestral gene.

from them (and presumably expressed only briefly or in low amounts).

From the organization of globin genes in a variety of species, we should be able to trace the evolution of present globin gene clusters from a single ancestral globin gene. Our present view of the evolutionary descent is pictured in **Figure 6.6**.

The leghemoglobin gene of plants, which is related to the globin genes, may represent the ancestral form. The farthest back that we can trace a globin gene in modern form is provided by the sequence of the single chain of mammalian myoglobin, which diverged from the globin line of descent ~800 million years ago. The myoglobin gene has the same organization as globin genes, so we may take the three-exon structure to represent their common ancestor.

Some "primitive" fish have only a single type of globin chain, so they must have diverged from the line of evolution before the ancestral globin gene was duplicated to give rise to the α and β variants. This appears to have occurred ~500 million years ago, during the evolution of the bony fish.

The next stage of evolution is represented by the state of the globin genes in the frog *Xenopus laevis*, which has two globin clusters. However, each cluster contains *both* α and β genes, of both larval and adult types. The cluster must therefore have evolved by duplication of a linked α-β pair, followed by divergence between the individual copies. Later the entire cluster was duplicated.

The amphibians separated from the mammalian/avian line ~350 million years ago, so the separation of the α- and β-globin genes must have resulted from a transposition in the mammalian/avian forerunner after this time. This probably occurred in the period of early vertebrate evolution. Because there are separate clusters for α and β globins in both birds and mammals, the α and β genes must have been physically separated before the mammals and birds diverged from their common ancestor, probably ~270 million years ago.

Changes have occurred within the separate α and β clusters in more recent times, as we see from the description of the divergence of the individual genes in the next section.

6.4 Sequence Divergence Is the Basis for the Evolutionary Clock

Key Terms

- **Divergence** is the percent difference in nucleotide sequence between two related DNA sequences or in amino acid sequences between two proteins.
- **Replacement sites** in a gene are those at which mutations alter the amino acid that is coded.
- A **silent site** in a coding region is a site where mutation does not change the sequence of the protein.
- The **evolutionary clock** is defined by the rate at which mutations accumulate in a given gene.

Key Concepts

- The sequences of homologous genes in different species vary at replacement sites (where mutation causes amino acid substitutions) and silent sites (where mutation does not affect the protein sequence).
- Mutations accumulate at silent sites ~10× faster than at replacement sites.
- The evolutionary divergence between two proteins is measured by the percent of positions at which the corresponding amino acids are different.
- Mutations accumulate at a more or less even speed after genes separate, so that the divergence between any pair of globin sequences is proportional to the time since their genes separated.

Most changes in protein sequences occur by small mutations that accumulate slowly with time. Point mutations and small insertions and deletions occur by chance, probably with more or less equal probability in all regions of the genome, except for hotspots at which mutations occur much more frequently. Most mutations that change the amino acid sequence are deleterious and are eliminated by natural selection.

When a species separates into two new species, each now constitutes an independent pool for evolution. By comparing the corresponding proteins in two species, we see the differences that have accumulated between them *since the time when their ancestors ceased to interbreed.* Some proteins are highly conserved, showing little or no change from species to species. This indicates that almost any change is deleterious and therefore selected against.

Other proteins show significant differences. We describe the difference between two proteins, or between the genes that code for them, as their **divergence**, the percent of positions at which there are differences.

The divergence between proteins can be different from the divergence between the corresponding nucleic acid sequences. The source of this difference is the representation of each amino acid in a three-base codon, in which often the third base has no effect on the meaning.

We may divide the nucleotide sequence of a coding region into potential **replacement sites** and **silent sites**:

- At replacement sites, a mutation alters the amino acid that is coded. The effect of the mutation (deleterious, neutral, or advantageous) depends on the result of the amino acid replacement.
- At silent sites, mutation only substitutes one synonym codon for another, so there is no change in the protein. (However, they can affect gene expression by changing the properties of the RNA.)
- Usually the replacement sites account for 75% of a coding sequence and the silent sites for 25%. This means that a nucleic acid divergence of 0.45% at replacement sites corresponds to an amino acid divergence of 1%.

In the example of the human β- and δ-globin chains, there are 10 differences in 146 residues, a divergence of 6.9%. The DNA sequence has 31 changes in 441 residues. However, these changes are distributed very differently in the replacement and silent sites. There are 11 changes in the 330 replacement sites, but 20 changes in only 111 silent sites. This gives (corrected) rates of divergence of 3.7% in the replacement sites and 32% in the silent sites, an order of magnitude in difference.

The striking difference in the divergence of replacement and silent sites demonstrates the existence of much greater constraints on nucleotide positions that influence protein constitution relative to those that do not.

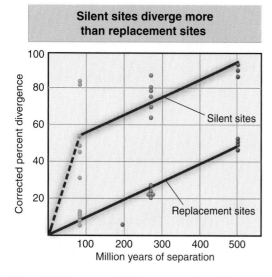

Silent sites diverge more than replacement sites

Figure 6.7 Divergence of DNA sequences depends on evolutionary separation. Each point on the graph represents a pairwise comparison.

Suppose we take the rate of mutation at silent sites to indicate the underlying rate of mutation. Then over the period since the β and δ genes diverged, there should have been changes at 32% of the 330 replacement sites, a total of 105. All but 11 of them have been eliminated, which means that ~90% of the mutations did not survive.

The divergence between any pair of globin sequences is (more or less) proportional to the time since they separated. This provides an **evolutionary clock** that measures the accumulation of mutations at an apparently even rate during the evolution of a given protein.

The rate of divergence can be measured as the percent difference per million years, or as its reciprocal, the unit evolutionary period (UEP), the time in millions of years that it takes for 1% divergence to develop. Once the clock has been established by pairwise comparisons between species (remembering the practical difficulties in establishing the actual time of speciation), it can be applied to related genes *within* a species. From their divergence, we can calculate how much time has passed since the duplication that generated them.

By comparing the sequences of homologous genes in different species, the rate of divergence at both replacement and silent sites can be determined, as plotted in **Figure 6.7**. In pairwise comparisons, there is an average divergence of ~0.096% per million years (or a UEP of 10.4). Considering the uncertainties in estimating the times at which the species diverged, the results lend good support to the idea that there is a linear clock for evolution of globin.

The rate at silent sites is not linear with regard to time. *If we assume that there must be zero divergence at zero years of separation*, we see that the rate of silent site divergence is much greater for the first ~100 million years of separation. One interpretation is that a fraction of roughly half of the silent sites is rapidly (within 100 million years) saturated by mutations; this fraction behaves as neutral sites (where mutation has no effect; see *6.5 The Rate of Neutral Substitution Can Be Measured from Divergence of Repeated Sequences*). The other fraction of the silent sites accumulates mutations more slowly, at a rate approximately the same as that of the replacement sites; this fraction identifies sites that are silent with regard to the protein, but that come under selective pressure for some other reason.

Now we can reverse the calculation of divergence rates to estimate the times since genes within a species have been apart. The difference between the human β and δ genes is 3.7% for replacement sites. At a UEP of 10.4, these genes must have diverged 10.4 × 3.7 = 40 million years ago—about the time of the separation of the lines leading to New World monkeys, Old World monkeys, great apes, and man. All of these higher primates have both β and δ genes, which suggests that the gene divergence commenced just before this point in evolution.

Proceeding farther back, the divergence between the replacement sites of γ and ε genes is 10%, which corresponds to a time of separation ~100 million years ago. The separation between embryonic and fetal globin genes therefore may have just preceded or accompanied the mammalian radiation.

An evolutionary tree for the human globin genes is constructed in **Figure 6.8**. Features that evolved before the mammalian radiation—such as the separation of β/δ from γ—should be found in all mammals. Features that evolved afterward—such as the separation of β- and δ-globin genes—should be found in individual lines of mammals.

In each species, there have been comparatively recent changes in the structures of the clusters, since we see differences in gene number (one adult β-globin gene in man, two in mouse) or in type (most often concerning whether there are separate embryonic and fetal genes).

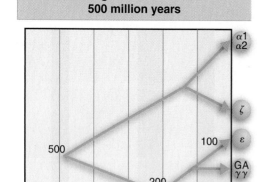

Globin genes evolved over 500 million years

Figure 6.8 Replacement site divergences between pairs of β-globin genes allow the history of the human cluster to be reconstructed. This tree accounts for the separation of classes of globin genes.

When sufficient data have been collected on the sequences of a particular gene, the arguments can be reversed, and comparisons between genes in different species can be used to assess taxonomic relationships.

6.5 The Rate of Neutral Substitution Can Be Measured from Divergence of Repeated Sequences

Key Concept

• The rate of substitution per year at neutral sites is greater in the mouse genome than in the human genome.

Key Term

• A **neutral mutation** has no significant effect on evolutionary fitness and usually has no effect on the phenotype.

Mutations that have no effect are called **neutral mutations**; the accumulation of mutations at neutral sites gives us our best insight into the underlying rate of mutation. We can estimate the rate of substitution at neutral sites by examining sequences that do not code for protein. (We use the term *neutral* here rather than *silent*, because there is no coding potential.) An informative comparison can be made of a common repetitive family in the human and mouse genomes.

The principle of the analysis is summarized in **Figure 6.9**. We start with a family of related sequences that have evolved by duplication and substitution. We assume that the common ancestral sequence can be deduced by taking the base that is most common at each position. Then we can calculate the divergence of each individual family member as the proportion of bases that differ from the deduced ancestral sequence. In this example, individual members diverge from 0.13 to 0.18, and the average is 0.16.

One gene family used for this analysis in the human and mouse genomes derives from a sequence that is thought to have ceased to be active at about the time of the divergence between man and rodents (the LINES family; see *22.9 Retroposons Fall into Three Classes*). This means that the family has been diverging without any selective pressure for the same length of time in both species. Its average divergence in man is ~0.17 substitutions per site, corresponding to a rate of 2.2×10^{-9} substitutions per base per year over the 75 million years since the separation. In the mouse genome, however, neutral substitutions have occurred at twice this rate, corresponding to 0.34 substitutions per site in the family, or a rate of 4.5×10^{-9}. However, note that if we calculated the rate per generation instead of per year, it would be greater in man than in mouse (~2.2×10^{-8} as opposed to ~10^{-9}). The difference between the species demonstrates that each species has systems that operate with a characteristic efficiency.

Comparing the mouse and human genomes allows us to assess whether syntenic (corresponding) sequences show signs of conservation or have differed at the rate expected from accumulation of neutral substitutions. About 5% of sites in these regions show signs of selection. This is much higher than the proportion that codes for protein or RNA (~1%). It implies that the genome includes many more stretches whose sequence is important for noncoding functions than for coding functions. Known regulatory elements are likely to compose only a small part of this proportion. This number also suggests that most (i.e., the rest) of the genome sequences do not have any function that depends on the exact sequence.

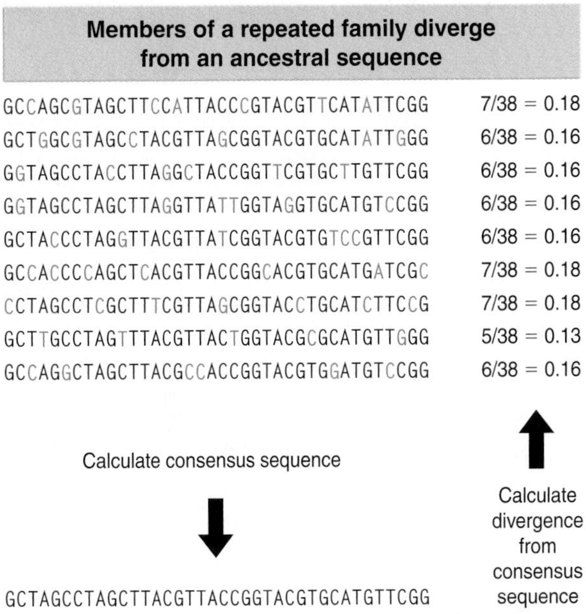

Members of a repeated family diverge from an ancestral sequence

GCCAGCGTAGCTTCCATTACCCGTACGTTCATATTCGG	7/38 = 0.18
GCTGGCGTAGCCTACGTTAGCGGTACGTGCATATTGGG	6/38 = 0.16
GGTAGCCTACCTTAGGCTACCGGTTCGTGCTTGTTCGG	6/38 = 0.16
GGTAGCCTAGCTTAGGTTATTGGTAGGTGCATGTCCGG	6/38 = 0.16
GCTACCCTAGGTTACGTTATCGGTACGTGTCCGTTCGG	6/38 = 0.16
GCCACCCCAGCTCACGTTACCGGCACGTGCATGATCGC	7/38 = 0.18
CCTAGCCTCGCTTTCGTTAGCGGTACCTGCATCTTCCG	7/38 = 0.18
GCTTGCCTAGTTTACGTTACTGGTACGCGCATGTTGGG	5/38 = 0.13
GCCAGGCTAGCTTACGCCACCGGTACGTGGATGTCCGG	6/38 = 0.16

Calculate consensus sequence

GCTAGCCTAGCTTACGTTACCGGTACGTGCATGTTCGG

Calculate divergence from consensus sequence

Figure 6.9 An ancestral consensus sequence for a family is calculated by taking the most common base at each position. The divergence of each existing current member of the family is calculated as the proportion of bases at which it differs from the ancestral sequence.

6.6 Unequal Crossing-Over Rearranges Gene Clusters

Key Terms

- **Thalassemia** is a disease of red blood cells resulting from lack of either α or β globin.

- **HbH** disease results from a condition in which there is a disproportionate amount of the abnormal tetramer β_4 relative to the amount of normal hemoglobin ($\alpha_2\beta_2$).

- **Hydrops fetalis** is a fatal disease resulting from the absence of the hemoglobin α gene.

- **Hb Lepore** is an unusual globin protein that results from unequal crossing-over between the β and δ genes. The genes become fused together to produce a single β-like chain that consists of the N-terminal sequence of δ joined to the C-terminal sequence of β.

- **Hb anti-Lepore** is a fusion gene produced by unequal crossing-over that has the N-terminal part of β globin and the C-terminal part of δ globin.

Key Concepts

- Unequal crossing-over between nonallelic members of a gene cluster creates new genes, each of which has the N-terminal part of one of the parental genes and the C-terminal part of the other parental gene.

- Different thalassemias are caused by various deletions that eliminate α- or β-globin genes. The severity of the disease depends on the individual deletion.

There are frequent opportunities for rearrangement in a cluster of related or identical genes. We can see the results by comparing the mammalian β clusters included in Figure 6.5. Although the clusters serve the same function, and all have the same general organization, each is different in size, there is variation in the total number and types of β-globin genes, and the numbers and structures of pseudogenes are different. All of these changes must have occurred since the mammalian radiation, ~85 million years ago (the last point in evolution common to all mammals).

The comparison makes the general point that gene duplication, rearrangement, and variation is as important a factor in evolution as the slow accumulation of point mutations in individual genes. What types of mechanisms are responsible for gene reorganization?

Unequal crossing-over (also known as nonreciprocal recombination) can occur as the result of pairing between two sites that are *not* homologous. Usually, recombination involves corresponding sequences of DNA held in exact alignment between the two homologous chromosomes. However, when there are two copies of a gene on each chromosome, an occasional misalignment allows pairing between them. (This requires some of the adjacent regions to go unpaired.) This can happen in a region of short repeats (see Figure 6.1) or in a gene cluster. **Figure 6.10** shows that unequal crossing-over in a gene cluster can have two consequences, quantitative and qualitative:

- The number of repeats increases in one chromosome and decreases in the other. In effect, one recombinant chromosome has a deletion and the other has an insertion. This happens irrespective of the exact location of the crossover. In the figure, the first recombinant has an increase in the number of gene copies from 2 to 3, while the second has a decrease from 2 to 1.

- If the recombination event occurs within a gene (rather than between genes), the result depends on whether the recombining genes are identical or only related. If the noncorresponding gene copies 1 and 2 are entirely homologous, there is no change in the sequence of either gene. However, unequal crossing-over also can occur when the adjacent genes are well related (although

Unequal crossing-over creates a duplication and a deletion

Normal crossing-over

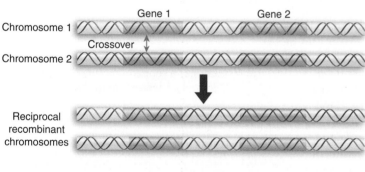

Unequal crossing-over

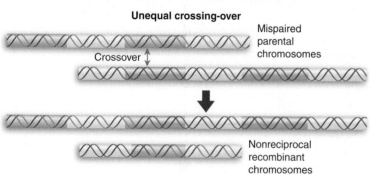

Figure 6.10 Gene number can be changed by unequal crossing-over. If gene 1 of one chromosome pairs with gene 2 of the other chromosome, the other gene copies are excluded from pairing. Recombination between the mispaired genes produces one chromosome with a single (recombinant) copy of the gene and one chromosome with three copies of the gene (one from each parent and one recombinant).

the probability is less than when they are identical). In this case, each of the recombinant genes has a sequence that is different from either parent.

Whether the chromosome has a selective advantage or disadvantage will depend on the consequence of any change in the sequence of the gene product as well as on the change in the number of gene copies.

Thalassemias result from mutations that reduce or prevent synthesis of either α or β globin. The occurrence of unequal crossing-over in the human globin gene clusters is revealed by the nature of certain thalassemias.

Figure 6.11 summarizes the deletions that cause the α-thalassemias. α-thal-1 deletions are long, varying in the location of the left end, with the positions of the right ends located beyond the known genes. They eliminate both α genes. The α-thal-2 deletions are short and eliminate only one of the two α genes. The L deletion removes 4.2 kb of DNA, including the α2 gene. It probably results from unequal crossing-over, because the ends of the deletion lie in homologous regions, just to the right of the ψα and α2 genes, respectively. The R deletion results from the removal of exactly 3.7 kb of DNA, the precise distance between the α1 and α2 genes. It appears to have been generated by unequal crossing-over between the α1 and α2 genes themselves. This is precisely the situation depicted in Figure 6.10.

Depending on the diploid combination of thalassemic chromosomes, an affected individual may have any number of α chains from zero to three. Individuals with three α genes do not display phenotypic differences from wild type. But with only one α gene, the excess β chains form the unusual tetramer β_4, which causes **HbH** disease. The complete absence of α genes results in **hydrops fetalis**, which is fatal at or before birth.

The **Hb Lepore** type provided the classic evidence that deletion can result from unequal crossing-over between linked genes. The β and δ genes differ only ~7% in sequence. Unequal recombination deletes the material between the genes, thus fusing them together (see Figure 6.10). The fused gene produces a single β-like chain that consists of the N-terminal sequence of δ joined to the C-terminal sequence of β.

Several types of Hb Lepore now are known, the difference between them lying in the point of transition from δ to β sequences. So when the δ and β genes pair for unequal crossing-over, the exact point of recombination determines the position at which the switch from δ to β sequence occurs in the amino acid chain.

The reciprocal of this event has been found in the form of **Hb anti-Lepore**, which is produced by a gene that has the N-terminal part of β and the C-terminal part of δ. The fusion gene lies between normal δ and β genes.

From the differences between the globin gene clusters of various mammals, we see that duplication followed (sometimes) by variation has been an important feature in the evolution of each cluster. The human thalassemic deletions demonstrate that unequal crossing-over continues to occur in both globin gene clusters. Each such event generates a duplication as well as the deletion, and we must account for the fate of both recombinant loci in the population. Deletions can also occur (in principle) by recombination between homologous sequences lying on the *same* chromosome. This does not generate a corresponding duplication.

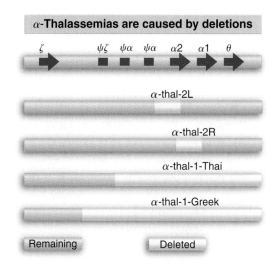

Figure 6.11 α-Thalassemias result from various deletions in the α-globin gene cluster.

6.7 Genes for rRNA Form Tandem Repeats Including an Invariant Transcription Unit

Key Terms

- **Ribosomal DNA (rDNA)** is usually a tandemly repeated series of genes coding for a precursor to the two large rRNAs.
- The **nucleolus** (plural, *nucleoli*) is a discrete region of the nucleus where ribosomes are produced.
- The **nucleolar organizer** is the region of a chromosome carrying genes coding for rRNA.
- The **nontranscribed spacer** is the region between transcription units in a tandem gene cluster.

Key Concepts

- Ribosomal RNA is coded by a large number of identical genes that are tandemly repeated to form one or more clusters.
- Each rDNA cluster is organized so that transcription units giving a joint precursor to the major rRNAs alternate with nontranscribed spacers.
- The nontranscribed spacers consist of shorter repeating units whose number varies so that the lengths of individual spacers differ.

In the cases we have discussed so far, there are differences between the individual members of a gene cluster that allow selective pressure to act independently upon each gene. A contrast is provided by two large gene clusters that contain many identical copies of the same gene or genes. Most organisms contain multiple copies of the genes for the histone proteins that are a major component of the chromosomes; and there are almost always multiple copies of the genes that code for the ribosomal RNAs. These large gene clusters pose some interesting evolutionary questions.

Ribosomal RNA is the predominant product of transcription, constituting some 80–90% of the total mass of cellular RNA in both eukaryotes and prokaryotes. The number of major rRNA genes varies from 7 in *E. coli*, 100–200 in lower eukaryotes, to several hundred in higher eukaryotes. The genes for the large and small rRNA (found respectively in the large and small subunits of the ribosome) usually form a tandem pair. (The sole exception is the yeast mitochondrion.)

The lack of any detectable variation in the sequences of the rRNA molecules implies that all the copies of each gene must be identical, or at least must have differences below the level of detection in rRNA (~1%). A point of major interest is what mechanism(s) prevent variations from accruing in the individual sequences.

In bacteria, the multiple rRNA gene pairs are dispersed. In most eukaryotic nuclei, the rRNA genes are in a tandem cluster or clusters. Sometimes these regions are called **rDNA**. (In some cases, the proportion of rDNA in the total DNA, together with its atypical base composition, is great enough to allow its isolation as a separate fraction directly from sheared genomic DNA.) An important diagnostic feature of a tandem cluster is that it generates a circular restriction map, as shown in **Figure 6.12**.

Suppose that each repeat unit has three restriction sites. When we map these fragments by conventional means, we find that A is next to B, which is next to C, which is next to A, generating the circular map. If the cluster is large, the internal fragments (A, B, C) will be present in much greater quantities than the terminal fragments (X, Y) which connect the cluster to adjacent

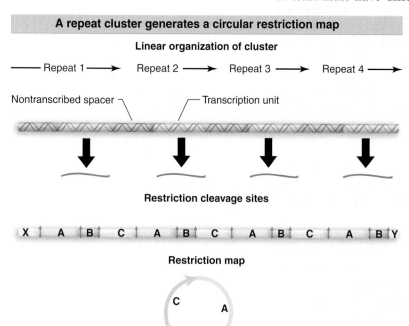

Figure 6.12 A tandem gene cluster has an alternation of transcription units and nontranscribed spacers and generates a circular restriction map.

DNA. In a cluster of 100 repeats, X and Y would be present at 1% of the level of A, B, C. This can make it difficult to obtain the ends of a gene cluster for mapping purposes.

The region of the nucleus where rRNA synthesis occurs has a characteristic appearance, with a core of fibrillar nature surrounded by a granular cortex. The fibrillar core is where the rRNA is transcribed from the DNA template; and the granular cortex is formed by the ribonucleoprotein particles into which the rRNA is assembled. The whole area is called the **nucleolus**. Its characteristic morphology is evident in **Figure 6.13**.

The particular chromosomal regions associated with a nucleolus are called **nucleolar organizers**. Each nucleolar organizer corresponds to a cluster of tandemly repeated rRNA genes on one chromosome. The concentration of the tandemly repeated rRNA genes, together with their very intensive transcription, is responsible for creating the characteristic morphology of the nucleoli.

The rRNAs for the large and small subunits are transcribed as a single precursor in both bacteria and eukaryotic nuclei. Following transcription, the precursor is cleaved to release the individual rRNA molecules. The transcription unit is shortest in bacteria and is longest in mammals (where it is known as 45S RNA, according to its rate of sedimentation). An rDNA cluster contains many transcription units, each separated from the next by a **nontranscribed spacer**. The alternation of transcription unit and nontranscribed spacer can be seen directly in electron micrographs. The example shown in **Figure 6.14** is taken from the newt *N. viridescens*, in which each transcription unit is intensively expressed, so that many RNA polymerases are simultaneously engaged in transcription on one repeating unit. The polymerases are so closely packed that the RNA transcripts form a characteristic matrix displaying increasing length along the transcription unit.

In some organisms, including *D. melanogaster* and *X. laevis*, all of the repeating units are present as a single tandem cluster on one chromosome. In mammals, the genes lie in several dispersed clusters—in man and mouse residing on five and six chromosomes, respectively. One interesting (but unanswered) question is how the corrective mechanisms that presumably function within a single cluster to ensure constancy of rRNA sequence are able to work when there are several clusters.

The sequence of the transcription unit is highly conserved, but the length of the nontranscribed spacer in a single gene cluster may vary widely. In mammals, the transcription unit

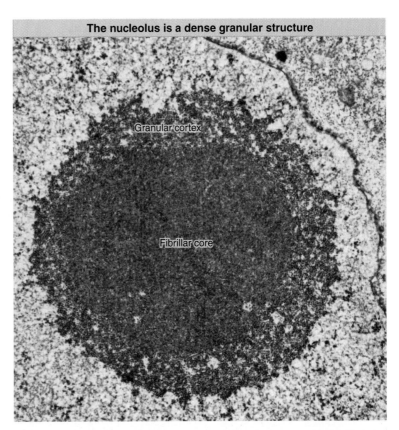

The nucleolus is a dense granular structure

Granular cortex

Fibrillar core

Figure 6.13 The nucleolar core identifies rDNA under transcription, and the surrounding granular cortex consists of assembling ribosomal subunits. This thin section shows the nucleolus of the newt *Notopthalmus viridescens*. Photograph kindly provided by Oscar L. Miller, Department of Biology, University of Virginia.

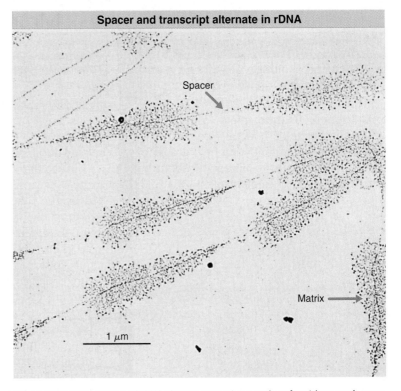

Spacer and transcript alternate in rDNA

Spacer

Matrix

1 μm

Figure 6.14 Transcription of rDNA clusters generates a series of matrices, each corresponding to one transcription unit and separated from the next by the nontranscribed spacer. Photograph kindly provided by Oscar L. Miller, Department of Biology, University of Virginia.

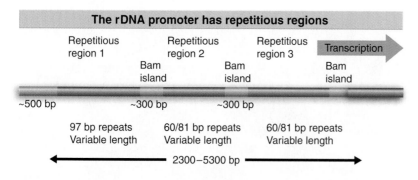

Figure 6.15 The nontranscribed spacer of *X. laevis* rDNA has an internally repetitious structure that is responsible for its varying length. The Bam islands are short, constant sequences that separate the repetitious regions.

is ~13 kb and the nontranscribed spacer can be up to ~30 kb. In spite of the variation in length, the sequences of longer nontranscribed spacers remain homologous with those of the shorter nontranscribed spacers. This implies that each nontranscribed spacer is *internally repetitious*, so that the variation in length results from changes in the number of repeats of some subunit.

The general nature of the nontranscribed spacer is illustrated by the example of *X. laevis* as diagrammed in **Figure 6.15**. Regions that are fixed in length alternate with regions that vary. Each of the three repetitious regions comprises a variable number of repeats of a rather short sequence. One type of repetitious region has repeats of a 97 bp sequence; the other, which occurs in two locations, has a repeating unit found in two forms, 60 bp and 81 bp long. The variation in the number of repeating units in the repetitious regions accounts for the overall variation in spacer length.

The contrast between the properties of the transcription unit and the nontranscribed spacer is revealing. The invariance of the transcription unit shows that it is highly selected. But the variation in the nontranscribed spacer lengths suggests that they are not subject to sequence selection, and shows that there is frequent unequal crossing-over. Crossing-over events in the nontranscribed space change the length of the cluster, but do not alter the properties of the individual transcription units. We see further in the next section that frequent crossing-over is the key mechanism that allows continuous contraction and expansion of a cluster to homogenize its copies.

6.8 Crossover Fixation Could Maintain Identical Repeats

Key Terms

- **Concerted evolution** describes the ability of two related genes to evolve together as though constituting a single locus.
- **Crossover fixation** refers to a possible consequence of unequal crossing-over that allows a mutation in one member of a tandem cluster to spread through the whole cluster (or to be eliminated).

Key Concepts

- Unequal crossing-over changes the size of a cluster of tandem repeats.
- Individual repeating units can be eliminated or can spread through a cluster.

When the duplication of a gene creates two identical copies, a change in the sequence of either one will not deprive the organism of a functional protein because the original amino acid sequence continues to be coded by the other copy. This relaxes the selective pressure on the genes, until one of them mutates sufficiently away from its original function to refocus all the selective pressure on the other.

Yet there are instances where duplicated genes retain the same function, coding for identical or nearly identical proteins. Identical proteins are coded by the two human α-globin genes, and there is only a single amino acid difference between the two γ-globin proteins. How is selective pressure exerted to maintain their sequence identity?

The most obvious possibility is that the two genes do not actually have identical functions, but differ in some (undetected) property, such as time or place of expression. Another possibility is that the need for two copies is quantitative, because neither by itself produces a sufficient amount of protein.

In more extreme cases of repetition, however, it is impossible to avoid the conclusion that no single copy of the gene is essential. When there are many copies of a gene, the immediate effects of mutation in any one copy must be very slight. The consequences of an individual mutation are diluted by the large number of copies of the gene that retain the wild-type sequence. Many mutant copies could accumulate before a lethal effect is generated.

Lethality becomes quantitative, a conclusion reinforced by the observation that half of the units of the rDNA cluster of *X. laevis* or *D. melanogaster* can be deleted without ill effect. So how are these units prevented from gradually accumulating deleterious mutations? And what chance is there for the rare favorable mutation to display its advantages in the cluster?

The basic principle of models to explain the maintenance of identity among repeated copies is to suppose that nonallelic genes are not independently inherited, but must be continually regenerated from *one* of the copies of a preceding generation. In the simplest case of two identical genes, when a mutation occurs in one copy, either it is by chance eliminated (because the sequence of the other copy takes over), or it is spread to both duplicates (because the mutant copy becomes the dominant version). Spreading exposes a mutation to selection. The result is that the two genes evolve together as though only a single locus existed. This is called **coincidental evolution** or **concerted evolution** (occasionally **coevolution**). It can be applied to a pair of identical genes or (with further assumptions) to a cluster containing many genes.

The **crossover fixation** model supposes that an entire cluster is subject to continual rearrangement by the mechanism of unequal crossing-over. Such events can explain the concerted evolution of multiple genes if unequal crossing-over causes all the copies to be regenerated physically from one copy.

Following the sort of event depicted in Figure 6.10, for example, the chromosome carrying a triple locus could suffer deletion of one of the genes. Of the two remaining genes, $1\frac{1}{2}$ represent the sequence of one of the original copies; only $\frac{1}{2}$ of the sequence of the other original copy has survived. Any mutation in the first region now exists in both genes and is subject to selective pressure.

Tandem clustering provides frequent opportunities for "mispairing" of genes whose sequences are the same, but that lie in different positions in their clusters. By continually expanding and contracting the number of units via unequal crossing-over, it is possible for all the units in one cluster to be derived from a small proportion of the units in an ancestral cluster. The variable lengths of the spacers are consistent with the idea that unequal crossing-over events take place in spacers that are internally mispaired. This can explain the homogeneity of the genes compared with the variability of the spacers. The genes are exposed to selection when individual repeating units are amplified within the cluster; but the spacers are irrelevant and can accumulate changes.

In a region of nonrepetitive DNA, recombination occurs between precisely matching points on the two homologous chromosomes, generating reciprocal recombinants. The basis for this precision is the ability of two duplex DNA sequences to align exactly. We know that unequal recombination can occur when there are multiple copies of genes whose exons are related, even though their flanking and intervening sequences may differ. This happens because of the mispairing between corresponding exons in *nonallelic* genes.

Imagine how much more frequently misalignment must occur in a tandem cluster of identical or nearly identical repeats. Except at the very ends of the cluster, the close relationship between successive repeats makes it impossible even to define the exactly corresponding repeats! This has two consequences: there is continual adjustment of the size of the cluster; and there is homogenization of the repeating unit.

Consider a sequence consisting of a repeating unit "ab" with ends "x" and "y." If we represent one chromosome in black and the other in color, the exact alignment between "allelic" sequences would be:

xabababababababababababababababababy

xabababababababababababababababababy

But probably *any* sequence *ab* in one chromosome could pair with *any* sequence *ab* in the other chromosome. In a misalignment such as:

xabababababababababababababababababy

xabababababababababababababababababababy

the region of pairing is no less stable than in the perfectly aligned pair, although it is shorter. We do not know very much about how pairing is initiated prior to recombination, but very likely it starts between short corresponding regions and then spreads. If it starts within satellite DNA, it is more likely than not to involve repeating units that do not have exactly corresponding locations in their clusters.

Now suppose that a recombination event occurs within the unevenly paired region. The recombinants will have different numbers of repeating units. In one case, the cluster becomes longer; in the other, it becomes shorter,

xabababababababababababababababababy

×

xabababababababababababababababababababy

↓

xababababababababababababababababababababy

+

xababababababababababababababababay

where "×" indicates the site of the crossover.

If this type of event is common, clusters of tandem repeats will undergo continual expansion and contraction. This can cause a particular repeating unit to spread through the cluster, as illustrated in **Figure 6.16**. Suppose that the cluster consists initially of a sequence *abcde*, where each letter represents a repeating unit. The different repeating units are closely enough related to one another to mispair for recombination. Then by a series of unequal recombination events, the size of the repetitive region increases or decreases, and also one unit spreads to replace all the others.

The crossover fixation model predicts that *any sequence of DNA that is not under selective pressure will be taken over by a series of identical tandem repeats generated in this way*. The critical assumption is that the process of crossover fixation is fairly rapid relative to mutation, so that new mutations either are eliminated (their repeats are lost) or come to take over the entire cluster. In the rDNA cluster, of course, a further factor is imposed by selection for an effective transcribed sequence.

6.9 Satellite DNAs Often Lie in Heterochromatin

Key Terms

- **Satellite DNA (simple sequence DNA)** consists of many tandem repeats (identical or related) of a short basic repeating unit. It is also known as highly repetitive DNA.

- **Buoyant density** measures the ability of a substance to float in some standard fluid, for example, CsCl.

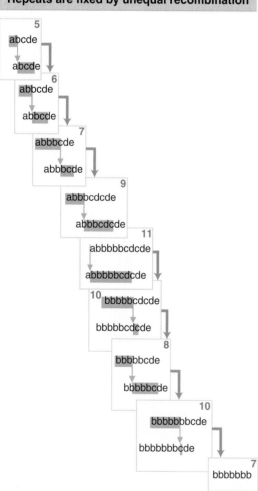

Repeats are fixed by unequal recombination

Figure 6.16 Unequal recombination allows one particular repeating unit to occupy the entire cluster. The numbers indicate the length of the repeating unit at each stage.

- A **density gradient** is used to separate macromolecules on the basis of differences in their density. It is prepared from a heavy soluble compound such as CsCl.
- A **cryptic satellite** is a satellite DNA sequence not identified as such by a separate peak on a density gradient; that is, it remains present in main-band DNA.
- **Constitutive heterochromatin** describes the inert state of permanently nonexpressed sequences, usually satellite DNA.
- *In situ* **hybridization (cytological hybridization)** is performed by denaturing the DNA of cells squashed on a microscope slide so that reaction is possible with an added single-stranded RNA or DNA; the added preparation is radioactively labeled and its hybridization is followed by autoradiography.

Key Concepts

- Satellite DNA has a very short repeating sequence and no coding function.
- It occurs in large blocks that can have distinct physical properties.
- It is often the major constituent of centromeric heterochromatin.

The tandem repetition of a short sequence often creates a fraction with distinctive physical properties that can be used to isolate it. In some cases, the repetitive sequence has a base composition distinct from the genome average, which allows it to form a separate fraction by virtue of its distinct buoyant density. A fraction of this sort is called **satellite DNA**. The term satellite DNA is essentially synonymous with simple sequence DNA. Consistent with its simple sequence, this DNA is not transcribed or translated.

Tandemly repeated sequences are especially liable to undergo misalignments during chromosome pairing, so the sizes of tandem clusters tend to be highly polymorphic, with wide variations between individuals. In fact, the smaller clusters of such sequences can be used to characterize individual genomes in the technique of "DNA fingerprinting" (see *6.12 Minisatellites Are Useful for Genetic Mapping*).

One feature of a simple repeating sequence is that its G · C content may differ from the genome average. Because the G · C content determines the **buoyant density**, it is often possible to separate such sequences by centrifuging DNA through a **density gradient** of CsCl. The DNA forms a band at the position corresponding to its own density. Fractions of DNA differing in G · C content by >5% can usually be separated on a density gradient.

When eukaryotic DNA is centrifuged on a density gradient, two types of material may be distinguished:

- Most of the genome forms a continuum of fragments that appears as a rather broad peak centered on the buoyant density corresponding to the average G · C content of the genome. This is called the main band.
- Sometimes an additional, smaller peak (or peaks) is seen at a different value. This material is the satellite DNA.

Satellites are present in many eukaryotic genomes. They may be either heavier or lighter than the main band; but it is uncommon for them to represent >5% of the total DNA. A clear example is provided by mouse DNA, shown in **Figure 6.17**. The graph is a quantitative scan of the bands formed when mouse DNA is centrifuged through a CsCl density gradient. The main band contains 92% of the genome and is centered on a buoyant density of 1.701 g-cm^{-3} (corresponding to its average G · C of 42%, typical for a mammal). The smaller peak represents 8% of the genome and has a distinct buoyant density of 1.690 g-cm^{-3}. It

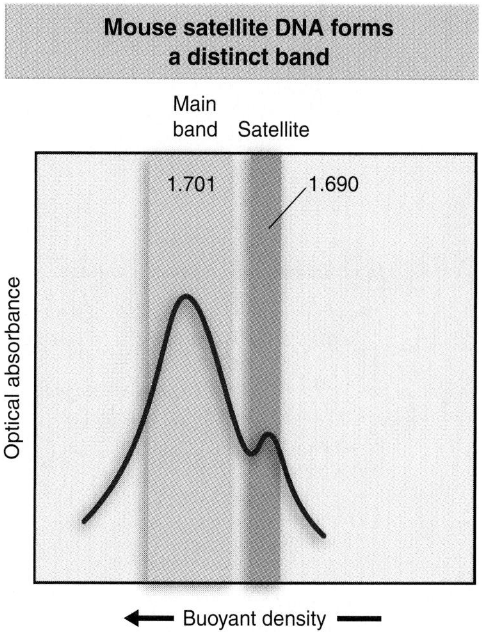

Mouse satellite DNA forms a distinct band

Figure 6.17 Mouse DNA is separated into a main band and a satellite by centrifugation through a density gradient of CsCl.

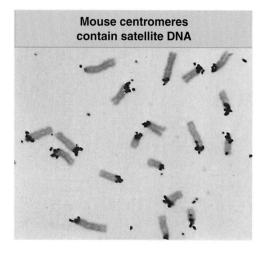

Mouse centromeres contain satellite DNA

Figure 6.18 Cytological hybridization shows that mouse satellite DNA is located at the centromeres. Photograph kindly provided by Mary Lou Pardue and Joseph G. Gall.

contains the mouse satellite DNA, whose G·C content (30%) is much lower than any other part of the genome.

Often most of the highly repetitive DNA of a genome can be isolated as satellites. When a highly repetitive DNA component does not separate as a satellite, on isolation its properties often prove to be similar to those of satellite DNA. That is to say, it consists of multiple tandem repeats but its behavior under centrifugation is anomalous. Material isolated in this manner is sometimes referred to as a **cryptic satellite**. Together the cryptic and apparent satellites usually account for all the large tandemly repeated blocks of highly repetitive DNA. When a genome has more than one type of highly repetitive DNA, each exists in its own satellite block (although sometimes different blocks are adjacent).

Satellite DNAs are often found in regions of **constitutive heterochromatin**. *Heterochromatin* is the term used to describe regions of chromosomes that are tightly coiled up and inert, in contrast with the *euchromatin* that represents most of the genome (see *28.5 Chromatin Is Divided into Euchromatin and Heterochromatin*). Chromatin may be facultative (meaning that its condition is reversible) or constitutive (meaning that its condition is permanent). Constitutive heterochromatin is commonly found at centromeres (the regions where the kinetochores form at mitosis and meiosis for controlling chromosome movement). The centromeric location of satellite DNA suggests that it has some structural function in the chromosome. This function could be connected with the process of chromosome segregation.

An example of the localization of satellite DNA for the mouse chromosomal complement is shown in **Figure 6.18**. This uses the technique of *in situ* **hybridization**, in which a radioactively labeled probe carrying a satellite sequence is directly hybridized with metaphase chromosomes. In this case, one end of each chromosome is labeled, because this is where the centromeres are located in *Mus musculus* chromosomes.

6.10 Arthropod Satellites Have Very Short Identical Repeats

Key Concept

- The repeating units of arthropod satellite DNAs are only a few nucleotides long. Most of the copies of the sequence are identical.

In the arthropods, as typified by insects and crabs, each satellite DNA appears to be rather homogeneous. Usually, a single very short repeating unit accounts for >90% of the satellite. This makes it relatively straightforward to determine the sequence.

Drosophila virilis has three major satellites and also a cryptic satellite, together representing >40% of the genome. The sequences of the satellites are summarized in **Figure 6.19**. The three major satellites have closely related sequences. A single base substitution is sufficient to generate either satellite II or III from the sequence of satellite I.

The satellite I sequence is present in other species of *Drosophila* related to *virilis*, and so may have preceded speciation. The sequences of satellites II and III seem to be specific to *D. virilis*, and so may have evolved from satellite I after speciation.

The main feature of these satellites is their very short repeating unit: only 7 bp. Similar satellites are found in other species. *D. melanogaster* has a variety of satellites, several of which have very short repeating units (5, 7, 10, or 12 bp). Comparable satellites are found in the crabs.

The close sequence relationship found among the *D. virilis* satellites is not necessarily a feature of other genomes, where the satellites may have

D. virilis has four related satellites

Satellite	Predominant sequence	Total length	Genome proportion
I	ACAAACT TGTTTGA	1.1×10^7	25%
II	ATAAACT TATTTGA	3.6×10^6	8%
III	ACAAATT TGTTTAA	3.6×10^6	8%
Cryptic	AATATAG TTATATC		

Figure 6.19 Satellite DNAs of *D. virilis* are related. More than 95% of each satellite consists of a tandem repetition of the predominant sequence.

unrelated sequences. *Each satellite has arisen by a lateral amplification of a very short sequence.* This sequence may represent a variant of a previously existing satellite (as in *D. virilis*), or could have some other origin.

Satellites are continually generated and lost from genomes. This makes it difficult to ascertain evolutionary relationships, since a current satellite could have evolved from some previous satellite that has since been lost. The important feature of these satellites is that *they represent very long stretches of DNA of very low sequence complexity, within which constancy of sequence can be maintained.*

6.11 Mammalian Satellites Consist of Hierarchical Repeats

Key Concept

- Mouse satellite DNA has evolved by duplication and mutation of a short repeating unit to give a basic repeating unit of 234 bp in which the original half-, quarter-, and eighth-repeats can be recognized.

In mammals, as typified by various rodents, the sequences constituting each satellite show appreciable divergence between tandem repeats. Common short sequences can be recognized by their preponderance among the oligonucleotide fragments released by chemical or enzymatic treatment. However, the predominant short sequence usually accounts for only a small minority of the copies. The other short sequences are related to the predominant sequence by a variety of substitutions, deletions, and insertions.

But a series of these variants of the short unit can constitute a longer repeating unit that is itself repeated in tandem with some variation. So mammalian satellite DNAs are constructed from a hierarchy of repeating units.

When any satellite DNA is digested with an enzyme that has a recognition site in its repeating unit, one fragment will be obtained for every repeating unit in which the site occurs. In fact, when the DNA of a eukaryotic genome is digested with a restriction enzyme, most of it gives a general smear, due to the random distribution of cleavage sites. But satellite DNA generates sharp bands, because a large number of fragments of identical or almost identical size are created by cleavage at restriction sites that lie a regular distance apart.

The satellite DNA of the mouse *M. musculus* is cleaved by the enzyme EcoRII into a series of bands, including a predominant monomeric fragment of 234 bp. This sequence must be repeated with few variations throughout the 60–70% of the satellite that is cleaved into the monomeric band. We may analyze this sequence in terms of its successively smaller constituent repeating units.

Figure 6.20 depicts the sequence in terms of two half-repeats. By writing the 234 bp sequence so that the first 117 bp are aligned with the second 117 bp, we see that the two halves are quite well related. They differ at 22 positions, corresponding to 19% divergence. This means

Figure 6.20 The repeating unit of mouse satellite DNA contains two half-repeats, which are aligned to show the identities (in red).

that the current 234 bp repeating unit must have been generated at some time in the past by duplicating a 117 bp repeating unit, after which differences accumulated between the duplicates.

Within the 117 bp unit, we can recognize two further subunits. Each of these is a quarter-repeat relative to the whole satellite. The four quarter-repeats are aligned in **Figure 6.21**. The upper two lines represent the first half-repeat of Figure 6.20; the lower two lines represent the second half-repeat. We see that the divergence between the four quarter-repeats has increased to 23 out of 58 positions, or 40%. The first three quarter-repeats are somewhat better related, and a large proportion of the divergence is due to changes in the fourth quarter-repeat.

Looking within the quarter-repeats, we find that each consists of two related subunits (one-eighth–repeats), shown as the α and β sequences in **Figure 6.22**. The α sequences all have an insertion of a C, and the β sequences all have an insertion of a trinucleotide, relative to a consensus sequence. This suggests that the quarter-repeat originated by the duplication of a sequence like the consensus sequence, after which additional changes generated the components we now see as α and β. Further changes then took place between tandemly repeated αβ sequences to generate the individual quarter- and half-repeats that exist today. Among the one-eighth–repeats, the present divergence is 19/31 = 61%.

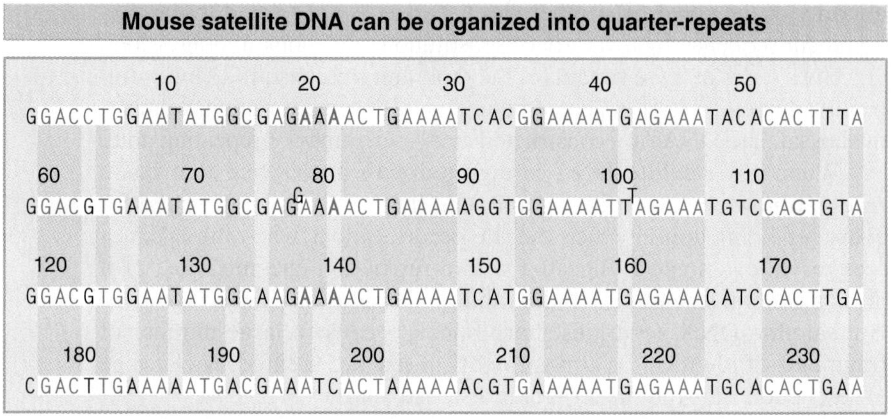

Figure 6.21 The alignment of quarter-repeats identifies homologies between the first and second half of each half-repeat. Positions that are the same in all four quarter-repeats are shown in color; identities that extend only through three quarter-repeats are indicated by black letters in the pink area.

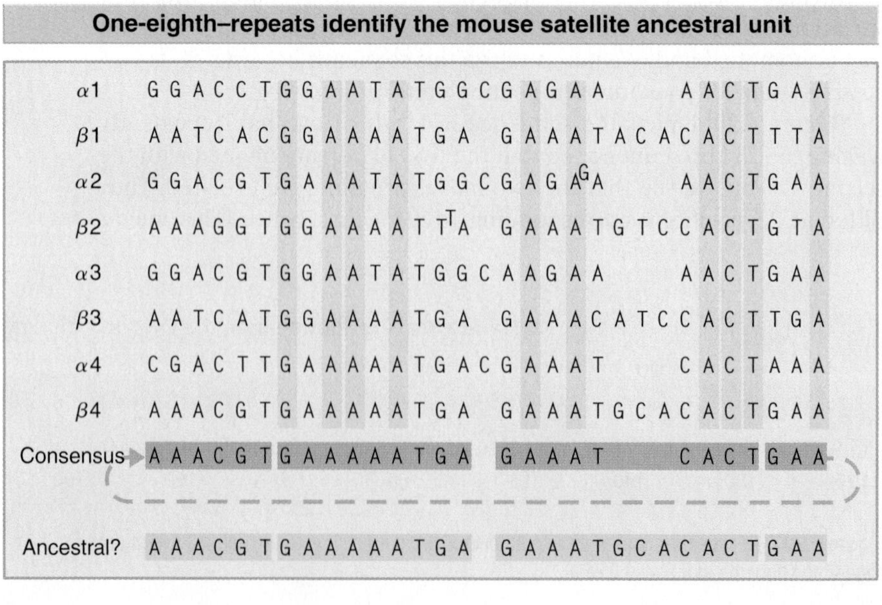

Mouse satellite DNA can be organized into quarter-repeats

```
            10        20        30        40        50
GGACCTGGAATATGGCGAGAAAACTGAAAATCACGGAAAATGAGAAATACACACTTTA

 60        70        80        90       100       110
                          G                    T
GGACGTGAAATATGGCGAGAAAACTGAAAAAGGTGGAAAATTAGAAATGTCCACTGTA

120       130       140       150       160       170
GGACGTGGAATATGGCAAGAAAACTGAAAATCATGGAAAATGAGAAACATCCACTTGA

  180       190       200       210       220       230
CGACTTGAAAAATGACGAAATCACTAAAAACGTGAAAAATGAGAAATGCACACTGAA
```

Figure 6.22 The alignment of one-eighth–repeats shows that each quarter-repeat consists of an α and a β half. The consensus sequence gives the most common base at each position. The "ancestral" sequence shows a sequence very closely related to the consensus sequence, which could have been the predecessor to the α and β units. (The satellite sequence is continuous, so that for the purposes of deducing the consensus sequence, we can treat it as a circular permutation, as indicated by joining the last GAA triplet to the first 6 bp.)

One-eighth–repeats identify the mouse satellite ancestral unit

```
α1          GGACCTGGAATATGGCGAGAA         AACTGAA
β1          AATCACGGAAAATGA   GAAATACACACTTTA
α2          GGACGTGAAATATGGCGAGAᴳA        AACTGAA
β2          AAAGGTGGAAAATTᵀA  GAAATGTCCACTGTA
α3          GGACGTGGAATATGGCAAGAA         AACTGAA
β3          AATCATGGAAAATGA   GAAACATCCACTTGA
α4          CGACTTGAAAAATGACGAAAT          CACTAAA
β4          AAACGTGAAAAATGA   GAAATGCACACTGAA
Consensus ▶ AAACGTGAAAAATGA   GAAAT    CACTGAA

Ancestral?  AAACGTGAAAAATGA   GAAATGCACACTGAA
```

The consensus sequence is analyzed directly in **Figure 6.23**, which demonstrates that the current satellite sequence can be treated as derivatives of a 9 bp sequence. We can recognize three variants of this sequence in the satellite, as indicated at the bottom of Figure 6.23. If in one of the repeats we take the next most frequent base at two positions instead of the most frequent, we obtain three well-related 9 bp sequences.

G A A A A A C G T

G A A A A A T G A

G A A A A A A C T

The origin of the satellite could well lie in an amplification of one of these three nonamers. The overall consensus sequence of the present satellite is GAAAAA$^{AG}_{TC}$T, which is effectively an amalgam of the three 9 bp repeats.

The average sequence of the monomeric fragment of the mouse satellite DNA explains its properties. The longest repeating unit of 234 bp is identified by the restriction cleavage. When single strands of denatured satellite DNA are renatured, the unit of recognition is probably the 117 bp half-repeat, because the 234 bp fragments can anneal both in register and in half-register (in the latter case, the first half-repeat of one strand renatures with the second half-repeat of the other).

6.12 Minisatellites Are Useful for Genetic Mapping

Key Terms

- **Microsatellite** DNAs consist of repetitions of extremely short (typically <10 bp) units.
- **Minisatellite** DNAs consist of ~10 copies of a short repeating sequence. The length of the repeating unit is measured in 10s of base pairs. The number of repeats varies between individual genomes.
- **VNTR** (variable number tandem repeat) regions describe very short repeated sequences, including microsatellites and minisatellites.
- **DNA fingerprinting** analyzes the differences between individuals of the fragments generated by using restriction enzymes to cleave regions that contain short repeated sequences. Because the lengths of the repeated regions are unique to every individual, the presence of a particular subset in any two individuals can be used to define their common inheritance (e.g., a parent–child relationship).

Key Concept

- The variation between microsatellites or minisatellites in individual genomes can be used to identify heredity unequivocally by showing that 50% of the bands in an individual are derived from a particular parent.

Sequences that resemble satellites in consisting of tandem repeats of a short unit, but that overall are much shorter, consisting of (for example) 5–50 repeats, are common in mammalian genomes. They were discovered by chance as fragments whose size is extremely variable in genomic libraries of human DNA. The variability is seen when a population contains fragments of many different sizes that represent

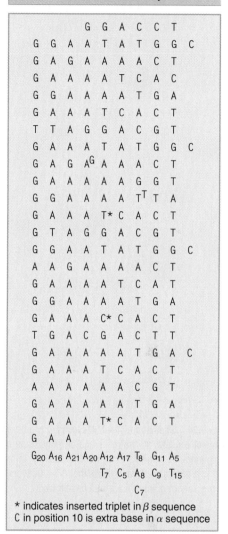

Figure 6.23 The existence of an overall consensus sequence is shown by writing the satellite sequence in terms of a 9 bp repeat.

the same genomic region; when individuals are examined, it turns out that there is extensive polymorphism, and that many different alleles can be found.

The name **microsatellite** is usually used when the length of the repeating unit is <10 bp, and the name **minisatellite** is used when the length of the repeating unit is ~10–100 bp, but the terminology is not precisely defined. These types of sequences are also called **VNTR** (variable number tandem repeat) regions.

The cause of the variation between individual genomes at microsatellites or minisatellites is that individual alleles have different numbers of the repeating unit. For example, one minisatellite has a repeat length of 64 bp, and is found in the population with the following distribution:

7%   18 repeats

11% 16 repeats

43% 14 repeats

36% 13 repeats

4%   10 repeats

The rate of genetic exchange at minisatellite sequences is ~10× greater than the rate of homologous recombination at meiosis, that is, in any random DNA sequence. This high rate of recombination makes minisatellites especially useful for genomic mapping, because there is a high probability that individuals will vary in their alleles at such a locus. An example of mapping by minisatellites is illustrated in **Figure 6.24**. This shows an extreme case in which two individuals both are het-

Figure 6.24 Alleles at a minisatellite locus may differ in the number of repeats, so that cleavage on either side of the locus generates restriction fragments that differ in length. By using a minisatellite with alleles that differ between parents, the pattern of inheritance can be followed.

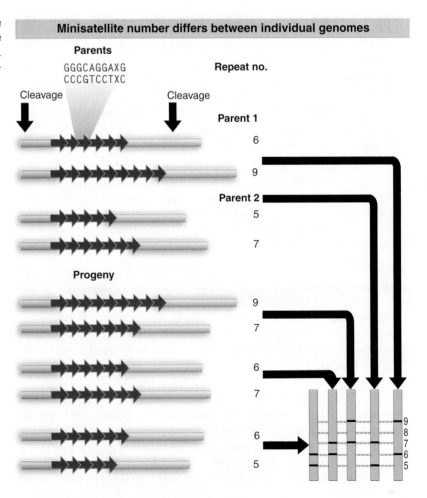

erozygous at a minisatellite locus, and in fact all four alleles are different. All progeny gain one allele from each parent in the usual way, and it is possible unambiguously to determine the source of every allele in the progeny. In the terminology of human genetics, the meioses described in this figure are highly informative, because of the variation between alleles.

Consider the situation shown in Figure 6.24, but multiplied many times by the existence of many such sequences. The effect of the variation at individual loci is to create a unique pattern for every individual. This makes it possible to assign heredity unambiguously between parents and progeny, by showing that 50% of the bands in any individual are derived from a particular parent. This is the basis of the technique known as **DNA fingerprinting**.

Both microsatellites and minisatellites are unstable, although for different reasons. Microsatellites undergo intrastrand mispairing, when slippage during replication leads to expansion of the repeat, as shown in **Figure 6.25**. Systems that repair damage to DNA, in particular those that recognize mismatched base pairs, are important in reversing such changes, as shown by a large increase in frequency when repair genes are inactivated. Because mutations in repair systems are an important contributory factor in the development of cancer, tumor cells often display variations in microsatellite sequences (see *20.11 Defects in Repair Systems Cause Mutations to Accumulate in Tumors*).

Minisatellites undergo the same sort of unequal crossing-over that we have discussed for satellites (see Figure 6.1). It is not clear at what repeating length the cause of the variation shifts from replication slippage to recombination.

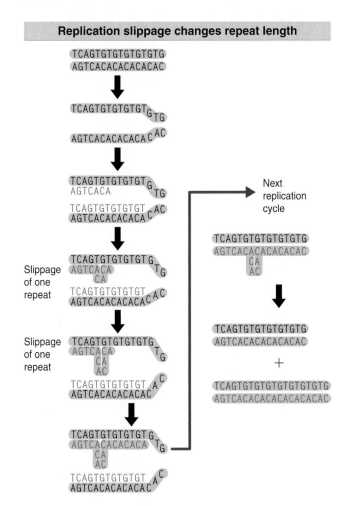

Replication slippage changes repeat length

Slippage of one repeat

Slippage of one repeat

Next replication cycle

Figure 6.25 Replication slippage occurs when the daughter strand slips back one repeating unit in pairing with the template strand. Each slippage event adds one repeating unit to the daughter strand. The extra repeats are extruded as a single strand loop. Replication of this daughter strand in the next cycle generates a duplex DNA with an increased number of repeats.

6.13 SUMMARY

Many genes belong to families, defined by the possession of related sequences in the exons of individual members. Families evolve by the duplication of a gene (or genes), followed by divergence between the copies. Some copies suffer inactivating mutations and become pseudogenes that no longer have any function. Pseudogenes also may be generated as DNA copies of the mRNA sequences.

An evolving set of genes may remain together in a cluster or may be dispersed to new locations by chromosomal rearrangement. The organization of existing clusters can sometimes be used to infer the series of events that has occurred. These events act with regard to sequence rather than function, and therefore include pseudogenes as well as active genes.

Mutations accumulate more rapidly in silent sites than in replacement sites (which affect the amino acid sequence). The rate of divergence at replacement sites can be used to establish a clock, calibrated in percent divergence per million years. The clock can then be used to calculate the time of divergence between any two members of the family.

A tandem cluster consists of many copies of a repeating unit that includes the transcribed sequence(s) and a nontranscribed spacer(s). rRNA gene clusters code only for a single rRNA precursor. Maintenance of active genes in clusters depends on mechanisms such as gene conversion or unequal crossing-over that cause muta-

tions to spread through the cluster, so that they become exposed to evolutionary pressure.

Satellite DNA consists of very short sequences repeated many times in tandem. Its distinct centrifugation properties reflect its biased base composition. Satellite DNA is concentrated in centromeric heterochromatin, but its function (if any) is unknown. The individual repeating units of arthropod satellites are identical. Those of mammalian satellites are related, and can be organized into a hierarchy reflecting the evolution of the satellite by the amplification and divergence of randomly chosen sequences.

Unequal crossing-over appears to have been a major determinant of satellite DNA organization. Crossover fixation explains the ability of variants to spread through a cluster.

Minisatellites and microsatellites consist of even shorter repeating sequences than satellites, <10 bp for microsatellites and 10–50 bp for minisatellites. The number of repeating units is usually 5–50. There is high variation in the repeat number between individual genomes. Microsatellite repeat number varies as the result of slippage during replication; the frequency is affected by systems that recognize and repair damage in DNA. Minisatellite repeat number varies as the result of recombination-like events. Variations in repeat number can be used to determine hereditary relationships by the technique known as DNA fingerprinting.

Messenger RNA

7.1 Introduction

Key Terms

- **Messenger RNA (mRNA)** is the intermediate that represents one strand of a gene coding for protein. Its coding region is related to the protein sequence by the triplet genetic code.

- **Transfer RNA (tRNA)** is the intermediate in protein synthesis that interprets the genetic code. Each tRNA can be linked to an amino acid. The tRNA has an anticodon sequence that is complementary to a triplet codon representing the amino acid.

- **Ribosomal RNA (rRNA)** is a major component of the ribosome. Each of the two subunits of the ribosome has a major rRNA as well as many proteins.

RNA is a central player in gene expression. Its function was first characterized as an intermediate in protein synthesis, but since then RNAs that play structural or functional roles at other stages of gene expression have been discovered. The involvement of RNA in many functions concerned with gene expression supports the general view that the entire process may have evolved in an "RNA world" in which RNA was originally the active component in maintaining and expressing genetic information. Many of these functions were subsequently assisted or taken over by proteins, with a consequent increase in versatility and efficiency.

As summarized in **Figure 7.1**, three types of RNA are directly involved in the production of proteins:

- Messenger RNA (**mRNA**) is an intermediate that carries the copy of a DNA sequence that represents protein.

- Transfer RNAs (**tRNA**) are small RNAs that provide amino acids corresponding to each particular codon in mRNA.

113

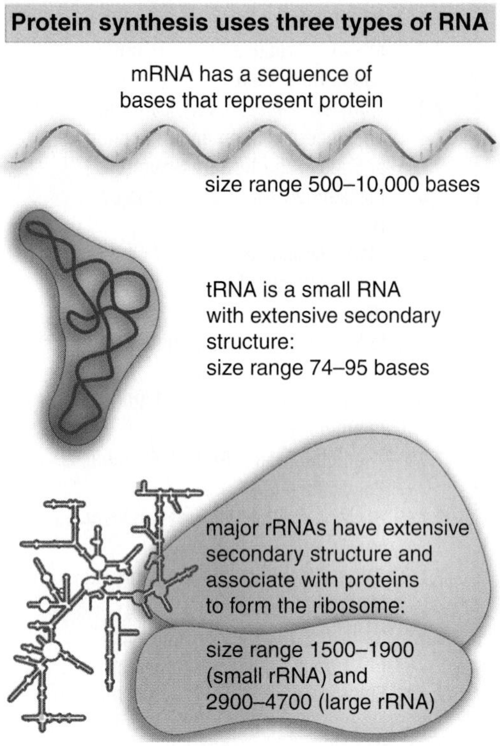

Protein synthesis uses three types of RNA

mRNA has a sequence of
bases that represent protein

size range 500–10,000 bases

tRNA is a small RNA
with extensive secondary
structure:
size range 74–95 bases

major rRNAs have extensive
secondary structure and
associate with proteins
to form the ribosome:

size range 1500–1900
(small rRNA) and
2900–4700 (large rRNA)

Figure 7.1 The three types of RNA universally required for gene expression are mRNA (carries the coding sequence), tRNA (provides the amino acid corresponding to each codon), and rRNA (a major component of the ribosome that provides the environment for protein synthesis).

• Ribosomal RNAs (**rRNA**) are components of the ribosome, a large ribonucleoprotein complex that contains many proteins as well as RNA components, and which provides the apparatus for actually polymerizing amino acids into a polypeptide chain.

The role that RNA plays in each of these cases is distinct.

For messenger RNA, its sequence is the important feature: each codon within the coding region of the mRNA represents an amino acid in the corresponding protein. However, the structure of the mRNA, in particular the sequences on either side of the coding region, can play an important role in controlling its activity, and therefore the amount of protein that is produced from it.

In tRNA, we see two of the common themes governing the use of RNA: its three-dimensional structure is important; and it has the ability to base pair with another RNA (mRNA). The three-dimensional structure is recognized first by an enzyme as providing a target that is appropriate for linkage to a specific amino acid. The linkage creates an aminoacyl-tRNA, which is the structure used for protein synthesis. The specificity of an aminoacyl-tRNA in protein synthesis is achieved by base pairing when its triplet sequence (the anticodon) pairs with the codon representing an amino acid.

With rRNA, we see another type of activity. One role of rRNA is structural, in providing a framework to which ribosomal proteins attach. But it also participates directly in the activities of the ribosome. One of the crucial activities of the ribosome is the ability to catalyze the formation of a peptide bond by which an amino acid is incorporated into protein. This activity resides in one of the rRNAs.

The important thing about this background, as we consider protein synthesis, is that we have to view RNA as a component that plays an active role and that can be a target for regulation by proteins or by other RNAs, and we should remember that the RNAs may have been the basis for the original apparatus. The theme that runs through all of the activities of RNA, in both protein synthesis and elsewhere, is that its functions depend critically upon base pairing, both to form its secondary structure, and to interact specifically with other RNA molecules. The coding function of mRNA is unique, but tRNA and rRNA are examples of a much broader class of noncoding RNAs with a variety of functions in gene expression.

7.2 mRNA Is Produced by Transcription and Is Translated

Key Terms

• **Transcription** is synthesis of RNA on a DNA template.
• **Translation** is synthesis of protein on an mRNA template.
• A **coding region** is a part of the gene that represents a protein sequence.
• A coding region consists of a series of **codons**; each codon is a triplet of nucleotides that represents an amino acid or a termination signal.
• The **template strand (antisense strand)** of DNA is complementary to the coding strand and is the one that acts as the template for synthesis of mRNA.
• The **coding strand (sense strand)** of DNA has the same sequence as the mRNA and is related by the genetic code to the protein sequence that it represents.

Key Concept

- Only one of the two strands of DNA is transcribed into mRNA and translated into protein.

Gene expression occurs by a two-stage process.

- **Transcription** generates a single-stranded RNA identical in sequence with one of the strands of the duplex DNA.
- **Translation** converts the nucleotide sequence of mRNA into the sequence of amino acids constituting a protein. The entire length of an mRNA is not translated, but each mRNA contains at least one **coding region** that is related to a protein sequence by the genetic code. The coding region consists of a continuous series of **codons**. Each codon is a nucleotide triplet that represents one amino acid.

Only one strand of a DNA duplex is transcribed into a messenger RNA. We distinguish the two strands of DNA as depicted in **Figure 7.2**:

- The strand of DNA that directs synthesis of the mRNA via complementary base pairing is called the **template strand** or **antisense strand**. (*Antisense* is used as a general term to describe a sequence of DNA or RNA that is complementary to mRNA.)
- The other DNA strand bears the *same* sequence as the mRNA (except for possessing T instead of U), and is called the **coding strand** or **sense strand**.

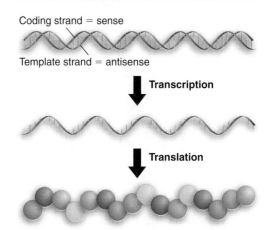

Figure 7.2 Transcription generates an RNA that is complementary to the DNA template strand and has the same sequence as the DNA coding strand. Translation reads each triplet of bases into one amino acid. Three turns of the DNA double helix contain 30 bp, which code for 10 amino acids.

7.3 Transfer RNA Forms a Cloverleaf

Key Concepts

- A tRNA has a sequence of 74–95 bases that folds into a cloverleaf secondary structure with four constant arms (and an additional arm in the longer tRNAs).
- A mature tRNA is generated by processing a precursor; the 3′ end is generated by cleavage followed by trimming of the last few bases, followed by addition of the common terminal trinucleotide sequence CCA.
- tRNA is charged to form aminoacyl-tRNA by forming an ester link from the 2′- or 3′-OH group of the adenylic acid at the end of the acceptor arm to the COOH group of the amino acid.
- The sequence of the anticodon is solely responsible for the specificity of the aminoacyl-tRNA.

Key Terms

- The **anticodon** is a trinucleotide sequence in tRNA, complementary to the codon in mRNA, that enables the tRNA to place the appropriate amino acid in response to the codon.
- The **cloverleaf** is the structure of tRNA drawn in two dimensions, showing four distinct "arms."
- An **arm** of tRNA is one of the four (or five) stem-loop structures that make up the secondary structure. A **stem** is a base-paired segment, and a **loop** is a single-stranded region at the end of an arm.
- The **acceptor arm** of tRNA is a short duplex that terminates in the CCA sequence to which an amino acid is linked.
- The **anticodon arm** of tRNA is a stem-loop structure that exposes the anticodon triplet at one end.
- An **aminoacyl-tRNA** is a tRNA linked to an amino acid. The COOH group of the amino acid is linked to the 3′- or 2′-OH group of the terminal base of the tRNA.
- **Aminoacyl-tRNA synthetases** are enzymes responsible for covalently linking amino acids to the 2′- or 3′-OH position of tRNA.

Messenger RNA can be distinguished from the apparatus responsible for its translation by the use of *in vitro* cell-free systems to synthesize proteins. *A protein-synthesizing system from one cell type can translate the mRNA from another, demonstrating that both the genetic code and the translation apparatus are universal.*

Each triplet codon in the mRNA represents an amino acid. The "adaptor" that matches each codon to its particular amino acid is **transfer RNA (tRNA)**. A tRNA has two crucial properties: First, it represents a single amino acid, to which it is *covalently linked*. Second, it contains a trinucleotide sequence, the **anticodon**, which is *complementary to the codon representing its amino acid*. The anticodon enables the tRNA to recognize the codon via complementary base pairing.

All tRNAs have common secondary and tertiary structures. The tRNA secondary structure can be written in the form of a **cloverleaf**,

tRNA is an adaptor

A

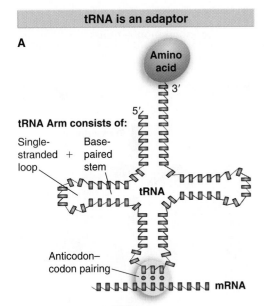

tRNA Arm consists of:

Single-stranded loop + Base-paired stem

Anticodon–codon pairing

B

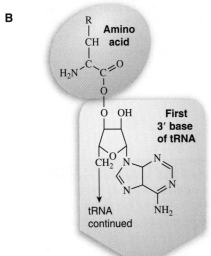

Figure 7.3 A tRNA has the dual properties of an adaptor that recognizes both the amino acid and codon. The 3′ adenosine is covalently linked to an amino acid. The anticodon pairs with the codon on mRNA.

illustrated in **Figure 7.3**, in which complementary base pairing forms **stems** for single-stranded **loops**. The stem-loop structures are called the **arms** of tRNA. Their sequences include "unusual" bases that are generated by modification of the four standard bases after synthesis of the polynucleotide chain.

The construction of the cloverleaf is illustrated in more detail in **Figure 7.4**. The four major arms are named for their structure or function:

- The **acceptor arm** has a base-paired stem that ends in an unpaired sequence whose free 2′- or 3′-OH group can be linked to an amino acid.
- The "TψC arm" is named for the characteristic triplet sequence in its loop. (ψ stands for pseudouridine, a modified base.)
- The **anticodon arm** always contains the anticodon triplet in the center of the loop.
- The "D arm" is named for its content of the base dihydrouridine (another of the modified bases in tRNA).
- Some tRNAs have a fifth arm, called the "extra arm"; it lies between the TψC and anticodon arms and varies from 3–21 bases.

The numbering system for tRNA illustrates the constancy of the structure. Positions in the most common tRNA structure, which has 76 residues, are numbered from 5′ to 3′. The overall range of tRNA lengths is 74–95 bases. The variation in length is caused by differences in the D arm and extra arm.

The base pairing that maintains the secondary structure is shown in Figure 7.4. Within a given tRNA, most of the base pairings are conventional partnerships of A·U and G·C, but occasional G·U, G·ψ, or A·ψ pairs are found. The unconventional base pairs are less stable than the conventional ones but still allow a double-helical structure to form.

When the sequences of tRNAs are compared, the bases found at some positions are invariant (or conserved); almost always a particular base is found at the position. Some positions are described as

tRNA secondary structure is a cloverleaf

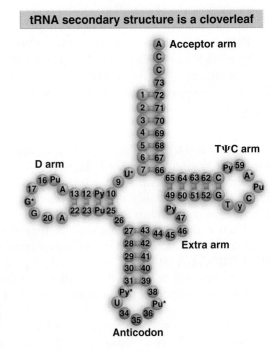

Figure 7.4 The tRNA cloverleaf has a conserved set of base pairing interactions.

semiinvariant (or semiconserved) because they are restricted to one type of base (purine versus pyrimidine), but either base of that type may be present.

tRNAs are commonly synthesized as precursor chains with additional material at one or both ends. The extra sequences are removed by combinations of endonucleolytic and exonucleolytic activities. One feature that is common to all tRNAs is that the three nucleotides at the 3′ terminus, always the triplet sequence CCA, are not coded in the genome but are added as part of tRNA processing. This is the result of an enzymatic process. The enzymes involved, and the process by which they determine the sequence of bases, is different in different organisms.

When a tRNA is *charged* with the amino acid corresponding to its anticodon, it is called **aminoacyl-tRNA**. The amino acid is linked by an ester bond from its carboxyl group to the 2′ or 3′ hydroxyl group of the ribose of the 3′ terminal base of the tRNA (which is always adenine). The process of charging a tRNA is catalyzed by a specific enzyme, **aminoacyl-tRNA synthetase**. There are 20 aminoacyl-tRNA synthetases. Each recognizes a single amino acid and all tRNAs onto which the amino acid can legitimately be placed.

There is at least one tRNA (but usually more) for each amino acid. A tRNA is named by using the three-letter abbreviation for the amino acid as a superscript. If there is more than one tRNA for the same amino acid, subscript numerals are used to distinguish them. So two tRNAs for tyrosine would be described as $tRNA_1^{Tyr}$ and $tRNA_2^{Tyr}$. A tRNA carrying an amino acid—that is, an aminoacyl-tRNA—is indicated by a prefix that identifies the amino acid. Ala-tRNA describes $tRNA^{Ala}$ carrying its amino acid.

Does the anticodon sequence alone allow aminoacyl-tRNA to recognize the correct codon? A classic experiment to test this question is illustrated in **Figure 7.5**. Reductive desulfuration converts the amino acid of cysteinyl-tRNA into alanine, generating alanyl-$tRNA^{Cys}$. The tRNA has an anticodon that responds to the codon UGU. Modification of the amino acid does not influence the specificity of the anticodon–codon interaction, so the alanine residue is incorporated into protein in place of cysteine. *Once a tRNA has been charged, the amino acid plays no further role in its specificity, which is determined exclusively by the anticodon.*

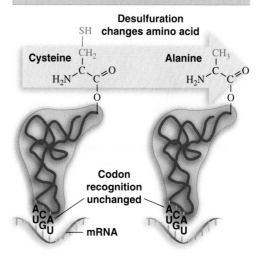

Figure 7.5 The meaning of tRNA is determined by its anticodon and not by its amino acid.

7.4 The Acceptor Stem and Anticodon Are at Ends of the Tertiary Structure

Key Concept

- The cloverleaf forms an L-shaped tertiary structure with the acceptor arm at one end and the anticodon arm at the other end.

The secondary structure of each tRNA folds into a compact L-shaped tertiary structure in which the 3′ end that binds the amino acid is distant from the anticodon that binds the mRNA. All tRNAs have the same general tertiary structure, although they are distinguished by individual variations.

The arrangement of the base paired double-helical stems of the secondary structure creates a tertiary structure with two double helices

All tRNAs share a tertiary structure

Cloverleaf has four arms

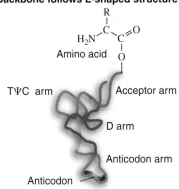

2D projection has 2 perpendicular duplexes

Backbone follows L-shaped structure

Amino acid

Figure 7.6 Transfer RNA folds into a compact L-shaped tertiary structure with the amino acid at one end and the anticodon at the other end.

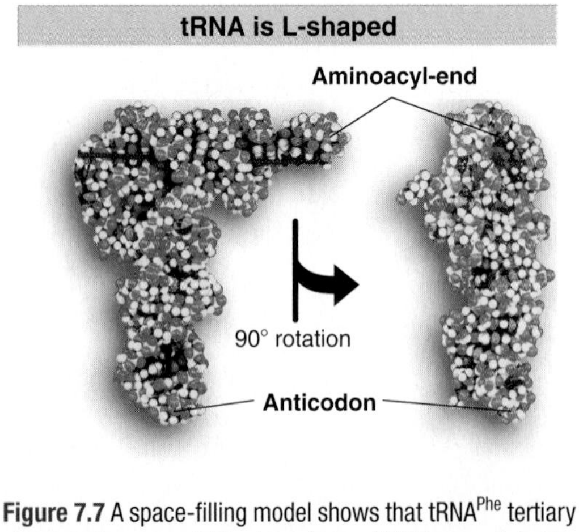

tRNA is L-shaped

Aminoacyl-end

90° rotation

Anticodon

Figure 7.7 A space-filling model shows that tRNA^{Phe} tertiary structure is compact. The two views of tRNA are rotated by 90°. Photograph kindly provided by Sung-Hou Kim, Department of Chemistry, University of California, Berkeley.

at right angles to each other, as illustrated in **Figure 7.6**. The acceptor stem and the TψC stem form one continuous double helix with a single gap; the D stem and anticodon stem form another continuous double helix, also with a gap. The region between the double helices, where the turn in the L-shape is made, contains the TψC loop and the D loop. So the amino acid resides at the extremity of one arm of the L-shape, and the anticodon loop forms the other end.

The tertiary structure is created by hydrogen bonding, mostly involving bases that are unpaired in the secondary structure. Many of the invariant and semiinvariant bases are involved in these H-bonds, which explains their conservation. Not every one of these interactions is universal, but probably they identify the *general* pattern for establishing tRNA structure.

A molecular model of the structure of yeast tRNA^{Phe} is shown in **Figure 7.7**. The left view corresponds with the bottom panel in Figure 7.6. Differences in the structure are found in other tRNAs, thus accommodating the dilemma that all tRNAs must have a similar shape, yet it must be possible to recognize differences between them. For example, in tRNA^{Asp}, the angle between the two axes is slightly greater, so the molecule has a slightly more open conformation.

The structure suggests a general conclusion about the function of tRNA. *Its sites for exercising particular functions are maximally separated.* The amino acid is as far distant from the anticodon as possible, which is consistent with their roles in protein synthesis.

7.5 Messenger RNA Is Translated by Ribosomes

Key Term

- The **ribosome** is a large assembly of RNA and proteins that synthesizes proteins under direction from an mRNA template. Bacterial ribosomes sediment at 70S, eukaryotic ribosomes at 80S. A ribosome can be dissociated into two subunits.

Key Concepts

- Ribosomes are characterized by their rate of sedimentation (70S for bacterial ribosomes and 80S for eukaryotic ribosomes).
- A ribosome consists of a large subunit (50S or 60S for bacteria and eukaryotes) and a small subunit (30S or 40S).
- The ribosome provides the environment in which aminoacyl-tRNAs add amino acids to the growing polypeptide chain in response to the corresponding triplet codons.
- A ribosome moves along an mRNA from 5′ to 3′.

Translation of an mRNA into a polypeptide chain is catalyzed by the **ribosome**. Ribosomes are traditionally described in terms of their (approximate) rate of sedimentation (measured in Svedbergs, in which a higher S value indicates a greater rate of sedimentation and a larger mass). Bacterial ribosomes generally sediment at ~70S. The ribosomes of the cytoplasm of higher eukaryotic cells are larger, usually sedimenting at ~80S.

The ribosome is a compact ribonucleoprotein particle consisting of two subunits. Each subunit has an RNA component, including one very large RNA molecule, and many proteins. The relationship between a ribosome and its subunits is depicted in **Figure 7.8**. The two subunits dissociate *in vitro* when the concentration of Mg^{2+} ions is reduced. In each case, the large subunit is about twice the mass of the small subunit. Bacterial (70S) ribosomes have subunits that sediment at 50S and 30S. The subunits of eukaryotic cytoplasmic (80S) ribosomes sediment at 60S and 40S. The two subunits work together as part of the complete ribosome, but each undertakes distinct reactions in protein synthesis.

All the ribosomes of a given cell compartment are identical. *They undertake the synthesis of different proteins by associating with the different mRNAs that provide the coding sequences.*

The ribosome provides the environment that controls the recognition between a codon of mRNA and the anticodon of tRNA. Reading the genetic code as a series of adjacent triplets, protein synthesis proceeds from the start of a coding region to the end. *A protein is assembled by the sequential addition of amino acids in the direction from the N-terminus to the C-terminus as a ribosome moves along the mRNA.*

A ribosome begins translation at the 5′ end of a coding region; it translates each triplet codon into an amino acid as it proceeds toward the 3′ end. At each codon, the appropriate aminoacyl-tRNA associates with the ribosome, donating its amino acid to the polypeptide chain. At any given moment, the ribosome can accommodate two aminoacyl-tRNAs corresponding to successive codons, making it possible for a peptide bond to form between the two amino acids. At each step, the growing polypeptide chain becomes longer by one amino acid.

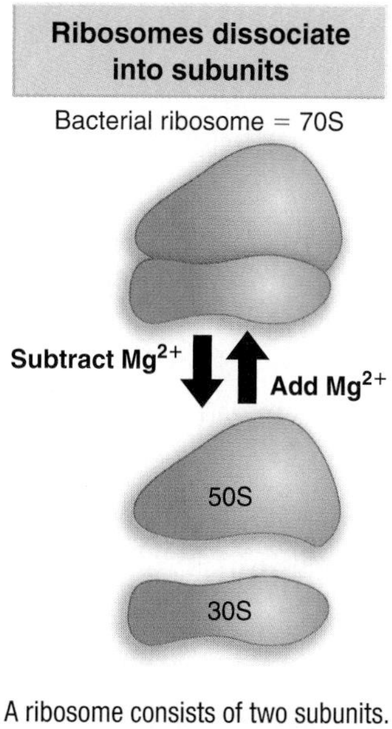

Ribosomes dissociate into subunits

Bacterial ribosome = 70S

Subtract Mg^{2+} Add Mg^{2+}

50S

30S

Figure 7.8 A ribosome consists of two subunits.

7.6 Many Ribosomes Bind to One mRNA

Key Term

- A **polyribosome (polysome)** is an mRNA that is simultaneously being translated by several ribosomes.

Key Concept

- An mRNA is simultaneously translated by several ribosomes. Each ribosome is at a different stage of progression along the mRNA.

When active ribosomes are isolated in a fraction associated with newly synthesized proteins, they are found in a complex consisting of an mRNA associated with several ribosomes. This is the **polyribosome**, or **polysome**. The 30S subunit of each ribosome is associated with the mRNA, and the 50S subunit carries the newly synthesized protein. The tRNA spans both subunits.

Each ribosome in the polysome independently synthesizes a single polypeptide during its traverse of the messenger sequence. Essentially the mRNA is pulled through the ribosome, and each codon is translated into an amino acid. So the mRNA has a series of ribosomes that carry increasing lengths of the protein product, moving from the 5′ to the 3′ end, as illustrated in **Figure 7.9**. A polypeptide chain in the process of synthesis is sometimes called a "nascent protein."

Each ribosome has a polypeptidyl-tRNA and an aminoacyl-tRNA

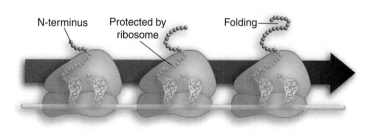

N-terminus   Protected by ribosome   Folding

Figure 7.9 A polyribosome consists of an mRNA being translated simultaneously by several ribosomes moving in the direction from 5′–3′. Each ribosome has two tRNA molecules, one carrying the nascent protein, the second carrying the next amino acid to be added.

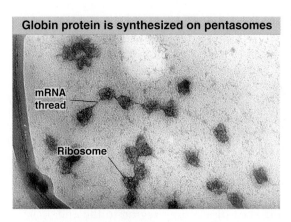

Globin protein is synthesized on pentasomes

mRNA
thread

Ribosome

Figure 7.10 Protein synthesis occurs on polysomes. Photograph kindly provided by Alexander Rich, Department of Biology, MIT.

Ribosomes recycle for translation

DNA

mRNA

Ribosomes translate mRNA and return to pool

mRNA is degraded

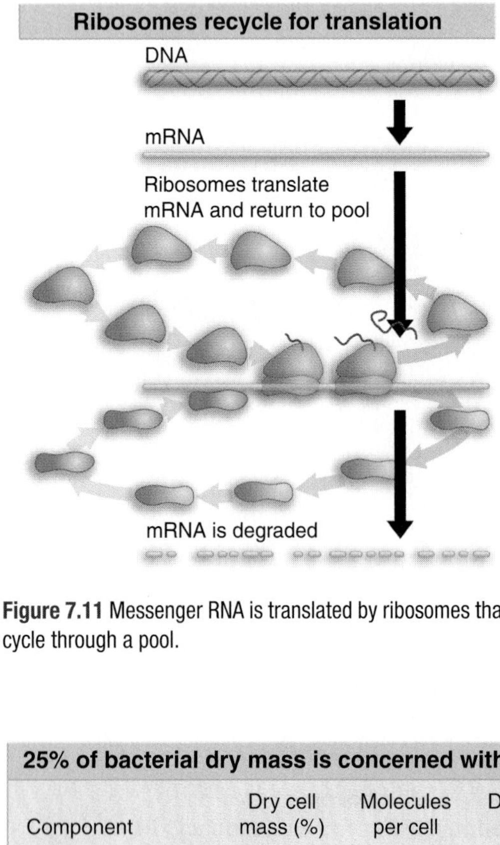

Figure 7.11 Messenger RNA is translated by ribosomes that cycle through a pool.

Roughly the 30–35 amino acids most recently added to a growing polypeptide chain are protected from the environment by the structure of the ribosome. Probably all of the preceding part of the polypeptide protrudes and is free to start folding into its proper conformation. So proteins can display parts of the mature conformation even before synthesis has been completed.

A classic characterization of polysomes is shown in the electron micrograph of **Figure 7.10**. Globin protein is synthesized by a set of five ribosomes attached to each mRNA. The ribosomes appear as squashed spherical objects of ~7 nm (70Å) in diameter, connected by a thread of mRNA. The ribosomes are located at various positions along the messenger. Those at one end have just started protein synthesis; those at the other end are about to complete production of a polypeptide chain.

The size of a polysome depends on several variables. In bacteria, it is very large, with tens of ribosomes simultaneously engaged in translation. Partly the size is due to the length of the mRNA (which usually codes for several proteins); partly it is due to the high efficiency with which the ribosomes attach to the mRNA.

Polysomes in the cytoplasm of a eukaryotic cell are likely to be smaller than those in bacteria; again, their size is a function both of the length of the mRNA (usually representing only a single protein in eukaryotes) and of the characteristic frequency with which ribosomes attach. An average eukaryotic mRNA probably has ~8 ribosomes attached at any one time.

Figure 7.11 illustrates the life cycle of the ribosome and mRNA. An mRNA is synthesized and immediately associates with ribosomes. The ribosomes are drawn from a pool (actually the pool consists of ribosomal subunits), used to translate an mRNA, and then return to the pool to be used in further cycles. The mRNA is degraded after being translated.

An overall view of the attention devoted to protein synthesis in the intact bacterium is given in **Figure 7.12**. The 20,000 or so ribosomes account for a quarter of the cell mass. There are >3000 copies of each tRNA, and altogether, the tRNA molecules outnumber the ribosomes by almost tenfold; most of them are present as aminoacyl-tRNAs—that is, ready to be used at once in protein synthesis. Because of their instability, it is difficult to calculate the number of mRNA molecules, but a reasonable guess would be ~1500, in

25% of bacterial dry mass is concerned with gene expression

Component	Dry cell mass (%)	Molecules per cell	Different types	Copies of each type
Wall	10	1	1	1
Membrane	10	2	2	1
DNA	1.5	1	1	1
mRNA	1	1,500	600	2–3
tRNA	3	200,000	60	>3,000
rRNA	16	38,000	2	19,000
Ribosomal proteins	9	10^6	52	19,000
Soluble proteins	46	2.0×10^6	1,850	>1,000
Small molecules	3	7.5×10^6	800	

Figure 7.12 Considering *E. coli* in terms of its macromolecular components.

varying states of synthesis and decomposition. There are ~600 different types of mRNA in a bacterium. This suggests that there are usually only 2–3 copies of each mRNA per bacterium. On average, each probably codes for ~3 proteins. If there are 1850 different soluble proteins, there must on average be >1000 copies of each protein in a bacterium.

7.7 The Life Cycle of Bacterial Messenger RNA

Key Terms

- **Nascent RNA** is a ribonucleotide chain that is still being synthesized, so that its 3′ end is paired with DNA where RNA polymerase is elongating.
- **Monocistronic mRNA** codes for one protein.
- **Polycistronic mRNA** includes coding regions representing more than one gene.
- A **coding region** is a part of the gene that represents a protein sequence.
- The **leader (5′ UTR)** of an mRNA is the untranslated region at the 5′ end that precedes the initiation codon.
- A **trailer (3′ UTR)** is the untranslated region at the 3′ end of an mRNA that follows the termination codon.
- The **intercistronic region** is the distance between the termination codon of one gene and the initiation codon of the next gene.

Key Concepts

- Transcription and translation occur simultaneously in bacteria, as ribosomes begin translating an mRNA before its synthesis has been completed.
- Bacterial mRNA is unstable and has a half-life of only a few minutes.
- A polycistronic mRNA has several coding regions that represent different genes.

Messenger RNA has the same function in all cells, but there are important differences in the details of the synthesis and structure of prokaryotic and eukaryotic mRNA.

A major difference in the production of mRNA depends on the locations where transcription and translation occur:

- In bacteria, mRNA is transcribed and translated in the single cellular compartment; and the two processes are so closely linked that they occur simultaneously. Because ribosomes attach to bacterial mRNA even before its transcription has been completed, the polysome is likely still to be attached to DNA. Bacterial mRNA usually is unstable, and is therefore translated into proteins for only a few minutes.

- In a eukaryotic cell, synthesis and maturation of mRNA occur exclusively in the nucleus. Only after these events are completed is the mRNA exported to the cytoplasm, where it is translated by ribosomes. Eukaryotic mRNA is relatively stable and continues to be translated for several hours.

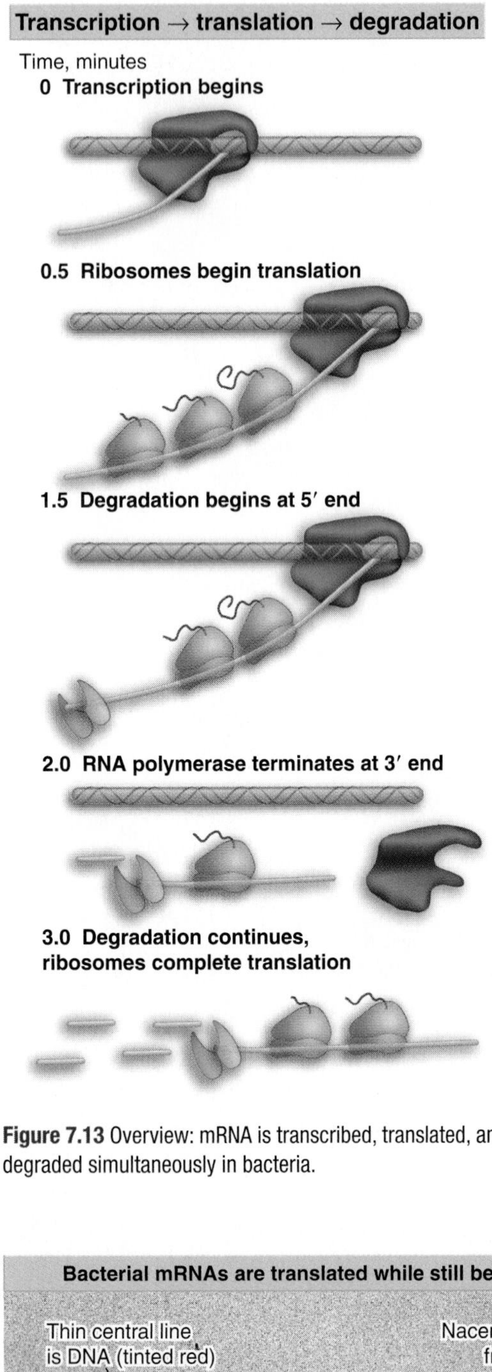

Transcription → translation → degradation

Time, minutes
0 Transcription begins

0.5 Ribosomes begin translation

1.5 Degradation begins at 5′ end

2.0 RNA polymerase terminates at 3′ end

3.0 Degradation continues, ribosomes complete translation

Figure 7.13 Overview: mRNA is transcribed, translated, and degraded simultaneously in bacteria.

Figure 7.13 shows the close relationship between transcription and translation in bacteria. Transcription begins when the enzyme RNA polymerase binds to DNA and then moves along, making a copy of one strand. Ribosomes immediately attach to the 5′ end of the mRNA and start translation, even before the rest of the message has been synthesized. Bacterial translation is very efficient, and most mRNAs are translated by a large number of tightly packed ribosomes that move along the mRNA while it is being synthesized. The 3′ end of the mRNA is generated when transcription terminates. Ribosomes continue to translate the mRNA while it survives, but it is degraded quite rapidly. An individual molecule of mRNA survives for only a matter of minutes or even less.

Bacterial transcription and translation take place at similar rates. At 37°C, mRNA is transcribed at ~40 nucleotides/second. This is very close to the rate of protein synthesis, roughly 15 amino acids/second. It therefore takes ~2 minutes to transcribe and translate an mRNA of 5000 bp, corresponding to 180 kD of protein. When expression of a new gene is initiated, its mRNA typically appears in the cell within ~2.5 minutes. The corresponding protein appears within perhaps another 0.5 minute.

The instability of most bacterial mRNAs is striking. Degradation of mRNA closely follows its translation. Probably degradation begins within 1 minute of the start of transcription, so that the 5′ end of the mRNA starts to degrade before the 3′ end has been transcribed or translated.

The stability of mRNA has a major influence on the amount of protein that is produced. It is usually expressed in terms of the half-life. The mRNA representing any particular gene has a characteristic half-life, but the average is ~2 minutes in bacteria.

This series of events is only possible, of course, because transcription, translation, and degradation all proceed in the same direction. The dynamics of gene expression have been caught *in flagrante delicto* in the electron micrograph of **Figure 7.14**. In these (unknown) transcription units, several mRNAs are under synthesis simultaneously; and each carries many ribosomes engaged in translation. (This corresponds to the stage shown in the second panel in Figure 7.13.) An RNA whose synthesis has not yet been completed is often called a **nascent RNA**.

Bacterial mRNAs vary greatly in the number of proteins for which they code. Some mRNAs represent only a single gene: they are **monocistronic**. Others (the majority) carry sequences coding for several proteins: they are **polycistronic**. In these cases, a single mRNA is transcribed from a group of adjacent genes. (Such a cluster of genes constitutes an operon that is controlled as a single genetic unit; see *Chapter 12 The Operon*.)

All mRNAs contain two types of regions. The **coding region** consists of a series of codons representing the amino acid sequence of the protein, starting (usually) with an initiation codon and ending with a termination codon. But the mRNA is always longer than the coding region; extra regions are present at both ends. An additional sequence at the 5′ end, preceding the start of the

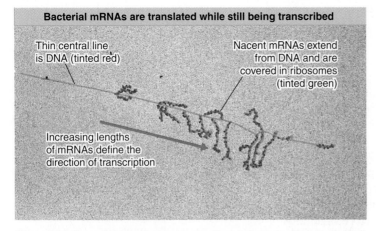

Bacterial mRNAs are translated while still being transcribed

Thin central line is DNA (tinted red)

Nacent mRNAs extend from DNA and are covered in ribosomes (tinted green)

Increasing lengths of mRNAs define the direction of transcription

Figure 7.14 Transcription units can be visualized in bacteria. Photograph kindly provided by Oscar L. Miller, Department of Biology, University of Virginia.

coding region, is described as the **leader**, or **5′ UTR** (untranslated region). An additional sequence following the termination signal, forming the 3′ end, is called the **trailer**, or **3′ UTR**. Although part of the transcription unit, these sequences do not code for protein.

A polycistronic mRNA also contains **intercistronic regions**, as illustrated in **Figure 7.15**. They vary greatly in length. They may be as long as 30 nucleotides in bacterial mRNAs (and even longer in phage RNAs), but they can also be very short, with as few as one or two nucleotides separating the termination codon for one protein from the initiation codon for the next. In an extreme case, two genes actually overlap, so that the last base of one coding region is also the first base of the next coding region.

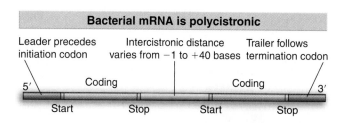

Figure 7.15 Bacterial mRNA includes untranslated as well as translated regions. Each coding region has its own initiation and termination signals. A typical bacterial mRNA has several coding regions.

7.8 Eukaryotic mRNA Is Modified During or After Its Transcription

Key Term

- **Poly(A)** is a stretch of ~200 bases of adenylic acid that is added to the 3′ end of mRNA following its synthesis.

Key Concepts

- A eukaryotic mRNA transcript is modified in the nucleus during or shortly after transcription.
- The modifications include the addition of a methylated cap at the 5′ end and a sequence of poly(A) at the 3′ end.
- The mRNA is exported from the nucleus to the cytoplasm only after all modifications have been completed.

The production of eukaryotic mRNA has additional stages after transcription. Transcription occurs in the usual way, initiating a transcript with a 5′ triphosphate end. However, the 3′ end is generated by cleaving the transcript, rather than by terminating transcription at a fixed site. Those RNAs that are derived from interrupted genes require splicing to remove the introns, generating a smaller mRNA that contains an intact coding sequence.

Figure 7.16 shows that both ends of the transcript are modified by additions of nucleotides (requiring additional enzyme systems). The 5′ end of the RNA is modified by addition of a "cap" virtually as soon as it appears. This replaces the triphosphate of the initial transcript with a nucleotide in reverse (3′ → 5′) orientation, thus "sealing" the end. The 3′ end is modified by addition of adenylic acid nucleotides [polyadenylic acid or **poly(A)**] immediately after its cleavage. Only after the completion of all modification and processing events can the eukaryotic mRNA be exported from the nucleus to the cytoplasm. The average time until departure for the cytoplasm is ~20 minutes. Once the mRNA has entered the cytoplasm, it is recognized by ribosomes and translated.

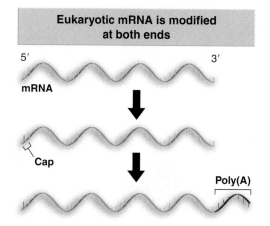

Figure 7.16 Eukaryotic mRNA is modified by addition of a cap to the 5′ end and poly(A) to the 3′ end.

Eukaryotic mRNA is modified and exported

Time, minutes

<1 Transcription begins: 5′ end is modified

6.0 3′ end of mRNA is released by cleavage

20.0 3′ end is polyadenylated

AAAAA

25.0 mRNA is transported to cytoplasm

Nucleus

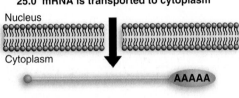

Cytoplasm

AAAAA

>240.0 Ribosomes translate mRNA

AAAAA

Figure 7.17 Overview: expression of mRNA in animal cells requires transcription, modification, processing, nucleocytoplasmic transport, and translation.

Figure 7.17 shows that the life cycle of eukaryotic mRNA is more protracted than that of bacterial mRNA. Transcription in animal cells occurs at about the same speed as in bacteria, ~40 nucleotides per second. Many eukaryotic genes are large; a gene of 10,000 bp takes ~5 minutes to transcribe.

Eukaryotic mRNA constitutes only a small proportion of the total cellular RNA (~3% of the mass). Half-lives are relatively short in yeast, ranging from 1–60 minutes. There is a substantial increase in stability in higher eukaryotes; animal cell mRNA is relatively stable, with half-lives ranging from 4–24 hours.

Eukaryotic polysomes are reasonably stable. The modifications at both ends of the mRNA contribute to the stability.

7.9 The 5′ End of Eukaryotic mRNA Is Capped

Key Terms

- A **cap** is the structure at the 5′ end of eukaryotic mRNA, added after transcription by linking the terminal phosphate of 5′ GTP to the terminal base of the mRNA. The added G (and sometimes some other bases) are methylated, giving a structure of the form ^{7}MeG5′ppp5′Np...
- A **cap 0** at the 5′ end of mRNA has only a methyl group on 7-guanine.
- A **cap 1** at the 5′ end of mRNA has methyl groups on the terminal 7-guanine and the 2′-O position of the next base.
- A **cap 2** has three methyl groups (7-guanine, 2′-O position of next base, and N^6 adenine) at the 5′ end of mRNA.

Key Concepts

- A guanylyl transferase generates a 5′ cap by adding a G to the terminal base of the transcript via a 5′–5′ link.
- 1–3 methyl groups are added to the base or ribose of the new terminal guanosine.

Transcription starts with a nucleoside triphosphate (usually a purine, A or G). The first nucleotide retains its 5′ triphosphate group and makes the usual phosphodiester bond from its 3′ position to the 5′ position of the next nucleotide. The initial sequence of the transcript can be represented as:

$$5'ppp^A_GpNpNpNp\ldots$$

The 5′ triphosphate end is modified immediately following the initiation of transcription by addition of a 5′ terminal G. The reaction is catalyzed by a nuclear enzyme, guanylyl transferase, and generates a unique structure that consists of two nucleotides, connected by a 5′–5′ triphosphate linkage. The overall reaction can be represented as a condensation between GTP and the original 5′ triphosphate terminus of the RNA:

$$\begin{array}{cc} 5' & 5' \\ \text{Gppp} + \text{pppApNpNp}\ldots \end{array}$$

$$\downarrow$$

$$5'-5'$$

$$\text{GpppApNpNp}\ldots + \text{pp} + \text{p}$$

The new G residue added to the end of the RNA is in the reverse orientation from all the other nucleotides.

The added structure is called a **cap**. It is a substrate for several methylation events. **Figure 7.18** shows the full structure of a cap after all possible methyl groups have been added. Types of caps are distinguished by the number of methylations:

- The first methylation occurs in all eukaryotes and consists of the addition of a methyl group to the 7 position of the terminal guanine. A cap that possesses this single methyl group is known as a **cap 0**. This is as far as methylation proceeds in unicellular eukaryotes. The enzyme responsible for this modification is called guanine-7-methyltransferase.

- The next step is to add another methyl group, to the 2′-O position of the penultimate base (which was actually the original first base of the transcript before any modifications were made). This reaction is catalyzed by another enzyme (2′-O-methyl-transferase). A cap with the two methyl groups is called **cap 1**. This is the predominant type of cap in all eukaryotes except unicellular organisms.

- In a small minority of higher eukaryotes, another methyl group is added to the second base. This happens only when the position is occupied by adenine; the reaction adds a methyl group at the N^6 position. The enzyme responsible acts only on an adenosine substrate that already has the methyl group in the 2′-O position.

- In some species, a methyl group is added to the third base of the capped mRNA. The substrate for this reaction is the cap 1 mRNA that already possesses two methyl groups. The third-base modification is always a 2′-O ribose methylation. This creates the **cap 2** type. This cap usually represents less than 10–15% of the total capped population.

In a population of eukaryotic mRNAs, every molecule is capped. The proportions of the different types of cap are characteristic for a particular organism. We do not know whether the structure of a particular mRNA is invariant or can have more than one type of cap.

In addition to the methylation in capping, a low frequency of internal methylation occurs in the mRNA only of higher eukaryotes. This is accomplished by the generation of N^6 methyladenine residues at a frequency of about one modification per 1000 bases. There are 1–2 methyladenines in a typical higher eukaryotic mRNA, although their presence is not obligatory, since some mRNAs do not have any.

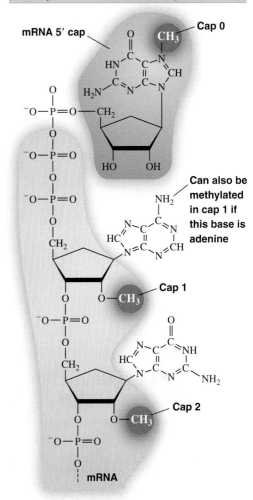

Eukaryotic mRNA has a methylated 5′ cap

Figure 7.18 The cap blocks the 5′ end of mRNA and may be methylated at several positions.

7.10 The Eukaryotic mRNA 3′ Terminus Is Polyadenylated

Key Concepts

- A length of poly(A) ~200 nucleotides long is added to a nuclear transcript after transcription.
- The poly(A) is bound by a specific protein (PABP).
- The poly(A) stabilizes the mRNA against degradation.

The 3′ terminus of almost all eukaryotic mRNA is modified by the addition of extra A residues to the sequence that is coded in the gene. The additional sequence is called the poly(A) tail, and mRNA with this feature is denoted **poly(A)$^+$**. The addition of poly(A) is catalyzed by the

Key Terms

- **Poly(A)$^+$ mRNA** is mRNA that has a 3′ terminal stretch of poly(A).

- **Poly(A) polymerase** is the enzyme that adds the stretch of polyadenylic acid to the 3′ of eukaryotic mRNA. It does not use a template.

- **Poly(A)-binding protein (PABP)** is the protein that binds to the 3′ stretch of poly(A) on a eukaryotic mRNA.

enzyme **poly(A) polymerase**, which adds ~200 A residues to the free 3'-OH end of the mRNA.

The poly(A) tract of both nuclear RNA and mRNA is associated with a protein, the **poly(A)-binding protein (PABP)**. Related forms of this protein are found in many eukaryotes. One PABP monomer of ~70 kD is bound every 10–20 bases of the poly(A) tail. So a common feature in many or most eukaryotes is that the 3' end of the mRNA consists of a stretch of poly(A) bound to a large mass of protein. Poly(A) is added as part of a reaction in which the 3' end of the mRNA is generated and modified by a complex of enzymes (see *26.14 The 3' Ends of mRNAs Are Generated by Cleavage and Polyadenylation*).

The presence of poly(A) has a direct effect on the structure of the 3' end of mRNA. However, it also has an indirect effect on the 5' end. The PABP binds to a protein (the eIF4G initiation factor for protein synthesis), which is bound at the 5' end. The reaction generates a closed loop, in which the 5' and 3' ends of the mRNA find themselves held in the same protein complex (see Figure 8.16). The formation of this complex may be responsible for some of the effects of poly(A) on the properties of mRNA. The most general effect is that poly(A) usually stabilizes mRNA against degradation; this requires binding of the PABP.

There are many examples in early embryonic development where polyadenylation of a particular mRNA controls its translation. In some cases, mRNAs are stored in a nonpolyadenylated form, and poly(A) is added when their translation is required; in other cases, poly(A)$^+$ mRNAs are deadenylated, and their translation is reduced.

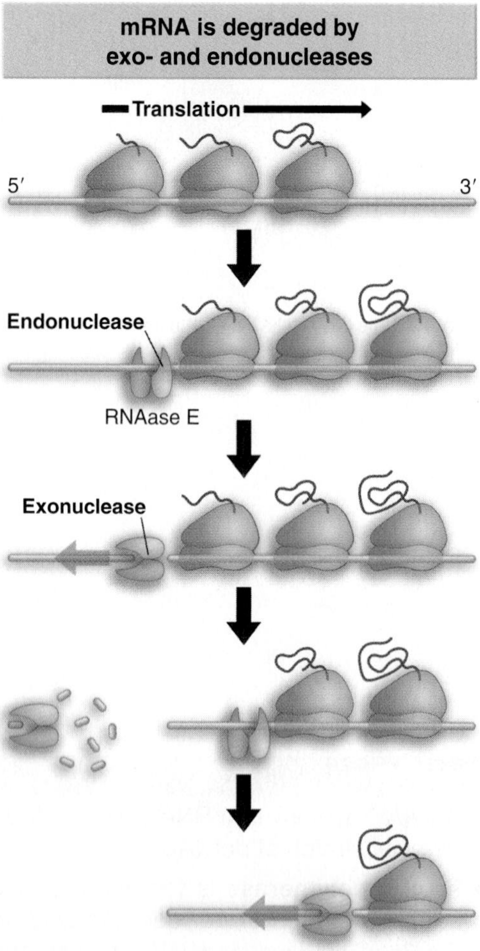

7.11 Bacterial mRNA Degradation Involves Multiple Enzymes

Key Term

- The **degradosome** is a complex of bacterial enzymes, including RNAase and helicase activities, which may be involved in degrading mRNA.

Key Concepts

- The overall direction of degradation of bacterial mRNA is 5'–3'.
- Degradation results from the combination of exonucleolytic cleavages followed by endonucleolytic degradation of the fragment from 3'–5'.

Bacterial mRNA is constantly degraded by a combination of endonucleases and exonucleases. Endonucleases cleave an RNA at internal sites. Exonucleases are involved in trimming reactions in which the extra residues are whittled away, base by base from the end. Bacterial exonucleases that act on single-stranded RNA proceed along the nucleic acid chain from the 3' end.

The way the two types of enzymes work together to degrade an mRNA is shown in **Figure 7.19**. Degradation of a bacterial mRNA is initiated by an endonucleolytic attack. Several 3' ends may be generated by endonucleolytic cleavages within the mRNA. The overall direction of degradation (as measured by loss of ability to synthesize proteins) is 5'–3'. This probably results from a succession of endonucleolytic cleavages behind the last ribosome. Degradation of the released fragments of mRNA into nucleotides then proceeds by exonucleolytic attack from the free 3'-OH end toward the 5' terminus (that is, in the

mRNA is degraded by exo- and endonucleases

Figure 7.19 Degradation of bacterial mRNA is a two-stage process. Endonucleolytic cleavages proceed 5'–3' behind the ribosomes. The released fragments are degraded by exonucleases that move 3'–5'.

opposite direction from transcription). The stability of each mRNA is determined by the susceptibility of its particular sequence to both endo- and exonucleolytic cleavages.

There are ~12 ribonucleases in *E. coli*. Mutants for the endoribonucleases (except ribonuclease I, which is without effect) accumulate unprocessed precursors to rRNA and tRNA, but are viable. Mutants for the exonucleases often have apparently unaltered phenotypes, which suggests that one enzyme can substitute for the absence of another. Mutants lacking multiple enzymes sometimes are inviable.

The process of degradation is catalyzed by a multienzyme complex (sometimes called the **degradosome**) that includes ribonuclease E, PNPase, and a helicase. RNAase E plays dual roles. Its N-terminal domain provides the endonuclease activity that makes the initial cut in mRNA. (It also is responsible for specific processing events that release rRNAs from a precursor RNA.) The C-terminal domain of RNAase E provides a scaffold that holds together the other components. The helicase unwinds the substrate RNA to make it available to PNPase (an exonuclease). According to this model, RNAase E makes the initial cut and then passes the fragments to the other components of the complex for processing. Other exonucleases can be used to complete the processing of cleaved RNAs, as shown by the lack of effects of mutation in any individual exonuclease.

7.12 Two Pathways Degrade Eukaryotic mRNA

Key Term

- The **exosome** is a complex of several exonucleases involved in degrading RNA.

Key Concepts

- The modifications at both ends of mRNA protect it against degradation by exonucleases.
- Specific sequences within an mRNA may have stabilizing or destabilizing effects.
- Destabilization may be triggered by loss of poly(A).
- The deadenylase of animal cells may bind directly to the 5′ cap.
- One yeast pathway involves exonucleolytic degradation from 5′–3′.
- Another yeast pathway uses a complex of several exonucleases that work in the 3′–5′ direction.

The major features of eukaryotic mRNA that affect its stability are summarized in **Figure 7.20**. Both structure and sequence are important. The 5′ and 3′ terminal structures protect against degradation, and

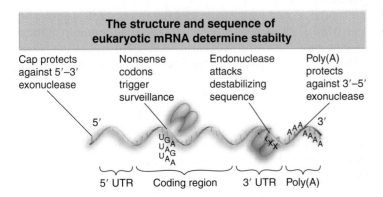

Figure 7.20 The terminal modifications of mRNA protect it against degradation. Internal sequences may activate degradation systems.

Proteins can stabilize mRNA

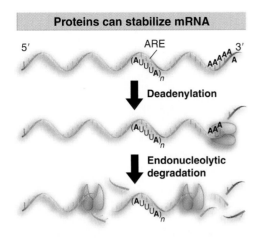

Figure 7.21 An ARE in a 3′ nontranslated region initiates degradation of mRNA.

Decapping leads to 5′–3′ degradation

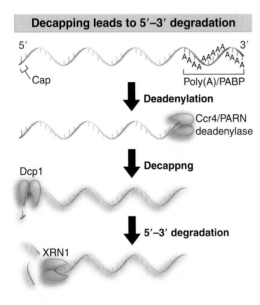

Figure 7.22 Deadenylation allows decapping to occur, which leads to endonucleolytic cleavage from the 5′ end.

The 3′–5′ pathway has three stages

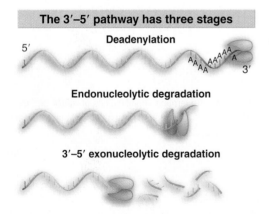

Figure 7.23 Deadenylation may lead directly to endonucleolytic cleavage and exonucleolytic cleavage from 3′ end(s).

specific sequences within the mRNA may either serve as targets to trigger degradation or may protect against degradation:

- The modifications at the 5′ and 3′ ends of mRNA play an important role in preventing exonuclease attack. The cap prevents 5′–3′ exonucleases from attacking the 5′ end, and the poly(A) prevents 3′–5′ exonucleases from attacking the 3′ end.
- Specific sequence elements within the mRNA may stabilize or destabilize it. The most common location for destabilizing elements is within the 3′ untranslated region. The presence of such an element shortens the lifetime of the mRNA.
- Within the coding region, mutations that create termination codons trigger a surveillance system that degrades the mRNA (see next section).

A common feature in some unstable mRNAs is the presence of an AU-rich sequence of ~50 bases (called the ARE) that is found in the 3′ trailer region. The consensus sequence in the ARE is the pentanucleotide AUUUA, repeated several times. **Figure 7.21** shows that the ARE triggers destabilization by a two-stage process: first the mRNA is deadenylated; then it is degraded. The deadenylation is probably needed because it causes loss of the poly(A)-binding protein, whose presence stabilizes the 3′ region.

We know most about degradation pathways in yeast. There are basically two pathways. Both start with removal of the poly(A) tail. This is catalyzed by a specific deadenylase that probably functions as part of a large protein complex. The catalytic subunit is related to bacterial RNAase D. The enzyme action is processive—once it has started to degrade a particular mRNA substrate, it continues to whittle away that mRNA, base by base.

The major degradation pathway is summarized in **Figure 7.22**. Deadenylation at the 3′ end triggers decapping at the 5′ end. The basis for this relationship is that the presence of the PABP [poly(A)-binding protein] on the poly(A) prevents the decapping enzyme from binding to the 5′ end. PABP is released when the length of poly(A) falls below 10–15 residues. The decapping reaction occurs by cleavage 1–2 bases from the 5′ end.

Each end of the mRNA influences events that occur at the other end. This is explained by the fact that the two ends of the mRNA are held together by the factors involved in protein synthesis (see Figure 8.16). The effect of PABP on decapping allows the 3′ end to have an effect in stabilizing the 5′ end. There is also a connection between the structure at the 5′ end and degradation at the 3′ end. The deadenylase directly binds to the 5′ cap, and this interaction is in fact needed for its exonucleolytic attack on the poly(A).

Removal of the cap triggers the 5′–3′ degradation pathway in which the mRNA is degraded rapidly from the 5′ end, by the 5′–3′ exonuclease XRN1.

In the second pathway, deadenylated yeast mRNAs can be degraded by the 3′–5′ exonuclease activity of the **exosome**, a complex of >9 exonucleases. The exosome is also involved in processing precursors for rRNAs. The aggregation of the individual exonucleases into the exosome complex may enable 3′–5′ exonucleolytic activities to be coordinately controlled. The exosome may also degrade fragments of mRNA released by endonucleolytic cleavage. **Figure 7.23** shows that the 3′–5′ degradation pathway may actually involve combinations of endonucleolytic and exonucleolytic action. The exosome is also found in the nucleus, where it degrades unspliced precursors to mRNA.

7.13 Nonsense Mutations Trigger a Eukaryotic Surveillance System

Key Terms

- **Nonsense-mediated mRNA decay** is a pathway that degrades an mRNA that has a nonsense mutation prior to the last exon.
- **Surveillance systems** check nucleic acids for errors. The term is used in several different contexts. One example is the system that degrades mRNAs that have nonsense mutations. Another is the set of systems that react to damage in the double helix. The common feature is that the system recognizes an invalid sequence or structure and triggers a response.

Key Concepts

- Nonsense mutations cause mRNA to be degraded.
- Genes coding for the nonsense-mediated mRNA decay system have been found in yeast and worms.

Another eukaryotic pathway for mRNA degradation is called **nonsense-mediated mRNA decay**. **Figure 7.24** shows that the introduction of a nonsense mutation (a mutation that terminates protein synthesis prematurely) often leads to increased degradation of the mRNA. The degradation event is linked to the termination event and occurs in the cytoplasm. It provides a quality control or **surveillance system** for removing nonfunctional mRNAs.

The surveillance system has been studied best in yeast and *C. elegans* but may also be important in animal cells. For example, during the formation of immunoglobulins and T cell receptors in cells of the immune system, genes are modified by somatic recombination and mutation (see *Chapter 23 Recombination in the Immune System*). This generates a significant number of nonfunctional genes, whose RNA products are disposed of by a surveillance system.

Genes that are required for the process have been identified in *S. cerevisiae* (*UPF* loci) and *C. elegans* (*smg* loci) by identifying suppressors of nonsense-mediated degradation. Mutations in these genes stabilize aberrant mRNAs, but do not affect the stability of most wild-type transcripts. One of these genes is conserved in eukaryotes (*UPF1/smg2*). It codes for an ATP-dependent helicase (an enzyme that unwinds double-stranded nucleic acids into single strands). This implies that recognition of the mRNA as an appropriate target for degradation requires a change in its structure.

Upf1 interacts with the protein factors (eRF1 and eRF3) that catalyze termination of protein synthesis, which is probably how it recognizes the termination event. It may then "scan" the mRNA by moving toward the 3′ end to look for downstream sequence elements (called *DSE*) at which the actual degradation is initiated.

In mammalian cells, the surveillance system appears to work only on mutations located prior to the last exon—in other words, there must be an intron after the site of mutation. This suggests that the system requires some event to occur in the nucleus, before the introns are removed by splicing. One possibility is that proteins attach to the mRNA in the nucleus at the exon–exon boundary when a splicing event occurs. **Figure 7.25** shows a general model for the operation of such a system. This is similar to the way

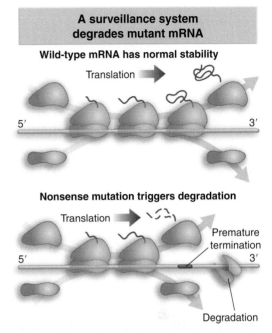

Figure 7.24 Nonsense mutations may cause mRNA to be degraded.

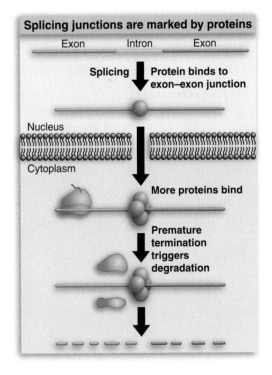

Figure 7.25 A surveillance system could have two parts. Protein(s) must bind in the nucleus to mark the result of a splicing event. Other proteins could bind to the mark either in the nucleus or cytoplasm. They are triggered to degrade the mRNA when ribosomes terminate prematurely.

in which an mRNA may be marked for export from the nucleus (see *26.9 Splicing Is Connected to Export of mRNA*). Attachment of a protein to the exon–exon junction creates a mark of the event that persists into the cytoplasm. Human homologues of the yeast Upf2,3 proteins may be involved in such a system. They bind specifically to mRNA that has been spliced.

7.14 Eukaryotic RNAs Are Transported

Key Concepts

- RNA is transported through a membrane as a ribonucleoprotein particle.
- All eukaryotic RNAs that function in the cytoplasm must be exported from the nucleus.
- tRNAs and the RNA component of a ribonuclease are imported into mitochondria.
- mRNAs can travel long distances between plant cells.

A bacterium consists of only a single compartment, so all the RNAs function in the same environment in which they are synthesized. This is most striking with mRNA, on which translation occurs simultaneously with transcription (see *7.7 The Life Cycle of Bacterial Messenger RNA*).

In eukaryotic cells, RNA is transported through membranes in the variety of instances summarized in **Figure 7.26**. Transporting a highly negative RNA through a hydrophobic membrane poses a significant thermodynamic problem, and the solution is to transport the RNA complexed with proteins.

All eukaryotic cells synthesize RNA in the nucleus; but mRNA, tRNA, and rRNA all function in the cytosol. Each type of RNA must be transported into the cytoplasm to assemble the apparatus for translation. The rRNA assembles with ribosomal proteins into immature ribosome subunits that are the substrates for the transport system. tRNA is transported by a specific protein system. mRNA is transported as a ribonucleoprotein, which forms on the RNA transcript in the nucleus (see *Chapter 26 RNA Splicing and Processing*).

One particular issue with the export of mRNA is how to distinguish the final, processed mRNA from precursors that are not fully processed (for example, which retain some introns). This may be accomplished by a system that uses some of the same methods as the surveillance system, in which a protein that binds to the RNA when it is spliced is required for components of the export apparatus to bind to the mRNA (see *26.9 Splicing Is Connected to Export of mRNA*).

Some RNAs are made in the nucleus, exported to the cytosol, and then imported into mitochondria. The mitochondria of some organisms do not code for all of the tRNAs that are required for protein synthesis. In these cases, the additional tRNAs must be imported from the cytosol. The enzyme ribonuclease P, which contains both RNA and protein subunits, is coded by nuclear genes but is found in mitochondria as well as in the nucleus. This means that the RNA must be imported into the mitochondria.

We know of some situations in which mRNA is even transported between cells. During development of the oocyte in *Drosophila*, certain mRNAs are transported into the egg from the nurse cells that surround it. The nurse cells have specialized junctions with the oocyte that allow

Eukaryotic RNA can be transported between cell compartments			
RNA	Transport	Location	
All RNA	Nucleus → cytoplasm	All cells	
tRNA	Nucleus → mitochondrion	Many cells	
mRNA	Nurse cell → oocyte	Fly embryogenesis	
mRNA	Anterior → posterior oocyte	Fly embryogenesis	
mRNA	Cell → cell	Plant phloem	

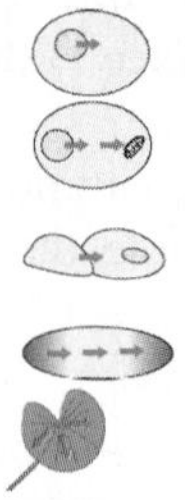

Figure 7.26 RNAs are transported through membranes in a variety of systems.

passage of material needed for early development. This material includes certain mRNAs. Once in the egg, these mRNAs take up specific locations. Some simply diffuse from the anterior end, through which they entered, but others are transported the full length of the egg to the posterior end by a motor attached to microtubules (as described in the next section).

The most striking mRNA transport has been found in plants. Movement of individual nucleic acids over long distances was first discovered in plants, in which viral movement proteins help propagate viral infection by transporting an RNA virus genome through the plasmodesmata (connections between cells). Plants also have a defense system that causes cells to silence an infecting virus, and this too may involve the spread of components including RNA over long distances between cells. Now it has turned out that similar systems may transport mRNAs between plant cells.

7.15 mRNAs Can Be Localized Within a Cell

Key Concepts

- Yeast ASH1 mRNA forms a ribonucleoprotein that binds to a myosin motor that transports it along actin filaments into the daughter bud.
- The mRNA is anchored and translated in the bud, so that the protein is found only in the bud.
- The mRNAs that establish the anterior and posterior systems in the fly oocyte are transcribed in nurse cells and transported through cytoplasmic bridges into oocytes.
- *bicoid* mRNA is localized close to the point of entry, but *oskar* and *nanos* mRNAs are transported the length of the oocyte to the posterior end by a motor attached to microtubules.

The cytosol is a crowded place, occupied by a high concentration of proteins. It is not clear how freely a polysome can diffuse within the cytosol, and most mRNAs are probably translated in random locations, determined by their point of entry into the cytosol, and the distance that they may have moved away from it. However, some mRNAs are translated at specific sites. This may be accomplished by several mechanisms:

- An mRNA may be specifically transported to a site where it is translated.
- It may be universally distributed but degraded at all sites except the site of translation.
- It may be freely diffusible but become trapped at the site of translation.

One of the best characterized cases of localization within a cell is that of ASH1 in yeast. ASH1 functions only in the budding daughter cell, because all the ASH1 mRNA is transported from the mother cell, where it is made, into the budding daughter cell. Mutations in any one of five genes, called *SHE1-5*, prevent the specific localization and cause ASH1 mRNA to be symmetrically distributed in both mother and daughter compartments. The proteins She1,2,3 bind ASH1 mRNA into a ribonucleoprotein particle that transports the mRNA into the daughter cell. **Figure 7.27** shows the functions of the proteins. She1 is a myosin (previously identified as Myo4), and She3 and She2 are proteins that connect the myosin to the mRNA. The myosin is a motor that moves the mRNA along actin filaments.

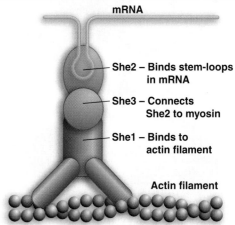

ASH1 mRNA is connected to a motor

mRNA

She2 – Binds stem-loops in mRNA

She3 – Connects She2 to myosin

She1 – Binds to actin filament

Actin filament

Figure 7.27 ASH1 mRNA forms a ribonucleoprotein containing a myosin motor that moves it along an actin filament.

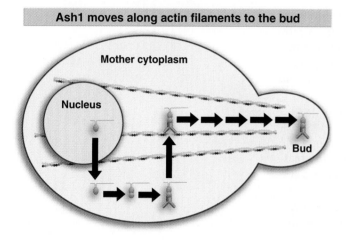

Ash1 moves along actin filaments to the bud

Figure 7.28 ASH1 mRNA is exported from the nucleus into the cytoplasm where it is assembled into a complex with the She proteins. The complex transports it along actin filaments to the bud.

mRNAs can be transported into the egg

oskar mRNA moves down tracks to posterior end; *bicoid* mRNA localizes at anterior end

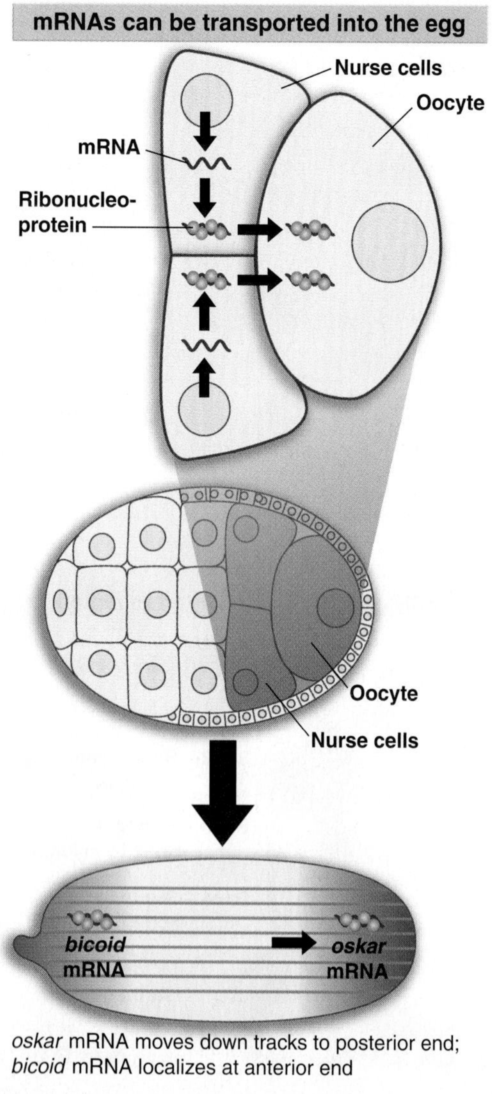

Figure 7.29 Some mRNAs are transported into the *Drosophila* egg as ribonucleoprotein particles. They move to their final sites of localization by association with microtubules.

Figure 7.28 summarizes the overall process. ASH1 mRNA is exported from the nucleus in the form of a ribonucleoprotein. In the cytoplasm it is first bound by She2, which recognizes some stem-loop secondary structures within the mRNA. Then She3 binds to She2, after which the myosin She1 binds. Then the particle hooks onto an actin filament and moves to the bud. When ASH1 mRNA reaches the bud, it is anchored there, probably by proteins that bind specifically to the mRNA.

Similar principles govern other cases in which mRNAs are transported to specific sites. The mRNA is recognized by means of *cis*-acting sequences, which usually are regions of secondary structure in the 3′ untranslated region. The mRNA is packaged into a ribonucleoprotein particle. In some cases, the transported mRNA can be visualized in very large particles, called mRNA granules. The particles are large enough (several times the size of a ribosome) to contain many protein and RNA components.

The early development of the *Drosophila* embryo depends on the localization of specific mRNAs at one end of the egg. How do these mRNAs reach the appropriate locations and what is responsible for maintaining them there? The egg is connected to surrounding nurse cells by cytoplasmic bridges. Several genes are transcribed in nurse cells, and their mRNAs are then transported through the cytoplasmic bridges into the oocyte as shown in **Figure 7.29**. Different mRNAs find their way to different locations in the oocyte. *bicoid* mRNA remains at the anterior end, but *oskar* mRNA is transported the length of the oocyte to the posterior end.

The typical means by which an mRNA is transported to a specific location in a cell involves movement along "tracks," which can be either actin filaments or microtubules. The mRNA is attached to the tracks by a motor protein that uses hydrolysis of ATP to drive movement. In the example of the *Drosophila* egg, microtubules are the tracks used to transport these and other mRNAs. In fact, the microtubules form a continuous network that connects the oocyte to the nurse cells through the cytoplasmic bridges. The motor dynein is used to move the mRNA particle along the microtubules.

Genes whose products are needed to transport these mRNAs are identified by mutants in which the mRNAs are not properly localized. The most typical disruption of the pattern is for the mRNAs simply to be distributed throughout the egg. The best characterized of these transport genes are *exuperantia* (*exu*) and *swallow* (*swa*). **Exu** protein is part of a large ribonucleoprotein complex. This complex assembles in the nurse cell, where it uses microtubule tracks to move to the cytoplasmic bridge. Then it passes across the bridge into the oocyte in a way that is independent of microtubules. In the oocyte, it attaches to microtubules to move to its location.

The properties of *exu* and *swa* mutants show that there are common components for the transport and localization of different mRNAs. We do not yet know what differences exist between the complexes that transport different mRNAs. However, we assume that there must be a component of each complex that is responsible for targeting it to the right location. It seems that whatever mRNA is being moved, the complex is first transported to the anterior end of the oocyte, where it aggregates. Then a decision is made on further localization, and the complex is transported to the appropriate site.

We know that the localization of the *bicoid* RNA to the anterior end of the oocyte depends upon sequences in the 3′ untranslated region. This is a common theme, and localization of *oskar* and *nanos* mRNAs is controlled in the same way. We assume that the 3′ sequences provide binding sites for specific protein(s) that are involved in localization. Corresponding sequences in each mRNA provide binding sites for the proteins that target the RNA to the appropriate sites in the oocyte. However, we are still missing the identification of the crucial protein that binds to the localizing sequence in the mRNA.

7.16 SUMMARY

Genetic information carried by DNA is expressed in two stages: transcription of DNA into mRNA, and translation of the mRNA into protein. Messenger RNA is transcribed from one strand of DNA and is complementary to this (noncoding) strand and identical with the other (coding) strand. The sequence of mRNA, in triplet codons 5′–3′, is related to the amino acid sequence of protein, N- to C-terminal.

The adaptor that interprets the meaning of a codon is transfer RNA, which has a compact L-shaped tertiary structure; one end of the tRNA has an anticodon that is complementary to the codon, and the other end can be covalently linked to the specific amino acid that corresponds to the target codon. A tRNA carrying an amino acid is called an aminoacyl-tRNA.

The ribosome provides the apparatus that allows aminoacyl-tRNAs to bind to their codons on mRNA. The small subunit of the ribosome is bound to mRNA; the large subunit carries the nascent polypeptide. A ribosome moves along mRNA from an initiation site in the 5′ region to a termination site in the 3′ region, and the appropriate aminoacyl-tRNAs respond to their codons, unloading their amino acids, so that the growing polypeptide chain extends by one residue for each codon traversed.

The translational apparatus is not specific for tissue or organism; an mRNA from one source can be translated by the ribosomes and tRNAs from another source. The number of times any mRNA is translated is a function of the affinity of its initiation site(s) for ribosomes and of its stability. There are some cases in which translation of groups of mRNA or individual mRNAs is specifically prevented: this is called translational control.

A typical mRNA contains both coding region(s) and nontranslated regions (5′ leader and 3′ trailer). Bacterial mRNA is usually polycistronic, with nontranslated regions between the cistrons. Each cistron is represented by a coding region that starts with a specific initiation site and ends with a termination site. Ribosome subunits associate at the initiation site and dissociate at the termination site of each coding region.

A growing *E. coli* bacterium has ~20,000 ribosomes and ~200,000 tRNAs, mostly in the form of aminoacyl-tRNA. There are ~1500 mRNA molecules, representing 2–3 copies of each of 600 different messengers.

A single mRNA can be translated by many ribosomes simultaneously, generating a polyribosome (or polysome). Bacterial polysomes are large, typically with tens of ribosomes bound to a single mRNA. Eukaryotic polysomes are smaller, typically with fewer than 10 ribosomes; each mRNA carries only a single coding sequence.

Bacterial mRNA has an extremely short half-life, only a few minutes. The 5′ end starts translation even while the downstream sequences are being transcribed. Degradation is initiated by endonucleases that cut at discrete sites, following the ribosomes in the 5′–3′ direction, after which exonucleases reduce the fragments to nucleotides by degrading them from the released 3′ end toward the 5′ end. Individual sequences may promote or retard degradation in bacterial mRNAs.

Eukaryotic mRNA must be processed in the nucleus before it is transported to the cytoplasm for translation. A methylated cap is added to the 5′ end. It consists of a nucleotide added to the original end by a 5′–5′ bond, after which methyl groups are added. Most eukaryotic mRNA has an ~200 base sequence of poly(A) added to its 3′ terminus in the nucleus after transcription. Eukaryotic mRNA exists as a ribonucleoprotein particle. Eukaryotic mRNAs are usually stable for several hours. They may have multiple sequences that initiate degradation; examples are known in which the process is regulated.

Yeast mRNA is degraded by (at least) two pathways. Both start with removal of poly(A) from the 3′ end, causing loss of poly(A)-binding protein, which in turn leads to removal of the methylated cap from the 5′ end. One pathway degrades the mRNA from the 5′ end by an exonuclease. Another pathway degrades from the 3′ end by the exosome, a complex containing several exonucleases.

Nonsense-mediated degradation leads to the destruction of mRNAs that have a termination (nonsense) codon prior to the last exon. The *UPF* loci in yeast and the *smg* loci in worms are required for the process. They include a helicase activity to unwind mRNA and a protein that interacts with the factors that terminate protein synthesis. The features of the process in mammalian cells suggest that some of the proteins attach to the mRNA in the nucleus when RNA splicing removes introns.

mRNAs can be transported to specific locations within a cell (especially in embryonic development). In the Ash1 system in yeast, mRNA is transported from the mother cell into the daughter cell by a myosin motor that moves on actin filaments. In plants, mRNAs can be transported long distances between cells.

Protein Synthesis

8.1 Introduction

An mRNA contains a series of codons that interact with the anticodons of aminoacyl-tRNAs so that a corresponding series of amino acids is incorporated into a polypeptide chain. The ribosome provides the environment for controlling the interaction between mRNA and aminoacyl-tRNA. The ribosome behaves like a small migrating factory that travels along the template engaging in rapid cycles of peptide bond synthesis. Aminoacyl-tRNAs shoot in and out of the particle at a fearsome rate, depositing amino acids; and elongation factors cyclically associate with and dissociate from the ribosome. Together with its accessory factors, the ribosome provides the full range of activities required for all the steps of protein synthesis.

Figure 8.1 shows the relative dimensions of the components of the protein synthetic apparatus. The ribosome consists of two subunits that have specific roles in protein synthesis. Messenger RNA associates with the small subunit; ~30 bases of the mRNA are bound at any time. The mRNA threads its way along the surface close to the junction of the subunits. Two tRNA molecules are active in protein synthesis at any moment; so polypeptide elongation involves reactions taking place at just two of the (roughly) 10 codons covered by the ribosome. The two tRNAs are inserted into internal sites that stretch across the subunits. A third tRNA may remain present on the ribosome after it has been used in protein synthesis, before being recycled.

The basic form of the ribosome has been conserved in evolution, but there are appreciable variations in the overall size and proportions of RNA and protein in the ribosomes of bacteria, eukaryotic cytoplasm, and organelles. **Figure 8.2** compares the components of bacterial and mammalian ribosomes.

Each of the ribosomal subunits contains a major rRNA and a number of small proteins (known as *r-proteins*). The large subunit may also contain smaller RNA(s). In *E. coli*, the small (30S) subunit consists of the 16S rRNA and 21 r-proteins. The large (50S) subunit contains 23S rRNA, the small 5S RNA, and 31 proteins. The major RNAs constitute the major part of the mass of the bacterial ribosome. Their presence is pervasive, and probably most or all of the ribosomal proteins actually contact rRNA. So the major rRNAs form what is sometimes thought of as the backbone of each subunit, a continuous thread whose presence dominates the structure, and which determines the positions of the ribosomal proteins. The large ribosomal subunit also contains a molecule of a 120 base *5S RNA* (in all ribosomes except those of mitochondria), which displays a highly base-paired structure.

The ribosomes of higher eukaryotic cytoplasm are larger than those of bacteria. The total content of both RNA and protein is greater; the major RNA molecules are longer (called 18S and 28S rRNAs), and there are more proteins. Most or all of the proteins are present in stoichiometric amounts. RNA is still the predominant component by mass. In eukaryotic cytosolic ribosomes, another small RNA is present in the large subunit. This is the *5.8S RNA*. Its sequence corresponds to the 5′ end of the prokaryotic 23S rRNA.

A feature of the primary structure of rRNA is the presence of methylated residues. There are ~10 methyl groups in 16S rRNA (located mostly toward the 3′ end of the molecule) and ~20 in 23S rRNA. In mammalian cells, the 18S and 28S rRNAs carry 43 and 74 methyl groups, respectively, so ~2% of the nucleotides are methylated (about three times the proportion methylated in bacteria).

The ribosome possesses several active centers, each of which is constructed from a group of proteins associated with a region of ribosomal RNA. The active centers require the direct participation of rRNA in a structural or even catalytic role. Some catalytic functions require individual proteins, but none of the activities can be reproduced by isolated proteins or groups of proteins; they function only in the context of the ribosome.

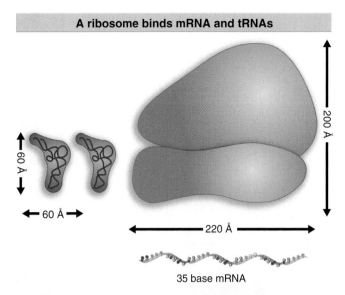

A ribosome binds mRNA and tRNAs

60 Å   60 Å   200 Å   220 Å   35 base mRNA

Figure 8.1 Size comparisons show that the ribosome is large enough to bind tRNAs and mRNA.

Figure 8.2 Ribosomes are large ribonucleoprotein particles that contain more RNA than protein and dissociate into large and small subunits.

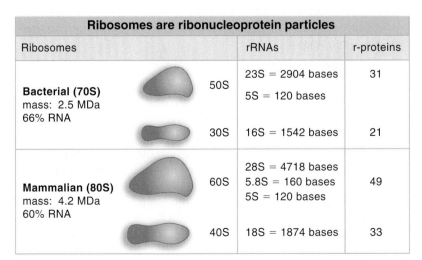

Ribosomes are ribonucleoprotein particles			
Ribosomes		rRNAs	r-proteins
Bacterial (70S) mass: 2.5 MDa 66% RNA	50S	23S = 2904 bases 5S = 120 bases	31
	30S	16S = 1542 bases	21
Mammalian (80S) mass: 4.2 MDa 60% RNA	60S	28S = 4718 bases 5.8S = 160 bases 5S = 120 bases	49
	40S	18S = 1874 bases	33

8.2 Protein Synthesis Occurs by Initiation, Elongation, and Termination

Key Terms

- The **A site** of the ribosome is the site that an aminoacyl-tRNA enters to base pair with the codon.
- The **P site** of the ribosome is the site that is occupied by peptidyl-tRNA, the tRNA carrying the nascent polypeptide chain, still paired with the codon to which it bound in the A site.
- **Peptidyl-tRNA** is the tRNA to which the nascent polypeptide chain has been transferred following peptide bond synthesis.
- **Deacylated tRNA** has no amino acid or polypeptide chain attached because it has completed its role in protein synthesis and is ready to be released from the ribosome.
- **Translocation** is the movement of the ribosome one codon along mRNA after the addition of each amino acid to the polypeptide chain.
- **Initiation** is the stage preceding the incorporation of subunits into a macromolecule. It consists of binding of the necessary components to the site where the reaction will start. For protein synthesis, initiation requires ribosomal subunits to bind to a site on mRNA.
- **Elongation** is the stage in a macromolecular synthesis reaction (replication, transcription, or translation) when the nucleotide or polypeptide chain is being extended by the addition of individual subunits.
- **Termination** is a separate reaction that ends a macromolecular synthesis reaction (replication, transcription, or translation), by stopping the addition of subunits, and (typically) causing disassembly of the synthetic apparatus.

Key Concepts

- The ribosome has three tRNA-binding sites.
- An aminoacyl-tRNA enters the A site.
- Peptidyl-tRNA is bound in the P site.
- Deacylated tRNA exits via the E site.
- An amino acid is added to the polypeptide chain by transferring the polypeptide from peptidyl-tRNA in the P site to aminoacyl-tRNA in the A site.

An amino acid is brought to the ribosome by an aminoacyl-tRNA. It is added to the growing protein chain by an interaction with the tRNA that brought the previous amino acid. Each of these tRNAs lies in a distinct site on the ribosome. **Figure 8.3** shows that the two sites have different features:

- The **A site** is where an incoming aminoacyl-tRNA enters. It presents the codon representing the next amino acid that will be added to the chain.
- The codon representing the previous amino acid to have been added to the nascent polypeptide chain lies in the **P site**. This site is occupied by **peptidyl-tRNA**, a tRNA carrying the nascent polypeptide chain.

Figure 8.4 shows that the aminoacyl end of the tRNA is located on the large subunit, while the anticodon at the other end interacts with

---

Aminoacylated tRNAs occupy the P and A sites

Codon "*n*" (P site) holds peptidyl-tRNA
Codon "*n* + 1" (A site) holds aminoacyl-tRNA

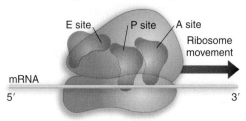

1. Before peptide bond formation peptidyl-tRNA occupies P site; aminoacyl-tRNA occupies A site

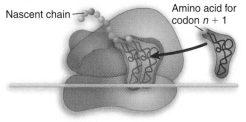

2. In peptide bond formation peptide is transferred from peptidyl-tRNA in P site to aminoacyl-tRNA in A site

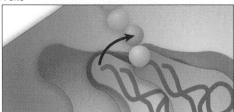

3. Translocation moves ribosome one codon; places peptidyl-tRNA in P site; deacylated tRNA leaves via E site; A site is empty for next aminoacyl-tRNA

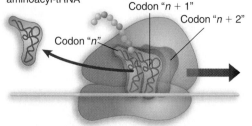

Repeat steps one through three until stop codon is reached

Figure 8.3 The ribosome has two sites for binding charged tRNAs.

the mRNA bound by the small subunit. So the P and A sites each extend across both ribosomal subunits.

For a ribosome to synthesize a peptide bond, it must be in the state shown in step 1 in Figure 8.3, when peptidyl-tRNA is in the P site and aminoacyl-tRNA is in the A site. Then a peptide bond forms when the polypeptide carried by the peptidyl-tRNA is transferred to the amino acid carried by the aminoacyl-tRNA. This reaction is catalyzed by the large subunit of the ribosome.

Transfer of the polypeptide generates the ribosome shown in step 2, in which the **deacylated tRNA**, lacking any amino acid, lies in the P site, while a new peptidyl-tRNA has been created in the A site. This peptidyl-tRNA is one amino acid residue longer than the peptidyl-tRNA that had been in the P site in step 1.

Then the ribosome moves one triplet along the messenger. This stage is called **translocation**. The movement transfers the deacylated tRNA out of the P site, and moves the peptidyl-tRNA into the P site (see step 3). The next codon to be translated now lies in the A site, ready for a new aminoacyl-tRNA to enter, when the cycle will be repeated. **Figure 8.5** summarizes the interaction between tRNAs and the ribosome.

The deacylated tRNA leaves the ribosome via another tRNA-binding site, the E site. This site is transiently occupied by the tRNA en route between leaving the P site and being released from the ribosome into the cytosol. So the flow of tRNA is into the A site, through the P site, and out through the E site. **Figure 8.6** compares the movement of

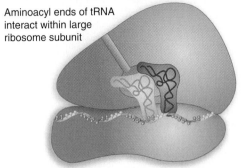

tRNA-binding sites extend across both subunits

Aminoacyl ends of tRNA interact within large ribosome subunit

Anticodons are bound to adjacent triplets on mRNA in small ribosome subunit

Figure 8.4 The P and A sites position the two interacting tRNAs across both ribosomal subunits.

Peptide bond synthesis involves transfer of polypeptide to aminoacyl-tRNA

Aminoacyl-tRNA enters the A site

Polypeptide is transferred to aminoacyl-tRNA

Translocation moves peptidyl-tRNA into P site

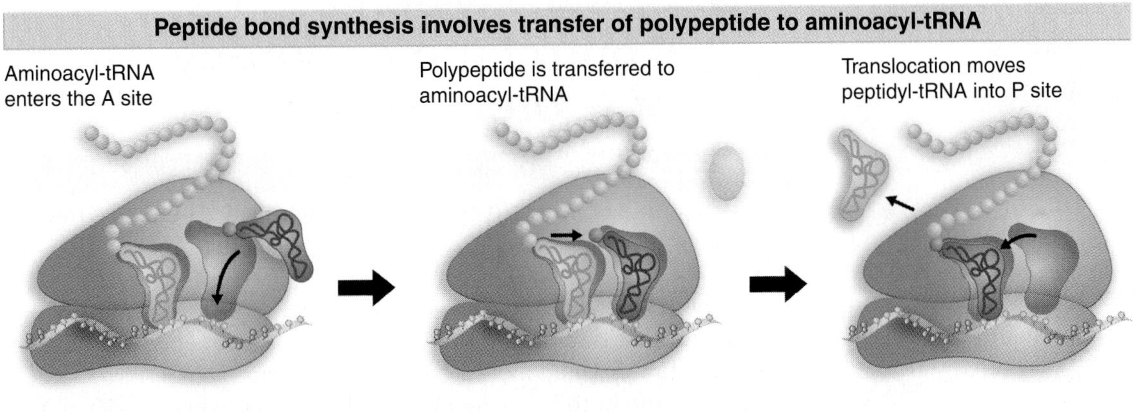

Figure 8.5 Aminoacyl-tRNA enters the A site, receives the polypeptide chain from peptidyl-tRNA, and is transferred into the P site for the next cycle of elongation.

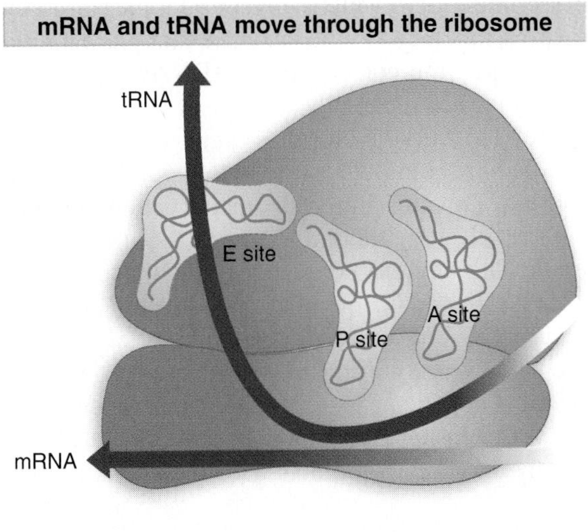

mRNA and tRNA move through the ribosome

tRNA

E site

A site

P site

mRNA

Figure 8.6 tRNA and mRNA move through the ribosome in the same direction.

Protein synthesis has three stages

Initiation 30S subunit on mRNA binding site is joined by 50S subunit and aminoacyl-tRNA binds

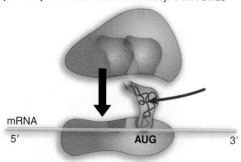

mRNA
5'   AUG   3'

Elongation Ribosome moves along mRNA, extending protein by transfer from peptidyl-tRNA to aminoacyl-tRNA

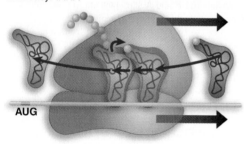

AUG

Termination Polypeptide chain is released from tRNA, and ribosome dissociates from mRNA

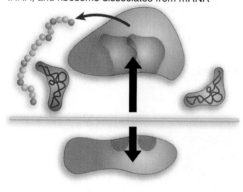

Figure 8.7 Protein synthesis falls into three stages.

tRNA and that of mRNA, which may be thought of as a sort of ratchet in which the reaction is driven by the codon–anticodon interaction.

Special reactions are needed to start and to terminate protein synthesis, so the process can be divided into the three stages shown in **Figure 8.7**:

- **Initiation** involves the reactions that precede formation of the peptide bond between the first two amino acids of the protein. The two ribosomal subunits join together to form the complete ribosome during the initiation reaction, and the first aminoacyl-tRNA is bound. This is a relatively slow step in protein synthesis, and usually determines the rate at which an mRNA is translated.
- **Elongation** includes all the reactions from synthesis of the first peptide bond to addition of the last amino acid. Amino acids are added to the chain one at a time; the addition of an amino acid is the most rapid step in protein synthesis. During elongation, the mRNA moves through the ribosome and is translated in triplets.
- **Termination** encompasses the steps that are needed to release the completed polypeptide chain; at the same time, the ribosome dissociates from the mRNA (and separates into subunits).

Different sets of accessory factors assist the ribosome at each stage. Energy is provided at various stages by the hydrolysis of GTP. A common theme is that the accessory factor has the GTPase activity, but can only exercise it when bound to a special site on the ribosome.

8.3 Special Mechanisms Control the Accuracy of Protein Synthesis

Key Concept

- The accuracy of protein synthesis is controlled by specific mechanisms at each stage.

We know that protein synthesis is generally accurate, because of the consistency that is found when we determine the sequence of a protein. There are few detailed measurements of the error rate *in vivo*, but it is generally thought to lie in the range of 1 error for every 10^4–10^5 amino acids incorporated. This means that the error rate is too low to have any effect on the phenotype of the cell.

It is not immediately obvious how such a low error rate is achieved. In fact, the nature of discriminatory events is a general issue raised by several steps in gene expression. How do the enzymes that place amino acids on tRNAs recognize just the corresponding tRNAs and amino acids? How does a ribosome recognize only the tRNA corresponding to the codon in the A site? How do the enzymes that synthesize DNA or RNA recognize only the base complementary to the template? Each case poses a similar problem: how to distinguish one particular member from the entire set, all of which share the same general features.

Probably any member initially can contact the active center by a random-hit process, but then the wrong members are rejected and only the appropriate one is accepted. The appropriate member is always in a minority (1 of 20 amino acids, 1 of ~40 tRNAs, 1 of 4 bases), so the criteria for discrimination must be strict. The point is that the system must have some mechanism for increasing discrimination from the level that would be achieved merely by making contacts with the available surfaces of the substrates.

Figure 8.8 summarizes the error rates at the steps that can affect the accuracy of protein synthesis.

Errors in transcribing mRNA are rare—probably $<10^{-6}$. This is an important stage to control, because a single mRNA molecule is translated into many protein copies. We do not know very much about the control mechanisms.

The ribosome can make two types of errors in protein synthesis. It may cause a frameshift by skipping a base when it reads the mRNA (or in the reverse direction by reading a base twice, once as the last base of one codon and then again as the first base of the next codon). These errors are rare, $\sim 10^{-5}$. Or it may allow an incorrect aminoacyl-tRNA to (mis)pair with a codon, so that the wrong amino acid is incorporated. This is probably the most common error in protein synthesis, $\sim 5 \times 10^{-4}$. It is controlled by ribosome structure and velocity.

A tRNA synthetase can make two types of errors. It can place the wrong amino acid on its tRNA; or it can charge its amino acid with the wrong tRNA. The incorporation of the wrong amino acid is more common, probably because the tRNA offers a larger surface with which the enzyme can make many more contacts to ensure specificity. Aminoacyl-tRNA synthetases have specific mechanisms to correct errors before a mischarged tRNA is released.

As we consider the processes involved in protein synthesis, we need to appreciate the mechanisms that control accuracy at each stage.

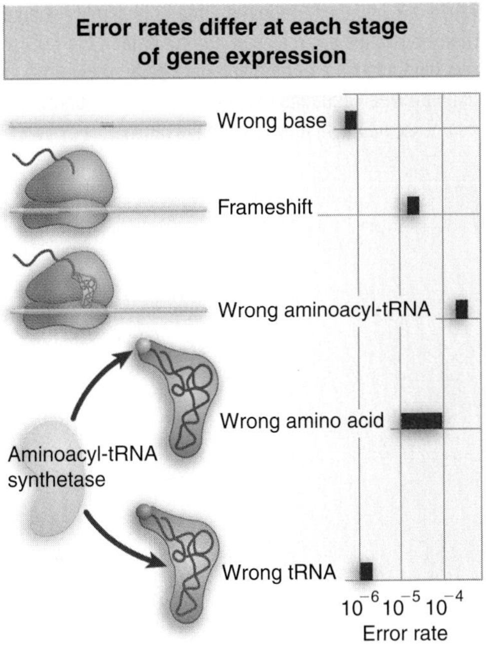

Figure 8.8 Errors occur at rates from 10^{-6} to 5×10^{-4} at different stages of protein synthesis.

8.4 Initiation in Bacteria Needs 30S Subunits and Accessory Factors

Key Concepts

- Initiation of protein synthesis requires separate 30S and 50S ribosomal subunits.
- Initiation factors (IF-1,2,3), which bind to 30S subunits, are also required.
- A 30S subunit carrying initiation factors binds to an initiation site on mRNA to form an initiation complex.
- IF-3 must be released to allow the 50S subunit to join the 30S-mRNA complex.

Key Terms

- A **ribosome-binding site** is a sequence on bacterial mRNA that includes an initiation codon that is bound by a 30S subunit in the initiation phase of protein synthesis.
- An **initiation complex** in bacterial protein synthesis contains a small ribosomal subunit, initiation factors, and initiator aminoacyl-tRNA bound to mRNA at an AUG initiation codon.
- **Initiation factors (IF)** (IF in prokaryotes, eIF in eukaryotes) are proteins that associate with the small subunit of the ribosome specifically at the stage of initiation of protein synthesis.
- **IF-1** is a bacterial initiation factor that stabilizes the initiation complex.
- **IF-2** is a bacterial initiation factor that binds the initiator tRNA to the initiation complex.
- **IF-3** is a bacterial initiation factor required for 30S subunits to bind to initiation sites in mRNA. It also prevents 30S subunits from binding to 50S subunits.

Bacterial ribosomes engaged in protein synthesis exist as 70S particles. At termination, they are released from the mRNA as free ribosomes. In growing bacteria, the majority of ribosomes are synthesizing proteins; the free pool is likely to contain ~20% of the ribosomes.

Ribosomes in the free pool can dissociate into separate subunits; so 70S ribosomes are in dynamic equilibrium with 30S and 50S subunits. *Initiation of protein synthesis is not a function of intact ribosomes, but is undertaken by the separate subunits*, which reassociate during the initiation reaction. **Figure 8.9** summarizes the ribosomal subunit cycle during protein synthesis in bacteria.

Initiation occurs at a special sequence on mRNA called the **ribosome-binding site**. This is a short sequence of bases that precedes the coding region. The small and large subunits associate at the ribosome-binding site to form an intact ribosome. The reaction occurs in two steps:

- Recognition of mRNA occurs when a small subunit binds to form an **initiation complex** at the ribosome-binding site.
- Then a large subunit joins the complex to generate a complete ribosome.

Figure 8.9 Initiation requires free ribosomal subunits. When ribosomes are released at termination, the 30S subunits bind initiation factors, and dissociate to generate free subunits. When subunits reassociate to give a functional ribosome at initiation, they release the factors.

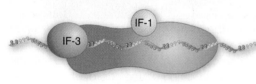
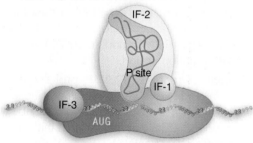
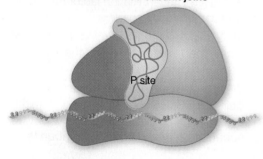

Ribosome subunits recycle

Figure 8.10 Initiation factors stabilize free 30S subunits and bind initiator tRNA to the 30S-mRNA complex.

Initiation requires factors and free subunits

1. 30S subunit binds to mRNA

2. IF-2 brings tRNA to P site

3. IFs are released and 50S subunit joins

Although the 30S subunit is involved in initiation, it is not by itself competent to undertake the reactions of binding mRNA and tRNA. It requires additional proteins called **initiation factors (IF)**. These factors are found only on 30S subunits, and they are released when the 30S subunits associate with 50S subunits to generate 70S ribosomes. This behavior distinguishes initiation factors from the structural proteins of the ribosome. The initiation factors are concerned solely with formation of the initiation complex; they are absent from 70S ribosomes, and they play no part in the stages of elongation. **Figure 8.10** summarizes the stages of initiation.

Bacteria use three initiation factors, numbered **IF-1**, **IF-2**, and **IF-3**. They are needed for both mRNA and tRNA to enter the initiation complex:

- IF-3 is needed for the 30S subunit to bind specifically to the initiation site in mRNA.
- IF-2 binds a special initiator tRNA and controls its entry into the ribosome.
- IF-1 binds to the 30S subunit only as a part of the initiation complex. It binds to the A site and prevents aminoacyl-tRNA from entering. Its location also may impede the 30S subunit from binding to the 50S subunit.

IF-3 has multiple functions: it is needed first to stabilize (free) 30S subunits; then it enables them to bind to mRNA; and, as part of the 30S-mRNA complex, it checks the accuracy of recognition of the first aminoacyl-tRNA.

The first function of IF-3 controls the equilibrium between ribosomal states, as shown in **Figure 8.11**. IF-3 binds to free 30S subunits that are released from the pool of 70S ribosomes. The presence of IF-3 prevents the 30S subunit from reassociating with a 50S subunit. The reaction between IF-3 and the 30S subunit is stoichiometric: one molecule of IF-3 binds per subunit. There is a relatively small amount of IF-3, so its availability determines the number of free 30S subunits.

Small subunits must have IF-3 in order to form initiation complexes with mRNA, but IF-3 must then be released from the 30S-mRNA complex in order to enable the 50S subunit to join. On its release, IF-3 immediately recycles by finding another 30S subunit.

IF-3 controls the ribosome-subunit equilibrium

Figure 8.11 Initiation requires 30S subunits that carry IF-3.

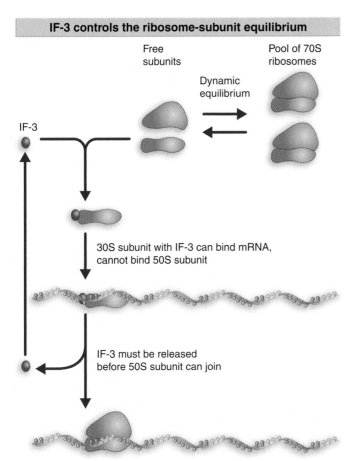

Free subunits

Pool of 70S ribosomes

Dynamic equilibrium

IF-3

30S subunit with IF-3 can bind mRNA, cannot bind 50S subunit

IF-3 must be released before 50S subunit can join

8.5 A Special Initiator tRNA Starts the Polypeptide Chain

Key Concepts

- Protein synthesis starts with a methionine amino acid usually coded by AUG.
- Different methionine tRNAs are involved in initiation and elongation.
- The initiator tRNA has unique structural features that distinguish it from all other tRNAs.
- The NH_2 group of the methionine bound to bacterial initiator tRNA is formylated.
- Eukaryotic initiator tRNA is a Met-tRNA that is different from the Met-tRNA used in elongation, but the methionine is not formylated.

Key Terms

- The **initiation codon** is a special codon (usually AUG) used to start synthesis of a protein.
- **tRNA$_f$^{Met}** is a special RNA that initiates protein synthesis in bacteria. It mostly uses AUG, but can also respond to GUG and UUG.
- **tRNA$_m$^{Met}** inserts methionine at internal AUG codons.
- **tRNA$_i$^{Met}** is a special tRNA that responds to initiation codons in eukaryotes.

Synthesis of all proteins starts with the same amino acid: methionine. The signal for initiating a polypeptide chain is a special **initiation codon** that marks the start of the reading frame. Usually the initiation codon is the triplet AUG, but in bacteria, GUG or UUG are also used.

The AUG codon represents methionine, and two types of tRNA can carry this amino acid. One is used for initiation, the other for recognizing AUG codons during elongation.

In bacteria and in eukaryotic organelles, the initiator tRNA carries a methionine residue that has been formylated on its amino group, forming a molecule of *N-formyl-methionyl-tRNA*. The tRNA is known as **tRNA$_f$^{Met}**. The name of the aminoacyl-tRNA is usually written fMet-tRNA$_f$.

Initiator Met-tRNA is formylated

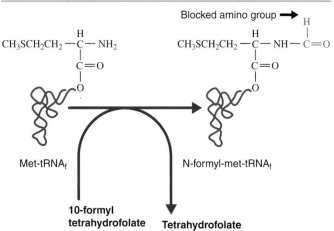

Figure 8.12 The initiator N-formyl-methionyl-tRNA (fMet-tRNA$_f$) is generated by formylation of methionyl-tRNA, using formyl-tetrahydrofolate as cofactor.

The initiator tRNA gains its modified amino acid in a two-stage reaction. First, it is charged with the amino acid to generate Met-tRNA$_f$; then the formylation reaction shown in **Figure 8.12** blocks the free NH$_2$ group. Although the blocked amino acid group would prevent the initiator from participating in chain elongation, it does not interfere with the ability to initiate a protein.

This tRNA is used only for initiation. It recognizes the codons AUG or GUG (occasionally UUG). The alternative codons are not recognized equally well; the extent of initiation declines by about half when AUG is replaced by GUG, and declines by about half again when UUG is employed.

The tRNA responsible for recognizing AUG codons in internal locations is **tRNA$_m^{Met}$**. This tRNA responds only to internal AUG codons. Its methionine cannot be formylated.

The differences between the two types of tRNA^{Met} are determined by their sequences. Although formylation of the methionine improves the efficiency with which the initiator is used, it is not absolutely necessary.

The formyl residue on the initiator methionine is removed by a specific deformylase enzyme to generate a normal NH$_2$ terminus. If methionine is to be the N-terminal amino acid of the protein, this is the only necessary step. In about half the proteins, the methionine at the terminus is removed by an aminopeptidase, creating a new terminus from R$_2$ (originally the second amino acid incorporated into the chain).

Initiation in eukaryotes has the same general features as in bacteria. Initiation in eukaryotic cytoplasm uses AUG as the initiator. The initiator tRNA is a distinct species, but its methionine does not become formylated. It is called **tRNA$_i^{Met}$**. So the difference between the initiating and elongating Met-tRNAs lies solely in the tRNA moiety, with Met-tRNA$_i$ used for initiation and Met-tRNA$_m$ used for elongation. The two tRNAs are distinguished by their tertiary structures and also by a phosphorylation of the 2′ ribose position on base 64 of the initiator.

8.6 mRNA Binds a 30S Subunit to Create the Binding Site for a Complex of IF-2 and fMet-tRNA$_f$

Key Terms

- The **context** of a codon in mRNA refers to the fact that neighboring sequences may change the efficiency with which a codon is recognized by its aminoacyl-tRNA or is used to terminate protein synthesis.

- A **ribosome-binding site** is a sequence on bacterial mRNA that includes an initiation codon that is bound by a 30S subunit in the initiation phase of protein synthesis.

- The **Shine-Dalgarno** sequence is the polypurine sequence AGGAGG centered about 10 bp before the AUG initiation codon on bacterial mRNA. It is complementary to the sequence at the 3′ end of 16S rRNA.

The meaning of the AUG and GUG codons depends on their **context**. When the AUG codon is used for initiation, it is read as formyl-methionine; when used within the coding region, it represents methionine. The meaning of the GUG codon is even more dependent on its location. When present as the first codon, it is read via the initiation reaction as formyl-methionine. Yet when present within a gene, it is read by Val-tRNA, one of the regular members of the tRNA set, to provide valine as required by the genetic code.

The initiation reaction involves binding of a 30S subunit to a **ribosome-binding site** on the mRNA. The two features of a bacterial ribosome-binding site are the AUG initiation codon and a polypurine sequence preceding it by ~10 bases that correspond to the hexamer:

$$5' \ldots A\,G\,G\,A\,G\,G \ldots 3'$$

This polypurine stretch is known as the **Shine-Dalgarno** sequence. It is complementary to a highly conserved sequence close to the 3′ end of 16S rRNA. Written in reverse direction, the rRNA sequence is the hexamer:

$$3' \ldots U\,C\,C\,U\,C\,C \ldots 5'$$

The Shine-Dalgarno sequence pairs with its complement in rRNA during mRNA-ribosome binding. Mutations of either partner in this reaction prevent an mRNA from being translated. The interaction is specific for bacterial ribosomes. This is a significant difference in the mechanism of initiation between prokaryotes and eukaryotes.

Figure 8.13 shows how an AUG initiation codon is distinguished from an AUG codon within a coding region. When an initiation complex forms at a ribosome-binding site, the initiation codon lies within the part of the P site carried by the small subunit. The only aminoacyl-tRNA that can become part of the initiation complex is the initiator, which has the unique property of being able to enter directly into the partial P site to recognize its codon.

When the large subunit joins the complex, the partial tRNA-binding sites are converted into the intact P and A sites. The initiator fMet-tRNA_f occupies the P site, and the A site is available for entry of the aminoacyl-tRNA complementary to the second codon of the gene. The first peptide bond forms between the initiator and the next aminoacyl-tRNA.

Initiation occurs when an AUG (or GUG) codon lies within a ribosome-binding site, because only the initiator tRNA can enter the partial P site in the 30S subunit. Internal reading prevails subsequently, when the codons are encountered by a ribosome that is continuing to translate an mRNA, because only the regular aminoacyl-tRNAs can enter the (complete) A site in the 70S ribosome.

30S subunits initiate; ribosomes elongate

Only fMet-tRNA_f enters partial P site on 30S subunit bound to mRNA

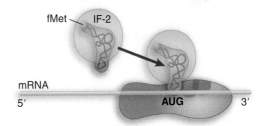

Only *aa*-tRNA enters A site on complete 70S ribosome

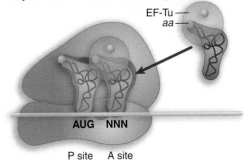

Figure 8.13 Only fMet-tRNA_f can be used for initiation by 30S subunits; only other aminoacyl-tRNAs (aa-tRNAs) can be used for elongation by 70S ribosomes.

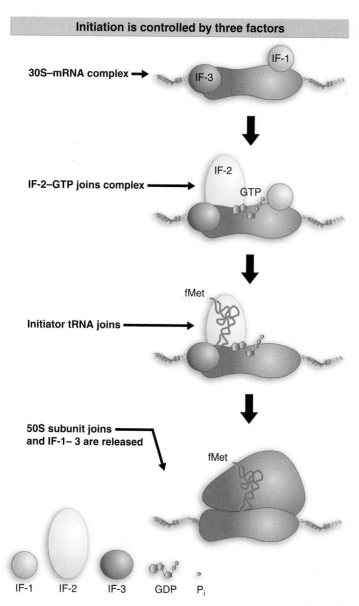

Initiation is controlled by three factors

30S–mRNA complex →

IF-1
IF-3

IF-2–GTP joins complex →

IF-2
GTP

Initiator tRNA joins →

fMet

50S subunit joins
and IF-1– 3 are released →

fMet

IF-1   IF-2   IF-3   GDP   P$_i$

Figure 8.14 IF-2 is needed to bind fMet-tRNA$_f$ to the 30S-mRNA complex. After 50S binding, all IF factors are released and GTP is cleaved.

Accessory factors are critical in controlling the usage of aminoacyl-tRNAs. All aminoacyl-tRNAs associate with the ribosome by binding to an accessory factor. The factor used in initiation is IF-2. (See *8.4 Initiation in Bacteria Needs 30S Subunits and Accessory Factors*. The accessory factor used at elongation is discussed in *8.8 Elongation Factor Tu Loads Aminoacyl-tRNA into the A Site*.)

The initiation factor IF-2 places the initiator tRNA into the P site. By forming a complex specifically with fMet-tRNA$_f$, IF-2 ensures that only the initiator tRNA, and none of the regular aminoacyl-tRNAs, participates in the initiation reaction.

Conversely, the accessory factor that places aminoacyl-tRNAs in the A site cannot bind fMet-tRNA$_f$, which is therefore excluded from use during elongation.

The accuracy of initiation is also assisted by IF-3, which stabilizes binding of the initiator tRNA by recognizing correct base pairing with the second and third bases of the AUG initiation codon.

Figure 8.14 details the series of events by which IF-2 places the fMet-tRNA$_f$ initiator in the P site. IF-2, bound to GTP, associates with the P site of the 30S subunit. At this point, the 30S subunit carries all the initiation factors. fMet-tRNA$_f$ then binds to the IF-2 on the 30S subunit. IF-2 then transfers the tRNA into the partial P site.

IF-2 has a ribosome-dependent GTPase activity that is triggered when the 50S subunit joins to generate a complete ribosome. This probably provides the energy required for conformational changes associated with the joining of the subunits.

8.7 Small Eukaryotic Subunits Scan for Initiation Sites on mRNA

Key Concepts

- Eukaryotic 40S ribosomal subunits bind to the 5′ end of mRNA and scan the mRNA until they reach an initiation site.
- A eukaryotic initiation site consists of a 10 nucleotide sequence that includes an AUG codon.
- Initiation factors are required for all stages of initiation, including binding the initiator tRNA, 40S subunit attachment to mRNA, movement along the mRNA, and joining of the 60S subunit.
- eIF2 and eIF3 bind the initiator Met-tRNA$_i$ and GTP, and the complex binds to the 40S subunit before it associates with mRNA.
- A 60S ribosomal subunit joins the complex at the initiation site.

Initiation of protein synthesis in eukaryotic cytoplasm resembles the process in bacteria, but the order of events is different, and the number of accessory factors is greater. Some of the differences in initiation are related to the requirement for a eukaryotic 40S subunit to scan the mRNA in order to reach the initiation codon.

Virtually all eukaryotic mRNAs are monocistronic, but each mRNA usually is substantially longer than necessary just to code for its protein. The average mRNA in eukaryotic cytoplasm is 1000–2000 bases long. It has a methylated cap at the 5′ terminus, and a relatively short (usually <100 base) nontranslated 5′ leader precedes the single coding region. The nontranslated 3′ trailer is often rather long, sometimes ~1000 bases, and carries 100–200 bases of poly(A) at the 3′ terminus.

The first feature to be recognized during translation of a eukaryotic mRNA is the methylated cap that marks the 5′ end. Binding of 40S subunits to mRNA requires several initiation factors, including proteins that recognize the structure of the cap. In some mRNAs, the AUG initiation codon lies within 40 bases of the 5′ terminus of the mRNA, so that both the cap and AUG lie within the span of ribosome binding. But in many other mRNAs, the cap and AUG are farther apart, in extreme cases ~1000 bases distant. How does the 40S subunit get from the cap to the initiation site?

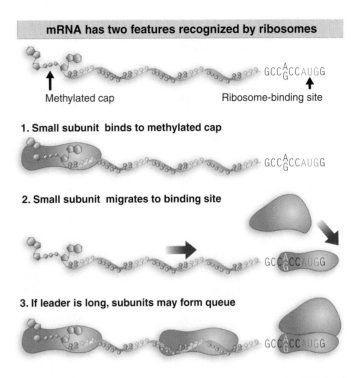

Figure 8.15 Eukaryotic ribosomes migrate from the 5′ end of mRNA to the ribosome-binding site, which includes an AUG initiation codon.

Figure 8.15 illustrates the scanning model, which supposes that the 40S subunit initially recognizes the 5′ cap and then migrates along the mRNA. Scanning from the 5′ end is a linear process, and the initiation factors help the ribosome to melt regions of the secondary structure that would block its movement.

Migration stops when the 40S subunit encounters the AUG initiation codon. Usually, although not always, the first AUG triplet sequence to be encountered is the initiation codon. However, the AUG triplet by itself is not sufficient to halt the migration; it is recognized efficiently as an initiation codon only when it is in the right context. An initiation codon may be recognized in the sequence NNNPuNNAUGG. The most important determinants of context are the bases in positions −4 and +1: the purine (A or G) 3 bases before the AUG codon, and the G immediately following it, can influence the efficiency of translation by 10×. When the leader sequence is long, additional 40S subunits can recognize the 5′ end before the first has left the initiation site, creating a line of subunits proceeding along the leader to the initiation site.

The vast majority of eukaryotic initiation events involve scanning from the 5′ cap, but there is an alternative means of initiation in which a 40S subunit associates directly with an internal site called an IRES. (This entirely bypasses any AUG codons that may be in the 5′ nontranslated region.) There are few sequence homologies between known IRES elements. The most common type of IRES element includes the AUG initiation codon at its left boundary. The 40S subunit binds directly to it, using a subset of the same factors that are required for initiation at 5′ ends.

Modification at the 5′ end occurs to almost all cellular and viral mRNAs, and is essential for their translation in eukaryotic cytosol. The sole exception to this rule is provided by a few viral mRNAs (such as poliovirus) that are not capped. They use the IRES pathway. This is especially important in picornavirus infection, where it was first discovered, because the infection destroys cap structures and inhibits the

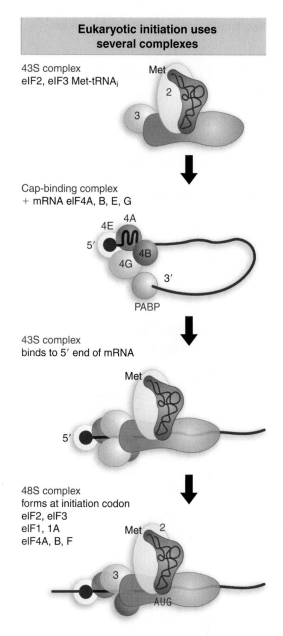

Eukaryotic initiation uses several complexes

43S complex
eIF2, eIF3 Met-tRNA_i

Met

Cap-binding complex
+ mRNA eIF4A, B, E, G

4E 4A
5′
4B
4G
3′
PABP

43S complex
binds to 5′ end of mRNA

Met
5′

48S complex
forms at initiation codon
eIF2, eIF3
eIF1, 1A
eIF4A, B, F

Met

AUG

Figure 8.16 Some initiation factors bind to the 40S ribosome subunit to form the 43S complex; others bind to mRNA. When the 43S complex binds to mRNA, it scans for the initiation codon and can be isolated as the 48S complex.

initiation factors that bind them. This prevents the translation of host mRNAs. Viral mRNAs can be translated because they use the IRES.

Eukaryotic cells have more initiation factors than bacteria—the current list includes 12 factors that are directly or indirectly required for initiation. The factors are named similarly to those in bacteria, sometimes by analogy with the bacterial factors, and are given the prefix "e" to indicate their eukaryotic identity. They act at all stages of the process, including:

- forming an initiation complex with the 5′ end of mRNA;
- forming a complex with Met-tRNA_i;
- binding the mRNA-factor complex to the Met-tRNA_i-factor complex;
- enabling the ribosome to scan mRNA from the 5′ end to the first AUG;
- detecting binding of initiator tRNA to AUG at the start site;
- mediating joining of the 60S subunit.

Figure 8.16 summarizes the stages of eukaryotic initiation, and shows which initiation factors are involved at each stage. eIF4A, eIF4B, eIF4F, and eIF4G bind to the 5′ mRNA cap. eIF2 and eIF3 form a ternary complex with the initiator tRNA that binds to the 40S ribosome subunit to form a 43S complex. Then the ternary complex scans mRNA looking for an initiation codon. One role of the factors during scanning is to help unwind any base-paired regions in the mRNA. eIF1 and eIF1A bind to the ribosome subunit-mRNA complex.

Note the circular arrangement of the mRNA associated with the cap-binding complex in Figure 8.16, when the interaction between the PABP and eIF4G brings the 5′ and 3′ ends of the mRNA into proximity.

8.8 Elongation Factor Tu Loads Aminoacyl-tRNA into the A Site

Key Terms

- **Elongation factors** (EF in prokaryotes, eEF in eukaryotes) are proteins that associate with ribosomes cyclically, during addition of each amino acid to the polypeptide chain.
- **EF-Tu** is the elongation factor that binds aminoacyl-tRNA and places it into the A site of a bacterial ribosome.

Key Concepts

- EF-Tu·GTP binds aminoacyl-tRNA and places it in the ribosome A site.
- The hydrolysis of GTP releases EF-Tu after the aminoacyl-tRNA has paired with its codon.

Once the complete ribosome has formed at the initiation codon, with initiator Met-tRNA in the P site, the A site is ready to accept an aminoacyl-tRNA. Any aminoacyl-tRNA except the initiator can enter the A site. Its entry is mediated by an **elongation factor** (**EF-Tu** in bacteria). The process is similar in eukaryotes. EF-Tu is a highly conserved protein throughout bacteria and mitochondria, and is homologous to its eukaryotic counterpart. The active form of EF-Tu carries a GTP nucleotide.

Just like its counterpart in initiation (IF-2), EF-Tu is associated with the ribosome only during the process of aminoacyl-tRNA entry. Once the aminoacyl-tRNA is in place, EF-Tu leaves the ribosome, to work again with another aminoacyl-tRNA.

Figure 8.17 shows the role of EF-Tu in bringing aminoacyl-tRNA to the A site. The binary complex of EF-Tu·GTP binds aminoacyl-tRNA to form a ternary complex of aminoacyl-tRNA·EF-Tu·GTP. The ternary complex binds only to the A site of ribosomes whose P site

EF-Tu recycles between GTP-bound and GDP-bound forms

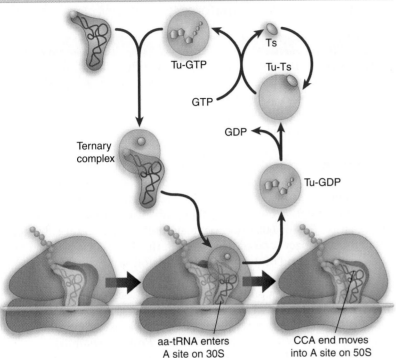

Ts

Tu-GTP   Tu-Ts

GTP

GDP

Ternary
complex

Tu-GDP

aa-tRNA enters
A site on 30S

CCA end moves
into A site on 50S

Figure 8.17 EF-Tu · GTP places aminoacyl-tRNA on the ribosome and then is released as EF-Tu · GDP. The reaction consumes GTP and releases GDP. EF-Ts is required to mediate the replacement of GDP by GTP.

is already occupied by peptidyl-tRNA. This is the critical reaction in ensuring that the aminoacyl-tRNA and peptidyl-tRNA are correctly positioned for peptide bond formation.

Aminoacyl-tRNA is loaded into the A site in two stages. First the anticodon end binds to the A site of the 30S subunit. Then codon-anticodon recognition triggers a change in the conformation of the ribosome. This stabilizes tRNA binding and causes EF-Tu to hydrolyze its GTP. The binary complex EF-Tu · GDP is released, and the 3′ end of the tRNA now moves into the A site on the 50S subunit. EF-Tu · GDP is inactive and does not bind aminoacyl-tRNA effectively. Another factor, EF-Ts, mediates the regeneration of the used form, EF-Tu · GDP, into the active form, EF-Tu · GTP.

The presence of EF-Tu prevents the aminoacyl end of aminoacyl-tRNA from entering the A site on the 50S subunit (see Figure 8.22). So the release of EF-Tu · GDP is needed for the ribosome to undertake peptide bond formation. The same principle is seen at other stages of protein synthesis: one reaction must be completed properly before the next can proceed.

In eukaryotes, the factor eEF1α is responsible for bringing aminoacyl-tRNA to the ribosome, again in a reaction that involves cleavage of a high-energy bond in GTP. Like its prokaryotic homologue (EF-Tu), it is an abundant protein. After hydrolysis of GTP, the active form is regenerated by the factor eEF1βγ, a counterpart to EF-Ts.

8.9 The Polypeptide Chain Is Transferred to Aminoacyl-tRNA

Key Terms

- **Peptidyl transferase** is the activity of the ribosomal 50S subunit that synthesizes a peptide bond when an amino acid is added to a growing polypeptide chain. The actual catalytic activity is a property of the rRNA.

Nascent polypeptide is transferred to aa-tRNA

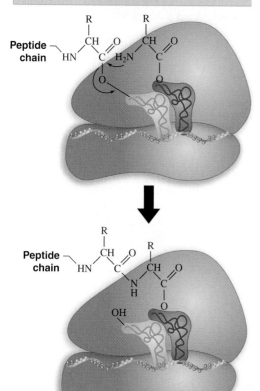

Figure 8.18 Peptide bond formation takes place by reaction between the polypeptide of peptidyl-tRNA in the P site and the amino acid of aminoacyl-tRNA in the A site.

Puromycin resembles aminoacyl-tRNA

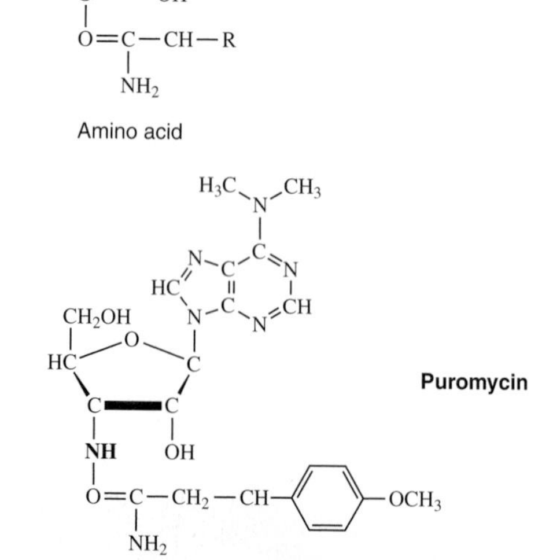

Figure 8.19 Puromycin mimics aminoacyl-tRNA because it resembles an aromatic amino acid linked to a sugar-base moiety.

- **Puromycin** is an antibiotic that terminates protein synthesis by mimicking a tRNA and becoming linked to the nascent protein chain.

Key Concepts

- The peptidyl transferase activity of the 50S subunit transfers the nascent polypeptide chain from peptidyl-tRNA in the P site to aminoacyl-tRNA in the A site.
- Peptide bond synthesis generates deacylated tRNA in the P site and peptidyl-tRNA in the A site.

A new peptide bond is synthesized by transferring the polypeptide attached to the tRNA in the P site to the aminoacyl-tRNA in the A site. The reaction is shown in **Figure 8.18**. The activity responsible for synthesis of the peptide bond is called **peptidyl transferase**. It is a function of the large (50S or 60S) ribosomal subunit. The reaction is triggered when the aminoacyl end of the tRNA in the A site swings into a location close to the end of the peptidyl-tRNA following the release of EF-Tu. This site has a peptidyl transferase activity that essentially ensures a rapid transfer of the peptide chain to the aminoacyl-tRNA. Both rRNA and 50S subunit proteins are necessary for this activity, but the actual act of catalysis is a property of the ribosomal RNA of the 50S subunit (see *8.16 Both rRNAs Play Active Roles in Protein Synthesis*).

The nature of the transfer reaction is revealed by the ability of the antibiotic **puromycin** to inhibit protein synthesis. Puromycin resembles an amino acid attached to the terminal adenosine of tRNA. **Figure 8.19** shows that puromycin has an N instead of the O that joins an amino acid to tRNA. The antibiotic is treated by the ribosome as though it were an incoming aminoacyl-tRNA. Then the polypeptide attached to peptidyl-tRNA is transferred to the NH_2 group of the puromycin.

Because the puromycin moiety is not anchored to the A site of the ribosome, the polypeptidyl-puromycin adduct is released from the ribosome in the form of polypeptidyl-puromycin. This premature termination of protein synthesis is responsible for the lethal action of the antibiotic.

8.10 Translocation Moves the Ribosome

Key Term

- **Translocation** is the movement of the ribosome one codon along mRNA after the addition of an amino acid to the polypeptide chain.

Key Concepts

- Ribosomal translocation moves the mRNA through the ribosome by 3 bases, moving deacylated tRNA into the E site, moving peptidyl-tRNA into the P site, and emptying the A site.
- The hybrid state model proposes that translocation occurs in two stages, in which the 50S moves relative to the 30S, and then the 30S moves along mRNA to restore the original conformation.

The cycle of addition of amino acids to the growing polypeptide chain is completed by **translocation**, when the ribosome advances three nucleotides along the mRNA. **Figure 8.20** shows that translocation expels the uncharged tRNA from the P site, so that the new peptidyl-tRNA can enter. The ribosome then has an empty A site ready for entry of the aminoacyl-tRNA corresponding to the next codon. As the figure shows, in bacteria the discharged tRNA is transferred from the P site to the E site (from which it is then expelled into the cytoplasm). In eukaryotes it is expelled directly into the cytosol.

Most thinking about translocation follows the hybrid-state model, which proposes that translocation occurs in two stages. **Figure 8.21** shows that first there is a shift of the 50S subunit relative to the 30S subunit; then a second shift occurs when the 30S subunit moves along mRNA to restore the original conformation.

tRNA moves through three ribosome sites	Translocation occurs in two stages

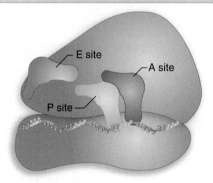

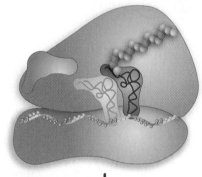

Pre-translocation:
Peptidyl-tRNA is in P site;
Aminoacyl-tRNA enters A site

50s subunit moves
relative to 30S

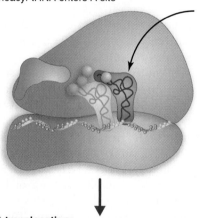

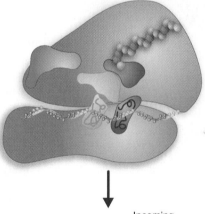

Post-translocation:
Deacylated tRNA moves to E site;
peptidyl-tRNA moves to P site

Incoming
aa-tRNA

Discharged tRNA
leaves via E site

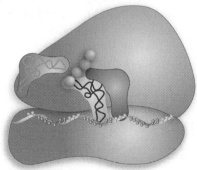

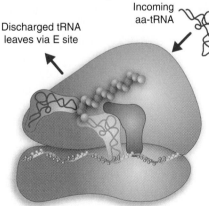

Figure 8.20 A bacterial ribosome has three tRNA-binding sites. Aminoacyl-tRNA enters the A site of a ribosome that has peptidyl-tRNA in the P site. Peptide bond synthesis deacylates the P site tRNA and generates peptidyl-tRNA in the A site. Translocation moves the deacylated tRNA into the E site and moves peptidyl-tRNA into the P site.

Figure 8.21 In the first stage of translocation, at peptide bond formation the aminoacyl end of the tRNA in the A site becomes relocated in the P site. In the second stage, the anticodon end of the tRNA becomes relocated in the P site.

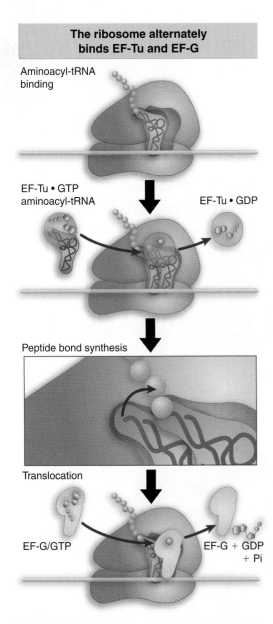

The ribosome alternately binds EF-Tu and EF-G

Aminoacyl-tRNA binding

EF-Tu • GTP aminoacyl-tRNA

EF-Tu • GDP

Peptide bond synthesis

Translocation

EF-G/GTP

EF-G + GDP + Pi

Figure 8.22 Binding of factors EF-Tu and EF-G alternates as a ribosome accepts a new aminoacyl-tRNA, forms a peptide bond, and translocates.

First the aminoacyl ends of the tRNAs (located in the 50S subunit) move into the new sites (while the anticodon ends remain bound to their anticodons in the 30S subunit). At this stage, the tRNAs are effectively bound in hybrid sites, consisting of the 50S E/ 30S P and the 50S P/ 30S A sites. Then movement is extended to the 30S subunit, so that the anticodon-codon pairing region finds itself in the right site. The most likely means of creating the hybrid state is by a movement of one ribosomal subunit relative to the other, so that translocation in effect involves two stages, the normal structure of the ribosome being restored by the second stage.

8.11 Elongation Factors Bind Alternately to the Ribosome

Key Term

- **EF-G** is an elongation factor needed for translocation in bacterial protein synthesis.

Key Concepts

- EF-G requires GTP to function in translocation, and has a structure resembling the aminoacyl-tRNA · EF-Tu · GTP complex.
- Binding of EF-Tu and EF-G to the ribosome is mutually exclusive.
- Translocation requires GTP hydrolysis, which triggers a change in EF-G, which in turn triggers a change in ribosome structure.

Translocation requires GTP and another elongation factor, **EF-G**. This factor is a major constituent of the cell; it is present at a level of ~1 copy per ribosome (~20,000 molecules per cell).

Ribosomes cannot bind EF-Tu and EF-G simultaneously, so protein synthesis follows the cycle illustrated in **Figure 8.22**, in which the factors are alternately bound to, and released from, the ribosome. So EF-Tu · GDP must be released before EF-G can bind; and then EF-G must be released before aminoacyl-tRNA · EF-Tu · GTP can bind.

Figure 8.23 shows an extraordinary similarity between the structures of the ternary complex of aminoacyl-tRNA · EF-Tu · GDP and EF-G. The structure of EF-G mimics the overall structure of EF-Tu bound to the amino acceptor stem of aminoacyl-tRNA. This suggests that they compete for the same binding site on the ribosome. The need for each factor to be released before the other can bind ensures that the events of protein synthesis proceed in an orderly manner.

Both elongation factors are monomeric GTP-binding proteins that are active when bound to GTP but inactive when bound to GDP. The triphosphate form is required for binding to the ribosome, which ensures that each factor obtains access to the ribosome only in the company of the GTP needed to fulfill its function.

8.12 Uncharged tRNA Causes the Ribosome to Trigger the Stringent Response

Key Terms

- **Stringent response** refers to the ability of a bacterium to shut down synthesis of tRNA and ribosomes in a poor-growth medium.

- An **alarmone** is a small molecule in bacteria that is produced as a result of stress and that acts to alter the state of gene expression. The unusual nucleotides ppGpp and pppGpp are examples.

- **ppGpp** is guanosine tetraphosphate. Diphosphate groups are attached to both the 5′ and 3′ positions.

- **pppGpp** is a guanosine pentaphosphate, with a triphosphate attached to the 5′ position and a diphosphate attached to the 3′ position.

- **Relaxed mutants** of *E. coli* do not display the stringent response to starvation for amino acids (or other nutritional deprivation).

- The **stringent factor** is the protein RelA, which is associated with ribosomes. It synthesizes ppGpp and pppGpp when uncharged aminoacyl-tRNA enters the A site.

- The **idling reaction** results in the production of pppGpp and ppGpp by ribosomes when an uncharged tRNA is present in the A site; this triggers the stringent response.

Key Concepts

- Poor growth conditions cause bacteria to produce the small-molecule regulators ppGpp and pppGpp.

- The trigger for the reaction is the entry of uncharged tRNA into the ribosomal A site, which activates the (p)ppGpp synthetase of the stringent factor RelA.

- One (p)ppGpp is produced every time an uncharged tRNA enters the A site.

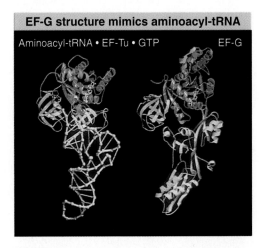

EF-G structure mimics aminoacyl-tRNA

Aminoacyl-tRNA · EF-Tu · GTP   EF-G

Figure 8.23 The structure of the ternary complex of aminoacyl-tRNA · EF-Tu · GTP (left) resembles the structure of EF-G (right). Structurally conserved domains of EF-Tu and EF-G are red and green; the tRNA and the domain resembling it in EF-G are purple. Photograph kindly provided by Poul Nissen.

The ribosome is not merely the key complex that synthesizes proteins, but it also triggers several types of regulatory response. The initial stimulus for these responses is the absence of amino acids (resulting from poor growth conditions), which in turn leads to a depletion of aminoacyl-tRNA. The ribosome is in a key position to detect a deficiency in aminoacyl-tRNA, because this deficiency stops it from functioning in protein synthesis. Lack of aminoacyl-tRNAs in general is used to trigger an alarm response; also, the absence of a specific aminoacyl-tRNA can be used to regulate the metabolic systems that produce the corresponding amino acid (see *13.5 Attenuation Can Be Controlled by Translation*).

When bacteria find themselves in such poor growth conditions that they lack a sufficient supply of amino acids to sustain protein synthesis, they shut down a wide range of activities. This is called the **stringent response**. We can view it as a mechanism for surviving hard times: the bacterium husbands its resources by engaging in only the minimum of activities until nutrient conditions improve.

The stringent response causes a massive (10–20×) reduction in the synthesis of rRNA and tRNA. This alone is sufficient to reduce the total amount of RNA synthesis to ∼5–10% of its previous level. The synthesis of certain mRNAs is reduced, leading to an overall reduction of ∼3× in mRNA synthesis. The rate of protein degradation is increased. Many metabolic adjustments occur, as seen in reduced synthesis of nucleotides, carbohydrates, lipids, etc.

The stringent response is controlled by the accumulation of two unusual nucleotides (sometimes called **alarmones**). **ppGpp** is guanosine tetraphosphate, with diphosphates attached to both 5′ and 3′ positions. **pppGpp** is guanosine pentaphosphate, with a 5′ triphosphate group and a 3′ diphosphate. These nucleotides are typical small-molecule effectors

RelA produces pppGpp

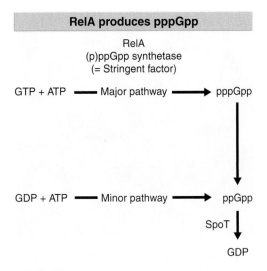

Figure 8.24 Stringent factor catalyzes the synthesis of pppGpp and ppGpp; ribosomal proteins can dephosphorylate pppGpp to ppGpp. ppGpp is degraded when it is no longer needed.

that function by binding to target proteins to alter their activities. Sometimes they are known collectively as (p)ppGpp.

Deprivation of any one amino acid, or mutation to inactivate any aminoacyl-tRNA synthetase, is sufficient to initiate the stringent response. The trigger that sets the entire series of events in train is *the presence of uncharged tRNA in the A site of the ribosome.* Under normal conditions, of course, only aminoacyl-tRNA is placed in the A site by EF-Tu (see *8.8 Elongation Factor Tu Loads Aminoacyl-tRNA into the A Site*). But when there is no aminoacyl-tRNA available to respond to a particular codon, the uncharged tRNA becomes able to gain entry.

Bacterial mutants that cannot produce the stringent reponse are called **relaxed mutants.** The most common site of relaxed mutation lies in the gene *relA*, which codes for a protein called the **stringent factor.** This factor is associated with ribosomes, although the amount is rather low—say, <1 molecule for every 200 ribosomes. So perhaps only a minority of the ribosomes are able to produce the stringent response.

The presence of uncharged tRNA in the A site blocks protein synthesis and it triggers an **idling reaction** by wild-type ribosomes. Provided that the A site is occupied by an uncharged tRNA *specifically responding to the codon,* the RelA protein catalyzes a reaction in which ATP donates a pyrophosphate group to the 3' position of either GTP or GDP. The formal name for this activity is (p)ppGpp synthetase.

Figure 8.24 shows the pathways for synthesis of (p)ppGpp. The RelA enzyme uses GTP as substrate more frequently, so that pppGpp is the predominant product. However, pppGpp is converted to ppGpp by several enzymes; among those able to perform this dephosphorylation are the translation factors EF-Tu and EF-G. The production of ppGpp via pppGpp is the most common route, *and ppGpp is the usual effector of the stringent response.*

The response of the ribosome to entry of uncharged tRNA is compared with normal protein synthesis in **Figure 8.25.** When EF-Tu places aminoacyl-tRNA in the A site, peptide bond synthesis is followed by ribosomal movement. But when uncharged tRNA is paired with the codon in the A site, the ribosome remains stationary and engages in the idling reaction. The connection that activates RelA may be a ribosomal protein, L11 of the 50S subunit, which is is located in the vicinity of the A and P sites, in a position to respond to the presence of a properly paired but uncharged tRNA in the A site. Relaxed mutations can occur in the gene for L11.

ppGpp is an effector for controlling several reactions, including the inhibition of transcription. In particular, it specifically inhibits transcription at the promoters of operons coding for rRNA. More generally, it causes the rate of transcription to decrease. Together these effects account for the ability of the stringent response to greatly reduce the energy that the cell spends on gene expression.

RelA responds to the state of tRNA

Aminoacyl-tRNA is substrate for peptide

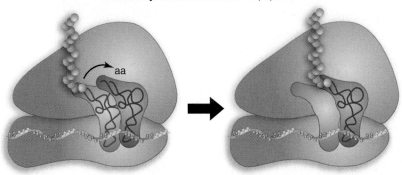

Uncharged tRNA triggers idling

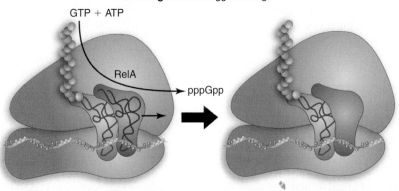

Figure 8.25 In normal protein synthesis, the presence of aminoacyl-tRNA in the A site is a signal for peptidyl transferase to transfer the polypeptide chain, followed by movement catalyzed by EF-G; but under stringent conditions, the presence of uncharged tRNA causes RelA protein to synthesize (p)ppGpp and to expel the tRNA.

How is ppGpp removed when conditions return to normal? A gene called *spoT* codes for an enzyme that provides the major catalyst for ppGpp degradation, as shown in Figure 8.24. The activity of this enzyme causes ppGpp to be rapidly degraded, with a half-life of ~20 sec; so the stringent response is reversed rapidly when synthesis of (p)ppGpp ceases.

8.13 Three Codons Terminate Protein Synthesis and Are Recognized by Protein Factors

Key Concepts

- The three termination codons are used in bacteria with relative frequencies UAA > UGA > UAG.
- Termination codons are recognized by protein release factors, not by aminoacyl-tRNAs.
- The structures of the class 1 release factors (RF1 and RF2 in *E. coli*) resemble aminoacyl-tRNA · EF-Tu and EF-G.
- The class 1 release factors respond to specific termination codons and hydrolyze the polypeptide-tRNA linkage.
- The class 1 release factors are assisted by class 2 release factors (such as RF3) that depend on GTP.
- The mechanism is similar in bacteria (which have two types of class 1 release factors) and eukaryotes (which have only one class 1 release factor).

Key Terms

- A **stop codon (termination codon)** is one of three triplets (UAG, UAA, UGA) that causes protein synthesis to terminate. They are also known historically as *nonsense codons*. The UAA codon is called **ochre**, and the UAG codon is called **amber**, after the names of the nonsense mutations by which they were originally identified.
- A **release factor (RF)** is required to terminate protein synthesis, causing release of the completed polypeptide chain and the ribosome from mRNA. Individual factors are numbered. Eukaryotic factors are called eRF.
- **RF1** is the bacterial release factor that recognizes UAA and UAG as signals to terminate protein synthesis.
- **RF2** is the bacterial release factor that recognizes UAA and UGA as signals to terminate protein synthesis.
- **RF3** is a protein synthesis termination factor related to the elongation factor EF-G. It functions to release the factors RF1 or RF2 from the ribosome after they act to terminate protein synthesis.

Only 61 triplets are assigned to amino acids. The other three triplets are **termination codons** (or **stop codons**) that end protein synthesis. They have casual names from the history of their discovery. The UAG triplet is called the **amber** codon; UAA is the **ochre** codon; and UGA is the **opal** codon. The UAG, UAA, and UGA triplet sequences are necessary and sufficient to end protein synthesis, whether occurring naturally at the end of a gene or created by mutation within a coding sequence.

(Sometimes the term *nonsense codon* is used to describe the termination triplets. "Nonsense" is really a misnomer, since the codons do have meaning, albeit a disruptive one in a mutant gene. A better term is **stop codon**.)

In bacterial genes, UAA is the most commonly used termination codon. UGA is used more heavily than UAG, although there appear to be more errors reading UGA. (An error in reading a termination codon, when an aminoacyl-tRNA improperly responds to it, results in the continuation of protein synthesis until another termination codon is encountered.)

There are two stages in ending translation. The *termination reaction* itself releases the protein chain from the last tRNA. The *post-termination reaction* releases the tRNA and mRNA and breaks the ribosome down into its subunits.

None of the termination codons is represented by a tRNA. They function in an entirely different manner from other codons, and are recognized directly by protein factors. (Because the reaction does not depend on codon–anticodon recognition, there seems to be no particular reason why it should require a triplet sequence. Presumably this reflects the evolution of the genetic code.)

Several factors have similar shapes

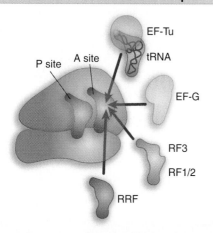

Figure 8.26 Molecular mimicry enables the elongation factor Tu-tRNA complex, the translocation factor EF-G, and the release factors RF1/2-RF3 to bind to the same ribosomal site, as does also RRF, whose role is shown in Figure 8.28.

eRF1 mimics tRNA

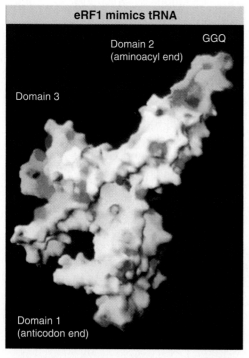

Figure 8.27 The eukaryotic termination factor eRF1 has a structure that mimics tRNA. The motif GGQ at the tip of domain 2 is essential for hydrolyzing the polypeptide chain from tRNA. Photograph kindly provided by David Barford.

Termination codons are recognized by class 1 **release factors (RF)**. In *E. coli* two class 1 release factors are specific for different sequences. **RF1** recognizes UAA and UAG; **RF2** recognizes UGA and UAA. The factors act at the ribosomal A site and require polypeptidyl-tRNA in the P site. They activate the ribosome to hydrolyze the peptidyl tRNA. Cleavage of polypeptide from tRNA takes place by a reaction analogous to the usual peptidyl transfer, except that the acceptor is H_2O instead of aminoacyl-tRNA. The release factors are present at much lower levels than initiation or elongation factors; there are ~600 molecules of each per cell, equivalent to 1 RF per 10 ribosomes. In eukaryotes, there is only a single class 1 release factor, called eRF.

The class 1 release factors are assisted by class 2 release factors, which are not codon-specific. In *E. coli*, the role of the class 2 factor, **RF3**, is to release the class 1 factor from the ribosome. RF3 is a GTP-binding protein that is related to the elongation factors.

RF3 resembles the GTP-binding domains of EF-Tu and EF-G, and RF1 and 2 resemble the C-terminal domain of EF-G, which mimics tRNA. This suggests that the release factors utilize the same site that is used by the elongation factors. **Figure 8.26** illustrates the idea that these factors all have the same general shape and bind to the ribosome successively at the same site (the A site or a region extensively overlapping with it).

The eukaryotic class 1 release factor, eRF1, is a single protein that recognizes all three termination codons. The structure of eRF1 follows a familiar theme: **Figure 8.27** shows that it consists of three domains that mimic the structure of tRNA.

The termination reaction releases the completed polypeptide, but leaves a deacylated tRNA and the mRNA still associated with the ribosome. **Figure 8.28** shows that the dissociation of the remaining components (tRNA, mRNA, 30S and 50S subunits) requires the factor RRF, the ribosome recycling factor. This acts together with EF-G in a reaction

Termination requires several protein factors

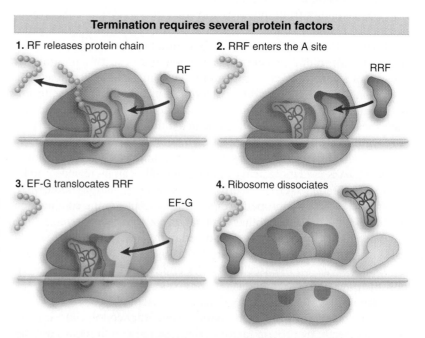

1. RF releases protein chain

2. RRF enters the A site

3. EF-G translocates RRF

4. Ribosome dissociates

Figure 8.28 The RF (release factor) terminates protein synthesis by releasing the protein chain. The RRF (ribosome recycling factor) releases the last tRNA, and EF-G releases RRF, causing the ribosome to dissociate.

that uses hydrolysis of GTP. Like the other factors involved in release, RRF has a structure that mimics tRNA, except that it lacks an equivalent for the 3′ amino acid-binding region. IF-3 is also required. RRF acts on the 50S subunit, and IF-3 removes deacylated tRNA from the 30S subunit. Once the subunits have separated, IF-3 remains necessary, of course, to prevent their reassociation.

8.14 Ribosomal RNA Pervades Both Ribosomal Subunits

Key Concepts

- Each rRNA has several distinct domains that fold independently.
- Virtually all ribosomal proteins are in contact with rRNA.
- Most of the contacts between ribosomal subunits are made between the 16S and 23S rRNAs.

Two-thirds of the mass of the bacterial ribosome is made up of rRNA. The most penetrating approach to analyzing the secondary structure of large RNAs is to compare the sequences of corresponding rRNAs in related organisms. Those regions that are important in the secondary structure retain the ability to interact by base pairing. So if a base pair is required, it can form at the same relative position in each rRNA. This approach has enabled detailed models to be constructed for both 16S and 23S rRNA.

Each of the major rRNAs can be drawn in a secondary structure with several discrete domains. 16S rRNA forms four general domains, in which just under half of the sequence is base paired (see Figure 8.38). 23S rRNA forms six general domains. The individual double-helical regions tend to be short (<8 bp). Often the duplex regions are not perfect, but contain bulges of unpaired bases. Comparable models have been drawn for mitochondrial rRNAs (which are shorter and have fewer domains) and for eukaryotic cytosolic rRNAs (which are longer and have more domains). The increase in length in eukaryotic rRNAs is due largely to the acquisition of sequences representing additional domains. The crystal structure of the ribosome shows that in each subunit the domains of the major rRNA fold independently and have discrete locations.

The 70S ribosome has an asymmetric construction. **Figure 8.29** shows a schematic of the structure of the 30S subunit, which is divided into four regions: the head, neck, body, and platform. **Figure 8.30** shows a similar representation of the 50S subunit, where two prominent features are the central protuberance (where 5S rRNA is located) and the stalk (made of multiple copies of protein L7). **Figure 8.31** shows that the platform of the small subunit fits into the notch of the large subunit. There is a cavity between the subunits that contains some of the important sites.

The structure of the 30S subunit follows the organization of 16S rRNA, with each structural feature corresponding to a domain of the rRNA. The body is based on the 5′ domain, the platform on the central domain, and the head on the 3′ region. **Figure 8.32** shows that the 30S subunit has an asymmetrical distribution of RNA and protein. One important feature is that the platform of the 30S subunit that provides the interface with the 50S subunit is composed almost entirely of RNA. Only two proteins (a small part of S7 and possibly part of S12) lie near the interface. This means that the association

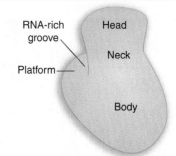

The 30S subunit has a platform

Figure 8.29 The 30S subunit has a head separated by a neck from the body, with a protruding platform.

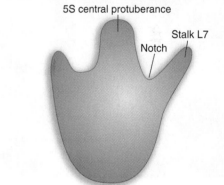

50S subunits have three features

Figure 8.30 The 50S subunit has a central protuberance where 5S rRNA is located, separated by a notch from a stalk made of copies of the protein L7.

30S + 50S = 70S

Figure 8.31 The platform of the 30S subunit fits into the notch of the 50S subunit to form the 70S ribosome.

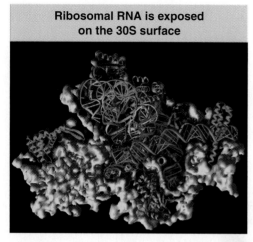

Figure 8.32 The 30S ribosomal subunit is a ribonucleoprotein particle. Proteins are in yellow. Photograph kindly provided by Venkitaraman Ramakrishnan.

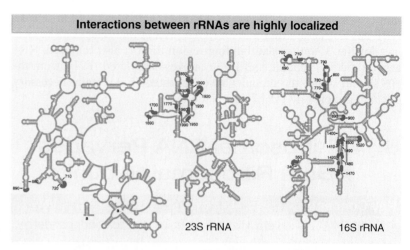

Figure 8.33 Contact points between the rRNAs are located in two domains of 16S rRNA and one domain of 23S rRNA. Photograph kindly provided by Harry Noller.

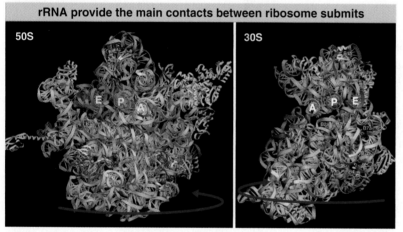

Figure 8.34 Contacts between the ribosomal subunits are mostly made by RNA (shown in purple). Contacts involving proteins are shown in yellow. The two subunits are rotated away from one another to show the faces where contacts are made; from a plane of contact perpendicular to the page, the 50S subunit is rotated 90° counter-clockwise, and the 30S is rotated 90° clockwise (this shows it in the reverse of the usual orientation). Photograph kindly provided by Harry Noller.

and dissociation of ribosomal subunits must depend on interactions with the 16S rRNA. This behavior supports the idea that the evolutionary origin of the ribosome may have been as a particle consisting of RNA rather than protein.

The 50S subunit has a more even distribution of components than the 30S, with long rods of double-stranded RNA crisscrossing the structure. The RNA forms a mass of tightly packed helices. The exterior surface largely consists of protein, except for the peptidyl transferase center. Almost all segments of the 23S rRNA interact with protein, but many of the proteins are relatively unstructured.

The junction of subunits in the 70S ribosome involves contacts between 16S rRNA (many in the platform region) with 23S rRNA. There are also some interactions between rRNA of each subunit with proteins in the other, and a few protein–protein contacts. **Figure 8.33** identifies the contact points on the rRNA structures. **Figure 8.34** opens out the structure (imagine the 50S subunit rotated counterclockwise and the 30S subunit rotated clockwise around the axis shown in the figure) to show the locations of the contact points on the face of each subunit.

8.15 Ribosomes Have Several Active Centers

Key Concepts

- Interactions involving rRNA are a key part of ribosome function.
- The environment of the tRNA-binding sites is largely determined by rRNA.

The basic message to remember about the ribosome is that it is a cooperative structure that depends on changes in the relationships among its active sites during protein synthesis. The active sites are not small, discrete regions like the active centers of enzymes. They are large regions whose construction and activities may depend just as much on the rRNA as on the ribosomal proteins. The crystal structures of the individual subunits and bacterial ribosomes give us a good impression of the overall organization and emphasize the role of the rRNA. The most recently described structure, at 5.5 Å resolution, clearly identifies the locations of the tRNAs and the functional sites. We can now account for many functions of the ribosome in terms of its structure.

Ribosomal functions are centered on the interaction with tRNAs. **Figure 8.35** shows the 70S ribosome with the positions of tRNAs in the three binding sites. The tRNAs in the A and P sites are nearly parallel to one another. All three tRNAs are aligned with their anticodon loops bound to the RNA-groove on the 30S subunit. The rest of each tRNA is bound to the 50S subunit. The environment surrounding each tRNA is mostly provided by rRNA. In each site, the rRNA contacts the tRNA at parts of the structure that are universally conserved.

It has always been a big puzzle to understand how two bulky tRNAs can fit next to one another in reading adjacent codons. The crystal structure shows a 45° kink in the mRNA between the P and A sites, which allows the tRNAs to fit as shown in the expansion of **Figure 8.36**. The tRNAs in the P and A sites are angled at 26° relative to each other at their anticodons. The closest approach between the backbones of the tRNAs occurs at the 3′ ends, where they converge to within 5 Å (perpendicular to the plane of the page). This allows the peptide chain to be transferred from the peptidyl-tRNA in the P site to the aminoacyl-tRNA in the A site.

Translocation involves large movements in the positions of the tRNAs within the ribosome. The anticodon end of tRNA moves ~28 Å from the A site to the P site, and then a further 20 Å from the P site to the E site. Because of the angle of each tRNA relative to the anticodon, the bulk of the tRNA moves much larger distances, 40 Å from A to P site, and 55 Å from P site to E site. This suggests that translocation requires a major reorganization of structure.

Much of the structure of the ribosome is occupied by its active centers. The schematic view of the ribosomal sites in **Figure 8.37** shows they occupy about two-thirds of the ribosomal structure. A tRNA enters the A site, is transferred by translocation into the P site, and then leaves the (bacterial) ribosome by the E site. The A and P sites extend across both ribosomal subunits; tRNA is paired with mRNA in the 30S subunit, but peptide transfer takes place in the 50S subunit. The A and P sites are adjacent, enabling translocation to move the tRNA from one site into the other. The E site is located near the P site (representing a position *en route* to the surface of the 50S subunit). The peptidyl transferase center is located on the 50S subunit, close to the aminoacyl ends of the tRNAs in the A and P sites (see next section).

All of the GTP-binding proteins that function in protein synthesis (EF-Tu, EF-G, IF-2, RF1,2,3) bind to a factor-binding site (sometimes called the GTPase center), which probably triggers their hydrolysis of GTP. It is located at the base of the stalk of the large subunit, which consists of the proteins L7/L12. (L7 is a modification of L12, and has an acetyl group on the N-terminus.) In addition to this region, the complex of protein L11 with a 58 base stretch of 23S rRNA provides the binding site for some antibiotics that affect GTPase activity. Neither of these ribosomal structures actually possesses GTPase activity, but they are both necessary for it. The role of the ribosome is to trigger GTP hydrolysis by factors bound in the factor-binding site.

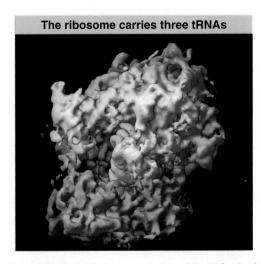

The ribosome carries three tRNAs

Figure 8.35 The 70S ribosome consists of the 50S subunit (gray) and the 30S subunit (blue) with three tRNAs located superficially: yellow in the A site, blue in the P site, and green in the E site. Photograph kindly provided by Harry Noller.

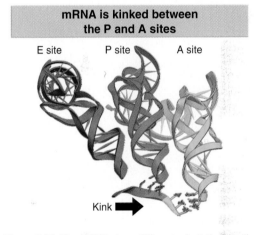

mRNA is kinked between the P and A sites

Figure 8.36 Three tRNAs have different orientations on the ribosome. mRNA turns between the P and A sites to allow aminoacyl-tRNAs to bind adjacent codons. Photograph kindly provided by Harry Noller.

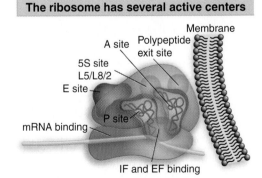

The ribosome has several active centers

Figure 8.37 The ribosome has several active centers. It may be associated with a membrane. mRNA takes a turn as it passes through the A and P sites, which are angled with regard to each other. The E site lies beyond the P site. The peptidyl transferase site (not shown) stretches across the tops of the A and P sites. The elongation factor-binding site lies at the base of the A and P sites.

Initial binding of 30S subunits to mRNA requires protein S1, which has a strong affinity for single-stranded nucleic acid. It is responsible for maintaining the single-stranded state in mRNA that is bound to the 30S subunit. This action is necessary to prevent the mRNA from taking up a base-paired conformation that would be unsuitable for translation. S1 has an extremely elongated structure and associates with S18 and S21. The three proteins constitute a domain that is involved in the initial binding of mRNA and in binding initiator tRNA. This locates the mRNA-binding site in the vicinity of the cleft of the small subunit (see Figure 8.31). The 3′ end of rRNA, which pairs with the mRNA initiation site, is located in this region.

A nascent protein debouches through the ribosome, away from the active sites, into the region in which ribosomes may be attached to membranes. A polypeptide chain emerges from the ribosome through an exit channel, which leads from the peptidyl transferase site to the surface of the 50S subunit. The tunnel is composed mostly of rRNA. It is quite narrow, only 1–2 nm wide, and ~10 nm long. The nascent polypeptide emerges from the ribosome ~15 Å away from the peptidyl transferase site. The tunnel can hold ~50 amino acids, and probably constrains the polypeptide chain so that it cannot fold until it leaves the exit domain.

8.16 Both rRNAs Play Active Roles in Protein Synthesis

Key Concepts

- 16S rRNA plays an active role in the functions of the 30S subunit. It interacts directly with mRNA, with the 50S subunit, and with the anticodons of tRNAs in the P and A sites.
- Peptidyl transferase activity resides exclusively in the 23S rRNA.

The ribosome was originally viewed as a collection of proteins with various catalytic activities, held together by protein–protein interactions and by binding to rRNA. But the discovery of RNA molecules with catalytic activities (see *Chapter 26 RNA Splicing and Processing*) immediately suggests that rRNA might play a more active role in ribosome function. There is now evidence that rRNA interacts with mRNA or tRNA at each stage of translation, and that the proteins are necessary to maintain the rRNA in a structure in which it can perform the catalytic functions. Several interactions involve specific regions of rRNA:

- The 3′ terminus of the rRNA interacts directly with mRNA at initiation.
- Specific regions of 16S rRNA interact directly with the anticodon regions of tRNAs in both the A site and the P site. Similarly, 23S rRNA interacts with the CCA terminus of peptidyl-tRNA in both the P site and A site.
- Subunit interaction involves interactions between 16S and 23S rRNAs (see *8.14 Ribosomal RNA Pervades Both Ribosomal Subunits*).

Much information about the individual steps of bacterial protein synthesis has been obtained by using antibiotics that inhibit the process at particular stages. The target for the antibiotic can be identified by the component in which resistant mutations occur. Some antibiotics act on individual ribosomal proteins, but several act on rRNA, which suggests that the rRNA is involved with many or even all of the functions of the ribosome.

The functions of rRNA have been investigated by two types of approaches. Structural studies show that particular regions of rRNA are located in important sites of the ribosome, and that chemical modifications of the bases in these regions impede particular ribosomal functions. And mutations identify bases in rRNA that are required for particular ribosomal functions. **Figure 8.38** summarizes the sites in 16S rRNA that have been identified by these means.

The tRNA makes contacts with the 23S rRNA in both the P and A sites. The classic criterion for proving the importance of the interaction is satisfied: a mutation in tRNA can be compensated by a mutation in rRNA.

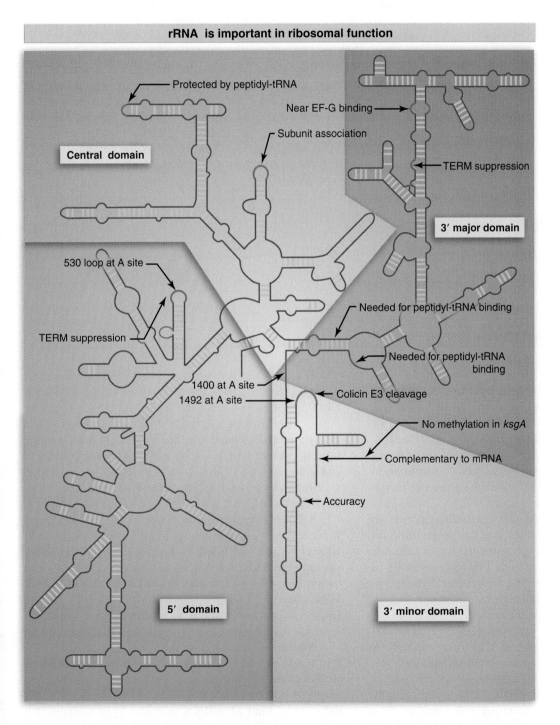

Figure 8.38 Some sites in 16S rRNA are protected from chemical probes when 50S subunits join 30S subunits or when aminoacyl-tRNA binds to the A site. Others are the sites of mutations that affect protein synthesis. TERM suppression sites may affect termination at termination codons. The large colored blocks indicate the four domains of the rRNA.

So there is a close role for rRNA in both the tRNA-binding sites. Indeed, we are moving toward describing the movements of tRNA between the A and P sites in terms of making and breaking contacts with rRNA.

What is the nature of the site on the 50S subunit that provides peptidyl transferase function? A long search for ribosomal proteins that might possess the catalytic activity was unsuccessful and led to the discovery that 23S rRNA can catalyze the formation of a peptide bond between peptidyl-tRNA and aminoacyl-tRNA. Activity is abolished by mutations in domain V of the rRNA, which lies in the P site. The crystal structure of an archaeal 50S subunit shows that the peptidyl transferase site basically consists of 23S rRNA. There is no protein within 18 Å of the active site where the transfer reaction occurs between peptidyl-tRNA and aminoacyl-tRNA!

The catalytic activity of isolated rRNA is quite low, and proteins that are bound to the 23S rRNA outside of the peptidyl transfer region are almost certainly required to enable the rRNA to form the proper structure *in vivo*. The idea that rRNA is the catalytic component is consistent with the results discussed in *Chapter 26 RNA Splicing and Processing* that identify catalytic properties in RNA that are involved with several RNA processing reactions. That rRNA is the catalytic component also fits with the notion that the ribosome evolved from functions originally possessed by RNA.

8.17 SUMMARY

A codon in mRNA is recognized by an aminoacyl-tRNA, which has an anticodon complementary to the codon and carries the amino acid corresponding to the codon. A special initiator tRNA (fMet-tRNA$_f$ in prokaryotes or Met-tRNA$_i$ in eukaryotes) recognizes the AUG codon, which is used to start the coding sequences. In prokaryotes, GUG and UUG are also used. Only the termination (nonsense) codons UAA, UAG, and UGA are not recognized by aminoacyl-tRNAs.

Ribosomes are released from protein synthesis to enter a pool of free ribosomes that are in equilibrium with separate small and large subunits. Small subunits bind to mRNA and then are joined by large subunits to generate an intact ribosome that undertakes protein synthesis. Recognition of a prokaryotic initiation site involves binding of a sequence at the 3' end of rRNA to the Shine-Dalgarno motif that precedes the AUG (or GUG) codon in the mRNA. Recognition of a eukaryotic mRNA involves binding to the 5' cap; the small subunit then migrates to the initiation site by scanning for AUG codons. When it recognizes an appropriate AUG codon (usually but not always the first it encounters), it is joined by a large subunit.

A ribosome can carry two aminoacyl-tRNAs simultaneously: its P site is occupied by a polypeptidyl-tRNA, which carries the polypeptide chain synthesized so far, while the A site is used for entry by an aminoacyl-tRNA carrying the next amino acid to be added to the chain. Bacterial ribosomes also have an E site, through which deacylated tRNA passes before it is released after being used in protein synthesis. The polypeptide chain in the P site is transferred to the aminoacyl-tRNA in the A site, creating a deacylated tRNA in the P site and a peptidyl-tRNA in the A site.

Following peptide bond synthesis, the ribosome translocates one codon along the mRNA, moving deacylated tRNA into the E site, and peptidyl tRNA from the A site into the P site. Translocation is catalyzed by the elongation factor EF-G, and, like several other stages of ribosome function, requires hydrolysis of GTP. During translocation, the ribosome passes through a hybrid stage in which the 50S subunit moves relative to the 30S subunit.

Additional factors are required at each stage of protein synthesis. They are defined by their cyclic association with, and dissociation from, the ribosome. IF factors are involved in prokaryotic initiation. IF-3 is needed for 30S subunits to bind to mRNA and also is responsible for maintaining the 30S subunit in a free form. IF-2 is needed for fMet-tRNA$_f$ to bind to the 30S subunit and is responsible for excluding other aminoacyl-tRNAs from the initiation reaction. GTP is hydrolyzed after the initiator tRNA has been bound to the initiation complex. The initiation factors must be released before a large subunit can join the initiation complex.

Eukaryotic initiation involves a greater number of factors. Some of them participate in the initial binding of the 40S subunit to the capped 5′ end of the mRNA. Then the initiator tRNA is bound by another group of factors. After this initial binding, the small subunit scans the mRNA until it recognizes the correct AUG codon. At this point, initiation factors are released and the 60S subunit joins the complex.

Prokaryotic EF factors are involved in elongation. EF-Tu binds aminoacyl-tRNA to the 70S ribosome. GTP is hydrolyzed when EF-Tu is released, and EF-Ts is required to regenerate the active form of EF-Tu. EF-G is required for translocation. Binding of the EF-Tu and EF-G factors to ribosomes is mutually exclusive, which ensures that each step must be completed before the next can be started.

The level of protein synthesis itself provides an important coordinating signal. Deficiency in aminoacyl-tRNAs causes an idling reaction on the ribosome, which leads to the synthesis of the unusual nucleotide ppGpp. This is an effector that inhibits initiation of transcription at certain promoters; it also has a general effect in inhibiting elongation on all templates.

Termination occurs at any one of the three special codons, UAA, UAG, or UGA. Class 1 RF factors that specifically recognize the termination codons activate the ribosome to hydrolyze the peptidyl-tRNA. A class 2 RF factor is required to release the class 1 RF factor from the ribosome. The GTP-binding factors IF-2, EF-Tu, EF-G, and RF3 all have similar structures, and bind to the same ribosomal site, the factor-binding site.

Ribosomes are ribonucleoprotein particles in which a majority of the mass is provided by rRNA. The shapes of all ribosomes are generally similar, but only those of bacteria (70S) have been characterized in detail. The small (30S) subunit has a squashed shape, with a "body" containing about two-thirds of the mass divided from the "head" by a cleft. The large (50S) subunit is more spherical, with a prominent "stalk" on the right and a "central protuberance." Locations of all proteins are known approximately in the small subunit.

Each subunit contains a single major rRNA, 16S and 23S in prokaryotes, 18S and 28S in eukaryotic cytosol.

There are also minor rRNAs, most notably 5S rRNA in the large subunit. Both major rRNAs have extensive base pairing, mostly in the form of short, imperfectly paired duplex stems with single-stranded loops. Conserved features in the rRNA can be identified by comparing sequences and the secondary structures that can be drawn for rRNA of a variety of organisms. The 16S rRNA has four distinct domains; the 23S rRNA has six distinct domains. Eukaryotic rRNAs have additional domains.

The crystal structure shows that the 30S subunit has an asymmetrical distribution of RNA and protein. RNA is concentrated at the interface with the 50S subunit. The 50S subunit has a surface of protein, with long rods of double-stranded RNA crisscrossing the structure. 30S–50S joining involves contacts between 16S rRNA and 23S rRNA.

Each subunit has several active centers, concentrated in the translational domain of the ribosome where proteins are synthesized. Proteins leave the ribosome through the exit domain, which can associate with a membrane. The major active sites are the P and A sites, the E site, the EF-Tu and EF-G binding sites, peptidyl transferase, and mRNA-binding site. Ribosome conformation may change at stages during protein synthesis; differences in the accessibility of particular regions of the major rRNAs have been detected.

The tRNAs in the A and P sites are parallel to one another. The anticodon loops are bound to mRNA in a groove on the 30S subunit. The rest of each tRNA is bound to the 50S subunit. A conformational shift of tRNA within the A site is required to bring its aminoacyl end into juxtaposition with the end of the peptidyl-tRNA in the P site. The peptidyl transferase site that links the P and A binding sites is made of 23S rRNA, which has the peptidyl transferase catalytic activity, although proteins are probably needed to acquire the right structure.

An active role for the rRNAs in protein synthesis is indicated by mutations that affect ribosomal function, interactions with mRNA or tRNA that can be detected by chemical crosslinking, and the requirement to maintain individual base pairing interactions with the tRNA or mRNA. The 3′ terminal region of the rRNA base pairs with mRNA at initiation. Internal regions make individual contacts with the tRNAs in both the P and A sites. Ribosomal RNA is the target for some antibiotics or other agents that inhibit protein synthesis.

Using the Genetic Code

9.1 Introduction

The sequence of a coding strand of DNA, read in the direction from 5′ to 3′, consists of nucleotide triplets (codons) corresponding to the amino acid sequence of a protein read from N-terminus to C-terminus. Sequencing of DNA and proteins makes it possible to compare corresponding nucleotide and amino acid sequences directly. There are 64 codons (each of four possible nucleotides can occupy each of the three positions of the codon, making $4^3 = 64$ possible trinucleotide sequences).

The breaking of the genetic code originally showed that genetic information is stored in the form of nucleotide triplets, but did not reveal how each codon specifies its corresponding amino acid. Before the advent of sequencing, codon assignments were deduced on the basis of two types of *in vitro* studies. A system using the translation of synthetic polynucleotides was introduced in 1961, when Nirenberg showed that polyuridylic acid [poly(U)] directs the assembly of phenylalanine into polyphenylalanine. This result means that UUU must be a codon for phenylalanine. A second system was later introduced in which a trinucleotide was used to mimic a codon, thus causing the corresponding aminoacyl-tRNA to bind to a ribosome. By identifying the amino acid component of the aminoacyl-tRNA, the meaning of the codon can be found. The two techniques together assigned meaning to all of the codons that represent amino acids.

Sixty-one of the 64 codons represent amino acids. The other three cause termination of protein synthesis. The assignment of amino acids to codons is not random, but shows relationships in which the third base has less effect on codon meaning; also, related amino acids are often represented by related codons. The meaning of a codon that represents an amino acid is determined by the tRNA that corresponds to it; the meaning of the termination codons is determined directly by protein factors.

9.2 Related Codons Represent Related Amino Acids

The code is summarized in **Figure 9.1**. Because the genetic code is actually read on the mRNA, usually it is described in terms of the four bases present in RNA: U, C, A, and G. Because there are more codons (61) than there are amino acids (20), almost all amino acids are represented by more than one codon. The only exceptions are methionine and tryptophan. Codons that have the same meaning are called **synonyms**.

Codons representing the same or related amino acids tend to be similar in sequence. Often the base in the third position of a codon is not significant, because the four codons differing only in the third base represent the same amino acid. Sometimes a distinction is made only between a purine versus a pyrimidine in this position. The reduced specificity at the last position is known as **third-base degeneracy**.

The tendency for similar amino acids to be represented by related codons minimizes the effects of mutations. It increases the probability that a single random base change will result in no amino acid substitution or in one involving amino acids of similar character. For example, a mutation of CUC to CUG has no effect, since both codons represent leucine; and a mutation of CUU to AUU results in replacement of leucine with isoleucine, a closely related amino acid.

Figure 9.2 plots the number of codons representing each amino acid against the frequency with which the amino acid is used in proteins (in *E. coli*). There is only a slight tendency for amino acids that are more common to be represented by more codons, and therefore it does not seem that the genetic code has been optimized with regard to the utilization of amino acids.

The genetic code is triplet			
U	**C**	**A**	**G**
UUU } Phe UUC UUA } Leu UUG	UCU UCC } Ser UCA UCG	UAU } Tyr UAC UAA } STOP UAG	UGU } Cys UGC UGA STOP UGG Trp
CUU CUC CUA } Leu CUG	CCU CCC } Pro CCA CCG	CAU } His CAC CAA } Gln CAG	CGU CGC } Arg CGA CGG
AUU AUC } Ile AUA AUG Met	ACU ACC } Thr ACA ACG	AAU } Asn AAC AAA } Lys AAG	AGU } Ser AGC AGA } Arg AGG
GUU GUC } Val GUA GUG	GCU GCC } Ala GCA GCG	GAU } Asp GAC GAA } Glu GAG	GGU GGC } Gly GGA GGG

First base (rows: U, C, A, G) — Second base

Figure 9.1 All the triplet codons have meaning: 61 represent amino acids, and three cause termination (STOP).

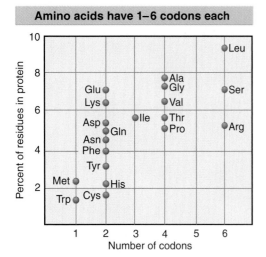

Figure 9.2 The number of codons for each amino acid does not correlate closely with its frequency of use in proteins.

Third bases have the least meaning

UUU	UCU	UAU	UGU
UUC	UCC	UAC	UGC
UUA	UCA	UAA	UGA
UUG	UCG	UAG	UGG
CUU	CCU	CAU	CGU
CUC	CCC	CAC	CGC
CUA	CCA	CAA	CGA
CUG	CCG	CAG	CGG
AUU	ACU	AAU	AGU
AUC	ACC	AAC	AGC
AUA	ACA	AAA	AGA
AUG	ACG	AAG	AGG
GUU	GCU	GAU	GGU
GUC	GCC	GAC	GGC
GUA	GCA	GAA	GGA
GUG	GCG	GAG	GGG

Third base relationship		Third base with same meaning	Codon number
Third base irrelevant		U, C, A, G	32
		U, C, A	3
Purines differ from pyrimidines		A or G	10
		U or C	14
unique		G only	2

Figure 9.3 Third bases have the least influence on codon meanings. Boxes indicate groups of codons within which third-base degeneracy ensures that the meaning is the same.

The three codons (UAA, UAG, and UGA) that do not represent amino acids are used specifically to terminate protein synthesis. One of these **stop codons** marks the end of every gene.

With a few rare exceptions, the genetic code is the same in all living organisms. This universality argues that the code must have been established very early in evolution. Perhaps the code started in a primitive form in which a small number of codons were used to represent comparatively few amino acids, possibly even with one codon corresponding to any member of a group of amino acids. More precise codon meanings and additional amino acids could have been introduced later. One possibility is that at first only two of the three bases in each codon were used; discrimination at the third position could have evolved later.

Evolution of the code could have become "frozen" at a point at which the system had become so complex that any changes in codon meaning would disrupt existing proteins by substituting unacceptable amino acids. Its universality implies that this must have happened at such an early stage that all living organisms are descended from a single pool of primitive cells in which this occurred.

9.3 Codon–Anticodon Recognition Involves Wobbling

Key Term

• The **wobble hypothesis** accounts for the ability of a tRNA to recognize more than one codon by unusual (non-G·C, non-A·T) pairing with the third base of a codon.

Key Concepts

• Multiple codons that represent the same amino acid most often differ at the third base position.

• The wobble in pairing between the first base of the anticodon and the third base of the codon results from the structure of the anticodon loop.

The function of tRNA in protein synthesis is fulfilled when it recognizes the codon in the ribosomal A site. The interaction between anticodon and codon takes place by base pairing, but under rules that extend pairing beyond the usual G·C and A·U partnerships. We can deduce the rules governing the interaction from the sequences of the anticodons that correspond to particular codons.

The genetic code itself yields some important clues about the process of codon recognition. The pattern of third-base degeneracy is drawn in **Figure 9.3**, which shows that in almost all cases either the third base is irrelevant or it makes a distinction only between purines and pyrimidines.

There are eight codon families in which all four codons sharing the same first two bases have the same meaning and one family in which three codons do; in these 35 codons, the third base has no role at all in specifying the amino acid. There are seven codon pairs in which the meaning is the same whichever pyrimidine is present at the third position; and there are five codon pairs in which either purine may be present without changing the amino acid that is coded.

There are only three cases in which a unique meaning is conferred by the presence of a particular base at the third position: AUG (for methionine), UGG (for tryptophan), and UGA (termination). So C and U never have a unique meaning in the third position, and A never signifies a unique amino acid.

Because the anticodon is complementary to the codon, it is the first base in the anticodon sequence written conventionally in the direction from 5' to 3' that pairs with the third base in the codon sequence written by the same convention. So the combination

Codon	5' A C G 3'
Anticodon	3' U G C 5'

is usually written as codon ACG/anticodon CGU, where the anticodon sequence must be read backward for complementarity with the codon.

To avoid confusion, we shall retain the usual convention in which all sequences are written 5'–3', but indicate anticodon sequences with a backward arrow as a reminder of the relationship with the codon. So the codon/anticodon pair shown above will be written as ACG and CGU←.

Does each triplet codon demand its own tRNA with a complementary anticodon? Or can a single tRNA respond to both members of a codon pair and to all (or at least some) of the three or four members of a codon family?

Often one tRNA can recognize more than one codon. This means that the base in the first position of the anticodon must be able to partner alternative bases in the corresponding third position of the codon. Base pairing at this position cannot be limited to the usual G · C and A · U partnerships.

The rules governing the recognition patterns are summarized in the **wobble hypothesis**, which states that the pairing between codon and anticodon at the first two codon positions always follows the usual rules, but that exceptional "wobbles" occur at the third position. Wobbling occurs because the conformation of the tRNA anticodon loop permits flexibility at the first base of the anticodon. **Figure 9.4** shows that G · U pairs can form in addition to the usual pairs.

This single addition creates a pattern of base pairing in which A can no longer have a unique meaning in the codon (because the U that recognizes it must also recognize G). Similarly, C also no longer has a unique meaning (because the G that recognizes it also must recognize U). **Figure 9.5** summarizes the pattern of recognition.

9.4 tRNA Contains Modified Bases

Key Term

- **Modification** of DNA or RNA includes all changes made to the nucleotides after their initial incorporation into the polynucleotide chain.

Key Concepts

- tRNAs contain >50 modified bases.
- Modification usually occurs by direct alteration of the primary bases in tRNA, but there are some exceptions in which a base is removed and replaced by another base.

Transfer RNA is unique among nucleic acids in its number of bases that are produced by **modification** of a base after it has been incorporated into the polyribonucleotide chain.

All classes of RNA display some degree of modification, but in all cases except tRNA this is confined to rather simple events, such as the addition of methyl groups. In tRNA, there is a vast range of modifications,

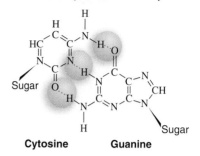

G-U pairs form at the third codon base

Standard base pairs occur at all positions

Cytosine   Guanine

Uracil   Adenine

G • U wobble pairing occurs only at third codon position

Uracil   Guanine

Figure 9.4 Wobble in base pairing allows G-U pairs to form between the third base of the codon and the first base of the anticodon.

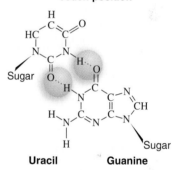

The third codon base wobbles	
Base in first position of anticodon	Base(s) recognized in third position of codon
U	A or G
C	G only
A	U only
G	C or U

Figure 9.5 Codon–anticodon pairing involves wobbling at the third position.

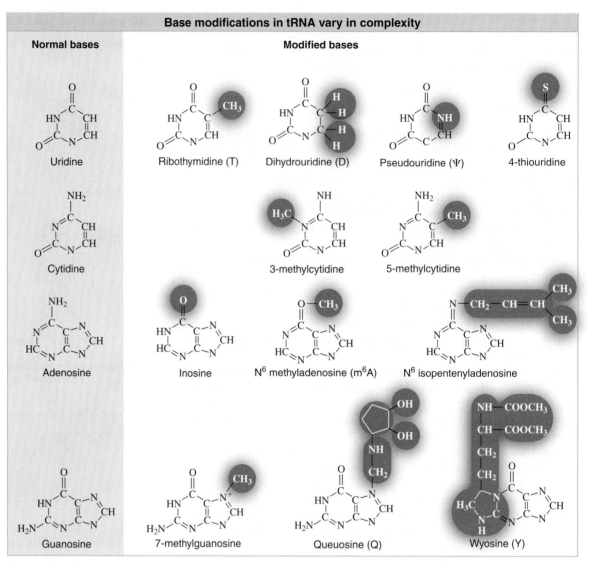

Base modifications in tRNA vary in complexity

Normal bases

Uridine

Cytidine

Adenosine

Guanosine

Modified bases

Ribothymidine (T)

Dihydrouridine (D)

Pseudouridine (Ψ)

4-thiouridine

3-methylcytidine

5-methylcytidine

Inosine

N⁶ methyladenosine (m⁶A)

N⁶ isopentenyladenosine

7-methylguanosine

Queuosine (Q)

Wyosine (Y)

Figure 9.6 All of the four bases in tRNA can be modified.

ranging from simple methylation to wholesale restructuring of the purine ring. All parts of the tRNA molecule may be modified. There are >50 different types of modified bases in tRNA.

Figure 9.6 shows some of the more common modified bases. Modifications of pyrimidines (C and U) are less complex than those of purines (A and G). In addition to the modifications of the bases themselves, there may also be methylation at the 2′-O position of the ribose ring.

The most common modifications of uridine are straightforward. Methylation at position 5 creates ribothymidine (T). The base is the same commonly found in DNA; but here it is attached to ribose, not deoxyribose. In RNA, thymine constitutes an unusual base, originating by modification of U.

Dihydrouridine (D) is generated by the saturation of a double bond, changing the ring structure. Pseudouridine (ψ) interchanges the positions of N and C atoms (see Figure 26.31). And 4-thiouridine has sulfur substituted for oxygen.

The nucleoside inosine is found normally in the cell as an intermediate in the purine biosynthetic pathway. However, it is not incorporated directly into RNA, where instead its existence depends on modification of A to create I. Other modifications of A include the addition of complex groups.

Some modifications are constant features of all tRNA molecules—for example, the D residues that give rise to the name of the D arm, and the ψ found in the TψC sequence. On the 3′ side of the anticodon there is always a modified purine, although the modification varies widely.

Other modifications are specific for particular tRNAs or groups of tRNAs. For example, wyosine bases have complicated modifications of guanosine and are characteristic of tRNA^Phe in bacteria, yeast, and mammals. There are also some species-specific patterns.

The modified nucleosides are synthesized by specific tRNA-modifying enzymes. There are many such enzymes (~60 in yeast), and they vary greatly in specificity. In some cases, a single enzyme acts to make a particular modification at a single position. In other cases, an enzyme can modify bases at several different target positions. Some enzymes undertake single reactions with individual tRNAs; others have a range of substrate molecules. The features recognized by the tRNA-modifying enzymes are unknown, but probably involve recognition of structural features surrounding the site of modification. Some modifications require the successive actions of more than one enzyme.

9.5 Modified Bases Affect Anticodon–Codon Pairing

Key Concept

- Modifications in the anticodon affect the pattern of wobble pairing and therefore are important in determining tRNA specificity.

The most direct effect of modification is seen in the anticodon, where change of sequence influences the ability to pair with the codon, thus determining the meaning of the tRNA. Modifications elsewhere in the vicinity of the anticodon also influence its pairing.

When bases in the anticodon are modified, further pairing patterns become possible in addition to those of the regular and wobble pairing of A, C, U, and G. **Figure 9.7** shows the patterns for inosine (I), which is often present at the first position of the anticodon. Inosine can pair with any one of three bases, U, C, and A.

This ability is especially important in the isoleucine codons, where AUA codes for isoleucine, while AUG codes for methionine. Because with the usual bases it is not possible to recognize A alone in the third position, any tRNA with U starting its anticodon would have to recognize AUG as well as AUA. So AUA must be read together with AUU and AUC, a problem that is solved by the existence of tRNA with I in the anticodon.

Some modifications create preferential readings of some codons with respect to others. Anticodons with uridine-5-oxyacetic acid or 5-methoxyuridine in the first position recognize A and G efficiently as third bases of the codon, but recognize U less efficiently. Another case in which multiple pairings can occur, but with some preferred to others, is provided by the series of queuosine and its derivatives. These modified G bases recognize both C and U, but pair with U more readily.

A restriction not allowed by the usual rules can be achieved by the employment of 2-thiouridine in the anticodon. **Figure 9.8** shows that its modification allows the base to continue to pair with A, but prevents it from indulging in wobble pairing with G.

These and other pairing relationships make the general point that there are multiple ways to construct a set of tRNAs able to recognize all the 61 codons representing amino acids. No particular pattern predominates in any given organism, although the absence of a certain pathway for modification can prevent the use of some recognition patterns. So a

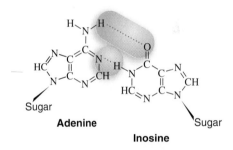

Inosine pairs with three bases

Cytosine · Inosine

Uracil · Inosine

Adenine · Inosine

Figure 9.7 Inosine can pair with any of U, C, and A.

2-thiouracil has restricted pairing

One bond is not enough

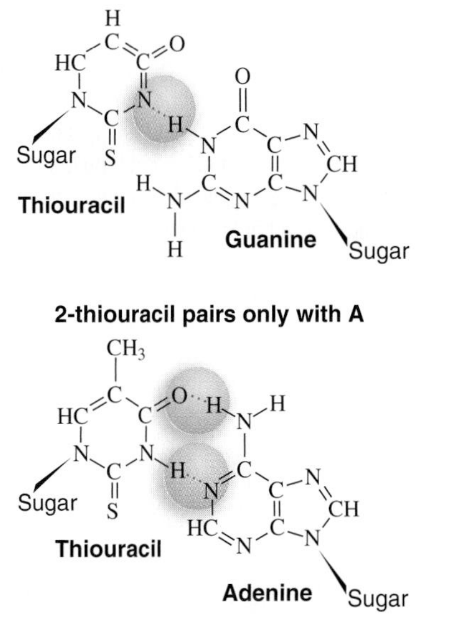

Thiouracil · Guanine

2-thiouracil pairs only with A

Thiouracil · Adenine

Figure 9.8 Modification to 2-thiouridine restricts pairing to A alone because only one H bond can form with G.

particular codon family is read by tRNAs with different anticodons in different organisms.

Often the tRNAs have overlapping responses, so that a particular codon is read by more than one tRNA. In such cases there may be differences in the efficiencies of the alternative recognition reactions. (As a general rule, codons that are commonly used tend to be more efficiently read.) And in addition to the construction of a set of tRNAs able to recognize all the codons, there may be multiple tRNAs that respond to the same codons.

The predictions of wobble pairing accord very well with the observed abilities of almost all tRNAs. But there are exceptions in which the codons recognized by a tRNA differ from those predicted by the wobble rules. Such effects probably result from the influence of neighboring bases and/or the conformation of the anticodon loop in the overall tertiary structure of the tRNA. Indeed, the importance of the structure of the anticodon loop is inherent in the idea of the wobble hypothesis itself. Further support for the influence of the surrounding structure is provided by the isolation of occasional mutants in which a change in a base in some other region of the molecule alters the ability of the anticodon to recognize codons.

9.6 There Are Sporadic Alterations of the Universal Code

Key Concepts

- Changes in the universal genetic code have occurred in some species.
- Changes in the genetic code are more common in mitochondrial genomes, where a phylogenetic tree can be constructed for the changes.
- In nuclear genomes, changes in the genetic code are sporadic and usually affect only termination codons.

The universality of the genetic code is striking, but some exceptions exist. They tend to affect the codons involved in initiation or termination and result from the production (or absence) of tRNAs representing certain codons. The changes found in bacterial or nuclear genomes are summarized in **Figure 9.9**.

Almost all of the changes that allow a codon to represent an amino acid affect termination codons. The most common change is that UGA is

Figure 9.9 Changes in the genetic code in eukaryotic nuclear or bacterial genomes usually assign amino acids to stop codons or change a codon so that it no longer specifies an amino acid (a NONE signal). A change in meaning from one amino acid to another is unusual.

Changes in the genetic code usually involve STOP/NONE signals

UUU UUC **Phe**	UCU UCC **Ser**	UAU UAC **Tyr**	UGU UGC **Cys**
UUA UUG **Leu**	UCA UCG	UAA UAG **STOP → Gln**	UGA **STOP → Trp, Cys, Sel** UGG
CUU CUC **Leu**	CCU CCC **Pro**	CAU CAC **His**	CGU CGC **Arg**
CUA CUG **Leu → Ser**	CCA CCG	CAA CAG **Gln**	CGA CGG **Arg → NONE**
AUU AUC **Ile**	ACU ACC **Thr**	AAU AAC **Asn**	AGU AGC **Ser**
AUA **Ile → NONE** AUG **Met**	ACA ACG	AAA AAG **Lys**	AGA **Arg → NONE** AGG **Arg**
GUU GUC **Val**	GCU GCC **Ala**	GAU GAC **Asp**	GGU GGC **Gly**
GUA GUG	GCA GCG	GAA GAG **Glu**	GGA GGG

not used for termination but instead codes for tryptophan. Some ciliates (unicellular protozoa) read UAA and UAG as glutamine instead of termination signals. The only substitution in coding for amino acids occurs in a yeast (*Candida*), where CUG means serine instead of leucine (and UAG is used as a sense codon).

Acquisition of a coding function by a termination codon requires two types of change: a tRNA must be mutated so as to recognize the codon; and the class 1 release factor must be mutated so that it does not terminate at this codon.

The other common type of change is loss of the tRNA that responds to a codon, so that the codon no longer specifies any amino acid. What happens at such a codon will depend on whether the termination factor evolves to recognize it.

All of these changes are sporadic, which is to say that they appear to have occurred independently in specific lines of evolution. They may be concentrated on termination codons, because these changes do not involve substitution of one amino acid for another. Once the genetic code was established, early in evolution, any general change in the meaning of a codon would cause a substitution in all the proteins that contain that amino acid. It seems likely that the change would be deleterious in at least some of these proteins, with the result that it would be strongly selected against. The divergent uses of the termination codons could represent their "capture" for normal coding purposes. If some termination codons were used only rarely, they could be recruited to coding purposes by changes that allowed tRNAs to recognize them.

Exceptions to the universal genetic code also occur in the mitochondria from several species. **Figure 9.10** constructs a phylogeny for the changes. It suggests that there was a universal code that was changed at various points in mitochondrial evolution. The earliest change was the employment of UGA to code for tryptophan, which is common to all (nonplant) mitochondria.

Why have changes been able to evolve in the mitochondrial code? Because the mitochondrion synthesizes only a small number of proteins (~10), the problem of disruption by changes in meaning is much less severe. Probably the codons that are altered were not used extensively in locations where amino acid substitutions would have been deleterious. The variety of changes found in mitochondria of different species suggests that they have evolved separately, and not by common descent from an ancestral mitochondrial code.

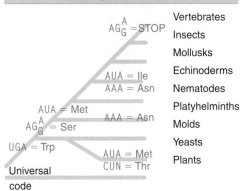

Figure 9.10 Changes in the genetic code in mitochondria can be traced in phylogeny. The minimum number of independent changes is generated by supposing that the AUA → Met and the AAA → Asn changes each occurred independently twice, and that the early AUA → Met change was reversed in echinoderms.

9.7 Novel Amino Acids Can Be Inserted at Certain Stop Codons

Key Concepts

- Changes in the reading of specific codons can occur in individual genes.
- The insertion of seleno-Cys-tRNA at certain UGA codons requires several proteins to modify the Cys-tRNA and insert it into the ribosome.
- Pyrrolysine can be inserted at certain UAG codons.

Specific changes in reading the code occur in individual genes. The specificity of such changes implies that the reading of the particular codon must be influenced by the surrounding bases.

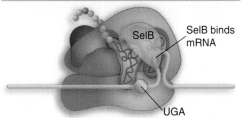

Figure 9.11 SelB is an elongation factor that specifically binds seleno-Cys-tRNA to a UGA codon that is followed by a stem-loop structure in mRNA.

A striking example is the incorporation of the modified amino acid seleno-cysteine at certain UGA codons within the genes that code for selenoproteins in both prokaryotes and eukaryotes. Usually these proteins catalyze oxidation-reduction reactions, and contain a single seleno-cysteine residue, which forms part of the active site. The most is known about the use of the UGA codons in three *E. coli* genes coding for formate dehydrogenase isozymes. The internal UGA codon is read by a seleno-Cys-tRNA.

The system in *E. coli* was identified by mutation in four genes that create a deficiency in selenoprotein synthesis. *selC* codes for tRNA (with the anticodon UCA$^{\leftarrow}$) that is charged with serine. *selA* and *selD* are required to modify the serine to seleno-cysteine. SelB is an alternative elongation factor. It is a guanine nucleotide-binding protein that acts as a specific translation factor for entry of seleno-Cys-tRNA into the A site of the ribosome; it thus provides (for this single tRNA) a replacement for factor EF-Tu. The sequence of SelB is related to both EF-Tu and IF-2.

Why is seleno-Cys-tRNA inserted only at certain UGA codons? These codons are followed by a stem-loop structure in the mRNA downstream of the UGA codon. **Figure 9.11** shows that the stem of this structure is recognized by an additional domain in SelB (one that is not present in EF-Tu or IF-2). A similar mechanism interprets some UGA codons in mammalian cells, except that two proteins are required to identify the appropriate UGA codons. One protein (SBP2) binds a stem-loop structure far downstream from the UGA codon, while the counterpart of SelB (called SECIS) binds to SBP2 and simultaneously binds the tRNA to the UGA codon.

Another example of the insertion of a special amino acid is the placement of pyrrolysine at a UAG codon. This happens in both an archaea and a bacterium. The mechanism is probably similar to the insertion of seleno-cysteine. An unusual tRNA is charged with lysine, which is presumably then modified. The tRNA has a CUA anticodon, which responds to UAG. There must be other components of the system that restrict its response to the appropriate UAG codons.

9.8 tRNAs Are Charged with Amino Acids by Synthetases

Key Term

- **Isoaccepting tRNAs (cognate tRNAs)** are the RNAs recognized by a particular aminoacyl-tRNA synthetase. They all are charged with the same amino acid.

Key Concepts

- Aminoacyl-tRNA synthetases are enzymes that charge tRNA with an amino acid to generate aminoacyl-tRNA in a two-stage reaction that uses energy from ATP.
- There are 20 aminoacyl-tRNA synthetases in each cell. Each charges all the tRNAs that represent a particular amino acid.
- Recognition of a tRNA is based on a small number of points of contact in the tRNA sequence.

It is necessary for tRNAs to have certain characteristics in common, yet be distinguished by others. The crucial feature that confers this capacity is the ability of tRNA to fold into a specific tertiary structure. Changes in the details of this structure, such as the angle of the two arms of the "L" or the protrusion of individual bases, may distinguish the individual tRNAs.

All tRNAs can fit in the P and A sites of the ribosome, where at one end they are associated with mRNA via codon–anticodon pairing, while at the other end the polypeptide is being transferred. Similarly, all tRNAs (except the initiator) share the ability to be recognized by the translation factors (EF-Tu or eEF1) for binding to the ribosome. The initiator tRNA is recognized instead by IF-2 or eIF2. So the tRNA set must possess common features for interaction with elongation factors, but the initiator tRNA can be distinguished.

Amino acids enter the protein synthesis pathway through the aminoacyl-tRNA synthetases, which provide the interface for connection with nucleic acid. All synthetases function by the two-step mechanism depicted in **Figure 9.12**:

- First, the amino acid reacts with ATP to form aminoacyl~adenylate, releasing pyrophosphate. Energy for the reaction is provided by cleaving the high-energy bond of the ATP.
- Then the activated amino acid is transferred to the tRNA, releasing AMP.

The synthetases sort the tRNAs and amino acids into corresponding sets. Each synthetase recognizes a single amino acid and all the tRNAs that can carry it. Usually, each amino acid is represented by more than one tRNA. Several tRNAs may be needed to respond to synonym codons, and sometimes there are multiple species of tRNA reacting with the same codon. Multiple tRNAs representing the same amino acid are called **isoaccepting tRNAs**; because they are all recognized by the same synthetase, they are also described as its **cognate tRNAs**.

A group of isoaccepting tRNAs must be charged only by the single aminoacyl-tRNA synthetase specific for their amino acid. So isoaccepting tRNAs must share some common feature(s) enabling the enzyme to distinguish them from the other tRNAs. The entire complement of tRNAs is divided into 20 isoaccepting groups; each group is able to identify itself to its particular synthetase.

Many attempts to deduce similarities in sequence between cognate tRNAs, or to induce chemical alterations that affect their charging, have shown that the basis for recognition is different for different tRNAs, and does not necessarily lie in some feature of primary or secondary structure alone. tRNAs are identified by their synthetases by contacts that recognize a small number of bases, typically from 1–5. Three types of features commonly are used:

- Usually (but not always), at least one base of the anticodon is recognized. Sometimes all the positions of the anticodon are important.
- Often one of the last three base pairs in the acceptor stem is recognized. An extreme case is represented by alanine tRNA, which is identified by a single unique base pair in the acceptor stem.
- The so-called discriminator base, which lies between the acceptor stem and the CCA terminus, is always invariant among isoacceptor tRNAs.

No one of these features constitutes a unique means of distinguishing 20 sets of tRNAs, or provides sufficient specificity, so it appears that recognition of tRNAs is idiosyncratic, each following its own rules. Recognition depends on an interaction between a few points of contact in the tRNA, concentrated at the extremities, and a few amino acids constituting the active site in the protein. The relative importance of the roles played by the acceptor stem and anticodon is different for each tRNA · synthetase interaction.

The charging reaction uses ATP

Synthetase enzyme has three binding sites

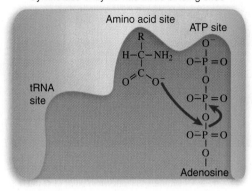

Amino acid and ATP form aminoacyl-AMP

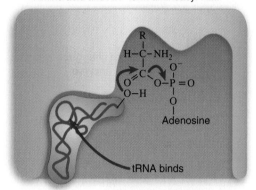

tRNA is charged with amino acid

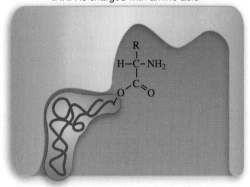

Figure 9.12 An aminoacyl-tRNA synthetase charges tRNA with an amino acid.

9.9 Aminoacyl-tRNA Synthetases Fall into Two Groups

Key Concept

- Aminoacyl-tRNA synthetases are divided into the class I and class II groups by sequence and structural similarities.

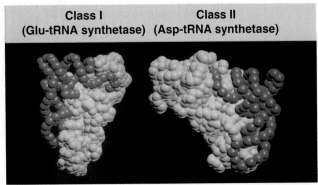

Class I (Glu-tRNA synthetase)   **Class II** (Asp-tRNA synthetase)

Figure 9.13 Crystal structures show that class I and class II aminoacyl-tRNA synthetases (shown in blue) bind the opposite faces of their tRNA substrates (red). Photographs kindly provided by Dr. Dino Moras.

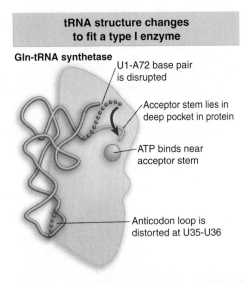

tRNA structure changes to fit a type I enzyme

Gln-tRNA synthetase

U1-A72 base pair is disrupted

Acceptor stem lies in deep pocket in protein

ATP binds near acceptor stem

Anticodon loop is distorted at U35-U36

Figure 9.14 A class I tRNA synthetase contacts tRNA at the minor groove of the acceptor stem and at the anticodon.

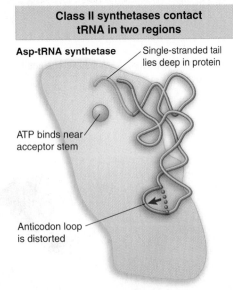

Class II synthetases contact tRNA in two regions

Asp-tRNA synthetase

Single-stranded tail lies deep in protein

ATP binds near acceptor stem

Anticodon loop is distorted

Figure 9.15 A class II aminoacyl-tRNA synthetase contacts tRNA at the major groove of the acceptor helix and at the anticodon loop.

In spite of their common function, synthetases are a rather diverse group of proteins. The individual subunits vary from 40–110 kD, and the enzymes may be monomeric, dimeric, or tetrameric. Synthetases have been divided into two general groups, each containing 10 enzymes, on the basis of the structure of the domain that contains the active site. The two groups show no relationship. Perhaps they evolved independently of one another. This makes it seem possible that an early form of life could have existed with proteins that were made up of just the 10 amino acids coded by one type or the other.

A general model for synthetase·tRNA binding suggests that the protein binds the tRNA along the "side" of the L-shaped molecule. The same general principle applies for all synthetase·tRNA binding: the tRNA is bound principally at its two extremities, and most of the tRNA sequence is not involved in recognition by a synthetase. However, the detailed nature of the interaction is different between class I and class II enzymes, as can be seen from the models of **Figure 9.13**, which are based on crystal structures. The two types of enzyme approach the tRNA from opposite sides, with the result that the tRNA-protein models look almost like mirror images of one another.

A class I enzyme (Gln-tRNA synthetase) approaches the D-loop side of the tRNA. It recognizes the minor groove of the acceptor stem at one end of the binding site, and interacts with the anticodon loop at the other end. **Figure 9.14** is a diagrammatic representation of the crystal structure of the tRNA^{Gln}·synthetase complex. A revealing feature of the structure is that contacts with the enzyme change the structure of the tRNA at two important points. These can be seen by comparing the dotted and solid lines in the anticodon loop and acceptor stem:

- Bases U35 and U36 in the anticodon loop are pulled farther out of the tRNA into the protein.
- The end of the acceptor stem is seriously distorted, with the result that base pairing between U1 and A72 is disrupted. The single-stranded end of the stem pokes into a deep pocket in the synthetase protein, which also contains the binding site for ATP.

This structure explains why changes in U35, G73, or the U1-A72 base pair affect the recognition of the tRNA by its synthetase. At all these positions, hydrogen bonding occurs between the protein and tRNA.

A class II enzyme (Asp-tRNA synthetase) approaches the tRNA from the other side, and recognizes the variable loop, and the major groove of the acceptor stem, as drawn in **Figure 9.15**. The acceptor stem remains in its regular helical conformation. ATP is probably bound near to the terminal adenine. At the other end of the binding site, there is a tight contact with the anticodon loop, which has a change in conformation that allows the anticodon to be in close contact with the protein.

9.10 Synthetases Use Proofreading to Improve Accuracy

Key Terms

- **Kinetic proofreading** describes a proofreading mechanism that depends on incorrect events proceeding more slowly than correct events, so that incorrect events are reversed before a subunit is added to a polymeric chain.

- **Chemical proofreading** describes a proofreading mechanism in which the correction event occurs after addition of an incorrect subunit to a polymeric chain, by reversal of the addition reaction.

Key Concept

- Specificity of recognition of both amino acid and tRNA is controlled by aminoacyl-tRNA synthetases by proofreading reactions that abort or reverse the catalytic reaction if the wrong component has been incorporated.

Aminoacyl-tRNA synthetases have a difficult job. Each synthetase must distinguish 1 out of 20 amino acids, and must differentiate its cognate tRNAs (typically 1–3) from the total set (perhaps 100 in all).

Many amino acids are closely related to one another, and all amino acids are related to the metabolic intermediates in their particular synthetic pathway. It is especially difficult to distinguish between two amino acids that differ only in the length of the carbon backbone (that is, by one CH_2 group). Intrinsic discrimination based on relative energies of binding two such amino acids would be only ~1/5. The synthetase enzymes improve this ratio ~1000 fold.

Intrinsic discrimination between tRNAs is better, because the tRNA offers a larger surface with which to make more contacts, but it is still true that all tRNAs conform to the same general structure, and there may be a quite limited set of features that distinguish the cognate tRNAs from the noncognate tRNAs.

Synthetases use proofreading mechanisms to control the recognition of both types of substrates. This means that they check the results of the reaction at some or all stages and reverse it if the wrong tRNA or amino acid has been used. They improve significantly on the intrinsic differences among amino acids or among tRNAs, but consistent with the intrinsic differences in each group, make more mistakes in selecting amino acids (error rates are 10^{-4}–10^{-5}) than in selecting tRNAs (error rates are ~10^{-6}) (see Figure 8.8).

Synthetases use **kinetic proofreading** to improve their discrimination of tRNAs. This relies on the greater intrinsic affinity of a cognate tRNA for its synthetase. A correctly bound tRNA makes contacts with the enzyme surface that allow the enzyme to aminoacylate it rapidly. An incorrectly bound tRNA makes fewer contacts, so the aminoacylation is not triggered so quickly, and this allows more time for the tRNA to dissociate from the enzyme before it is trapped by the reaction.

Specificity for amino acids varies among the synthetases. Some are highly specific for initially binding a single amino acid, but others can also activate amino acids closely related to the proper substrate. Although the analog amino acid can sometimes be converted to the adenylate form, in none of these cases is an incorrectly activated amino acid actually used to form a stable aminoacyl-tRNA.

There are two stages during formation of aminoacyl-tRNA at which proofreading of an incorrect aminoacyl-adenylate may occur. **Figure 9.16** shows that both use **chemical proofreading**, in which the catalytic reaction is reversed. The extent to which one pathway or the other predominates varies with the individual synthetase. The presence of the cognate tRNA usually is needed to trigger proofreading:

- The noncognate aminoacyl-adenylate may be hydrolyzed when the cognate tRNA binds. This mechanism is used predominantly by several synthetases, including those for methionine, isoleucine, and valine.

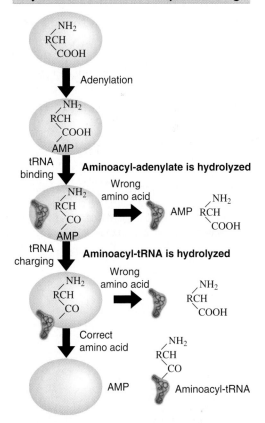

Synthetases use chemical proofreading

Figure 9.16 When a synthetase binds the incorrect amino acid, proofreading requires binding of the cognate tRNA. It may take place either by a conformation change that causes hydrolysis of the incorrect aminoacyl-adenylate, or by transfer of the amino acid to tRNA, followed by hydrolysis.

Errors are controlled at each stage	
Step	Frequency of Error
Activation of valine to Val-AMP^{lle}	1/225
Release of Val-tRNA	1/270
Overall rate of error	1/225 × 1/270 = 1/60,000

Figure 9.17 The accuracy of charging tRNA^{lle} by its synthetase depends on error control at two stages.

The double sieve model uses two active sites

The editing site is smaller than the synthetic site

Synthetic site   Editing site

Leu is too large to fit in the synthetic site

Ile fits in the synthetic site but not the editing site

Val passes from the synthetic site to editing site

Figure 9.18 Ile-tRNA synthetase has two active sites. Amino acids larger than Ile cannot be activated because they do not fit in the synthetic site. Amino acids smaller than Ile are removed because they are able to enter the editing site.

- Some synthetases use chemical proofreading at a later stage. The wrong amino acid is actually transferred to tRNA, is then recognized as incorrect by its structure in the tRNA-binding site, and so is hydrolyzed and released. The process requires a continual cycle of linkage and hydrolysis until the correct amino acid is transferred to the tRNA.

A classic example in which discrimination between amino acids depends on the presence of tRNA is provided by the Ile-tRNA synthetase of *E. coli*. The enzyme can charge valine with AMP, but hydrolyzes the valyl-adenylate when tRNA^{lle} is added. The overall error rate depends on the specificities of the individual steps, as summarized in **Figure 9.17**. The overall error rate of 1.5×10^{-5} is less than the measured rate at which valine is substituted for isoleucine (in rabbit globin), which is $2-5 \times 10^{-4}$. So mischarging probably provides only a small fraction of the errors that actually occur in protein synthesis.

Ile-tRNA synthetase uses size as a basis for discrimination among amino acids. **Figure 9.18** shows that it has two active sites: the synthetic (or activation) site and the editing (or hydrolytic) site. The crystal structure of the enzyme shows that the synthetic site is too small to allow leucine (a close analog of isoleucine) to enter. All amino acids larger than isoleucine are excluded from activation because they cannot enter the synthetic site. An amino acid that can enter the synthetic site is placed on tRNA. Then the enzyme tries to transfer it to the editing site. Isoleucine is safe from editing because it is too large to enter the editing site. However, valine can enter this site, and as a result an incorrect Val-tRNA^{lle} is hydrolyzed. Essentially the enzyme provides a double molecular sieve, in which size of the amino acid is used to discriminate between closely related species.

9.11 Suppressor tRNAs Have Mutated Anticodons that Read New Codons

Key Terms

- A **suppressor** is a second mutation that compensates for or alters the effects of a primary mutation.

- A **nonsense suppressor** is a gene coding for a mutant tRNA able to respond to one or more of the termination codons and insert an amino acid at that site.

- A **missense suppressor** codes for a tRNA that has been mutated to recognize a different codon. By inserting a different amino acid at a mutant codon, the tRNA suppresses the effect of the original mutation.

- **Readthrough** occurs during transcription or translation when RNA polymerase or the ribosome, respectively, ignores a termination signal because of a mutation of the template or the behavior of an accessory factor.

Key Concepts

- A suppressor tRNA typically has a mutation in the anticodon that changes the codons to which it responds.

- Each type of nonsense codon can be suppressed by a tRNA (or more than one) with a mutant anticodon that recognizes that codon.

- When a mutant anticodon corresponds to a termination codon, an amino acid is inserted and the polypeptide chain is extended beyond the termination codon. This results in nonsense suppression at a site of nonsense mutation or in readthrough at a natural termination codon.
- Suppressor tRNAs compete with wild-type tRNAs that have the same anticodon to read the corresponding codon(s) or with release factors that recognize termination codons.
- Efficient suppression is deleterious because it results in readthrough past normal termination codons.
- Missense suppression occurs when a tRNA recognizes a different codon from usual, so that one amino acid is substituted for another.

Isolation of mutant tRNAs has been one of the most potent tools for analyzing the ability of a tRNA to respond to its codon(s) in mRNA, and for determining the effects that different parts of the tRNA molecule have on codon–anticodon recognition.

Mutant tRNAs are isolated by virtue of their ability to overcome the effects of mutations in genes coding for proteins. In general genetic terminology, a mutation that is able to overcome the effects of another mutation is called a **suppressor**.

In tRNA suppressor systems, the primary mutation changes a codon in an mRNA so that the protein product is no longer functional. The secondary, suppressor mutation changes the anticodon of a tRNA, so that it recognizes the mutant codon instead of (or as well as) its original target codon. The amino acid that is now inserted restores protein function. The suppressors are described as **nonsense suppressors** or **missense suppressors**, depending on the nature of the original mutation.

In a cell that does not have a suppressor mutation, a nonsense mutation is recognized only by a release factor, terminating protein synthesis. The suppressor mutation creates an aminoacyl-tRNA that can recognize the termination codon; by inserting an amino acid, it allows protein synthesis to continue beyond the site of nonsense mutation. This new capacity of the translation system allows a full-length protein to be synthesized, as illustrated in **Figure 9.19**. If the amino acid inserted by suppression is different from the amino acid that was originally present at this site in the wild-type protein, the activity of the protein may be altered. Each type of nonsense suppressor is specific for one of the three nonsense codons; that is, its anticodon recognizes one of the codons UAA, UAG, or UGA.

A nonsense suppressor is isolated by its ability to respond to a mutant nonsense codon. But the same triplet sequence constitutes one of the normal termination signals of the cell! The mutant tRNA that suppresses the nonsense mutation can also suppress natural termination at the corresponding codon. **Figure 9.20** shows that this **readthrough** results in the synthesis of a longer protein, with additional C-terminal material. The extended protein ends at the next termination triplet sequence found in the reading frame. Any extensive suppression of termination is likely to be deleterious to the cell by producing extended proteins whose functions are thereby altered.

Amber codons are used relatively infrequently to terminate protein synthesis in *E. coli*, and as a result amber suppressors tend to be relatively efficient (10–50%). The ochre codon is used most frequently as a natural termination signal, and as a result ochre suppressors are difficult to isolate and are always much less efficient (<10%). UGA is the least efficient of the termination codons in its natural function; it is misread by Trp-tRNA as frequently as 1–3% in wild-type cells. In spite of this deficiency, however, it is used more commonly than the amber triplet to terminate bacterial genes.

Suppressor tRNA recognizes a nonsense codon

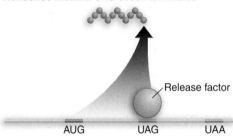

Wild type: UUG codon is read by Leu-tRNA

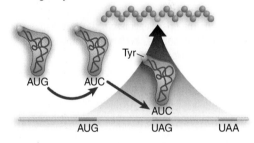

Nonsense mutant: UAG codon terminates

Suppressor mutation: changes Tyr-tRNA anticodon

Figure 9.19 Nonsense mutations can be suppressed by a tRNA with a mutant anticodon that inserts an amino acid at the mutant codon, producing a full-length protein in which the original Leu residue has been replaced by Tyr.

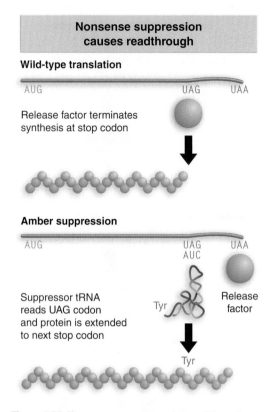

Figure 9.20 Nonsense suppressors also read through natural termination codons, synthesizing proteins that are longer than wild type.

Missense mutations change a codon representing one amino acid into a codon representing another amino acid, one that cannot function in the protein in place of the original residue. (Formally, any substitution of amino acids constitutes a missense mutation, but in practice a substitution is detected only if it changes the activity of the protein.) The mutation can be suppressed by the insertion either of the original amino acid or of some other amino acid that can function in the protein.

Figure 9.21 demonstrates that missense suppression can be accomplished in the same way as nonsense suppression, by mutating the anticodon of a tRNA carrying an acceptable amino acid so that it responds to the mutant codon. So missense suppression involves a change in the meaning of the codon from one amino acid to another.

As with nonsense suppressors, an event that suppresses a missense mutation at one site may replace the wild-type amino acid at another site with a new amino acid. The change may inhibit normal protein function. The absence of any strong missense suppressors is therefore explained by the damaging effects that would be caused by a general and efficient substitution of amino acids.

Suppression is most often considered in the context of a mutation that changes the reading of a codon. However, there are some situations in which a stop codon coded for in wild-type DNA is read, at a low frequency, as an amino acid. The first example to be discovered was the coat protein gene of the RNA phage Qβ. The formation of infective Qβ particles requires that the stop codon at the end of this gene is suppressed at a low frequency to generate a small proportion of coat proteins with a C-terminal extension. In effect, this stop codon is leaky. The reason is that Trp-tRNA recognizes the stop codon at a low frequency.

Readthrough past stop codons occurs also in eukaryotes, where it is employed most often by RNA viruses. This may involve the suppression of UAG/UAA by Tyr-tRNA, Gln-tRNA, or Leu-tRNA, or the suppression of UGA by Trp-tRNA or Arg-tRNA. The extent of partial suppression is dictated by the context surrounding the codon.

9.12 Recoding Changes Codon Meanings

Key Term

- In **recoding** events, the meaning of a codon or series of codons is changed from that predicted by the genetic code. Recoding may involve altered interactions between aminoacyl-tRNA and mRNA that are influenced by the ribosome.

Key Concepts

- Changes in codon meaning can be caused by mutant tRNAs or by tRNAs with special properties.
- The reading frame can be changed by frameshifting or bypassing, both of which depend on properties of the mRNA.

The reading frame of a messenger usually is invariant. Translation starts at an AUG codon and continues in triplets to a termination codon. Reading takes no notice of sense: insertion or deletion of a base causes a frameshift mutation, in which the reading frame is changed beyond the site of mutation. Ribosomes and tRNAs continue ineluctably in triplets, synthesizing an entirely different series of amino acids.

There are some exceptions to the usual pattern of translation that enable a reading frame with an interruption of some sort—such as a nonsense codon or frameshift—to be translated into a full-length protein. **Recoding** events are responsible for making exceptions to the usual rules.

In one type of recoding, changing the meaning of a single codon allows one amino acid to be substituted in place of another, or for an amino acid to be inserted at a termination codon. **Figure 9.22** shows that these changes rely on the properties of an individual tRNA that responds to the codon:

- Suppression involves recognition of a codon by a (mutant) tRNA that usually would respond to a different codon (see previous section).
- Redefinition of the meaning of a codon occurs when an aminoacyl-tRNA is modified (see previous section).

Changing the reading frame, another type of recoding, occurs in two types of situation:

- Frameshifting typically involves changing the reading frame when aminoacyl-tRNA slips by one base (+1 forward or −1 backward, as discussed in the next section). The result shown in **Figure 9.23** is that translation continues past a termination codon.
- Bypassing involves a movement of the ribosome to change the codon that is paired with the peptidyl-tRNA in the P site. The sequence between the two codons is not represented in protein. As shown in **Figure 9.24**, this allows translation to continue past any termination codons in the intervening region.

9.13 Frameshifting Occurs at Slippery Sequences

Key Term

- **Programmed frameshifting** is required for expression of the protein sequences coded beyond a specific site at which a +1 or −1 frameshift occurs at some typical frequency.

Key Concepts

- The reading frame may be influenced by the sequence of mRNA and the ribosomal environment.
- Slippery sequences allow a tRNA to shift by one base after it has paired with its anticodon, thereby changing the reading frame.
- Translation of some genes depends upon the regular occurrence of programmed frameshifting.

Frameshifting is associated with specific tRNAs in two circumstances: First, some mutant tRNA suppressors recognize a "codon" for four bases instead of the usual three bases. Second, certain "slippery" sequences allow a tRNA to move a base up or down mRNA in the A site.

The simplest type of external frameshift suppressor corrects the reading frame when a mutation has been caused by inserting an additional base within a stretch of identical residues. For example, a G may be inserted in a run of several contiguous G bases. The frameshift suppressor is a tRNA^{Gly} that has an extra base inserted in its anticodon loop, converting the anticodon from the usual triplet sequence CCC$^{\leftarrow}$ to the quadruplet sequence CCCC$^{\leftarrow}$. The suppressor tRNA recognizes a 4-base "codon."

Situations in which frameshifting is a normal event are presented by phages and viruses. Such events may affect the continuation or termination of protein synthesis, and result from the intrinsic properties of the mRNA.

In retroviruses, translation of the first gene is terminated by a nonsense codon in phase with the reading frame. The second gene lies in a different reading frame, and (in some viruses) is translated by a frameshift that changes into the second reading frame and therefore bypasses the termination codon (see Figure 9.23 and *22.3 Retroviral Genes*

Missense suppressors compete with wild type

Wild type: GGA codon is read by Gly-tRNA

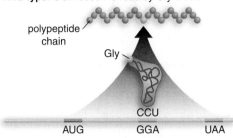

Missense: AGA is read by Arg-tRNA

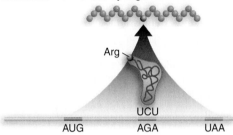

Suppression: AGA is read by mutant Gly-tRNA

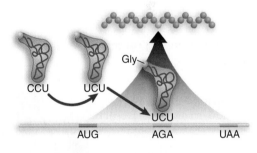

Figure 9.21 Missense suppression occurs when the anticodon of tRNA is mutated so that it responds to the wrong codon. The suppression is only partial because both the wild-type tRNA and the suppressor tRNA can respond to AGA.

Special or mutant tRNAs change meaning

Suppression is caused by mutated anticodon

Special factor + tRNA recognizes codon

Figure 9.22 A mutation in an individual tRNA (usually in the anticodon) can suppress the usual meaning of that codon. In a special case, a specific tRNA bound by an unusual elongation factor becomes able to recognize a termination codon adjacent to a hairpin loop.

Frameshifts can suppress termination

−1 frameshift in HIV retrovirus

N N N N U U U U U U A G G N N N N N N N N

Last codon read in initial reading frame

First codon read in new reading frame

Reading without frameshift

N N N N U U U U U U A G G N N N N N N N N

Reading after frameshift

N N N N U U U U U U A G G N N N N N N N N

Figure 9.23 A tRNA that slips one base in pairing with a codon causes a frameshift that can suppress termination. The efficiency is usually ~5%.

Bypassing skips between identical codons

60 nucleotide bypass in phage T4 gene *60*

GAUGGAUGAC . . . . AUUGGAUUA

Last codon in original reading frame

First codon in new reading frame

Reading without bypass

GAUGGAUGAC . . . . AUUGGAUUA

Reading after bypass

GAUGGAUGAC . . . . AUUGGAUUA

Figure 9.24 Bypassing occurs when the ribosome moves along mRNA so that the peptidyl-tRNA in the P site is released from pairing with its codon and then pairs instead with another codon farther along.

Frameshifting controls translation

Alternative modes of translation give Tya or Tya-Tyb

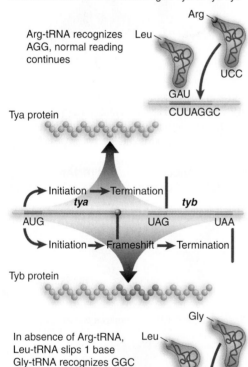

Arg-tRNA recognizes AGG, normal reading continues

Arg

Leu

UCC

GAU

CUUAGGC

Tya protein

Initiation → Termination

tya

AUG   UAG   UAA

Initiation → Frameshift → Termination

tyb

Tyb protein

In absence of Arg-tRNA, Leu-tRNA slips 1 base Gly-tRNA recognizes GGC

Gly

Leu

CCG

GAU

Frameshift

CUUAGGC

Figure 9.25 A +1 frameshift is required for expression of the *tyb* gene of the yeast Ty element. The shift occurs at a 7-base sequence at which two Leu codon(s) are followed by a scarce Arg codon.

Code for Polyproteins). The efficiency of the frameshift is low, typically ~5%. In fact, this is important in the biology of the virus; an increase in efficiency can be damaging. **Figure 9.25** illustrates the similar situation of the yeast Ty element, in which the termination codon of *tya* must be bypassed by a frameshift in order to read the subsequent *tyb* gene.

Such situations make the important point that the rare (but predictable) occurrence of "misreading" events can be relied on as a necessary step in natural translation. This is called **programmed frameshifting**. It occurs at particular sites at frequencies that are 100–1000× greater than the rate at which errors are made at nonprogrammed sites (~3×10^{-5} per codon).

There are two common features in this type of frameshifting:

- A "slippery" sequence allows an aminoacyl-tRNA to pair with its codon and then to move +1 (rare) or −1 base (more common) to pair with an overlapping triplet sequence that can also pair with its anticodon.

- The ribosome is delayed at the frameshifting site to allow time for the aminoacyl-tRNA to rearrange its pairing. The cause of the delay can be an adjacent codon that requires a scarce aminoacyl-tRNA, a termination codon that is recognized slowly by its release factor, or a structural impediment in mRNA (for example, a "pseudoknot," a particular conformation of RNA) that impedes the ribosome.

Slippery events can be movement in either direction; a −1 frameshift is caused when the tRNA moves backwards, and a +1 frameshift is caused when it moves forwards. In either case, the result is to expose an out-of-phase triplet in the A site for the next aminoacyl-tRNA. The frameshifting event precedes peptide bond synthesis. In the most common type of case, in which the frameshift is triggered by a slippery sequence in conjunction with a downstream hairpin in mRNA, the surrounding sequences influence its efficiency.

The frameshifting in Figure 9.25 shows the behavior of a typical slippery sequence. The 7-nucleotide sequence CUUAGGC is usually recognized by Leu-tRNA at CUU followed by Arg-tRNA at AGG. However, the Arg-tRNA is scarce, and when its scarcity results in a delay, the Leu-tRNA slips from the CUU codon to the overlapping UUA triplet. This causes a frameshift, because the next triplet in phase with the new pairing (GGC) is read by Gly-tRNA. Slippage usually occurs in the P site (when the Leu-tRNA actually has become peptidyl-tRNA, carrying the nascent chain).

9.14 Bypassing Involves Ribosome Movement

Key Concept

- When a ribosome encounters a GGA codon adjacent to a stop codon in a specific stem-loop structure, it moves directly to a specific GGA downstream without adding amino acids to the polypeptide.

Certain sequences trigger a bypass event, when a ribosome stops translation, slides along mRNA with peptidyl-tRNA remaining in the P site, and then resumes translation. This is a rather rare phenomenon, with only about three authenticated examples. The most dramatic example of bypassing is in gene *60* of phage T4, where the ribosome moves 60 nucleotides along the mRNA, as shown in Figure 9.24.

The key to the bypass system is that there are identical (or synonymous) codons at either end of the sequence that is skipped. They are sometimes referred to as the "take-off" and "landing" sites. Before bypass, the ribosome is positioned with a peptidyl-tRNA paired with the take-off codon in the P site, with an empty A site waiting for an aminoacyl-tRNA to enter. **Figure 9.26** shows that the ribosome slides along mRNA in this condition until the peptidyl-tRNA can become paired with the codon in the landing site. A remarkable feature of the system is its high efficiency, ~50%.

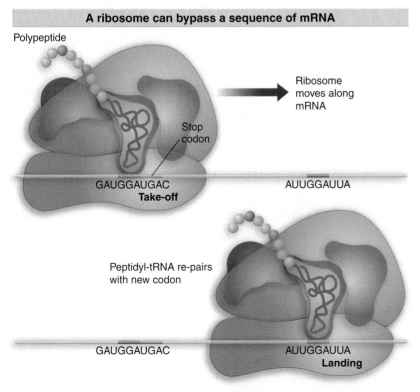

A ribosome can bypass a sequence of mRNA

Polypeptide

Ribosome moves along mRNA

Stop codon

GAUGGAUGAC
Take-off

AUUGGAUUA

Peptidyl-tRNA re-pairs with new codon

GAUGGAUGAC

AUUGGAUUA
Landing

Figure 9.26 In bypass mode, a ribosome with its P site occupied can stop translation. It slides along mRNA to a site where the peptidyl-tRNA pairs with a new codon in the P site. Then protein synthesis resumes.

The sequence of the mRNA triggers the bypass. The important features are the two GGA codons for take-off and landing, the spacing between them, a stem-loop structure that includes the take-off codon, and the stop codon adjacent to the take-off codon. The protein under synthesis is also involved.

The take-off stage requires the peptidyl-tRNA to unpair from its codon. This is followed by a movement of the mRNA that prevents it from re-pairing. Then the ribosome scans the mRNA until the peptidyl-tRNA can re-pair with the codon in the landing reaction. This is followed by the resumption of protein synthesis when aminoacyl-tRNA enters the A site in the usual way.

Like frameshifting, the bypass reaction depends on a pause by the ribosome. The probability that peptidyl-tRNA will dissociate from its codon in the P site is increased by delays in the entry of aminoacyl-tRNA into the A site. Starvation for an amino acid can trigger bypassing in bacterial genes because of the delay when there is no aminoacyl-tRNA available to enter the A site. In phage T4 gene *60*, one role of mRNA structure may be to reduce the efficiency of termination, thus creating the delay that is needed for the take-off reaction.

9.15 SUMMARY

The sequence of mRNA read in triplets $5' \rightarrow 3'$ is related by the genetic code to the amino acid sequence of protein read from N- to C-terminus. Of the 64 triplets, 61 code for amino acids and three provide termination signals. Synonym codons that represent the same amino acids are related, often differing only in the third base of the codon. This third-base degeneracy, coupled with a pattern in which related amino acids tend to be coded by related codons, minimizes the effects of mutations. The genetic code is universal, and must have been established very early in evolution. Changes in the code in nuclear genomes are rare, but some changes have occurred during mitochondrial evolution.

Multiple tRNAs may respond to a particular codon. The set of tRNAs responding to the various codons for each amino acid is distinctive for each organism. Codon–anticodon recognition involves wobbling at the first position of the anticodon (third position of the codon), which allows some tRNAs to recognize multiple codons. All tRNAs have modified bases, introduced by enzymes that recognize target bases in the tRNA structure. Codon–anticodon pairing is influenced by modifications of the anticodon itself and also by the context of adjacent bases, especially on the 3′ side of the anticodon.

Each amino acid is recognized by a particular aminoacyl-tRNA synthetase, which also recognizes all of the tRNAs coding for that amino acid. Aminoacyl-tRNA synthetases vary widely, but fall into two general groups according to the structure of the catalytic domain. Synthetases of each group bind the tRNA from the side, making contacts principally with the extremities of the acceptor stem and the anticodon stem-loop; the two types of synthetases bind tRNA from opposite sides. The relative importance of the acceptor stem and the anticodon region for specific recognition varies with the individual tRNA. Aminoacyl-tRNA synthetases have proofreading functions that scrutinize the aminoacyl-tRNA products and hydrolyze incorrectly joined aminoacyl-tRNAs.

Mutations may allow a tRNA to read different codons; most commonly, such mutations are in the anticodon itself. Alteration of its specificity may allow a tRNA to suppress a mutation in a gene coding for protein. A tRNA that recognizes a termination codon provides a nonsense suppressor; one that changes the amino acid responding to a codon is a missense suppressor. Suppressors of UAG and UGA codons are more efficient than those of UAA codons, which is explained by the fact that UAA is the most commonly used natural termination codon. But the efficiency of all suppressors depends on the context of the individual target codon.

Frameshifts of the +1 type may be caused by aberrant tRNAs that read "codons" of four bases. Frameshifts of either +1 or −1 may be caused by slippery sequences in mRNA that allow a peptidyl-tRNA to slip from its codon to an overlapping sequence that can also pair with its anticodon. This frameshifting also requires another sequence that causes the ribosome to delay. Frameshifts determined by the mRNA sequence may be required for expression of natural genes. Bypassing occurs when a ribosome stops translation and moves along mRNA with its peptidyl-tRNA in the P site until the peptidyl-tRNA pairs with an appropriate codon; then translation resumes.

10

Protein Localization Requires Special Signals

10.1 Introduction

Key Terms

- The **endoplasmic reticulum (ER)** is an organelle involved in the synthesis of lipids, membrane proteins, and secretory proteins. It is a single compartment that extends from the outer layer of the nuclear envelope into the cytoplasm. It has subdomains, such as the rough ER and smooth ER.

- The **Golgi apparatus** is an organelle that receives newly synthesized proteins from the endoplasmic reticulum and processes them for subsequent delivery to other destinations. It is composed of several flattened membrane disks arranged in a stack.

Key Concepts

- Proteins that are imported into cytoplasmic organelles are synthesized on free ribosomes in the cytosol.
- Proteins that are imported into the ER–Golgi system are synthesized on ribosomes that are associated with the ER.

All proteins are synthesized on ribosomes in one or the other of two locations: The vast majority of proteins are synthesized in the cytosol. A small minority are synthesized within organelles (mitochondria or chloroplasts).

Proteins synthesized in the cytosol can be divided into two general classes with regard to the ribosomes that synthesize them. The basic distinction is whether the ribosomes are "free" or "membrane-associated." **Figure 10.1** maps the cell to show the possible ultimate destinations for a newly synthesized protein and the systems that transport it.

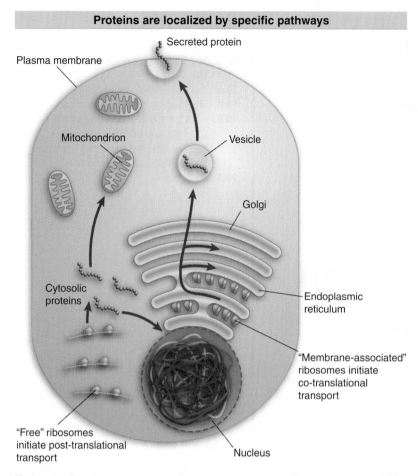

Proteins are localized by specific pathways

Figure 10.1 Proteins synthesized on free ribosomes are released into the cytosol. Some have signals that target the proteins for transport to organelles such as the nucleus or mitochondria. Membrane-associated ribosomes transfer newly synthesized proteins directly into the endoplasmic reticulum, and may be transported farther along the system to the Golgi, plasma membrane, or exterior of the cell.

Proteins synthesized on free ribosomes, which are not associated with membranes, are released into the cytosol when their synthesis is completed. We can divide them into three groups according to how the protein is localized after its release:

- Some of these proteins remain free in the cytosol in quasi-soluble form.
- Some associate with macromolecular cytosolic structures, such as filaments, microtubules, or centrioles. This association depends on protein–protein interactions between the newly synthesized polypeptide and the existing structure.
- Some proteins are transported into the nucleus or into other organelles such as mitochondria (or chloroplasts in plant cells). All the proteins of the nucleus are provided by cytosolic synthesis, as are most of the proteins of the mitochondrion.

Eukaryotic cells have a complex series of membranes called the **endoplasmic reticulum (ER)** that extends from the outer membrane of the nucleus. Close to the ER are the membrane-enclosed stacks of the **Golgi apparatus**, which extend toward the plasma membrane. All proteins that associate with the ER, Golgi, or plasma membrane are synthesized by ribosomes that associate with the ER. These ribosomes are sometimes described as being membrane-associated.

10.2 Protein Translocation May Be Post-Translational or Co-Translational

Key Terms

- Protein **translocation** describes the movement of a protein across a membrane. This occurs across the membranes of organelles in eukaryotes, or across the plasma membrane in bacteria. Each membrane across which proteins are translocated has a channel specialized for the purpose.

- **Post-translational translocation** is the movement of a protein across a membrane after the protein has been synthesized and released from the ribosome.

- **Co-translational translocation** describes the movement of a protein across a membrane as the protein is being synthesized. The term is usually restricted to cases in which the ribosome binds to the channel. This form of translocation may be restricted to the endoplasmic reticulum.

- The **leader** of a protein is a short N-terminal sequence responsible for initiating passage into or through a membrane.

- A protein to be imported into an organelle or secreted from bacteria is called a **preprotein** until its signal sequence has been removed.

Key Concepts

- Proteins associate with membranes by means of specific amino acid sequences called signal sequences.

- Signal sequences are most often leaders that are located at the N-terminus.

- N-terminal signal sequences are usually cleaved off the protein during the insertion process.

In order to be associated with a cellular organelle (such as a membrane), a protein requires a specific signal. The same principle for associating with membranes applies to proteins synthesized by free and by membrane-bound ribosomes. The signal for interacting with the target membrane is provided by a short sequence of amino acids in the protein that is recognized by a receptor associated with the organelle. The way in which the signal works is somewhat different for the two classes of ribosomes.

Special arrangements are required for proteins to associate with membranes. The protein presents a hydrophilic surface, but the membrane is hydrophobic. Like oil and water, the two would prefer not to mix. The task is accomplished by a special structure consisting of a proteinaceous channel in the membrane through which the protein can pass. The term **translocation** describes the process of inserting into or passing through a membrane.

Post-translational translocation describes the process for proteins that are released from ribosomes and subsequently become associated with membranes. The protein has a sequence that interacts with a receptor that introduces it into the channel or conveys it through. **Figure 10.2** summarizes some signals used by proteins released from cytosolic ribosomes. Mitochondrial and chloroplast proteins that are synthesized on cytosolic ribosomes associate with the organelle membranes after they have been released from the ribosomes; they have N-terminal sequences of ~25 amino acids in length that are recognized by receptors on the organelle envelope. (A similar principle is used also for importing proteins into the nucleus. Nuclear proteins have "nuclear localization signals" that enable them to pass through nuclear pores.) One type of signal that determines transport to the peroxisome (a membrane-enclosed organelle) is a very short C-terminal sequence.

The signal works in a different way for proteins that enter the ER-Golgi system from ribosomes. **Figure 10.3** shows that the nascent protein passes into the ER directly from the ribosome while it is being synthesized. Then it may be transported farther along the membrane network to the Golgi or plasma membrane, or secreted from the cell. Because the protein associates with the membrane during protein synthesis, the process is described as **co-translational translocation**.

In both post-translational and co-translational translocation, the association with the membrane is determined by the protein. A common feature is found in proteins that use N-terminal sequences to be transported co-translationally to the ER or post-translationally to mitochondria or chloroplasts. The N-terminal sequence is a **leader** that is not part of the mature protein. The protein carrying this leader is called a **preprotein**, and is a transient precursor to the mature protein. The leader is cleaved from the protein during protein translocation.

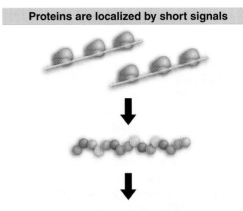

Proteins are localized by short signals

Organelle	Signal location	Type	Signal length
Mitochondrion	N-terminal	Amphipathic helix	12–30
Chloroplast	N-terminal	Charged	>25
Nucleus	Internal	Basic or bipartite	4–9
Peroxisome	C-terminal	Short peptide	3–4

Figure 10.2 Proteins synthesized on free ribosomes in the cytosol are directed after their release to specific destinations by short signal motifs.

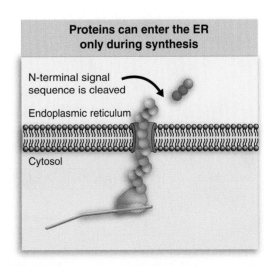

Proteins can enter the ER only during synthesis

N-terminal signal sequence is cleaved

Endoplasmic reticulum

Cytosol

Figure 10.3 Proteins can enter the ER–Golgi pathway only by associating with the endoplasmic reticulum while they are being synthesized.

10.3 The Signal Sequence Interacts with the SRP

Key Terms

- A **signal sequence** is a short region of a protein that directs it to the endoplasmic reticulum for co-translational translocation.

- The **signal recognition particle (SRP)** is a ribonucleoprotein complex that recognizes signal sequences during translation and guides the ribosome to the translocation channel. SRPs from different organisms may have different compositions, but all contain related proteins and RNAs.

- **Signal peptidase** is an enzyme within the membrane of the ER that specifically removes the signal sequences from proteins as they are translocated. Analogous activities are present in bacteria, archaea, and in each organelle in a eukaryotic cell into which proteins are targeted and translocated by means of removable targeting sequences. Signal peptidase is one component of a larger protein complex.

Key Concepts

- The signal sequence binds to the SRP (signal recognition particle).
- Signal-SRP binding causes protein synthesis to pause.
- Protein synthesis resumes when the SRP binds to the SRP receptor in the membrane.
- The signal sequence is cleaved from the translocating protein by the signal peptidase located on the "inside" face of the membrane.

Protein translocation into the endoplasmic reticulum can be divided into two general stages: first ribosomes carrying nascent polypeptides associate with the membranes; and then the nascent chain is transferred to the channel and translocates through it. The process is initiated by the leader sequence of the nascent protein, which is also known as the **signal sequence**. Usually this is a sequence of 15–30 N-terminal amino acids that is cleaved from the protein during translocation. At or close to the N-terminus are several polar residues, and within the leader is a hydrophobic core consisting exclusively or very largely of hydrophobic amino acids. There is no other conservation of sequence. **Figure 10.4** gives an example.

The attachment of ribosomes to membranes requires the **signal recognition particle (SRP)**. The SRP has two important abilities:

- It can bind to the signal sequence of a nascent secretory protein.
- It can bind to a protein (the SRP receptor) located in the membrane.

The SRP and SRP receptor function catalytically to transfer a ribosome carrying a nascent protein to the membrane. The first step is the recognition of the signal sequence by the SRP. Then the SRP binds to the SRP receptor, and the ribosome binds to the membrane. The stages of translation of membrane proteins are summarized in **Figure 10.5**.

The role of the SRP receptor in protein translocation is transient. When the SRP binds to the signal sequence, it arrests translation. This usually happens when ~70 amino acids have been incorporated into the polypeptide chain (at this point the 25 residue leader has become exposed, with the next ~40 amino acids still buried in the ribosome).

Then when the SRP binds to the SRP receptor, the SRP releases the signal sequence. The ribosome becomes bound by another component of the membrane. At this point, translation can resume. When the

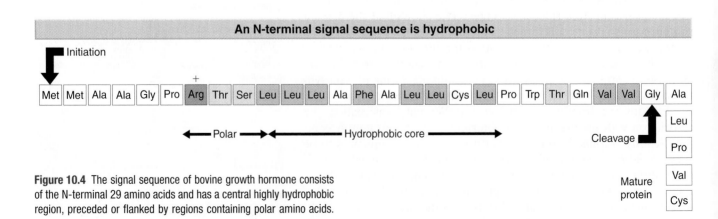

An N-terminal signal sequence is hydrophobic

Figure 10.4 The signal sequence of bovine growth hormone consists of the N-terminal 29 amino acids and has a central highly hydrophobic region, preceded or flanked by regions containing polar amino acids.

The signal sequence initiates membrane entry

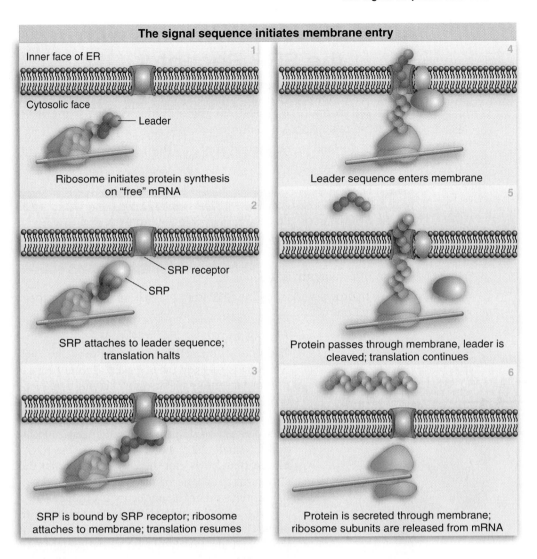

Figure 10.5 Ribosomes synthesizing secretory proteins are attached to the membrane via the signal sequence on the nascent polypeptide.

ribosome has been passed on to the membrane, the role of SRP and SRP receptor has been played. They now recycle, and are free to sponsor the association of another nascent polypeptide with the membrane.

This process may be needed to control the conformation of the protein. If the nascent protein were released into the cytoplasm, it could take up a conformation in which it might be unable to traverse the membrane. The ability of the SRP to inhibit translation while the ribosome is being handed over to the membrane is therefore important in preventing the protein from being released into the aqueous environment.

The signal peptide is cleaved from a translocating protein by a complex of five proteins called the **signal peptidase**. The complex is several times more abundant than the SRP and SRP receptor. Its amount is equivalent roughly to the amount of bound ribosomes, suggesting that it functions in a structural capacity. It is located on the lumenal face of the ER membrane, which implies that the entire signal sequence must cross the membrane before the cleavage event occurs. Homologous signal peptidases can be recognized in eubacteria, archaea, and eukaryotes.

10.4 The SRP Interacts with the SRP Receptor

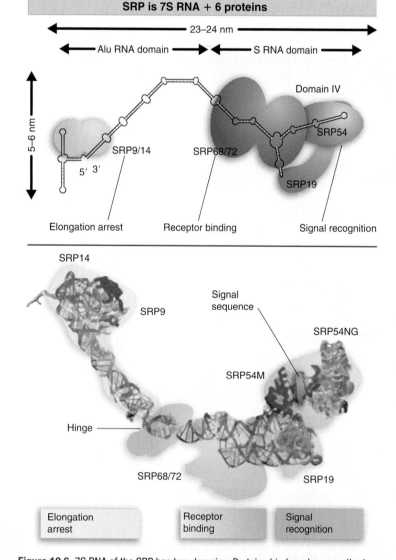

Figure 10.6 7S RNA of the SRP has two domains. Proteins bind as shown on the two-dimensional diagram above to form the crystal structure shown below. Each function of the SRP is associated with a discrete part of the structure.

The interaction between the SRP and the SRP receptor is the key event in eukaryotic translation in targetting a ribosome carrying a nascent protein to the membrane.

The SRP is an 11S ribonucleoprotein complex, containing six proteins (total mass 240 kD) and a small (305 base, 100 kD) 7S RNA. **Figure 10.6** shows that the 7S RNA provides the structural backbone of the particle; the individual proteins do not assemble in its absence.

The 7S RNA of the SRP is divided into two parts. The 100 bases at the 5′ end and 45 bases at the 3′ end are closely related to the sequence of Alu RNA, a common mammalian small RNA. They therefore define the **Alu domain**. The remaining part of the RNA constitutes the **S domain**.

Different parts of the SRP structure depicted in Figure 10.6 have separate functions in protein targeting. SRP54 is the most important subunit. It is located at one end of the RNA structure, and is directly responsible for recognizing the substrate protein by binding to the signal sequence. It also binds to the SRP receptor in conjunction with the SRP68-SRP72 dimer that is located at the central region of the RNA. The SRP9-SRP14 dimer is located at the other end of the molecule; it is responsible for elongation arrest.

The SRP is a flexible structure. In its unengaged form (not bound to signal sequence), it is quite extended, as can be seen from the crystal structure of Figure 10.6. **Figure 10.7** shows that binding to a signal sequence triggers a change of conformation, and the protein bends at a hinge to allow the SRP54 end to contact the ribosome at the protein exit site, while the SRP19 swings around to contact the ribosome at the elongation factor binding site. This enables it to cause the elongation arrest that gives time for targeting to the translocation site on the membrane.

The SRP receptor is a dimer containing subunits SRα (72 kD) and SRβ (30 kD). The β subunit is an integral membrane protein. The amino-terminal end of the large α subunit is anchored by the β subunit. The bulk of the α protein protrudes into the cytosol. A large part of the sequence of the cytoplasmic region of the protein resembles a nucleic acid-binding protein, with many positive residues. This suggests the possibility that the SRP receptor recognizes the 7S RNA in the SRP.

There is a counterpart to SRP in bacteria, although it contains fewer components. *E. coli* contains a 4.5S RNA that associates with ribosomes and is homologous to the 7S RNA of the SRP. It associates with two proteins: Ffh is homologous to SRP54; FtsY is homologous to the α subunit of the SRP receptor. In fact, FtsY replaces the functions of both the α and β SRP subunits; its N-terminal domain substitutes for SRPβ in membrane targeting, and the C-terminal domain interacts with the target protein.

The role of the bacterial 4.5S RNA-complex is more limited than that of SRP-SRP receptor. It probably plays the role of keeping the nascent protein in an appropriate conformation until it interacts with other components of the secretory apparatus. It is needed for the translocation of integral membrane proteins. The basis for differential selection of substrates is that the *E. coli* SRP recognizes a potential transmembrane sequence in the protein. Chloroplasts have counterparts to the Ffh and FtsY proteins, but do not require an RNA component.

Why should the SRP have an RNA component? The answer must lie in the evolution of the SRP: it must have originated very early in evolution, in an RNA-dominated world, presumably in conjunction with a ribosome whose functions were mostly carried out by RNA. Bacterial 4.5S/Ffh could represent the original connection between protein synthesis and secretion; in eukaryotes the SRP has acquired the additional roles of causing translational arrest and targeting to the membrane.

The crystal structure of the complex between the protein-binding domain of 4.5S RNA and the RNA-binding domain of Ffh suggests that RNA continues to play a role in the function of SRP. The 4.5S RNA has a region (domain IV) that is very similar to domain IV in 7S RNA (see Figure 10.6). Ffh consists of three domains (N, G, and M). The M domain (named for a high content of methionines) performs the key binding functions. It has a helix-loop-helix motif that is typical of DNA-binding proteins (see *14.10 Repressor Uses a Helix-Turn-Helix Motif to Bind DNA*). This motif binds to a duplex region of the 4.5S RNA in domain IV. Next to it, a hydrophobic pocket created by the methionine side chains binds the signal sequence of a target protein. The juxtaposition raises the possibility that a signal sequence actually binds to both the protein and RNA components of the SRP.

GTP hydrolysis plays an important role in inserting the signal sequence into the membrane. Both the SRP and the SRP receptor have GTPase capability. The signal-binding subunit of the SRP, SRP54, is a GTPase. And both subunits of the SRP receptor are GTPases. All of the GTPase activities are necessary for a nascent protein to be transferred to the membrane. **Figure 10.8** shows that the SRP starts out with GDP when it binds to the signal sequence. The ribosome then stimulates replacement of the GDP with GTP. The signal sequence inhibits hydrolysis of the GTP. This ensures that the complex has GTP bound when it encounters the SRP receptor.

For the nascent protein to be transferred from the SRP to the membrane, the SRP must be released from the SRP receptor. **Figure 10.9** shows that this requires hydrolysis of the GTPs of both the SRP and the SRP receptor. The reaction has been characterized in the bacterial system, where it has the unusual feature that Ffh activates hydrolysis by FtsY, and FtsY reciprocally activates hydrolysis by Ffh.

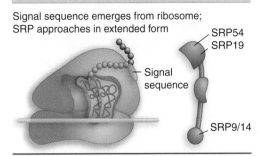

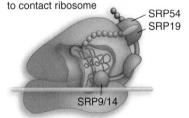

Figure 10.7 SRP binds to a signal sequence as it emerges from the ribosome. The binding causes the SRP to change conformation by bending at a "hinge," allowing SRP54 to contact the ribosome at the protein exit site while SRP19 makes a second set of contacts.

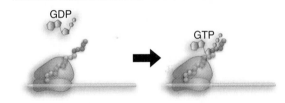

Figure 10.8 The SRP carries GDP when it binds the signal sequence. The ribosome causes the GDP to be replaced with GTP.

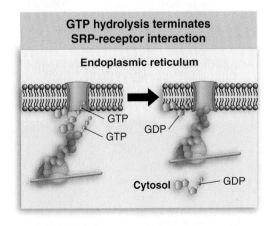

Figure 10.9 The SRP and SRP receptor both hydrolyze GTP when the signal sequence is transferred to the membrane.

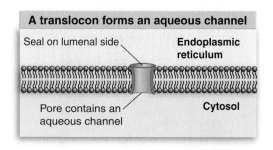

A translocon forms an aqueous channel

Seal on lumenal side

Endoplasmic reticulum

Pore contains an aqueous channel

Cytosol

Figure 10.10 The translocon is a trimer of Sec61 that forms a channel through the membrane. It is sealed on the lumenal (ER) side.

10.5 The Translocon Forms a Pore

Key Term

• A **translocon** is a discrete structure in a membrane that forms a channel through which (hydrophilic) proteins may pass.

Key Concepts

• The Sec61 trimeric complex provides the channel for proteins to pass through a membrane.

• A translocating protein passes directly from the ribosome to the translocon without exposure to the cytosol.

The basic problem in passing a (largely) hydrophilic protein through a hydrophobic membrane is that the energetics of the interaction between the charged protein and the hydrophobic lipids are highly unfavorable. However, a protein in the process of translocation across an ER membrane remains in an aqueous environment created by a channel through the bilayer. **Figure 10.10** shows that the channel is formed by proteins that are part of the ER membrane. The translocating protein travels directly from an enclosed tunnel in the ribosome into the aqueous channel, interacting with the resident proteins rather than with the lipid bilayer.

The channel through the membrane is called the **translocon**. It consists of a cylinder formed from the heterotrimeric protein Sec61. An aggregate of 3–4 heterotrimers of the Sec61 complex (consisting of three transmembrane proteins: Sec61α,β,γ) forms cylindrical oligomers with a diameter of ~85 Å and a central pore of ~20 Å. The complex is well conserved in evolution, and is called SecY in bacteria and archaea (see *10.7 Bacteria Use Both Co-translational and Post-translational Translocation*).

A similar trimeric structure for the channel is found in all organisms. The Sec61 α subunit (or the corresponding SecY subunit in bacteria and archaea) provides the pore through which the protein passes, and is the best conserved in sequence. The active pore is an oligomer containing dimers of the trimeric complex, organized back to back so that the α subunits are fused into a single channel.

Access to the pore is controlled (or "gated") on *both* sides of the membrane. Before attachment of the ribosome, the pore is closed on the lumenal side to stop free transfer of ions between the ER and the cytosol. **Figure 10.11** shows that when the ribosome attaches, it seals the pore on the cytosolic side. When the nascent protein extends fully across the channel, the pore opens on the lumenal side. The translocating protein fills the channel completely, so ions cannot pass through during translocation. So at all times, the pore is closed on one side or the other, maintaining the ionic integrities of the separate compartments.

The translocon is versatile, and can be used by translocating proteins in several ways:

• It is the means by which nascent proteins are transferred from cytosolic ribosomes to the lumen of endoplasmic reticulum.

• It is also the route by which integral membrane proteins of the ER system are transferred to the membrane; this requires the channel to open or disaggregate in some unknown way so that the protein can move laterally into the lipid bilayer.

• Proteins can also be transferred from the ER back to the cytosol to be degraded there; This is known as reverse translocation.

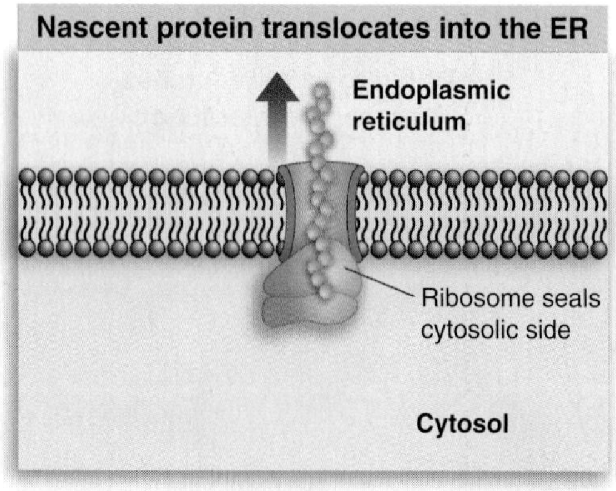

Nascent protein translocates into the ER

Endoplasmic reticulum

Ribosome seals cytosolic side

Cytosol

Figure 10.11 A nascent protein is transferred directly from the ribosome to the translocon. The ribosome seals the channel on the cytosolic side.

10.6 Post-Translational Membrane Insertion Depends on Leader Sequences

Key Terms

- The **TOM complex** resides in the outer membrane of the mitochondrion and is responsible for importing proteins from the cytosol into the space between the membranes.
- The **TIM complex** resides in the inner membrane of mitochondria and is responsible for transporting proteins from the intermembrane space into the interior of the organelle.

Key Concepts

- N-terminal leader sequences provide the information that allows proteins to associate with mitochondrial or chloroplast membranes.
- The N-terminal part of a leader sequence targets a protein to the mitochondrial matrix or chloroplast lumen.
- An adjacent sequence can control further targeting, to a membrane or the intermembrane spaces.
- The sequences are cleaved successively from the protein.
- Transport through the outer and inner mitochondrial membranes uses different receptor complexes.
- The TOM (outer membrane) complex is a large complex in which substrate proteins are directed to the Tom40 channel by one of two subcomplexes.
- Different TIM (inner membrane) complexes are used depending on whether the substrate protein is targeted to the inner membrane or to the lumen.
- Proteins pass directly from the TOM to the TIM complex.

Mitochondria and chloroplasts synthesize only some of their proteins. Mitochondria synthesize only ~10 organelle proteins; chloroplasts synthesize ~50 proteins. The majority of organelle proteins are synthesized in the cytosol by the same pool of free ribosomes that synthesize cytosolic proteins. They must then be imported into the organelle.

Many proteins that enter mitochondria or chloroplasts by a post-translational process have leader sequences that are responsible for primary recognition of the outer membrane of the organelle. As shown in the simplified diagram of **Figure 10.12**, the leader sequence initiates the interaction between the precursor and the organelle membrane. The protein passes through the membrane, and the leader is cleaved by a protease on the organelle side.

The leaders of proteins imported into mitochondria and chloroplasts usually have both hydrophobic and basic amino acids. They consist of stretches of uncharged amino acids interrupted by basic amino acids, and they lack acidic amino acids. There is little other homology. An example is given in **Figure 10.13**. Recognition of the leader does not depend on its exact sequence, but rather on its ability to form an amphipathic helix, in which one face has hydrophobic amino acids, and the other face presents the basic amino acids.

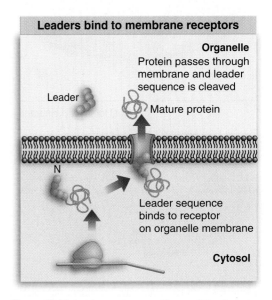

Leaders bind to membrane receptors

Organelle
Protein passes through membrane and leader sequence is cleaved

Leader

Mature protein

N

Leader sequence binds to receptor on organelle membrane

Cytosol

Figure 10.12 Leader sequences allow proteins to recognize mitochondrial or chloroplast surfaces by a post-translational process.

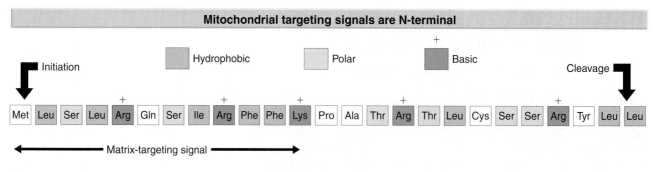

Mitochondrial targeting signals are N-terminal

Initiation | Hydrophobic | Polar | Basic | Cleavage

Met Leu Ser Leu Arg Gln Ser Ile Arg Phe Phe Lys Pro Ala Thr Arg Thr Leu Cys Ser Ser Arg Tyr Leu Leu

◄─────── Matrix-targeting signal ───────►

Figure 10.13 The leader sequence of yeast cytochrome *c* oxidase subunit IV consists of 25 neutral and basic amino acids. The first 12 amino acids are sufficient to transport any attached polypeptide into the mitochondrial matrix.

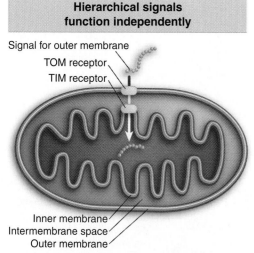

Hierarchical signals function independently

Signal for outer membrane
TOM receptor
TIM receptor

Inner membrane
Intermembrane space
Outer membrane

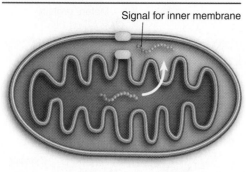

Signal for inner membrane

Figure 10.14 Mitochondria have receptors for protein transport in the outer and inner membranes. Recognition at the outer membrane may lead to transport through both receptors into the matrix, where the leader is cleaved. If it has a membrane-targeting signal, it may be re-exported.

The mitochondrion is surrounded by an envelope consisting of two membranes. Proteins imported into mitochondria may be located in the outer membrane, the intermembrane space, the inner membrane, or the matrix. A protein that is a component of one of the membranes may be oriented so that it faces one side or the other.

What is responsible for directing a mitochondrial protein to the appropriate compartment? The leader sequence contains all the information needed to localize an organelle protein. The "default" pathway for a protein imported into a mitochondrion is to move through both membranes into the matrix. This property is conferred by the N-terminal part of the leader sequence. A protein that is localized within the intermembrane space or in the inner membrane itself requires an additional signal, which specifies its destination within the organelle. A multipart leader contains signals that function in a hierarchical manner, as summarized in **Figure 10.14**. The first part of the leader targets the protein to the organelle, and the second part is required if its destination is elsewhere than the matrix. The two parts of the leader are removed by successive cleavages.

The two parts of a leader that contains both types of signal have different compositions. As indicated in **Figure 10.15**, the 35 N-terminal amino acids resemble other organelle leader sequences in the high content of uncharged amino acids, punctuated by basic amino acids. The next 19 amino acids, however, constitute an uninterrupted stretch of uncharged amino acids, long enough to span a lipid bilayer. This membrane-targeting signal resembles the sequences that help translocate proteins into membranes of the endoplasmic reticulum.

Cleavage of the matrix-targeting signal is the sole processing event required for proteins that reside in the matrix. This signal must also be cleaved from proteins that reside in the intermembrane space; but following this cleavage, the membrane-targeting signal (which is now the N-terminal sequence of the protein) directs the protein to its destination in the outer membrane, intermembrane space, or inner membrane. Then it in turn is cleaved.

There are different receptors for transport through each membrane in the chloroplast and mitochondrion. In the chloroplast they are called TOC and TIC, and in the mitochondrion they are called the **TOM complex** and the **TIM complex**, referring to the outer and inner membranes, respectively.

When a protein is translocated through the TOM complex, it passes from a state in which it is exposed to the cytosol into a state in which it is exposed to the intermembrane space. However, it is not usually released, but instead is transferred directly to the TIM complex. There are two TIM complexes in the inner membrane, which are used for transporting different classes of proteins.

A mitochondrial protein folds under different conditions before and after its passage through the membrane. Ionic conditions and the chaperones that are present are different in the cytosol and in the mitochondrial matrix. It is possible that a mitochondrial protein can attain its mature conformation *only* in the mitochondrion.

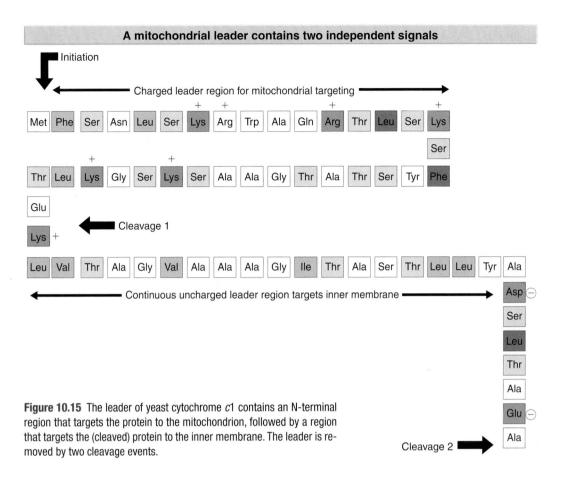

A mitochondrial leader contains two independent signals

Figure 10.15 The leader of yeast cytochrome *c*1 contains an N-terminal region that targets the protein to the mitochondrion, followed by a region that targets the (cleaved) protein to the inner membrane. The leader is removed by two cleavage events.

10.7 Bacteria Use Both Co-Translational and Post-Translational Translocation

Key Term

- The **periplasm** (or periplasmic space) is the region between the inner and outer membranes in the bacterial envelope.

Key Concepts

- Bacterial proteins that are exported to or through membranes use both post-translational and co-translational mechanisms.

- The bacterial SecYEG translocon in the inner membrane is related to the eukaryotic Sec61 translocon.

- Various chaperones are involved in directing secreted proteins to the translocon.

The bacterial envelope consists of two membrane layers. The space between them is called the **periplasm**. Proteins are exported from the cytoplasm to reside in the envelope or to be secreted from the cell. The mechanisms of secretion from bacteria are similar to those characterized for eukaryotic cells, and we can recognize some related components. **Figure 10.16** shows that proteins that are exported from the cytoplasm have one of four fates:

- to be inserted into the inner membrane;
- to be translocated through the inner membrane to rest in the periplasm;

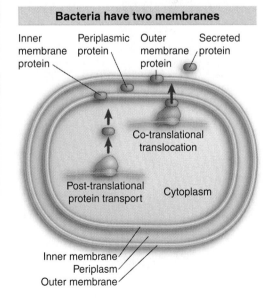

Bacteria have two membranes

Figure 10.16 Bacterial proteins may be exported either post-translationally or co-translationally, and may be located within either membrane or the periplasmic space, or may be secreted.

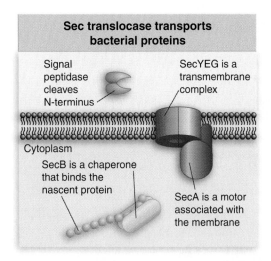

Figure 10.17 The Sec system has the SecYEG translocon embedded in the membrane, the SecA associated protein that pushes proteins through the channel, the SecB chaperone that transfers nascent proteins to SecA, and the signal peptidase that cleaves the N-terminal signal from the translocated protein.

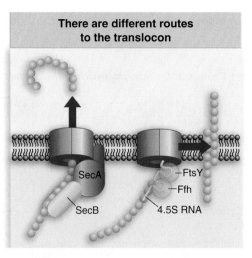

Figure 10.18 SecB/SecA transfers proteins that pass through the membrane. 4.4S RNA transfers proteins that enter the membrane.

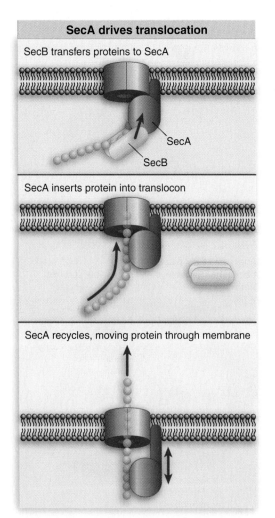

Figure 10.19 SecB transfers a nascent protein to SecA, which inserts the protein into the channel. Translocation requires hydrolysis of ATP and a protonmotive force. SecA undergoes cycles of association and dissociation with the channel and provides the motive force to push the protein through.

- to be inserted into the outer membrane; or
- to be translocated through the outer membrane into the medium.

Different protein complexes in the inner membrane are responsible for transport of proteins depending on whether their fate is to pass through or stay within the inner membrane. This resembles the situation in mitochondria, where different complexes in each of the inner and outer membranes handle different subsets of protein substrates depending on their destinations (see previous section). A difference from import into organelles is that transfer in *E. coli* may be either co- or post-translational. Some proteins are secreted both co-translationally and post-translationally, and the relative kinetics of translation versus secretion through the membrane could determine the balance.

There are several systems for transport through the inner membrane. The best characterized is the Sec system, whose components are shown in **Figure 10.17**. The translocon that is embedded in the membrane consists of three subunits that are related to the components of eukaryotic Sec61. Each of the subunits is an integral transmembrane protein. The functional translocon is a trimer with one copy of each subunit. The major pathway for directing proteins to the translocon consists of SecB and SecA. SecB binds to the nascent protein to control its folding, and then transfers the protein to SecA, which in turn transfers it to the translocon.

Figure 10.18 shows that there are two predominant ways of directing proteins to the Sec channel:

- the SecB chaperone; and
- the 4.5S RNA-based SRP.

SecB binds to a nascent protein to retard folding, the purpose being to inhibit improper folding of the newly synthesized protein. Second, it has an affinity for the protein SecA. This allows it to target a precursor protein to the membrane. The SecB-SecYEG pathway is used for translocation of proteins that are secreted into the periplasm and is summarized in **Figure 10.19**. The SRP is used for proteins that are inserted into the membrane.

10.8 SUMMARY

Synthesis of proteins in the cytosol starts on free ribosomes. Proteins that are secreted from the cell or that are inserted into membranes of the endoplasmic reticulum start with an N-terminal signal sequence that causes the ribosome to become attached to the membrane of the endoplasmic reticulum. The protein is translocated through the membrane by co-translational transfer. The process starts when a signal sequence is recognized by the SRP (a ribonucleoprotein particle), which interrupts translation. The SRP binds to the SRP receptor in the ER membrane, and transfers the signal sequence to the Sec61 receptor in the membrane. Synthesis resumes, and the protein is translocated through the membrane while it is being synthesized, although there is no energetic connection between the processes. The channel through the membrane provides a hydrophilic environment, and is largely made of the protein Sec61. A secreted protein passes completely through the membrane into the ER lumen. Integral membrane proteins are transferred from the channel to the membrane. The signal sequences of secreted proteins are usually N-terminal, and are cleaved from the protein when it passes to the lumenal side of the membrane; some integral membrane proteins have internal signal sequences. A protein usually passes through the channel in an unfolded form, and association with other proteins (chaperones) when it emerges is necessary for folding into the correct conformation.

Proteins that are imported into mitochondria or chloroplasts are released from cytoplasmic ribosomes and then associate with receptors on the organelle membrane. Mitochondria and chloroplasts have separate receptor complexes that create channels through each of the outer and inner membranes. All imported proteins pass directly from the TOM complex in the outer membrane to a TIM complex in the inner membrane. Proteins that reside in the intermembrane space or in the outer membrane are reexported from the TIM complex after entering the matrix. The TOM complex uses different receptors for imported proteins depending on whether they have N-terminal or internal signal sequences, and directs both types into the TOM40 channel. There are two TIM receptors in the inner membrane, one used for proteins whose ultimate destination is the inner matrix, the other used for proteins that are re-exported to the intermembrane space or outer membrane. The signals that determine which receptors are recognized are usually located at the N-terminus of the protein.

Transcription

11.1 Introduction

Key Terms

- The **coding strand (sense strand)** of DNA has the same sequence as the mRNA and is related by the genetic code to the protein sequence that it represents.

- The **template strand** of DNA is the strand that is copied by RNA polymerase to generate a complementary single-stranded RNA.

- **RNA polymerases** are enzymes that synthesize RNA using a DNA template (formally described as DNA-dependent RNA polymerases).

- A **promoter** is a region of DNA where RNA polymerase binds to initiate transcription.

- **Startpoint (startsite)** refers to the position on DNA corresponding to the first base incorporated into RNA.

- A **terminator** is a sequence of DNA that causes RNA polymerase to terminate transcription.

- A **transcription unit** is the sequence between sites of initiation and termination by RNA polymerase; may include more than one gene.

- **Upstream** identifies sequences in the opposite direction from expression; for example, the bacterial promoter is upstream of the transcription unit, the initiation codon is upstream of the coding region.

- **Downstream** identifies sequences farther in the direction of expression; for example, the coding region is downstream from the initiation codon.

- A **primary transcript** is the original unmodified RNA product corresponding to a transcription unit.

Transcription produces an RNA chain representing one strand of a DNA duplex. **Figure 11.1** shows that the RNA is *identical in sequence* with one strand of the DNA, which is called the **coding strand**. The RNA is *complementary* to the other strand, which provides the **template strand** for its synthesis.

RNA synthesis is catalyzed by the enzyme **RNA polymerase**. Transcription starts when RNA polymerase binds to a special region, the **promoter**, at the start of the gene. The promoter surrounds the first base pair that is transcribed into RNA, the **startpoint**. From this point, RNA polymerase moves along the template, synthesizing RNA, until it reaches a **terminator** sequence. This action defines a **transcription unit** that extends from the promoter to the terminator. The critical feature of the transcription unit, depicted in **Figure 11.2**, is that it constitutes a stretch of DNA *expressed via the production of a single RNA molecule*. A transcription unit may include more than one gene.

Sequences prior to the startpoint are described as **upstream** of it; those after the startpoint (within the transcribed sequence) are **downstream** of it. Sequences are conventionally written so that transcription proceeds from left (upstream) to right (downstream). This corresponds to writing the mRNA in the usual $5' \rightarrow 3'$ direction.

Often the DNA sequence is written to show only the coding strand, which has the same sequence as the RNA. Base positions are numbered in both directions away from the startpoint, which is assigned the value +1; numbers are increased going downstream. The base before the startpoint is numbered −1, and the negative numbers increase going upstream. (There is no base assigned the number 0.)

The immediate product of transcription is called the **primary transcript**. It would consist of an RNA extending from the promoter to the terminator, possessing the original 5′ and 3′ ends. However, the primary transcript is almost always immediately modified. In prokaryotes, it is rapidly degraded simultaneously with translation (mRNA) or cleaved to give mature products (rRNA and tRNA). In eukaryotes, it is modified at the ends (mRNA) and/or cleaved to give mature products (all RNA).

Transcription is the first stage in gene expression, and the principal step at which it is controlled. Regulatory proteins determine whether a particular gene is available to be transcribed by RNA polymerase. The initial (and often the only) step in regulation is the decision on whether or not to transcribe a gene. Most regulatory events occur at the initiation of transcription, although subsequent stages in transcription (or other stages of gene expression) are sometimes regulated.

Within this context, there are two basic questions in gene expression:

- How does RNA polymerase find promoters on DNA? This is a particular example of a more general question: how do proteins distinguish their specific binding sites in DNA from other sequences?
- How do regulatory proteins interact with RNA polymerase (and with one another) to activate or to repress specific steps in the initiation, elongation, or termination of transcription?

In this chapter, we analyze the interactions of bacterial RNA polymerase with DNA, from its initial contact with a gene, through the act of transcription, culminating in its release when the transcript has been completed. *Chapter 12 The Operon* discusses the various means by which regulatory proteins can assist or prevent bacterial RNA polymerase from recognizing a particular gene for transcription. *Chapter 13*

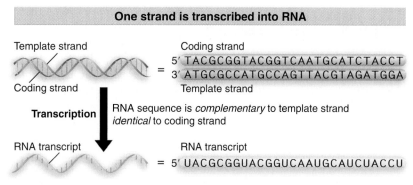

One strand is transcribed into RNA

Template strand

Coding strand

Coding strand
= 5′ TACGCGGTACGGTCAATGCATCTACCT
  3′ ATGCGCCATGCCAGTTACGTAGATGGA
Template strand

Transcription   RNA sequence is *complementary* to template strand *identical* to coding strand

RNA transcript

RNA transcript
= 5′ UACGCGGUACGGUCAAUGCAUCUACCU

Figure 11.1 RNA is transcribed from the template strand of duplex DNA.

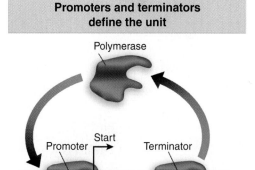

Promoters and terminators define the unit

Polymerase

Start
Promoter   Terminator

−35   −10−1+1   +10

Proximal   Distal

Upstream Downstream

Figure 11.2 A transcription unit is a sequence of DNA transcribed into a single RNA, starting at the promoter and ending at the terminator.

RNA synthesis occurs in the transcription bubble

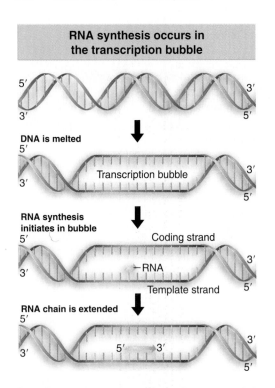

Figure 11.3 DNA strands separate to form a transcription bubble. RNA is synthesized by complementary base pairing with one of the DNA strands.

The transcription bubble moves along DNA

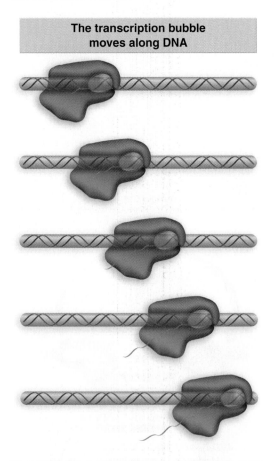

Figure 11.4 Transcription takes place in a bubble, in which RNA is synthesized by base pairing with one strand of DNA in the transiently unwound region. As the bubble progresses, the DNA duplex reforms behind it, displacing the RNA in the form of a single polynucleotide chain.

Regulatory RNA discusses other means of regulation, including the use of small RNAs. In *Chapter 14 Phage Strategies* we consider how individual regulatory interactions can be connected into more complex networks. In *Chapter 24 Promoters and Enhancers* and *Chapter 25 Regulating Eukaryotic Transcription*, we consider the analogous reactions between eukaryotic RNA polymerases and their templates.

11.2 Transcription Occurs by Base Pairing in a "Bubble" of Unpaired DNA

Key Concepts

- RNA polymerase separates the two strands of DNA in a transient "bubble" and uses one strand as a template to direct synthesis of a complementary sequence of RNA.
- The length of the bubble is ~12–14 bp, and the length of RNA–DNA hybrid within it is ~8–9 bp.

Transcription takes place by the usual process of complementary base pairing. **Figure 11.3** illustrates the general principle. RNA synthesis takes place within a "transcription bubble," in which DNA is transiently separated into its single strands, and the template strand is used to direct synthesis of the RNA strand.

The RNA chain is synthesized from the 5′ end toward the 3′ end. The 3′—OH group of the last nucleotide added to the chain reacts with an incoming nucleotide 5′ triphosphate. The incoming nucleotide loses its terminal two phosphate groups (γ and β); its α group is used in the phosphodiester bond linking it to the chain. The overall reaction rate is ~40 nucleotides/second at 37°C (for the bacterial RNA polymerase); this is about the same as the rate of translation (15 amino acids/sec), but much slower than the rate of DNA replication (800 bp/sec).

RNA polymerase creates the transcription bubble when it binds to a promoter. **Figure 11.4** shows that as RNA polymerase moves along the DNA, the bubble moves with it, and the RNA chain grows longer. The process of base pairing and base addition within the bubble is catalyzed and scrutinized by the enzyme.

The structure of the bubble within RNA polymerase is shown in the expanded view of **Figure 11.5**. As RNA polymerase moves along the DNA template, it unwinds the duplex at the front of the bubble (the unwinding point), and rewinds the DNA at the back (the rewinding point). The length of the transcription bubble is ~12–14 bp, but the length of the RNA–DNA hybrid region within it is ~8–9 bp. As the enzyme moves on, the DNA duplex reforms, and the RNA is displaced as a free polynucleotide chain. About the last 25 ribonucleotides added to a growing chain are complexed with DNA and/or enzyme at any moment.

11.3 The Transcription Reaction Has Three Stages

Key Terms

- **Initiation** describes the stages of transcription up to synthesis of the first bond in RNA. This includes binding of RNA polymerase to the promoter and melting a short region of DNA into single strands.

- **Elongation** is the stage in transcription in which the nucleotide chain is being extended by the addition of individual nucleotides.
- **Termination** is a separate reaction that ends transcription by stopping the addition of nucleotides, and causes dissociation of RNA polymerase from DNA.

Key Concepts

- RNA polymerase initiates transcription after binding to a promoter site on DNA.
- During elongation the transcription bubble moves along DNA and the RNA chain is extended in the 5′–3′ direction.
- Transcription stops, the DNA duplex reforms and RNA polymerase dissociates at a terminator site.

The transcription reaction can be divided into the stages illustrated in **Figure 11.6**, in which a bubble is created, RNA synthesis begins, the bubble moves along the DNA, and finally the reaction is terminated.

Template recognition begins with the binding of RNA polymerase to the double-stranded DNA at a promoter to form a "closed complex," in which the DNA remains duplex. Then the strands of DNA are separated to form the "open complex" that makes the template strand available for base pairing with ribonucleotides. This occurs when the transcription bubble is created by a local unwinding that begins at the site bound by RNA polymerase.

Initiation describes the synthesis of the first nucleotide bonds in RNA. The enzyme remains at the promoter while it synthesizes the first ~9 nucleotide bonds. The initiation phase is protracted by the occurrence of abortive events, in which the enzyme makes short transcripts, releases them, and then starts synthesis of RNA again. The initiation phase ends when the enzyme succeeds in extending the chain and clears the promoter. Abortive initiation probably involves synthesizing an RNA chain that fills the active site. If the RNA is released, the initiation is aborted and must start again. Initiation is accomplished when the enzyme moves the next region of the DNA into the active site. *The sequence of DNA needed for RNA polymerase to bind to the template and accomplish the initiation reaction defines the promoter.*

During **elongation** the enzyme moves along the DNA and extends the growing RNA chain. As the enzyme moves, it unwinds the DNA helix to expose a new segment of the template as a single strand. Nucleotides are covalently added to the 3′ end of the growing RNA chain, forming an RNA–DNA hybrid in the unwound region. Behind the unwound region, the DNA template strand pairs again with the coding strand to reform the double helix. The RNA emerges as a free single strand. *Elongation involves the movement of the transcription bubble by a disruption of DNA structure in which the template strand of the transiently unwound region is paired with the nascent RNA at the growing point.*

Termination involves recognition of the point at which no further bases should be added to the chain. To terminate transcription, the formation of phosphodiester bonds must cease, and the transcription complex must come apart. When the last base is added to the RNA chain, the transcription bubble collapses as the RNA–DNA hybrid is disrupted, the DNA reforms in duplex state, and the enzyme and RNA are both released. *The sequence of DNA required for these reactions defines the terminator.*

RNA polymerase surrounds the bubble

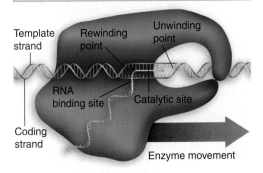

Figure 11.5 During transcription, the bubble is maintained within bacterial RNA polymerase, which unwinds and rewinds DNA and synthesizes RNA.

RNA polymerase catalyzes transcription

Template recognition: RNA polymerase binds to duplex DNA

DNA is unwound at promoter

Initiation: Very short chains are synthesized and released

Elongation: Polymerase synthesizes RNA

Termination: RNA polymerase and RNA are released

Figure 11.6 Transcription has several stages. RNA polymerase binds to the promoter and melts DNA, remains stationary during initiation, moves along the template during elongation, and dissociates at termination.

RNA polymerase can recover from pausing

RNA polymerase during elongation

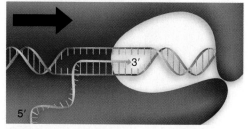

RNA polymerase is stalled and backtracks

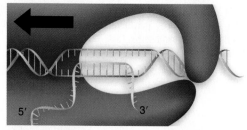

3' region of RNA is cleaved

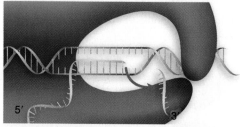

New 3' end is located in the catalytic site

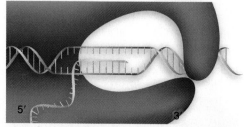

Catalytic site resumes elongation

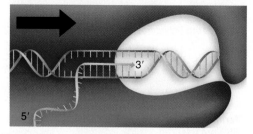

Figure 11.7 A stalled RNA polymerase can be released by cleaving the 3' end of the transcript.

In addition to synthesizing RNA, RNA polymerase must be able to handle situations when transcription is blocked. This can happen, for example, when DNA is damaged. A model system for such situations is provided by arresting elongation *in vitro* by omitting one of the necessary precursor nucleotides. When the missing nucleotide is restored, the enzyme can overcome the block by cleaving the 3' end of the RNA, to create a new 3' terminus for chain elongation. The cleavage involves accessory factors in addition to the enzyme itself, although the catalytic site of RNA polymerase undertakes the actual cleavage. The model shown in **Figure 11.7** suggests that the enzyme "backtracks" on the DNA. The 3' terminus of the RNA is exposed in single-stranded form. Cleavage restores a normal elongation complex.

11.4 A Model for Enzyme Movement Is Suggested by the Crystal Structure

Key Concepts

- DNA moves through a groove in yeast RNA polymerase that makes a sharp turn at the active site.
- A protein bridge changes conformation to control the entry of nucleotides to the active site.

We now have much information about the structure and function of RNA polymerase from the crystal structures of the bacterial and yeast enzymes. The bacterial and eukaryotic enzymes share a common type of structure, in which there is a "channel" ~25 Å wide that could be the path for DNA. The length of the channel could hold 16 bp in the bacterial enzyme, and ~25 bp in the eukaryotic enzyme, but this represents only part of the total length of DNA bound during transcription.

The views of the crystal structure in **Figure 11.8** and **Figure 11.9** show how the yeast RNA polymerase enzyme surrounds the DNA. A catalytic Mg^{2+} ion is found at the active site. **Figure 11.10** shows that DNA is forced to take a turn at the entrance to the active site, because of an adjacent wall of protein. The length of the RNA hybrid is limited by another protein obstruction, called the rudder. Nucleotides probably enter the active site from below, via pores through the structure.

A side view of RNA polymerase II

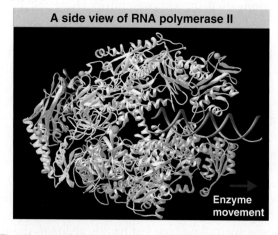

Enzyme movement

Figure 11.8 The side view of the crystal structure of RNA polymerase II from yeast shows that DNA is held downstream by a pair of jaws and is clamped in position in the active site, which contains an Mg^{2+} ion. Photograph kindly provided by Roger Kornberg, Dept. of Structural Biology, Stanford University School of Medicine.

An end view of RNA polymerase II

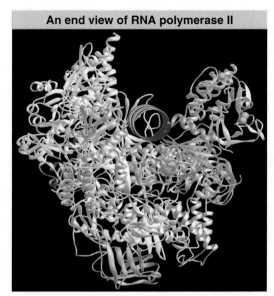

Figure 11.9 The end view of the crystal structure of RNA polymerase II from yeast shows that DNA is surrounded by ∼270° of protein. Photograph kindly provided by Roger Kornberg, Dept. of Structural Biology, Stanford University School of Medicine.

The active site holds the transcription bubble

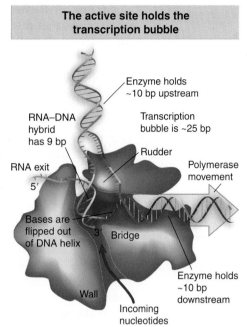

Enzyme holds ∼10 bp upstream

Transcription bubble is ∼25 bp

RNA–DNA hybrid has 9 bp

Rudder

RNA exit

5′

Polymerase movement

Bases are flipped out of DNA helix

3′   Bridge

Wall

Incoming nucleotides

Enzyme holds ∼10 bp downstream

Figure 11.10 DNA is forced to make a turn at the active site by a wall of protein. The RNA–DNA hybrid has bases flipped out of the helix. Its length is limited by the rudder. Nucleotides enter through a pore in the protein.

The transcription bubble includes 9 bp of DNA–RNA hybrid. Where the DNA makes its turn, the bases downstream are flipped out of the DNA helix. As the enzyme moves along DNA, the base in the template strand at the start of the turn will be flipped to face the nucleotide entry site. The RNA–DNA hybrid is 9 bp long, and the 5′ end of the RNA is forced to leave the DNA when it hits the protein rudder.

Once DNA has been melted, the individual strands have a flexible structure in the transcription bubble. This enables DNA to take its turn in the active site. But before transcription starts, the DNA double helix is a relatively rigid straight structure. How does this structure enter the polymerase without being blocked by the wall? The answer is that the enzyme makes a large conformational shift. Adjacent to the wall is a clamp. In the free form of RNA polymerase, this clamp swings away from the wall to allow DNA to follow a straight path through the enzyme. After DNA has been melted to create the transcription bubble, the clamp swings back into position against the wall.

One of the dilemmas of any nucleic acid polymerase is that the enzyme must make tight contacts with the nucleic acid substrate and product, but must break these contacts and remake them with each cycle of nucleotide addition. Consider the situation illustrated in **Figure 11.11**. A polymerase makes a series of specific contacts with the bases at particular positions. For example, contact "1" is made with the base at the end of the growing chain, and contact "2" is made with the base in the template strand that is complementary to the next base to be added. But the bases that occupy these locations in the nucleic acid chains change every time a nucleotide is added!

The top and bottom panels of Figure 11.11 show the same situation: a base is about to be added to the growing chain. The difference is that the growing chain has been extended by one base in the bottom panel. The geometry of both complexes is exactly the same, but contacts "1" and "2" in the bottom panel are made to bases in the nucleic acid

Polymerases must make and break bonds

Polymerase contacts nucleic acid

Template

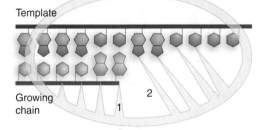

Growing chain

2

1

New base is added and bonds are broken

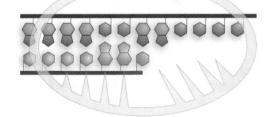

Enzyme moves and remakes bonds

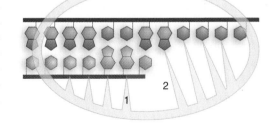

2

1

Figure 11.11 Movement of a nucleic acid polymerase requires breaking and remaking bonds to the nucleotides at fixed positions relative to the enzyme structure (two contact positions are indicated by numbers). The nucleotides in all the fixed positions change each time the enzyme moves a base along the template.

The bridge controls nucleotide entry

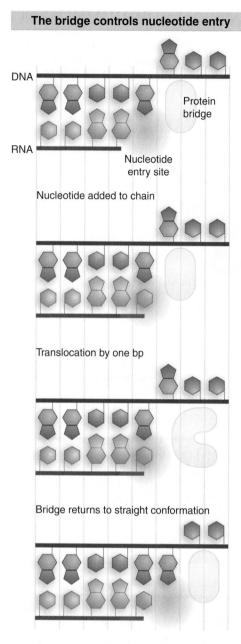

DNA

Protein
bridge

RNA

Nucleotide
entry site

Nucleotide added to chain

Translocation by one bp

Bridge returns to straight conformation

Figure 11.12 The RNA polymerase elongation cycle starts with a straight bridge adjacent to the nucleotide entry site. After nucleotide addition, the enzyme moves one base pair and the bridge bends as it retains contact with the newly added nucleotide. When the bridge is released, the cycle can start again.

chains that are located one position farther along the chain. The middle panel shows that this must mean that, after the base is added, and before the enzyme moves relative to the nucleic acid, the contacts made to specific positions must be broken; then later they are remade to bases that occupy those positions after the movement.

The RNA polymerase structure suggests an insight into how the enzyme retains contact with its substrate while breaking and remaking bonds. A "bridge" in the protein is adjacent to the active site (see Figure 11.10). **Figure 11.12** suggests that the change in conformation of the bridge structure is closely related to translocation of the enzyme along the nucleic acid.

At the start of the cycle of translocation, the bridge has a straight conformation adjacent to the nucleotide entry site. This allows the next nucleotide to bind at the nucleotide entry site. The bridge is in contact with the newly added nucleotide. Then the enzyme moves one base pair along the substrate. The bridge changes its conformation, bending to keep contact with the newly added nucleotide. In this conformation, the bridge obscures the nucleotide entry site. To end the cycle, the bridge returns to its straight conformation, allowing access again to the nucleotide entry site. The bridge acts as a ratchet that releases the DNA and RNA strands for translocation while holding onto the end of the growing chain.

11.5 RNA Polymerase Consists of the Core Enzyme and Sigma Factor

Key Terms

- The **holoenzyme (complete enzyme)** is the complex of five subunits—the core enzyme ($\alpha_2\beta\beta'$) plus the sigma (σ) factor—that is competent to initiate bacterial transcription.
- The **core enzyme** is the complex of RNA polymerase subunits needed for elongation. It does not include additional subunits or factors that may be needed for initiation or termination.
- **Sigma factor** is the subunit of bacterial RNA polymerase needed for initiation; it is the major influence on selection of promoters.
- A **loose binding site** is any random sequence of DNA that is bound by the core RNA polymerase when it is not engaged in transcription.

Key Concepts

- Bacterial RNA polymerase can be divided into the $\alpha_2\beta\beta'$ core enzyme that catalyzes transcription and the sigma subunit that is required only for initiation.
- The channel that binds DNA lies at the interface of the β and β' subunits.
- Sigma factor changes the DNA-binding properties of RNA polymerase so that its affinity for general DNA is reduced and its affinity for promoters is increased.
- Binding constants of RNA polymerase for different promoters vary over 6 orders of magnitude, corresponding to the frequency with which transcription is initiated at each promoter.

The best characterized RNA polymerases are those of eubacteria, for which *E. coli* is a typical case. *A single type of RNA polymerase appears* to be responsible for almost all synthesis of mRNA, rRNA, and tRNA in a eubacterium. About 7000 RNA polymerase molecules are present in an *E. coli* cell. Many of them are engaged in transcription; probably 2000–5000 enzymes are synthesizing RNA at any one time, the number depending on the growth conditions.

The **holoenzyme**, or **complete enzyme**, in *E. coli* consists of five subunits, and has a molecular weight of ~465 kD. The holoenzyme ($\alpha_2\beta\beta'\sigma$) can be separated into two components, the **core enzyme** ($\alpha_2\beta\beta'$) and the **sigma factor** (the σ polypeptide), which is concerned specifically with promoter recognition.

The β and β' subunits together make up the catalytic center. Their sequences are related to those of the largest subunits of eukaryotic RNA polymerases (see *24.2 Eukaryotic RNA Polymerases Consist of Many Subunits*), suggesting that there are common features to the actions of all RNA polymerases. The α subunit is required for assembly of the enzyme, and also plays a role in the interaction of RNA polymerase with some regulatory factors.

The crystal structure of the bacterial enzyme shows that the channel for DNA lies at the interface of the β and β' subunits. The DNA is unwound at the active site, where an RNA chain is being synthesized. **Figure 11.13** shows that the β and β' subunits contact DNA at many points downstream of the active site. They stabilize the separated single strands by making several contacts with the coding strand in the region of the transcription bubble. The RNA is contacted largely in the region of the transcription bubble.

The drug rifampicin (a member of the rifamycin antibiotic family) blocks transcription by bacterial RNA polymerase. It is a major drug used against tuberculosis. The crystal structure of RNA polymerase bound to rifampicin explains its action: it binds in a pocket of the β subunit, >12 Å away from the active site, but in a position where it blocks the path of the elongating RNA. By preventing the RNA chain from extending beyond 2–3 nucleotides, this blocks transcription.

Only the holoenzyme can initiate transcription. The sigma factor ensures that the holoenzyme binds in a stable manner to DNA only at promoters. Sigma is not needed for elongation, and may be released after initiation.

Core enzyme has the ability to synthesize RNA on a DNA template, but cannot initiate transcription at the proper sites. It has a general affinity for DNA, in which electrostatic attraction between the basic protein and the acidic nucleic acid plays a major role. Any (random) sequence of DNA that is bound by core enzyme in this general binding reaction is described as a **loose binding site**. No change occurs in the DNA, which remains duplex. The complex at such a site is stable, with a half-life for dissociation of the enzyme from DNA of ~60 minutes. *Core enzyme does not distinguish between promoters and other sequences of DNA.*

Figure 11.14 shows that sigma factor introduces a major change in the affinity of RNA polymerase for DNA. *In comparison with core enzyme, the holoenzyme has a drastically reduced ability to recognize loose binding sites*—that is, to bind to any general sequence of DNA. The half-life of such a complex is reduced to <1 second. So sigma factor destabilizes the general binding ability very considerably.

But sigma factor also *confers the ability to recognize specific binding sites.* The holoenzyme binds to promoters very tightly, with an association constant increased from that of core enzyme by (on average) 1000× and with a half-life of several hours.

The specificity of holoenzyme for promoters compared to other sequences is ~10^7, but this is only an average, because there is wide variation in the rate at which the holoenzyme binds to different promoter sequences. This is an important parameter in determining the efficiency of an individual promoter in initiating transcription.

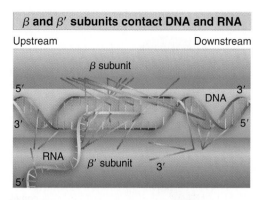

β and β' subunits contact DNA and RNA

Figure 11.13 Both the template and coding strands of DNA are contacted by the β and β' subunits largely in the region of the transcription bubble and downstream. The RNA is contacted mostly in the transcription bubble.

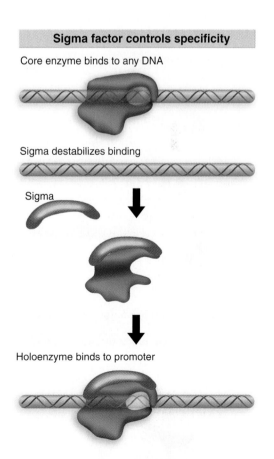

Sigma factor controls specificity

Core enzyme binds to any DNA

Sigma destabilizes binding

Sigma

Holoenzyme binds to promoter

Figure 11.14 Core enzyme binds indiscriminately to any DNA. Sigma factor reduces the affinity for sequence-independent binding, and confers specificity for promoters.

All RNA polymerase is bound to DNA

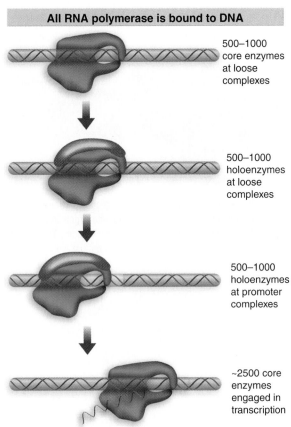

500–1000 core enzymes at loose complexes

500–1000 holoenzymes at loose complexes

500–1000 holoenzymes at promoter complexes

~2500 core enzymes engaged in transcription

Figure 11.15 Core enzyme and holoenzyme are distributed on DNA, and very little RNA polymerase is free.

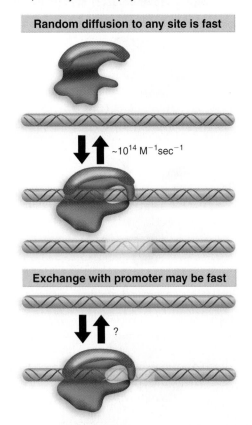

Random diffusion to any site is fast

~10^{14} M^{-1}sec^{-1}

Exchange with promoter may be fast

?

Figure 11.16 RNA polymerase binds very rapidly to random DNA sequences and could find a promoter by direct displacement of the bound DNA sequence.

11.6 How Does RNA Polymerase Find Promoter Sequences?

Key Concepts

- The rate at which RNA polymerase binds to promoters is too fast to be accounted for by random diffusion.
- RNA polymerase probably binds to random sites on DNA and exchanges them for other sites very rapidly until a promoter is found.

How is RNA polymerase distributed in the cell? The main feature is that virtually all of it is bound to DNA, with no free core enzyme or holoenzyme. A (somewhat speculative) picture of the enzyme's distribution is depicted in **Figure 11.15**:

- Excess core enzyme exists largely as closed loose complexes, because the enzyme enters into them rapidly and leaves them slowly.
- There is enough sigma factor for about one third of the polymerases to exist as holoenzymes, and they are distributed between loose complexes at nonspecific sites and binary complexes (mostly closed) at promoters.
- About half of the RNA polymerases consist of core enzymes engaged in transcription.

RNA polymerase must find promoters within the context of the genome. Suppose that a promoter is a stretch of ~60 bp; how is it distinguished from the 4×10^6 bp that constitute the *E. coli* genome?

The simplest model is to suppose that RNA polymerase moves by random diffusion. Holoenzyme very rapidly associates with, and dissociates from, loose binding sites. So it could continue to make and break a series of closed complexes until (by chance) it encounters a promoter. Movement from one site to another is limited by the speed of diffusion through the medium. However, the actual speed with which RNA polymerase finds some promoters is greater than would be possible by diffusion. This excludes random diffusion as the means by which RNA polymerase finds its promoter.

RNA polymerase must therefore use some other means to seek its binding sites. **Figure 11.16** shows that the process could be speeded up if the initial target for RNA polymerase is the whole genome, not just a specific promoter sequence. By increasing the target size, the rate constant for diffusion to DNA is correspondingly increased, and is no longer limiting.

If this idea is correct, a free RNA polymerase binds DNA and then remains in contact with it. How does the enzyme move from a random (loose) binding site on DNA to a promoter? The most likely model is to suppose that the bound sequence is directly displaced by another sequence. Having taken hold of DNA, the enzyme exchanges this sequence for another sequence very rapidly, and continues to exchange sequences until a promoter is found. Then the enzyme forms a stable, open complex, after which initiation occurs. The search process becomes much faster because association and dissociation are virtually simultaneous, and time is not spent commuting between sites. Direct displacement can give a "directed walk," in which the enzyme moves preferentially from a weak site to a stronger site.

11.7 Sigma Factor Controls Binding to DNA

Key Terms

- An **open complex** describes the stage of initiation of transcription in which RNA polymerase causes the two strands of DNA to separate to form the "transcription bubble."
- **Tight binding** of RNA polymerase to DNA describes the formation of an open complex (when the strands of DNA have separated).
- The **ternary complex** in initiation of transcription consists of RNA polymerase and DNA and a dinucleotide that represents the first two bases in the RNA product.
- **Abortive initiation** describes a process in which RNA polymerase starts transcription but terminates before it has left the promoter. It then reinitiates. Several cycles may occur before the elongation stage begins.

Key Concepts

- When RNA polymerase binds to a promoter, it separates the DNA strands to form a transcription bubble and incorporates up to nine nucleotides into RNA.
- There may be a cycle of abortive initiations before the enzyme moves to the next phase.
- Sigma factor may be released from RNA polymerase when the nascent RNA chain reaches 8–9 bases in length.
- The release of sigma factor changes binding affinity for DNA so that core enzyme can move along DNA.

We can now describe the stages of transcription in terms of the interactions between different forms of RNA polymerase and the DNA template. The initiation reaction can be described by the parameters that are summarized in **Figure 11.17**:

- The holoenzyme·promoter reaction starts by forming a closed binary complex. "Closed" means that the DNA remains duplex.
- The closed complex is converted into an **open complex** by "melting" of a short region of DNA within the sequence bound by the enzyme. The series of events leading to formation of an open complex is called **tight binding**. This reaction is fast. Sigma factor is involved in the melting reaction (see *11.11 Substitution of Sigma Factors May Control Initiation*).
- The next step is to incorporate the first two nucleotides; then a phosphodiester bond forms between them. This generates a **ternary complex** that contains RNA as well as DNA and enzyme. Further nucleotides can be added without any enzyme movement to generate an RNA chain of up to nine bases. After each base is added, there is a certain probability that the enzyme will release the chain, making the reaction an **abortive initiation**, after which the enzyme begins again with the first base. A cycle of abortive initiations usually occurs, generating a series of very short oligonucleotides.
- When initiation succeeds, sigma is no longer necessary, and the enzyme makes the transition to the elongation ternary complex of core enzyme·DNA·nascent RNA. The critical parameter here is *how long it takes for the polymerase to leave the promoter so another polymerase can initiate*. This parameter is the promoter clearance time; its minimum value of 1–2 seconds establishes the maximum frequency of initiation as ~1 event per second. The enzyme then moves along the template, and the RNA chain extends beyond 10 bases.

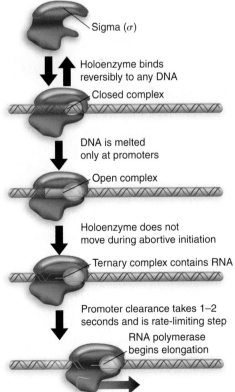

There are several stages in initiation

Sigma (σ)

Holoenzyme binds reversibly to any DNA

Closed complex

DNA is melted only at promoters

Open complex

Holoenzyme does not move during abortive initiation

Ternary complex contains RNA

Promoter clearance takes 1–2 seconds and is rate-limiting step

RNA polymerase begins elongation

Figure 11.17 RNA polymerase passes through several steps prior to elongation. A closed binary complex is converted to an open form and then into a ternary complex.

Sigma and core enzyme must dissociate

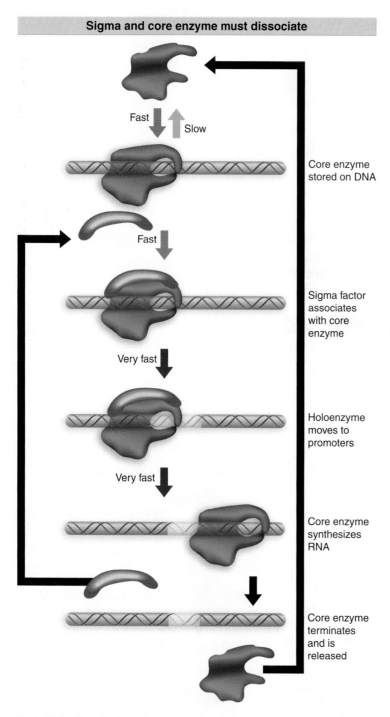

Fast   Slow

Core enzyme stored on DNA

Fast

Sigma factor associates with core enzyme

Very fast

Holoenzyme moves to promoters

Very fast

Core enzyme synthesizes RNA

Core enzyme terminates and is released

Figure 11.18 Sigma factor and core enzyme recycle at different points in transcription.

It has been a tenet of transcription since soon after the discovery of sigma factor that the factor is released after initiation. However, this may not be strictly true. Direct measurements of elongating RNA polymerase complexes show that ~70% of them retain sigma factor. Since a third of elongating polymerases lack sigma, the original conclusion is certainly correct that it is not necessary for elongation. In those cases where it remains associated with core enzyme, the sigma factor is likely to be associated in a different way from that of the initiation complex.

RNA polymerase encounters a dilemma in reconciling its needs for initiation with those for elongation. Initiation requires tight binding *only* to particular sequences (promoters), while elongation requires close association with *all* sequences that the enzyme encounters during transcription. **Figure 11.18** illustrates how the dilemma is solved by the reversible association between sigma factor and core enzyme.

Sigma factor is involved only in initiation. It becomes unnecessary when abortive initiation is concluded and RNA synthesis has been successfully initiated. At this point, the core enzyme in the ternary complex is bound very tightly to DNA. It is essentially "locked in" until elongation has been completed. When transcription terminates, the core enzyme is released. It is then "stored" by binding to a loose site on DNA.

Core enzyme has a high intrinsic affinity for DNA, which is increased by the presence of nascent RNA. But its affinity for loose binding sites is too high to allow the enzyme to distinguish promoters efficiently from other sequences. By reducing the stability of the loose complexes, sigma allows the process to occur much more rapidly; and by stabilizing the association at tight binding sites, the factor drives the reaction irreversibly into the formation of open complexes. The enzyme releases sigma (or changes the nature of its association with it) after initiation, and then reverts to a general affinity for all DNA, irrespective of sequence, that allows it to continue transcription.

Sigma factor enables the holoenzyme to bind specifically to promoters. As an independent polypeptide, sigma does not bind to DNA, but when holoenzyme forms a tight binding complex, sigma contacts the DNA in the region upstream of the startpoint. This difference is due to a change in the conformation of sigma factor when it binds to core enzyme. The N-terminal region of free sigma factor suppresses the activity of the DNA-binding region; when sigma binds to core enzyme, this inhibition is released, and it becomes able to bind specifically to promoter sequences (see also Figure 11.30 in Section 11.12). The inability of free sigma factor to recognize promoter sequences may be important: if sigma could freely bind to promoters, it might block holoenzyme from initiating transcription.

11.8 Promoter Recognition Depends on Consensus Sequences

Key Terms

- **Conserved sequences** are identified when many examples of a particular nucleic acid or protein are compared and the same individual bases or amino acids are always found at particular locations.
- A **consensus sequence** is an idealized sequence in which each position represents the base most often found when many actual sequences are compared.
- The **−10 sequence** is the consensus sequence centered about 10 bp before the startpoint of a bacterial gene. It is involved in melting DNA during the initiation reaction.
- The **−35 sequence** is the consensus sequence centered about 35 bp before the startpoint of a bacterial gene. It is involved in initial recognition by RNA polymerase.

Key Concepts

- A promoter is defined by the presence of short consensus sequences at specific locations.
- The bacterial promoter consensus sequences consist of a purine at the startpoint, the hexamer TATAAT centered at −10, and another hexamer centered at −35.
- Individual promoters usually differ from the consensus at one or more positions.
- The consensus sequences at −35 and −10 provide most of the contact points for RNA polymerase in the promoter.
- The points of contact lie on one face of the DNA.

As a sequence of DNA whose function is to be *recognized by proteins*, a promoter differs from sequences whose role is to be transcribed or translated. The information for promoter function is provided directly by the DNA sequence: its structure is the signal. This is a classic example of a *cis*-acting site, as defined previously in Figure 2.15 and Figure 2.16. By contrast, expressed regions gain their meaning only after the information is transferred into the form of some other nucleic acid or protein.

All promoters share certain sequence features that are recognized by RNA polymerase. In the bacterial genome, the minimum length of DNA that could distinguish a promoter is 12 bp. (Any shorter sequence is likely to occur—just by chance—a sufficient number of additional times to provide false signals. The minimum length required for unique recognition increases with the size of genome.) The 12 bp sequence need not be contiguous. If a specific number of base pairs separates two constant shorter sequences, their combined length could be less than 12 bp, since the *distance* of separation itself provides a part of the signal (even if the intermediate *sequence* is itself irrelevant).

Attempts to identify the features in DNA that are necessary for RNA polymerase binding started by comparing the sequences of different promoters. Any essential nucleotide sequence should be present in all the promoters. Such a sequence is said to be a **conserved sequence**. However, a conserved sequence need not necessarily be conserved at every single position; some variation is permitted. How do we analyze a sequence of DNA to determine whether it is sufficiently conserved to constitute a recognizable signal?

Putative DNA recognition sites can be defined in terms of an idealized sequence that represents the base most often present at each position. A **consensus sequence** is defined by aligning all known examples so as to maximize their homology. For a sequence to be accepted as a consensus, each particular base must be reasonably predominant at its position, and most of the actual examples must be related to the consensus by rather few substitutions—say, no more than 1–2.

The striking feature in the sequence of promoters in *E. coli* is the *lack of any extensive conservation of sequence* over the 60 bp associated with RNA polymerase. The sequence of much of the binding site is irrelevant. But some short stretches within the promoter are conserved, and they are critical for its function. *Conservation of only very short consensus sequences is a typical feature of regulatory sites (such as promoters) in both prokaryotic and eukaryotic genomes.*

The conserved features in a bacterial promoter are: the startpoint; the −10 sequence; the −35 sequence; and the separation between the −10 and −35 sequences:

- The startpoint is usually (>90% of the time) a purine. It is common for the startpoint to be the central base in the sequence CAT, but the conservation of this triplet is not great enough to regard it as an obligatory signal.
- Just upstream of the startpoint, a 6 bp region is recognizable in almost all promoters. The center of the hexamer generally is close to 10 bp upstream of the startpoint; the distance varies in known promoters from position −18 to −9. Named for its location, the hexamer is often called the **−10 sequence**. Its consensus is *TATAAT*, and can be summarized in the form

$$T_{80}A_{95}T_{45}A_{60}A_{50}T_{96}$$

where the subscript denotes the percent occurrence of the most frequently found base, varying from 45–96%. (A position at which there is no discernible preference for any base would be indicated by N.) If the frequency indicates likely importance in binding RNA polymerase, we would expect the initial highly conserved TA and the final almost completely conserved T in the −10 sequence to be the most important bases.
- Another conserved hexamer is centered ~35 bp upstream of the startpoint. This is called the **−35 sequence**. The consensus is *TTGACA*; in more detailed form, the conservation is

$$T_{82}T_{84}G_{78}A_{65}C_{54}A_{45}$$

- The distance separating the −35 and −10 sites is between 16–18 bp in 90% of promoters; in the exceptions, it is as little as 15 or as great as 20 bp. *Although the actual sequence in the intervening region is unimportant, the distance is critical in holding the two sites at the appropriate separation for the geometry of RNA polymerase.*
- Some promoters have an A·T–rich sequence located farther upstream. This is called the UP element. It interacts with the α subunit of the RNA polymerase. It is typically found in promoters that are highly expressed, such as the promoters for rRNA genes.

The optimal promoter is a sequence consisting of the −35 hexamer, separated by 17 bp from the −10 hexamer, lying 7 bp upstream of the startpoint. **Figure 11.19** summarizes the permitted range of variation from the optimum.

RNA polymerase initially binds the promoter region from −50 to +20. **Figure 11.20** shows the contacts that RNA polymerase makes with DNA at a typical promoter. The regions at −35 and −10 contain most of the contact points for the enzyme.

The promoter has three components

Figure 11.19 A typical promoter has three components, consisting of consensus sequences at −35 and −10 and the startpoint.

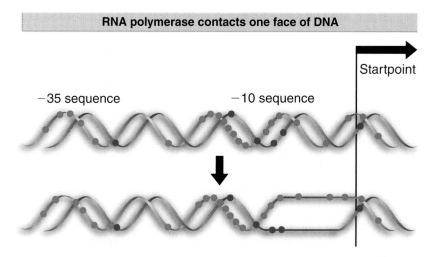

RNA polymerase contacts one face of DNA

Figure 11.20 One face of the promoter contains most of the contact points for RNA (shown by circles on the DNA strands). The initial region of unwinding extends from within the −10 sequence to past the startpoint.

Viewed in three dimensions, the points of contact upstream of the −10 sequence all lie on one face of DNA. Most lie on the coding strand. These bases are probably recognized in the initial formation of a closed binary complex. This would make it possible for RNA polymerase to approach DNA from one side and recognize that face of the DNA. As DNA unwinding commences, further sites that originally lay on the other face of DNA can be recognized and bound.

The region of DNA that is unwound in the binary complex can be identified directly by chemical changes in its availability. When the strands of DNA are separated, the unpaired bases become susceptible to reagents that cannot reach them in the double helix. Experiments with single DNA strands implicate positions between −9 and +3 in the initial melting reaction. The region unwound during initiation therefore includes the right end of the −10 sequence and extends just past the startpoint.

11.9 Promoter Efficiencies Can Be Increased or Decreased by Mutation

Key Concepts

- Down mutations, which decrease promoter efficiency, usually decrease conformance to consensus sequences, whereas up mutations increase conformance to consensus sequences.
- Mutations in the −35 sequence usually affect initial binding of RNA polymerase.
- Mutations in the −10 sequence usually affect the melting reaction that converts a closed to an open complex.

Key Terms

- A **down mutation** in a promoter decreases the rate of transcription.
- An **up mutation** in a promoter increases the rate of transcription.

Mutations in promoters affect the level of expression of the gene(s) they control, without altering the gene products themselves. Most bacterial mutants with such mutations have lost the ability to transcribe genes adjacent to the promotor, completely or in part. Their mutations are known as **down mutations**. Less often, mutants are found in which there is increased transcription from the promoter. They have **up mutations**.

Contact → binding → melting

−35 −10 Start

1. RNA polymerase initially contacts −35 sequence

2. Closed complex forms over promoter region

3. Melting at −10 region converts complex to open form

Figure 11.21 The −35 sequence is used for initial recognition, and the −10 sequence is used for the melting reaction that converts a closed complex to an open complex.

Is the most effective promoter one that has the actual consensus sequences? This expectation is borne out by the simple rule that up mutations usually increase homology with one of the consensus sequences or bring the distance between them closer to 17 bp. Down mutations usually decrease the resemblance of either site with the consensus or make the distance between them longer than 17 bp. Down mutations tend to be concentrated in the most highly conserved positions, which confirms that these positions are the main determinant of promoter efficiency. However, there are occasional exceptions to these rules.

There is ~100-fold variation in the rate at which RNA polymerase binds to different promoters *in vitro*, which correlates well with the frequencies of transcription when their genes are expressed *in vivo*. By measuring the kinetic constants for formation of a closed complex and its conversion to an open complex, we can dissect the two stages of the initiation reaction. Mutations in the two consensus regions have different effects:

- Down mutations in the −35 sequence reduce the rate of closed complex formation, but do not inhibit the conversion to an open complex.
- Down mutations in the −10 sequence do not affect the initial formation of a closed complex, but they slow its conversion to the open form.

These results suggest the model shown in **Figure 11.21**. The function of the −35 sequence is to provide the signal for recognition by RNA polymerase, while the −10 sequence allows the complex to convert from closed to open form. We might view the −35 sequence as constituting a "recognition domain," while the −10 sequence constitutes an "unwinding domain" of the promoter.

The consensus sequence of the −10 site consists exclusively of A · T base pairs, which assists the melting of DNA into single strands. The lower energy needed to disrupt A · T pairs compared with G · C pairs means that a stretch of A · T pairs demands the minimum amount of energy for strand separation.

The sequence immediately around the startpoint influences the initiation event. The initial transcribed region (from +1 to +30) influences the rate at which RNA polymerase clears the promoter, and therefore has an effect upon promoter strength. So the overall strength of a promoter cannot be predicted entirely from its −35 and −10 consensus sequences.

A "typical" promoter relies upon its −35 and −10 sequences to be recognized by RNA polymerase, but one or the other of these sequences can be absent from some (exceptional) promoters. In at least some of these cases, the promoter cannot be recognized by RNA polymerase alone, and the reaction requires ancillary proteins, which overcome the deficiency in intrinsic interaction between RNA polymerase and the promoter.

11.10 Supercoiling Is an Important Feature of Transcription

Key Concepts

- Negative supercoiling increases the efficiency of some promoters by assisting the melting reaction.
- Transcription generates positive supercoils ahead of the enzyme and negative supercoils behind it, and these must be removed by gyrase and topoisomerase.

Supercoiling has an important influence on transcription at both initiation and elongation. When DNA is negatively supercoiled, the two strands are less tightly wound around one another; at high enough negative supercoiling density they may separate (see *1.7 Supercoiling Affects the Structure of DNA*).

Negative supercoiling may assist initiation by making it easier for the strands of DNA to be separated. Indeed, the efficiency of some bacterial promoters is influenced by the degree of supercoiling. However, it is puzzling that some promoters are influenced by the extent of supercoiling while others are not. One possible explanation is that the dependence of a promoter on supercoiling is determined by its sequence. This would predict that some promoters have sequences that are easier to melt (and are therefore less dependent on supercoiling), while others have more difficult sequences (and have a greater need to be supercoiled). An alternative is that the location of the promoter might be important if different regions of the bacterial chromosome have different degrees of supercoiling.

Supercoiling also has a continuing involvement with transcription. As RNA polymerase transcribes DNA, the DNA is unwound and rewound. The consequences are illustrated in the *twin domain* model for transcription, as shown in **Figure 11.22**. As RNA polymerase pushes forward along the double helix, it generates positive supercoils (more tightly wound DNA) ahead and leaves negative supercoils (partially unwound DNA) behind. For each helical turn traversed by RNA polymerase, $+1$ turn is generated ahead and -1 turn behind.

Transcription therefore has a significant effect on the local structure of DNA. As a result, two enzymes, gyrase (which introduces negative supercoils) and topoisomerase I (which removes negative supercoils) are required to rectify the situation in front of and behind the polymerase, respectively. Blocking the activities of gyrase and topoisomerase causes major changes in the supercoiling of DNA. For example, in yeast lacking an enzyme that removes negative supercoils, the density of negative supercoiling doubles in a transcribed region. This makes it seem likely that transcription is responsible for generating a significant proportion of the supercoiling in the cell.

There is a similar situation during replication, when DNA must be unwound at a moving replication fork, so that the individual single strands can be used as templates to synthesize daughter strands.

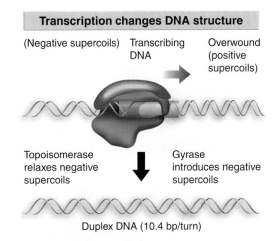

Transcription changes DNA structure

(Negative supercoils)   Transcribing DNA   Overwound (positive supercoils)

Topoisomerase relaxes negative supercoils

Gyrase introduces negative supercoils

Duplex DNA (10.4 bp/turn)

Figure 11.22 Transcription generates more tightly wound (positively supercoiled) DNA ahead of RNA polymerase, while the DNA behind becomes less tightly wound (negatively supercoiled).

11.11 Substitution of Sigma Factors May Control Initiation

Key Terms

- **Early genes** are transcribed before the replication of phage DNA. They code for regulators and other proteins needed for later stages of infection.

- **Middle genes** are phage genes that are regulated by the proteins coded by early genes. Some proteins coded by middle genes catalyze replication of the phage DNA; others regulate the expression of a later set of genes.

- **Late genes** are transcribed when phage DNA is being replicated. They code for components of the phage particle.

- **Sporulation** is the generation of a spore by a bacterium (by morphological conversion) or by a yeast cell (as the product of meiosis).

- The **vegetative phase** describes the period of normal growth and division of a bacterium. For a bacterium that can sporulate, this contrasts with the sporulation phase, when spores form.

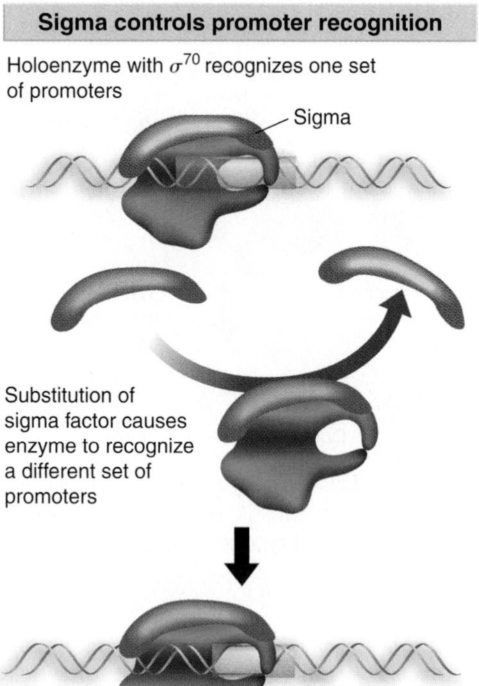

Sigma controls promoter recognition

Holoenzyme with σ^{70} recognizes one set of promoters

Substitution of sigma factor causes enzyme to recognize a different set of promoters

Figure 11.23 The sigma factor associated with core enzyme determines the set of promoters at which transcription is initiated.

E. coli has several sigma factors

Gene	Factor	Use
rpoD	σ^{70}	General
rpoS	σ^{S}	Stress
rpoH	σ^{32}	Heat shock
rpoE	σ^{E}	Heat shock
rpoN	σ^{54}	Nitrogen starvation
fliA	$\sigma^{28}(\sigma^{F})$	Flagellar synthesis

Figure 11.24 In addition to σ^{70}, *E. coli* has several sigma factors that are induced by particular environmental conditions.

Key Concepts

- *E. coli* has several sigma factors, each of which causes RNA polymerase to initiate at a set of promoters defined by specific −35 and −10 sequences.
- σ^{70} is used for general transcription, and the other sigma factors are activated by special conditions.
- A cascade of sigma factors is created when one sigma factor is required to transcribe the gene coding for the next sigma factor.
- Sigma factor cascades are used to control transcription in some bacteriophage infections and during bacterial sporulation.

Because sigma factor determines the choice of sites for initiation, it is possible to control the specificity of transcription by changing the sigma factor. **Figure 11.23** illustrates the basic idea. When the sigma factor is changed, RNA polymerase behaves in exactly the same way as before, except that it binds to a different promoter sequence.

E. coli uses alternative sigma factors to respond to general environmental changes. They are listed in **Figure 11.24**, and are named either by molecular weight of the product or for the gene. The general factor, responsible for transcription of most genes under normal conditions, is σ^{70}. The alternative sigma factors σ^{S}, σ^{32}, σ^{E}, and σ^{54} are activated in response to environmental changes; σ^{28} is used for expression of flagellar genes during normal growth, but its level of expression responds to changes in the environment. All the sigma factors except σ^{54} belong to the same protein family and function in the same general manner.

Temperature fluctuation is a common type of environmental challenge. Many organisms, both prokaryotic and eukaryotic, respond in a similar way. Upon an increase in temperature, synthesis of the proteins currently being made is turned off or down, and a new set of proteins is synthesized. The new proteins are the products of the *heat shock genes*. They play a role in protecting the cell against environmental stress, and are synthesized in response to other conditions as well as heat shock. In *E. coli*, the expression of 17 heat shock proteins is triggered by changes at transcription. The gene *rpoH* is a regulator needed to switch on the heat shock response. Its product is σ^{32}, which functions as an alternative sigma factor that causes transcription of the heat shock genes.

The heat shock response is accomplished by increasing the amount of σ^{32} when the temperature increases. The basic signal that induces production of σ^{32} is the accumulation of unfolded (partially denatured) proteins that results from increase in temperature. The σ^{32} protein is unstable, which is important in allowing its quantity to be increased or decreased rapidly. σ^{70} and σ^{32} can compete for the available core enzyme, so that the set of genes transcribed during heat shock depends on the balance between them.

More complex regulatory circuits can be constructed by forming cascades of sigma factors; in such a cascade, a set of genes activated by one sigma factor includes a gene coding for a sigma factor that in turn activates another set of genes. **Figure 11.25** shows how regulatory genes in bacteriophage SPO1 (which infects the bacterium *B. subtilis*) are joined into such a cascade.

Phage infection passes through three stages of gene expression. Immediately on infection, the **early genes** of the phage are transcribed by the holoenzyme of the host bacterium. They are essentially indistinguishable from host genes whose promoters have the intrinsic ability to be recognized by the RNA polymerase $\alpha_2\beta\beta'\sigma^{43}$. (The major sigma factor in *B. subtilis* is σ^{43}.) One of the early phage genes is called *28*. It codes for a sigma factor (gp28) that replaces the host sigma factor and causes the **middle genes** to be transcribed. Two of the middle genes (called *33* and *34*) code for the subunits of another protein that replaces gp28; this causes RNA polymerase to transcribe the **late genes**.

The successive replacements of sigma factor have dual consequences. Each time sigma changes, the RNA polymerase becomes able to recognize a new class of genes, *and* it no longer recognizes the previous class. These switches therefore constitute global changes in the activity of RNA polymerase. Probably all, or virtually all, of the core enzyme becomes associated with the sigma factor of the moment.

Perhaps the most extensive example of switches in sigma factors is provided by **sporulation**, an alternative lifestyle available to some bacteria. At the end of the **vegetative phase** in a bacterial culture, logarithmic growth ceases because nutrients in the medium become depleted. This triggers sporulation, in which DNA replication is followed by segregation of the daughter chromosomes between the mother cell and a spore. Sporulation drastically changes the biosynthetic activities of the bacterium, with many genes being involved. The basic level of control lies at transcription. Some of the genes that functioned in the vegetative phase are turned off during sporulation, but most continue to be expressed. In addition, the genes specific for sporulation are expressed only during this period. At the end of sporulation, ~40% of the bacterial mRNA is sporulation-specific.

New forms of the RNA polymerase become active in sporulating cells; they contain the same core enzyme as vegetative cells, but have different proteins in place of the vegetative σ^{43}. The changes in transcriptional specificity occur in both the mother cell and the spore. The principle is that in each compartment the existing sigma factor is successively displaced by a new factor that causes transcription of a different set of genes.

Each sigma factor causes RNA polymerase to initiate at a particular set of promoters. By analyzing the sequences of these promoters, we can show that each set is identified by unique sequence elements. Indeed, the sequence of each type of promoter ensures that it is recognized only by RNA polymerase directed by the appropriate sigma factor.

We can deduce the general rules for promoter recognition from the identification of the genes responding to the sigma factors found in *E. coli* and those involved in sporulation in *B. subtilis*. A significant feature of the promoters for each enzyme is that *they have the same size and location relative to the startpoint, and they show conserved sequences only around the usual centers of −35 and −10.*

As summarized in **Figure 11.26**, the consensus sequences for each set of promoters are different from

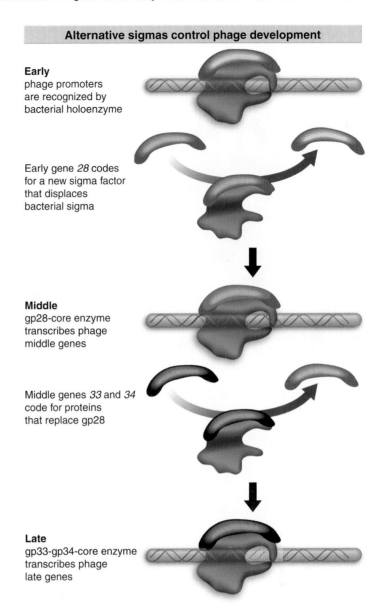

Alternative sigmas control phage development

Early
phage promoters are recognized by bacterial holoenzyme

Early gene *28* codes for a new sigma factor that displaces bacterial sigma

Middle
gp28-core enzyme transcribes phage middle genes

Middle genes *33* and *34* code for proteins that replace gp28

Late
gp33-gp34-core enzyme transcribes phage late genes

Figure 11.25 Transcription of phage SPO1 genes is controlled by two successive substitutions of the sigma factor that change the initiation specificity.

Sigma factors recognize promoters by consensus sequences				
Gene	Factor	−35 Sequence	Separation	−10 Sequence
rpoD	σ^{70}	TTGACA	16–18 bp	TATAAT
rpoH	σ^{32}	CCCTTGAA	13–15 bp	CCCGATNT
rpoN	σ^{54}	CTGGNA	6 bp	TTGCA
fliA	$\sigma^{28}(\sigma^{F})$	CTAAA	15 bp	GCCGATAA
sigH	σ^{H}	AGGANPuPu	11–12 bp	GCTGAATCA

Figure 11.26 *E. coli* sigma factors recognize promoters with different consensus sequences.

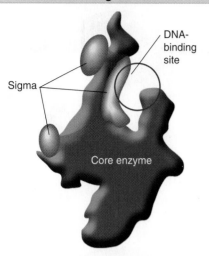

Figure 11.27 Sigma factor has an elongated structure that extends along the surface of the core subunits when the holoenzyme is formed.

one another at either or both of the −35 and −10 positions. This means that an enzyme containing a particular sigma factor can recognize only its own set of promoters, so that transcription of the different groups is mutually exclusive. Substitution of one sigma factor by another therefore turns off transcription of the old set of genes as well as turning on transcription of a new set of genes. (Some genes are expressed by RNA polymerases with alternative sigma factors because they have more than one promoter, each with a different set of consensus sequences.)

11.12 Sigma Factors Directly Contact DNA

Key Concepts

- σ^{70} changes its structure to release its DNA-binding regions when it associates with core enzyme.
- σ^{70} binds both the −35 and −10 sequences.

The influence of sigma factor on promoter recognition suggests that the sigma subunit must itself contact the promoter sequences. This requires different sigma factors all to bind to the same core enzyme in the same way, so that they are positioned to make critical contacts with the promoter sequences in the vicinity of −35 and −10.

Comparisons of the crystal structures of the core enzyme and holoenzyme show that sigma factor lies largely on the surface of the core enzyme. **Figure 11.27** shows that it has an elongated structure that extends past the DNA-binding site. This places it in a position to contact DNA during the initial binding. The DNA helix has to move 16 Å from the initial position to enter the active site. **Figure 11.28** illustrates this movement, looking in cross section down the helical axis of the DNA.

Direct evidence that sigma contacts the promoter directly at both the −35 and −10 consensus sequences is provided by mutations in sigma that suppress mutations in the consensus sequences. When a mutation at a particular position in the promoter prevents recognition by RNA polymerase, and a compensating mutation in sigma factor allows the polymerase to use the mutant promoter, the most likely explanation is that the relevant base pair in DNA is contacted by the amino acid that has been substituted.

Comparisons of the sequences of several bacterial sigma factors identify regions that have been conserved. Among them are regions that are especially important for the interaction with the DNA of the promoter. Two short regions (named 2.4 and 4.2) contact bases in the −10 and −35 elements, respectively. Both of these regions form short stretches of α-helix in the protein. They contact bases on the coding strand, and continue to hold these contacts after the DNA has been unwound in this region. This suggests that sigma factor could be important in the melting reaction.

The use of α-helical motifs in proteins to recognize duplex DNA sequences is common (see *14.10 Repressor Uses a Helix-Turn-Helix Motif to Bind DNA*). Amino acids separated by 3–4 positions lie on the same face of an α-helix and are therefore in a position to contact adjacent base pairs. **Figure 11.29** shows that amino acids lying along one face of the 2.4 region α-helix contact the bases at positions −12 to −10 of the −10 promoter sequence.

The N-terminal region of σ^{70} has important regulatory functions. If it is removed, the shortened protein becomes able to bind independently to promoter sequences. This

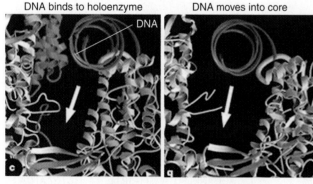

Figure 11.28 DNA initially contacts sigma factor (pink) and core enzyme (gray). It moves deeper into the core enzyme to make contacts at the −10 sequence. When sigma is released, the width of the passage containing DNA increases. Photograph courtesy of Roger Komberg, Dept. of Structural Biology, Stanford University School of Medicine.

suggests that the N-terminal region behaves as an autoinhibition domain. It occludes the DNA-binding domains when σ^{70} is free. Association with core enzyme changes the conformation of sigma so that the inhibition is released, and the DNA-binding domains can contact DNA.

Figure 11.30 schematizes the conformational change in sigma at open complex formation. When sigma binds to the core polymerase, the N-terminal domain swings ~20 Å away from the DNA-binding domains, and the DNA-binding domains separate from one another by ~15 Å, taking up a more elongated conformation. Mutations in either the −10 or −35 sequences prevent an (N-terminal-deleted) σ^{70} from binding to DNA, which suggests that σ^{70} contacts both sequences simultaneously.

In the free holoenzyme, the N-terminal domain is located in the active site of the core enzyme components, essentially mimicking the location that DNA will occupy when a transcription complex is formed. When the holoenzyme forms an open complex on DNA, the N-terminal sigma domain is displaced from the active site. Its relationship with the rest of the protein is therefore very flexible, and changes when sigma binds to core enzyme, and again when the holoenzyme binds to DNA.

11.13 There Are Two Types of Terminators in *E. coli*

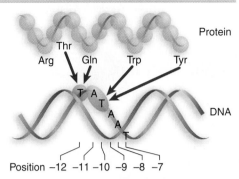

Figure 11.29 Amino acids in the 2.4 α-helix of σ^{70} contact specific bases in the coding strand of the −10 promoter sequence.

Key Terms

- A **terminator** is a sequence of DNA that causes RNA polymerase to terminate transcription.
- **Intrinsic terminators** are able to terminate transcription by bacterial RNA polymerase in the absence of any additional factors.
- **Rho-dependent terminators** are sequences that terminate transcription by bacterial RNA polymerase in the presence of the rho factor.
- **Rho (ρ) factor** is a protein that assists *E. coli* RNA polymerase to terminate transcription at certain terminators (called rho-dependent terminators).
- **Antitermination** is a mechanism of transcriptional control in which termination is prevented at a specific terminator site, allowing RNA polymerase to read into the genes beyond it.
- **Readthrough** at transcription or translation occurs when RNA polymerase or the ribosome, respectively, ignores a termination signal because of a mutation of the template or the behavior of an accessory factor.

Key Concept

- Termination may require both recognition of the terminator sequence in DNA and the formation of a hairpin structure in the RNA product.

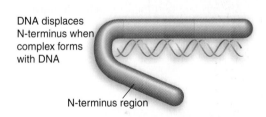

Figure 11.30 The N-terminus of sigma blocks the DNA-binding regions from binding to DNA. When an open complex forms, the N-terminus swings 20 Å away, and the two DNA-binding regions separate by 15 Å.

Once RNA polymerase has started transcription, it moves along the template, synthesizing RNA, until it meets a **terminator** (*t*) sequence. At this point, the enzyme stops adding nucleotides to the growing RNA chain, releases the completed product, and dissociates from the DNA template. Termination requires that all hydrogen bonds holding the RNA–DNA hybrid together must be broken, after which the DNA duplex re-forms.

There are two types of terminators in *E. coli*, and they are distinguished according to whether RNA polymerase requires any additional factors to terminate *in vitro*:

- Core enzyme can terminate at certain sites in the absence of any other factor. These sites are called **intrinsic terminators**.

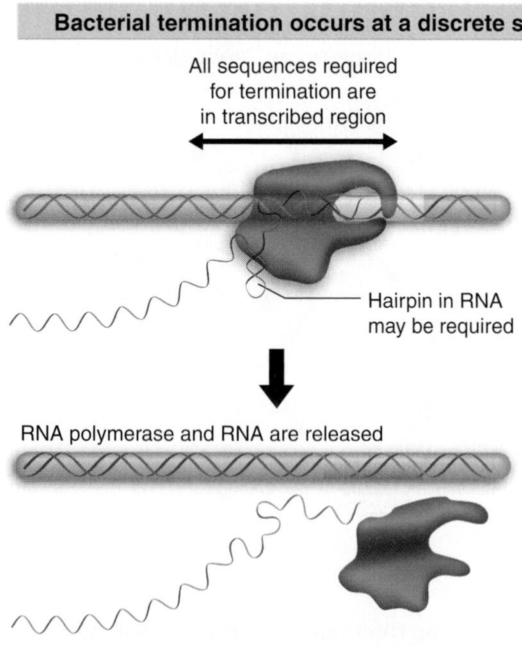

Bacterial termination occurs at a discrete site

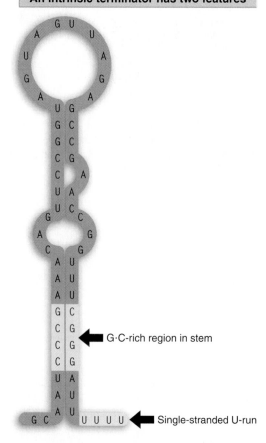

All sequences required
for termination are
in transcribed region

Hairpin in RNA
may be required

RNA polymerase and RNA are released

Figure 11.31 The DNA sequences required for termination are upstream of the terminator sequence. Formation of a hairpin in the RNA may be necessary.

An intrinsic terminator has two features

G·C-rich region in stem

Single-stranded U-run

Figure 11.32 Intrinsic terminators include palindromic regions that form hairpins varying in length from 7–20 bp. The stem-loop structure includes a G · C-rich region and is followed by a run of U residues.

- **Rho-dependent terminators** are defined by the need for addition of **rho (ρ) factor.**

Both types of terminators share the characteristics summarized in **Figure 11.31**:

- Terminator sequences are located before the point at which the last base is added to the RNA. The responsibility for termination lies with the *sequences already transcribed* by RNA polymerase. So termination relies on scrutiny of the template or product that the polymerase is currently transcribing.
- Terminators often require a hairpin to form in the secondary structure of the RNA being transcribed. This causes RNA polymerase to pause at the terminator sequence, allowing time for the termination event to occur.

Terminators vary widely in their efficiencies of termination. At some terminators, the termination event can be *prevented* by specific ancillary factors that interact with RNA polymerase. **Antitermination** causes the enzyme to continue transcription past the terminator sequence, an event called **readthrough** (the same term is used to describe a ribosome's suppression of termination codons).

In approaching the termination event, we must regard it not simply as a mechanism for generating the 3′ end of the RNA molecule, but as an opportunity to control gene expression. So the stages when RNA polymerase associates with DNA (initiation) or dissociates from it (termination) both are subject to specific control. There are interesting parallels between the systems employed in initiation and termination. Both require breaking of hydrogen bonds (initial melting of DNA at initiation, RNA–DNA dissociation at termination); and both require additional proteins to interact with the core enzyme.

11.14 Intrinsic Termination Requires a Hairpin and U-Rich Region

Key Concept

- Intrinsic terminators consist of a G·C-rich hairpin in the RNA product followed by a U-rich region in which termination occurs.

Intrinsic terminators have the two structural features evident in **Figure 11.32**: a hairpin in the secondary structure; and a region that is rich in U residues at the very end of the unit. Both features are needed for termination. The hairpin usually contains a G · C-rich region near the base of the stem. The typical distance between the hairpin and the U-rich region is 7–9 bases. About half of the genes in *E. coli* have intrinsic terminators.

The role of the hairpin in RNA is probably to cause RNA polymerase to slow, thus creating an opportunity for termination to occur. Pausing also occurs at sites that resemble terminators but have a greater separation (typically 10–11 bases) between the hairpin and the U-run. But if the pause site does not correspond to a terminator, usually the enzyme moves on again to continue transcription. The length of the pause varies, but at a typical terminator lasts ~60 seconds.

A downstream U-rich region destabilizes the RNA–DNA hybrid when RNA polymerase pauses at the terminator hairpin. The rU · dA RNA–DNA hybrid has an unusually weak base-paired structure; it requires the least energy of any RNA–DNA hybrid to break the association between the two strands. When the polymerase pauses, the RNA–DNA hybrid unravels from the weakly bonded rU · dA terminal region. Often the actual termination event takes place at any one of several positions toward or at the end of the U-rich region, as though the enzyme "stutters" during termination. The U-rich region in RNA corresponds to an A · T-rich region in DNA, so we see that A · T-rich regions are important in intrinsic termination as well as initiation.

Both the sequence of the hairpin and the length of the U-run influence the efficiency of termination, but termination efficiency (at least *in vitro*) varies 2–90%, and does not correlate in any simple way with the constitution of the hairpin or the number of U residues in the U-rich region. The hairpin and U-region are therefore necessary, but not sufficient, and additional parameters influence the interaction with RNA polymerase. In particular, the sequences both upstream and downstream of the intrinsic terminator influence its efficiency.

Less is known about the signals and ancillary factors of termination for eukaryotic polymerases. Each class of polymerase uses a different mechanism (see *Chapter 26 RNA Splicing and Processing*).

11.15 How Does Rho Factor Work?

Key Concept

- The termination factor *rho* is a helicase that binds to the C-rich, G-poor sequence of a *rut* site in nascent RNA, and then tracks along RNA to reach RNA polymerase, where it releases the RNA from the DNA template.

Rho factor is an essential protein in *E. coli* that functions solely at the stage of termination. It acts at rho-dependent terminators, which account for about half of *E. coli* terminators.

Figure 11.33 shows how rho functions. First it binds to a sequence within the transcript upstream of the site of termination. This sequence is called a *rut* site (an acronym for *rho utilization*). Then rho tracks along the RNA until it catches up to RNA polymerase. When the RNA polymerase reaches the termination site, rho acts on the RNA–DNA hybrid in the enzyme to cause release of the RNA. Pausing by the polymerase at the site of termination allows time for rho factor to translocate to the hybrid stretch, and is an important feature of termination.

We see an important general principle here. When we know the site on DNA at which some protein exercises its effect, we cannot assume that this coincides with the DNA sequence that it initially recognizes. They can be separate, and there need not be a fixed relationship between them. In fact, *rut* sites in different transcription units are found at varying distances preceding the sites of termination. A similar distinction is made by antitermination factors (see next section).

The common feature of *rut* sites is that the sequence is rich in C residues and poor in G residues, and has no secondary structure. An example is given in **Figure 11.34**; C is by far the most common base (41%) and G is the least common

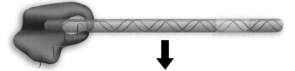

Rho terminates transcription

RNA polymerase transcribes DNA

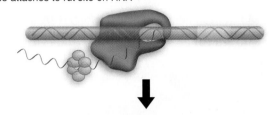

Rho attaches to *rut* site on RNA

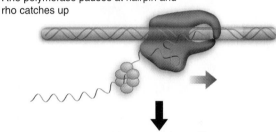

Rho polymerase pauses at hairpin and rho catches up

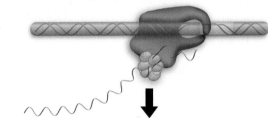

Rho translocates along RNA

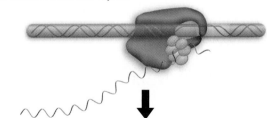

Rho unwinds DNA–RNA hybrid

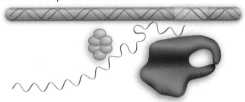

Termination: all components released

Figure 11.33 Rho factor binds to RNA at a *rut* site and translocates along RNA until it reaches the RNA–DNA hybrid in RNA polymerase, where it releases the RNA from the DNA.

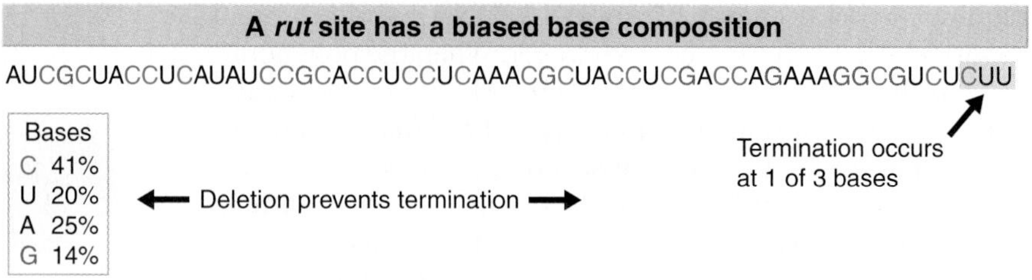

Figure 11.34 A sequence rich in C and poor in G, upstream of the termination site, is called a *rut* site and is discussed below. The sequence is at the 3′ end of the RNA.

A *rut* site has a biased base composition

AUCGCUACCUCAUAUCCGCACCUCCUCAAACGCUACCUCGACCAGAAAGGCGUCUCUU

Bases	
C	41%
U	20%
A	25%
G	14%

← Deletion prevents termination →

Termination occurs at 1 of 3 bases

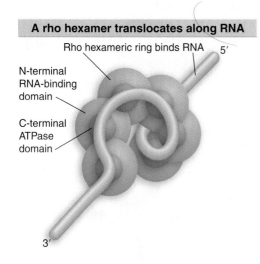

A rho hexamer translocates along RNA

Rho hexameric ring binds RNA 5′

N-terminal RNA-binding domain

C-terminal ATPase domain

3′

Figure 11.35 Rho has an N-terminal RNA-binding domain and a C-terminal ATPase domain. A hexamer in the form of a gapped ring binds RNA along the exterior of the N-terminal domains. The 5′ end of the RNA is bound by a secondary binding site in the interior of the hexamer.

base (14%). Rho is a member of the family of hexameric ATP-dependent helicases. The subunit has an RNA-binding domain and an ATP hydrolysis domain. The hexamer functions by passing nucleic acid through the hole in the middle of the assembly formed from the RNA-binding domains of the subunits.

Figure 11.35 shows that rho binds the RNA on the surface of the N-terminal domains from one end, pushing the other end of the bound stretch into the interior of the hexamer, where it binds to the C-terminal domains. When it reaches the RNA–DNA hybrid region at the point of transcription, rho uses its helicase activity to unwind the duplex structure and to release the RNA.

Some rho mutations can be suppressed by mutations in other genes. Studying the mutants is an excellent way to identify proteins that interact with rho, and implicates the β subunit of RNA polymerase as interacting with the factor.

11.16 Antitermination Is a Regulatory Event

Key Terms

- **Antitermination** is a mechanism of transcriptional control in which termination is prevented at a specific terminator site, allowing RNA polymerase to read into the genes beyond it.
- **Antitermination proteins** allow RNA polymerase to transcribe through certain terminator sites.

Key Concepts

- Termination is prevented when antitermination proteins act on RNA polymerase to cause it to read through a specific terminator or terminators.
- Phage lambda has two antitermination proteins, pN and pQ, that act on different transcription units.
- The site where an antiterminator protein acts is upstream of the terminator site in the transcription unit.
- The location of the antiterminator site varies in different cases, and can be in the promoter or within the transcription unit.

Antitermination is a control mechanism used in both phage regulatory circuits and bacterial operons. **Figure 11.36** shows that antitermination controls the ability of the enzyme to read past a terminator into genes lying beyond. In the example shown in the figure, the default pathway is for RNA polymerase to terminate at the end of region 1. But antitermination allows it to continue transcription through region 2. Because the promoter does not change, both situations produce an RNA with the same 5′ sequences; the difference is that after antitermination the RNA is extended to include new sequences at the 3′ end.

Termination of transcript can be regulated

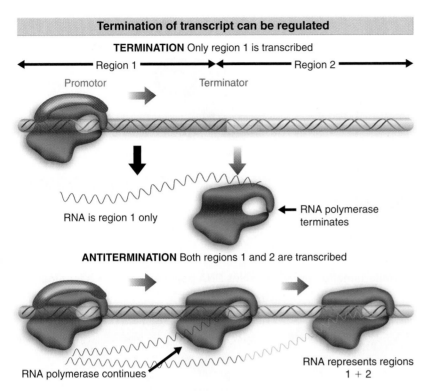

TERMINATION Only region 1 is transcribed

Region 1   Region 2

Promotor   Terminator

RNA is region 1 only

RNA polymerase terminates

ANTITERMINATION Both regions 1 and 2 are transcribed

RNA polymerase continues

RNA represents regions 1 + 2

Figure 11.36 Antitermination can control transcription by determining whether RNA polymerase terminates or reads through a particular terminator into the following region.

Antitermination was discovered in bacteriophage infections. A common feature in the control of phage infection is that very few of the phage genes (the "early" genes) can be transcribed by the bacterial host RNA polymerase. Among these genes, however, are regulator(s) whose product(s) allow the next set of phage genes to be expressed (see *14.4 Two Types of Regulatory Event Control the Lytic Cascade*). One of these types of regulator is an **antitermination protein**. As **Figure 11.37** shows, this protein enables RNA polymerase to read through a terminator, extending the RNA transcript. In the absence of the antitermination protein, RNA polymerase terminates at the terminator (top panel). When the antitermination protein is present, it continues past the terminator (lower panel).

The best characterized example of antitermination is provided by phage lambda, with which the phenomenon was discovered. It is used at two stages of phage expression. The antitermination protein produced at each stage is specific for the particular transcription units that are expressed at that stage, as summarized in the table of Figure 11.37.

The host RNA polymerase initially transcribes two genes, which are called the *immediate early* genes. The transition to the next stage of expression is controlled by preventing termination at the ends of the immediate early genes, with the result that the *delayed early* genes are expressed. (We discuss the overall regulation of lambda development in *Chapter 14 Phage Strategies*.) The antitermination protein pN acts specifically on the immediate early transcription units. Later during infection, another antitermination protein, pQ, acts specifically on the late transcription unit, to allow its transcription to continue past a terminator sequence.

The different specificities of pN and pQ establish an important general principle: *RNA polymerase interacts with transcription units in such a way that an ancillary factor can sponsor antitermination specifically*

Figure 11.37 An antitermination protein can act on RNA polymerase to enable it to read through a specific terminator.

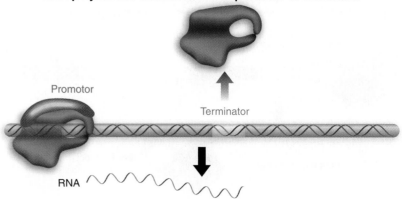

Antitermination extends the transcription unit

RNA polymerase transcribes from promoter to terminator

Promotor

Terminator

RNA

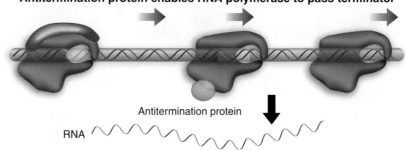

Antitermination protein enables RNA polymerase to pass terminator

Antitermination protein

RNA

Antitermination proteins act on specific terminators

Transcription unit	Promotor	Terminator	Antitermination protein
Immediate early	P_L	t_L	pN
Immediate early	P_{R1}	t_{R1}	pN
Late	$P_{R'}$	$t_{R'}$	pQ

for some transcripts. Termination can be controlled with the same sort of precision as initiation.

What sites are involved in controlling the specificity of antitermination? The antitermination activity of pN is highly specific, but *the antitermination event is not determined by the terminators (t_{L1} and t_{R1}); the recognition site needed for antitermination lies upstream in the transcription unit—that is, at a different place from the terminator site at which the action eventually is accomplished.*

The recognition sites required for pN action are called *nut* (for *N utilization*). The sites responsible for determining leftward and rightward antitermination are described as *nutL* and *nutR*, respectively. **Figure 11.38** shows that their locations relative to each transcription unit are quite different. *nutL* lies between the startpoint of P_L and the beginning of the *N* coding region. By contrast, *nutR* lies between the end of the *cro* gene and t_{R1}. So *nutL* is near the promoter, but *nutR* is near the terminator. (*qut* is different yet again, and lies within the promoter.)

How does antitermination occur? When pN recognizes the *nut* site, it must act on RNA polymerase to ensure that the enzyme can no longer respond to the terminator. The variable locations of the *nut* sites indicate that this event is linked neither to initiation nor to termination, but can occur to RNA polymerase as it elongates the RNA chain past the *nut* site. As illustrated in **Figure 11.39**, the polymerase then becomes a juggernaut that continues past the terminator, heedless of its signal.

Is the ability of pN to recognize a short sequence within the transcription unit an example of a more widely used mechanism for antitermination? Other phages, related to lambda, have different *N* genes and different antitermination specificities. The region of the phage genome in which the *nut* sites lie has a different sequence in each of these phages, and each phage must therefore have characteristic *nut* sites recognized specifically by its own pN. Each of these pN products must have the same general ability to interact with the transcription apparatus in an antitermination capacity, but has a different specificity for the sequence of DNA that activates the mechanism.

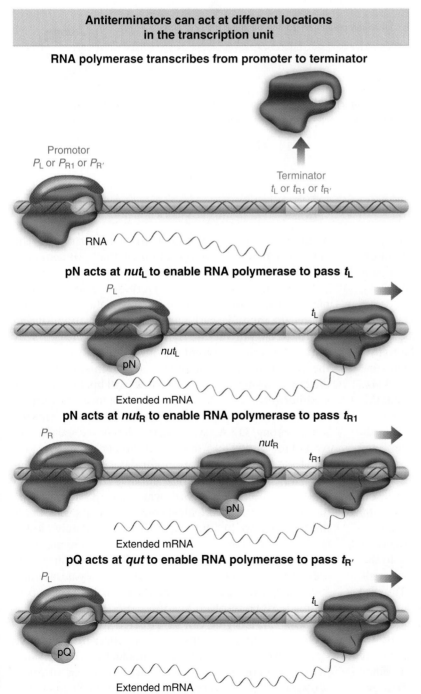

Antiterminators can act at different locations in the transcription unit

RNA polymerase transcribes from promoter to terminator

Promotor
P_L or P_{R1} or $P_{R'}$

Terminator
t_L or t_{R1} or $t_{R'}$

RNA

pN acts at *nut*$_L$ to enable RNA polymerase to pass t_L

P_L

t_L

nut$_L$

pN

Extended mRNA

pN acts at *nut*$_R$ to enable RNA polymerase to pass t_{R1}

P_R

nut$_R$

t_{R1}

pN

Extended mRNA

pQ acts at *qut* to enable RNA polymerase to pass $t_{R'}$

P_L

t_L

pQ

Extended mRNA

Figure 11.38 Host RNA polymerase transcribes lambda genes and terminates at *t* sites. pN allows it to read through terminators in the L and R1 units; pQ allows it to read through the R′ terminator. The sites at which pN acts (*nut*) and at which pQ acts (*qut*) are located at different relative positions in the transcription units.

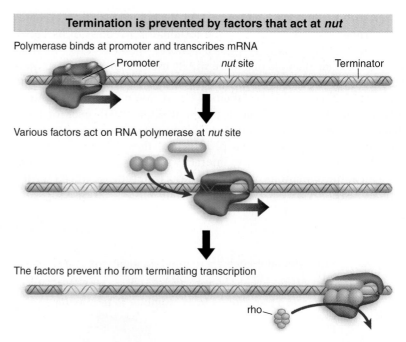

Termination is prevented by factors that act at *nut*

Polymerase binds at promoter and transcribes mRNA

Promoter   *nut* site   Terminator

Various factors act on RNA polymerase at *nut* site

The factors prevent rho from terminating transcription

rho

Figure 11.39 Ancillary factors bind to RNA polymerase as it passes the *nut* site. They prevent rho from causing termination when the polymerase reaches the terminator.

11.17 SUMMARY

A transcription unit comprises the DNA between a promoter, where transcription initiates, and a terminator, where it ends. One strand of the DNA in this region serves as a template for synthesis of a complementary strand of RNA. The RNA–DNA hybrid region is short and transient, contained in a transcription "bubble" that moves along DNA. The RNA polymerase holoenzyme that synthesizes bacterial RNA can be separated into two components. Core enzyme is a multimer of structure $\alpha_2\beta\beta'$ that is responsible for elongating the RNA chain. Sigma factor (σ) is a single subunit that is required at the stage of initiation for recognizing the promoter.

Core enzyme has a general affinity for DNA. The addition of sigma factor reduces the affinity of the enzyme for nonspecific binding to DNA, but increases its affinity for promoters. The rate at which RNA polymerase finds its promoters is too great to be accounted for by diffusion and random contacts with DNA; direct exchange of DNA sequences held by the enzyme may be involved.

Bacterial promoters are identified by two short conserved sequences centered at -35 and -10 relative to the startpoint. Most promoters have sequences that are well related to the consensus sequences at these sites. The distance separating the consensus sequences is 16–18 bp. RNA polymerase initially "touches down" at the -35 sequence and then extends its contacts over the -10 region. The enzyme covers ~77 bp of DNA. The initial "closed" binary complex is converted to an "open" binary complex by melting of a sequence of ~12 bp that ex-

tends from the -10 region to the startpoint. The A·T-rich base pair composition of the -10 sequence may be important for the melting reaction.

The binary complex is converted to a ternary complex by the incorporation of ribonucleotide precursors. There are multiple cycles of abortive initiation, during which RNA polymerase synthesizes and releases very short RNA chains without moving from the promoter. At the end of this stage, there is a change in structure, and the core enzyme contracts to cover ~50 bp. Sigma factor is either released (30% of cases) or changes its form of association with the core enzyme. Then core enzyme moves along DNA, synthesizing RNA. A locally unwound region of DNA moves with the enzyme.

The "strength" of a promoter describes the frequency at which RNA polymerase initiates transcription; it is related to the closeness with which its -35 and -10 sequences conform to the ideal consensus sequences, but is influenced also by the sequences immediately downstream of the startpoint. Negative supercoiling increases the strength of certain promoters. Transcription generates positive supercoils ahead of RNA polymerase and leaves negative supercoils behind the enzyme. The supercoiling must be resolved by topoisomerases.

The core enzyme can be directed to recognize promoters with different consensus sequences by alternative sigma factors. In *E. coli*, these sigma factors are activated by adverse conditions, such as heat shock or nitrogen starvation. *B. subtilis* contains a single major sigma factor with the same specificity as the *E. coli* sigma factor, and also contains a variety of minor sigma factors. Another series of factors is activated when sporulation is initiated; sporulation is regulated by two cascades in which sigma factor replacements occur in the daughter cell and mother cell. A cascade for regulating transcription by substitution of sigma factors is also used by phage SPO1.

The geometry of RNA polymerase-promoter recognition is similar for holoenzymes containing all sigma factors (except σ^{54}). Each sigma factor causes RNA polymerase to initiate transcription at a promoter that conforms to a particular consensus at -35 and -10. Direct contacts between sigma and DNA at these sites have been demonstrated for *E. coli* σ^{70}. The σ^{70} factor of *E. coli* has an N-terminal autoinhibitory domain that prevents the DNA-binding regions from recognizing DNA. The autoinhibitory region is displaced by DNA when the holoenzyme forms an open complex.

Bacterial RNA polymerase terminates transcription at two types of sites. Intrinsic terminators contain a G·C-rich hairpin followed by a U-rich region. They are recognized by core enzyme alone. Rho-dependent terminators require rho factor. Rho binds to *rut* sites that are rich in C and poor in G residues and that precede the actual site of termination. Rho provides a hexameric ATP-dependent helicase activity that translocates along the RNA until it reaches the RNA–DNA hybrid region in the transcription bubble of RNA polymerase, where it dissociates the RNA from DNA. In both types of termination, pausing by RNA polymerase is important in providing time for the actual termination event to occur.

Antitermination is used by some phages to regulate progression from one stage of gene expression to the next. The lambda gene *N* codes for an antitermination protein (pN) that is necessary to allow RNA polymerase to read through the terminators located at the ends of the immediate early genes. Another antitermination protein, pQ, is required later in phage infection. pN and pQ (*nut* and *qut* respectively) act on RNA polymerase as it passes specific sites. These sites are located at different relative positions in their respective transcription units.

The Operon

12.1 Introduction

Key Terms

- A *trans*-acting product can function on any copy of its target DNA (or RNA). This implies that it is a diffusible protein or RNA.
- A *cis*-acting site affects the activity only of sequences on its own molecule of DNA (or RNA). This property usually implies that the site does not code for protein.
- A **structural gene** codes for any RNA or protein product other than a regulator.
- A **regulator gene** codes for a product (typically protein) that controls the expression of other genes (usually at the level of transcription).
- Genes that are subject to **negative control** are turned on unless a specific event occurs to turn them off.
- A **repressor** is a protein that inhibits expression of a gene. It may act to prevent transcription by binding to an operator site in DNA, or to prevent translation by binding to RNA.
- The **operator** is the site on DNA at which a repressor protein binds to prevent transcription from initiating at the adjacent promoter.
- **Positive control** describes a system in which a gene is not expressed unless some action occurs to turn it on.
- A **transcription factor** is required for RNA polymerase to initiate transcription at specific promoter(s), but is not itself part of the enzyme.

The basic concept for how transcription is controlled in bacteria was provided by the classic formulation of the model for control of gene expression by Jacob and Monod in 1961. They distinguished between two types of sequences in DNA: sequences that code for ***trans*-acting** products; and ***cis*-acting** sequences that function exclusively within the DNA. Gene activity is regulated by the specific interactions of the *trans*-acting products (usually proteins) with the *cis*-acting sequences (usually sites in DNA). In more formal terms:

- A gene is a sequence of DNA that codes for a diffusible product. This product may be protein (as in the case of the majority of genes) or may be RNA (as in the case of genes that code for tRNA and rRNA). *The crucial feature is that the product diffuses away from its site of synthesis to act elsewhere.* Any gene product that is free to diffuse away from the gene that made it to function elsewhere is described as *trans*-acting.
- The description *cis*-acting applies to any sequence of DNA that is not converted into any other form, but that functions exclusively as a DNA sequence *in situ*, affecting only the DNA to which it is physically linked. (In some cases, a *cis*-acting sequence functions in an RNA rather than in a DNA molecule.)

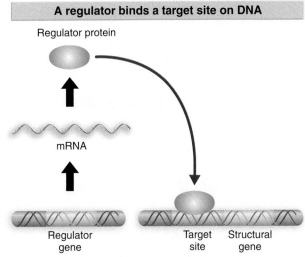

A regulator binds a target site on DNA

Regulator protein

mRNA

Regulator gene   Target site   Structural gene

Figure 12.1 A regulator gene codes for a protein that acts at a target site on DNA.

To help distinguish between the components of regulatory circuits and the genes they regulate, we sometimes use the terms *structural gene* and *regulator gene*. A **structural gene** is simply any gene that codes for a protein (or RNA) product. Structural genes represent an enormous variety of protein structures and functions, including structural proteins, enzymes with catalytic activities, and regulatory proteins. A **regulator gene** is a gene that codes for a protein (or an RNA) involved in regulating the expression of other genes.

The simplest form of the regulatory model is illustrated in **Figure 12.1**: *a regulator gene codes for a protein that controls transcription by binding to a particular site on DNA.* This interaction can regulate a target gene in either a positive manner (the interaction turns the gene on) or in a negative manner (the interaction turns the gene off). The sites on DNA are usually (but not exclusively) located just upstream of the target gene.

The sequences that mark the beginning and end of the transcription unit, the promoter and terminator, are examples of *cis*-acting sites. *A promoter serves to initiate transcription only of the gene or genes physically connected to it on the same stretch of DNA.* In the same way, a terminator can terminate transcription only by an RNA polymerase that has traversed the preceding gene(s).

About half the genes in bacteria are regulated by **negative control**, in which a **repressor** protein prevents a gene from being expressed. **Figure 12.2** shows that the "default state" for such a gene is to be expressed via the recognition of its promoter by RNA polymerase. Close to the promoter is another *cis*-acting site called the **operator**, which is the target for the repressor protein. When the repressor binds to the operator, RNA polymerase is prevented from initiating transcription, and *gene expression is therefore turned off.*

Positive control is used in bacteria with about equal frequency to negative control, and it is the most common mode of control in eukaryotes. A **transcription factor** is required to assist RNA polymerase in initiating at the promoter. **Figure 12.3** shows that the typical default state of a eukaryotic gene is inactive: RNA polymerase cannot by itself initiate transcription at the promoter. Several *trans*-acting factors have target sites in the vicinity of the promoter, and *binding of some or all of these factors enables RNA polymerase to initiate transcription.*

The unifying theme is that regulatory proteins are *trans*-acting factors that recognize *cis*-acting elements (usually) upstream of the gene.

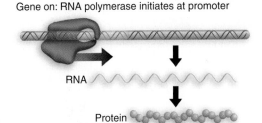

A repressor blocks RNA polymerase

cis-acting operator/promoter precedes structural gene(s)

Promoter Operator   Structural gene(s)

Gene on: RNA polymerase initiates at promoter

RNA

Protein

Gene is turned off when repressor binds to operator

Figure 12.2 In negative control, a *trans*-acting repressor binds to the *cis*-acting operator to turn off transcription.

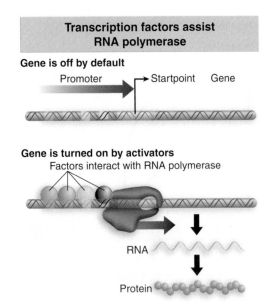

Transcription factors assist RNA polymerase

Gene is off by default

Promoter → Startpoint Gene

Gene is turned on by activators

Factors interact with RNA polymerase

RNA

Protein

Figure 12.3 In positive control, *trans*-acting factors must bind to *cis*-acting sites in order for RNA polymerase to initiate transcription at the promoter.

The consequences of this recognition are to activate or to repress the gene, depending on the type of regulatory protein. A typical feature is that regulator proteins function by recognizing very short sequences in DNA, usually <10 bp in length, although the protein actually binds over a somewhat greater distance of DNA. The bacterial promoter is an example: although RNA polymerase covers >70 bp of DNA at initiation, the crucial sequences that it recognizes are the hexamers (6 bp sequences) centered at −35 and −10.

Going beyond the interactions in which a protein or RNA regulates expression of a single gene, we find that bacteria have responses in which the expression of many genes is coordinated. The simplest form of regulating multiple genes occurs when they all have a copy of the same *cis*-acting regulatory elements. Genes that have a copy of the same operator sequence are coordinately repressed by a single repressor protein. Genes that have the same type of promoter are coordinately activated by a factor that causes RNA polymerase to use that promoter.

12.2 Structural Gene Clusters Are Coordinately Controlled

Key Term

- An **operon** is a unit of bacterial gene expression and regulation, including structural genes and control elements in DNA recognized by regulator gene product(s).

Key Concept

- Genes coding for proteins that function in the same pathway may be located adjacent to one another and controlled as a single unit that is transcribed into a polycistronic mRNA.

Bacterial structural genes are often organized into clusters that include genes coding for proteins whose functions are related. It is common for the genes coding for the enzymes of a metabolic pathway to be organized into such a cluster. In addition to the enzymes actually involved in the pathway, other related activities may be included in the unit of coordinate control; for example, the unit may include the gene for the protein that transports the small molecule substrate into the cell. Clustering of structural genes allows them to be coordinately controlled by interactions at a single promoter: as a result of these interactions, the entire set of genes is either transcribed or not transcribed.

The cluster of the three *lac* structural genes, *lacZYA*, is typical. **Figure 12.4** summarizes the organization of the structural genes, their associated *cis*-acting regulatory elements, and the *trans*-acting regulatory gene (the repressor). *The key feature is that the cluster is transcribed into a single polycistronic mRNA from a promoter where initiation of transcription is regulated.*

The protein products enable cells to take up and metabolize β-galactosides, such as lactose. The roles of the three structural genes are:

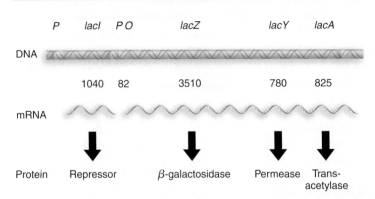

The *lac* operon includes *cis*-acting regulator elements and protein-coding structural genes

| P | lacI | P O | lacZ | lacY | lacA |

DNA

1040 82 3510 780 825

mRNA

Protein: Repressor β-galactosidase Permease Trans-acetylase

Figure 12.4 The *lac* operon occupies ~6000 bp of DNA. At the left the *lacI* gene has its own promoter and terminator. The end of the *lacI* region is adjacent to the promoter, *P*. The operator, *O*, occupies the first 26 bp of the transcription unit. The long *lacZ* gene starts at base 39, and is followed by the *lacY* and *lacA* genes and a terminator.

- *lacZ* codes for the enzyme β-galactosidase, whose active form is a tetramer of ~500 kD. The enzyme breaks a β-galactoside into its component sugars. For example, lactose is cleaved into glucose and galactose (which are then further metabolized).
- *lacY* codes for the β-galactoside permease, a 30 kD membrane-bound protein constituent of the transport system. This transports β-galactosides into the cell.
- *lacA* codes for β-galactoside transacetylase, an enzyme that transfers an acetyl group from acetyl-CoA to β-galactosides.

Mutations in either *lacZ* or *lacY* can create the *lac* genotype, in which cells cannot utilize lactose. (The genotypic description "*lac*" without a qualifier indicates loss-of-function.) The *lacZ* mutations abolish enzyme activity, directly preventing metabolism of lactose. The *lacY* mutants cannot take up lactose from the medium. The role of *lacA* is still not understood.

The entire system, including structural genes and the elements that control their expression, forms a common unit of regulation; this is called an **operon**. The activity of the operon is controlled by regulator gene(s), whose protein products interact with the *cis*-acting control elements.

12.3 The *lac* Genes Are Controlled by a Repressor

Key Concepts

- Transcription of the *lacZYA* gene cluster is controlled by a repressor protein that binds to an operator that overlaps the promoter at the start of the cluster.
- The repressor protein is a tetramer of identical subunits coded by the gene *lacI*.

Key Term

- The default state of genes that are controlled by **negative regulation** is to be expressed. A specific intervention is required to turn them off.

We can distinguish between structural genes and regulator genes by the effects of mutations. A deleterious mutation in a structural gene deprives the cell of the particular protein for which the gene codes. But a mutation in a regulator gene influences the expression of all the structural genes that it controls. The consequences of a regulatory mutation reveal the type of regulation.

Transcription of the *lacZYA* genes is controlled by a regulator protein synthesized by the *lacI* gene. It happens that *lacI* is located adjacent to the structural genes, but it comprises an independent transcription unit with its own promoter and terminator. Because *lacI* specifies a diffusible product, in principle it need not be located near the structural genes; it can function equally well if moved elsewhere, or carried on a separate DNA molecule (the classic test for a *trans*-acting regulator).

The *lac* genes are controlled by **negative regulation**: *they are transcribed unless turned off by the regulator protein.* A mutation that inactivates the regulator causes the structural genes to remain in the expressed condition. The product of *lacI* is called the *Lac repressor*, because its function is to prevent the expression of the structural genes.

The repressor functions by binding to an operator (formally denoted O_{lac}) at the start of the *lacZYA* cluster. The operator lies between the promoter (P_{lac}) and the structural genes (*lacZYA*). *When the repressor binds at the operator, it prevents RNA polymerase from initiating transcription at the promoter.* The repressor is a tetramer of identical subunits of 38 kD each, and there are ~10 tetramers in a wild-type cell.

Figure 12.5 expands our view of the region at the start of the *lac* structural genes. The operator extends from position −5 just upstream of the mRNA startpoint to position +21 within the transcription unit.

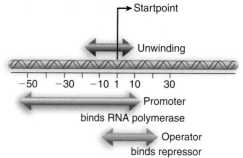

Figure 12.5 Repressor and RNA polymerase bind at sites that overlap around the transcription startpoint of the *lac* operon.

So it overlaps the right end of the promoter. We discuss the relationship between repressor and RNA polymerase in more detail in *12.10 Repressor Binds to Three Operators and Interacts with RNA Polymerase.*

12.4 The *lac* Operon Can Be Induced

Key Terms

- **Induction** refers to the ability of bacteria (or yeast) to synthesize certain enzymes only when their substrates are present; applied to gene expression, it refers to switching on transcription as a result of interaction of the inducer with the regulator protein.
- The level of response from a system in the absence of a stimulus is its **basal level**. The basal level of transcription of a gene is the level in the absence of any specific activation.
- **Repression** describes the ability of bacteria to prevent synthesis of certain enzymes when their products are present; more generally, it refers to inhibition of transcription (or translation) by binding of repressor protein to a specific site on DNA (or mRNA).
- An **inducer** is a small molecule that triggers gene transcription by binding to a regulator protein.
- A **corepressor** is a small molecule that triggers repression of transcription by binding to a regulator protein.

Key Concepts

- Small molecules that induce an operon are identical with or related to the substrate for its enzymes.
- β-galactosides are the substrates for the enzymes coded by *lacZYA*.
- In the absence of β-galactosides, the *lac* operon is expressed only at a very low (basal) level.
- Addition of β-galactosides induces transcription of all three genes of the operon.
- Because the *lac* mRNA is extremely unstable, induction can be rapidly reversed.
- The same types of systems that allow substrates to induce operons coding for metabolic enzymes can be used to allow end products to repress the operons that code for biosynthetic enzymes.

Bacteria need to respond swiftly to changes in their environment. The supply of nutrients can fluctuate at any time; survival depends on the ability to switch from metabolizing one substrate to another. Yet economy also is important because a bacterium that indulges in energetically expensive ways to meet the demands of the environment is likely to be at a disadvantage. So a bacterium avoids synthesizing the enzymes of a pathway in the absence of the substrate, but is ready to produce the enzymes if the substrate should appear.

The synthesis of enzymes in response to the appearance of a specific substrate is called **induction**. This type of regulation is widespread in bacteria, and occurs also in unicellular eukaryotes (such as yeasts). The *lactose* system of *E. coli* provides the paradigm for this sort of control mechanism.

When cells of *E. coli* are grown in the absence of a β-galactoside, there is no need for β-galactosidase, and the cells contain very few molecules of the enzyme—say, <5. When a suitable substrate is added, the enzyme activity appears very rapidly in the bacteria. Within 2–3 minutes some enzyme is present, and soon there are ~5000 molecules of enzyme per bacterium. (Under suitable conditions, β-galactosidase can account for 5–10% of the total soluble protein of the bacterium.) If the substrate is removed from the medium, the synthesis of enzyme stops as rapidly as it had originally started.

Figure 12.6 summarizes the essential features of induction. In the absence of inducer, the operon is transcribed at a very low **basal level**. Control of transcription of the *lac* genes responds very rapidly to the inducer, as shown in the upper part of the figure. Transcription is stimulated as soon as inducer is added; the amount of *lac* mRNA increases rapidly to an induced level that reflects a balance between synthesis and degradation of the mRNA.

The *lac* mRNA is extremely unstable, and decays with a half-life of only ~3 minutes. This feature allows induction to be reversed rapidly. Transcription ceases as soon as the inducer is removed; and in a very short time all the *lac* mRNA has been destroyed, and the cellular content has returned to the basal level.

The production of protein is followed in the lower part of the figure. Translation of the *lac* mRNA produces β-galactosidase (and the products of the other *lac* genes). There is a short lag between the appearance of *lac* mRNA and the appearance of the first completed enzyme molecules (it is ~2 minutes after the rise of mRNA from basal level before protein begins to increase). There is a similar lag between reaching maximal induced levels of mRNA and protein. When inducer is removed, synthesis of enzyme ceases almost immediately (as the mRNA is degraded), but the β-galactosidase in the cell is more stable than the mRNA, so the enzyme activity remains at the induced level for longer.

This type of rapid response to changes in nutrient supply not only provides the ability to metabolize new substrates, but also is used to shut off endogenous synthesis of compounds that suddenly appear in the medium. For example, *E. coli* synthesizes the amino acid tryptophan through the action of the enzyme tryptophan synthetase. But if tryptophan is provided in the medium on which the bacteria are growing, the production of the enzyme is immediately halted. This effect is called **repression**. It allows the bacterium to avoid spending resources on unnecessary synthetic activities.

Induction and repression represent the same phenomenon. In one case the bacterium adjusts its ability to use a given substrate (such as lactose) for growth; in the other it adjusts its ability to synthesize a particular metabolic intermediate (such as an essential amino acid). The trigger in the two cases is the small molecule that is respectively the substrate for the enzyme or the product of the enzyme activity. Small molecules that cause the production of enzymes able to metabolize them are called **inducers**. Those that prevent the production of enzymes able to synthesize them are called **corepressors**.

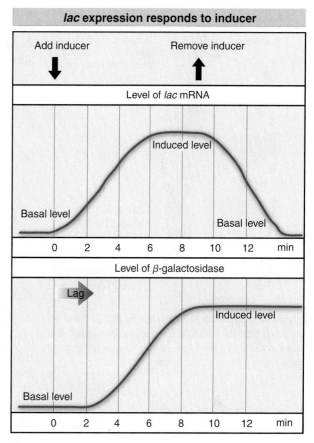

Figure 12.6 Addition of inducer results in rapid induction of *lac* mRNA, and is followed after a short lag by synthesis of the enzymes; removal of inducer is followed by rapid cessation of synthesis.

12.5 Repressor Is Controlled by a Small Molecule Inducer

Key Terms

- **Gratuitous inducers** resemble authentic inducers of transcription but are not substrates for the induced enzymes.
- **Allosteric regulation** describes the ability of a protein to change its conformation (and therefore activity) at one site as the result of binding a small molecule to a second site located elsewhere on the protein.
- **Coordinate regulation** refers to the common control of a group of genes.

Key Concepts

- An inducer functions by converting the repressor protein into an inactive form.
- Repressor protein has two binding sites, one for the operator and another for the inducer.
- Repressor is inactivated by an allosteric interaction in which binding of inducer at its site changes the properties of the DNA-binding site.

A repressor tetramer binds the operator

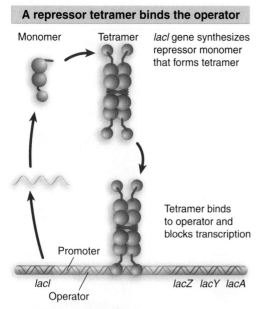

Monomer Tetramer *lacI* gene synthesizes repressor monomer that forms tetramer

Tetramer binds to operator and blocks transcription

Promoter

lacI *lacZ lacY lacA*

Operator

Figure 12.7 Repressor maintains the *lac* operon in the inactive condition by binding to the operator. The shape of the repressor is represented as a series of connected domains as revealed by its crystal structure.

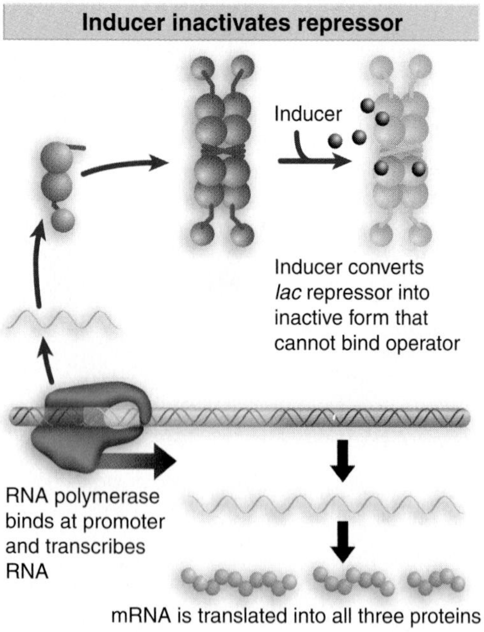

Inducer inactivates repressor

Inducer

Inducer converts *lac* repressor into inactive form that cannot bind operator

RNA polymerase binds at promoter and transcribes RNA

mRNA is translated into all three proteins

Figure 12.8 Addition of inducer converts repressor to an inactive form that cannot bind the operator. This allows RNA polymerase to initiate transcription.

The ability to act as inducer or corepressor is highly specific. Only the substrate/product or a closely related molecule can serve. *But the activity of the small molecule does not depend on its interaction with the target enzyme.* Some inducers resemble the natural inducers of the *lac* operon, but cannot be metabolized by the enzyme. These **gratuitous inducers** are extremely useful because they remain in the cell in their original form. (A real inducer would be metabolized, interfering with study of the system.) The gratuitous inducer used in the *lac* system is isopropylthiogalactoside (IPTG).

The component that responds to the inducer is the repressor protein coded by *lacI*. The *lacZYA* structural genes are transcribed into a single mRNA from a promoter just upstream of *lacZ*. The state of the repressor determines whether this promoter is turned off or on:

- **Figure 12.7** shows that in the absence of an inducer, the genes are not transcribed, because repressor protein is in an active form that is bound to the operator.
- **Figure 12.8** shows that when an inducer is added, the repressor is converted into an inactive form that leaves the operator. Then transcription starts at the promoter and proceeds through the genes to a terminator located beyond the 3′ end of *lacA*.

The crucial features of the control circuit reside in the dual properties of the repressor: it can prevent transcription; and it can recognize the small-molecule inducer. The repressor has two binding sites, one for the operator and one for the inducer. When the inducer binds at its site, it changes the conformation of the protein in such a way as to influence the activity of the operator-binding site. The ability of one site in the protein to control the activity of another is called **allosteric regulation**.

Induction accomplishes a **coordinate regulation**: *all the genes are expressed (or not expressed) in unison.* The mRNA is translated sequentially from its 5′ end, which explains why induction always causes the appearance of β-galactosidase, β-galactoside permease, and β-galactoside transacetylase, in that order. Translation of a common mRNA ensures that all of the necessary enzymes are available together, and explains why the relative amounts of the three enzymes always remain the same under varying conditions of induction.

12.6 *cis*-Acting Constitutive Mutations Identify the Operator

Key Terms

- An **uninducible** mutant is one in which the affected gene(s) cannot be expressed.
- A **constitutive** process is one that occurs all the time, unchanged by any form of stimulus or external condition.

Key Concepts

- Mutations in the operator can cause constitutive expression of all three *lac* structural genes.
- Operator mutations are *cis*-acting and affect only those genes on the contiguous stretch of DNA.

Mutations in the regulatory circuit may either abolish expression of the operon or cause unregulated expression. Mutants that cannot be expressed at all are called **uninducible**. The continued functioning of a gene that does not respond to regulation is called **constitutive** gene expression.

Components of the regulatory circuit of the operon can be identified by mutations that *affect the expression of all the structural genes and map outside them*. They fall into two classes. The promoter and the operator are identified as targets for the regulatory proteins (RNA polymerase and repressor, respectively) by *cis*-acting mutations. And the locus *lacI* is identified as the gene that codes for the repressor protein by mutations that eliminate the *trans*-acting product.

The operator was originally identified by constitutive mutations, denoted O^c, whose distinctive properties provided the first evidence for *an element that functions without being represented in a diffusible product*.

The structural genes contiguous with an O^c mutation are expressed constitutively because the mutation changes the operator so that the repressor no longer binds to it. So the repressor cannot prevent RNA polymerase from initiating transcription. The operon is transcribed constitutively, as illustrated in **Figure 12.9**.

The operator can control only the lac *genes that are adjacent to it*. This property defines the operator as a typical *cis*-acting site, whose function depends upon recognition of its DNA sequence by some *trans*-acting factor. *cis*-dominance is a characteristic of any site that is *physically contiguous with the sequences it controls*. If a control site functions as part of a polycistronic mRNA, mutations in it will display *exactly the same pattern* of *cis*-dominance as they would if functioning in DNA. The critical feature is that the control site cannot be physically separated from the genes that it regulates. From the genetic point of view, it does not matter whether the site and genes are together on DNA or on RNA.

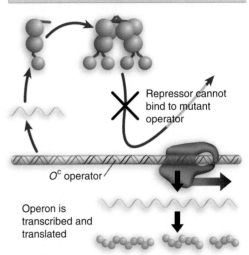

Constitutive operator mutant cannot bind repressor

Repressor cannot bind to mutant operator

O^c operator

Operon is transcribed and translated

Figure 12.9 Operator mutations are constitutive because the operator is unable to bind repressor protein; this allows RNA polymerase to have unrestrained access to the promoter. The O^c mutations are *cis*-acting, because they affect only the contiguous set of structural genes.

12.7 *trans*-Acting Mutations Identify the Regulator Gene

Key Terms

- The **DNA-binding site** of a protein is the region that binds to DNA. Several types of motifs are known for DNA-binding sites. In regulatory proteins, their activities may be controlled by changes in conformation that are triggered by a small molecule binding elsewhere on the protein.

- The **inducer-binding site** of a repressor or activator is the discrete site on the protein at which the small molecule inducer binds. It affects the structure of the DNA-binding site by an allosteric interaction.

- **Interallelic complementation (intragenic complementation)** describes the change in the properties of a heteromultimeric protein brought about by the interaction of subunits coded by two different mutant alleles; the mixed protein may be more or less active than the protein consisting of subunits only of one or the other type.

- **Negative complementation** occurs when interallelic complementation allows a mutant subunit to suppress the activity of a wild-type subunit in a multimeric protein.

- A **dominant negative** mutation results in a mutant gene product that prevents the function of the wild-type gene product, causing loss or reduction of gene activity in cells containing both the mutant and wild-type alleles. This usually happens because the protein is a multimer whose function can be inhibited by one mutant subunit.

Key Concepts

- Mutations in the *lacI* gene are *trans*-acting and affect expression of all *lacZYA* clusters in the bacterium.

- Mutations that eliminate *lacI* function cause constitutive expression and are recessive.

- Mutations in the DNA-binding site of the repressor are constitutive because the repressor cannot bind the operator.

- Mutations in the inducer-binding site of the repressor prevent it from being inactivated and cause uninducibility.

- Active repressor is a tetramer of identical subunits.

- When mutant and wild-type repressor subunits are present, a single *lacI*$^{-d}$ mutant subunit can inactivate a tetramer whose other subunits are wild type.

- *lacI*$^{-d}$ mutations occur in the DNA-binding site. Their effect is explained by the fact that repressor activity requires all DNA-binding sites in the tetramer to be active.

Mutations in the *lacI* gene can cause the operon to be uninducible (it cannot be turned on in any circumstances) or constitutively active (it is permanently turned on, irrespective of circumstances). These two types of mutations identify different active sites in the protein.

Constitutive transcription is caused by mutations of the *lacI*$^{-}$ type, which are caused by loss of function (including deletions of the gene). When the repressor is inactive or absent, transcription can initiate at the promoter. **Figure 12.10** shows that the *lacI*$^{-}$ mutants express the structural genes all the time (constitutively), *irrespective of whether the inducer is present or absent*, because the repressor is inactive. One important subset of *lacI*$^{-}$ mutations (called *lacI*$^{-d}$; see below) is localized in the **DNA-binding site** of the repressor. They abolish the ability to turn off the gene by damaging the site that the protein uses to contact the operator.

Uninducible mutants are caused by mutations that abolish the ability of repressor to bind the inducer. They are described as *lacI*^{s}. The repressor is "locked in" to the active form that recognizes the operator and prevents transcription. These mutations identify the **inducer-binding site**. The addition of inducer has no effect because its binding site is absent, and therefore it is impossible to convert the repressor to the inactive form. The mutant repressor binds to all *lac* operators in the cell to prevent their transcription, and cannot be pried off, irrespective of the properties of any wild-type repressor protein that is present.

An important feature of the repressor is that it is multimeric. Repressor subunits associate at random in the cell to form the active protein tetramer. When two different alleles of the *lacI* gene are present,

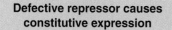

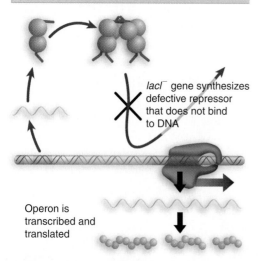

Defective repressor causes constitutive expression

lacI^{-} gene synthesizes defective repressor that does not bind to DNA

Operon is transcribed and translated

Figure 12.10 Mutations that inactivate the *lacI* gene cause the operon to be constitutively expressed, because the mutant repressor protein cannot bind to the operator.

the subunits made by each can associate to form a heterotetramer, whose properties can differ from those of either homotetramer. This type of interaction between subunits is a characteristic feature of multimeric proteins and is described as **interallelic complementation**.

Combinations of certain repressor mutants display a form of interallelic complementation called **negative complementation**. The *lacI*⁻ᵈ mutation alone results in the production of a repressor that cannot bind the operator, and is therefore constitutive like the *lacI*⁻ alleles. Because the *lacI*⁻ type of mutation inactivates the repressor, it is usually recessive to the wild type. However, the −d notation indicates that this variant of the negative type is dominant when paired with a wild-type allele. Such mutations are called **dominant negative**. The reason for their behavior is that one mutant subunit in a tetramer can antagonize the function of the wild-type subunits as discussed in the next section.

12.8 Repressor Is a Tetramer Made of Two Dimers

Key Concepts

- A single repressor subunit can be divided into the N-terminal DNA-binding domain, a hinge, and the core of the protein.
- The DNA-binding domain contains two short α-helical regions that bind the major groove of DNA.
- The inducer-binding site and the regions responsible for multimerization are located in the core.
- Monomers form a dimer by making contacts between core domain 2 and between the oligomerization helices.
- Dimers form a tetramer by interactions between the oligomerization helices.
- Different types of mutations occur in different domains of the repressor subunit.

The repressor has several domains, as shown in the crystal structure illustrated in **Figure 12.11**. A major feature is that the DNA-binding domain is separate from the rest of the protein.

The DNA-binding domain occupies residues 1–59. It consists of two α-helices separated by a turn. This is a common DNA-binding motif, known as the HTH (helix-turn-helix); the two α-helices fit into the major groove of DNA, where they make contacts with specific bases (see *14.10 Repressor Uses a Helix-turn-helix Motif to Bind DNA*). This region is connected by a *hinge* to the main body of the protein. In the DNA-binding form of repressor, the hinge forms a small α-helix (as shown in Figure 12.11); but when the repressor is not bound to DNA, this region is disordered. The HTH and hinge together correspond to the **headpiece**.

The bulk of the core consists of two regions with similar structures (core domains 1 and 2). Each has a six-stranded parallel β-sheet sandwiched between two α-helices on either side. The inducer binds in a cleft between the two regions.

At the C-terminus, there is an α-helix that contains two leucine heptad repeats. This is the oligomerization domain. The oligomerization helices of four monomers associate to maintain the tetrameric structure.

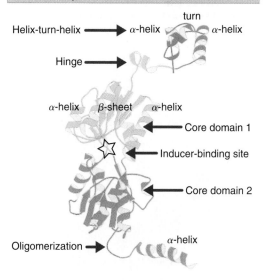

Lac repressor has several domains

Helix-turn-helix → α-helix   turn   α-helix

Hinge →

α-helix   β-sheet   α-helix

Core domain 1

Inducer-binding site

Core domain 2

Oligomerization →   α-helix

Figure 12.11 The structure of a monomer of Lac repressor identifies several independent domains. Photograph kindly provided by Mitchell Lewis, Dept. of Biochemistry & Biophysics, University of Pennsylvania.

A regulator binds a target site on DNA

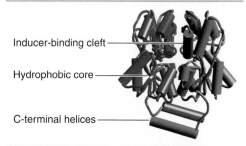

Inducer-binding cleft

Hydrophobic core

C-terminal helices

Two dimers make a tetramer

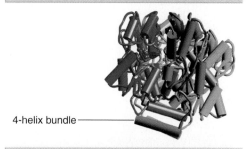

4-helix bundle

Mutations identify functional sites

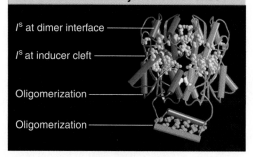

I^s at dimer interface

I^s at inducer cleft

Oligomerization

Oligomerization

Figure 12.12 The crystal structure of the core region of Lac repressor identifies the interactions between monomers in the tetramer. Each monomer is identified by a different color. Mutations are colored as: dimer interface—yellow; inducer-binding—blue; oligomerization—white and purple. Photographs kindly provided by Alan Friedman.

Repressor is a tetramer of two dimers

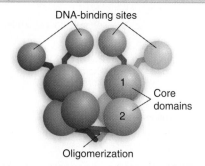

DNA-binding sites

1

Core domains

2

Oligomerization

Figure 12.13 The repressor tetramer consists of two dimers. Dimers are held together by contacts involving core domain 2 as well as by the oligomerization helix. The dimers are linked into the tetramer by the oligomerization interface.

Figure 12.12 shows the structure of the tetrameric core (using a different modeling system from Figure 12.11). It consists in effect of two dimers. The body of the dimer contains a loose interface between the N-terminal regions of the core monomers, a cleft at which inducer binds, and a hydrophobic core (top). The C-terminal regions of each monomer protrude as parallel helices. (The headpiece would join on to the N-terminal regions at the top.) Together the dimers interact to form a tetramer (center) that is held together by a C-terminal bundle of four helices.

Sites of mutations are shown by beads on the structure in the bottom panel of Figure 12.12. $lacI^s$ mutations make the repressor unresponsive to the inducer, so that the operon is uninducible. They map in two groups: blue shows those in the inducer-binding cleft, and yellow shows those that affect the dimer interface. The first group abolishes the inducer binding site; the second group prevents the effects of inducer binding from being transmitted to the DNA-binding site. $lacI^-$ mutations that affect oligomerization map in two groups. White shows mutations in core domain 2 that prevent dimer formation. Purple shows those in the oligomerization helix that prevent the dimers from forming tetramers.

From these data we can derive the schematic of **Figure 12.13**, which shows how the monomers are organized into the tetramer. Two monomers form a dimer by means of contacts at core domain 2 and in the oligomerization helix. The dimer has two DNA-binding domains at one end of the structure, and the oligomerization helices at the other end. Two dimers then form a tetramer by interactions at the oligomerization interface. We can map the types of mutations onto this structure as summarized in the scheme of **Figure 12.14**.

The special class of dominant-negative $lacI^{-d}$ mutations lie in the DNA-binding site of the repressor subunit. This explains their ability to prevent mixed tetramers from binding to the operator; a reduction in the number of binding sites reduces the specific affinity for the operator. The role of the N-terminal region in specifically binding DNA is shown also by its location as the site of occurrence of "tight binding" mutations. These increase the affinity of the repressor for the operator, sometimes so much that it cannot be released by inducer. They are rare.

Uninducible $lacI^s$ mutations map in a region of the core domain 1 extending from the inducer-binding site to the hinge. One group lies in amino acids that contact the inducer, and these mutations function by preventing binding of inducer. The remaining mutations lie at sites that must be involved in transmitting the allosteric change in conformation to the hinge when inducer binds.

12.9 Repressor Binding to the Operator Is Regulated by an Allosteric Change in Conformation

Key Term

- A **palindrome** is a DNA sequence that reads the same on each strand of DNA when the strand is read in the 5′ to 3′ direction. It consists of adjacent inverted repeats.

Key Concepts

- Repressor protein binds to the double-stranded DNA sequence of the operator.
- The operator is a palindromic sequence of 26 bp.
- Each inverted repeat of the operator binds to the DNA-binding site of one repressor subunit.
- The DNA-binding domain of a monomer inserts into the major groove of DNA.
- Active repressor has a conformation in which the two DNA-binding domains of a dimer can insert into successive turns of the double helix.
- Inducer binding changes the conformation so that the two DNA-binding sites are not in the right geometry to make simultaneous contacts.

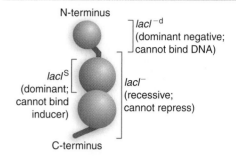

Mutations identify repressor domains

Figure 12.14 The locations of three types of mutations in lactose repressor are mapped on the domain structure of the protein. Recessive *lacI*⁻ mutants that cannot repress can map anywhere in the protein. Dominant negative *lacI*⁻ᵈ mutants that cannot repress map to the DNA-binding domain. Dominant *lacI*ˢ mutants that cannot induce because they do not bind inducer map to core domain 1.

How does the repressor recognize the specific sequence of operator DNA? The operator has a feature common to many recognition sites for bacterial regulator proteins: it is a **palindrome**. The inverted repeats are highlighted in **Figure 12.15**. Each repeat can be regarded as a half-site of the operator.

The importance of particular bases within the operator sequence can be determined by identifying those that contact the repressor protein or in which mutations change the binding of repressor. The region of DNA contacted by protein extends for 26 bp, and within this region are eight sites at which constitutive mutations occur. This emphasizes the same point made by promoter mutations: *A small number of essential specific contacts within a larger region can be responsible for sequence-specific association of DNA with protein.*

The symmetry of the DNA sequence reflects the symmetry in the protein. Each of the identical subunits in a repressor tetramer has a DNA-binding site. Two of these sites contact the operator in such a way that each inverted repeat (half site) of the operator makes the same pattern of contacts with a repressor monomer. This is shown by symmetry in the contacts that repressor makes with the operator (the pattern between +1 and +6 is identical with that between +21 and +16) and by matching constitutive mutations in each inverted repeat.

Early work suggested a model in which the headpiece is relatively independent of the core. It can bind to operator DNA by making the same pattern of contacts with a half site as an intact repressor. However, its affinity for DNA is many orders of magnitude less than that of the intact repressor. The reason for the difference is that the dimeric form of the intact repressor allows two headpieces to contact the operator simultaneously, each binding to one half site. **Figure 12.16** shows that the two DNA-binding domains in a dimeric unit contact DNA by inserting into successive turns of the major groove. This enormously increases affinity for the operator.

Binding of inducer causes an immediate conformational change in the repressor protein. Binding of two molecules of inducer to the repressor tetramer is adequate to release repression. Binding of inducer changes the orientation of the headpieces relative to the core, with the result that the two headpieces in a dimer can no longer bind

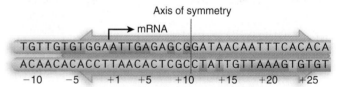

The *lac* operator has dyad symmetry

Axis of symmetry

→ mRNA

```
TGTTGTGTGGAATTGAGAGCGGATAACAATTTCACACA
ACAACACACCTTAACACTCGCCTATTGTTAAAGTGTGT
-10    -5   +1  +5   +10   +15   +20  +25
```

Figure 12.15 The *lac* operator has a symmetrical sequence. The sequence is numbered relative to the startpoint for transcription at +1. The pink arrows to left and right identify the two dyad repeats. The green blocks indicate the positions of identity.

Inducer controls repressor conformation

Headpieces bind successive turns in major groove

Inducer binding changes conformation at the hinge, so that headpieces cannot fit into major groove

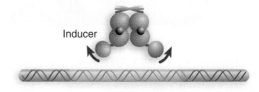

Inducer

Figure 12.16 Inducer changes the structure of the core so that the headpieces of a repressor dimer are no longer in an orientation that permits binding to DNA.

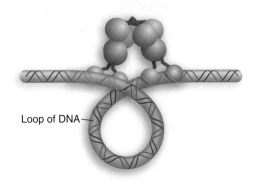

Figure 12.17 If both dimers in a repressor tetramer bind to DNA, the DNA between the two binding sites is held in a loop.

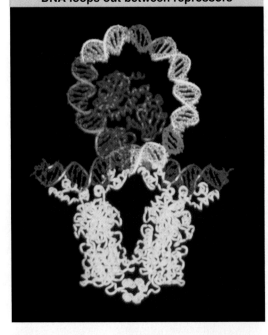

Figure 12.18 When a repressor tetramer binds to two operators, the stretch of DNA between them is forced into a tight loop. (The blue structure in the center of the looped DNA represents CAP, another regulator protein that binds in this region.) Photograph kindly provided by Mitchell Lewis, Dept. of Biochemistry & Biophysics, University of Pennsylvania.

DNA simultaneously. This eliminates the advantage of the multimeric repressor, and reduces the affinity for the operator.

An inducer reduces the affinity of the repressor for the operator below the threshold that is needed to ensure that the operators are bound. When the inducer enters the cell, it binds to the repressor. After repressor binding, any repressor that is bound to an operator is released; any repressor that is free has too low an affinity for the operator to bind to one.

12.10 Repressor Binds to Three Operators and Interacts with RNA Polymerase

Key Concepts

- Each dimer in a repressor tetramer can bind an operator, so that the tetramer can bind two operators simultaneously.
- Full repression requires the repressor to bind to an additional operator downstream or upstream as well as to the operon at *lacZ*.
- Binding of repressor at the operator stimulates binding of RNA polymerase at the promoter.

The allosteric transition that results from binding of the inducer occurs in the repressor dimer. So why is a tetramer required to establish full repression?

Each dimer can bind an operator sequence. This enables the intact repressor to bind to two operator sites simultaneously. In fact, there are two further operator sites in the initial region of the *lac* operon. The original operator, *O1*, is located just at the start of the *lacZ* gene. It has the strongest affinity for the repressor. Weaker operator sequences (sometimes called pseudo-operators) are located on either side; *O2* is 410 bp downstream of the startpoint, and *O3* is 83 bp upstream of it.

Figure 12.17 shows what happens when a DNA-binding protein can bind simultaneously to two separated sites on DNA. The DNA between the two sites forms a loop from a base where the protein has bound the two sites. The length of the loop depends on the distance between the two binding sites. When Lac repressor binds simultaneously to *O1* and to one of the other operators, it causes the DNA between them to form a rather short loop, significantly constraining the DNA structure. A scale model for binding of the tetrameric repressor to two operators is shown in **Figure 12.18**.

Binding at the additional operators affects the level of repression. Elimination of either the downstream operator (*O2*) or the upstream operator (*O3*) reduces the efficiency of repression by 2–4×. However, if *both O2 and O3 are eliminated, repression is reduced 100×. This suggests that the ability of the repressor to bind to one of the two other operators as well as to O1 is important for establishing repression.* We do not know how and why this simultaneous binding increases repression.

We know most about the direct effects of binding of the repressor to the operator (*O1*). It was originally thought that repressor binding would occlude RNA polymerase from binding to the promoter. However, we now know that the two proteins may be bound to DNA simultaneously, and *the binding of repressor actually enhances the binding of RNA polymerase!* But the bound enzyme is prevented from initiating transcription. The repressor in effect causes RNA polymerase to be stored at the promoter. When inducer is added, the repressor is released, and RNA polymerase can initiate transcription immediately. The overall effect of repressor has been to speed up the induction process.

Does this model apply to other systems? The interaction between RNA polymerase, repressor, and the promoter/operator region is distinct in each system, because the operator does not always overlap with the same region of the promoter (see Figure 12.22). For example, in phage lambda, the operator lies in the upstream region of the promoter, and binding of the repressor occludes the binding of RNA polymerase (see *Chapter 14 Phage Strategies*). So a bound repressor does not interact with RNA polymerase in the same way in all systems.

12.11 The Operator Competes with Low-Affinity Sites to Bind Repressor

Key Concepts

- Proteins that have a high affinity for a specific DNA sequence also have a low affinity for other DNA sequences.

- Every base pair in the bacterial genome is the start of a low-affinity binding site for repressor.

- The large number of low-affinity sites ensures that all repressor protein is bound to DNA.

- Repressor binds to the operator by moving there from a low-affinity site.

- In the absence of inducer, the operator has an affinity for repressor that is $10^7 \times$ that of a low-affinity site.

- The level of 10 repressor tetramers per cell ensures that the operator is bound by repressor 96% of the time.

- Induction reduces the affinity for the operator to $10^4 \times$ that of low-affinity sites, so that only 3% of operators are bound.

- Induction causes repressor to move from the operator to a low-affinity site by direct displacement.

Probably all proteins that have a high affinity for a specific sequence also possess a low affinity for any (random) DNA sequence. A large number of low-affinity sites will compete just as well for a repressor tetramer as a small number of high-affinity sites. There is only one high-affinity site in the *E. coli* genome: the operator. The remainder of the DNA provides low-affinity binding sites. Every base pair in the genome starts a new low-affinity site. (Just moving one base pair along the genome, out of phase with the operator itself, creates a low-affinity site!) So there are 4.2×10^6 low-affinity sites.

The large number of low-affinity sites means that, even in the absence of a specific binding site, all or virtually all of the repressors are bound to DNA; no repressors are free in solution. *All but 0.01% of repressors are bound to (random) DNA.* Since there are ~10 molecules of repressor per cell, this is tantamount to saying that there is no free repressor protein. This has an important implication for the interaction of repressor with the operator: it means that we are concerned with the *partitioning* of the repressor on DNA, in which the single high-affinity site of the operator *competes* with the large number of low-affinity sites.

Repressor specifically binds operator DNA		
DNA	Repressor	Repressor + inducer
Operator	2×10^{13}	2×10^{10}
Other DNA	2×10^{6}	2×10^{6}
Specificity	10^{7}	10^{4}
Operators bound	96%	3%
Operon is:	Repressed	Induced

Figure 12.19 Lac repressor binds strongly and specifically to its operator, but is released by inducer. All equilibrium constants are in M^{-1}.

The efficiency of repression therefore depends on the relative affinity of the repressor for its operator compared with other (random) DNA sequences. The affinity must be great enough to overcome the large number of random sites. We can see how this works by comparing the equilibrium constants for *lac* repressor/operator binding with repressor/general DNA binding. **Figure 12.19** shows that the ratio is 10^{7} for an active repressor, enough to ensure that the operator is bound by repressor 96% of the time, so the operon is effectively repressed. However, when inducer is added, the ratio is reduced to 10^{4}. At this level, only 3% of the operators are bound, and the operon is effectively induced.

The consequence of these affinities is that in an uninduced cell, one tetramer of repressor usually is bound to the operator. All or almost all of the remaining tetramers are bound at random to other regions of DNA, as illustrated in **Figure 12.20**. There are likely to be very few or no repressor tetramers free within the cell.

The addition of inducer abolishes the ability of repressor to bind specifically at the operator. Those repressors bound at the operator are released, and bind to random (low-affinity) sites. So in an induced cell, the repressor tetramers are "stored" on random DNA sites. In a noninduced cell, a tetramer is bound at the operator, while the remaining repressor molecules are bound to nonspecific sites. *The effect of induction is therefore to change the distribution of repressor on DNA, rather than to generate free repressor.* In the same way that RNA polymerase probably moves between promoters and other DNA by swapping one sequence for another, the repressor also may directly displace one bound DNA sequence with another in order to move between sites.

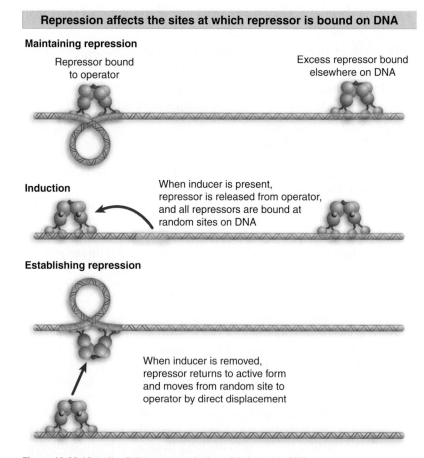

Repression affects the sites at which repressor is bound on DNA

Maintaining repression

Repressor bound to operator

Excess repressor bound elsewhere on DNA

Induction

When inducer is present, repressor is released from operator, and all repressors are bound at random sites on DNA

Establishing repression

When inducer is removed, repressor returns to active form and moves from random site to operator by direct displacement

Figure 12.20 Virtually all the repressor in the cell is bound to DNA.

We can define the parameters that influence the ability of a regulator protein to saturate its target site by comparing the equilibrium equations for specific and nonspecific binding. As might be expected intuitively, the important parameters are:

- The size of the genome dilutes the ability of a protein to bind specific target sites.
- The specificity of the protein counters the effect of the mass of DNA.
- The amount of protein that is required increases with the total amount of DNA in the genome and decreases with the specificity.
- The amount of protein also must be in reasonable excess of the total number of specific target sites, so we expect regulators with many targets to be found in greater quantities than regulators with fewer targets.

12.12 Repression Can Occur at Multiple Loci

Key Concept

- A repressor acts on all loci that have a copy of its target operator sequence.

The *lac* repressor acts only on the operator of the *lacZYA* cluster. However, some repressors control dispersed structural genes by binding at more than one operator. An example is the *trp* repressor, which controls three unlinked sets of genes:

- An operator at the cluster of structural genes *trpEDBCA* controls coordinate synthesis of the enzymes that synthesize tryptophan from chorismic acid.
- An operator at another locus controls the *aroH* gene, which codes for one of the three enzymes that catalyze the initial reaction in the common pathway of aromatic amino acid biosynthesis.
- The *trpR* regulator gene is repressed by its own product, the *trp* repressor. So the repressor protein acts to reduce its own synthesis. This circuit is an example of *autogenous* control. Such circuits are quite common in regulatory genes, and may be either negative or positive.

A related 21 bp operator sequence is present at each of the three loci at which the *trp* repressor acts. The conservation of sequence is indicated in **Figure 12.21**. Each operator contains appreciable (but not identical) dyad symmetry. The features conserved at all three operators

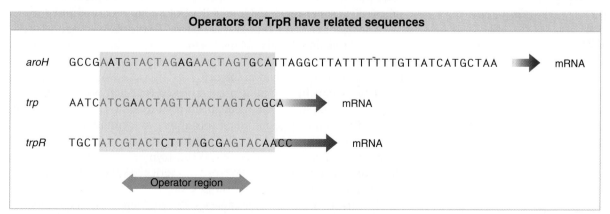

Figure 12.21 The *trp* repressor recognizes operators at three loci. Conserved bases are shown in red. The location of the startpoint and mRNA varies, as indicated by the blue arrows.

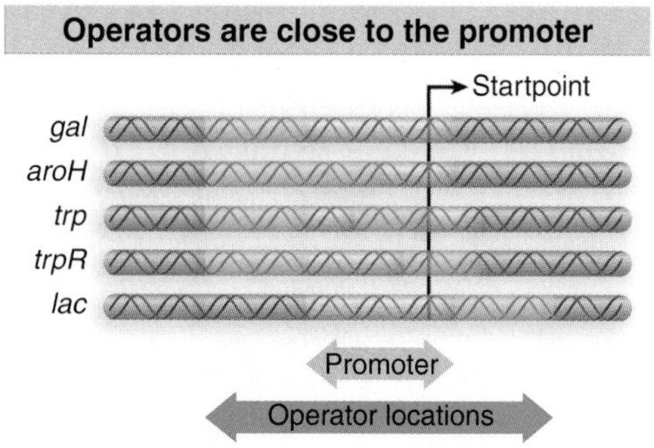

Operators are close to the promoter

gal
aroH
trp
trpR
lac

→ Startpoint

Promoter

Operator locations

Figure 12.22 Operators may lie at various positions relative to the promoter.

include the important points of contact for *trp* repressor. This explains how one repressor protein acts on several loci: *each locus has a copy of a specific DNA-binding sequence recognized by the repressor* (just as each promoter shares consensus sequences with other promoters).

Figure 12.22 summarizes the variety of relationships between operators and promoters. A notable feature of the dispersed operators recognized by TrpR is their presence at different locations within the promoter in each locus. In *trpR* the operator lies between positions -12 and $+9$, while in the *trp* operon it occupies positions -23 to -3, but in the *aroH* locus it lies farther upstream, between -49 and -29. In other cases, the operator lies downstream from the promoter (as in *lac*), or apparently just upstream of the promoter (as in *gal*, where the nature of the repressive effect is not quite clear). The ability of the repressors to act at operators whose positions are different in each target promoter suggests that there could be differences in the exact mode of repression, the common feature being that RNA polymerase is prevented from initiating transcription at the promoter.

12.13 Operons May Be Repressed or Induced

Key Terms

- The **derepressed** state describes a gene that is turned on because a small molecule corepressor is absent. It has the same effect as the induced state that is produced by a small molecule inducer for a gene that is regulated by induction. In describing the effect of a mutation, *derepressed* and *constitutive* have the same meaning.

- **Super-repressed** is a mutant condition in which a repressible operon cannot be derepressed, so it is always turned off. (This corresponds to the uninducible condition for an operon that is regulated by induction.)

Key Concept

- An operon is repressed when a small-molecule corepressor is added if the corepressor activates a repressor protein.

The terminology used for repressible systems describes the active state of the operon as **derepressed**; this has the same meaning as *induced*. The condition in which a (mutant) operon cannot be derepressed is sometimes called **super-repressed**; this is the exact counterpart of *uninducible*.

The *trp* operon is a repressible system. Tryptophan is the end product of the reactions catalyzed by a series of biosynthetic enzymes. Both the activity and the synthesis of the tryptophan enzymes are controlled by the level of tryptophan in the cell.

Tryptophan functions as a corepressor that activates a repressor protein. This is the classic mechanism for repression, as illustrated in **Figure 12.23**. In conditions when the supply of tryptophan is plentiful, the operon is repressed because the repressor protein·corepressor complex is bound at the operator. When tryptophan is in short supply, the corepressor is inactive, therefore has reduced specificity for the operator, and is stored elsewhere on DNA.

Deprivation of repressor causes ~70-fold increase in the frequency of initiation events at the *trp* promoter. Even under repressing conditions, the structural genes continue to be expressed at a low basal level

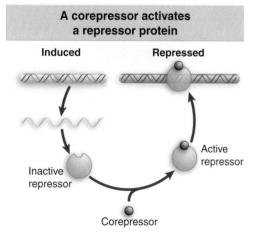

A corepressor activates a repressor protein

Induced   Repressed

Active repressor

Inactive repressor

Corepressor

Figure 12.23 An operon is repressed when a corepressor converts the repressor protein to an active form that can bind the operator.

(sometimes also called the repressed level). The efficiency of repression at the operator is much lower than in the *lac* operon (where the basal level is only ~1/1000 of the induced level).

We have treated both induction and repression as phenomena that rely upon allosteric changes induced in regulator proteins by small molecules. Other types of interactions also can be used to control the activities of regulator proteins. One example is OxyR, a transcriptional activator of genes induced by hydrogen peroxide. The OxyR protein is directly activated by oxidation, so it provides a sensitive measure of oxidative stress. Another common type of signal is phosphorylation of a regulator protein.

12.14 Cyclic AMP Is an Inducer That Activates CRP to Act at Many Operons

Key Terms

- **CRP activator (CAP activator)** is a positive regulator protein activated by cyclic AMP. It is needed for RNA polymerase to initiate transcription of many operons of *E. coli*.
- **Adenylate cyclase** is an enzyme that uses ATP as a substrate to generate cyclic AMP, in which 5′ and 3′ positions of the sugar ring are connected via a phosphate group.

Key Concepts

- A dimer of the activator protein CRP is activated by a single molecule of cyclic AMP to bind to its target sequence at a promoter.
- CRP-binding sites lie at highly variable locations relative to the promoter.
- CRP interacts with RNA polymerase, but the details of the interaction depend on the relative locations of the CRP-binding site and the promoter.

Many promoters are recognized by RNA polymerase unless some regulatory action occurs to prevent initiation. But there are some promoters where initiation requires additional proteins. Proteins that are required for initiation of transcription are called activators. Like repressors, they function to control the ability of RNA polymerase to initiate at a specific promoter(s), but their role is to assist the RNA polymerase instead of antagonizing it. Typically the activator overcomes a deficiency in the promoter, for example, a poor consensus sequence at −35 or −10.

One of the most widely acting activators is a protein called **CRP activator** that controls the activity of a large set of operons in *E. coli*. The protein is a positive control factor whose presence is necessary to initiate transcription at dependent promoters. CRP is active *only in the presence of cyclic AMP*. **Figure 12.24** shows the regulatory circuit.

Cyclic AMP is synthesized by the enzyme **adenylate cyclase**. The reaction uses ATP as substrate and introduces a 3′–5′ link via phosphodiester bonds, generating the structure drawn in **Figure 12.25**.

The CRP factor binds to DNA, and complexes of cyclic AMP·CRP·DNA can be isolated at each promoter at which it functions. The factor is a dimer of two identical subunits of 22.5 kD, which can be activated by a single molecule of cyclic AMP. A CRP monomer contains a DNA-binding region and a transcription-activating region.

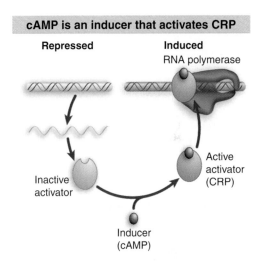

cAMP is an inducer that activates CRP

Figure 12.24 A small molecule inducer (cyclic AMP) converts an activator protein (CRP) to a form that binds the promoter and assists RNA polymerase to initiate transcription.

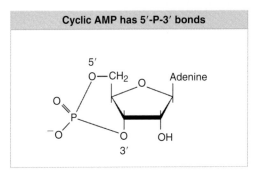

Cyclic AMP has 5′-P-3′ bonds

Figure 12.25 Cyclic AMP has a single phosphate group connected to both the 3′ and 5′ positions of the sugar ring.

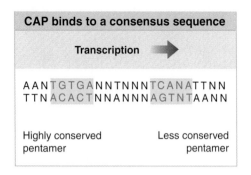

Figure 12.26 The consensus sequence for CRP contains the well-conserved pentamer TGTGA and (sometimes) an inversion of this sequence (TCANA).

A CRP dimer binds to a site of ~22 bp at a responsive promoter. The binding sites include variations of the consensus sequence given in **Figure 12.26**. Mutations preventing CRP action usually are located within the well-conserved pentamer $\frac{\text{TGTGA}}{\text{ACACT}}$ which appears to be the essential element in recognition. CRP binds most strongly to sites that contain two (inverted) versions of the pentamer, because this enables both subunits of the dimer to bind to the DNA. Many binding sites lack the second pentamer, however, and in these the second subunit must bind a different sequence (if it binds to DNA). The hierarchy of binding affinities for CRP helps to explain why different genes are activated by different levels of cyclic AMP.

12.15 Translation Can Be Regulated

Key Term

- **Autogenous control** describes the action of a gene product that either inhibits (negative autogenous control) or activates (positive autogenous control) expression of the gene coding for it.

Key Concepts

- A repressor protein can regulate translation by preventing a ribosome from binding to an initiation codon.
- Translation of an r-protein operon can be controlled by a product of the operon that binds to a site on the polycistronic mRNA.
- Autogenous control is often used for proteins that are incorporated into macromolecular assemblies.

One mechanism for controlling gene expression at the level of translation is an exact parallel to the use of a repressor that inhibits transcription by binding to DNA to prevent RNA polymerase from utilizing a promoter. *Repressor function is provided by a protein that binds to a target region on mRNA to prevent ribosomes from recognizing the initiation region.* **Figure 12.27** illustrates the most common form of this interaction, in which the regulator protein binds directly to a sequence that includes the AUG initiation codon, thereby preventing the ribosome from binding.

Some examples of translational repressors and their targets are summarized in **Figure 12.28**. A classic example is the coat protein of the RNA phage R17, which binds to a hairpin that encompasses the ribosome-binding site in the phage mRNA. Similarly, the T4 RegA protein binds to a consensus sequence that includes the AUG initiation codon in several T4 early mRNAs; and T4 DNA polymerase binds to a

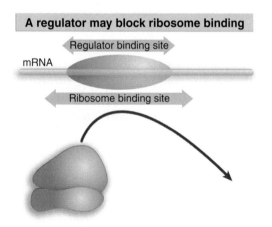

Figure 12.27 A regulator protein may block translation by binding to a site on mRNA that overlaps the ribosome-binding site at the initiation codon.

Translational repressors bind to mRNA		
Repressor	Target gene	Site of action
R17 coat protein	R17 replicase	Hairpin that includes ribosome binding site
T4 RegA	Early T4 mRNAs	Various sequences including initiation codon
T4 DNA polymerase	T4 DNA polymerase	Shine-Dalgarno sequence
T4 p32	Gene 32	Single-stranded 5′ leader

Figure 12.28 Proteins that bind to sequences within the initiation regions of mRNAs may function as translational repressors.

sequence in its own mRNA that includes the Shine–Dalgarno element needed for ribosome binding.

Autogenous control occurs whenever a protein (or RNA) regulates its own production. The usual type of effect is that accumulation of the protein prevents production of more protein at the level of either transcription or translation. Sometimes this applies to an individual product, such as T4 p32, where the protein prevents translation of its own RNA. In other cases, accumulation of one product of an operon may turn off the synthesis of all products of the operon, as in the inhibition of translation of the operons coding for ribosomal proteins.

The genes coding for the ribosomal proteins (r-proteins) are grouped into several operons. In each of these operons, accumulation of one protein inhibits further synthesis of itself and of some of the other gene products. Each of the regulators is a ribosomal protein that binds directly to rRNA. Its effect on translation is a result of its ability also to bind to its own mRNA. The sites on mRNA at which these proteins bind either overlap the sequence where translation is initiated or lie nearby and probably influence the accessibility of the initiation site by inducing conformational changes.

The use of r-proteins that bind rRNA to establish autogenous regulation immediately suggests that this provides a mechanism to link r-protein synthesis to rRNA synthesis. A generalized model is depicted in **Figure 12.29**. Suppose that the binding sites for the autogenous regulator r-proteins on rRNA are much stronger than those on the mRNAs. Then so long as any free rRNA is available, the newly synthesized r-proteins will associate with it to start ribosome assembly. There will be no free r-protein available to bind to the mRNA, so its translation will continue. But as soon as the synthesis of rRNA slows or stops, free r-proteins begin to accumulate. Then they are available to bind their mRNAs, repressing further translation. This circuit ensures that each r-protein operon responds in the same way to the level of rRNA: as soon as there is an excess of r-protein relative to rRNA, synthesis of the protein is repressed.

Autogenous control is a common type of regulation among proteins that are incorporated into macromolecular assemblies. The assembled particle itself may be unsuitable as a regulator, because it is too large, too numerous, or too restricted in its location. But the need for synthesis of its components may be reflected in the pool of free precursor subunits. If the assembly pathway is blocked for any reason, free subunits accumulate and shut off the unnecessary synthesis of further components.

rRNA controls the level of free r-proteins

When rRNA is available, the r-proteins associate with it. Translation of mRNA continues.

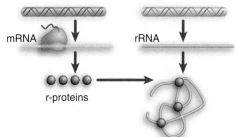

mRNA   rRNA

r-proteins

When no rRNA is available, r-proteins accumulate. An r-protein binds to mRNA and prevents translation.

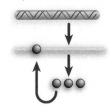

Figure 12.29 Translation of the r-protein operons is autogenously controlled and responds to the level of rRNA.

12.16 SUMMARY

Transcription is regulated by the interaction between *trans*-acting factors and *cis*-acting sites. A *trans*-acting factor is the product of a regulator gene. It is usually protein but can be RNA. Because it diffuses in the cell, it can act on any appropriate target gene. A *cis*-acting site in DNA (or RNA) is a sequence that functions by being recognized *in situ*. It has no coding function and can regulate only those sequences that are physically contiguous with it. Bacterial genes coding for proteins whose functions are related, such as successive enzymes in a path-

way, may be organized in a cluster that is transcribed into a polycistronic mRNA from a single promoter. Control of this promoter regulates expression of the entire pathway. The unit of regulation, containing structural genes and *cis*-acting elements, is called the operon.

Initiation of transcription is regulated by interactions in the vicinity of the promoter. The ability of RNA polymerase to initiate at the promoter is prevented or activated by other proteins. Genes that are active unless they are turned off are said to be under negative control. Genes that are active only when specifically turned on are said to be under positive control. The type of control can be determined by the dominance relationships between wild type and mutants that are constitutive/derepressed (permanently on) or uninducible/super-repressed (permanently off).

A repressor protein prevents RNA polymerase either from binding to the promoter or from activating transcription. The repressor binds to a target sequence, the operator, that usually is located around or upstream of the startpoint. Operator sequences are short and often are palindromic. The repressor is often a homomultimer whose symmetry reflects that of its target.

The ability of the repressor protein to bind to its operator is regulated by a small molecule. An inducer prevents a repressor from binding; a corepressor activates it. Binding of the inducer or corepressor to its site produces a change in the structure of the DNA-binding site of the repressor. This allosteric reaction occurs both in free repressor proteins and directly in repressor proteins already bound to DNA.

The lactose pathway operates by induction, when an inducer β-galactoside prevents the repressor from binding its operator; transcription and translation of the *lacZ* gene then produce β-galactosidase, the enzyme that metabolizes β-galactosides. The tryptophan pathway operates by repression; the corepressor (tryptophan) activates the repressor protein, so that it binds to the operator and prevents expression of the genes that code for the enzymes that biosynthesize tryptophan. A repressor can control multiple targets that have copies of an operator consensus sequence.

A protein with a high affinity for a particular target sequence in DNA has a lower affinity for all DNA. The ratio defines the specificity of the protein. Because there are many more nonspecific sites (any DNA sequence) than specific target sites in a genome, a DNA-binding protein such as a repressor or RNA polymerase is "stored" on DNA; probably none or very little is free. The specificity for the target sequence must be great enough to counterbalance the excess of nonspecific sites over specific sites. The balance for bacterial proteins is adjusted so that the amount of protein and its specificity allow specific recognition of the target in "on" conditions, but allow almost complete release of the target in "off" conditions.

Some promoters cannot be recognized by RNA polymerase (or are recognized only poorly) unless a specific activator protein is present. Activator proteins also may be regulated by small molecules. The CRP activator becomes able to bind to target sequences in the presence of cyclic AMP. All promoters that respond to CRP have at least one copy of the target sequence. Direct contact between one subunit of CRP and RNA polymerase is required to activate transcription.

A common means for controlling translation is for a regulator protein to bind to a site on the mRNA that overlaps the ribosome-binding site at the initiation codon. This prevents ribosomes from initiating translation. RegA of T4 is a general regulator that functions on several target mRNAs at the level of translation. Most proteins that repress translation possess this capacity in addition to other functional roles; in particular, translation is controlled in some cases of autogenous regulation, when a gene product regulates expression of the operon containing its own gene.

13

Regulatory RNA

13.1 Introduction

Key Concept

- RNA functions as a regulator by forming a region of secondary structure (either inter- or intra-molecular) that changes the properties of a target sequence.

The basic principle of regulation in bacteria is that gene expression is controlled by a regulator that interacts with a specific sequence or structure in DNA or mRNA at some stage prior to the synthesis of protein. The stage of expression that is controlled can be transcription, when the target for regulation is DNA; or it can be at translation, when the target for regulation is RNA. When control is during transcription, it can be at initiation or at termination. The regulator can be a protein or an RNA. "Controlled" can mean that the regulator turns off (represses) the target or that it turns on (activates) the target. Expression of many genes can be coordinately controlled by a single regulator gene on the principle that each target contains a copy of the sequence or structure that the regulator recognizes. Regulators may themselves be regulated, most typically in response to small molecules whose supply responds to environmental conditions. Regulators may be controlled by other regulators to make complex circuits. Let's compare the ways that different types of regulators work.

Protein regulators work on the principle of allostery. The protein has two binding sites, one for a nucleic acid target, the other for a small molecule. Binding of the small molecule to its site changes the conformation in such a way as to alter the affinity of the other site for the nucleic acid. The way in which this happens is known in detail for the Lac repressor (see *12.9 Repressor Binding to the Operator Is Regulated by an Allosteric Change in Conformation*). Protein regulators are often multimeric, with a symmetrical organization that allows two subunits to

243

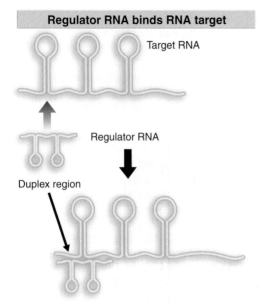

Regulator RNA binds RNA target

Target RNA

Regulator RNA

Duplex region

Figure 13.1 A regulator RNA is a small RNA with a single-stranded region that can pair with a single-stranded region in a target RNA.

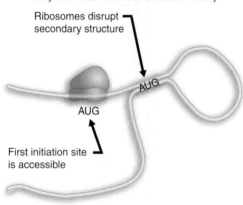

Ribosome movement can control transition

Only one initiation site is available initially

Ribosomes disrupt secondary structure

AUG

AUG

First initiation site is accessible

Transition exposes second initiation site

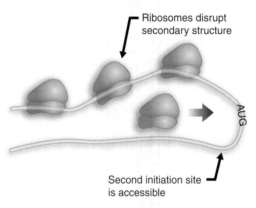

Ribosomes disrupt secondary structure

AUG

Second initiation site is accessible

Figure 13.2 Secondary structure can control initiation. Only one initiation site is available in the phage RNA, but translation of the first cistron changes the conformation of the RNA so that another initiation site become available.

contact a palindromic target on DNA. This can generate cooperative binding effects that create a more sensitive response to regulation.

Regulation via RNA uses changes in secondary structure as the guiding principle. The ability of an RNA to shift between different conformations with regulatory consequences is the nucleic acid's alternative to the allosteric changes of protein conformation. The changes in structure may result from either intramolecular or intermolecular interactions.

The most common role for intramolecular changes is for an RNA molecule to assume alternative secondary structures by utilizing different schemes for base pairing. The properties of the alternative conformations may be different. Changes in secondary structure of an mRNA can result in a change in its ability to be translated. Secondary structure also is used to regulate the termination of transcription, when the alternative structures differ in whether they permit termination.

In intermolecular interactions, an RNA regulator recognizes its target by the familiar principle of complementary base pairing. **Figure 13.1** shows that the regulator is usually a small RNA molecule with extensive secondary structure, but with a single-stranded region(s) that is complementary to a single-stranded region in its target. The formation of a double helical region between regulator and target can have two types of consequence:

- Formation of the double helical structure may itself be sufficient. In some cases, protein(s) can bind only to the single-stranded form of the target sequence, and are therefore prevented from acting by duplex formation. In other cases, the duplex region becomes a target for binding, for example, by nucleases that degrade the RNA and therefore prevent its expression.
- Duplex formation may be important because it sequesters a region of the target RNA that would otherwise participate in some alternative secondary structure.

13.2 Alternative Secondary Structures Can Affect Translation or Transcription

Key Terms

- **Attenuation** describes the regulation of bacterial operons by controlling termination of transcription at a site located before the first structural gene.
- An **attenuator** is a terminator sequence at which attenuation occurs.

Key Concepts

- Translation of one cistron in a polycistronic mRNA may control the translation of another cistron via changes in secondary structure.
- Termination of transcription can be attenuated by controlling formation of the necessary hairpin structure in RNA.
- The most direct mechanisms for attenuation involve proteins that either stabilize or destabilize the hairpin.

Secondary structure is used to control translation indirectly when translation of one cistron requires changes in secondary structure that depend on translation of a preceding cistron. This happens during translation of the RNA phages, whose cistrons always are expressed in a set order. **Figure 13.2** shows that the phage RNA takes up a secondary structure in which only one initiation sequence is accessible; the second cannot be recognized by ribosomes because it is base paired with other regions of the RNA. However, translation of the first cistron disrupts the secondary structure, allowing ribosomes to bind to the initiation site of the next cistron. In this mRNA, secondary structure controls translatability.

Several operons are regulated by **attenuation**, a mechanism that controls the ability of RNA polymerase to read through an **attenuator**, which is an intrinsic terminator located at the beginning of a transcription unit. *The principle of attenuation is that some external event controls the formation of the hairpin needed for intrinsic termination.* If the hairpin forms, termination prevents RNA polymerase from transcribing the structural genes. If the hairpin is prevented from forming, RNA polymerase elongates through the terminator, and the genes are expressed. Different types of mechanisms are used in different systems for controlling the structure of the RNA.

Attenuation may be regulated by proteins that bind to RNA, either to stabilize or to destabilize formation of the hairpin required for termination. **Figure 13.3** shows an example in which a protein prevents formation of the terminator hairpin. The activity of such a protein may be intrinsic or may respond to a small molecule in the same manner as a repressor protein responds to corepressor.

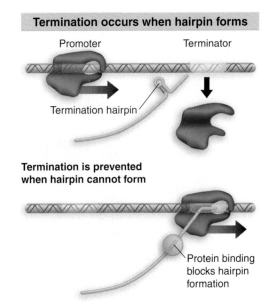

Termination occurs when hairpin forms

Termination is prevented when hairpin cannot form

Figure 13.3 Attenuation occurs when a terminator hairpin in RNA is prevented from forming.

13.3 Termination of *B. subtilis trp* Genes Is Controlled by Tryptophan and by tRNA^{Trp}

Key Concepts

- A terminator protein called TRAP is activated by tryptophan to prevent transcription of *trp* genes.
- Activity of TRAP is (indirectly) inhibited by uncharged tRNA^{Trp}.

The circuitry that controls transcription via termination can use both direct and indirect means to respond to the level of small-molecule products or substrates. The basic principle is that when the product of a pathway is available, it causes transcription to be terminated to stop production of the enzymes responsible for producing it. Such systems have been extensively characterized for the control of tryptophan production.

In *B. subtilis*, a protein called TRAP (formerly called MtrB) is activated by tryptophan to bind to a sequence in the leader of the nascent transcript. TRAP forms a multimer of 11 subunits. Each subunit binds a single tryptophan amino acid and a trinucleotide (GAG or UAG) of RNA. The RNA is wound in a circle around the protein. **Figure 13.4** shows that the result is to ensure the availability of the regions that are required to form the terminator hairpin. The termination of transcription then prevents production of the tryptophan biosynthetic enzymes. In effect, TRAP is a terminator protein that responds to the level of tryptophan. In the absence of TRAP, an alternative secondary structure precludes the formation of the terminator hairpin.

However, the TRAP protein in turn is also controlled by tRNA^{Trp}. **Figure 13.5** shows that uncharged tRNA^{Trp} binds to the mRNA for a protein called anti-TRAP. This is necessary to suppress formation of a termination hairpin in the mRNA. The result is the synthesis of anti-TRAP, which binds to TRAP, and prevents it from repressing the tryptophan operon. By this complex series of events, the absence of tryptophan generates the uncharged tRNA, which causes synthesis of anti-TRAP, which prevents function of TRAP, which causes expression of tryptophan genes.

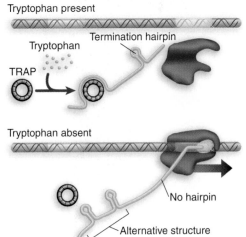

TRAP controls the *B. subtilis trp* operon

Tryptophan present

Tryptophan

Termination hairpin

TRAP

Tryptophan absent

No hairpin

Alternative structure

Figure 13.4 TRAP is activated by tryptophan and binds to *trp* mRNA. This allows the termination hairpin to form, with the result that RNA polymerase terminates, and the genes are not expressed. In the absence of tryptophan, TRAP does not bind, and the mRNA assumes a structure that prevents the terminator hairpin from forming.

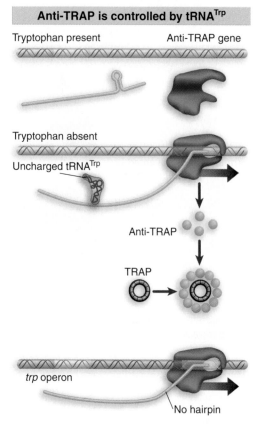

Anti-TRAP is controlled by tRNA^Trp

Tryptophan present

Anti-TRAP gene

Tryptophan absent

Uncharged tRNA^Trp

Anti-TRAP

TRAP

trp operon

No hairpin

Figure 13.5 Under normal conditions (in the presence of tryptophan) transcription terminates before the anti-TRAP gene. When tryptophan is absent, uncharged tRNA^Trp base pairs with the anti-TRAP mRNA, preventing formation of the terminator hairpin, thus causing expression of anti-TRAP.

Expression of the *B. subtilis trp* genes is therefore controlled by both tryptophan and tRNA^Trp. When tryptophan is present, there is no need for it to be synthesized. This is accomplished when tryptophan activates TRAP and therefore inhibits expression of the enzymes that synthesize tryptophan. The presence of uncharged tRNA^Trp indicates that there is a shortage of tryptophan. The uncharged tRNA activates the anti-TRAP, and thereby activates transcription of the *trp* genes.

13.4 The *E. coli tryptophan* Operon Is Controlled by Attenuation

Key Concepts

- An attenuator (intrinsic terminator) is located between the promoter and the first gene of the *trp* cluster.
- The absence of tryptophan suppresses termination and results in a 10× increase in transcription.

A complex regulatory system is used in the *E. coli trp* operon (where attenuation was originally discovered). The changes in secondary structure that control attenuation are determined by the position of the ribosome on mRNA. **Figure 13.6** shows that termination requires that *the ribosome can translate a leader segment that precedes the trp genes in the mRNA*. When the ribosome translates the leader region, a termination hairpin forms at terminator 1. But when the ribosome is prevented

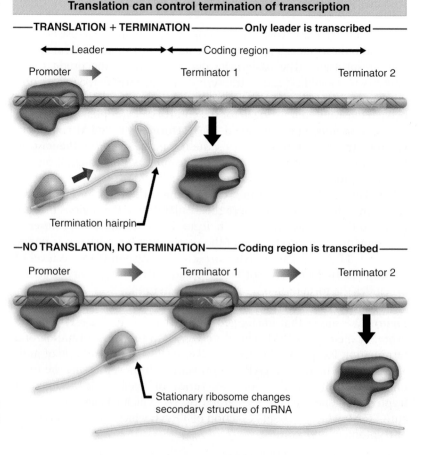

Translation can control termination of transcription

——TRANSLATION + TERMINATION——— Only leader is transcribed ———

←— Leader —→ ←— Coding region —→

Promoter   Terminator 1   Terminator 2

Termination hairpin

—NO TRANSLATION, NO TERMINATION——— Coding region is transcribed ———

Promoter   Terminator 1   Terminator 2

Stationary ribosome changes secondary structure of mRNA

Figure 13.6 Termination can be controlled via changes in RNA secondary structure that are determined by ribosome movement.

from translating the leader, the termination hairpin does not form, and RNA polymerase transcribes the coding region. *This mechanism of antitermination therefore depends upon the ability of external circumstances to influence ribosome movement in the leader region.*

The *trp* operon consists of five structural genes, coding for the three enzymes that convert chorismic acid to tryptophan. Expression of the operon is controlled by two separate mechanisms. Repression of expression is exercised by a repressor protein (coded by the unlinked gene *trpR*) that binds to an operator that is adjacent to the promoter. Attenuation controls the progress of RNA polymerase into the operon by regulating whether termination occurs at a site preceding the first structural gene. An attenuator (intrinsic terminator) is located between the promoter and the *trpE* gene. It provides a barrier to transcription into the structural genes. RNA polymerase terminates there to produce a 140-base transcript.

Termination at the attenuator responds to the level of tryptophan, as illustrated in **Figure 13.7**. In the presence of adequate amounts of tryptophan, termination is efficient. But in the absence of tryptophan, RNA polymerase can continue into the structural genes.

Repression and attenuation respond in the same way to the level of tryptophan. When tryptophan is present, the operon is repressed; and most of the RNA polymerases that escape from the promoter then terminate at the attenuator. When tryptophan is removed, RNA polymerase has free access to the promoter, and also is no longer compelled to terminate prematurely.

Attenuation has ~10× effect on transcription. When tryptophan is present, termination is effective, and the attenuator allows only ~10% of the RNA polymerases to proceed. In the absence of tryptophan, attenuation allows virtually all of the polymerases to proceed. Together with the ~70× increase in initiation of transcription that results from the release of repression, this allows an ~700-fold range of regulation of the operon.

13.5 Attenuation Can Be Controlled by Translation

Key Terms

- The **leader peptide** is the product that would result from translation of a short coding sequence used to regulate transcription of the tryptophan operon by controlling ribosome movement.
- **Ribosome stalling** describes the inhibition of movement that occurs when a ribosome reaches a codon for which there is no corresponding charged aminoacyl-tRNA.

Key Concepts

- The leader region of the *trp* operon has a 14-codon open reading frame that includes two codons for tryptophan.
- The structure of RNA at the attenuator depends on whether this reading frame is translated.
- In the presence of tryptophan, the leader is translated, and the attenuator is able to form the hairpin that causes termination.
- In the absence of tryptophan, the ribosome stalls at the tryptophan codons and an alternative secondary structure prevents formation of the hairpin, so that transcription continues.

How can termination of transcription at the attenuator respond to the level of tryptophan? The sequence of the leader region suggests a mechanism. **Figure 13.8** shows that it has a short coding sequence for a

Transcription is controlled by translation

Transcription of leader region

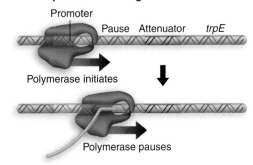

Polymerase initiates

Polymerase pauses

When tryptophan is not present transcription continues into operon

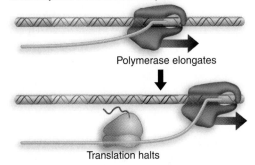

Polymerase elongates

Translation halts

When tryptophan is present transcription terminates at attenuator

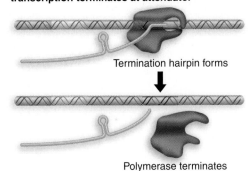

Termination hairpin forms

Polymerase terminates

Figure 13.7 An attenuator controls the progression of RNA polymerase into the *trp* genes. RNA polymerase initiates at the promoter and then proceeds to position 90, where it pauses before proceeding to the attenuator at position 140. In the absence of tryptophan, the polymerase continues into the structural genes (*trpE* starts at +163). In the presence of tryptophan there is ~90% probability of termination to release the 140-base leader RNA.

Figure 13.8 The *trp* operon has a short sequence coding for a leader peptide that is located between the operator and the attenuator.

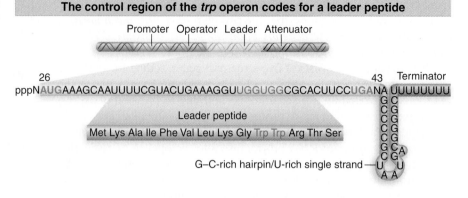

The control region of the *trp* operon codes for a leader peptide

leader peptide of 14 amino acids. The sequence contains two successive codons for tryptophan. Ribosomes that are translating the leader peptide are forced to stop at the Trp codons when the cell runs out of tryptophan. The sequence of the mRNA suggests that this **ribosome stalling** influences termination at the attenuator.

The leader sequence can be written in alternative base-paired structures. The ability of the ribosome to proceed through the leader region controls transitions between these structures. The structure determines whether the mRNA can provide the features needed for termination.

Figure 13.9 draws these structures. In the first, region 1 pairs with region 2; and region 3 pairs with region 4. The pairing of regions 3 and 4 generates the hairpin that precedes the U_8 sequence: this is the essential signal for intrinsic termination. Probably the RNA would take up this structure in lieu of any outside intervention.

A different structure forms if region 1 is prevented from pairing with region 2. In this case, region 2 is free to pair with region 3. Then region 4 has no available pairing partner; so it is compelled to remain single-stranded. So the terminator hairpin cannot be formed.

Figure 13.10 shows that the position of the ribosome can determine which structure is formed, in such a way that termination is attenuated only in the absence of tryptophan. The crucial feature is the position of the Trp codons in the leader peptide-coding sequence.

Figure 13.9 The *trp* leader region can exist in alternative base-paired conformations. The center shows the four regions that can base pair. Region 1 is complementary to region 2, which is complementary to region 3, which is complementary to region 4. On the left is the conformation produced when region 1 pairs with region 2, and region 3 pairs with region 4. On the right is the conformation when region 2 pairs with region 3, leaving regions 1 and 4 unpaired.

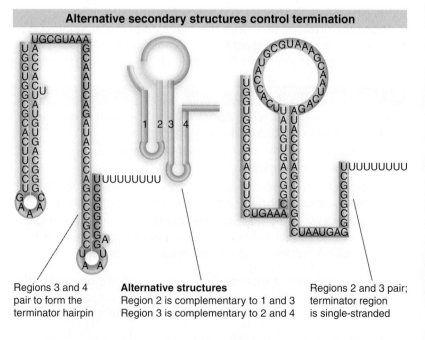

Alternative secondary structures control termination

Regions 3 and 4 pair to form the terminator hairpin

Alternative structures
Region 2 is complementary to 1 and 3
Region 3 is complementary to 2 and 4

Regions 2 and 3 pair; terminator region is single-stranded

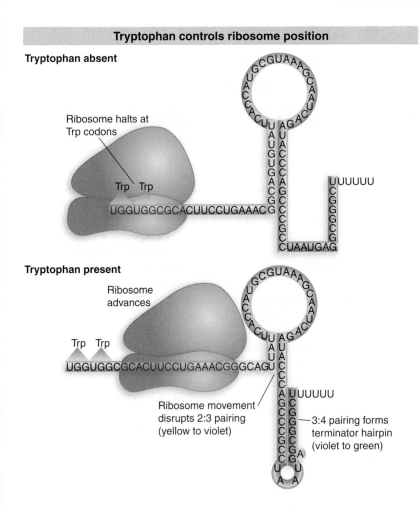

Tryptophan controls ribosome position

Tryptophan absent

Ribosome halts at Trp codons

Trp Trp

UGGUGGCGCACUUCCUGAAACG

UUUUUU

Tryptophan present

Ribosome advances

Trp Trp

UGGUGGCGCACUUCCUGAAACGGGCAGU

Ribosome movement disrupts 2:3 pairing (yellow to violet)

UUUUUU

3:4 pairing forms terminator hairpin (violet to green)

Figure 13.10 The alternatives for RNA polymerase at the attenuator depend on the location of the ribosome, which determines whether regions 3 and 4 can pair to form the terminator hairpin.

When tryptophan is present, ribosomes are able to synthesize the leader peptide. They continue along the leader section of the mRNA to the UGA codon, which lies between regions 1 and 2. As shown in the lower part of Figure 13.10, by progressing to this point, the ribosomes extend over region 2 and prevent it from base pairing. The result is that region 3 is available to base pair with region 4, generating the terminator hairpin. Under these conditions, therefore, RNA polymerase terminates at the attenuator.

When there is no tryptophan, ribosomes stall at the Trp codons, which are part of region 1, as shown in the upper part of the figure. So region 1 is sequestered within the ribosome and cannot base pair with region 2. This means that regions 2 and 3 become base paired before region 4 has been transcribed. This compels region 4 to remain in a single-stranded form. In the absence of the terminator hairpin, RNA polymerase continues transcription past the attenuator.

Figure 13.11 summarizes the role of Trp-tRNA in controlling expression of the operon. *By providing a mechanism to sense the inadequacy of the supply of Trp-tRNA, attenuation responds directly to the need of the cell for tryptophan in protein synthesis.*

E. coli and *B. subtilis* therefore use the same types of mechanisms, involving control of mRNA structure in response to the presence or absence of a tRNA, but they have combined the individual interactions in different ways. The end result is the same: to inhibit production of the enzymes when there is an excess supply of the amino acid, and to activate production when a shortage is indicated by the accumulation of uncharged tRNA^{Trp}.

Trp-tRNA controls the *E. coli* trp operon directly

Tryptophan present

Trp

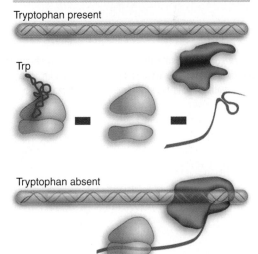

Tryptophan absent

Figure 13.11 In the presence of tryptophan tRNA, ribosomes translate the leader peptide and are released. This allows hairpin formation, so that RNA polymerase terminates. In the absence of tryptophan tRNA, the ribosome is blocked, the termination hairpin cannot form, and RNA polymerase continues.

13.6 Antisense RNA Can Be Used to Inactivate Gene Expression

Key Term

- An **antisense gene** codes for an (antisense) RNA that has a complementary sequence to an RNA that is its target.

Key Concept

- Antisense genes block expression of their targets when introduced into eukaryotic cells.

Base pairing offers a powerful means for one RNA to control the activity of another. There are many cases in both prokaryotes and eukaryotes in which a (usually rather short) single-stranded RNA base pairs with a complementary region of an mRNA, and as a result prevents expression of the mRNA. One of the early illustrations of this effect was provided by an artificial situation, in which **antisense genes** were introduced into eukaryotic cells.

Antisense genes are constructed by reversing the orientation of a gene with regard to its promoter, so that the "antisense" strand is transcribed, as illustrated in **Figure 13.12**. An antisense RNA is in effect a synthetic RNA regulator, and can inactivate a target RNA in either prokaryotic or eukaryotic cells.

At what level does the antisense RNA inhibit expression? It could in principle prevent transcription of the authentic gene, processing of its RNA product, or translation of the messenger. Results with different systems show that the inhibition depends on formation of RNA · RNA duplex molecules, but this can occur either in the nucleus or in the cytoplasm. In the case of an antisense gene stably carried by a cultured cell, sense-antisense RNA duplexes form in the nucleus, preventing normal processing and/or transport of the sense RNA. In another case, injection of antisense RNA into the cytoplasm inhibits translation by forming duplex RNA in the 5′ region of the mRNA.

This technique offered an early approach for turning off genes at will; for example, the function of a regulatory gene can be investigated by introducing an antisense version. Using the same principle, now we have more sophisticated possibilities as the result of the discovery that small RNAs can inhibit expression of complementary sequences (see *13.10 RNA Interference Is Related to Gene Silencing*).

Figure 13.12 Antisense RNA can be generated by reversing the orientation of a gene with respect to its promoter, and can anneal with the wild-type transcript to form duplex RNA.

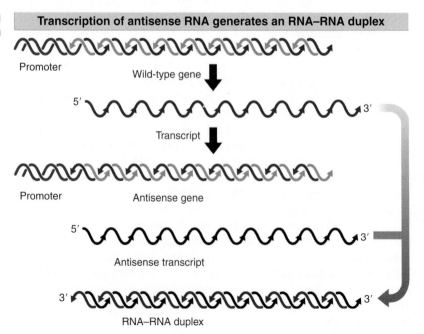

Transcription of antisense RNA generates an RNA–RNA duplex

Promoter

Wild-type gene

5′   3′

Transcript

Promoter   Antisense gene

5′   3′

Antisense transcript

3′   3′

RNA–RNA duplex

13.7 Small RNA Molecules Can Regulate Translation

Key Term

- A **riboswitch** is a catalytic RNA whose activity responds to a small ligand.

Key Concepts

- A regulator RNA functions by forming a duplex region with a target RNA.
- The duplex may block initiation of translation, cause termination of transcription, or create a target for an endonuclease.

Repressors and activators are *trans*-acting proteins. Yet the formal circuitry of a regulatory network could equally well be constructed by using an RNA as regulator. In fact, the original model for the operon left open the question of whether the regulator might be RNA or protein. Indeed, the construction of synthetic antisense RNAs turns out to mimic a class of RNA regulators that is becoming of increasing importance.

Like a protein regulator, a small regulator RNA is an independently synthesized molecule that diffuses to a target site consisting of a specific nucleotide sequence. The target for a regulator RNA is a single-stranded nucleic acid sequence. The regulator RNA functions by complementarity with its target, at which it can form a double-stranded region.

We can imagine two general mechanisms for the action of a regulator RNA:

- Formation of a duplex region with the target nucleic acid directly prevents its ability to function, by forming or sequestering a specific site. **Figure 13.13** illustrates the situation in which a protein that acts by binding to single-stranded RNA is prevented from binding because a duplex has formed. **Figure 13.14** shows the opposite type of relationship in which the formation of a duplex creates a target site for an endonuclease that destroys the RNA target.
- Formation of a duplex region in one part of the target molecule changes the conformation of another region, thus indirectly affecting its function. **Figure 13.15** shows an example. The mechanism is

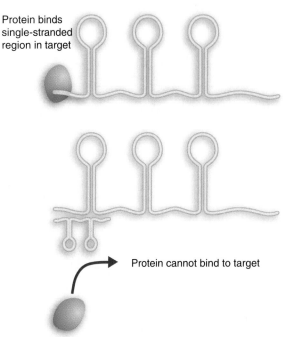

Regulator excludes protein binding

Protein binds single-stranded region in target

Protein cannot bind to target

Figure 13.13 A protein that binds to a single-stranded region in a target RNA could be excluded by a regulator RNA that forms a duplex in this region.

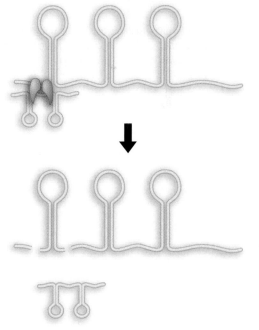

Endonuclease cleaves duplex target

Figure 13.14 By binding to a target RNA to form a duplex region, a regulator RNA may create a site that is attacked by a nuclease.

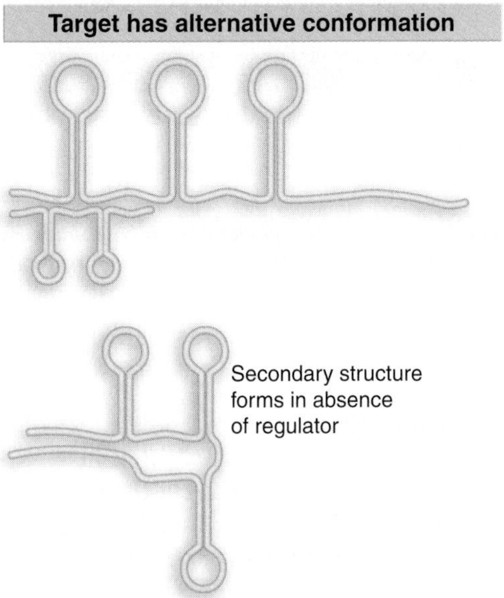

Target has alternative conformation

Secondary structure forms in absence of regulator

Figure 13.15 The secondary structure formed by base pairing between two regions of the target RNA may be prevented from forming by base pairing with a regulator RNA. In this example, the ability of the 3′ end of the RNA to pair with the 5′ end is prevented by the regulator.

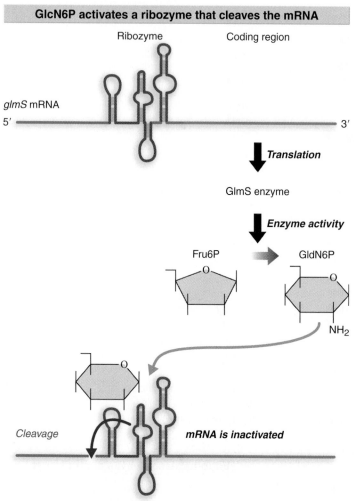

GlcN6P activates a ribozyme that cleaves the mRNA

Figure 13.16 The 5′ untranslated region of the mRNA for the enzyme that synthesizes GlcN6P contains a ribozyme that is activated by the metabolic product. The ribozyme inactivates the mRNA by cleaving it.

essentially similar to the use of secondary structure in attenuation (see *13.2 Alternative Secondary Structures Can Affect Translation or Transcription*), except that the interacting regions are on different RNA molecules instead of being part of the same RNA molecule.

The feature common to both types of RNA-mediated regulation is that changes in the secondary structure of the target control its activity.

A small RNA regulator typically can be turned on by controlling transcription of its gene, or turned off by an enzyme that degrades the RNA regulator product. Usually it is not possible otherwise to regulate the activity of an RNA regulator. In fact, it used to be thought that it would not be possible for an RNA to have allosteric properties; unlike repressor proteins that control operons, an RNA usually cannot respond to small molecules by changing its ability to recognize its target.

The discovery of the **riboswitch** provides an exception to this rule. **Figure 13.16** summarizes the regulation of the system that produces the metabolite GlcN6P. The gene *glmS* codes for an enzyme that synthesizes GlcN6P (glucosamine-6-phosphate) from fructose-6-phosphate and glutamine. The mRNA contains a long 5′ untranslated region before the coding frame. Within the UTR is a ribozyme—a sequence of RNA that has a catalytic activity (see *27.4 Ribozymes Have Various Catalytic Activities*). In this case, the catalytic activity is that of an endonuclease that cleaves its own RNA. The endonuclease is activated by binding of the metabolite product, GlcN6P, to the ribozyme. So the accumulation of GlcN6P activates the ribozyme, which cleaves the mRNA, preventing further translation. This is an exact parallel to allosteric control of a repressor protein by the end product of a metabolic pathway. There are several examples of such riboswitches in bacteria.

13.8 Bacteria Contain Regulator RNAs

Key Term

- An **sRNA** is a small bacterial RNA that functions as a regulator of gene expression.

Key Concepts

- *E. coli* has several sRNAs that coordinate control of expression of many target genes.
- The *oxyS* sRNA activates or represses expression of >10 loci at the post-transcriptional level.

In bacteria, regulator RNAs are short molecules, collectively known as **sRNAs**: *E. coli* contains at least 17 different sRNAs. Some of the sRNAs are general regulators that affect many target genes. Oxidative stress provides an interesting example of a general control system in which RNA is the regulator. When exposed to reactive oxygen species, bacteria respond by inducing antioxidant defense genes. Hydrogen peroxide activates the transcription activator OxyR, which controls the expression of several inducible genes. One of these genes is *oxyS*, which codes for a small RNA.

Figure 13.17 shows two salient features of the control of *oxyS* expression. In a wild-type bacterium under normal conditions, it is not expressed. The pair of gels on the left side of the figure shows that it is expressed at high levels in a mutant bacterium with a constitutively active *oxyR* gene. This identifies *oxyS* as a target for activation by *oxyR*. The pair of gels on the right side of the figure shows that *oxyS* RNA is transcribed within 1 minute of exposure to hydrogen peroxide.

The *oxyS* RNA is a short sequence that does not code for protein. It is a *trans*-acting regulator that affects gene expression at posttranscriptional levels. It has >10 target loci; at some of them, it activates expression, at others it represses expression. **Figure 13.18** shows the mechanism of repression of one target, the *flhA* mRNA. Three stem-loop structures protrude in the secondary structure of *oxyR* mRNA, and the loop close to the 3′ terminus is complementary to a sequence just preceding the initiation codon of *flhA* mRNA. Base pairing between *oxyS* RNA and *flhA* RNA prevents the ribosome from binding to the initiation codon, and therefore represses translation. There is also a second pairing interaction that involves a sequence within the coding region of *flhA*.

Another target for *oxyS* is *rpoS*, the gene coding for an alternative sigma factor (which activates a general stress response). By inhibiting production of the sigma factor, *oxyS* ensures that the specific response to oxidative stress does not trigger the response that is appropriate for other stress conditions. The *rpoS* gene is also regulated by two other sRNAs (d*srA* and r*prA*), which activate it. These three sRNAs appear to be global regulators that coordinate responses to various environmental conditions.

The actions of all three sRNAs are assisted by an RNA-binding protein called Hfq. The Hfq protein was originally identified as a bacterial host factor needed for replication of the RNA bacteriophage Qβ. It is related to the Sm proteins of eukaryotes that bind to many of the snRNAs (small nuclear RNAs) that have regulatory roles in gene expression (see *26.5 snRNAs Are Required for Splicing*). Mutations in the gene coding Hfq have many effects, identifying it as a pleiotropic protein. Hfq binds to many of the sRNAs of *E. coli*. It increases the effectiveness of *oxyS* RNA by enhancing its ability to bind to its target mRNAs. The effect of Hfq is probably mediated by causing a small change in the secondary structure of *oxyS* RNA that improves the exposure of the single-stranded sequences that pair with the target mRNAs.

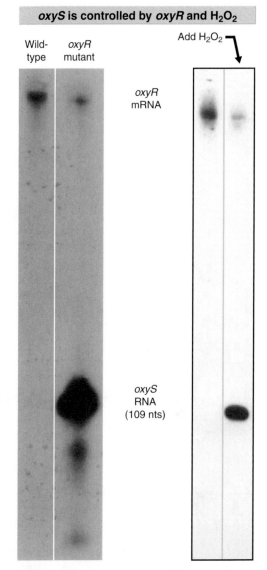

oxyS is controlled by oxyR and H₂O₂

Wild-type   oxyR mutant

Add H₂O₂

oxyR mRNA

oxyS RNA (109 nts)

Figure 13.17 The gels on the left show that *oxyS* RNA is induced in an *oxyR* constitutive mutant. The gels on the right show that *oxyS* RNA is induced within 1 minute of adding hydrogen peroxide to a wild-type culture. Photograph kindly provided by Gisela Storz.

Figure 13.18 *oxyS* RNA inhibits translation of *flhA* mRNA by base pairing with a sequence just upstream of the AUG initiation codon.

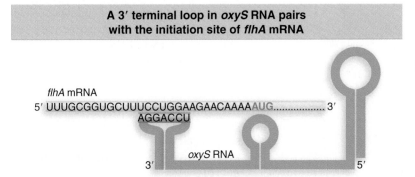

A 3′ terminal loop in oxyS RNA pairs with the initiation site of flhA mRNA

flhA mRNA
5′ UUUGCGGUGCUUUCCUGGAAGAACAAAAAUG................ 3′
AGGACCU

oxyS RNA
3′   5′

13.9 MicroRNAs Are Regulators in Many Eukaryotes

Key Term

- **MicroRNAs** are very short RNAs that may regulate gene expression.

Key Concepts

- Animal and plant genomes code for many short (~22 base) RNA molecules, called microRNAs.
- MicroRNAs regulate gene expression by base pairing with complementary sequences in target mRNAs.

Very small RNAs are gene regulators in many eukaryotes. The first example was discovered in the nematode *C. elegans* as the result of the interaction between the regulator gene *lin4* and its target gene, *lin14*. **Figure 13.19** illustrates the behavior of this regulatory system. The *lin14* target gene regulates larval development. Expression of *lin14* is controlled by *lin4*, which codes for a small transcript of 22 nucleotides. The *lin4* transcripts are complementary to a 10 base sequence that is repeated seven times in the 3′ nontranslated region of *lin14*. Expression of *lin4* represses expression of *lin14* post-transcriptionally, most likely because the base pairing reaction between the two RNAs leads to degradation of the mRNA.

The *lin4* RNA is an example of a **microRNA**. There are ~55 genes in the *C. elegans* genome coding for microRNAs of 21–24 nucleotide length. They have varying patterns of expression during development and are likely to be regulators of gene expression. Many of the microRNAs of *C. elegans* are contained in a large (15S) ribonucleoprotein particle.

Many of the *C. elegans* microRNAs have homologues in mammals, so the mechanism may be widespread. They are also found in plants. Of 16 microRNAs in *Arabidopsis*, eight are completely conserved in rice, suggesting widespread conservation of this regulatory mechanism.

The mechanism of production of the microRNAs is also widely conserved. In the example of *lin4*, the gene is transcribed into a transcript that forms a double-stranded region that becomes a target for a nuclease called Dicer. This has an N-terminal helicase activity, enabling it to unwind the double-stranded region, and two nuclease domains that are related to the bacterial ribonuclease III. Related enzymes are found in flies, worms, and plants. Active microRNA is generated when a precursor RNA transcript is cleaved. Interfering with the enzyme activity blocks the production of microRNAs.

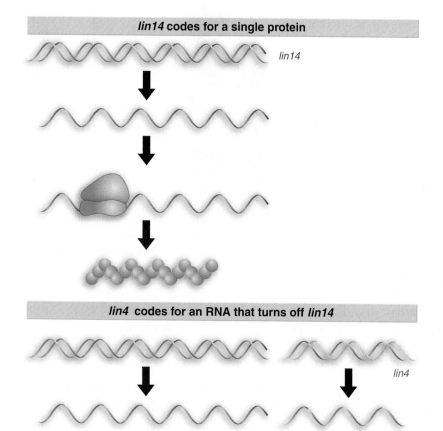

lin14 codes for a single protein

lin14

lin4 codes for an RNA that turns off _lin14_

lin4

No protein

Figure 13.19 *lin4* RNA regulates expression of *lin14* by binding to the 3′ nontranslated region.

13.10 RNA Interference Is Related to Gene Silencing

Key Terms

- **RNA interference (RNAi)** describes the technique in which double-stranded RNA is introduced into cells to eliminate or reduce the activity of a target gene; the introduced RNA sequences are complementary to the target gene. The technique works because of the ability of double-stranded RNA sequences to trigger degradation of the mRNA of the gene.
- **RNA silencing** describes the ability of a double-stranded RNA to suppress expression of the corresponding gene systemically in a plant.
- **Cosuppression** describes the ability of a transgene (usually in plants) to inhibit expression of the corresponding endogenous gene.

Key Concepts

- RNA interference triggers degradation of mRNAs complementary to either strand of a short double-stranded RNA.
- Double-stranded RNA may cause silencing of host genes.

The regulation of mRNAs by microRNAs is mimicked by the phenomenon of **RNA interference (RNAi)**. This was discovered when it was observed that antisense and sense RNAs can be equally effective in inhibiting gene expression. The reason is that preparations of either type of (supposedly) single-stranded RNA are actually contaminated by small amounts of double-stranded RNA.

The double-stranded RNA (dsRNA) is degraded by ATP-dependent cleavage to give oligonucleotides of 21–23 bases. The short RNA is sometimes called siRNA (short interfering RNA). **Figure 13.20** shows that the cleavage makes breaks in a long dsRNA to generate siRNA fragments with short (2 base) protruding 3' ends. The same enzyme (Dicer) that generates microRNAs is responsible for the cleavage.

RNAi occurs post-transcriptionally when an siRNA induces degradation of a complementary mRNA. **Figure 13.21** suggests that the siRNA may provide a template that directs a nuclease to degrade mRNAs that are complementary to one or both strands, perhaps by a process in which the mRNA pairs with the fragments. It is likely that a helicase is required to assist the pairing reaction. The siRNA directs cleavage of the mRNA in the middle of the paired segment. These reactions occur within a ribonucleoprotein complex called RISC (RNA-induced silencing complex).

RNAi has become a powerful technique for ablating the expression of a specific target gene in invertebrate cells, especially in *C. elegans* and *D. melanogaster*. However, the technique has been limited in mammalian cells, which have the more generalized response to dsRNA of shutting down protein synthesis and degrading mRNA. **Figure 13.22** shows that this happens because of two reactions. The dsRNA activates the enzyme PKR, which inactivates the translation initiation factor eIF2a by phosphorylating it. And it activates 2'5' oligoadenylate synthetase, whose product activates RNAase L, which degrades all mRNAs. However, it turns out that these reactions require dsRNA that is

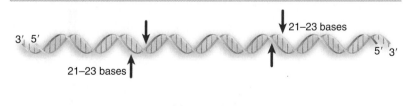

dsRNA is cleaved ~22 bases from the 3' ends to generate siRNA

21–23 bases

21–23 bases

21–23 base siRNA with protruding 3' ends

Figure 13.20 siRNA that mediates RNA interference is generated by cleaving dsRNA into smaller fragments. The cleavage reaction occurs 21–23 nucleotides from a 3' end. The siRNA product has protruding bases on its 3' ends.

RNAi works by generating siRNA

Nuclease cleaves dsRNA to siRNA

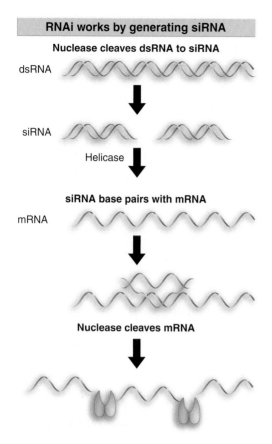

Figure 13.21 RNAi is generated when a dsRNA is cleaved into fragments that direct cleavage of the corresponding mRNA.

longer than 26 nucleotides. If shorter dsRNA (21–23 nucleotides) is introduced into mammalian cells, it triggers the specific degradation of complementary RNAs just as with the RNAi technique in worms and flies. With this advance, it seems likely that RNAi will become the universal mechanism of choice for turning off the expression of a specific gene.

RNA interference is related to natural processes in which gene expression is silenced. Plants and fungi show **RNA silencing** (sometimes called post-transcriptional gene silencing) in which dsRNA inhibits expression of a gene. The most common source of the RNA is a replicating virus. This mechanism may have evolved as a defense against viral infection. When a virus infects a plant cell, the formation of dsRNA triggers the suppression of expression from the plant genome. RNA silencing has the further remarkable feature that it is not limited to the cell in which the viral infection occurs: it can spread throughout the plant systemically. Presumably the propagation of the signal involves passage of RNA or fragments of RNA. It may require some of the same features that are involved in movement of the virus itself. It is possible that RNA silencing involves an amplification of the signal by an RNA-dependent RNA synthesis process in which a novel polymerase uses the siRNA as a primer to synthesize more RNA on a template of complementary RNA.

A related process is the phenomenon of **cosuppression**, in which introduction of a transgene causes the corresponding endogenous gene to be silenced. This has been largely characterized in plants. The implication is that the transgene must make antisense as well as sense RNA copies, and this inhibits expression of the endogenous gene.

Silencing takes place by RNA–RNA interactions. It is also possible that dsRNA may inhibit gene expression by interacting with the DNA. If a DNA copy of a viroid RNA sequence is inserted into a plant genome, it becomes methylated when the viroid RNA replicates. This suggests that the RNA sequence could be inducing methylation of the DNA sequence. Similar targeting of methylation of DNA corresponding to sequences represented in dsRNA has been detected in plant cells. Methylation of DNA is associated with repression of transcription, so this could be another means of silencing genes represented in dsRNA. Nothing is known about the mechanism.

Figure 13.22 dsRNA inhibits protein synthesis and triggers degradation of all mRNA in mammalian cells as well as having sequence-specific effects.

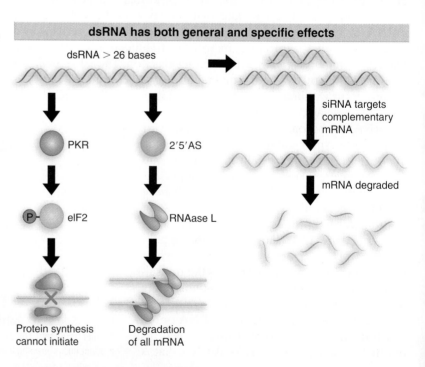

13.11 SUMMARY

Gene expression can be regulated positively by factors that activate a gene or negatively by factors that repress a gene. The first and most common level of control is at the initiation of transcription, but termination of transcription may also be controlled. Translation may be controlled by regulators that interact with mRNA. The regulatory products may be proteins, which often are controlled by allosteric interactions in response to the environment, or RNAs, which function by base pairing with the target RNA to change its secondary structure. Regulatory networks can be created by linking regulators so that the production or activity of one regulator is controlled by another.

Attenuation is a mechanism that relies on regulation of termination to control transcription through bacterial operons. It is commonly used in operons that code for enzymes involved in biosynthesis of an amino acid. The polycistronic mRNA of the operon starts with a sequence that can form alternative secondary structures. One of the structures has a hairpin loop that provides an intrinsic terminator upstream of the structural genes; the alternative structure lacks the hairpin. Various types of interaction determine whether the hairpin forms. One is for a protein to bind to the mRNA to prevent formation of the alternative structure. In the *trp* operon of *B. subtilis*, the TRAP protein has this function; it is controlled by the anti-TRAP protein, whose production in turn is controlled by the level of uncharged aminoacyl-tRNA^{Trp}. In the *trp* (and other) operons of *E. coli*, the choice of which structure forms is controlled by the progress of translation through a short leader sequence that includes codons for the amino acid(s) that are the product of the system. In the presence of aminoacyl-tRNA bearing such amino acid(s), ribosomes translate the leader peptide, allowing a secondary structure to form that supports termination. In the absence of this aminoacyl-tRNA, the ribosome stalls, resulting in a new secondary structure in which the hairpin needed for termination cannot form. The supply of aminoacyl-tRNA therefore (inversely) controls amino acid biosynthesis.

Small regulator RNAs are found in both bacteria and eukaryotes. *E. coli* has ~17 sRNA species. The *oxyS* sRNA controls about 10 target loci at the post-transcriptional level; some of them are repressed, and others are activated. Repression is caused when the sRNA binds to a target mRNA to form a duplex region that includes the ribosome-binding site. MicroRNAs are ~22 bases long and are produced in many eukaryotes by cleavage of a longer transcript. They function by base pairing with target mRNAs to form duplex regions that are susceptible to cleavage by endonucleases. The degradation of the mRNA prevents its expression. The technique of RNA interference is becoming the method of choice for inactivating eukaryotic genes. It uses the introduction of short dsRNA sequences with one strand complementary to the target RNA, and it works by inducing degradation of the targets. This may be related to a natural defense system in plants called RNA silencing.

Phage Strategies

14.1 Introduction

Key Terms

- **Bacteriophages (phages)** are viruses that infect bacteria.
- **Lytic infection** of a bacterium by a phage ends in the destruction of the bacterium with release of progeny phage.
- **Lysis** describes the death of a bacterium at the end of a phage infective cycle when the bacterium bursts open, releasing the progeny of the infecting phage (because phage enzymes disrupt the bacterium's cytoplasmic membrane or cell wall). The same term also applies to eukaryotic cells; for example, when infected cells are attacked by the immune system.
- **Lysogeny** describes the ability of a phage to survive in a bacterium as a stable prophage component of a bacterial genome.
- **Prophage** is a phage genome covalently integrated as a linear part of a bacterial chromosome.
- **Integration** of a viral or another DNA sequence describes its insertion into a host genome as a region covalently linked on either side to the host sequences.
- **Induction** of a prophage describes its entry into the lytic (infective) cycle as a result of destruction of the lysogenic repressor, which leads to excision of free phage DNA from the bacterial chromosome.
- The **excision** of a phage or episome or other sequence describes its release from the host chromosome as an autonomous DNA molecule.

A virus consists of a nucleic acid genome contained in a protein coat. In order to reproduce, the virus must infect a host cell. The typical pattern of an infection is to subvert the functions of the host cell to the purpose of

14.3 Lytic Development Is Controlled by a Cascade

- A **cascade** is a sequence of events, each of which is stimulated by the previous one. In transcriptional regulation, as seen in sporulation and phage lytic development, it means that regulation is divided into stages, and at each stage, one of the genes that are expressed codes for a regulator needed to express the genes of the next stage.

- **Early genes** are transcribed before the replication of phage DNA. They code for regulators and other proteins needed for later stages of infection.

- **Immediate early genes** in phage lambda are equivalent to the early class of other phages. They are transcribed immediately upon infection by the host RNA polymerase.

- **Delayed early genes** in phage lambda are equivalent to the middle genes of other phages. They cannot be transcribed until regulator protein(s) coded by the immediate early genes have been synthesized.

- **Middle genes** are phage genes that are regulated by the proteins coded by early genes. Some proteins coded by middle genes catalyze replication of the phage DNA; others regulate the expression of a later set of genes.

- **Late genes** are transcribed when phage DNA is being replicated. They code for components of the phage particle.

Key Concepts

- The early genes transcribed by host RNA polymerase following infection include regulators required for expression of the middle genes.

- The middle genes include regulators to transcribe the late genes.

- The cascade results in the ordered expression of groups of genes during phage infection.

The organization of the phage genetic map often reflects the sequence of lytic development. The concept of the operon is taken to somewhat of an extreme, in which the genes coding for proteins with related functions are clustered to allow their control with the maximum economy. This allows the pathway of lytic development to be controlled with a small number of regulatory switches.

The lytic cycle is under positive control, so that each group of phage genes can be expressed only when an appropriate signal is given. **Figure 14.3** shows that the regulatory genes function in a **cascade**, in which a gene expressed at one stage is necessary for synthesis of the genes that are expressed at the next stage.

The first stage of gene expression necessarily relies on the transcription apparatus of the host cell. Usually only a few genes are expressed at this stage. Their promoters are indistinguishable from those of host genes. The name of the first-stage genes depends on the phage. In most cases, they are known as the **early genes**. In phage lambda, they are given the evocative description of **immediate early genes**. Irrespective of the name, they constitute only a preliminary, representing just the initial part of the early period. Sometimes they are exclusively occupied with the transition to the next period. At all events, *one of these genes always codes for a protein that is necessary for transcription of the next group of genes.*

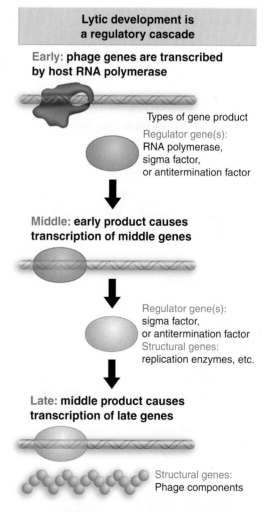

Figure 14.3 Phage lytic development proceeds by a regulatory cascade, in which a gene product at each stage is needed for expression of the genes at the next stage.

This second group of genes is known variously as the **delayed early** or **middle gene** group. Its expression typically starts as soon as the regulator protein coded by the early gene(s) is available. Depending on the nature of the control circuit, the initial set of early genes may or may not continue to be expressed at this stage. Often the expression of host genes is reduced. Together the two sets of early genes account for all necessary phage functions except those needed to assemble the particle coat itself and to lyse the cell.

When the replication of phage DNA begins, it is time for the **late genes** to be expressed. Their transcription at this stage usually is arranged by embedding a further regulator gene within the previous (delayed early or middle) set of genes. This regulator may be another antitermination factor (as in lambda) or it may be another sigma factor (as in SPO1).

A lytic infection often falls into three stages, as shown in Figure 14.3. The first stage consists of early genes transcribed by host RNA polymerase (sometimes the regulators are the only products at this stage). The second stage consists of genes transcribed under direction of the regulator produced in the first stage (most of these genes code for enzymes needed for replication of phage DNA). The final stage consists of genes for phage components, transcribed under direction of a regulator synthesized in the second stage.

The use of these successive controls, in which each set of genes contains a regulator that is necessary for expression of the next set, creates a cascade in which groups of genes are turned on (and sometimes off) at particular times. The means used to construct each phage cascade are different, but the results are similar.

14.4 Two Types of Regulatory Event Control the Lytic Cascade

Key Concept

- Regulator proteins used in phage cascades may sponsor initiation at new (phage) promoters or cause the host polymerase to read through transcription terminators.

At every stage of phage expression, one or more of the active genes is a regulator that is needed for the subsequent stage. The regulator may take the form of a new RNA polymerase, a sigma factor that redirects the specificity of the host RNA polymerase, or an antitermination factor that allows it to read a new group of genes (see *11.16 Antitermination Is a Regulatory Event*). Now let's compare the use of switching at initiation or termination to control gene expression.

Figure 14.4 shows that phages use two types of mechanisms for changing promoter recognition. One is to replace the sigma factor of the host enzyme with another factor that redirects its specificity in initiation. An alternative is to synthesize a new phage RNA polymerase. In either case, the critical feature that distinguishes the new set of genes is their possession of *different promoters from those originally recognized by host RNA polymerase*. **Figure 14.5** shows that the two sets of transcripts are independent; as a consequence, early gene expression can cease after the new sigma factor or polymerase has been produced.

Antitermination provides an alternative mechanism for phages to control the switch from early genes to the next stage of expression. The use of antitermination depends on a particular arrangement of genes. **Figure 14.6** shows that the early genes lie adjacent to the genes that are to

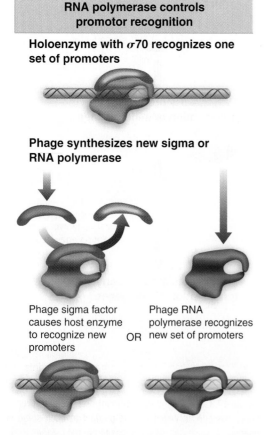

RNA polymerase controls promotor recognition

Holoenzyme with σ70 recognizes one set of promoters

Phage synthesizes new sigma or RNA polymerase

Phage sigma factor causes host enzyme to recognize new promoters

OR

Phage RNA polymerase recognizes new set of promoters

Figure 14.4 A phage may control transcription at initiation either by synthesizing a new sigma factor that replaces the host sigma or by synthesizing a new RNA polymerase.

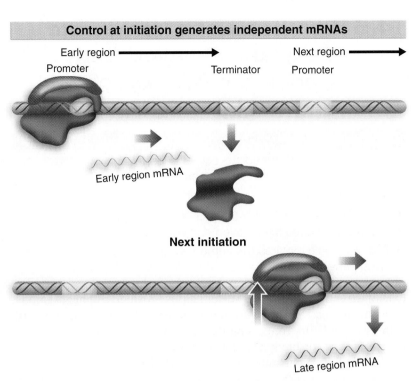

Control at initiation generates independent mRNAs

Early region ━━━▶

Promoter

Terminator

Next region ━━━▶

Promoter

Early region mRNA

Next initiation

Late region mRNA

Figure 14.5 Control at initiation utilizes independent transcription units, each with its own promoter and terminator, which produce independent mRNAs. The transcription units need not be located near one another.

be expressed next, but are separated from them by terminator sites. *If termination is prevented at these sites, the polymerase reads through into the genes on the other side.* So in antitermination, the *same promoters* continue to be recognized by RNA polymerase. The new genes are expressed only by extending the RNA chain to form molecules that contain the early gene sequences at the 5′ end and the new gene sequences at the 3′ end. Because the two types of sequence remain linked, early gene expression inevitably continues.

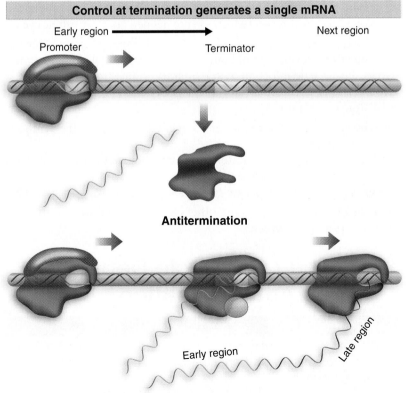

Control at termination generates a single mRNA

Early region ━━━▶

Promoter

Terminator

Next region

Early region mRNA

Antitermination

Early region   Late region

Figure 14.6 Control at termination requires adjacent units, so that transcription can read from the first gene into the next gene. This produces a single mRNA that contains both sets of genes.

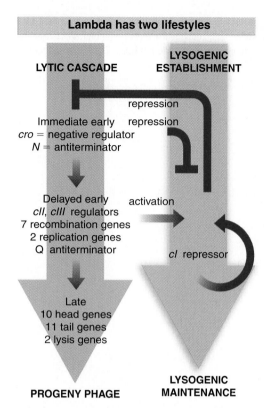

Lambda has two lifestyles

LYTIC CASCADE — LYSOGENIC ESTABLISHMENT

repression

Immediate early   repression
cro = negative regulator
N = antiterminator

Delayed early   activation
cII, cIII regulators
7 recombination genes
2 replication genes
Q antiterminator

cI repressor

Late
10 head genes
11 tail genes
2 lysis genes

PROGENY PHAGE   LYSOGENIC MAINTENANCE

Figure 14.7 The lambda lytic cascade is interlocked with the circuitry for lysogeny.

14.5 Lambda Uses Immediate Early and Delayed Early Genes for Both Lysogeny and the Lytic Cycle

Key Concepts

- Lambda has two immediate early genes, *N* and *cro*, which are transcribed by host RNA polymerase.
- *N* is required to express the delayed early genes.
- Three of the delayed early genes are regulators.
- Lysogeny requires the delayed early genes *cII–cIII*.
- The lytic cycle requires the immediate early gene *cro* and the delayed early gene *Q*.

One of the most intricate cascade circuits is provided by phage lambda. Actually, the cascade for lytic development itself is straightforward, with two regulators controlling the successive stages of development. But the circuit for the lytic cycle is interlocked with the circuit for establishing lysogeny, as summarized in **Figure 14.7**.

When lambda DNA enters a new host cell, the lytic and lysogenic pathways start off the same way. Both require expression of the immediate early and delayed early genes. But then they diverge: lytic development follows if the late genes are expressed; lysogeny ensues if synthesis of the repressor is established.

Lambda has only two immediate early genes, transcribed independently by host RNA polymerase:

- *N* codes for an antitermination factor whose action at the *nut* sites allows transcription to proceed into the delayed early genes (see *11.16 Antitermination Is a Regulatory Event*).
- *cro* has dual functions: it prevents synthesis of the repressor (a necessary action if the lytic cycle is to proceed); and it turns off expression of the early genes (which are not needed later in the lytic cycle).

The delayed early genes include two replication genes (needed for lytic infection), seven recombination genes (some involved in recombination during lytic infection, two necessary to integrate lambda DNA into the bacterial chromosome for lysogeny), and three regulators. The regulators have opposing functions:

- The *cII–cIII* pair of regulators is needed to establish the synthesis of repressor.
- The *Q* regulator is an antitermination factor that allows host RNA polymerase to transcribe the late genes.

So the delayed early genes serve two masters: some are needed for the phage to enter lysogeny, the others are concerned with controlling the order of the lytic cycle.

14.6 The Lytic Cycle Depends on Antitermination

Key Concepts

- pN is a lambda antitermination factor that allows RNA polymerase to continue transcription past the ends of the two immediate early genes.

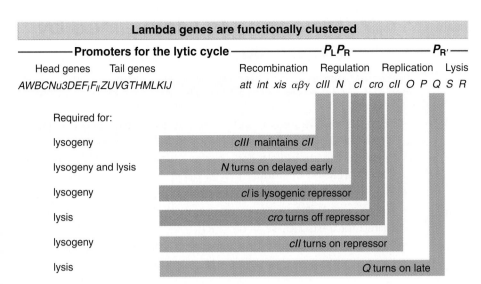

Lambda genes are functionally clustered

Figure 14.8 The lambda map shows clustering of related functions. The genome is 48,514 bp.

- pQ is the product of a lambda delayed early gene and is an antiterminator that allows RNA polymerase to transcribe the late genes.

- Because lambda DNA circularizes after infection, the late genes form a single transcription unit.

To disentangle the two pathways, let's first consider just the lytic cycle. **Figure 14.8** gives the map of lambda phage DNA. A group of genes concerned with regulation is surrounded by genes needed for recombination and replication. The genes coding for structural components of the phage are clustered. All the genes necessary for the lytic cycle are expressed in polycistronic transcripts from three promoters.

Figure 14.9 shows that the two immediate early genes, N and cro, are transcribed by host RNA polymerase. N is transcribed toward the left, and cro toward the right. Each transcript is terminated at the end of the gene. The protein pN is the regulator that allows transcription to continue into the delayed early genes. It is an antitermination factor that suppresses use of the terminators t_L and t_R. In the presence of pN, transcription continues to the left of N into the recombination genes, and to the right of cro into the replication genes.

The map in Figure 14.8 gives the organization of the lambda DNA as it exists in the phage particle. But shortly after infection, the ends of the DNA join to form a circle. **Figure 14.10** shows the true state of lambda DNA during infection. The late genes are welded into a single group, containing the lysis genes S–R from the right end of the linear DNA, and the head and tail genes A–J from the left end.

The late genes are expressed as a single transcription unit, starting from a promoter $P_{R'}$ that lies between Q and S. The late promoter is used constitutively. However, in the absence of the product of gene Q (which is the last gene in the rightward delayed early unit), late transcription terminates at a site t_{R3}. The transcript resulting from this termination event is 194 bases long; it is known as 6S RNA. When pQ becomes available, it suppresses termination at t_{R3} and the 6S RNA is extended, with the result that the late genes are expressed.

Similar controls apply to λ left and right transcription

Lambda has leftward and rightward transcription units

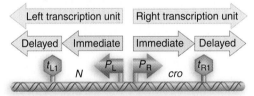

Immediate early Only N and cro are transcribed

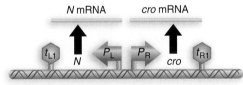

pN extends transcription into delayed early genes

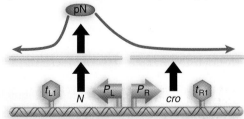

Figure 14.9 Phage lambda has two early transcription units; in the "leftward" unit, the "upper" strand is transcribed toward the left; in the "rightward" unit, the "lower" strand is transcribed toward the right. Genes N and cro are the immediate early functions, and are separated from the delayed early genes by the terminators. Synthesis of N protein allows RNA polymerase to pass the terminators t_{L1} to the left and t_{R1} to the right.

Lambda has three stages of development

State of lambda DNA	Stage and activity
	**Early** Host RNA polymerase transcribes *N* and *cro* from P_L and P_R
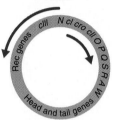	**Delayed early** pN permits transcription from same promoters to continue past *N* and *cro*
	**Late** Transcription initiates at $P_{R'}$ (between *Q* and *S*) and pQ permits it to continue through all late genes

Figure 14.10 Lambda DNA circularizes during infection, so that the late gene cluster is intact in one transcription unit.

14.7 Lysogeny Is Maintained by Repressor Protein

Key Concepts

- Mutants in the *cI* gene cannot maintain lysogeny.
- *cI* codes for a repressor protein that acts at the O_L and O_R operators to block transcription of the immediate early genes.
- Because the immediate early genes trigger a regulatory cascade, their repression prevents the lytic cycle from proceeding.
- Repressor binding at O_L blocks transcription of gene *N* from P_L.
- Repressor binding at O_R blocks transcription of *cro* but also is required for transcription of *cI*.
- Repressor binding to the operators therefore simultaneously blocks entry to the lytic cycle and promotes its own synthesis.

Looking at the lambda lytic cascade, we see that the entire program is set in train by initiating transcription at the two promoters P_L and P_R for the immediate early genes *N* and *cro*. Because lambda uses antitermination to proceed to the next stage of (delayed early) expression, the same two promoters continue to be used throughout the early period.

The expanded map of the regulatory region drawn in **Figure 14.11** shows that the promoters P_L and P_R lie on either side of the *cI* gene. Associated with each promoter is an operator (O_L, O_R) at which repressor protein binds to prevent RNA polymerase from initiating transcription. The sequence of each operator overlaps with the promoter that it controls; so often these are described as the P_L/O_L and P_R/O_R control regions.

Because of the sequential nature of the lytic cascade, the control regions provide a pressure point at which entry to the entire cycle can be controlled. *By denying RNA polymerase access to these promoters, a repressor protein prevents the phage genome from entering the lytic cycle.* The repressor functions in the same way as repressors of bacterial operons: it binds to specific operators.

The repressor protein is coded by the *cI* gene. Mutants in this gene cannot maintain lysogeny, but always enter the lytic cycle. Expression of the repressor protein has two consequences: it maintains the lysogenic state; and it also provides immunity for a lysogenic cell against superinfection by new phage lambda genomes.

The repressor binds independently to the two operators. Its ability to block transcription at the associated promoters is illustrated in **Figure 14.12**.

At O_L the repressor has the same sort of effect that we have already discussed for several other systems: it prevents RNA polymerase from initiating transcription at P_L. This stops the expression of gene *N*. Because P_L is used for all leftward early gene transcription, this action prevents expression of the entire leftward early transcription unit. *So the lytic cycle is blocked before it can proceed beyond the early stages.*

At O_R, repressor binding prevents the use of P_R. So *cro* and the other rightward early genes cannot be expressed. The repressor at O_2 also stimulates transcription of its own gene from P_{RM}.

The nature of this control circuit explains the biological features of lysogenic existence. Lysogeny is stable because the control circuit ensures that, so long as the

Lambda has a compact regulatory region

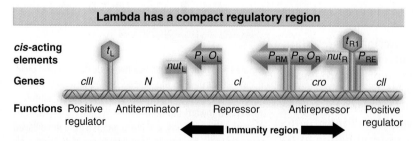

Figure 14.11 The lambda regulatory region contains a cluster of *trans*-acting functions and *cis*-acting elements.

level of repressor is adequate, there is continued expression of the *cI* gene. The result is that O_L and O_R remain occupied indefinitely. By repressing the entire lytic cascade, this action maintains the prophage in its inert form.

14.8 The Repressor and Its Operators Define the Immunity Region

Key Terms

- **Immunity** in phages refers to the ability of a prophage to prevent another phage of the same type from infecting a cell. It results from the synthesis of phage repressor by the prophage genome.
- **Virulent** phage mutants are unable to establish lysogeny.
- The **immunity region** is a segment of the phage genome that enables a prophage to inhibit additional phages of the same type from infecting the bacterium. This region has a gene that encodes for the repressor, as well as the sites to which the repressor binds.

Key Concepts

- Several lambdoid phages have different immunity regions.
- A lysogenic phage confers immunity to further infection by any other phage with the same immunity region.

The presence of repressor explains the phenomenon of **immunity**. If a second lambda phage DNA enters a lysogenic cell, repressor protein synthesized from the resident prophage genome immediately binds to O_L and O_R in the new genome. This prevents the second phage from entering the lytic cycle.

The operators were originally identified as the targets for repressor action by **virulent** phage mutations (λvir). These mutations prevent the repressor from binding at O_L or O_R, with the result that the phage inevitably proceeds into the lytic pathway when it infects a new host bacterium. And λvir mutants can reproduce in lysogenic cells because the virulent mutations in O_L and O_R allow the incoming phage to ignore the resident repressor and thus to enter the lytic cycle. Virulent mutations in phages are the equivalent of operator-constitutive mutations in bacterial operons.

A prophage is induced to enter the lytic cycle when the lysogenic circuit is broken. This happens when the repressor is inactivated, as discussed in the next section. The absence of repressor allows RNA polymerase to bind at P_L and P_R, starting the lytic cycle as shown in **Figure 14.13**.

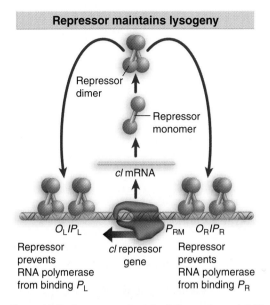

Repressor maintains lysogeny

Repressor dimer

Repressor monomer

cI mRNA

O_L/P_L

Repressor prevents RNA polymerase from binding P_L

cI repressor gene

P_{RM} O_R/P_R

Repressor prevents RNA polymerase from binding P_R

Figure 14.12 Repressor acts at the left operator and right operator to prevent transcription of the immediate early genes (*N* and *cro*). It also acts at the promoter P_{RM} to activate transcription by RNA polymerase of its own gene.

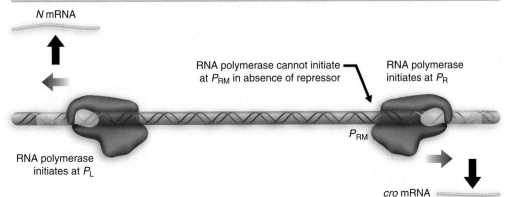

RNA polymerase initiates at P_L and P_R but not at P_{RM} during the lytic cycle

N mRNA

RNA polymerase cannot initiate at P_{RM} in absence of repressor

RNA polymerase initiates at P_R

P_{RM}

RNA polymerase initiates at P_L

cro mRNA

Figure 14.13 In the absence of repressor, RNA polymerase initiates at the left and right promoters. It cannot initiate at P_{RM} in the absence of repressor.

The region including the left and right operators, the *cI* gene, and the *cro* gene determines the immunity of the phage. Any phage that possesses this region has the same type of immunity, because *it specifies both the repressor protein and the sites on which the repressor acts.* Accordingly, this is called the **immunity region** (as marked in Figure 14.11). Each of the four lambdoid phages φ80, 21, 434, and λ has a unique immunity region. When we say that a lysogenic phage confers immunity to any other phage of the same type, we mean more precisely that the immunity is to any other phage that has the same immunity region (irrespective of differences in other regions).

14.9 The DNA-Binding Form of Repressor Is a Dimer

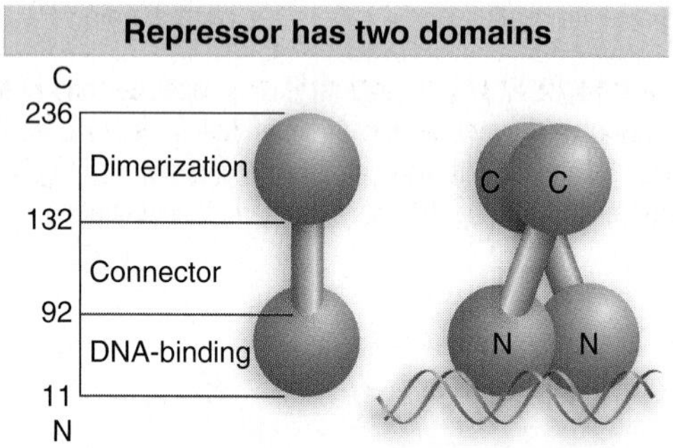

Key Concepts

- A repressor monomer has two distinct domains.
- The N-terminal domain contains the DNA-binding site.
- The C-terminal domain dimerizes.
- Binding to the operator requires the dimeric form so that two DNA-binding domains can contact the operator simultaneously.
- Cleavage of the repressor between the two domains reduces the affinity for the operator and induces a lytic cycle.

The lambda repressor subunit is a polypeptide of 27 kD with the two distinct domains summarized in **Figure 14.14**.

- The N-terminal domain, residues 1–92, provides the operator-binding site.
- The C-terminal domain, residues 132–236, is responsible for dimerization.

Each domain can exercise its function independently of the other. The C-terminal fragment can form oligomers. The N-terminal fragment can bind the operators, although with a lower affinity than the intact repressor. So the information for specifically contacting DNA is contained within the N-terminal domain, but the efficiency of the process is enhanced by the attachment of the C-terminal domain.

The dimeric structure of the repressor is crucial in maintaining lysogeny. The induction of a lysogenic prophage to enter the lytic cycle is caused by cleavage of the repressor subunit in the connector region. (This is a counterpart to the allosteric change in conformation that results when a small-molecule inducer inactivates the repressor of a bacterial operon, a capacity that the lysogenic repressor does not have.) Induction occurs under certain adverse conditions, such as exposure of lysogenic bacteria to UV irradiation, which leads to proteolytic inactivation of the repressor.

In the intact state, dimerization of the C-terminal domains ensures that when the repressor binds to DNA its two N-terminal domains each contact DNA simultaneously. But cleavage releases the C-terminal domains from the N-terminal domains. As illustrated in **Figure 14.15**, this means that the N-terminal domains can no longer dimerize; as a result, they do not have sufficient affinity for the operator to remain bound to DNA. Also, two dimers usually cooperate to bind at an operator, and the cleavage destabilizes this interaction.

Repressor has two domains

C
236
Dimerization
132
Connector
92
DNA-binding
11
N

Figure 14.14 The N-terminal and C-terminal regions of the repressor form separate domains. The C-terminal domains associate to form dimers; the N-terminal domains bind DNA.

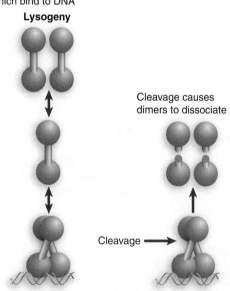

Repressor cleavage induces lytic cycle

Monomers are in equilibrium with dimers, which bind to DNA

Lysogeny

Cleavage causes dimers to dissociate

Cleavage →

Figure 14.15 Repressor dimers bind to the operator. The affinity of the N-terminal domains for DNA is controlled by the dimerization of the C-terminal domains.

14.10 Repressor Uses a Helix-Turn-Helix Motif to Bind DNA

Key Terms

- The **helix-turn-helix** motif describes an arrangement of two α-helices that form a site that binds to DNA, one fitting into the major groove of DNA and other lying across it.
- The **recognition helix** is the one of the two helices of the helix-turn-helix motif that makes contacts with DNA that are specific for particular bases. This determines the specificity of the DNA sequence that is bound.

Key Concepts

- Each DNA-binding region in the repressor contacts a half-site in the DNA.
- The DNA-binding site of repressor includes two short α-helical regions that fit into the successive turns of the major groove of DNA.
- A DNA-binding site is a (partially) palindromic sequence of 17 bp.
- The amino acid sequence of the recognition helix makes contacts with particular bases in the operator sequence that it recognizes.

A repressor dimer is the unit that binds to DNA. It recognizes a sequence of 17 bp displaying partial symmetry about an axis through the central base pair. **Figure 14.16** shows an example of a binding site. The sequence on each side of the central base pair is sometimes called a "half-site." Each individual N-terminal region contacts a half-site. Several DNA-binding proteins that regulate bacterial transcription share a similar mode of holding DNA, in which the active domain contains two short regions of α-helix that contact DNA. (Some transcription factors in eukaryotic cells use a similar motif; see *25.11 Homeodomains Bind Related Targets in DNA*.)

The N-terminal domain of lambda repressor contains several stretches of α-helix, arranged as illustrated diagrammatically in **Figure 14.17**. Two of the helical regions are responsible for binding DNA. The **helix-turn-helix** model for contact is illustrated in **Figure 14.18**. Looking at a single monomer, we see that α-helix-3 consists of 9 amino acids, lying at an angle to the preceding region of 7 amino acids that forms α-helix-2. In the dimer, the two apposed helix-3 regions lie 34 Å apart, enabling them to fit into successive major grooves of DNA. The helix-2 regions lie at an angle that would place them across the groove. The symmetrical binding of the dimer to the site means that each N-terminal domain of the dimer contacts a similar set of bases in its half-site.

Related forms of the α-helical motifs employed in the helix-loop-helix of the lambda repressor are found in several DNA-binding proteins, including CRP, the *lac* repressor, and several other phage repressors. By comparing the abilities of these proteins to bind DNA, we can define the roles of each helix:

- Contacts between helix-2 and helix-3 are maintained by interactions between hydrophobic amino acids.
- Contacts between helix-3 and DNA rely on hydrogen bonds between the amino acid side chains and the exposed positions of the base pairs. This helix is responsible for recognizing the specific target DNA sequence, and is therefore also known as the **recognition helix**. By comparing the contact patterns summarized in **Figure 14.19**, we see that repressor and Cro select different sequences in DNA as their most favored targets because they have different amino acids in the corresponding positions in helix-3.

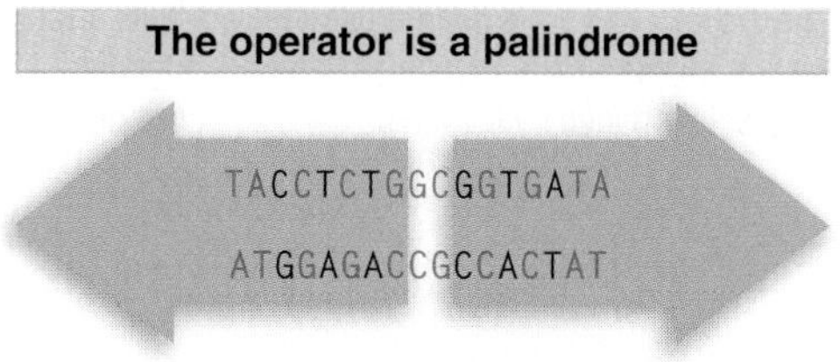

The operator is a palindrome

```
TACCTCTGGCGGTGATA
ATGGAGACCGCCACTAT
```

Figure 14.16 The operator is a 17 bp sequence with an axis of symmetry through the central base pair. Each half site is marked by the arrows. Base pairs that are identical in each operator half are in red.

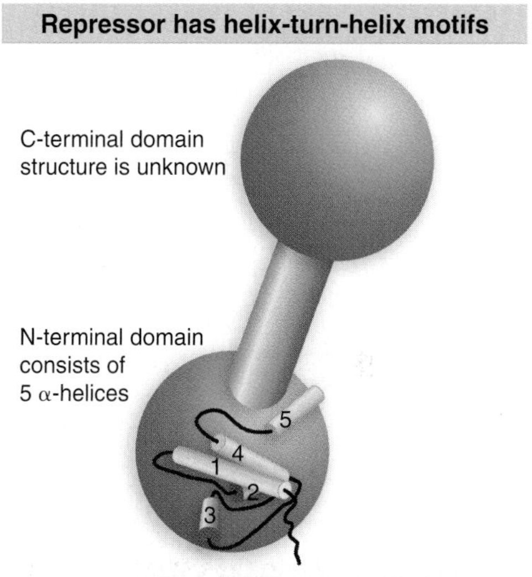

Repressor has helix-turn-helix motifs

C-terminal domain structure is unknown

N-terminal domain consists of 5 α-helices

Figure 14.17 Lambda repressor's N-terminal domain contains five stretches of α-helix; helices 2 and 3 bind DNA.

Repressor binds DNA via two α-helices

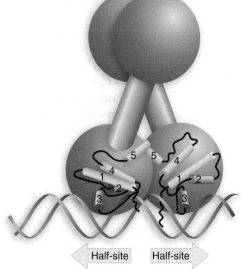

Half-site   Half-site

Figure 14.18 In the two-helix model for DNA binding, helix-3 of each monomer lies in the wide groove on the same face of DNA, and helix-2 lies across the groove.

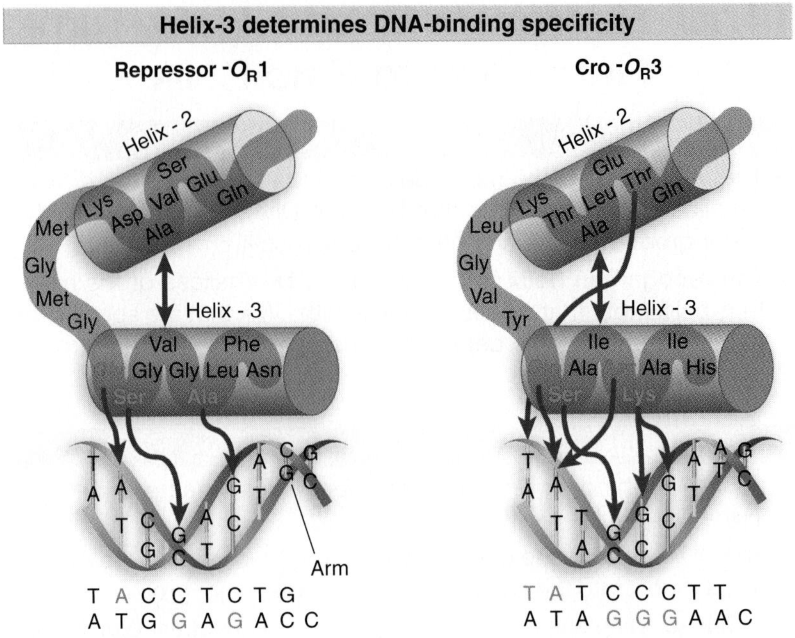

Figure 14.19 Two proteins that use the two-helix arrangement to contact DNA recognize lambda operators with affinities determined by the amino acid sequence of helix-3.

Helix-3 determines DNA-binding specificity

Repressor -O_R1   Cro -O_R3

Helix - 2   Helix - 3

Met Gly Met Gly Lys Ser Asp Val Glu Ala Gln

Val Phe Gly Gly Leu Asn Ser Ala

T A C C T C T G
A T G G A G A C C

Leu Gly Val Tyr Lys Glu Thr Leu Ala Gln

Ile Ile Ala Ala His

Arm

T A T C C T T
A T A G G G A A C

- Contacts from helix-2 to the DNA are hydrogen bonds connecting with the phosphate backbone. These interactions are necessary for binding, but do not control the specificity of target recognition. In addition to these contacts, a large part of the overall energy of interaction with DNA is provided by ionic interactions with the phosphate backbone.

What happens if we manipulate the coding sequence to construct a new protein by substituting the recognition helix in one repressor with the corresponding sequence from a closely related repressor? The specificity of the hybrid protein is that of its new recognition helix. *The amino acid sequence of this short region determines the sequence specificities of the individual proteins, and is able to act in conjunction with the rest of the polypeptide chain.*

The bases contacted by helix-3 lie on one face of DNA, as can be seen from the positions indicated on the helical diagram in Figure 14.19. However, repressor makes an additional contact with the other face of DNA. The last six N-terminal amino acids of the N-terminal domain form an "arm" extending around the back. **Figure 14.20** shows the view from the back. Lysine residues in the arm make contacts with G residues in the major groove, and also with the phosphate backbone. The interaction between the arm and DNA contributes heavily to DNA binding; the affinity of a mutant armless repressor for DNA is reduced by ~1000 fold.

N-terminal arm wraps around DNA

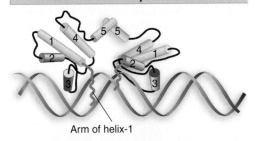

Arm of helix-1

Figure 14.20 A view from the back shows that the bulk of the repressor contacts one face of DNA, but its N-terminal arms reach around to the other face.

14.11 Repressor Dimers Bind Cooperatively to the Operator

Key Concepts

- Repressor binding to one operator increases the affinity for binding a second repressor dimer to the adjacent operator.

- The affinity is 10× greater for O_L1 and O_R1 than other operators, so they are bound first.

- Cooperativity allows repressor to bind the O_L2 and O_R2 sites at lower concentrations.

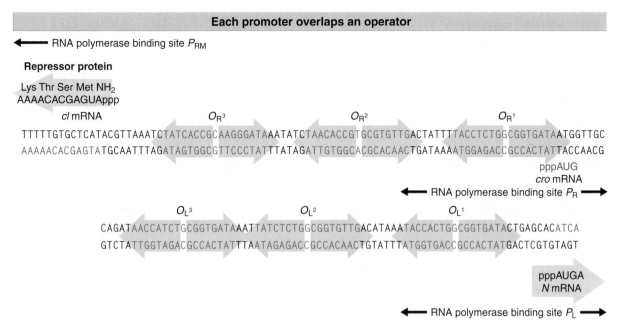

Figure 14.21 Each operator contains three repressor-binding sites, and overlaps with the promoter at which RNA polymerase binds. The orientation of O_L has been reversed from usual to facilitate comparison with O_R.

Each lambda operator contains three repressor-binding sites. As can be seen from **Figure 14.21**, no two of the six individual repressor-binding sites are identical, but they all conform with a consensus sequence. The binding sites within each operator are separated by spacers of 3–7 bp that are rich in A·T base pairs. The sites at each operator are numbered so that O_R consists of the series of binding sites O_R1-O_R2-O_R3, while O_L consists of the series O_L1-O_L2-O_L3. In each case, site 1 lies closest to the startpoint for transcription in the promoter, and sites 2 and 3 lie farther upstream.

Faced with a choice of three binding sites at each operator, how does repressor decide where to start binding? At each operator, site 1 has a greater affinity (roughly tenfold) than the other sites for the repressor. So the repressor always binds first to O_L1 and O_R1.

Lambda repressor binds to subsequent sites within each operator in a cooperative manner. The presence of a dimer at site 1 greatly increases the affinity with which a second dimer can bind to site 2. When both sites 1 and 2 are occupied, this interaction does *not* extend farther to site 3. At the concentrations of repressor usually found in a lysogenic cell, both sites 1 and 2 are filled at each operator, but site 3 is not occupied.

The C-terminal domain is responsible for the cooperative interaction between dimers as well as for the dimer formation between subunits. **Figure 14.22** shows that the cooperative interaction involves both subunits of each dimer; that is, each subunit contacts its counterpart in the other dimer, forming a tetrameric structure.

A result of cooperative binding is to increase the effective affinity of repressor for the operator at physiological concentrations. This enables a lower concentration of repressor to achieve occupancy of the operator. This is an important consideration in a system in which release of repression has irreversible consequences.

From the sequences shown in Figure 14.21, we see that O_L1 and O_R1 lie more or less in the center of the RNA polymerase binding sites of P_L and P_R, respectively. Occupancy of O_L1-O_L2 and O_R1-O_R2 thus physically blocks access of RNA polymerase to the corresponding promoters.

Lambda repressors bind DNA cooperatively

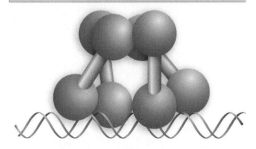

Figure 14.22 When two lambda repressor dimers bind cooperatively, each of the subunits of one dimer contacts a subunit in the other dimer.

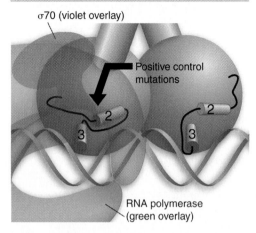

Helix-2 interacts with RNA polymerase

σ70 (violet overlay)

Positive control mutations

2

3

2

3

RNA polymerase (green overlay)

Figure 14.23 Positive control mutations identify a small region at helix-2 that interacts directly with RNA polymerase.

14.12 Repressor Maintains an Autogenous Circuit

Key Concepts

- The DNA-binding region of repressor at O_R2 contacts RNA polymerase and stabilizes its binding to P_{RM}.
- This is the basis for the autogenous control of repressor maintenance.

The cI gene in lambda is transcribed from the promoter P_{RM} that lies at its right end, close to P_R/O_R. (The subscript "RM" stands for repressor maintenance.) Transcription is terminated at the left end of the gene. The mRNA starts with the AUG initiation codon; because of the absence of the usual ribosome binding site, the mRNA is translated somewhat inefficiently, producing only a low level of repressor protein.

The presence of repressor at O_R has dual effects. It blocks expression from P_R. But it assists transcription from P_{RM}. *RNA polymerase can initiate efficiently at P_{RM} only when repressor is bound at O_R.* The repressor behaves as a positive regulator protein that is necessary for transcription of the cI gene. *Because the repressor is the product of cI, this interaction creates a positive autogenous circuit, in which the presence of repressor is necessary to support its own continued synthesis.*

The RNA polymerase binding site at P_{RM} is adjacent to O_R2. This explains how repressor autogenously regulates its own synthesis. When two dimers are bound at O_R1-O_R2, the amino terminal domain of the dimer at O_R2 interacts with RNA polymerase. The nature of the interaction is identified by mutations in the repressor that abolish positive control because they cannot stimulate RNA polymerase to transcribe from P_{RM}. They map within a small group of amino acids, located on the outside of helix-2 or in the turn between helix-2 and helix-3. The mutations reduce the negative charge of the region; conversely, mutations that increase the negative charge enhance the activation of RNA polymerase. This suggests that the group of amino acids constitutes an "acidic patch" that functions by an electrostatic interaction with a basic region on RNA polymerase.

The location of these "positive control mutations" in the repressor is indicated on **Figure 14.23**. They lie at a site on repressor that is close to a phosphate group on DNA that is also close to RNA polymerase. So the group of amino acids on the repressor that exerts positive control is in a position to contact the polymerase. The important principle is that *protein–protein interactions can release energy that is used to help to initiate transcription.*

The target site on RNA polymerase that repressor contacts is in the σ^{70} subunit, within the region that contacts the -35 region of the promoter. The interaction between repressor and polymerase is needed for the polymerase to initiate transcription.

This explains how low levels of repressor positively regulate its own synthesis. So long as enough repressor is available to fill O_R2, RNA polymerase will continue to transcribe the cI gene from P_{RM}.

14.13 Cooperative Interactions Increase the Sensitivity of Regulation

Key Concepts

- Repressor dimers bound at O_L1 and O_L2 interact with dimers bound at O_R1 and O_R2 to form octamers.
- These cooperative interactions increase the sensitivity of regulation.

Lambda repressor dimers interact cooperatively at both the left and right operators, so that their normal condition when occupied by repressor is to have dimers at both the 1 and 2 binding sites. In effect, each operator has a tetramer of repressor. However, this is not the end of the story. The two tetramers interact with one another to form an octamer, as depicted in **Figure 14.24**, which shows the distribution of repressors at the operator sites that are occupied in a lysogenic cell. Repressors are occupying O_L1, O_L2, O_R1, and O_R2; and the repressor at the last of these sites is interacting with RNA polymerase, which is initiating transcription at P_{RM}.

The interaction between the two operators has several consequences. It stabilizes repressor binding, and therefore makes it possible for repressor to occupy operators at lower concentrations. Because binding at O_R2 stabilizes RNA polymerase binding at P_{RM}, this enables low concentrations of repressor to autogenously stimulate their own production.

The DNA between the O_L and O_R sites (that is, the gene *cI*) forms a large loop, held together by the repressor octamer. The octamer brings the sites O_L3 and O_R3 into proximity. As a result, two repressor dimers can bind to each of these sites and interact with one another, as shown in **Figure 14.25**. The occupation of O_R3 prevents RNA polymerase from binding to P_{RM}, and therefore turns off expression of repressor.

This shows us how the expression of the *cI* gene becomes exquisitely sensitive to repressor concentration. At the lowest concentrations, it forms the octamer and activates RNA polymerase in a positive autogenous regulation. An increase in concentration allows binding to O_L3 and O_R3 and turns off transcription in a negative autogenous regulation. The threshold levels of repressor that are required for each of these events is reduced by the cooperative interactions, making the overall regulatory system much more sensitive. Any change in repressor level triggers the appropriate regulatory response to restore the lysogenic level.

And because the overall level of repressor has been reduced (about threefold from the level that would be required if there were no cooperative effects), there is less repressor that has to be eliminated when it becomes necessary to induce the prophage. This increases the efficiency of induction.

14.14 The *cII* and *cIII* Genes Are Needed to Establish Lysogeny

Key Concepts

- The delayed early gene products cII and cIII are necessary for RNA polymerase to initiate transcription at the promoter P_{RE}.
- cII acts directly at the promoter and cIII protects cII from degradation.
- Transcription from P_{RE} leads to synthesis of repressor and also blocks the transcription of *cro*.
- P_{RE} has atypical sequences at −10 and −35.
- RNA polymerase binds the P_{RE} promoter only in the presence of cII.
- cII binds to sequences close to the −35 region.

The control circuit for maintaining lysogeny presents a paradox. *The presence of repressor protein is necessary for its own synthesis.* This explains how the lysogenic condition is perpetuated. But how is the synthesis of repressor established in the first place?

When a lambda DNA enters a new host cell, RNA polymerase cannot transcribe *cI*, because there is no repressor present to aid its binding at

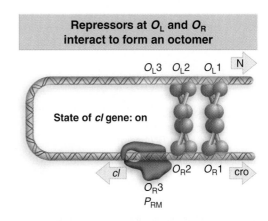

Figure 14.24 In the lysogenic state, the repressors bound at O_L1 and O_L2 interact with those bound at O_R1 and O_R2. RNA polymerase is bound at P_{RM} (which overlaps with O_R3) and interacts with the repressor bound at O_R2.

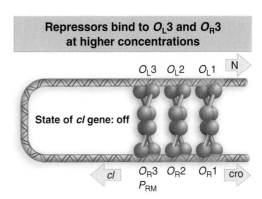

Figure 14.25 O_L3 and O_R3 are brought into proximity by formation of the repressor octamer, and an increase in repressor concentration allows dimers to bind at these sites and to interact.

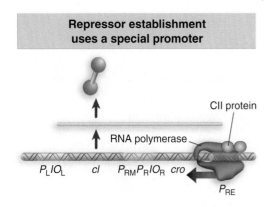

Figure 14.26 Repressor synthesis is established by the action of cII and RNA polymerase at P_{RE} to initiate transcription that extends from the antisense strand of *cro* through the *cI* gene.

P_{RM}. But this same absence of repressor means that P_R and P_L are available. So the first event when lambda DNA infects a bacterium is for genes *N* and *cro* to be transcribed. Then pN allows transcription to be extended farther. This allows *cIII* (and other genes) to be transcribed on the left, while *cII* (and other genes) are transcribed on the right (see Figure 14.11).

The *cII* and *cIII* genes share with *cI* the property that mutations in them affect the establishment of lysogeny. But there is a difference. The *cI* mutants can neither establish nor maintain lysogeny. The *cII* or *cIII* mutants have some difficulty in establishing lysogeny, but once established, they are able to maintain it by the *cI* autogenous circuit.

This implicates the *cII* and *cIII* genes as positive regulators whose products are needed for an alternative system for repressor synthesis. The system is needed only to *initiate* the expression of *cI* to circumvent the inability of the autogenous circuit to engage in *de novo* synthesis. They are not needed for continued expression.

The cII protein acts directly on gene expression. Between the *cro* and *cII* genes is another promoter, called P_{RE}. (The subscript "RE" stands for *repressor establishment*.) *This promoter can be recognized by RNA polymerase only in the presence of cII*, whose action is illustrated in **Figure 14.26**.

The P_{RE} promoter has a poor fit with the consensus at -10 and lacks a consensus sequence at -35. This deficiency explains its dependence on *cII*. The promoter cannot be transcribed by RNA polymerase alone, but can be transcribed when cII is added. cII is a classic activator: a protein that functions at the promoter to enable RNA polymerase to initiate transcription.

The cII protein is extremely unstable *in vivo*. Its degradation is the result of the action of the host *Hfl* genes. (*Hfl* stands for *High frequency lysogenization*, which is the phenotype of mutations that inactivate the degradation system.) The role of cIII is to protect cII against this degradation.

Transcription from P_{RE} promotes lysogeny in two ways. Its direct effect is that *cI* is translated into repressor protein. An indirect effect is that transcription proceeds through the *cro* gene in the "wrong" direction. So the 5′ part of the RNA corresponds to an antisense transcript of *cro*; in fact, it hybridizes to authentic *cro* mRNA, inhibiting its translation. This is important because *cro* expression is needed to enter the lytic cycle (see *14.16 The Cro Repressor Is Needed for Lytic Infection*).

The *cI* coding region on the P_{RE} transcript is very efficiently translated, in contrast with the weak translation of the P_{RM} transcript. In fact, repressor is synthesized $\sim$7–8 times more effectively via expression from P_{RE} than from P_{RM}. This reflects the fact that the P_{RE} transcript has an efficient ribosome-binding site, whereas the P_{RM} transcript has no ribosome-binding site and actually starts with the AUG initiation codon.

14.15 Lysogeny Requires Several Events

Key Concepts

- cII/cIII cause repressor synthesis to be established, and also trigger inhibition of late gene transcription.
- Establishment of repressor turns off immediate and delayed early gene expression.
- Repressor turns on the maintenance circuit for its own synthesis.
- Lambda DNA is integrated into the bacterial genome at the final stage in establishing lysogeny.

Now we can see how lysogeny is established during an infection by lambda. **Figure 14.27** recapitulates the early stages and shows what happens as the result of expression of *cIII* and *cII*. The presence of cII allows P_{RE} to be used for transcription extending through *cI*. Repressor protein is synthesized in high amounts from this transcript. Immediately it binds to O_L and O_R.

By directly inhibiting any further transcription from P_L and P_R, repressor binding turns off the expression of all phage genes. This halts the synthesis of cII and cIII, which are unstable; they decay rapidly, with the result that P_{RE} can no longer be used. So the synthesis of repressor via the establishment circuit is brought to a halt.

But repressor now is present at O_R. It switches on the maintenance circuit for expression from P_{RM}. Repressor continues to be synthesized, although at the lower level typical of P_{RM} function. So the establishment circuit starts off repressor synthesis at a high level; then repressor turns off all other functions, while at the same time turning on the maintenance circuit, which functions at the low level adequate to sustain lysogeny.

We shall not now deal in detail with the other functions needed to establish lysogeny, but we can just briefly remark that the infecting lambda DNA must be inserted into the bacterial genome (see *19.12 Specialized Recombination in Phage Lambda Involves Specific Sites*). The insertion requires the product of gene *int*, which is expressed from its own promoter P_I, at which cII also is necessary. The functions necessary for establishing the lysogenic control circuit are therefore under the same control as the function needed to integrate the phage DNA into the bacterial genome. So the establishment of lysogeny is under a control that ensures all the necessary events occur with the same timing.

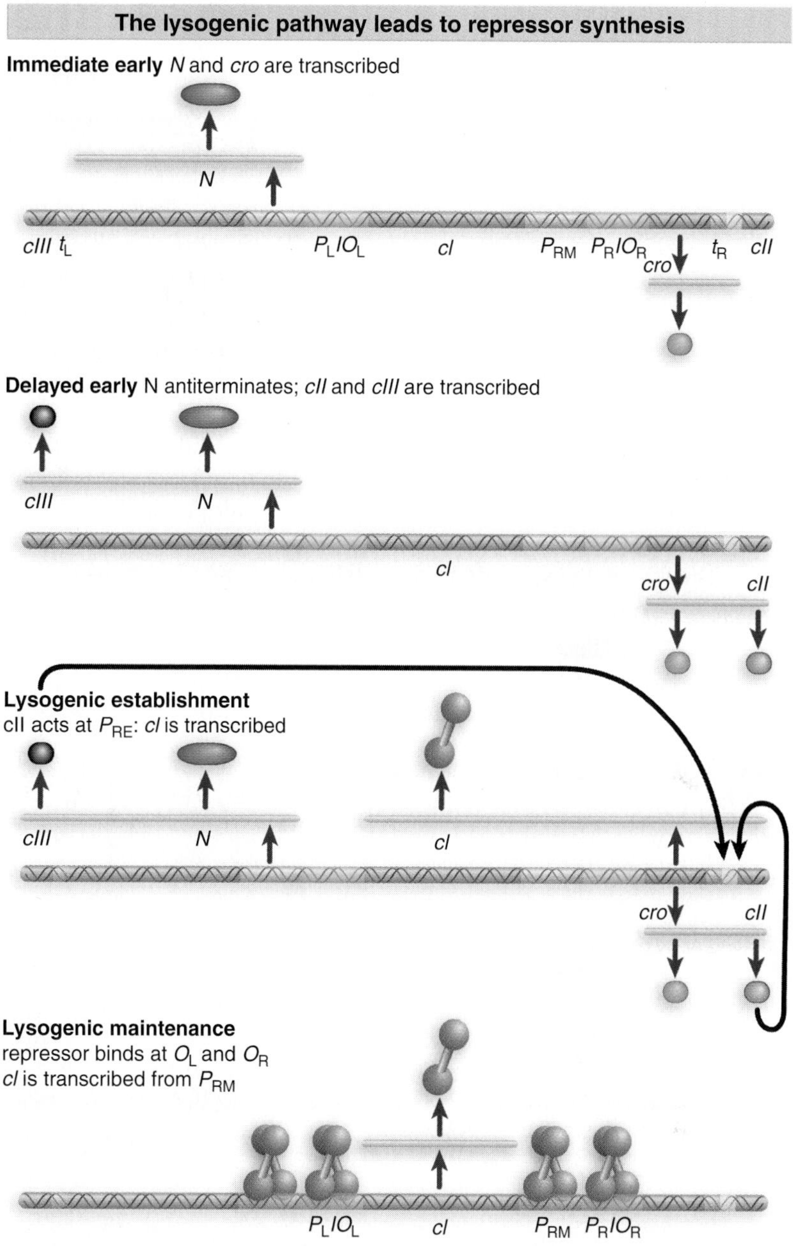

Emphasizing the tricky quality of lambda's intricate cascade, we now know that cII promotes lysogeny in another, indirect manner. It sponsors transcription from a promoter called P_{anti-Q}, which is located within the Q gene. This transcript is an antisense version of the Q region, and it hybridizes with Q mRNA to prevent translation of Q protein, whose synthesis is essential for lytic development. So the same mechanisms that directly promote lysogeny by causing transcription of the *cI* repressor gene also indirectly help lysogeny by inhibiting the expression of *cro* (see above) and Q, the regulator genes needed for the antagonistic lytic pathway.

The lysogenic pathway leads to repressor synthesis

Immediate early *N* and *cro* are transcribed

Delayed early N antiterminates; *cII* and *cIII* are transcribed

Lysogenic establishment
cII acts at P_{RE}: *cI* is transcribed

Lysogenic maintenance
repressor binds at O_L and O_R
cI is transcribed from P_{RM}

Figure 14.27 A cascade is needed to establish lysogeny, but then this circuit is switched off and replaced by the autogenous repressor-maintenance circuit.

14.16 The Cro Repressor Is Needed for Lytic Infection

Key Concepts

- Cro binds to the same operators as repressor but with different affinities.
- When Cro binds to O_R3, it prevents RNA polymerase from binding to P_{RM}, and blocks maintenance of repressor.
- When Cro binds to other operators at O_R or O_L, it prevents RNA polymerase from expressing immediate early genes, which (indirectly) blocks repressor establishment.

Lambda has the alternatives of entering lysogeny or starting a lytic infection. Lysogeny is initiated by establishing an autogenous maintenance circuit that inhibits the entire lytic cascade through applying pressure at two points. The program for establishing lysogeny proceeds through some of the same events that are required for the lytic cascade (expression of delayed early genes via expression of N is needed). We now face a problem. How does the phage enter the lytic cycle?

The key influence on the lytic cycle is the role of gene *cro*, which codes for another repressor. *Cro is responsible for preventing the synthesis of the cI repressor protein;* this action shuts off the possibility of establishing lysogeny. *Cro* mutants usually establish lysogeny rather than entering the lytic pathway, because they lack the ability to switch events away from the expression of repressor.

Cro forms a small dimer that acts within the immunity region. It has two effects:

- It prevents the synthesis of repressor via the maintenance circuit; that is, it prevents transcription via P_{RM}.
- It also inhibits the expression of early genes from both P_L and P_R.

This means that, when a phage enters the lytic pathway, Cro has responsibility both for preventing the synthesis of repressor and (subsequently) for turning down the expression of the early genes.

Cro achieves its function by binding to the same operators as the repressor protein. How can two proteins have the same sites of action, yet have such opposite effects? The answer lies in the different affinities that each protein has for the individual binding sites within the operators. Let us just consider O_R, where more is known, and where Cro exerts both its effects. The series of events is illustrated in **Figure 14.28**. (Note that the first two stages are identical to those of the lysogenic circuit shown in Figure 14.27.)

The lytic pathway leads to expression of *cro* and late genes

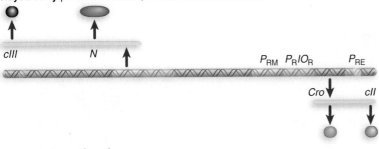

Immediate early *N* and *cro* are transcribed

$cIII$ t_L P_L/O_L cI P_{RM} P_R/O_R t_R cII

Cro

Delayed early pN antiterminates; *cII* and *cIII* are transcribed

$cIII$ N P_{RM} P_R/O_R P_{RE}

Cro   cII

Delayed early continuation Cro binds to O_L and O_R

$cIII$ N P_{RM} P_R/O_R

Cro   cII

Late expression Cro represses *cI* and all early genes; pQ activates late expression

$P_{R'}$

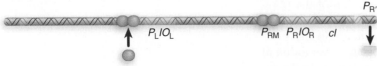

P_L/O_L P_{RM} P_R/O_R cI

Figure 14.28 The lytic cascade requires Cro protein, which directly prevents repressor maintenance via P_{RM}, as well as turning off delayed early gene expression, indirectly preventing repressor establishment.

The affinity of Cro for O_R3 is greater than its affinity for O_R2 or O_R1. So it binds first to O_R3. This inhibits RNA polymerase from binding to P_{RM}. So Cro's first action is to prevent the maintenance circuit for lysogeny from coming into play.

Then Cro binds to O_R2 or O_R1. Its affinity for these sites is similar, and there is no cooperative effect. Its presence at either site is sufficient to prevent RNA polymerase from using P_R. This in turn stops the production of the early functions (including Cro itself). Because cII is unstable, any use of P_{RE} is brought to a halt. So the two actions of Cro together block *all* production of repressor.

So far as the lytic cycle is concerned, Cro turns down (although it does not completely eliminate) the expression of the early genes. Its incomplete effect is explained by its affinity for O_R1 and O_R2, which is about eight times lower than that of repressor. This effect of Cro does not occur until the early genes have become more or less superfluous, because pQ is present; by this time, the phage has started late gene expression, and is concentrating on the production of progeny phage particles.

14.17 What Determines the Balance Between Lysogeny and the Lytic Cycle?

Key Concepts

- The delayed early stage when both Cro and repressor are being expressed is common to lysogeny and the lytic cycle.
- The critical event is whether cII causes sufficient synthesis of repressor to overcome the action of Cro.

The programs for the lysogenic and lytic pathways in lambda are so intimately related that it is impossible to predict the fate of an individual phage genome when it enters a new host bacterium. Will the antagonism between repressor and Cro be resolved by establishing the autogenous maintenance circuit shown in Figure 14.27, or by turning off repressor synthesis and entering the late stage of development shown in Figure 14.28?

The same pathway is followed in both cases right up to the brink of decision. Both involve the expression of the immediate early genes and extension into the delayed early genes. The difference between them comes down to the question of whether repressor or Cro obtains occupancy of the two operators.

The early phase during which the decision is taken is limited in duration in either case. No matter which pathway the phage follows, expression of all early genes will be prevented as P_L and P_R are repressed; and, as a consequence of the disappearance of cII and cIII, production of repressor via P_{RE} will cease.

The critical question comes down to whether the cessation of transcription from P_{RE} is followed by activation of P_{RM} and the establishment of lysogeny, or whether P_{RM} fails to become active and the pQ regulator commits the phage to lytic development. **Figure 14.29** shows the critical stage, at which both repressor and Cro are being synthesized.

The initial event in establishing lysogeny is the binding of repressor at O_L1 and O_R1. Binding at the first sites is rapidly succeeded by cooperative binding of further repressor dimers at O_L2 and O_R2. This shuts off the synthesis of Cro and starts up the synthesis of repressor via P_{RM}.

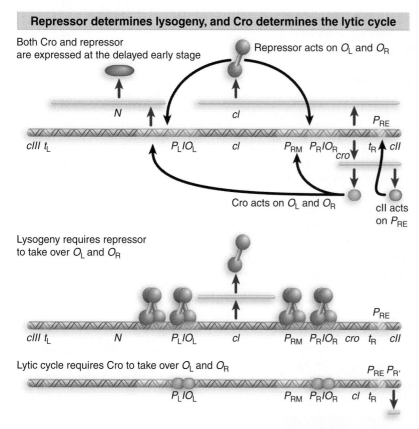

Repressor determines lysogeny, and Cro determines the lytic cycle

Both Cro and repressor are expressed at the delayed early stage

Repressor acts on O_L and O_R

N cl P_{RE}

$cIII\ t_L$ P_L/O_L cl $P_{RM}\ P_R/O_R$ t_R cII
cro

Cro acts on O_L and O_R

cII acts on P_{RE}

Lysogeny requires repressor to take over O_L and O_R

P_{RE}

$cIII\ t_L$ N P_L/O_L cl $P_{RM}\ P_R/O_R\ cro\ t_R$ cII

Lytic cycle requires Cro to take over O_L and O_R

$P_{RE}\ P_{R'}$

P_L/O_L $P_{RM}\ P_R/O_R\ cl\ t_R$

Figure 14.29 The critical stage in deciding between lysogeny and lysis is when delayed early genes are being expressed. If cII causes sufficient synthesis of repressor, lysogeny will result because repressor occupies the operators. Otherwise Cro occupies the operators, resulting in a lytic cycle.

The initial event in entering the lytic cycle is the binding of Cro at O_R3. This stops the lysogenic-maintenance circuit from starting up at P_{RM}. Then Cro must bind to O_R1 or O_R2, and to O_L1 or O_L2, to turn down early gene expression. By halting production of cII and cIII, this action leads to the cessation of repressor synthesis via P_{RE}. Repressor establishment is shut off when the unstable cII and cIII proteins decay.

The critical influence over the switch between lysogeny and lysis is cII. If cII is active, synthesis of repressor via the establishment promoter is effective; and, as a result, repressor gains occupancy of the operators. If cII is not active, repressor establishment fails, and Cro binds to the operators.

The level of cII protein under any particular set of circumstances determines the outcome of an infection. Mutations that increase the stability of cII increase the frequency of lysogenization. Such mutations occur in *cII* itself or in other genes. The cause of cII's instability is its susceptibility to degradation by host proteases. Its level in the cell is influenced by *cIII*, which helps to protect cII against degradation. Although the presence of cIII does not guarantee the survival of cII, in the absence of cIII, cII is virtually always inactivated.

Host gene products act on this pathway. Mutations in the host genes *hflA* and *hflB* increase lysogeny (*hfl* stands for *high frequency lysogenization*). The mutations stabilize cII because they inactivate host protease(s) that degrade it.

The influence of the host cell on the level of cII provides a route for the bacterium to interfere with the decision-taking process. For example, host proteases that degrade cII are activated by growth on rich medium, so lambda tends to lyse cells that are growing well, but is more likely to enter lysogeny on cells that are starving (and which lack components necessary for efficient lytic growth).

14.18 SUMMARY

Phages have a lytic life cycle, in which infection of a host bacterium is followed by production of a large number of phage particles, lysis of the cell, and release of the viruses. Some phages also can exist in a lysogenic form, in which the phage genome is integrated into the bacterial chromosome and is inherited in this inert, latent form like any other bacterial gene.

Lytic infection falls typically into three phases. In the first phase a small number of phage genes are transcribed by the host RNA polymerase. One or more of these genes is a regulator that controls expression of the group of genes expressed in the second phase. The pattern is repeated in the second phase, when one or more genes is a regulator needed for expression of the genes of the third phase. Genes of the first two phases code for enzymes needed to reproduce phage DNA; genes of the final phase code for structural components of the phage particle. It is common for the very early genes to be turned off during the later phases.

In phage lambda, the genes are organized into groups whose expression is controlled by individual regulatory events. The immediate early gene N codes for an antiterminator that allows transcription of the leftward and rightward groups of delayed early genes from the early promoters P_R and P_L. The delayed early gene Q has a similar antitermination function that allows transcription of all late genes from the promoter $P_{R'}$. The lytic cycle is repressed, and the lysogenic state maintained, by expression of the cI gene, whose product is a repressor protein that acts at the operators O_R and O_L to prevent use of the promoters P_R and P_L, respectively. A lysogenic phage genome expresses only the cI gene, from its promoter P_{RM}. Transcription from this promoter is controlled by positive autogenous regulation, in which repressor bound at O_R activates RNA polymerase at P_{RM}.

Each operator consists of three binding sites for repressor. Each site is palindromic, consisting of symmetrical half-sites. Repressor functions as a dimer. Each half binding site is contacted by a repressor monomer. The N-terminal domain of repressor contains a helix-turn-helix motif that contacts DNA. Helix-3 is the recognition helix, responsible for making specific contacts with base pairs in the operator. Helix-2 is involved in positioning helix-3; it is also involved in contacting RNA polymerase at P_{RM}. The C-terminal domain is required for dimerization. Induction is caused by cleavage between the N- and C-terminal domains, which prevents the DNA-binding regions from functioning in dimeric form, thereby reducing their affinity for DNA and making it impossible to maintain lysogeny. Repressor-operator binding is cooperative, so that once one dimer has bound to the first site, a second dimer binds more readily to the adjacent site.

The helix-turn-helix motif is used by other DNA-binding proteins, including lambda Cro, which binds to the same operators, but has a different affinity for the individual operator sites, determined by the sequence of helix-3. Cro binds individually to operator sites, starting with O_R3, in a noncooperative manner. It is needed for progression through the lytic cycle. Its binding to O_R3 first prevents synthesis of repressor from P_{RM}; then its binding to O_R2 and O_R1 prevents continued expression of early genes, an effect also seen in its binding to O_L1 and O_L2.

Establishment of repressor synthesis requires use of the promoter P_{RE}, which is activated by the product of the cII gene. The product of $cIII$ is required to stabilize the cII product against degradation. By turning off cII and $cIII$ expression, Cro acts to prevent lysogeny. By turning off all transcription except that of its own gene, repressor acts to prevent the lytic cycle. The choice between lysis and lysogeny depends on whether repressor or Cro gains occupancy of the operators in a particular infection. The stability of cII protein in the infected cell is a primary determinant of the outcome.

15

The Replicon

15.1 Introduction

Key Terms

- The **replicon** is a unit of the genome in which DNA is replicated. Each replicon contains an origin for initiation of replication.

- **Single-copy replication control** describes a system in which there is only one copy of a replicon per unit bacterium. The bacterial chromosome and some plasmids have this type of regulation.

- A plasmid is said to be under **multicopy control** when the control system allows the plasmid to exist in more than one copy per individual bacterial cell.

Replication of DNA is a key regulatory event in cell division. For every time a cell divides, its entire set of DNA sequences must be replicated once and only once. This is accomplished by controlling the initiation of replication, which occurs only at a unique site called the *origin*.

The origin defines the unit of DNA in which an individual act of replication occurs: a **replicon** is a length of DNA that contains an origin and that is replicated whenever an initiation event occurs at that origin. **Figure 15.1** illustrates the general nature of the relationship between replicons and genomes in prokaryotes and eukaryotes.

A genome in a prokaryotic cell (usually a circular molecule of DNA) constitutes a single replicon. This means that replication of the entire bacterial chromosome depends on a single initiation event that occurs at the unique origin. The initiation event occurs once for every cell division, and is known as **single-copy replication control**. The frequency of initiation at the bacterial origin is controlled by its state of methylation (see *15.4 Methylation of the Bacterial Origin Regulates Initiation*).

Bacteria may contain additional genetic information in the form of plasmids. *A plasmid is an autonomous circular DNA genome that constitutes a separate replicon*. A plasmid replicon may show single-copy control, which means that it replicates once every time the bacterial chromosome replicates. Or it may be under **multicopy control**, when it is present in a greater number of copies than the bacterial chromosome. Each phage or virus DNA also constitutes a replicon, able to initiate many times during an infectious cycle. Perhaps a better way to view the prokaryotic replicon, therefore, is to reverse the definition: *any DNA molecule that contains an origin can be replicated autonomously in the cell*.

A major difference in the organization of bacterial and eukaryotic genomes is seen in their replication. Each eukaryotic chromosome (usually a very long linear molecule of DNA) is divided into a large number of replicons. (We will see later why the origin is placed in the center of the replicon.) Like the single replicon of the bacterial chromosome, each of these replicons "fires" once and only once in each cell cycle. Eukaryotic origin usage is controlled by the ability of protein factors to bind to it (see *15.7 Licensing Factor Controls Rereplication and Consists of MCM Proteins*). The replicons of a eukaryotic genome are not active simultaneously, but are activated over a fairly protracted period, which is called S phase. This implies the existence of additional levels of control, to ensure that each replicon does fire once and does not fire a second time. And because many replicons are activated independently, another signal must exist to indicate when the entire process of replicating all replicons has been completed.

In contrast with nuclear chromosomes, which have a single-copy type of control, the DNA of mitochondria and chloroplasts may be regulated more like plasmids that exist in multiple copies per bacterium. There are multiple copies of each organelle DNA per cell, and the control of organelle DNA replication must be related to the cell cycle.

In all these systems, the key question is to define the sequences that function as origins and to determine how they are recognized by the appropriate proteins of the apparatus for replication. We start by considering the basic construction of replicons and the various forms that they take in bacteria and eukaryotic cells. In *Chapter 16 Extrachromosomal Replicons*, we consider autonomously replicating units in bacteria. In *Chapter 17 Bacterial Replication Is Connected to the Cell Cycle*, we turn to the question of how replication of the genome is coordinated with bacterial division, and what is responsible for segregating the genomes to daughter bacteria.

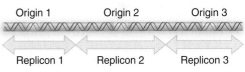

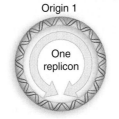

Figure 15.1 A bacterium usually has a circular chromosome that is replicated from a single origin, but a eukaryotic chromosome has many origins, each defining a separate replicon.

15.2 An Origin Usually Initiates Bidirectional Replication

Key Concepts

- A replication fork is initiated at the origin and then moves sequentially along DNA.
- Replication is bidirectional when an origin creates two replication forks that move in opposite directions.

Key Terms

- A **replication eye** is a region in which DNA has been replicated within a longer, unreplicated region.
- A **replication fork (growing point)** is the point at which strands of parental duplex DNA are separated so that replication can proceed. A complex of proteins including DNA polymerase is found at the fork.

Replication starts at an origin by separating the two strands of the DNA duplex. **Figure 15.2** shows that each of the parental strands then acts as a template to synthesize a complementary daughter strand.

A molecule of DNA engaged in replication has two types of regions. **Figure 15.3** shows that when replicating DNA is viewed by electron

DNA replication initiates at an origin

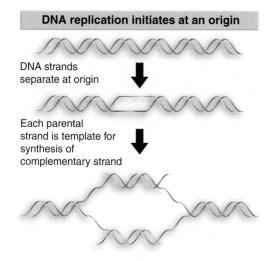

DNA strands separate at origin

Each parental strand is template for synthesis of complementary strand

Figure 15.2 An origin is a sequence of DNA at which replication is initiated by separating the parental strands and initiating synthesis of new DNA strands. Each new strand is complementary to the parental strand that acts as the template for its synthesis.

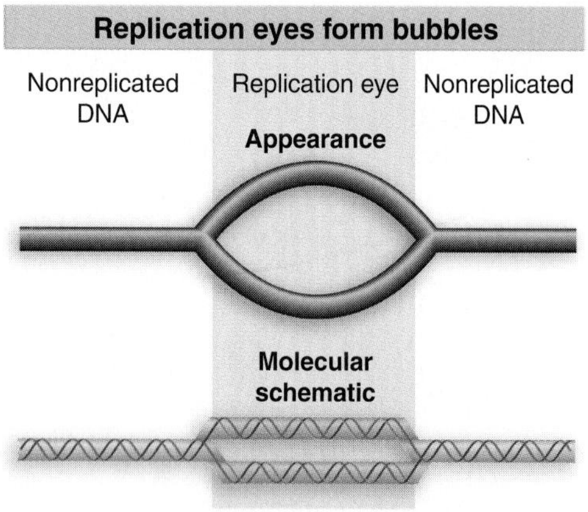

Replication eyes form bubbles

Nonreplicated DNA   Replication eye   Nonreplicated DNA

Appearance

Molecular schematic

Figure 15.3 Replicated DNA is seen as a replication eye flanked by nonreplicated DNA.

Replication is usually bidirectional

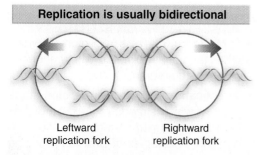

Leftward replication fork   Rightward replication fork

Figure 15.4 Replication is bidirectional when two replication forks are established at the origin, moving in opposite directions.

microscopy, the replicated region appears as a **replication eye** within the nonreplicated DNA. The nonreplicated region consists of the parental duplex; this opens into the replicated region where the two daughter duplexes have formed.

The point at which replication is occurring is called the **replication fork** (sometimes also known as the **growing point**). *A replication fork moves sequentially along the DNA, from its starting point at the origin.*

The most common form of replication is bidirectional, with two replication forks setting out from the origin in opposite directions, as shown in **Figure 15.4**. As the replication forks continue to move apart, the replication eye increases in size, and eventually it becomes larger than the nonreplicated region.

15.3 The Bacterial Genome Is a Single Circular Replicon

Key Concepts

- Bacterial replicons are usually circles that replicate bidirectionally from a single origin.
- The origin of *E. coli*, *oriC*, is 245 bp in length.
- The two replication forks usually meet halfway around the circle, but there are *ter* sites that cause termination if the replication forks go too far.

Prokaryotic replicons are usually circular, so that the DNA forms a closed circle with no free ends. Circular structures include the bacterial chromosome itself, all plasmids, and many bacteriophages, and are also common in chloroplast and mitochondrial DNAs. **Figure 15.5** summarizes

Replication requires DNA synthesis and chromosome separation

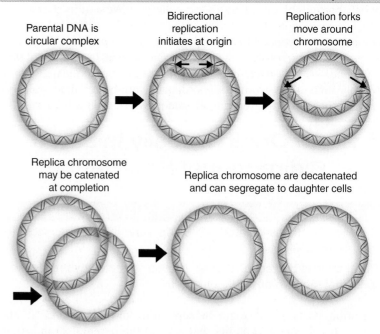

Parental DNA is circular complex

Bidirectional replication initiates at origin

Replication forks move around chromosome

Replica chromosome may be catenated at completion

Replica chromosome are decatenated and can segregate to daughter cells

Figure 15.5 Bidirectional replication of a circular bacterial chromosome is initiated at a single origin. The replication forks move around the chromosome. If the replica chromosomes are catenated, they must be disentangled before they can segregate to daughter cells.

the stages of replicating a circular chromosome. After replication has initiated at the origin, two replication forks proceed in opposite directions. The circular chromosome is sometimes described as a θ structure at this stage, because of its appearance. An important consequence of circularity is that the completion of the process can generate two chromosomes that are linked because one passes through the other (they are said to be *catenated*), and specific enzyme systems may be required to separate them (see *17.7 Chromosomal Segregation May Require Site-Specific Recombination*).

The genome of *E. coli* is replicated bidirectionally from a single origin, identified as the genetic locus *oriC*. Two replication forks initiate at *oriC* and move around the genome (at approximately the same speed) to a meeting point. Termination occurs in a discrete region. One interesting question is what ensures that the DNA is replicated right across the region where the forks meet.

Sequences that cause termination are called *ter* sites. A *ter* site contains a short (~23 bp) sequence that causes termination *in vitro*. The termination sequences function in only one orientation. The *ter* site is recognized by a protein (called Tus in *E. coli* and RTP in *B. subtilis*) that recognizes the consensus sequence and prevents the replication fork from proceeding. However, deletion of the *ter* sites does not prevent normal replication cycles from occurring, although it does affect segregation of the daughter chromosomes.

Termination in *E. coli* and *B. subtilis* has the interesting features reported in **Figure 15.6**. We know that the replication forks usually meet and halt replication at a point midway around the chromosome from the origin. But two termination regions (*terE,D,A* and *terC,B* in *E. coli*, and *terI, terII* and also some other sites in *B. subtilis*) have been identified, located ~100 kb on either side of this meeting point. Each contains multiple terminators. Each terminus is specific for one direction of fork movement, and they are arranged in such a way that each fork would have to pass the other to reach the terminus to which it is susceptible. This arrangement creates a "replication fork trap." If for some reason one fork is delayed, so that the forks fail to meet at the usual central position, the more rapid fork will be trapped at the *ter* region to wait for the arrival of the slow fork.

What happens when a replication fork encounters a protein bound to DNA? We assume that repressors (for example) are displaced and then rebind. A particularly interesting question is what happens when a replication fork encounters an RNA polymerase engaged in transcription. A replication fork moves >10× faster than RNA polymerase. If they are proceeding in the same direction, either the replication fork must displace the polymerase or it must slow down as it waits for the RNA polymerase to reach its terminator. It appears that a DNA polymerase moving in the same direction as an RNA polymerase can "bypass" it without disrupting transcription, but we do not understand how this happens.

A conflict arises when the replication fork meets an RNA polymerase traveling in the opposite direction, that is, toward it. Can it displace the RNA polymerase? Or do both replication and transcription come to a halt? An indication that these encounters cannot easily be resolved is provided by the organization of the *E. coli* chromosome. Almost all active transcription units are oriented so that they are expressed in the same direction as the replication fork that passes them. The exceptions all comprise small transcription units that are infrequently expressed. The difficulty of generating inversions containing highly expressed genes argues that head-on encounters between a replication fork and a series of transcribing RNA polymerases may be lethal.

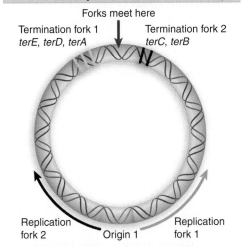

Forks usually meet before terminating

Figure 15.6 Replication termini in *E. coli* are located beyond the point at which the replication forks actually meet.

15.4 Methylation of the Bacterial Origin Regulates Initiation

Key Term

- **Hemimethylated DNA** is methylated on one strand of a target sequence that has a cytosine on each strand.

Key Concepts

- *oriC* contains 11 $\frac{GATC}{CTAG}$ repeats that are methylated on adenine on both strands.
- Replication generates hemimethylated DNA, which cannot initiate replication.
- There is a 13-minute delay before the $\frac{GATC}{CTAG}$ repeats are remethylated.

The bacterial origin contains sequences that are methylated, and which are in different states before and after replication. This difference is used as a mark to distinguish a replicated origin from a nonreplicated origin.

The *E. coli* origin *oriC* contains 11 copies of the sequence $\frac{GATC}{CTAG}$ which is a target for methylation at the N6 position of adenine by the Dam methylase. The reaction is illustrated in **Figure 15.7**.

Before replication, the palindromic target site is methylated on the adenines of each strand. Replication inserts the normal (nonmodified) bases into the daughter strands, generating **hemimethylated DNA**, in which one strand is methylated and one strand is unmethylated. So the replication event converts Dam target sites from fully methylated to hemimethylated condition.

What is the consequence for replication? The ability of a plasmid relying upon *oriC* to replicate in *dam⁻ E. coli* depends on its state of methylation. If the plasmid is methylated, it undergoes a single round of replication, to generate hemimethylated products, as described in **Figure 15.8**. The hemimethylated plasmids then accumulate, suggesting that a hemimethylated origin cannot be used to initiate a replication cycle.

Two explanations suggest themselves. Initiation may require full methylation of the Dam target sites in the origin. Or initiation may be inhibited by hemimethylation of these sites. The latter seems to be the case, because an origin of nonmethylated DNA can function effectively.

So hemimethylated origins cannot initiate again until the Dam methylase has converted them into fully methylated origins. The GATC sites at the origin remain hemimethylated for ~13 minutes after replication. This long period is unusual, because at typical GATC sites elsewhere in the genome, remethylation begins immediately (<1.5 min) following replication.

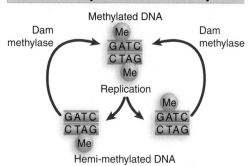

The Dam methylase maintains methylation

Figure 15.7 Replication of methylated DNA gives hemimethylated DNA, which maintains its state at GATC sites until the Dam methylase restores the fully methylated condition.

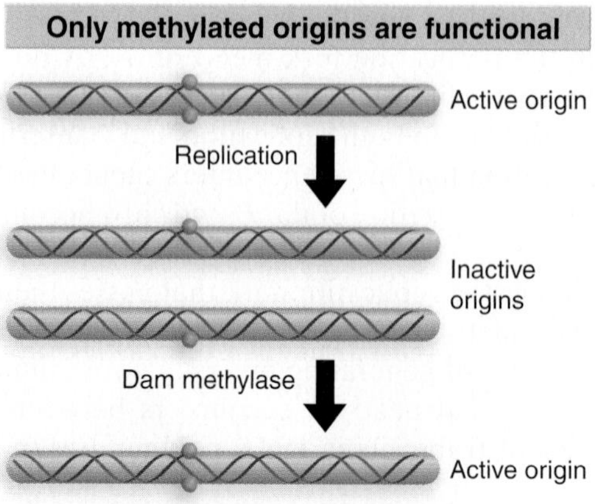

Only methylated origins are functional

Figure 15.8 Only fully methylated origins can initiate replication; hemimethylated daughter origins cannot be used again until they have been restored to the fully methylated state.

15.5 Each Eukaryotic Chromosome Contains Many Replicons

Key Concepts

- Eukaryotic replicons are 40–100 kb in length.
- A chromosome is divided into many replicons.
- Individual replicons are activated at characteristic times during S phase.
- Regional activation patterns suggest that replicons near one another are activated at the same time.

Key Term

- **S phase** is the restricted part of the eukaryotic cell cycle during which synthesis of DNA occurs.

In eukaryotic cells, the replication of DNA is confined to part of the cell cycle called **S phase**, which usually lasts a few hours in a higher eukaryotic cell. Replication of the large amount of DNA contained in a eukaryotic chromosome is accomplished by dividing it into many individual replicons. Only some of these replicons are engaged in replication at any point in S phase. Presumably each replicon is activated at a specific time during S phase, although the evidence on this issue is not decisive.

The start of S phase is signaled by the activation of the first replicons. Over the next few hours, initiation events occur at other replicons in an ordered manner. Chromosomal replicons usually display bidirectional replication.

Individual replicons in eukaryotic genomes are relatively small, typically ~40 kb in yeast or fly, ~100 kb in animal cells. However, they can vary >10-fold in length within a genome. The rate of replication is ~2000 bp/min, which is much slower than the 50,000 bp/min of bacterial replication fork movement.

From the speed of replication, it is evident that a mammalian genome could be replicated in ~1 hour if all replicons functioned simultaneously. But S phase actually lasts for >6 hours in a typical somatic cell, which implies that no more than 15% of the replicons are likely to be active at any given moment.

How are origins selected for initiation at different times during S phase? In *S. cerevisiae*, the default appears to be for origins to replicate early, but *cis*-acting sequences can cause origins linked to them to replicate at late times.

Available evidence suggests that chromosomal replicons do not have termini at which the replication forks cease movement and (presumably) dissociate from the DNA. It seems more likely that a replication fork continues from its origin until it meets a fork proceeding toward it from the adjacent replicon. We have already mentioned the potential topological problem of joining the newly synthesized DNA at the junction of the replication forks.

The propensity of replicons located in the same vicinity to be active at the same time could be explained by "regional" controls, in which groups of replicons are initiated more or less coordinately, as opposed to a mechanism in which individual replicons are activated one by one in dispersed areas of the genome. Two structural features suggest the possibility of large-scale organization. Quite large regions of the chromosome can be characterized as "early replicating" or "late replicating," implying that there is little interspersion of replicons that fire at early and late times. And visualization of replicating forks by labeling with DNA precursors shows 100–300 "foci" instead of uniform staining; each focus shown in **Figure 15.9** probably contains >300 replication forks. The foci could represent fixed structures through which replicating DNA must move.

Replication forks form foci

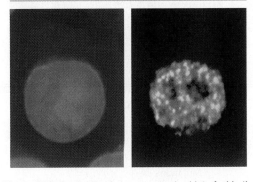

Figure 15.9 Replication forks are organized into foci in the nucleus. Cells were labeled with BrdU. The leftmost panel was stained with propidium iodide to identify bulk DNA. The right panel was stained using an antibody to BrdU to identify replicating DNA. Photographs kindly provided by A. D. Mills and Ron Laskey.

15.6 Replication Origins Bind the ORC

Because a eukaryotic chromosome contains many replicons, we cannot identify origins directly by mutations that prevent replication, as we can with a bacterial chromosome. However, any segment of DNA that has an origin should be able to replicate. This provided the first means to identify an origin, by testing sequences for their ability to support replication of artificially created independent DNA molecules. A sequence that confers the ability to replicate efficiently in yeast is called an **ARS** (autonomous replicating sequence). *ARS* elements are derived from origins of replication.

An *ARS* element consists of an A·T-rich region that contains discrete sites in which mutations affect origin function. Base composition rather than sequence may be important in the rest of the region. **Figure 15.10** shows a systematic mutational analysis along the length of an origin. Origin function is abolished completely by mutations in a "core" region, called the A-domain, that contains an 11 bp consensus sequence consisting of A·T base pairs. This consensus sequence (sometimes called the *ACS,* for *ARS* Consensus Sequence) is the only homology between known *ARS* elements.

Mutations in three adjacent elements, numbered B1–B3, reduce origin function. An origin can function effectively with any two of the B elements, so long as a functional A element is present. (Imperfect copies of the core consensus, typically conforming at 9–11 positions, are found close to, or overlapping with, each B element, but they do not appear to be necessary for origin function.)

The origin is a target for the ORC (origin recognition complex), a complex of six proteins with a mass of ~400 kD. ORC binds to the A and B1 elements on the A·T-rich strand, and is associated with *ARS* elements throughout the cell cycle. This means that initiation depends on changes in its condition rather than *de novo* association with an origin (discussed in the next section). By counting the number of sites to which ORC binds, we can estimate that there are about 400 origins of replication in the yeast genome. This means that the average length of a replicon is ~35,000 bp.

ORC was first found in *S. cerevisiae* (where it is called scORC), but similar complexes have now been characterized in *S. pombe* (spORC), *Drosophila* (DmORC), and *Xenopus* (XlORC). All of the ORC

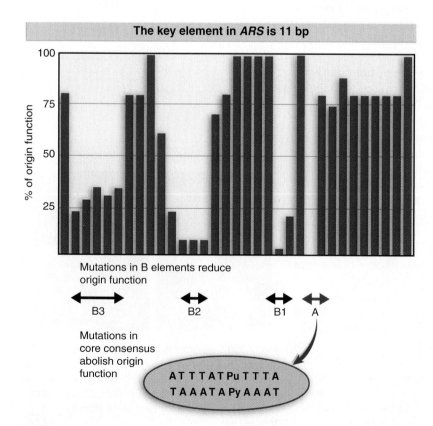

Figure 15.10 An *ARS* extends for ~50 bp and includes a consensus sequence (A) and additional elements (B1–B3).

complexes bind to DNA. Although none of the binding sites have been characterized in the same detail as in *S. cerevisiae*, in several cases they are at locations associated with the initiation of replication. It seems clear that ORC is an initiation complex whose binding identifies an origin of replication. Although the ORC is conserved, it has not so far been possible to identify origins of replication in higher eukarytic DNA that function like *ARS* elements.

15.7 Licensing Factor Controls Rereplication and Consists of MCM Proteins

Key Terms

- A **licensing factor** is located in the nucleus and is necessary for replication; it is inactivated or destroyed after one round of replication. New licensing factors must be provided for further rounds of replication.
- The **prereplication complex** is a protein–DNA complex at the origin in *S. cerevisiae* that is required for DNA replication. The complex contains the ORC complex, Cdc6, and the MCM proteins.
- The **postreplication complex** is a protein–DNA complex in *S. cerevisiae* that consists of the ORC complex bound to the origin.

Key Concepts

- Licensing factor is necessary for initiation of replication at each origin.
- Licensing factor is present in the nucleus prior to replication, but is inactivated or destroyed by replication.
- Initiation of another replication cycle becomes possible only after licensing factor reenters the nucleus after mitosis.
- The ORC proteins are associated with yeast origins throughout the cell cycle.
- Cdc6 protein is an unstable protein that is synthesized only in G1.
- Cdc6 binds to ORC and allows MCM proteins to bind.
- When replication is initiated, Cdc6 and MCM proteins are displaced. The degradation of Cdc6 prevents reinitiation.
- Some MCM proteins are in the nucleus throughout the cycle, but others may enter only after mitosis.

A eukaryotic genome is divided into multiple replicons, and the origin in each replicon is activated once and only once in a single division cycle. This could be achieved by providing some rate-limiting component that functions only once at an origin or by the presence of a repressor that prevents rereplication at origins that have been used. The critical questions about the nature of this regulatory system are how the system determines whether any particular origin has been replicated, and what protein components are involved.

Insights into the nature of the protein components have been provided by using a system in which a substrate DNA undergoes only one cycle of replication. *Xenopus* eggs have all the components needed to replicate DNA—in the first few hours after fertilization they undertake 11 division cycles without new gene expression—and they can replicate the DNA in a nucleus that is injected into the egg. **Figure 15.11** summarizes the features of this system.

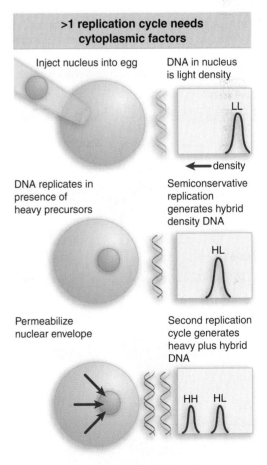

>1 replication cycle needs cytoplasmic factors

Inject nucleus into egg

DNA in nucleus is light density

LL

←density

DNA replicates in presence of heavy precursors

Semiconservative replication generates hybrid density DNA

HL

Permeabilize nuclear envelope

Second replication cycle generates heavy plus hybrid DNA

HH   HL

Figure 15.11 A nucleus injected into a *Xenopus* egg can replicate only once unless the nuclear membrane is permeabilized to allow subsequent replication cycles.

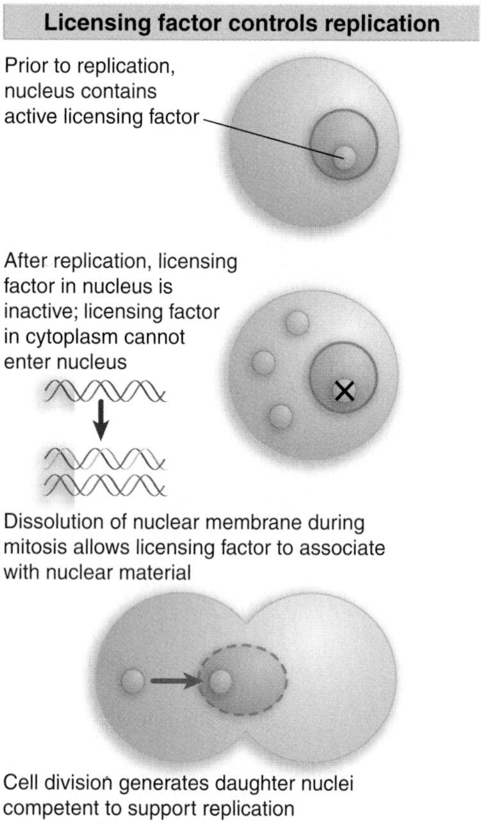

Licensing factor controls replication

Prior to replication, nucleus contains active licensing factor

After replication, licensing factor in nucleus is inactive; licensing factor in cytoplasm cannot enter nucleus

Dissolution of nuclear membrane during mitosis allows licensing factor to associate with nuclear material

Cell division generates daughter nuclei competent to support replication

Figure 15.12 Licensing factor in the nucleus is inactivated after replication. A new supply of licensing factor can enter only when the nuclear membrane breaks down at mitosis.

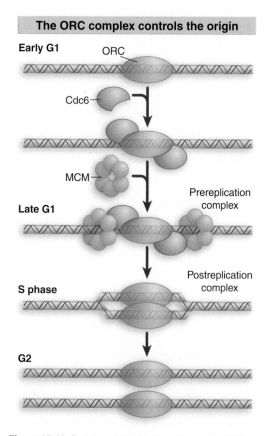

The ORC complex controls the origin

Early G1   ORC

Cdc6

MCM

Prereplication complex

Late G1

Postreplication complex

S phase

G2

Figure 15.13 Proteins at the origin control susceptibility to initiation.

When a sperm or interphase nucleus is injected into the egg, its DNA is replicated only once. If protein synthesis is blocked in the egg, the membrane around the injected material remains intact, and the DNA cannot replicate again. However, in the presence of protein synthesis, the nuclear membrane breaks down just as it would for a normal cell division, and in this case subsequent replication cycles can occur. The same result can be achieved by using agents that permeabilize the nuclear membrane. This suggests that the nucleus contains a protein (or more than one) needed for replication that is used up in some way by a replication cycle; although more of the protein is present in the egg cytoplasm, it can only enter the nucleus if the nuclear membrane breaks down.

Figure 15.12 explains the control of reinitiation by proposing that this protein is a **licensing factor**. It is present in the nucleus prior to replication. One round of replication either inactivates or destroys the factor, and another round cannot occur until further factor is provided. Factor in the cytoplasm can gain access to the nuclear material only at the subsequent mitosis when the nuclear envelope breaks down.

The key event in controlling replication is the behavior of the ORC complex at the origin. The striking feature is that ORC remains bound at the origin through the entire cell cycle. However, changes occur in the binding of other proteins to the ORC-origin complex. **Figure 15.13** summarizes the cycle of events at the origin.

At the end of the cell cycle, ORC is bound to the A–B1 elements of the origin, and protects the region against DNAase, but there is a site that is hypersensitive to the enzyme in the center of B1. There is change during G1, most strikingly by the loss of the hypersensitive site. This results from the binding of Cdc6 protein to the ORC. In yeast, Cdc6 is a highly unstable protein (half-life <5 minutes). It is synthesized during G1, and typically binds to the ORC between the exit from mitosis and late G1. Its rapid degradation means that no protein is available later in the cycle. In mammalian cells Cdc6 is controlled differently; it is phosphorylated during S phase, and as a result is exported from the nucleus. Cdc6 provides the connection between ORC and a complex of proteins that is involved in licensing and initiation. Cdc6 has an ATPase activity that is required for it to support initiation.

The licensing factor and the system that controls its availability in yeast are identified by two different types of mutations:

- The licensing factor is identified by mutations in *MCM2,3,5*, which prevent initiation of replication.
- Mutations that have the opposite effect, and allow the accumulation of excess quantities of DNA, are found in genes that code for components of the ubiquitination system that is responsible for degrading certain proteins. This suggests that licensing factor may be destroyed after the start of a replication cycle.

In yeast, MCM2,3,5 enter the nucleus only during mitosis. Homologues are found in animal cells, where MCM3 is bound to chromosomal material before replication, but is released after replication. The animal cell MCM2,3,5 complex remains in the nucleus throughout the cell cycle, suggesting that it may be one component of the licensing factor. Another component, able to enter only at mitosis, may be necessary for MCM2,3,5 to associate with chromosomal material.

The presence of Cdc6 at the yeast origin allows MCM proteins to bind to the complex. Their presence is necessary for initiation. The origin therefore enters S phase in the condition of a **prereplication complex**, containing ORC, Cdc6, and MCM proteins. When initiation occurs, Cdc6 and MCM are displaced, returning the origin to the state of the **postreplication complex**, which contains only ORC. Because Cdc6 is rapidly degraded during S phase, it is not available to support reloading of MCM proteins, and so the origin cannot be used for a second cycle of initiation during the S phase.

The MCM2–7 proteins form a six-member ring-shaped complex around DNA. Some of the ORC proteins have similarities to replication proteins that load DNA polymerase on to DNA. It is possible that ORC uses hydrolysis of ATP to load the MCM ring on to DNA. In *Xenopus* extracts, replication can be initiated if ORC is removed after it has loaded Cdc6 and MCM proteins. This shows that the major role of ORC is to identify the origin to the Cdc6 and MCM proteins that control initiation and licensing.

15.8 SUMMARY

Replicons in bacterial or eukaryotic chromosomes have a single unifying feature: replication is initiated at an origin, once and only once in each cell division cycle. The origin is located within the replicon, and replication typically is bidirectional, with replication forks proceeding away from the origin in both directions. Replication is not usually terminated at specific sequences, but continues until DNA polymerase meets another DNA polymerase, either halfway round a circular replicon, or at the junction between two linear replicons.

An origin consists of a discrete sequence at which replication of DNA is initiated. Origins of replication tend to be rich in A · T base pairs. A bacterial chromosome contains a single origin, which is responsible for initiating replication once every division cycle. The *oriC* origin in *E. coli* is a sequence of 245 bp. Any DNA molecule with this sequence can replicate in *E. coli*. Replication of the circular bacterial chromosome produces a θ structure, in which the replicated DNA starts out as a small replicating eye. Replication proceeds until the eye occupies the whole chromosome. The bacterial origin contains sequences that are methylated on both strands of DNA. Replication produces hemimethylated DNA, which cannot function as an origin. There is a delay before the hemimethylated origins are remethylated to convert them to the functional state, and this is responsible for preventing initiation from occurring too frequently.

A eukaryotic chromosome is divided into many individual replicons. Replication occurs during a discrete part of the cell cycle (S phase), but because not all replicons are active simultaneously, the process may take several hours. Eukaryotic replication is at least an order of magnitude slower than bacterial replication. Origins sponsor bidirectional replication, and are probably used in a fixed order during S phase. Each replicon is activated only once in each cycle. Origins of replication were isolated as *ARS* sequences in yeast by virtue of their ability to support replication of any sequence attached to them. The core of an *ARS* is an 11 bp A · T-rich sequence that is bound by the ORC protein complex, which remains bound throughout the cell cycle. Utilization of the origin is controlled by the MCM licensing factors that associate with the ORC.

16

Extrachromosomal Replicons

16.1 Introduction

Key Terms

- A **plasmid** is a circular, extrachromosomal DNA. It is autonomous and can replicate itself.
- **Lysogeny** describes the ability of a phage to survive in a bacterium as a stable prophage component of the bacterial genome.
- An **episome** is a plasmid able to integrate into bacterial DNA.
- **Immunity** describes the ability of an extrachromosomal element to prevent another of the same type from infecting a cell. It applies to both plasmids and phages, but has a different molecular basis in each case.

A bacterium may be a host for independently replicating genetic units in addition to its chromosome. These extrachromosomal genomes fall into two general types: plasmids and bacteriophages (phages). Some plasmids and all phages have the ability to transfer from a donor bacterium to a recipient by an infective process. An important distinction between them is that plasmids exist only as free DNA genomes, but bacteriophages are viruses that package a nucleic acid genome into a protein coat, and are released from the bacterium at the end of an infective cycle.

Plasmids are self-replicating circular molecules of DNA that are maintained in the cell in a stable and characteristic number of copies; that is, the number remains constant from generation to generation. *Single-copy plasmids* are maintained at the same relative quantity as the bacterial host chromosome, that is, one per unit bacterium. Like the host chromosome, they rely on a specific apparatus to be segregated equally at each bacterial division. *Multicopy plasmids* exist in many copies per unit bacterium and may be segregated to daughter bacteria stochastically (meaning that there are enough copies that each daughter cell always gains some by a random distribution).

Plasmids and phages are defined by their ability to reside in a bacterium as independent genetic units. However, certain plasmids and some phages can also exist as sequences within the bacterial genome. In this case, the same sequence that constitutes the independent plasmid or phage genome is found within the chromosome, and is inherited like any other bacterial gene. Phages that are found as part of the bacterial chromosome are said to show **lysogeny**; plasmids that have the ability to behave like this are called **episomes**. Related processes are used by phages and episomes to insert into and excise from the bacterial chromosome.

A parallel between lysogenic phages and plasmids and episomes is that they maintain a selfish possession of their bacterium and often make it impossible for another element of the same type to become established. This effect is called **immunity**, although the molecular basis for plasmid immunity is different from lysogenic immunity, and is a consequence of the replication control system.

Figure 16.1 summarizes the types of genetic units that can be propagated in bacteria as independent genomes. Lytic phages may have genomes of any type of nucleic acid; they transfer between cells by release of infective particles. Lysogenic phages have double-stranded DNA genomes, as do plasmids and episomes. Some plasmids transfer between cells by a conjugative process (with direct contact between donor and recipient cells). A feature of the transfer process in both cases is that on occasion some bacterial host genes are transferred with the phage or plasmid DNA, so these events play a role in allowing exchange of genetic information between bacteria.

The key feature in determining the behavior of each type of unit is how its origin is used. An origin in a bacterial or eukaryotic chromosome is used to initiate a single replication event that extends across the replicon. However, replicons can also be used to sponsor other forms of replication. The most common alternative is used by the small, independently replicating units of viruses. The objective of a viral replication cycle is to produce many copies of the viral genome before the host cell is lysed to release them. Some viruses replicate in the same way as a host genome, with an initiation event leading to production of duplicate copies, each of which then replicates again, and so on. However, others use a mode of replication in which many copies are produced as a tandem array following a single initiation event. A similar type of event is triggered by episomes when an integrated plasmid DNA ceases to be inert and initiates a replication cycle.

Many prokaryotic replicons are circular, and this indeed is a necessary feature for replication modes that produce multiple tandem copies. However, some extrachromosomal replicons are linear, and in such cases we have to account for the ability to replicate the end of the replicon. (Of course, eukaryotic chromosomes are linear, so the same problem applies to the replicons at each end, but they have a special system for resolving the problem.)

Phages and plasmids live in bacteria			
Type of unit	Genome structure	Mode of propagation	Consequences
Lytic phage	ds- or ss-DNA or RNA; linear or circular	Infects susceptible host	Usually kills host
Lysogenic phage	ds-DNA	Linear sequence in host chromosome	Immunity to infection
Plasmid	ds-DNA circle	Replicates at defined copy number; may be transmissible	Immunity to plasmids in same group
Episome	ds-DNA circle	Free circle or linear integrated	May transfer host DNA

Figure 16.1 Several types of independent genetic units exist in bacteria.

16.2 The Ends of Linear DNA Are a Problem for Replication

The ability of all known nucleic acid polymerases, DNA or RNA, to synthesize a new strand only in the 5′–3′ direction poses a problem for synthesizing DNA at the end of a linear replicon. Consider the two

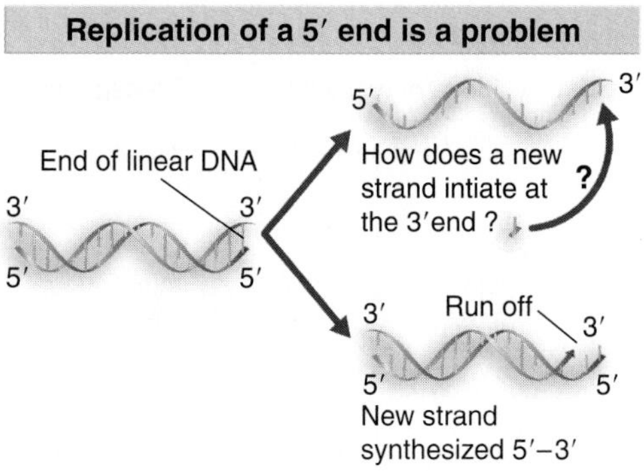

Replication of a 5' end is a problem

Figure 16.2 Replication could run off the 3' end of a newly synthesized linear strand, but could it initiate at a 5' end?

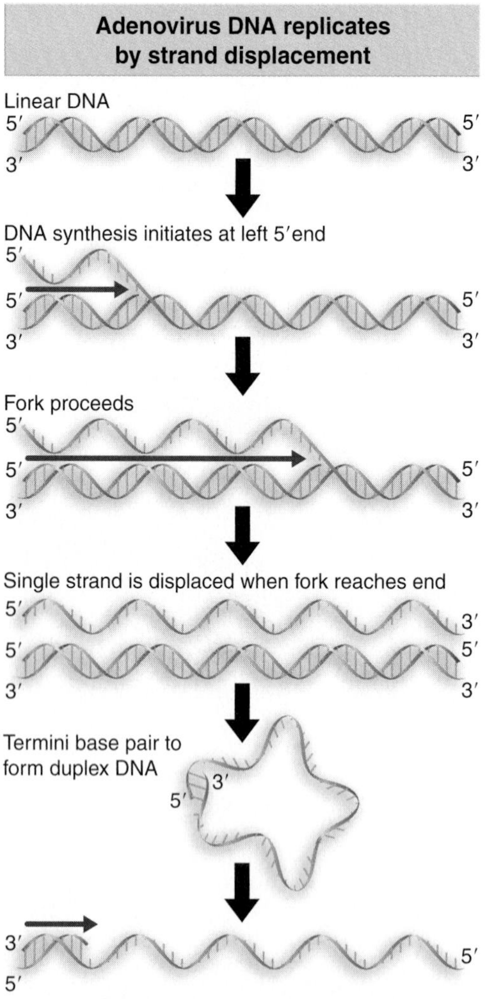

Adenovirus DNA replicates by strand displacement

Figure 16.3 Adenovirus DNA replication initiates separately at the two ends of the molecule and proceeds by strand displacement.

parental strands depicted in **Figure 16.2**. The lower strand presents no problem: it can act as a template to synthesize a daughter strand that runs right up to the end, where presumably the polymerase falls off. But to synthesize a complement at the end of the upper strand, synthesis must start right at the very last base (or else this strand would become shorter in successive cycles of replication).

We do not know whether initiation right at the end of a linear DNA is feasible. We usually think of a polymerase as binding at a site *surrounding* the position at which a base is to be incorporated. So a special mechanism must be employed for replication at the ends of linear replicons. Several types of solution may be imagined to accommodate the need to copy a terminus:

- The problem may be circumvented by converting a linear replicon into a circular or multimeric molecule. Phages such as T4 or lambda use such mechanisms (see *16.4 Rolling Circles Produce Multimers of a Replicon*).
- The DNA may form an unusual structure—for example, by creating a hairpin at the terminus, so that there is no free end. Formation of a crosslink is required in replication of the linear mitochondrial DNA of *Paramecium*.
- Instead of being precisely determined, the end may be variable. Eukaryotic chromosomes may adopt this solution, in which the number of copies of a short repeating unit at the end of the DNA changes (see *28.12 Telomeres Are Synthesized by a Ribonucleoprotein Enzyme*). A mechanism to add or remove units makes it unnecessary to replicate right up to the very end.
- A protein may intervene to make initiation possible at the actual terminus. Several linear viral nucleic acids have proteins that are *covalently linked to the 5' terminal base*. The best characterized examples are adenovirus DNA, phage φ29 DNA, and poliovirus RNA.

16.3 Terminal Proteins Enable Initiation at the Ends of Viral DNAs

Key Terms

- **Strand displacement** is a mode of replication of some viruses in which a new DNA strand grows by displacing the previous (homologous) strand of the duplex.
- A **terminal protein** allows replication of a linear phage genome to start at the very end. The protein attaches to the 5' end of the genome through a covalent bond, is associated with a DNA polymerase, and contains a cytosine residue that serves as a primer.

Key Concepts

- The dsDNA viruses adenovirus and φ29 have terminal proteins that initiate replication by generating a new 5' end.
- The newly synthesized strand displaces the corresponding strand of the original duplex.
- The strand that is released base pairs at the ends to form a duplex origin that initiates synthesis of a complementary strand.

An example of initiation at a linear end is provided by adenovirus and φ29 DNAs, which actually replicate from both ends, using the mechanism of **strand displacement** illustrated in **Figure 16.3**. The same events can occur independently at either end. Synthesis of a new strand starts at one end, displacing the homologous strand that was previously paired in the duplex. When the replication fork reaches the other end of the molecule, the displaced strand is released as a free single strand. It is then replicated independently; this requires the formation of a duplex origin by base pairing between some short complementary sequences at the ends of the molecule.

In several viruses that use such mechanisms, a protein is found covalently attached to each 5′ end. In the case of adenovirus, a **terminal protein** is linked to the mature viral DNA via a phosphodiester bond to serine, as indicated in **Figure 16.4**.

How does the attachment of the protein overcome the initiation problem? The terminal protein has a dual role: it carries a cytidine nucleotide that provides the primer; and it is associated with DNA polymerase. In fact, linkage of terminal protein to a nucleotide is undertaken by DNA polymerase in the presence of adenovirus DNA. This suggests the model illustrated in **Figure 16.5**. The complex of polymerase and terminal protein, bearing the priming C nucleotide, binds to the end of the adenovirus DNA. The free 3′—OH end of the C nucleotide is used to prime the elongation reaction by the DNA polymerase. This generates a new strand whose 5′ end is covalently linked to the initiating C nucleotide. (The reaction actually involves displacement of protein from DNA rather than binding *de novo*. The 5′ end of adenovirus DNA is bound to the terminal protein that was used in the previous replication cycle. The old terminal protein is displaced by the new terminal protein for each new replication cycle.)

Terminal protein binds to the region located between 9 and 18 bp from the end of the DNA. The adjacent region, between positions 17 and 48, is essential for the binding of a host protein, nuclear factor I, which is also required for the initiation reaction. The initiation complex may therefore form between positions 9 and 48, a fixed distance from the actual end of the DNA.

16.4 Rolling Circles Produce Multimers of a Replicon

Key Term

- The **rolling circle** is a mode of replication in which a replication fork proceeds around a circular template for an indefinite number of revolutions; the DNA strand newly synthesized in each revolution displaces the strand synthesized in the previous revolution, giving a tail containing a linear series of sequences complementary to the circular template strand.

Key Concept

- A rolling circle generates single-stranded multimers of the original sequence.

The structures generated by replication depend on the relationship between the template and the replication fork. The critical features are whether the template is circular or linear, and whether the replication fork is engaged in synthesizing both strands of DNA or only one.

Replication of only one strand is used to generate copies of some circular molecules. A nick opens one strand, and then the free 3′—OH end generated by the nick is extended by the DNA polymerase. The

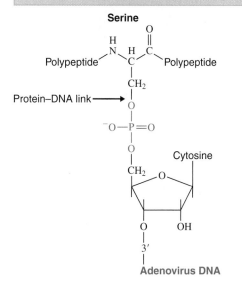

Figure 16.4 The 5′ terminal phosphate at each end of adenovirus DNA is covalently linked to serine in the 55 kD Ad-binding protein.

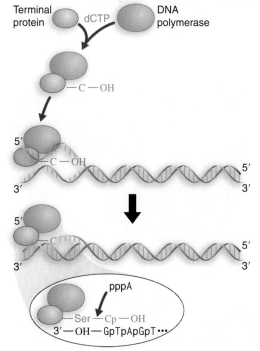

Figure 16.5 Adenovirus terminal protein binds to the 5′ end of DNA and provides a C—OH end to prime synthesis of a new DNA strand.

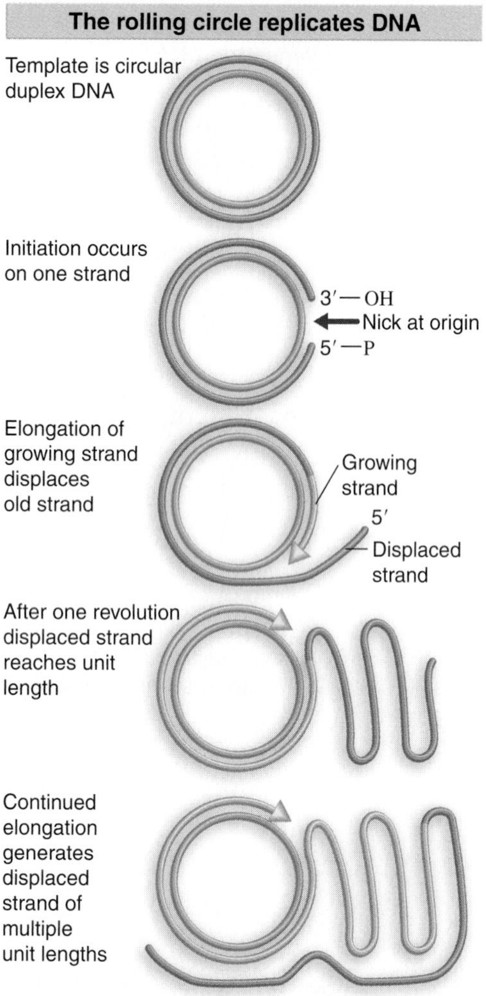

The rolling circle replicates DNA

Template is circular
duplex DNA

Initiation occurs
on one strand

3′ — OH
← Nick at origin
5′ — P

Elongation of
growing strand
displaces
old strand

Growing
strand
5′
Displaced
strand

After one revolution
displaced strand
reaches unit
length

Continued
elongation
generates
displaced
strand of
multiple
unit lengths

Figure 16.6 The rolling circle generates a multimeric single-stranded tail.

newly synthesized strand displaces the original parental strand. The ensuing events are depicted in **Figure 16.6**.

This type of structure is called a **rolling circle**, because the growing point can be envisaged as rolling around the circular template strand. It could in principle continue to do so indefinitely. As it moves, the replication fork extends the outer strand and displaces the previous partner. An example is shown in the electron micrograph of **Figure 16.7**.

Because the newly synthesized material is covalently linked to the original material, the displaced strand has the original unit genome at its 5′ end. The original unit is followed by any number of unit genomes, synthesized by continuing revolutions of the template. Each revolution displaces the material synthesized in the previous cycle.

The rolling circle is used in several ways *in vivo*. Some pathways that are used to replicate DNA are depicted in **Figure 16.8**.

Cleavage of a unit length tail generates a copy of the original circular replicon in linear form. The linear form may be maintained as a single strand or may be converted into a duplex by synthesis of the complementary strand (which is identical in sequence to the template strand of the original rolling circle).

The rolling circle provides a means for amplifying the original (unit) replicon. This mechanism is used to generate amplified rDNA in the *Xenopus* oocyte. The genes for rRNA are organized as a large number of contiguous repeats in the genome. A single repeating unit from the genome is converted into a rolling circle. The displaced tail, containing many units, is converted into duplex DNA; later it is cleaved from the circle so that the two ends can be joined together to generate a large circle of amplified rDNA. The amplified material therefore consists of a large number of identical repeating units.

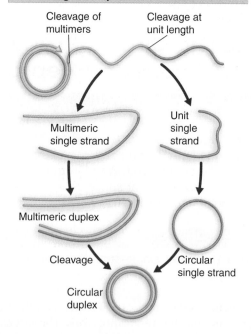

Rolling circle products are versatile

Cleavage of
multimers

Cleavage at
unit length

Multimeric
single strand

Unit
single
strand

Multimeric duplex

Cleavage

Circular
single strand

Circular
duplex

Circular
single strand

Figure 16.8 The fate of the displaced tail determines the types of products generated by rolling circles. Cleavage at unit length generates monomers, which can be converted to duplex and circular forms. Cleavage of multimers generates a series of tandemly repeated copies of the original unit. Note that the conversion to double-stranded form could occur earlier, before the tail is cleaved from the rolling circle.

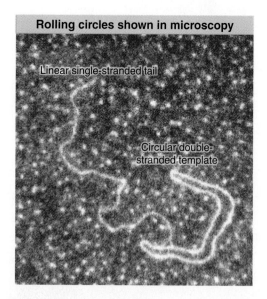

Rolling circles shown in microscopy

Linear single-stranded tail

Circular double-stranded template

Figure 16.7 A rolling circle appears as a circular molecule with a linear tail by electron microscopy. Photograph kindly provided by Dr. David Dressler, Harvard University.

16.5 Rolling Circles Are Used to Replicate Phage Genomes

Key Term

- A **relaxase** is an enzyme that cuts one strand of DNA, and binds to the free 5′ end.

Key Concept

- The φX A protein is a *cis*-acting relaxase that generates single-stranded circles from the tail produced by rolling circle replication.

Replication by rolling circles is common among bacteriophages. Unit genomes can be cleaved from the displaced tail, generating monomers that can be packaged into phage particles or used for further replication cycles. Phage φX174 consists of a single-stranded circular DNA, known as the plus (+) strand. A complementary strand, called the minus (−) strand, is synthesized as the first step in replication. This action generates the duplex circle shown at the top of **Figure 16.9**. Replication then proceeds by a rolling circle mechanism.

The duplex circle is converted to a covalently closed form, which becomes supercoiled. A protein coded by the phage genome, the A protein, nicks the (+) strand of the duplex DNA at a specific site that defines the origin for replication. After nicking the origin, the A protein remains connected to the 5′ end that it generates, while the 3′ end is extended by DNA polymerase.

The structure of the DNA plays an important role in this reaction, for the DNA can be nicked *only when it is negatively supercoiled* (wound about its axis in space in the opposite sense from the handedness of the double helix; see *1.7 Supercoiling Affects the Structure of DNA*). The A protein is able to bind to a single-stranded decamer fragment of DNA that surrounds the site of the nick. This suggests that the supercoiling is needed to assist the formation of a single-stranded region that provides the A protein with its binding site. (An enzymatic activity in which a protein cleaves duplex DNA and binds to a released 5′ end is sometimes called a **relaxase**.) The nick generates a 3′ — OH end and a 5′–phosphate end (covalently attached to the A protein), both of which have roles to play in φX174 replication.

Using the rolling circle, the 3′ — OH end of the nick is extended into a new chain. The chain is elongated around the circular (−) strand template, until it reaches the starting point and displaces the origin. Now the A protein functions again. It remains connected with the rolling circle as well as to the 5′ end of the displaced tail, and it is therefore in the vicinity as the growing point returns past the origin. So the same A protein is available again to recognize the origin and nick it, now attaching to the end generated by the new nick. The cycle can be repeated indefinitely.

Following this nicking event, the displaced single (+) strand is freed as a circle. The A protein is involved in the circularization. In fact, the joining of the 3′ and 5′ ends of the (+) strand product is accomplished by the A protein as part of the reaction by which it is released at the end of one cycle of replication, and starts another cycle.

The A protein has an unusual property that may be connected with these activities. It is *cis*-acting *in vivo*. (This behavior is not reproduced *in vitro*, as can be seen from its activity on any DNA template in a cell-free system.) *The implication is that in vivo the A protein synthesized by a particular genome can attach only to the DNA of that genome.* We do not know how this is accomplished. However, its activity *in vitro* shows how it remains associated with the same parental (−) strand template.

The A protein is *cis*-acting

A protein nicks the origin and binds to 5′ end

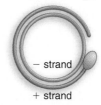

Rolling circle replication displaces plus strand

Replication fork passes origin, A protein nicks DNA, and binds to the new 5′ end

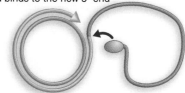

Released plus strand forms covalent circle

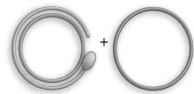

Cycle repeats

Figure 16.9 φX174 RF DNA is a template for synthesizing single-stranded viral circles. The A protein remains attached to the same genome through indefinite revolutions, each time nicking the origin on the viral (+) strand and transferring to the new 5′ end. At the same time, the released viral strand is circularized.

The A protein has two active sites; this may allow it to cleave the "new" origin while still retaining the "old" origin; then it ligates the displaced strand into a circle.

The displaced ($+$) strand may follow either of two fates after circularization. During the replication phase of viral infection, it may be used as a template to synthesize the complementary ($-$) strand. The duplex circle may then be used as a rolling circle to generate more progeny. During phage morphogenesis, the displaced ($+$) strand is packaged into the phage virion.

16.6 The F Plasmid Is Transferred by Conjugation Between Bacteria

Key Terms

- **Conjugation** is a process in which DNA is transferred from a donor to a recipient cell.
- The **F plasmid** is an episome that can be free or integrated in *E. coli*, and which in either form can sponsor conjugation.
- The **transfer region** is a segment on the F plasmid that is required for bacterial conjugation.
- A **pilus (pili)** is a surface appendage on a bacterium that allows the bacterium to attach to other bacterial cells. Its appearance is that of a short, thin, flexible rod. During conjugation, pili are used to transfer DNA from one bacterium to another.
- **Pilin** is the subunit that is polymerized into the pilus in bacteria.

Key Concepts

- A free F factor is a replicon that is maintained at the level of one plasmid per bacterial chromosome.
- An F factor can integrate into the bacterial chromosome, in which case its own replication system is suppressed.
- The F-pilus coded by the F factor enables an F-positive bacterium to contact an F-negative bacterium and to initiate conjugation.

An interesting example of a connection between replication and the propagation of a genetic unit is provided by bacterial **conjugation**, in which a plasmid genome or host chromosome is transferred from one bacterium to another.

Conjugation is mediated by the **F plasmid**, which is the classic example of an episome, an element that may exist as a free circular plasmid, or that may become integrated, like a lysogenic bacteriophage, into the bacterial chromosome as a linear sequence. The F plasmid is a large circular DNA, ~100 kb in length.

The F factor can integrate at several sites in the *E. coli* chromosome, often by a recombination event between certain sequences (called IS sequences; see *21.5 Transposons Cause Rearrangement of DNA*) that are present on both the host chromosome and F plasmid. In its free (plasmid) form, the F plasmid utilizes its own replication origin (*oriV*) and control system, and is maintained at a level of one copy per bacterial chromosome. When it is integrated into the bacterial chromosome, this system is suppressed, and F DNA is replicated as a part of the chromosome.

The presence of the F plasmid, whether free or integrated, has important consequences for the host bacterium. Bacteria that are F-positive are able to conjugate (or mate) with bacteria that are F-negative. Conjugation requires a contact between donor (F-positive) and recipient (F-negative) bacteria; contact is followed by transfer of the F factor. If the F factor exists as a free plasmid in the donor bacterium, it is transferred as a plasmid, and the transfer converts the F-negative recipient into an F-positive state. If the F factor is present in an integrated form in the donor, conjugation may also transfer some or all of the bacterial chromosome. Many plasmids have conjugation systems that operate in a generally similar manner, but the F factor was the first to be discovered, and remains the paradigm for this type of genetic transfer.

A large (~33 kb) region of the F plasmid, called the **transfer region**, is required for conjugation. It contains ~40 genes, named as *tra* and *trb* loci, that are required for the transmission of DNA. Only four of the *tra* genes in the major transcription unit are concerned directly with the transfer of DNA; most are concerned with the properties of the bacterial cell surface and with maintaining contacts between mating bacteria.

F-positive bacteria possess surface appendages called **pili** (singular **pilus**) that are coded by the F factor. The gene *traA* codes for the single subunit protein, **pilin**, that is polymerized into the pilus. At least 12 *tra* genes are required for the modification and assembly of pilin into the pilus. The F-pili are hairlike structures, 2–3 μm long, that protrude from the bacterial surface. A typical F-positive cell has 2–3 pili. The pilin subunits are polymerized into a hollow cylinder, ~8 nm in diameter, with a 2 nm axial hole.

Mating is initiated when the tip of the F-pilus contacts the surface of the recipient cell. **Figure 16.10** shows an example of *E. coli* cells beginning to mate. A donor cell does not contact other cells carrying the F factor, because the genes *traS* and *traT* code for "surface exclusion" proteins that make the cell a poor recipient in such contacts. This effectively restricts donor cells to mating with F-negative cells. (And the presence of F-pili has secondary consequences; they provide the sites to which RNA phages and some single-stranded DNA phages attach, so F-positive bacteria are susceptible to infection by these phages, whereas F-negative bacteria are resistant.)

The initial contact between donor and recipient cells is easily broken, but other *tra* genes act to stabilize the association, bringing the mating cells closer together. The F pili are essential for initiating pairing, but retract or disassemble as part of the process by which the mating cells are brought into close contact. There must be a channel through which DNA is transferred, but the pilus itself does not appear to provide it.

16.7 Conjugation Transfers Single-Stranded DNA

Key Term

- An **Hfr** cell is a bacterium that has an integrated F plasmid within its chromosome. Hfr stands for high frequency recombination, referring to the fact that chromosomal genes are transferred from an Hfr cell to an F⁻ cell much more frequently than from an F⁺ cell.

Key Concepts

- Transfer of an F factor is initiated when rolling circle replication begins at *oriT*.
- The free 5′ end initiates transfer into the recipient bacterium.
- The transferred DNA is converted into double-stranded form in the recipient bacterium.
- When an F factor is free, conjugation "infects" the recipient bacterium with a copy of the F factor.
- When an F factor is integrated, conjugation causes transfer of the bacterial chromosome until the process is interrupted by random breakage of the contact between donor and recipient bacteria.

Transfer of the F factor is initiated at a site called *oriT*, the origin of transfer, which is located at one end of the transfer region. The transfer process may be initiated when the protein TraM recognizes that a mating pair has formed. Then TraY binds near *oriT* and causes TraI to bind. TraI is a relaxase, like φX174 A protein. TraI nicks *oriT* at a unique site (called *nic*), and then forms a covalent link to the 5′ end that has been generated. TraI also catalyzes the unwinding of ~200 bp of DNA (this is a helicase activity). **Figure 16.11** shows that the freed 5′ end leads the

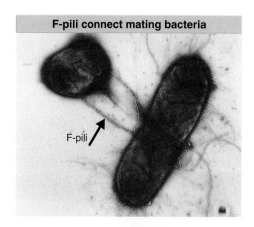

F-pili connect mating bacteria

F-pili

Figure 16.10 Mating bacteria are initially connected when donor F pili contact the recipient bacterium. Photograph kindly provided by Professor Ronald A. Skurray, School of Biological Sciences, University of Sydney.

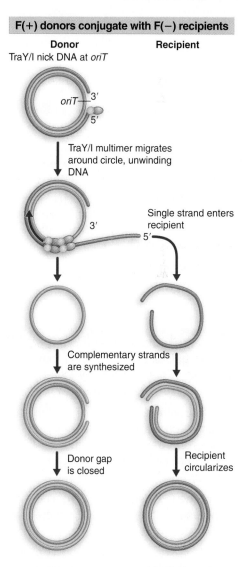

F(+) donors conjugate with F(−) recipients

Donor   Recipient

TraY/I nick DNA at *oriT*

oriT — 3′
5′

TraY/I multimer migrates around circle, unwinding DNA

3′   Single strand enters recipient
5′

Complementary strands are synthesized

Donor gap is closed   Recipient circularizes

Figure 16.11 Transfer of DNA occurs when the F factor is nicked at *oriT* and a single strand is led by the 5′ end into the recipient. Only one unit length is transferred. Complementary strands are synthesized to the single strand remaining in the donor and to the strand transferred into the recipient.

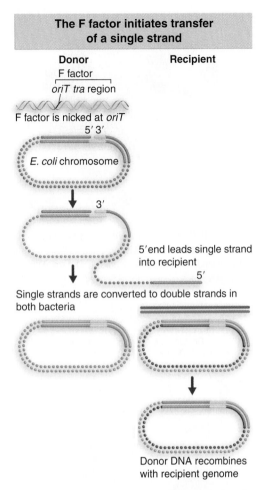

The F factor initiates transfer of a single strand

Donor   Recipient

F factor
oriT tra region

F factor is nicked at *oriT*

5' 3'

E. coli chromosome

3'

5'end leads single strand
into recipient

5'

Single strands are converted to double strands in
both bacteria

Donor DNA recombines
with recipient genome

Figure 16.12 Transfer of chromosomal DNA occurs when an integrated F factor is nicked at *oriT*. Transfer of DNA starts with a short sequence of F DNA and continues until prevented by loss of contact between the bacteria.

way into the recipient bacterium. A complement for the transferred single strand is synthesized in the recipient bacterium, which as a result is converted to the F-positive state.

A complementary strand must be synthesized in the donor bacterium to replace the strand that has been transferred. If this happens concomitantly with the transfer process, the state of the F plasmid resembles the rolling circle of Figure 16.6. Conjugating DNA usually has the appearance of a rolling circle, but replication as such is not necessary to provide the driving energy, and single-strand transfer is independent of DNA synthesis. Only a single unit length of the F factor is transferred to the recipient bacterium. This implies that some (unidentified) feature terminates the process after one revolution, after which the covalent integrity of the F plasmid is restored.

When an integrated F plasmid initiates conjugation, the orientation of transfer is directed away from the transfer region, into the bacterial chromosome. **Figure 16.12** shows that, following a sequence of F DNA, bacterial DNA is transferred. The process continues until it is interrupted by the breaking of contacts between the mating bacteria. It takes ~100 minutes to transfer the entire bacterial chromosome, and under standard conditions, contact is often broken before the completion of transfer.

Donor DNA that enters a recipient bacterium is converted to double-stranded form, and may recombine with the recipient chromosome. So conjugation affords a means to transfer genetic material between bacteria (a contrast with their usual asexual growth). A strain of *E. coli* with an integrated F factor supports such recombination at relatively high frequencies (compared to strains that lack integrated F factors); such strains are described as **Hfr** (for *high frequency recombination*). Each position of integration for the F factor gives rise to a different Hfr strain, with a characteristic pattern of transferring bacterial markers to a recipient chromosome.

16.8 The Ti Bacterial Plasmid Transfers Genes into Plant Cells

Key Terms

- **Crown gall disease** is a tumor that can be induced in many plants by infection with the bacterium *Agrobacterium tumefaciens*.

- The **Ti plasmid** is an episome of the bacterium *Agrobacterium tumefaciens* that carries the genes responsible for the induction of crown gall disease in infected plants.

- An **opine** is a derivative of arginine that is synthesized by plant cells infected with crown gall disease.

- **T-DNA** is the segment of the Ti plasmid of *Agrobacterium tumefaciens* that is transferred to the plant cell nucleus during infection. It carries genes that transform the plant cell.

Key Concepts

- Infection with the bacterium *A. tumefaciens* can transform plant cells into tumors.

- The infectious agent is a plasmid carried by the bacterium.

- The plasmid also carries genes for synthesizing and metabolizing opines (arginine derivatives) that are used by the tumor cell.

- Only part of the DNA of the Ti plasmid (T-DNA) is transferred to the plant cell nucleus, but the *vir* genes outside this region are required for the transfer process.

- The *vir* genes are induced by phenolic compounds released by plants in response to wounding.

- The membrane protein VirA is autophosphorylated on histidine when it binds an inducer and activates VirG by transferring the phosphate group to it.

The interaction between bacteria and certain plants involves the transfer of DNA from the bacterial genome to the plant genome. **Crown gall**

disease, shown in **Figure 16.13**, can be induced in most dicotyledonous plants by the soil bacterium *Agrobacterium tumefaciens*. The bacterium is a parasite that effects a genetic change in the eukaryotic host cell, with consequences for both parasite and host. It improves conditions for survival of the parasite. And it causes the plant cell to grow as a tumor.

Agrobacteria are required to induce tumor formation, but the tumor cells do not require the continued presence of bacteria. Like animal tumors, the plant cells have been transformed into a state in which new mechanisms govern growth and differentiation. Transformation is caused by the expression within the plant cell of genetic information transferred from the bacterium.

The tumor-inducing principle of *Agrobacterium* resides in the **Ti plasmid**, which is perpetuated as an independent replicon within the bacterium. The plasmid carries genes involved in various bacterial and plant cell activities, including those required to generate the transformed state, and a set of genes concerned with synthesis or utilization of **opines** (novel derivatives of arginine). Ti plasmids (and thus the *Agrobacteria* in which they reside) can be divided into four groups according to the types of opines that are made.

The interaction between *Agrobacterium* and a plant cell is illustrated in **Figure 16.14**. The bacterium does not enter the plant cell, but transfers part of the Ti plasmid to the plant nucleus. The transferred part of the Ti genome is called **T-DNA**. It becomes integrated into the plant genome, where it expresses the functions needed to synthesize opines and to transform the plant cell.

Transformation of plant cells requires three types of function carried in the *Agrobacterium*:

- Three loci on the *Agrobacterium* chromosome, *chvA*, *chvB*, *pscA*, are required for the initial stage of binding the bacterium to the plant cell. They are responsible for synthesizing a polysaccharide on the bacterial cell surface.
- The *vir* region carried by the Ti plasmid outside the T-DNA region is required to release and initiate transfer of the T-DNA.
- The T-DNA is required to transform the plant cell.

The T-region occupies ~23 kb of the ~200 kb Ti genome, and codes for the proteins necessary to maintain the plant cell in a transformed state. However, functions affecting oncogenicity—the ability to form tumors—are not confined to the T-region. Those genes located outside the T-region must be concerned with establishing the tumorigenic state, but their products are not needed to perpetuate it. They may be concerned with transfer of T-DNA into the plant nucleus or perhaps with subsidiary functions such as the balance of plant hormones in the infected tissue.

16.9 Transfer of T-DNA Resembles Bacterial Conjugation

Key Concepts

- T-DNA is generated when a nick at its right boundary creates a primer for synthesis of a new DNA strand.
- The preexisting single strand that is displaced by the new synthesis is transferred to the plant cell nucleus.

T plasmids induce plant teratomas

Figure 16.13 An *Agrobacterium* carrying a Ti plasmid of the nopaline type induces a teratoma, in which differentiated structures develop. Photograph kindly provided by Jeff Schell.

T-DNA integrates into the plant cell

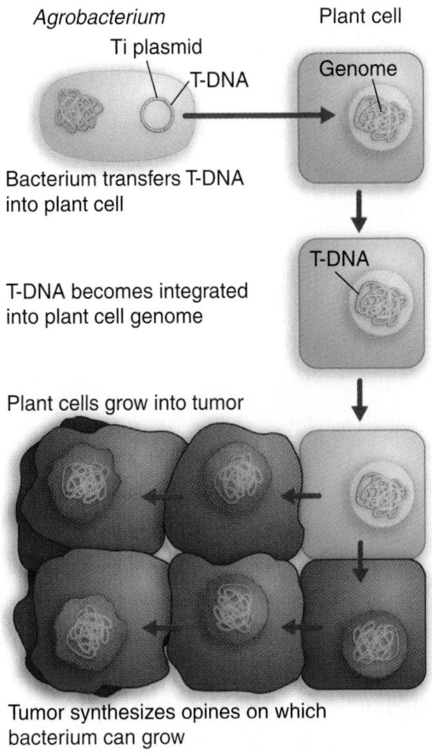

Agrobacterium   Plant cell
   Ti plasmid
              T-DNA   Genome

Bacterium transfers T-DNA
into plant cell

                                    T-DNA

T-DNA becomes integrated
into plant cell genome

Plant cells grow into tumor

Tumor synthesizes opines on which
bacterium can grow

Figure 16.14 T-DNA is transferred from *Agrobacterium* carrying a Ti plasmid into a plant cell, where it becomes integrated into the nuclear genome and expresses functions that transform the host cell.

- Transfer is terminated when DNA synthesis reaches a nick at the left boundary.
- The T-DNA is transferred as a complex of single-stranded DNA with the VirE2 single strand-binding protein.
- The single-stranded T-DNA is converted into double-stranded DNA and integrated into the plant genome.
- The mechanism of integration is not known. T-DNA can be used to transfer genes into a plant nucleus.

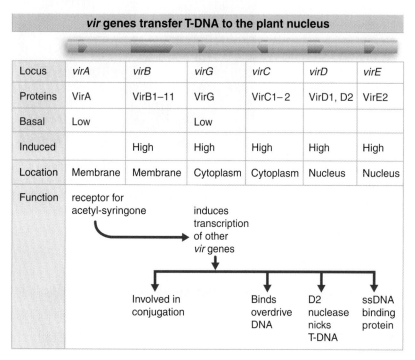

vir genes transfer T-DNA to the plant nucleus

Locus	virA	virB	virG	virC	virD	virE
Proteins	VirA	VirB1–11	VirG	VirC1–2	VirD1, D2	VirE2
Basal	Low		Low			
Induced		High	High	High	High	High
Location	Membrane	Membrane	Cytoplasm	Cytoplasm	Nucleus	Nucleus
Function	receptor for acetyl-syringone		induces transcription of other vir genes			
			Involved in conjugation	Binds overdrive DNA	D2 nuclease nicks T-DNA	ssDNA binding protein

Figure 16.15 The *vir* region of the Ti plasmid has six loci that are responsible for transferring T-DNA to an infected plant.

The transfer of T-DNA from Agrobacterium into a plant cell is coded by six virulence loci, *virA–G*, which reside in a 40 kb region outside the T-DNA. Their organization is summarized in **Figure 16.15**. Each locus is transcribed as an individual unit; some contain more than one open reading frame.

We may divide the transforming process into (at least) two stages:

- Agrobacterium contacts a plant cell, and the *vir* genes are induced.
- *vir* gene products cause T-DNA to be transferred to the plant cell nucleus, where it is integrated into the genome.

The *vir* genes fall into two groups, corresponding to these stages. Genes *virA* and *virG* are regulators that respond to a change in the plant by inducing the other genes. So mutations in *virA* and *virG* are avirulent; the mutants carrying them cannot express the remaining *vir* genes. Genes *virB,C,D,E* code for proteins involved in the transfer of DNA. Mutations in *virB* and *virD* are avirulent in all plants, but the effects of mutations in *virC* and *virE* vary with the type of host plant.

The genes *virA* and *virG* are expressed constitutively at a low level. The signal to which they respond is provided by phenolic compounds generated by plants as a response to wounding. **Figure 16.16** presents an example. *Nicotiana tabacum* (tobacco) generates the molecules acetosyringone and α-hydroxyacetosyringone. Exposure to these compounds activates *virA*, which acts on *virG*, which in turn induces the expression of *virB,C,D,E*. This reaction explains why *Agrobacterium* infection succeeds only on wounded plants.

VirA and VirG are an example of a classic type of bacterial system in which stimulation of a sensor protein causes autophosphorylation and transfer of the phosphate to the second protein. VirA forms a homodimer that is located in the inner membrane; it may respond to the presence of the phenolic compounds in the periplasmic space. Exposure to these compounds causes VirA to become autophosphorylated on histidine. The phosphate group is then transferred to an Asp residue in VirG. The phosphorylated VirG binds to promoters of the *virB,C,D,E* genes to activate transcription. When *virG* is activated, its transcription is induced from a new startpoint, different from that used for constitutive expression, with the result that the amount of VirG protein is increased.

Of the other *vir* loci, *virD* is the best characterized. The *virD* locus has four open reading frames. Two of the proteins coded at *virD*, VirD1

Wounding produces acetosyringone

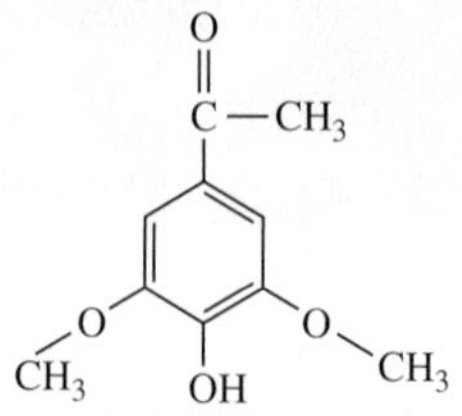

Figure 16.16 Acetosyringone (4-acetyl-2,6-dimethoxyphenol) is produced by *N. tabacum* upon wounding, and induces transfer of T-DNA from *Agrobacterium*.

and VirD2, provide an endonuclease that initiates the transfer process by nicking T-DNA at a specific site.

The transfer process actually selects the T-region for entry into the plant. **Figure 16.17** shows that the T-DNA of a nopaline plasmid is demarcated from the flanking regions in the Ti plasmid by repeats of 25 bp, which differ at only two positions between the left and right ends. When T-DNA is integrated into a plant genome, it has a well-defined right junction, which retains 1–2 bp of the right repeat. The left junction is variable; the boundary of T-DNA in the plant genome may be located at the 25 bp repeat or at one of a series of sites extending over ~100 bp within the T-DNA. Sometimes multiple tandem copies of T-DNA are integrated at a single site.

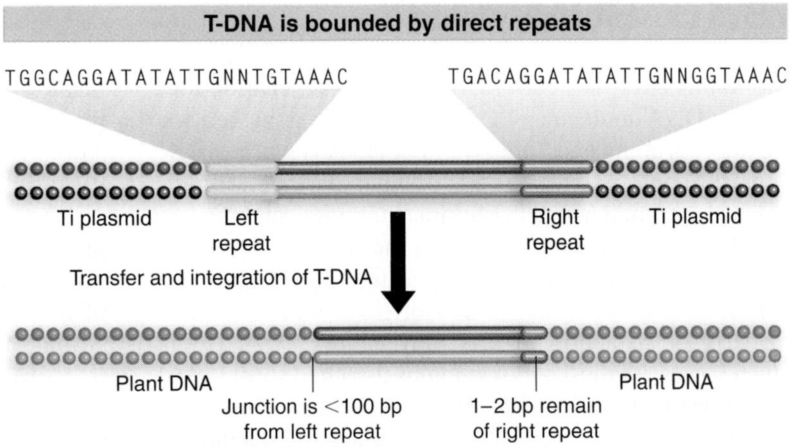

T-DNA is bounded by direct repeats

TGGCAGGATATATTGNNTGTAAAC   TGACAGGATATATTGNNGGTAAAC

Ti plasmid   Left repeat   Right repeat   Ti plasmid

Transfer and integration of T-DNA

Plant DNA   Plant DNA

Junction is <100 bp from left repeat   1–2 bp remain of right repeat

Figure 16.17 T-DNA has almost identical repeats of 25 bp at each end in the Ti plasmid. The right repeat is necessary for transfer and integration to a plant genome. T-DNA that is integrated in a plant genome has a precise junction that retains 1–2 bp of the right repeat, but the left junction varies and may be up to 100 bp short of the left repeat.

A model for transfer is illustrated in **Figure 16.18**. A nick is made at the right 25 bp repeat. It provides a priming end for synthesis of a DNA single strand. Synthesis of the new strand displaces the old strand, which is used in the transfer process. The process is terminated when DNA synthesis reaches a nick at the left repeat. This model explains why the right repeat is essential, and it accounts for the polarity of the process. If the left repeat fails to be nicked, production of DNA for transfer could continue farther along the Ti plasmid.

The single molecule of single-stranded DNA produced for transfer in the infecting bacterium is transferred as a DNA–protein complex, sometimes called the T-complex. The DNA is covered by the VirE2 single-strand binding protein, which has a nuclear localization signal and is responsible for transporting T-DNA into the plant cell nucleus. A single molecule of the D2 subunit of the endonuclease remains bound at the 5′ end. The *virB* operon codes for 11 products that are involved in the transfer reaction.

This model for transfer of T-DNA closely resembles the events of bacterial conjugation, by which the *E. coli* chromosome is transferred from one cell to another in single-stranded form. The genes of the *virB* operon are homologous to the *tra* genes of certain bacterial plasmids that are involved in conjugation (see *16.7 Conjugation Transfers Single-Stranded DNA*). A difference is that the transfer of T-DNA is (usually) limited by the boundary of the left repeat, whereas transfer of bacterial DNA is indefinite.

We do not know how the transferred DNA is integrated into the plant genome. At some stage, the newly generated single strand must be converted into duplex DNA. Circles of T-DNA that are found in infected plant cells appear to be generated by recombination between the left and right 25 bp repeats, but we do not know if they are intermediates. The actual event is likely to involve a nonhomologous recombination, because there is no homology between the T-DNA and the sites of integration.

As a practical matter, the ability of *Agrobacterium* to transfer T-DNA to the plant genome makes it possible to introduce new genes into plants. Because the transfer/integration and oncogenic functions are separate, it is possible to engineer new Ti plasmids in which the oncogenic functions have been replaced by other genes whose effect on the plant we wish to test. The existence of a natural system for delivering genes to the plant genome is widely used for genetic engineering of plants.

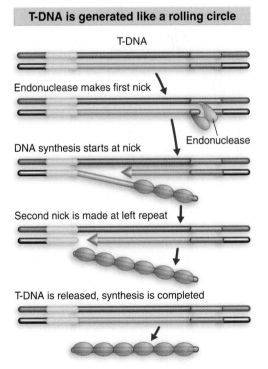

T-DNA is generated like a rolling circle

T-DNA

Endonuclease makes first nick

DNA synthesis starts at nick   Endonuclease

Second nick is made at left repeat

T-DNA is released, synthesis is completed

Figure 16.18 T-DNA is generated by displacement when DNA synthesis starts at a nick made at the right repeat. The reaction is terminated by a nick at the left repeat.

16.10 SUMMARY

The rolling circle is an alternative form of replication for circular DNA molecules in which an origin is nicked to provide a priming end. One strand of DNA is synthesized from this end, displacing the original partner strand, which is extruded as a tail. Multiple genomes can be produced by continuing revolutions of the circle.

Rolling circles are used to replicate some phages. The A protein that nicks the φX174 origin has the unusual property of *cis*-action. It acts only on the DNA from which it was synthesized. It remains attached to the displaced strand until an entire strand has been synthesized, and then nicks the origin again, releasing the displaced strand and starting another cycle of replication.

Rolling circles also characterize bacterial conjugation, when an F plasmid is transferred from a donor to a recipient cell, following the initiation of contact between the cells by means of the F-pili. A free F plasmid infects new cells by this means; an integrated F factor creates an Hfr strain that may transfer chromosomal DNA. In conjugation, replication is used to synthesize complements to the single strand remaining in the donor and to the single strand transferred to the recipient, but does not provide the motive power.

Agrobacteria induce tumor formation in wounded plant cells. The wounded cells secrete phenolic compounds that activate *vir* genes carried by the Ti plasmid of the bacterium. The *vir* gene products cause a single strand of DNA from the T-DNA region of the plasmid to be transferred to the plant cell nucleus. Transfer is initiated at one boundary of T-DNA, but ends at variable sites. The single strand is converted into a double strand and integrated into the plant genome. Genes within the T-DNA transform the plant cell, and cause it to produce particular opines (derivatives of arginine). Genes in the Ti plasmid allow Agrobacteria to metabolize the opines produced by the transformed plant cell. T-DNA has been used to develop vectors for transferring genes into plant cells.

17

Bacterial Replication Is Connected to the Cell Cycle

17.1 Introduction

Key Term

- The **unit cell** describes the state of an *E. coli* bacterium generated by a new division. It is 1.7 μm long and has a single replication origin.

The way in which replication is controlled and linked to the cell cycle is a major difference between prokaryotes and eukaryotes.

In eukaryotes, the chromosomes reside in the nucleus, each chromosome consists of many replicons, replication requires coordination of these replicons to reproduce DNA during a discrete period of the cell cycle, the decision on whether to replicate is taken by a complex pathway that regulates the cell cycle, and the duplicated chromosomes are segregated to daughter cells during mitosis by means of a special apparatus.

Figure 17.1 shows that in bacteria, replication is triggered at a single origin when the cell mass increases past a threshold level, and the segregation of the daughter chromosomes is accomplished by ensuring that they find themselves on opposite sides of the septum that grows to divide the bacterium into two.

How does the cell know when to initiate the replication cycle? The initiation event occurs at a constant ratio of cell mass to the number of chromosome origins. Cells growing more rapidly are larger and possess a greater number of origins. The growth of *E. coli* can be described in terms of the **unit cell**, an entity 1.7 μm long. A bacterium contains one origin per unit cell; a rapidly growing cell with two origins is 1.7–3.4 μm long.

How is cell mass titrated? An initiator protein could be synthesized continuously throughout the cell cycle; accumulation of a critical amount would trigger initiation. This explains why protein synthesis is

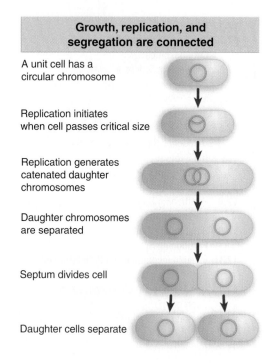

Growth, replication, and segregation are connected

A unit cell has a circular chromosome

Replication initiates when cell passes critical size

Replication generates catenated daughter chromosomes

Daughter chromosomes are separated

Septum divides cell

Daughter cells separate

Figure 17.1 Replication initiates at the bacterial origin when a cell passes a critical threshold of size. Completion of replication produces daughter chromosomes that may be linked by recombination or that may be catenated. They are separated and moved to opposite sides of the septum before the bacterium is divided into two.

needed for the initiation event. An alternative possibility is that an inhibitor protein might be synthesized at a fixed point, and diluted below an effective level by the increase in cell volume.

Some of the events in partitioning the daughter chromosomes are consequences of the circularity of the bacterial chromosome. Circular chromosomes are said to be *catenated* when one passes through another, connecting them. Topoisomerases are required to separate them. An alternative type of structure is formed when a recombination event occurs: a single recombination between two monomers converts them into a single dimer. This is resolved by a specialized recombination system that recreates the independent monomers. Essentially the partitioning process is handled by enzyme systems that act directly on discrete DNA sequences.

17.2 Bacteria Can Have Multiforked Chromosomes

Key Terms

- The **doubling time** is the period (usually measured in minutes) that it takes for a bacterial cell to reproduce.

- A **multiforked chromosome** (in a bacterium) has more than one set of replication forks because a second initiation has occurred before the first cycle of replication has been completed.

Key Concepts

- A fixed time of 40 minutes is required to replicate the *E. coli* chromosome, and 20 minutes is then required to divide the cell.

- When cells divide more often than once every 60 minutes, one cycle of replication may be initiated before the previous cycle has been completed.

The rate of bacterial growth is assessed by the **doubling time**, the period required for the number of cells to double. The shorter the doubling time, the faster the growth rate. *E. coli* cells can grow at rates ranging from doubling times as fast as 18 minutes to slower than 180 minutes. Because the bacterial chromosome is a single replicon, the frequency of replication cycles is controlled by the number of initiation events at the single origin. The replication cycle can be defined in terms of two constants:

- C is the fixed time of ~40 minutes required to replicate the entire bacterial chromosome. Its duration corresponds to a rate of replication fork movement of ~50,000 bp/minute. (The rate of DNA synthesis is more or less invariant at a constant temperature; it proceeds at the same speed unless and until the supply of precursors becomes limiting.)

- D is the fixed time of ~20 minutes that elapses between the completion of a round of replication and the cell division with which it is connected. This period may represent the time required to assemble the components needed for division.

(The constants C and D can be viewed as representing the maximum speed with which the bacterium is capable of completing these processes. They apply for all growth rates between doubling times of 18 and 60 minutes, but both constant phases become longer when the cell cycle occupies >60 minutes.)

A cycle of chromosome replication must be initiated a fixed time before a cell division, $C + D = 60$ minutes. For bacteria dividing more frequently than every 60 minutes, a cycle of replication must be initiated before the end of the preceding division cycle. You might say that a cell is born already pregnant with the next generation.

Consider the example of cells dividing every 35 minutes. The cycle of replication connected with a division must have been initiated 25 minutes before the preceding division. This situation is illustrated in **Figure 17.2**, which shows the chromosomal complement of a bacterial cell at 5-minute intervals throughout the cycle.

At division (35/0 minutes), the cell receives a partially replicated chromosome. The replication fork continues to advance. At 10 minutes, when this "old" replication fork has not yet reached the terminus, cell mass passes a threshold, and initiation occurs at both origins on the partially replicated chromosome. The start of these "new" replication forks creates a **multiforked chromosome**.

At 15 minutes—that is, at 20 minutes before the next division—the old replication fork reaches the terminus. Its arrival allows the two daughter chromosomes to separate; each of them has already been partially replicated by the new replication forks (which now are the only replication forks). These forks continue to advance.

At the point of division, the two partially replicated chromosomes segregate. This recreates the point at which we started. The single replication fork becomes "old," it terminates at 15 minutes, and 20 minutes later there is a division. We see that the initiation event occurs $1\,{}^{25}\!/_{35}$ cell cycles before the division event with which it is associated.

The general principle of the link between initiation and the cell cycle is that, as cells grow more rapidly (the cycle is shorter), the initiation event occurs an increasing number of cycles before the related division. There are correspondingly more chromosomes in the individual bacterium. This relationship can be viewed as the cell's response to its inability to reduce the periods of C and D to keep pace with the shorter cycle.

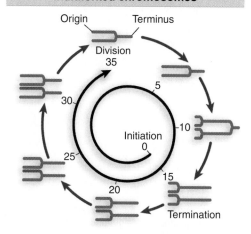

Figure 17.2 The fixed interval of 60 minutes between initiation of replication and cell division produces multiforked chromosomes in rapidly growing cells. Note that only the replication forks moving in one direction are shown; actually the chromosome is replicated symmetrically by two sets of forks moving in opposite directions on circular chromosomes.

17.3 The Septum Divides a Bacterium into Progeny Each Containing a Chromosome

Key Concepts

- Septum formation is initiated at the annulus, which is a ring around the cell where the structure of the envelope is altered.
- New annuli are initiated at 50% of the distance from the septum to each end of the bacterium.
- When the bacterium divides, each daughter has an annulus at the midcenter position.
- Septation starts when the cell reaches a fixed length.
- The septum consists of the same peptidoglycans that compose the bacterial envelope.

Key Terms

- The **nucleoid** is the structure in a prokaryotic cell that contains the genome. The DNA is bound to proteins and is not enclosed by a membrane.
- **Anucleate cells** lack a nucleoid, but are of similar shape to wild-type bacteria.
- A **septum** is the structure that forms in the center of a dividing bacterium, providing the site at which the daughter bacteria will separate.
- A **periseptal annulus** is a ring-like area where inner and outer membranes appear fused. Formed around the circumference of the bacterium, the periseptal annulus determines the location of the septum.

Chromosome segregation in bacteria is especially interesting because the DNA itself is involved in the mechanism for partition. Although the bacterial apparatus is much simpler than the complex mitotic apparatus of eukaryotes, it is quite accurate, as indicated by the fact that **anucleate cells**, which lack a **nucleoid** (bacterial chromosome), form <0.03% of a bacterial population.

The division of a bacterium into two daughter cells is accomplished by the formation of a **septum**, a structure that forms in the center of the

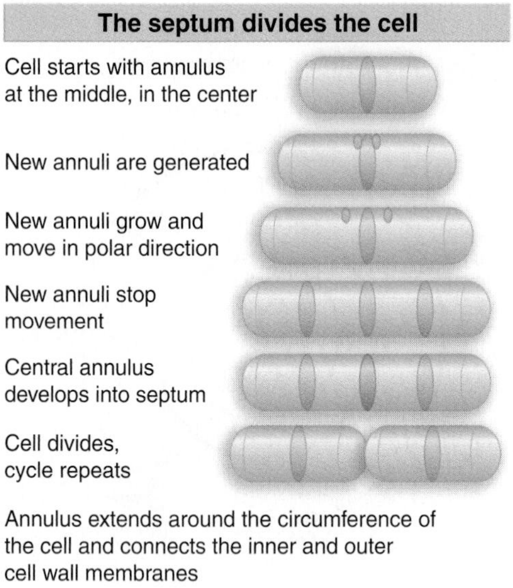

The septum divides the cell

Cell starts with annulus
at the middle, in the center

New annuli are generated

New annuli grow and
move in polar direction

New annuli stop
movement

Central annulus
develops into septum

Cell divides,
cycle repeats

Annulus extends around the circumference of
the cell and connects the inner and outer
cell wall membranes

Figure 17.3 Duplication and displacement of the periseptal annulus give rise to the formation of a septum that divides the cell.

cell as an invagination from the surrounding envelope. The septum forms an impenetrable barrier between the two parts of the cell and provides the site at which the two daughter cells eventually separate entirely. Two related questions address the role of the septum in division: what determines the location at which it forms; and what ensures that the daughter chromosomes lie on opposite sides of it?

The formation of the septum is preceded by the organization of the **periseptal annulus**. This is observed as a zone in *E. coli* or *S. typhimurium* in which the structure of the envelope is altered so that the inner membrane is connected more closely to the cell wall and outer membrane layer. As its name suggests, the annulus extends around the cell. **Figure 17.3** summarizes its development.

The annulus starts at a central position in a new cell. As the cell grows, two events occur. A septum forms at the midcell position defined by the annulus. And new annuli form on either side of the initial annulus. These new annuli are displaced from the center and move along the cell to positions at $^1/_4$ and $^3/_4$ of the cell length. They will become the midcell positions after the next division. The displacement of the periseptal annuli to the correct positions may be the crucial event that ensures the division of the cell into daughters of equal size. (The mechanism of movement is unknown.) Septation begins when the cell reaches a fixed length (2L), and the distance between the new annuli is always L. We do not know how the cell measures length, but the relevant parameter appears to be linear distance as such (not area or volume).

The septum consists of the same components as the cell envelope: there is a rigid layer of peptidoglycan in the periplasm, between the inner and outer membranes. The peptidoglycan is made by polymerization of tri- or pentapeptide-disaccharide units in a reaction involving connections between both types of subunit (transpeptidation and transglycosylation). The rodlike shape of the bacterium is maintained by a pair of activities, PBP2 and RodA. They are interacting proteins, coded by the same operon. RodA is a member of the SEDS family (SEDS stands for shape, elongation, division, and sporulation) that is present in all bacteria that have a peptidoglycan cell wall. Each SEDS protein functions together with a specific transpeptidase, which catalyzes the formation of the cross-links in the peptidoglycan. PBP2 (penicillin-binding protein 2) is the transpeptidase that interacts with RodA. Mutations in the gene for either protein cause the bacterium to lose its extended shape, becoming round. This demonstrates the important principle that shape and rigidity can be determined by the simple extension of a polymeric structure. Another enzyme is responsible for generating the peptidoglycan in the septum (see *17.5 FtsZ Is Necessary for Septum Formation*). The septum initially forms as a double layer of peptidoglycan, and the protein EnvA is required to split the covalent links between the layers, so that the daughter cells may separate.

17.4 Mutations in Division or Segregation Affect Cell Shape

Key Term

- A **minicell** is an anucleate bacterial (*E. coli*) cell produced by a division that generates a cytoplasm without a nucleoid.

Key Concepts

- *fts* mutants form long filaments because the septum fails to form to divide the daughter bacteria.
- Minicells form in mutants that produce too many septa; they are small and lack DNA.
- Anucleate cells of normal size are generated by partition mutants in which the duplicate chromosomes fail to separate.

A difficulty in isolating mutants that affect cell division is that mutations in the critical functions may be lethal and/or pleiotropic. For example, if formation of the annulus occurs at a site that is essential for overall growth of the envelope, it would be difficult to distinguish mutations that specifically interfere with annulus formation from those that inhibit envelope growth generally. Most mutations in the division apparatus have been identified as conditional mutants (whose division is affected under nonpermissive conditions; typically they are temperature sensitive). Mutations that affect cell division or chromosome segregation cause striking phenotypic changes. **Figure 17.4** and **Figure 17.5** illustrate the opposite consequences of failure in the division process and failure in segregation:

- Long *filament*s form when septum formation is inhibited, but chromosome replication is unaffected. The bacteria continue to grow, and even continue to segregate their daughter chromosomes, but septa do not form, so the cell consists of a very long filamentous structure, with the nucleoids regularly distributed along the length of the cell. This phenotype is displayed by *fts* mutants (named for temperature-sensitive filamentation), which identify defects that lie in the division process itself.

- **Minicells** form when septum formation occurs too frequently or in the wrong place, with the result that one of the new daughter cells lacks a chromosome. The minicell has a rather small size, and lacks DNA, but otherwise appears morphologically normal. Anucleate cells form when segregation is aberrant; like minicells, they lack a chromosome, but because septum formation is normal, their size is unaltered. This phenotype is caused by *par* (partition) mutants (named because they are defective in chromosome segregation).

17.5 FtsZ Is Necessary for Septum Formation

Key Term

- The **septal ring (Z-ring)** is a complex of several proteins coded by *fts* genes of *E. coli* that forms at the midpoint of the cell. It gives rise to the septum at cell division. The first of the proteins to be incorporated is FtsZ, which gave rise to the original name of the Z-ring.

Key Concepts

- The product of *ftsZ* is required for septum formation at preexisting sites.
- FtsZ is a GTPase that forms a ring on the inside of the bacterial envelope. It is connected to other cytoskeletal components.

The gene *ftsZ* plays a central role in division. Mutations in *ftsZ* block septum formation and generate filaments. Overexpression induces minicells, by causing an increased number of septation events per unit cell mass. The *ftsZ* mutants act at stages varying from the displacement of the periseptal annuli to septal morphogenesis. FtsZ is therefore required for usage of preexisting sites for septum formation, but does not itself affect the formation of the periseptal annuli or their localization.

FtsZ functions at an early stage of septum formation. Early in the division cycle, FtsZ is localized throughout the cytoplasm. As the cell elongates and begins to constrict in the middle, FtsZ becomes localized in a ring around the circumference. The structure is sometimes called

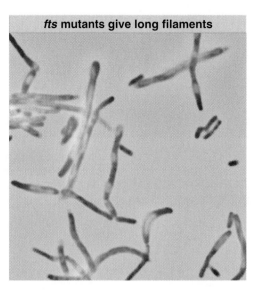

fts mutants give long filaments

Figure 17.4 Failure of cell division generates multinucleated filaments. Photograph kindly provided by Sota Hiraga.

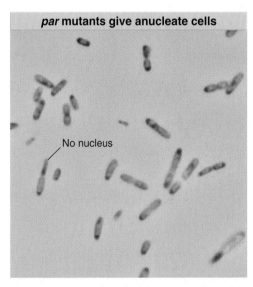

par mutants give anucleate cells

No nucleus

Figure 17.5 *E. coli* generate anucleate cells when chromosome segregation fails. Cells with chromosomes stain blue; daughter cells lacking chromosomes have no blue stain. This field shows cells of the *mukB* mutant; both normal and abnormal divisions can be seen. Photograph kindly provided by Sota Hiraga.

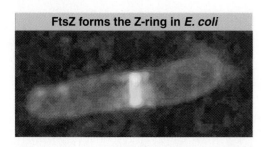

FtsZ forms the Z-ring in *E. coli*

Figure 17.6 Immunofluorescence with an antibody against FtsZ shows that it is localized at the midcell. Photograph kindly provided by William Margolin, Dept. of Microbiology & Molecular Genetics, University of Texas Medical School.

the **Z-ring**. **Figure 17.6** shows that it lies in the position of the midcenter annulus of Figure 17.3. The formation of the Z-ring is the rate-limiting step in septum formation. In a typical division cycle, it forms in the center of the cell 1–5 min after division, remains for 15 min, and then quickly constricts to pinch the cell into two.

The structure of FtsZ resembles tubulin (the subunit of the microtubule filaments of eukaryotic cells), suggesting that assembly of the ring could resemble the formation of microtubules. FtsZ has GTPase activity, and GTP cleavage is used to support the oligomerization of FtsZ monomers into the ring structure. The Z-ring is a dynamic structure, in which there is continuous exchange of subunits with a cytoplasmic pool.

Two other proteins needed for division, ZipA and FtsA, interact directly and independently with FtsZ. ZipA is an integral membrane protein, located in the inner bacterial membrane. It provides the means for linking FtsZ to the membrane. FtsA is a cytosolic protein, but is often found associated with the membrane. The Z-ring can form in the absence of either ZipA or FtsA, but cannot form if both are absent. This suggests that they have overlapping roles in stabilizing the Z-ring, and perhaps in linking it to the membrane.

The products of several other *fts* genes join the Z-ring in a defined order after FtsA has been incorporated. They are all transmembrane proteins. The final structure is sometimes called the **septal ring**. It consists of a multiprotein complex that is able to constrict the membrane. One of the last components to be incorporated into the septal ring is FtsW, which is a protein belonging to the SEDS family. *ftsW* is expressed as part of an operon with *ftsI*, which codes for a transpeptidase (also called PBP3 for *p*enicillin-*b*inding *p*rotein 3), a membrane-bound protein that has its catalytic site in the periplasm. FtsW is responsible for incorporating FtsI into the septal ring. This suggests a model for septum formation in which the transpeptidase activity of FtsI then causes the peptidoglycan to grow inward, thus pushing the inner membrane and pulling the outer membrane.

FtsZ is the major cytoskeletal component of septation. It is common in bacteria, and is found also in chloroplasts. Mitochondria, which also share an evolutionary origin with bacteria, usually do not have FtsZ. Instead, they use a variant of the protein dynamin, which is involved in pinching off vesicles from membranes of eukaryotic cytoplasm. This functions from the outside of the organelle, squeezing the membrane to generate a constriction.

The common feature, then, in the division of bacteria, chloroplasts, and mitochondria is the use of a cytoskeletal protein that forms a ring round the organelle, and either pulls or pushes the membrane to form a constriction.

17.6 *min* Genes Regulate the Location of the Septum

Key Concepts

- The location of the septum is controlled by *minC,D,E*.
- The number and location of septa is determined by the ratio of MinE/MinC,D.
- The septum forms where MinE is able to form a ring.
- At normal concentrations, MinC/D allows a midcenter ring, but prevents additional rings of MinE from forming at the poles.

Information about the localization of the septum is provided by mini-cell mutants. The original minicell mutation lies in the locus *minB*; deletion of *minB* generates minicells by allowing septation to occur at the poles as well as (or instead of) at midcell. This suggests that the cell possesses the ability to initiate septum formation either at midcell or at the poles; and the role of the wild-type *minB* locus is to suppress septation at the poles. In terms of the events depicted in Figure 17.3, this implies that a newborn cell has potential septation sites associated both with the annulus at midcenter and with the poles. One pole was formed from the septum of the previous division; the other pole represents the septum from the division before that. Perhaps the poles retain remnants of the annuli from which they were derived, which remain able to nucleate septation.

The *minB* locus consists of three genes, *minC,D,E*. Their roles are summarized in **Figure 17.7**. The products of *minC* and *minD* form a division inhibitor. MinD is required to activate MinC, which prevents FtsZ from polymerizing into the Z-ring.

Expression of MinCD in the absence of MinE, or overexpression even in the presence of MinE, causes a generalized inhibition of division. The resulting cells grow as long filaments without septa. Expression of MinE at levels comparable to MinCD confines the inhibition to the polar regions, so restoring normal growth. MinE protects the mid-cell sites from inhibition. Overexpression of MinE induces minicells, because the presence of excess MinE counteracts the inhibition at the poles as well as at midcell, allowing septa to form at both locations.

The determinant of septation at the proper (midcell) site is therefore the ratio of MinCD to MinE. The wild-type level prevents polar septation, while permitting midcell septation. The effects of MinC/D and MinE are inversely related; absence of MinCD or too much MinE causes indiscriminate septation, forming minicells; too much MinCD or absence of MinE inhibits midcell as well as polar sites, resulting in filamentation.

MinE forms a ring at the septal position. Its accumulation suppresses the action of MinCD in the vicinity, thus allowing formation of the septal ring (which includes FtsZ and ZipA).

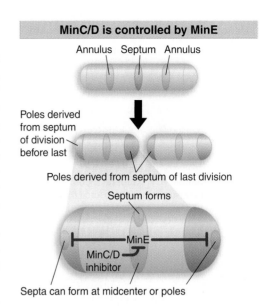

MinC/D is controlled by MinE

Annulus   Septum   Annulus

Poles derived from septum of division before last

Poles derived from septum of last division

Septum forms

MinE

MinC/D inhibitor

Septa can form at midcenter or poles

Figure 17.7 MinC/D is a division inhibitor, whose action is confined to the polar sites by MinE.

17.7 Chromosomal Segregation May Require Site-Specific Recombination

Key Term

- **Site-specific recombination (specialized recombination)** occurs between two specific sequences.

Key Concept

- The Xer site-specific recombination system acts on a target sequence near the chromosome terminus to recreate monomers if a generalized recombination event has converted the bacterial chromosome to a dimer.

After replication has created duplicate copies of a bacterial chromosome or plasmid, the copies can recombine. **Figure 17.8** demonstrates the consequences. A single intermolecular recombination event between two circles generates a dimeric circle; further recombination can generate higher multimeric forms. Such an event reduces the number of physically segregating units. In the extreme case of a single-copy plasmid that has just replicated,

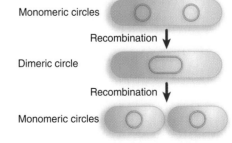

Plasmid genomes may recombine

Monomeric circles

Recombination

Dimeric circle

Recombination

Monomeric circles

Figure 17.8 Intermolecular recombination merges monomers into dimers, and intramolecular recombination releases individual units from oligomers.

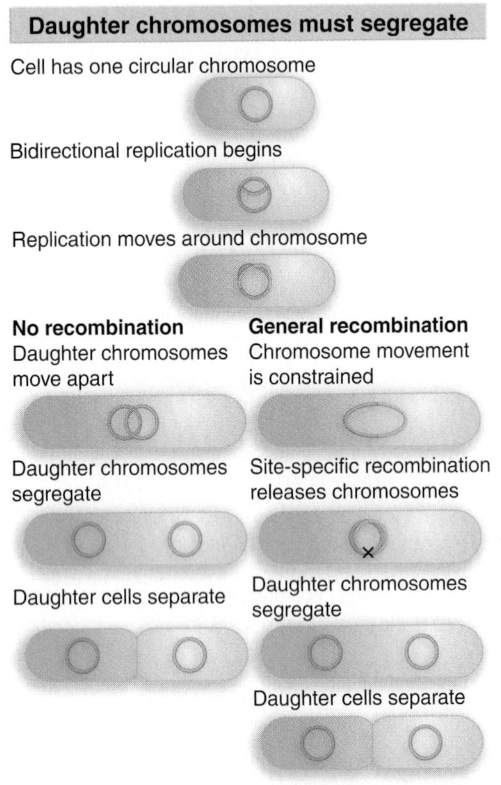

Daughter chromosomes must segregate

Cell has one circular chromosome

Bidirectional replication begins

Replication moves around chromosome

No recombination	**General recombination**
Daughter chromosomes move apart	Chromosome movement is constrained

Daughter chromosomes segregate	Site-specific recombination releases chromosomes

Daughter cells separate	Daughter chromosomes segregate

Daughter cells separate

Figure 17.9 A circular chromosome replicates to produce two monomeric daughters that segregate to daughter cells. However, a generalized recombination event generates a single dimeric molecule. This can be resolved into two monomers by a site-specific recombination.

formation of a dimer by recombination means that the cell only has one unit to segregate, and the plasmid therefore must inevitably be lost from one daughter cell. To counteract this effect, plasmids often have **site-specific recombination** systems that act upon particular sequences to sponsor an intramolecular recombination that restores the monomeric condition.

The same types of event can occur with the bacterial chromosome, and **Figure 17.9** shows how they affect its segregation. If no recombination occurs, there is no problem, and the separate daughter chromosomes can segregate to the daughter cells. But a dimer is produced when there is homologous recombination between the daughter chromosomes produced by a replication cycle. If there has been such a recombination event, the daughter chromosomes cannot separate. In this case, a second recombination is required to achieve resolution in the same way as for a plasmid dimer.

Most bacteria with circular chromosomes posses the Xer site-specific recombination system. In *E. coli*, this consists of two recombinases, XerC and XerD, which act on a 28 bp target site, called *dif*, that is located in the terminus region of the chromosome. The use of the Xer system is related in an interesting way to cell division. The relevant events are summarized in **Figure 17.10**. XerC can bind to a pair of *dif* sequences and form a recombination junction between them. The complex may form soon after the replication fork passes over the *dif* sequence, which explains how the two copies of the target sequence can find one another consistently. The *dif* target sequence must be located in a specific region of ~30 kb; if it is moved outside of this region, it cannot support the reaction.

Resolution of the junction to give recombinants occurs only in the presence of FtsK, a protein located in the septum that is required for chromosome segregation and cell division. FtsK is a large transmembrane protein. Its N-terminal domain is associated with the membrane, and causes it to be localized to the septum. Its C-terminal domain has two functions. One is to cause Xer to resolve a dimer into two monomers. It also has an ATPase activity, which it can use to translocate along DNA. This could be used to pump DNA through the septum. Bacteria that have the Xer system always have an FtsK homolog, and vice versa, which suggests that the system has evolved so that resolution is connected to the septum.

So there is a site-specific recombination available when the terminus sequence of the chromosome is close to the septum. But the bacterium wants to have a recombination only when there has already been a general recombination event to generate a dimer. (Otherwise the site-specific recombination would create the dimer!) How does the system know whether the daughter chromosomes exist as independent monomers or have been recombined into a dimer?

The answer may be that segregation of chromosomes starts soon after replication. If there has been no recombination, the two chromosomes move apart from one another. But the ability of the relevant sequences to move apart from one another may be constrained if a dimer has been formed. This forces them to remain in the vicinity of the septum, where they are exposed to the Xer system.

17.8 Partitioning Separates the Chromosomes

Key Concepts

- Replicon origins may be attached to the inner bacterial membrane.
- Chromosomes make abrupt movements from the midcenter to the $\frac{1}{4}$ and $\frac{3}{4}$ positions.

Partitioning is the process by which the two daughter chromosomes find themselves on either side of the position at which the septum forms. Two types of event are required for proper partitioning:

- The two daughter chromosomes must be released from one another so that they can segregate following termination. This requires disentangling of DNA regions that are coiled around each other in the vicinity of the terminus. Most mutations affecting partitioning map in genes coding for topoisomerases—enzymes with the ability to pass DNA strands through one another. The mutations prevent the daughter chromosomes from segregating, with the result that the DNA is located in a single large mass at midcell. Septum formation then releases an anucleate cell and a cell containing both daughter chromosomes. This tells us that the bacterium must be able to disentangle its chromosomes topologically in order to segregate them into different daughter cells.

- Mutations that affect the partition process itself are rare. We expect to find two classes. *cis*-acting mutations should occur in DNA sequences that are the targets for the partition process. *trans*-acting mutations should occur in genes that code for the proteins that cause segregation, which could include proteins that bind to DNA or activities that control the locations on the envelope to which DNA might be attached. Both types of mutation have been found in the systems responsible for partitioning plasmids, but only *trans*-acting functions have been found in the bacterial chromosome. In addition, mutations in plasmid site-specific recombination systems increase plasmid loss (because the dividing cell has only one dimer to partition instead of two monomers), and therefore have a phenotype that is similar to partition mutants.

Chromosome segregation requires that the daughter chromosomes find themselves on opposite sides of the septum when it divides the cell. **Figure 17.11** shows that the formation of a septum could segregate the chromosomes into the different daughter cells if the origins are connected to sites that lie on either side of the septum. The question now becomes what ensures that the daughter chromosomes are in such locations. Replicated chromosomes are capable of abrupt movements to their final positions at $1/4$ and $3/4$ cell length. If protein synthesis is inhibited before the termination of replication, the chromosomes fail to segregate and remain close to the midcell position. But when protein synthesis is allowed to resume, the chromosomes move to the quarter positions in the absence of any further envelope elongation. This suggests that an active process, requiring protein synthesis, may move the chromosomes to specific locations.

Segregation is interrupted by mutations of the *muk* class, which give rise to anucleate progeny at a much increased frequency: both daughter chromosomes remain on the same side of the septum instead of segregating. Mutations in the *muk* genes are not lethal, and may identify components of the apparatus that segregates the chromosomes. The gene *mukA* is identical with the gene for a known outer membrane protein (*tolC*), whose product may be involved with attaching the chromosome to the envelope. The gene *mukB* codes for a large (180 kD) globular protein, which has the same general type of organization as the two groups of SMC proteins that participate in condensing and holding together eukaryotic chromosomes (SMC stands for structural maintenance of chromosomes). SMC-like proteins have also been found in other bacteria.

The insight into the role of MukB was the discovery that some mutations in *mukB* can be suppressed by mutations in *topA*, the gene coding

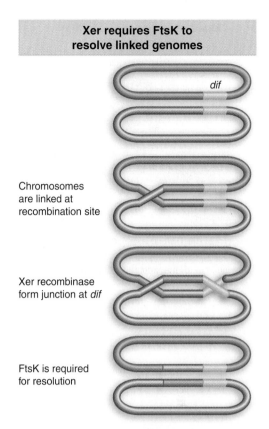

Xer requires FtsK to resolve linked genomes

Chromosomes are linked at recombination site

Xer recombinase form junction at *dif*

FtsK is required for resolution

Figure 17.10 A recombination event creates two linked chromosomes. Xer creates a Holliday junction at the *dif* site, but can resolve it only in the presence of FtsK.

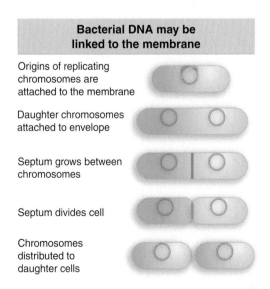

Bacterial DNA may be linked to the membrane

Origins of replicating chromosomes are attached to the membrane

Daughter chromosomes attached to envelope

Septum grows between chromosomes

Septum divides cell

Chromosomes distributed to daughter cells

Figure 17.11 Attachment of bacterial DNA to the membrane could provide a mechanism for segregation.

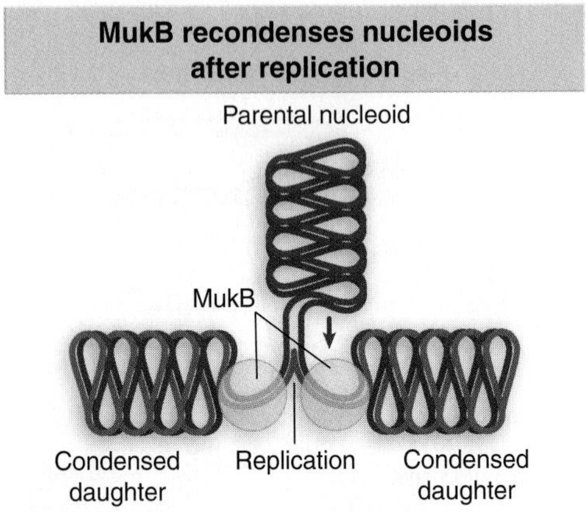

MukB recondenses nucleoids after replication

Parental nucleoid

MukB

Condensed daughter   Replication   Condensed daughter

Figure 17.12 The DNA of a single parental nucleoid becomes decondensed during replication. MukB is an essential component of the apparatus that recondenses the daughter nucleoids.

for topoisomerase I. This led to the model that the function of MukBEF proteins is to condense the nucleoid. A defect in this function is the cause of failure to segregate properly. The defect can be compensated by preventing topoisomerases from relaxing negative supercoils; the resulting increase in supercoil density helps to restore the proper state of condensation and thus to allow segregation.

We still do not understand how genomes are positioned in the cell, but the process may be connected with condensation. **Figure 17.12** shows a current model. The parental genome is centrally positioned. It must be decondensed to pass through the replication apparatus. The daughter chromosomes emerge from replication, are disentangled by topoisomerases, and then passed in an uncondensed state to MukBEF, which causes them to form condensed masses at the positions that will become the centers of the daughter cells.

There have been suspicions for years that a physical link exists between bacterial DNA and the membrane, but the evidence remains indirect. Bacterial DNA can be found in membrane fractions, which tend to be enriched in genetic markers near the origin, the replication fork, and the terminus. The proteins present in these membrane fractions may be affected by mutations that interfere with the initiation of replication. The growth site could be a structure on the membrane to which the origin must be attached for initiation.

17.9 Single-Copy Plasmids Have a Partitioning System

Key Term

- The **copy number** is the number of copies of a plasmid that is maintained in a bacterium (relative to the number of copies of the origin of the bacterial chromosome).

Key Concepts

- Single-copy plasmids exist at one plasmid copy per bacterial chromosome origin.
- Multicopy plasmids exist at >1 plasmid copy per bacterial chromosome origin.
- Homologous recombination between circular plasmids generates dimers and higher multimers.
- Plasmids have site-specific recombination systems that undertake intramolecular recombination to regenerate monomers.
- Partition systems ensure that duplicate plasmids are segregated to different daughter cells produced by a division.

The type of system that a plasmid uses to ensure that it is distributed to both daughter cells at division depends upon its type of replication system. Each type of plasmid is maintained in its bacterial host at a characteristic **copy number**:

- Single-copy control systems resemble that of the bacterial chromosome and result in one replication per cell division. A single-copy plasmid effectively maintains parity with the bacterial chromosome.
- Multicopy control systems allow multiple initiation events per cell cycle, with the result that there are several copies of the plasmid per bacterium. Multicopy plasmids exist in a characteristic number (typically 10–20) per bacterial chromosome.

Copy number is primarily a consequence of the type of replication control mechanism. The system responsible for initiating replication determines how many origins can be present in the bacterium. Since each plasmid consists of a single replicon, the number of origins is the same as the number of plasmid molecules.

Single-copy plasmids have a system for replication control whose consequences are similar to that governing the bacterial chromosome. A single origin can be replicated once; then the daughter origins are segregated to the different daughter cells.

Multicopy plasmids have a replication system that allows a pool of origins to exist. If the number is great enough (in practice >10 per bacterium), an active segregation system becomes unnecessary, because even a statistical distribution of plasmids to daughter cells will result in cells lacking plasmids at frequencies $<10^{-6}$.

Plasmids are maintained in bacterial populations with very low rates of loss ($<10^{-7}$ per cell division is typical, even for a single-copy plasmid). The systems that control plasmid segregation can be identified by mutations that increase the frequency of loss, but that do not act upon replication itself. Several types of mechanism are used to ensure the survival of a plasmid in a bacterial population. It is common for a plasmid to carry several systems, often of different types, all acting independently to ensure its survival.

Single-copy plasmids require partitioning systems to ensure that the duplicate copies find themselves on opposite sides of the septum at cell division, and are therefore segregated to a different daughter cell. In fact, functions involved in partitioning were first identified in plasmids. The components of a common system are summarized in **Figure 17.13**. Typically there are two *trans*-acting loci (*parA* and *parB*) and a *cis*-acting element (*parS*) located just downstream of the two genes. ParA is an ATPase. It binds to ParB, which binds to the *parS* site on DNA. Deletions of any of the three loci prevent proper partition of the plasmid. Systems of this type have been characterized for the plasmids F, P1, and R1. In spite of their overall similarities, there are no significant sequence homologies between the corresponding genes or *cis*-acting sites.

parS plays a role for the plasmid that is equivalent to the centromere in a eukaryotic chromosone. Binding of the ParB protein to it creates a structure that segregates the plasmid copies to opposite daughter cells. A bacterial protein, IHF, also binds at this site to form part of the structure. The complex of ParB and IHF with *parS* is called the partition complex. *parS* is a 34 bp sequence containing the IHF-binding site flanked on either side by sequences called *boxA* and *boxB* that are bound by ParB.

IHF is the integration host factor, named for the role in which it was first discovered (forming a structure necessary for the integration of phage lambda DNA into the host chromosome). IHF is a heterodimer with the capacity to form a large structure in which DNA is wrapped on the surface. The role of IHF is to bend the DNA so that ParB can bind simultaneously to the separated *boxA* and *boxB* sites, as indicated in **Figure 17.14**. Complex formation is initiated when *parS* is bound by a heterodimer of IHF together with a dimer of ParB. This enables additional dimers of ParB to bind cooperatively. The interaction of ParA with the partition complex structure is essential but transient.

The protein-DNA complex that assembles on IHF during phage lambda integration binds two DNA molecules to enable them to recombine. The role of the partition complex is different: to ensure that two DNA molecules segregate apart from one another. We do not know yet how the formation of the individual complex accomplishes this task. One possibility is that it attaches the DNA to some physical site—for example, on the membrane—and then the sites of attachment are segregated by growth of the septum.

Proteins related to ParA and ParB are found in several bacteria. In *B. subtilis*, they are called Soj and SpoOJ, respectively. Mutations in these loci prevent sporulation, because of a failure to segregate one daughter chromosome into the forespore. In sporulating cells, SpoOJ

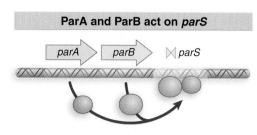

Figure 17.13 A common segregation system consists of genes *parA* and *parB* and the target site *parS*.

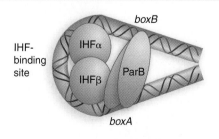

Figure 17.14 The partition complex is formed when IHF binds to DNA at *parS* and bends it so that ParB can bind to sites on either side. The complex is initiated by a heterodimer of IHF and a dimer of ParB, and then more ParB dimers bind.

localizes at the pole and may be responsible for localizing the origin there. It is possible that SpoOJ binds both old and newly synthesized origins, maintaining a status equivalent to chromosome pairing, until the chromosomes are segregated to the opposite poles. In *C. crescentus*, ParA and ParB localize to the poles of the bacterium, and ParB binds sequences close to the origin, thus localizing the origin to the pole. These results suggest that a specific apparatus is responsible for localizing the origin to the pole. The next stage of the analysis will be to identify the cellular components with which this apparatus interacts.

17.10 Plasmid Incompatibility Is Determined by the Replicon

Key Term	Key Concept
• A **compatibility group** of plasmids contains members unable to coexist in the same bacterial cell.	• Plasmids in a single compatibility group have origins that are regulated by a common control system.

The phenomenon of plasmid incompatibility is related to the regulation of plasmid copy number and segregation. A **compatibility group** is defined as a set of plasmids whose members are unable to coexist in the same bacterial cell. The reason for their incompatibility is that they cannot be distinguished from one another at some stage that is essential for plasmid maintenance. DNA replication and segregation are stages at which this may apply.

The negative control model for plasmid incompatibility follows the idea that copy number control is achieved by synthesizing a repressor that measures the concentration of origins. (Formally this is the same as the titration model for regulating replication of the bacterial chromosome.)

The introduction of a new origin in the form of a second plasmid of the same compatibility group mimics the result of replication of the resident plasmid; two origins now are present. So any further replication is prevented until after the two plasmids have been segregated to different cells to create the correct prereplication copy number as illustrated in **Figure 17.15**.

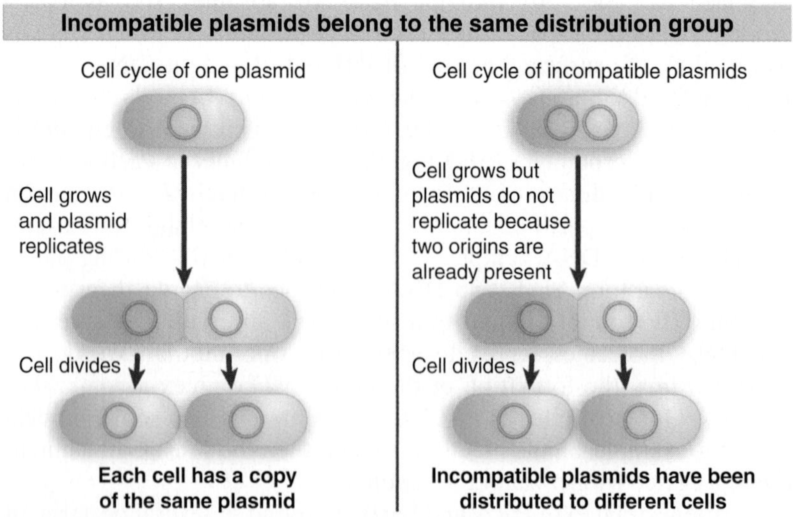

Incompatible plasmids belong to the same distribution group

Cell cycle of one plasmid

Cell cycle of incompatible plasmids

Cell grows and plasmid replicates

Cell grows but plasmids do not replicate because two origins are already present

Cell divides

Cell divides

Each cell has a copy of the same plasmid

Incompatible plasmids have been distributed to different cells

Figure 17.15 Two plasmids are incompatible (they belong to the same compatibility group) if their origins cannot be distinguished at the stage of initiation. The same model could apply to segregation.

A similar effect would be produced if the system for segregating the products to daughter cells could not distinguish between two plasmids. For example, if two plasmids have the same *cis*-acting partition sites, competition between them would ensure that they would be segregated to different cells, and therefore could not survive in the same line.

The presence of a member of one compatibility group does not directly affect the survival of a plasmid belonging to a different group. Only one replicon of a given compatibility group (of a single-copy plasmid) can be maintained in the bacterium, but it does not interact with replicons of other compatibility groups.

17.11 How Do Mitochondria Replicate and Segregate?

Key Concepts

- mtDNA replication and segregation to daughter mitochondria is stochastic.
- Mitochondrial segregation to daughter cells is also stochastic.

Mitochondria must be duplicated during the cell cycle and segregated to the daughter cells. We understand some of the mechanics of this process, but not its regulation.

At each stage in the duplication of mitochondria—DNA replication, DNA segregation to duplicate mitochondria, organelle segregation to daughter cells—the process appears to be stochastic, governed by a random distribution of each copy. The theory of distribution in this case is analogous that of multicopy bacterial plasmids, with the same conclusion that >10 copies are required to ensure that each daughter gains at least one copy (see *previous section*). When there are mtDNAs with allelic variations (either because of inheritance from different parents or because of mutation), the stochastic distribution may generate cells that have only one of the alleles.

Replication of mtDNA may be stochastic because there is no control over which particular copies are replicated, so that in any cycle some mtDNA molecules may replicate more times than others. The total number of copies of the genome may be controlled by titrating mass in a way similar to bacteria.

A mitochondrion divides by developing a ring around the organelle that constricts to pinch it into two halves. The mechanism is similar in principle to that for bacterial division. The apparatus that is used in plant cell mitochondria is similar to that of bacteria and uses a homologue of the bacterial protein FtsZ (see *17.5 FtsZ Is Necessary for Septum Formation*). The molecular apparatus is different in animal cell mitochondria, and uses the protein dynamin that is involved in formation of membranous vesicles. An individual organelle may have more than one copy of its genome.

We do not know whether there is a partitioning mechanism for segregating mtDNA molecules within the mitochondrion, or whether they are simply inherited by daughter mitochondria according to which half of the mitochondrion they happen to lie in. **Figure 17.16** shows that the combination of replication and segregation mechanisms can result in a stochastic assignment of DNA to each of the copies, so that the distribution of mitochondrial genomes to daughter mitochondria does not depend on their parental origins.

The assignment of mitochondria to daughter cells at mitosis also appears to be random. Indeed, it was the observation of somatic variation in plants that first suggested the existence of genes that could be lost

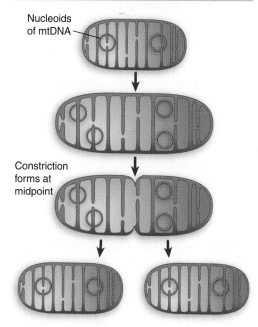

Figure 17.16 Mitochondrial DNA replicates by increasing the number of genomes in proportion to mitochondrial mass, but without ensuring that each genome replicates the same number of times. This can lead to changes in the representation of alleles in the daughter mitochondria.

from one of the daughter cells because they were not inherited according to Mendel's laws.

In some situations a mitochondrion has both paternal and maternal alleles. This has two requirements: that both parents provide alleles to the zygote (which of course is not the case when there is maternal inheritance; see *4.10 Organelles Have DNA*); and that the parental alleles are found in the same mitochondrion. For this to happen, parental mitochondria must have fused.

The size of the individual mitochondrion may not be precisely defined. Indeed, there is a continuing question as to whether an individual mitochondrion represents a unique and discrete copy of the organelle or whether it is in a dynamic flux in which it can fuse with other mitochondria. We know that mitochondria can fuse in yeast, because recombination between mtDNAs can occur after two haploid yeast strains have mated to produce a diploid strain. This implies that the two mtDNAs must have been exposed to one another in the same mitochondrial compartment. Attempts have been made to test for the occurrence of similar events in animal cells by looking for complementation between alleles after two cells have been fused, but the results are not clear.

17.12 SUMMARY

A fixed time of 40 min. is required to replicate the *E. coli* chromosome and a further 20 min. is required before the cell can divide. When cells divide more rapidly than every 60 min., a replication cycle is initiated before the end of the preceding division cycle. This generates multiforked chromosomes. The initiation event depends on titration of cell mass, probably by accumulating an initiator protein. Initiation may occur at the cell membrane, since the origin is associated with the membrane for a short period after initiation.

The septum that divides the cell grows at a location defined by the preexisting periseptal annulus; a locus of three genes (*minCDE*) codes for products that regulate whether the midcell periseptal annulus or the polar sites derived from previous annuli are used for septum formation. Absence of septum formation generates multinucleated filaments; excess of septum formation generates anucleate minicells.

Many transmembrane proteins interact to form the septum. ZipA is located in the inner bacterial membrane and binds to FtsZ, which is a tubulin–like protein that can polymerize into a filamentous structure called a Z-ring. FtsA is a cytosolic protein that binds to FtsZ. Several other *fts* products, all transmembrane proteins, join the Z-ring in an ordered process that generates a septal ring. The last proteins to bind are the SEDS protein FtsW and the transpeptidase ftsI (PBP3), which together function to produce the peptidoglycans of the septum. Chloroplasts use a related division mechanism that has an FtsZ-like

protein, but mitochondria use a different process in which the membrane is constricted by a dynamin-like protein.

Plasmids and bacteria have site-specific recombination systems that regenerate pairs of monomers by resolving dimers created by general recombination. The Xer system acts on a target sequence located in the terminus region of the chromosome. The system is active only in the presence of the FtsK protein of the septum, which may ensure that it acts only when a dimer needs to be resolved.

Plasmid partitioning involves the interaction of the ParB protein with the *parS* target site to build a structure that includes the IHF protein. This partition complex ensures that replica chromosomes segregate into different daughter cells. The mechanism of segregation may involve movement of DNA, possibly by the action of MukB in condensing chromosomes into masses at different locations as they emerge from replication.

Plasmids have a variety of systems that ensure or assist partition, and an individual plasmid may carry systems of several types. The copy number of a plasmid describes whether it is present at the same level as the bacterial chromosome (one per unit cell) or in greater numbers. Plasmid incompatibility can be a consequence of the mechanisms involved in either replication or partition (for single-copy plasmids). Two plasmids that share the same control system for replication are incompatible because the number of replication events ensures that there is only one plasmid for each bacterial genome.

18

DNA Replication

18.1 Introduction

Key Term

- The **replisome** is the multiprotein structure that assembles at the bacterial replication fork to undertake synthesis of DNA. It contains DNA polymerase and other enzymes.

Replication of duplex DNA requires different activities for the stages of initiation, elongation, and termination. **Figure 18.1** shows the first stages of the process.

- Initiation begins with recognition of an origin by a large protein complex. Before DNA can be synthesized, the parental strands must be separated and transiently stabilized in the single-stranded state. Then synthesis of daughter strands can be initiated at the replication fork.
- Elongation is undertaken by another complex of proteins. The **replisome** exists only as a protein complex associated with the particular structure that DNA takes at the replication fork. It does not preexist as an independent unit (for example, analogous to the ribosome), but assembles *de novo* at the origin for each replication cycle. As the replisome moves along DNA, the parental strands unwind and daughter strands are synthesized.
- At the end of the replicon, joining and/or termination reactions are necessary. Following termination, the duplicate chromosomes must be separated from one another, which requires manipulation of higher-order DNA structure.

Inability to replicate DNA is fatal for a growing cell. Mutants in replication must therefore be obtained as *conditional lethals*. These are able to accomplish replication under permissive conditions (provided by the normal temperature of incubation), but they are defective under

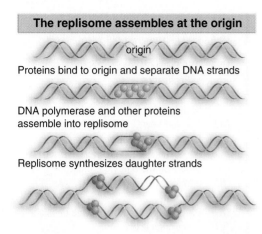

Figure 18.1 Replication initiates when a protein complex binds to the origin and melts the DNA there. Then the components of the replisome, including DNA polymerase, assemble. The replisome moves along DNA, synthesizing both new strands.

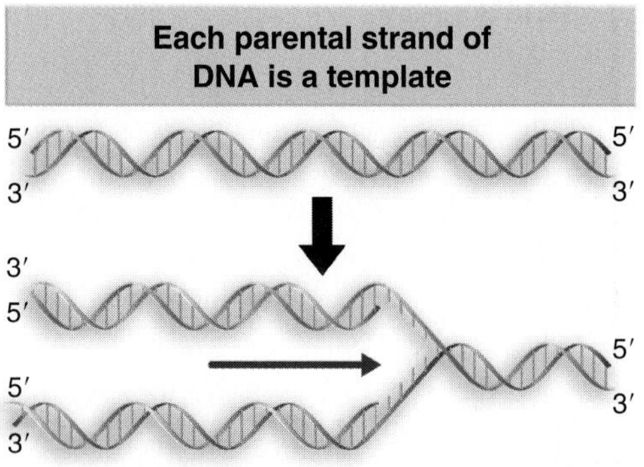

Each parental strand of DNA is a template

Figure 18.2 Semiconservative replication synthesizes two new strands of DNA.

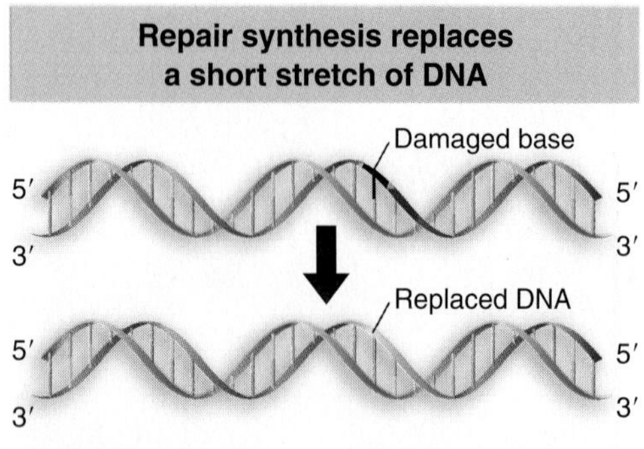

Repair synthesis replaces a short stretch of DNA

Damaged base

Replaced DNA

Figure 18.3 Repair synthesis replaces a short stretch of one strand of DNA containing a damaged base.

---

nonpermissive conditions (provided by the higher temperature of 42°C). A comprehensive series of such temperature-sensitive mutants in *E. coli* identified a set of loci called the *dna* genes.

18.2 DNA Polymerases Are the Enzymes that Make DNA

Key Terms

- **Replication** of duplex DNA takes place by synthesis of two new strands that are complementary to the parental strands. The parental duplex is replaced by two identical daughter duplexes, each of which has one parental strand and one newly synthesized strand. Replication is called semiconservative because the conserved units are the single strands of the parental duplex.
- **Repair** of damaged DNA can take place by repair synthesis, when a strand that has been damaged is excised and replaced by the synthesis of a new stretch. Repair can also take place by recombination reactions in which the duplex region containing the damaged strand is replaced by an undamaged region from another copy of the genome.
- A **DNA polymerase** is an enzyme that synthesizes a daughter strand(s) of DNA (under direction from a DNA template). A particular polymerase may function in repair or replication (or both).
- A **DNA replicase** is a DNA-synthesizing enzyme required specifically for replication.

Key Concepts

- DNA is synthesized in both semiconservative replication and repair reactions.
- A bacterium or eukaryotic cell has several different DNA polymerase enzymes.
- One bacterial DNA polymerase undertakes semiconservative replication; the others function in repair reactions.
- Eukaryotic nuclei, mitochondria, and chloroplasts each have a single unique DNA polymerase required for replication and other DNA polymerases that function in ancillary or repair activities.

There are two basic types of DNA synthesis.

Figure 18.2 shows the result of semiconservative **replication**. The two strands of the parental duplex are separated, and each serves as a template for synthesis of a new strand. The parental duplex is replaced with two daughter duplexes, each of which has one parental strand and one newly synthesized strand.

Figure 18.3 shows the consequences of a **repair** reaction. One strand of DNA has been damaged. It is excised and new material is synthesized to replace it.

An enzyme that can synthesize a new DNA strand on a template strand is called a **DNA polymerase**. Both prokaryotic and eukaryotic cells contain multiple DNA polymerase activities. Only some of these enzymes actually undertake replication; sometimes these are called **DNA replicases**. The others perform subsidiary roles in replication and/or participate in repair synthesis.

All prokaryotic and eukaryotic DNA polymerases share the same fundamental type of synthetic activity. Each can extend a DNA chain by adding nucleotides one at a time to a 3′—OH end, as illustrated diagrammatically in **Figure 18.4**. The choice of the nucleotide to add to the chain is dictated by base pairing with the template strand.

Some DNA polymerases function as independent enzymes, but others (most notably the replicases) are incorporated into large protein assemblies. The DNA-synthesizing subunit is only one of several functions of the replicase, which typically contains many other activities concerned with unwinding DNA, initiating new strand synthesis, and so on.

E. coli has five DNA polymerases. DNA polymerase III is the replicase. DNA polymerase I (coded by *polA*) is involved in the repair of damaged DNA and, in a subsidiary role, in semiconservative replication. DNA polymerase II is required to restart replication when a fork stalls at a site of damaged DNA. DNA polymerases IV and V participate in specific repair reactions, and are called translesion DNA polymerases because they can synthesize DNA on a template strand that contains damaged bases.

Several classes of eukaryotic DNA polymerases have been identified. DNA polymerases δ and ε are required for nuclear replication; DNA polymerase α is concerned with "priming" (initiating) replication. Other DNA polymerases participate in repairing damaged nuclear DNA (β and also ε) or in mitochondrial DNA replication (γ).

18.3 DNA Polymerases Control the Fidelity of Replication

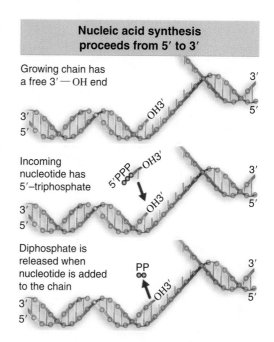

Figure 18.4 DNA is synthesized by adding nucleotides to the 3'—OH end of the growing chain, so that the new chain grows in the 5' → 3' direction. The precursor for DNA synthesis is a nucleoside triphosphate, which loses the terminal two phosphate groups in the reaction.

Key Terms

- **Processivity** describes the ability of an enzyme to perform multiple catalytic cycles with a single template instead of dissociating after each cycle.

- **Proofreading** refers to any mechanism for correcting errors in protein or nucleic acid synthesis that involves scrutiny of individual units *after* they have been added to the chain.

Key Concepts

- DNA polymerases often have a 3'–5' exonuclease activity that is used to excise incorrectly paired bases.

- The fidelity of replication is improved by proofreading by a factor of ~100.

The fidelity of replication poses the same sort of problem we have encountered already in considering (for example) the accuracy of translation. It relies on the specificity of base pairing. However, we would expect errors to occur with a frequency of ~10^{-3} per base pair replicated if specificity relied solely upon chemical interactions between the bases. The actual rate in bacteria seems to be ~10^{-8} to 10^{-10}. This corresponds to ~1 error per genome per 1000 bacterial replication cycles, or ~10^{-6} per gene per generation.

We can divide the errors that DNA polymerase makes during replication into two general classes.

- Frameshifts occur when an extra nucleotide is inserted or omitted. Fidelity with regard to frameshifts is affected by the **processivity** of the enzyme: its tendency to remain on a single template rather than to dissociate and reassociate. This is particularly important for the replication of a homopolymeric stretch, for example, a long sequence of $dT_n : dA_n$, in which "replication slippage" can change the length of the homopolymeric run. As a general rule, increased processivity reduces the likelihood of such events. In multimeric DNA

DNA polymerases have exonuclease activity

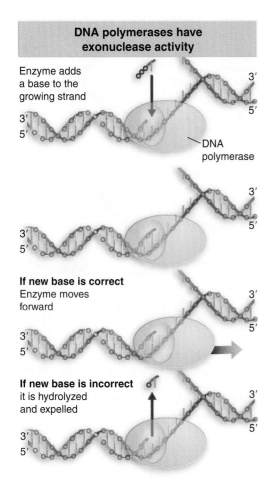

Enzyme adds a base to the growing strand

If new base is correct
Enzyme moves forward

If new base is incorrect
it is hydrolyzed and expelled

Figure 18.5 Bacterial DNA polymerases scrutinize the base pair at the end of the growing chain and excise the nucleotide added in the case of a misfit.

DNA polymerases have a common structure

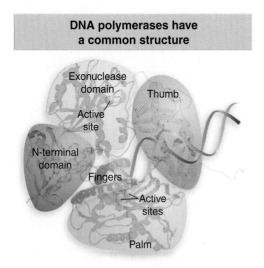

Figure 18.6 The common organization of DNA polymerases has a palm that contains the catalytic site, fingers that position the template, a thumb that binds DNA and is important in processivity, an exonuclease domain with its own active site, and an N-terminal domain.

polymerases, processivity is usually increased by a particular subunit that is not needed for catalytic activity *per se*.

- Substitutions occur when the wrong (improperly paired) nucleotide is incorporated. The error level is determined by the efficiency of **proofreading**, in which the enzyme scrutinizes the newly formed base pair and removes the nucleotide if it is mispaired.

All bacterial DNA polymerases possess a $3'-5'$ exonucleolytic activity that proceeds in the reverse direction from DNA synthesis. This provides the proofreading function illustrated diagrammatically in **Figure 18.5**. In the chain elongation step, a precursor nucleotide enters the position at the end of the growing chain. A bond is formed. The enzyme moves one base pair farther, ready for the next precursor nucleotide to enter. If a mistake has been made, however, the enzyme uses the exonucleolytic activity to excise the last base that was added.

Different DNA polymerases handle the relationship between the polymerizing and proofreading activities in different ways. In some cases, the activities are part of the same protein subunit, but in others they are contained in different subunits. Each DNA polymerase has a characteristic error rate that is reduced by its proofreading activity. Proofreading typically decreases the error rate in replication by $\sim 100\times$, from $\sim 10^{-5}$ to $\sim 10^{-7}$ per base pair replicated. Systems that recognize errors and correct them following replication then eliminate some of the errors, bringing the overall rate to $< 10^{-9}$ per base pair replicated (see *20.6 Controlling the Direction of Mismatch Repair*).

18.4 DNA Polymerases Have a Common Structure

Key Concepts

- Many DNA polymerases have a large cleft composed of three domains that resemble a hand.
- DNA lies across the "palm" in a groove created by the "fingers" and "thumb."

Figure 18.6 shows that all DNA polymerases share some common structural features. The enzyme structure can be divided into several independent domains, which are described by analogy with a human right hand. DNA binds in a large cleft composed of three domains. The "palm" domain has important conserved sequence motifs that provide the catalytic active site. The "fingers" position the template correctly at the active site. The "thumb" binds the DNA as it exits the enzyme, and is important in processivity. The most important conserved regions of each of these three domains converge to form a continuous surface at the catalytic site. The exonuclease activity resides in an independent domain with its own catalytic site. The N-terminal domain extends into the nuclease domain. DNA polymerases fall into five families based on sequence homologies; the palm is well conserved among them, but the thumb and fingers provide analogous secondary structure elements from different sequences.

The catalytic reaction in a DNA polymerase occurs at an active site in which a nucleotide triphosphate pairs with an (unpaired) single strand of DNA. The DNA lies across the palm in a groove that is created by the thumb and fingers. **Figure 18.7** shows the crystal structure of the T7 enzyme complexed with DNA (in the form of a primer annealed to a template strand) and an incoming nucleotide that is about to be added to the primer. The DNA is in the classic B-form duplex up to the last two base pairs at the $3'$ end of the primer, which are in the more open A-form.

A sharp turn in the DNA exposes the template base to the incoming nucleotide. The 3′ end of the primer (to which bases are added) is anchored by the fingers and palm. The DNA is held in position by contacts that are made principally with the phosphodiester backbone (thus enabling the polymerase to function with DNA of any sequence).

In structures of DNA polymerases of this family complexed only with DNA (that is, lacking the incoming nucleotide), the orientation of the fingers and thumb relative to the palm is more open, with the O helix (O, O1, O2; see Figure 18.7) rotated away from the palm. This suggests that the O helix rotates inward to grasp the incoming nucleotide and create the active catalytic site. When a nucleotide binds, the fingers domain rotates 60° toward the palm, with the tops of the fingers moving by 30 Å. The thumb domain also rotates toward the palm by 8°. These changes are cyclical: they are reversed when the nucleotide is incorporated into the DNA chain, which then translocates through the enzyme to recreate an empty site.

The exonuclease activity is responsible for removing mispaired bases. But the catalytic site of the exonuclease domain is distant from the active site of the catalytic domain. The enzyme alternates between polymerizing and editing modes, as determined by a competition between the two active sites for the 3′ primer end of the DNA. Amino acids in the active site contact the incoming base in such a way that the enzyme structure is affected by a mismatched base. When a mismatched base pair occupies the catalytic site, the fingers cannot rotate toward the palm to bind the incoming nucleotide. This leaves the 3′ end free to bind to the active site in the exonuclease domain, which is accomplished by a rotation of the DNA in the enzyme structure.

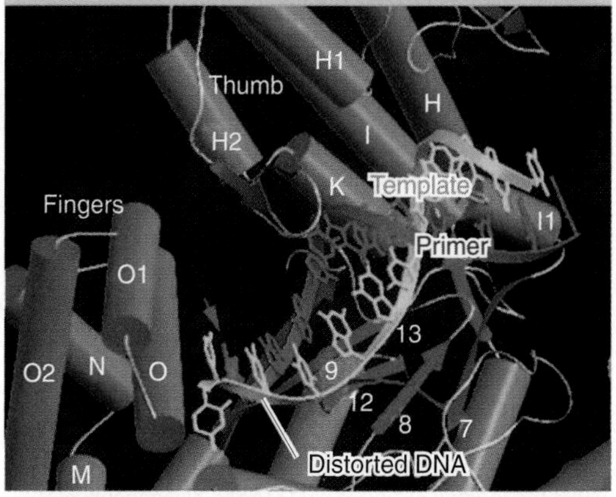

Figure 18.7 The crystal structure of phage T7 DNA polymerase shows that the template strand takes a sharp turn that exposes it to the incoming nucleotide. Photograph kindly provided by Charles Richardson and Tom Ellenberger, Dept. of Biological Chemistry and Molecular Pharmacology, Harvard Medical School.

18.5 The Two New DNA Strands Have Different Modes of Synthesis

Key Concept

- The DNA replicase advances continuously when it synthesizes the leading strand (5′–3′), but synthesizes the lagging strand by making short fragments that are subsequently joined together.

Key Terms

- The **leading strand** of DNA is synthesized continuously in the 5′–3′ direction.
- The **lagging strand** of DNA must grow overall in the 3′–5′ direction and is synthesized discontinuously in the form of short fragments (5′–3′) that are later connected covalently.
- **Okazaki fragments** are the short stretches of 1000–2000 bases produced during discontinuous replication; they are later joined into a covalently intact strand.

The antiparallel structure of the two strands of duplex DNA poses a problem for replication. As the replication fork advances, daughter strands must be synthesized on both of the exposed parental single strands. The fork moves in the direction from 5′–3′ on one strand, and in the direction from 3′–5′ on the other strand. Yet nucleic acids are synthesized only from a 5′ end toward a 3′ end. The problem is solved by synthesizing the strand that grows overall from 3′–5′ in a series of short fragments, each actually synthesized in the "backwards" direction—that is, with the customary 5′–3′ polarity.

Consider the region immediately behind the replication fork, as illustrated in **Figure 18.8**. We describe events in terms of the different properties of each of the newly synthesized strands:

- On the **leading strand** DNA synthesis can proceed continuously in the 5′ to 3′ direction as the parental duplex is unwound.
- On the **lagging strand** a stretch of single-stranded parental DNA must be exposed, and then a segment is synthesized in the reverse direction (relative to fork movement). A series of these fragments are synthesized, each 5′–3′; then they are joined together to create an intact lagging strand.

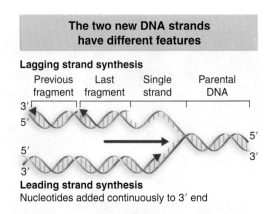

The two new DNA strands have different features

Lagging strand synthesis

Previous fragment | Last fragment | Single strand | Parental DNA

3′
5′

5′
3′

5′

3′

Leading strand synthesis
Nucleotides added continuously to 3′ end

Figure 18.8 The leading strand is synthesized continuously while the lagging strand is synthesized discontinuously.

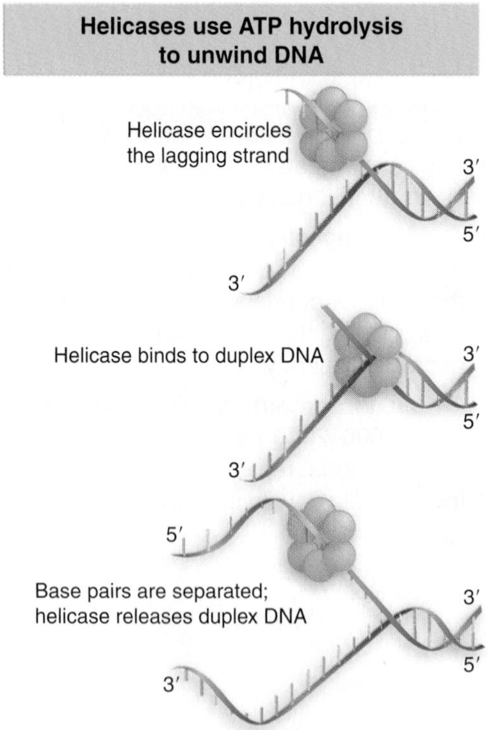

Helicases use ATP hydrolysis to unwind DNA

Helicase encircles the lagging strand

3′

5′

3′

Helicase binds to duplex DNA

3′

5′

3′

5′

Base pairs are separated; helicase releases duplex DNA

3′

5′

3′

Figure 18.9 A hexameric helicase moves along one strand of DNA. It probably changes conformation when it binds to the duplex, uses ATP hydrolysis to separate the strands, and then returns to the conformation it has when bound only to a single strand.

Discontinuous replication can be followed by the fate of a very brief label of radioactivity. The label enters newly synthesized DNA in the form of short fragments, sedimenting in the range of 7–11S, corresponding to ~1000–2000 bases in length. These **Okazaki fragments** are found in replicating DNA in both prokaryotes and eukaryotes. After longer periods of incubation, the label enters larger segments of DNA. The transition results from covalent linkages between Okazaki fragments.

18.6 Replication Requires a Helicase and Single-Strand Binding Protein

Key Terms

- A **helicase** is an enzyme that uses energy provided by ATP hydrolysis to separate the strands of a nucleic acid duplex.
- The **single-strand binding protein (SSB)** attaches to single-stranded DNA, thereby preventing the DNA from forming a duplex.

As the replication fork advances, it unwinds the duplex DNA. One of the template strands is rapidly converted to duplex DNA as the leading daughter strand is synthesized. The other remains single-stranded until a sufficient length has been exposed to initiate synthesis of an Okazaki fragment of the lagging strand in the backward direction. The generation and maintenance of single-stranded DNA is therefore a crucial aspect of replication. Two types of functions are needed to convert double-stranded DNA to the single-stranded state:

- A **helicase** is an enzyme that separates the strands of DNA, using the hydrolysis of ATP to provide the necessary energy.
- A **single-strand binding protein (SSB)** binds to the single-stranded DNA, preventing it from reforming the duplex state. The SSB binds as a monomer, but typically in a cooperative manner in which the binding of additional monomers to the existing complex is enhanced.

Helicases separate the strands of a duplex nucleic acid in a variety of situations, ranging from strand separation at the growing point of a replication fork to catalyzing migration of Holliday (recombination) junctions along DNA. There are 12 different helicases in *E. coli*. A helicase is generally multimeric. A common form of helicase is a hexamer. This typically translocates along DNA by using its multimeric structure to provide multiple DNA-binding sites.

Figure 18.9 shows a generalized schematic model for the action of a hexameric helicase. It is likely to have one conformation that binds to duplex DNA and another that binds to single-stranded DNA. Alternation between them drives the motor that melts the duplex, and requires ATP hydrolysis—typically 1 ATP is hydrolyzed for each base pair that is unwound. A helicase usually initiates unwinding at a single-stranded region adjacent to a duplex, and may function with a particular polarity, preferring single-stranded DNA with a 3′ end (3′–5′ helicase) or with a 5′ end (5′–3′ helicase).

Under normal circumstances *in vivo*, the unwinding, coating, and replication reactions proceed in tandem. The SSB binds to DNA as the replication fork advances, keeping the two parental strands separate so that they are in the appropriate condition to act as templates. SSB is needed in stoichiometric amounts at the replication fork. (Some phages, such as T4, use different SSB proteins; this shows that there may be specific interactions between components of the replication apparatus and the SSB.)

18.7 Priming Is Required to Start DNA Synthesis

Key Terms

- A **primer** is a short sequence (often of RNA) that is paired with one strand of DNA and provides a free 3'—OH end at which a DNA polymerase starts synthesis of a deoxyribonucleotide chain.
- The **primase** is a type of RNA polymerase that synthesizes short segments of RNA that will be used as primers for DNA replication.

Key Concepts

- All DNA polymerases require a 3'—OH priming end to initiate DNA synthesis.
- The priming end can be provided by an RNA primer, a nick in DNA, or a priming protein.
- For DNA replication, a special RNA polymerase called a primase synthesizes an RNA chain that provides the priming end.
- *E. coli* has two types of priming reaction, which occur at the bacterial origin (*oriC*) and the φX174 origin.
- Priming of replication on double-stranded DNA always requires a replicase, SSB, and primase.

A common feature of all DNA polymerases is that they cannot initiate synthesis of a chain of DNA *de novo*. **Figure 18.10** shows the features required for initiation. Synthesis of the new strand can only start from a pre-existing 3'—OH end; and the template strand must be converted to a single-stranded condition.

The 3'—OH end is called a **primer**. The primer can take various forms. Types of priming reaction are summarized in **Figure 18.11**:

- A sequence of RNA is synthesized on the template, so that the free 3'—OH end of the RNA chain is extended by the DNA polymerase. This is commonly used in replication of cellular DNA, and by some viruses.
- A preformed RNA pairs with the template, allowing its 3'—OH end to be used to prime DNA synthesis. This mechanism is used by retroviruses to prime reverse transcription of RNA (see *22.4 Viral DNA Is Generated by Reverse Transcription*).
- A primer terminus is generated within duplex DNA. The most common mechanism is the introduction of a nick, as used to initiate rolling circle replication. In this case, the preexisting strand is displaced by new synthesis (see Figure 16.6).
- A protein primes the reaction directly by presenting a nucleotide to the DNA polymerase. This reaction is used by certain viruses (see *16.2 The Ends of Linear DNA Are a Problem for Replication*).

Priming activity is required to provide 3'—OH ends to start off the DNA chains on both the leading and lagging strands. The leading strand requires only one such initiation event, which occurs at the origin. But there must be a series of initiation events on the lagging strand, since each Okazaki fragment requires its own start *de novo*. Each Okazaki fragment starts with a primer sequence of RNA, 11–12 bases long, that provides the 3'—OH end for extension by DNA polymerase.

A **primase** is required to catalyze the actual priming reaction. This is provided by a special RNA polymerase activity, the product of the *dnaG* gene. The enzyme is a single polypeptide of 60 kD (much smaller

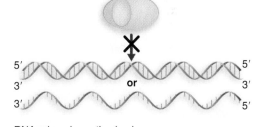

A free 3' end is required for priming

Priming terminus

Single-stranded template

Figure 18.10 A DNA polymerase requires a 3'—OH end to initiate replication.

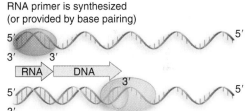

There are many ways to generate 3' ends

DNA polymerases cannot intiate DNA synthesis on duplex or single-stranded DNA without a primer

or

RNA primer is synthesized (or provided by base pairing)

RNA DNA

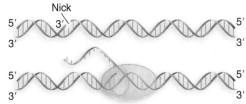

Duplex DNA is nicked to provide free end for DNA polymerase

Nick

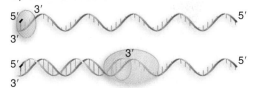

A priming nucleotide is provided by a protein that binds to DNA

Figure 18.11 There are several methods for providing the free 3'—OH end that DNA polymerases require to initiate DNA synthesis.

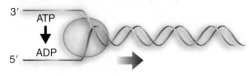

Priming requires helicase, SSB, and primase

Helicase DnaB 5′–3′ helicase (5′–3′)

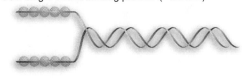

SSB single-strand binding protein (~60/fork)

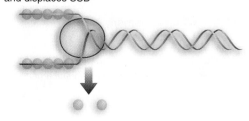

PriA (φX only)
recognizes primosome assembly site
and displaces SSB

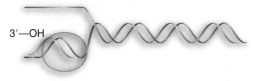

DnaG primase synthesizes RNA

Figure 18.12 Initiation requires several enzymatic activities, including helicases, single-strand binding proteins, and synthesis of the primer.

than RNA polymerase). The primase is an RNA polymerase that is used only under specific circumstances, that is, to synthesize short stretches of RNA that are used as primers for DNA synthesis. DnaG primase associates transiently with the replication complex, and typically synthesizes an 11–12 base primer. Primers start with the sequence pppAG, opposite the sequence 3′-GTC-5′ in the template.

There are two types of priming reaction in *E. coli*.

- The *oriC* system, named for the bacterial origin, uses the DnaG primase with the protein complex at the replication fork.
- The φX system, named for phage φX174, requires an initiation complex consisting of additional components, called the primosome (see *18.14 The Primosome Is Needed to Restart Replication*).

Sometimes replicons are referred to as being of the φX or *oriC* type. The types of activities in the initiation reaction are summarized in **Figure 18.12**. Although other replicons in *E. coli* may have alternatives for some of these particular proteins, the same general types of activity are required in every case. A helicase is required to generate single strands, a single-strand binding protein is required to maintain the single-stranded state, and the primase synthesizes the RNA primer.

18.8 DNA Polymerase Holoenzyme Consists of Subcomplexes

Key Terms

- The **clamp** is a protein complex that forms a circle around DNA; by connecting to DNA polymerase, it ensures that the enzyme action is processive.
- The **clamp loader** is a five subunit protein complex which is responsible for loading the β clamp onto DNA at the replication fork.

Key Concepts

- The *E. coli* replicase DNA polymerase III is a 900 kD complex with a dimeric structure.
- Each monomeric unit has a catalytic core, a dimerization subunit, and a processivity component (the β clamp).
- A clamp loader places the processivity subunits on DNA, and they form a circular clamp around the nucleic acid.
- One catalytic core is associated with each template strand.

We can now relate the subunit structure of *E. coli* DNA polymerase III to the activities required for DNA synthesis and propose a model for its action. The holoenzyme is a complex of 900 kD that contains 10 proteins organized into four types of subcomplex:

- There are two copies of the catalytic core. Each catalytic core contains the α subunit (the DNA polymerase activity), ε subunit (3′–5′ proofreading exonuclease), and θ subunit (stimulates exonuclease).
- There are two copies of the dimerizing subunit, τ, which link the two catalytic cores together.
- There are two copies of the **clamp**, which is responsible for holding catalytic cores on to their template strands. Each clamp consists of a homodimer of β subunits that binds around the DNA and ensures processivity.

- The γ complex is a group of five proteins, the **clamp loader**, that places the clamp on DNA.

A model for the assembly of DNA polymerase III is depicted in **Figure 18.13**. The holoenzyme assembles on DNA in three stages:

- First the clamp loader uses hydrolysis of ATP to bind β subunits to a template-primer complex.
- Binding to DNA changes the conformation of the β subunits to create a high affinity for the core polymerase. This enables core polymerase to bind, and this is the means by which the core polymerase is brought to DNA.
- A τ dimer binds to the core polymerase, and provides a dimerization function that binds a second core polymerase (associated with another β clamp). The holoenzyme is asymmetric, because it has only one clamp loader.

Each of the core complexes of the holoenzyme synthesizes one of the new strands of DNA. Because the clamp loader is also needed for unloading the β complex from DNA, the two cores have different abilities to dissociate from DNA. This corresponds to the need to synthesize a continuous leading strand (where polymerase remains associated with the template) and a discontinuous lagging strand (where polymerase repetitively dissociates and reassociates). The clamp loader is associated with the core polymerase that synthesizes the lagging strand, and plays a key role in the ability to synthesize individual Okazaki fragments.

18.9 The Clamp Controls Association of Core Enzyme with DNA

Key Concepts

- The core on the leading strand is processive because its clamp keeps it on the DNA.
- The clamp associated with the core on the lagging strand dissociates at the end of each Okazaki fragment and reassembles for the next fragment.
- The helicase DnaB is responsible for interacting with the primase DnaG to initiate each Okazaki fragment.

The β dimer makes the holoenzyme highly processive. β is strongly bound to DNA, but can slide along a duplex molecule. The crystal structure of β shows that it forms a ring-shaped dimer. The model in **Figure 18.14** shows the β ring in relationship to a DNA double helix. The ring has an external diameter of 80 Å and an internal cavity of 35 Å, almost twice the diameter of the DNA double helix (20 Å). The space between the protein ring and the DNA is filled by water. The β subunit consists of three globular domains with similar organization (although their sequences are different). As a result, the dimer has 6-fold symmetry, reflected in 12 α-helices that line the inside of the ring.

The dimer surrounds the duplex, providing the "sliding clamp" that allows the holoenzyme to slide along DNA. The structure explains the high processivity: there is no way for the enzyme to fall off! The α-helices on the inside have some positive charges that may interact with the DNA via the intermediate water molecules. Because the protein clamp does not directly contact the DNA, it may be able to "ice skate" along the DNA, making and breaking contacts via the water molecules.

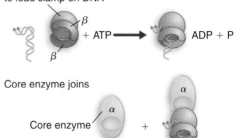

A dimer synthesizes lagging and leading strands

Clamp loader (χ,ψ,δε,γ,δ) cleaves ATP to load clamp on DNA

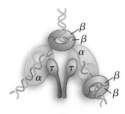

Core enzyme joins

tau + second core joins to give a symmetric dimer

Figure 18.13 DNA polymerase III holoenzyme assembles in stages, generating an enzyme complex that synthesizes the DNA of both new strands.

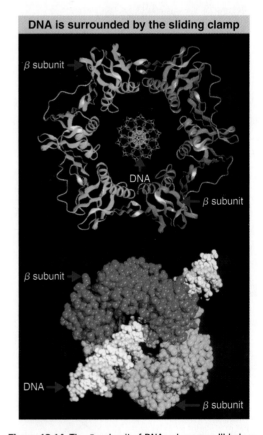

DNA is surrounded by the sliding clamp

β subunit

DNA

β subunit

β subunit

DNA

β subunit

Figure 18.14 The β subunit of DNA polymerase III holoenzyme consists of a head to tail dimer (the two monomers are shown in red and orange) that forms a ring completely surrounding a DNA duplex (shown in the center). Photograph kindly provided by Dr. John Kuriyan, University of California, Berkeley.

How does the clamp get on to the DNA? Because the clamp is a circle of subunits surrounding DNA, its assembly or removal requires the use of an energy-dependent process by the clamp loader. The γ clamp loader is a pentameric circular structure that binds an open form of the β ring preparatory to loading it on to DNA. In effect, the ring is opened at one of the interfaces between the two β subunits by the δ subunit of the clamp loader. The clamp loader uses hydrolysis of ATP to provide the energy to open the ring of the clamp and insert DNA into its central cavity.

The relationship between the β clamp and the γ clamp loader is a paradigm for similar systems used by DNA replicases ranging from bacteriophages to animal cells. The clamp is a heteromer (sometimes a dimer, sometimes a trimer) that forms a ring around DNA with a set of 12 α-helices forming 6-fold symmetry for the structure as a whole. The clamp loader has some subunits that hydrolyze ATP to provide energy for the clamp-loading reaction.

18.10 Coordinating the Synthesis of the Lagging and Leading Strands

Key Concepts

- Different catalytic units are required to synthesize the leading and lagging strands.
- In *E. coli* both units contain the same catalytic subunit (DnaE).
- In other organisms, different catalytic subunits may be required for each strand.

Each new DNA strand is synthesized by an individual enzyme catalytic unit. **Figure 18.15** shows that the behavior of these two units is different because the new DNA strands are growing in opposite directions. One enzyme unit is moving with the unwinding point and synthesizing the leading strand continuously. The other unit is moving "backwards," relative to the DNA, along the exposed single strand. Only short segments of template are exposed at any one time. When synthesis of one Okazaki fragment is completed, synthesis of the next Okazaki fragment is required to start at a new location approximately in the vicinity of the growing point for the leading strand. This requires a translocation relative to the DNA of the enzyme unit that is synthesizing the lagging strand.

In some cases, such as in *E. coli*, there is only a single type of DNA polymerase catalytic subunit used in replication, the DnaE protein. The active replicase is a dimer, and each half of the dimer contains DnaE as the catalytic subunit, supported by other proteins (which differ between the leading and lagging strands). In other cases, such as *B. subtilis*, there are two different types of catalytic subunits. PolC is the homologue to *E. coli*'s DnaE, and is responsible for synthesizing the leading strand. A related protein, DnaE$_{BS}$, is the catalytic subunit that synthesizes the lagging strand. Eukaryotic DNA polymerases have the same general structure as the bacterial enzymes, with separate catalytic units synthesizing the leading and lagging strands, but it is not clear whether the same or different types of catalytic subunits are used (see *18.12 Separate Eukaryotic DNA Polymerases Undertake Initiation and Elongation*).

The use of separate catalytic units to synthesize each new DNA strand raises a question: how is synthesis of the lagging strand coordinated with synthesis of the leading strand? **Figure 18.16** shows that the catalytic units behave differently. As the replisome moves along DNA, unwinding the parental strands, one catalytic unit elongates the leading strand pro-

Separate enzyme units synthesize lagging and leading strands

Lagging enzyme starts new fragments

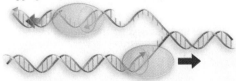

Leading enzyme elongates continuously

Figure 18.15 Leading and lagging strand polymerases move apart.

Leading and lagging catalytic units behave differently

Lagging catalytic subunit dissociates and reassociates

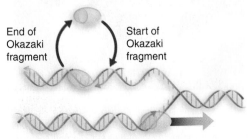

End of Okazaki fragment   Start of Okazaki fragment

Leading catalytic unit functions processively

Figure 18.16 A replicase contains separate catalytic units for synthesizing the leading and lagging strands.

cessively. The other catalytic unit synthesizes an Okazaki fragment, dissociates from DNA, and then reassociates closer to the growing point to start synthesis of the next Okazaki fragment where a primer has been formed.

The basic principle that is established by the dimeric polymerase model is that, while one polymerase subunit synthesizes the leading strand continuously, the other cyclically initiates and terminates the Okazaki fragments of the lagging strand within a large single-stranded loop formed by its template strand. **Figure 18.17** draws a generic model for the operation of such a replicase. The replication fork is created by a helicase, typically forming a hexameric ring, that translocates in the 5′–3′ direction on the template for the lagging strand. The helicase is connected to two DNA polymerase catalytic subunits, each of which is associated with a sliding clamp.

DnaB is the central component in both φX and *oriC* replicons. It provides the 5′–3′ helicase activity that unwinds DNA. Energy for the reaction is provided by cleavage of ATP. Basically DnaB is the active component of the growing point. In *oriC* replicons, DnaB is initially loaded at the origin as part of a large complex (see *18.13 Creating the Replication Forks at an Origin*). DnaB forms the growing point at which the DNA strands are separated as the replication fork advances.

What is responsible for recognizing the sites for initiating synthesis of Okazaki fragments? In *oriC* replicons, the connection between priming and the replication fork is provided by the dual properties of DnaB: it is the helicase that propels the replication fork; and it interacts with the DnaG primase at an appropriate site. Following primer synthesis, the primase is released. The length of the priming RNA is limited to 8–14 bases. Apparently DNA polymerase III is responsible for displacing the primase.

We can describe this model for DNA polymerase III in terms of the individual components of the enzyme complex, as illustrated in **Figure 18.18**. A catalytic core is associated with each template strand of DNA. The holoenzyme moves continuously along the template for the leading strand; the template for the lagging strand is "pulled through," creating a loop in the DNA. DnaB creates the unwinding point, and translocates along the DNA in the "forward" direction.

Synthesis of the leading strand creates a loop of single-stranded DNA that provides the template for lagging strand synthesis, and this loop becomes larger as the unwinding point advances. After initiation of an Okazaki fragment, the lagging strand core complex pulls the single-stranded template through the β clamp while synthesizing the new strand. The single-stranded template must extend for the length of at least one Okazaki fragment before the lagging polymerase completes one fragment and is ready to begin the next.

What happens when the Okazaki fragment is completed? All of the components of the replication apparatus function processively (that is, they remain associated with the DNA), except for the primase and the β clamp. **Figure 18.19** shows that they dissociate when the synthesis of each fragment is completed, releasing the loop. A new β clamp is then recruited by the clamp loader to initiate the next Okazaki fragment. The lagging strand polymerase transfers from one β clamp to the next in each cycle, without dissociating from the replicating complex.

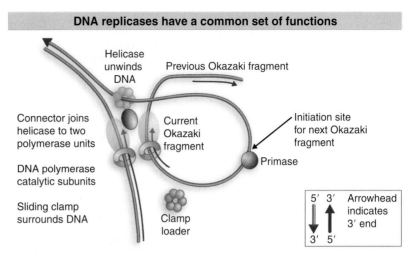

DNA replicases have a common set of functions

Helicase unwinds DNA

Previous Okazaki fragment

Connector joins helicase to two polymerase units

Current Okazaki fragment

Initiation site for next Okazaki fragment

Primase

DNA polymerase catalytic subunits

Sliding clamp surrounds DNA

Clamp loader

5′ 3′ / 3′ 5′ Arrowhead indicates 3′ end

Figure 18.17 The helicase creating the replication fork is connected to two DNA polymerase catalytic subunits, each of which is held onto DNA by a sliding clamp. The polymerase that synthesizes the leading strand moves continuously. The polymerase that synthesizes the lagging strand dissociates at the end of an Okazaki fragment and then reassociates with a primer in the single-stranded template loop to synthesize the next fragment.

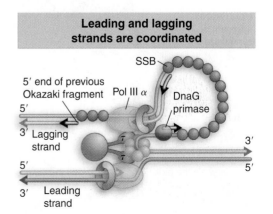

Leading and lagging strands are coordinated

SSB

5′ end of previous Okazaki fragment

Pol III α

DnaG primase

5′

3′ Lagging strand

3′

5′

5′

τ

τ

3′ Leading strand

Figure 18.18 Each catalytic core of Pol III synthesizes a daughter strand. DnaB is responsible for forward movement at the replication fork.

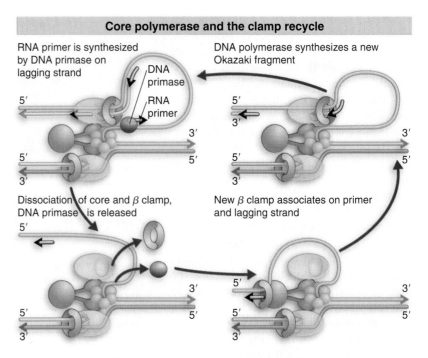

Figure 18.19 Core polymerase and the β clamp dissociate at completion of Okazaki fragment synthesis and reassociate at the beginning of the next Okazaki fragment.

18.11 Okazaki Fragments Are Linked by Ligase

Key Term

- **DNA ligase** makes a bond between an adjacent 3′—OH and 5′–phosphate end where there is a nick in one strand of duplex DNA.

Key Concepts

- Each Okazaki fragment starts with a primer and stops before the next fragment.
- DNA polymerase I removes the primer and replaces it with DNA.
- DNA ligase makes the bond that connects the 3′ end of one Okazaki fragment to the 5′ end of the next fragment.

We can now expand our view of the joining of Okazaki fragments, as illustrated in **Figure 18.20**. The series of events includes synthesis of an RNA primer, its extension with DNA, removal of the RNA primer, its replacement by a stretch of DNA, and the covalent linking of adjacent Okazaki fragments.

Synthesis of an Okazaki fragment terminates just before the start of the RNA primer of the preceding fragment. When the primer is removed, there will be a gap. The gap is filled by DNA polymerase I, which uses its 5′–3′ exonuclease activity to remove the RNA primer while simultaneously replacing it with a DNA sequence extended from the 3′—OH end of the next Okazaki fragment. In mammalian systems (where the DNA polymerase does not have a 5′–3′ exonuclease activity), RNA primers are removed by a two-step process. First RNAase H (an enzyme that is specific for a DNA–RNA hybrid substrate) makes an endonucleolytic cleavage; then a 5′–3′ exonuclease called FEN1 removes the RNA.

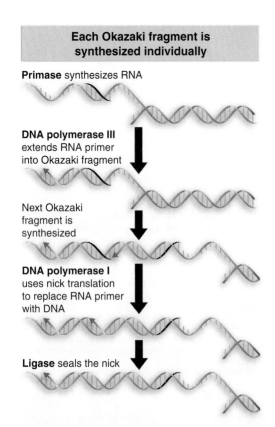

Figure 18.20 Synthesis of Okazaki fragments requires priming, extension, removal of RNA, gap filling, and nick ligation.

Once the RNA has been removed and replaced, the adjacent Okazaki fragments must be linked together. The $3' \mathrm{-OH}$ end of one fragment is adjacent to the $5'$–phosphate end of the previous fragment. The enzyme **DNA ligase** makes a bond by using a complex with AMP. **Figure 18.21** shows that the AMP of the enzyme complex becomes attached to the $5'$–phosphate of the nick; and then a phosphodiester bond is formed with the $3' \mathrm{-OH}$ terminus of the nick, releasing the enzyme and the AMP.

18.12 Separate Eukaryotic DNA Polymerases Undertake Initiation and Elongation

Key Concepts

- A replication fork has one complex of DNA polymerase α/primase and two complexes of DNA polymerase δ and/or ε.
- The DNA polymerase α/primase complex initiates the synthesis of both DNA strands.
- DNA polymerase δ elongates the leading strand and a second DNA polymerase δ or DNA polymerase ε elongates the lagging strand.

Eukaryotic cells have a large number of DNA polymerases. These polymerases can be broadly divided into those required for semiconservative replication and those involved in synthesizing material to repair damaged DNA. Nuclear DNA replication requires DNA polymerases α, δ, and ε, and mitochondrial replication requires DNA polymerase γ. All the other enzymes are concerned with synthesizing stretches of new DNA to replace damaged material. **Figure 18.22** shows that all of the nuclear replicases are large heterotetrameric enzymes. In each case, one of the subunits has the responsibility for catalysis, and the others are concerned with ancillary functions, such as priming, processivity, or proofreading. These enzymes all replicate DNA with high fidelity, as does the slightly less complex mitochondrial enzyme. The repair polymerases have much simpler structures, often consisting of a single monomeric subunit (although it may function in the context of a complex of other repair enzymes). Of the enzymes involved in repair, only DNA polymerase β has a fidelity approaching the replicases: all of the others have much greater error rates.

Each of the three nuclear DNA replicases has a different function:

- DNA polymerase α initiates the synthesis of new strands.
- DNA polymerase δ elongates the leading strand.
- DNA polymerase ε may be involved in lagging strand synthesis, but also has other roles.

The ability of DNA polymerase α to initiate a new strand is unusual. It is used to initiate both the leading and lagging strands. Reflecting its dual capacity to prime and extend chains, it is sometimes called pol α/primase. It binds to the initiation complex at the origin and synthesizes a short strand consisting of ~10 bases of RNA followed by 20–30 bases of DNA (sometimes called iDNA). Then it is replaced by an enzyme that will extend the chain. On the leading strand, this is DNA polymerase δ. The change of polymerase is called the *pol switch*. It is accomplished by interactions among several components of the initiation complex.

DNA polymerase δ is a highly processive enzyme that continuously synthesizes the leading strand. Its processivity results from its interaction with two other proteins, RF-C and PCNA.

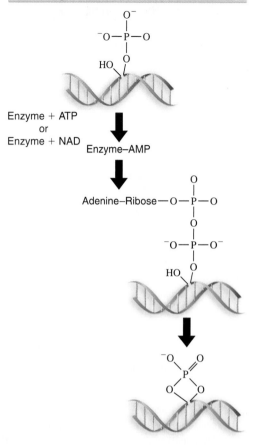

DNA ligase uses an AMP intermediate

Figure 18.21 DNA ligase seals nicks between adjacent nucleotides by employing an enzyme–AMP intermediate.

DNA polymerases undertake replication or repair

DNA polymerase	Function	Structure
High fidelity replicases		
α	Nuclear replication	350 kD tetramer
δ	"	250 kD tetramer
ε	"	350 kD tetramer
γ	Mitochondrial replication	200 kD dimer
High fidelity repair		
β	Base excision repair	39 kD monomer
Low fidelity repair		
ζ	Thymine dimer bypass	heteromer
η	Base damage repair	monomer
ι	Required in meiosis	monomer
κ	Deletion and base substitution	monomer

Figure 18.22 Eukaryotic cells have many DNA polymerases. The replicative enzymes operate with high fidelity. Except for the β enzyme, the repair enzymes all have low fidelity. Replicative enzymes have large structures, with separate subunits for different activities. Repair enzymes have much simpler structures.

The roles of RF-C and PCNA are analogous to the *E. coli* γ clamp loader and β processivity unit (see *18.9 The Clamp Controls Association of Core Enzyme with DNA*). RF-C is a clamp loader that catalyzes the loading of PCNA on to DNA. It binds to the 3' end of the iDNA and uses ATP-hydrolysis to open the ring of PCNA so that it can encircle the DNA. The processivity of DNA polymerase δ is maintained by PCNA, which tethers DNA polymerase δ to the template. (PCNA is called proliferating cell nuclear antigen for historical reasons.) The crystal structure of PCNA closely resembles the *E. coli* β subunit: a trimer forms a ring that surrounds the DNA. Although the sequence and subunit organization are different from the dimeric β clamp, the function is likely to be similar.

We are less certain about events on the lagging strand. One possibility is that DNA polymerase δ also elongates the lagging strand. It has the capability to dimerize, which suggests a model analogous to the behavior of *E. coli* replicase (see *18.8 DNA Polymerase Holoenzyme Consists of Subcomplexes*). However, there are some indications that DNA polymerase ε may elongate the lagging strand, although it also has been identified with other roles.

A general model suggests that a replication fork contains one complex of DNA polymerase α/primase and two other DNA polymerase complexes. One is DNA polymerase δ and the other is either a second DNA polymerase δ or may possibly be a DNA polymerase ε. The two complexes of DNA polymerase δ/ε behave in the same way as the two complexes of DNA polymerase III in the *E. coli* replisome: one synthesizes the leading strand, and the other synthesizes Okazaki fragments on the lagging strand. The exonuclease MF1 removes the RNA primers of Okazaki fragments. The enzyme DNA ligase I is specifically required to seal the nicks between the completed Okazaki fragments.

18.13 Creating the Replication Forks at an Origin

Key Concepts

- Initiation at *oriC* requires the sequential assembly of a large protein complex.
- DnaA binds to short repeated sequences and forms an oligomeric complex that melts DNA.
- Six DnaC monomers bind each hexamer of DnaB and this complex binds to the origin.
- A hexamer of DnaB forms the replication fork. Gyrase and SSB are also required.

Starting a cycle of replication of duplex DNA requires several successive activities:

- The two strands of DNA must be separated by a melting reaction over a short region.
- The replication fork is generated and then starts unwinding DNA.
- The first nucleotides of the new chain must be synthesized into the primer. This action is required only once for the leading strand, but is repeated at the start of each Okazaki fragment on the lagging strand.

Initiation of replication at *oriC* starts with formation of a complex that requires six proteins: DnaA, DnaB, DnaC, HU, Gyrase, and SSB. Of the six proteins involved in prepriming, DnaA draws our attention as the only one uniquely involved in initiation vis-à-vis elongation. DnaB/DnaC provides the "engine" of initiation at the origin.

The first stage in complex formation is binding to *oriC* by DnaA. The reaction involves action at two types of sequences: 9 bp and 13 bp repeats. Together the 9 bp and 13 bp repeats define the limits of the 245 bp minimal origin, as indicated in **Figure 18.23**. The four 9 bp consensus sequences on the right side of *oriC* provide the initial binding sites for DnaA. **Figure 18.24** shows that it binds cooperatively to form a central core around which *oriC* DNA is wrapped. Then DnaA acts at three A·T-rich 13 bp tandem repeats located in the left side of *oriC*. In the presence of ATP, DnaA melts the DNA strands at each of these sites to form an open complex. All three 13 bp repeats must be opened for the reaction to proceed to the next stage.

Altogether, 2–4 monomers of DnaA bind at the origin, and they recruit two "prepriming" complexes of DnaB–DnaC to bind, so that there is one for each of the two (bidirectional) replication forks. Each DnaB-DnaC complex consists of six DnaC monomers bound to a hexamer of DnaB. Each DnaB–DnaC complex transfers a hexamer of DnaB to an opposite strand of DNA. DnaC hydrolyzes ATP in order to release DnaB.

The region of strand separation in the open complex is large enough for both DnaB hexamers to bind, initiating the two replication forks. As DnaB binds, it displaces DnaA from the 13 bp repeats, and extends the length of the open region. Then it uses its helicase activity to extend the region of unwinding. Each DnaB activates a DnaG primase, in one case to initiate the leading strand, and in the other to initiate the first Okazaki fragment of the lagging strand.

Some further proteins are required to support the unwinding reaction. Gyrase provides a swivel that allows one strand to rotate around the other; without this reaction, unwinding would generate torsional strain in the DNA. The protein SSB stabilizes the single-stranded DNA as it is formed. The length of duplex DNA that usually is unwound to initiate replication is probably <60 bp. Also, the general DNA-binding protein HU, which can bend DNA, is involved in building the structure that leads to formation of the open complex.

Input of energy in the form of ATP is required at several stages for the prepriming reaction. It is required for unwinding DNA. The helicase action of DnaB depends on ATP hydrolysis, and the swivel action of gyrase requires ATP hydrolysis. ATP is also needed for the action of primase and to activate DNA polymerase III.

18.14 The Primosome Is Needed to Restart Replication

Key Concepts
- Initiation of φX replication requires the primosome complex to displace SSB from the origin.
- A replication fork stalls when it arrives at damaged DNA.
- After the damage has been repaired, the primosome is required to reinitiate replication.

Early work on replication made extensive use of phage φX174, and led to the discovery of a complex system for priming. The single-stranded DNA of φX174 is coated with SSB. A **primosome** assembles at a unique site, called the assembly site (*pas*). The *pas* is the equivalent of an origin for synthesis of the complementary strand of φX174. The primosome consists of six proteins: PriA, PriB, PriC, DnaT, DnaB, and DnaC. The key event in localizing the primosome is the ability of PriA to displace SSB from single-stranded DNA.

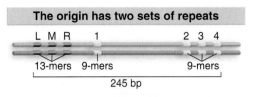
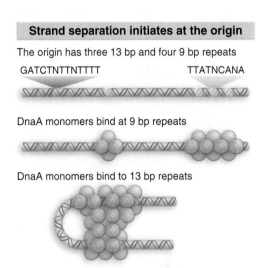
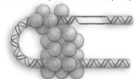
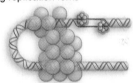

Figure 18.23 The minimal origin is defined by the distance between the outside members of the 13-mer and 9-mer repeats.

Figure 18.24 Prepriming involves formation of a complex by sequential association of proteins, leading to the separation of DNA strands.

Key Term
- The **primosome** is the complex of proteins involved in the priming action that initiates replication on φX-type origins. It is also involved in restarting stalled replication forks.

DNA damage can halt replication

Replication fork advances on normal DNA

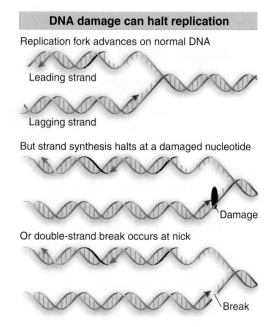

Leading strand

Lagging strand

But strand synthesis halts at a damaged nucleotide

Damage

Or double-strand break occurs at nick

Break

Figure 18.25 Replication is halted by a damaged base or nick in DNA.

Although the primosome forms initially at the *pas* on φX174 DNA, primers are initiated at a variety of sites. PriA translocates along the DNA, displacing SSB, to reach additional sites at which priming occurs. As in *oriC* replicons, DnaB plays a key role in unwinding and priming in φX replicons. The role of PriA is to load DnaB to form a replication fork.

It has always been puzzling that φX origins should use a complex structure that is not required to replicate the bacterial chromosome. Why does the bacterium provide this complex?

The answer is provided by the fate of stalled replication forks. **Figure 18.25** compares an advancing replication fork with what happens when there is damage to a base in the DNA or a nick in one strand. In either case, DNA synthesis is halted, and the replication fork is either stalled or disrupted. Replication fork stalling appears to be quite common; estimates for the frequency in *E. coli* suggest that 18–50% of bacteria encounter a problem during a replication cycle.

The damaged fork must be repaired, typically by a recombination event that excises and replaces the damage (see *20.8 Recombination Is an Important Mechanism to Recover from Replication Errors*). After the damage has been repaired, the replication fork must be restarted. **Figure 18.26** shows that this may be accomplished by assembly of the primosome, which in effect reloads DnaB so that helicase action can continue.

Replication fork reactivation is a common (and therefore important) reaction. It may be required in most chromosomal replication cycles. It is impeded by mutations in either the retrieval systems that replace the damaged DNA or in the components of the primosome.

A stalled replication fork can restart

Replication fork stalls at damaged nucleotide

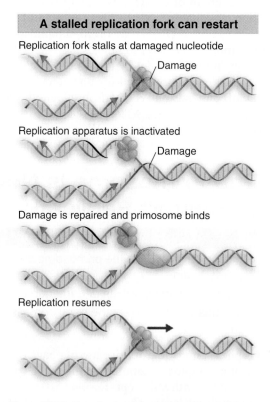

Damage

Replication apparatus is inactivated

Damage

Damage is repaired and primosome binds

Replication resumes

Figure 18.26 The primosome is required to restart a stalled replication fork after the DNA has been repaired.

18.15 SUMMARY

DNA synthesis occurs by semidiscontinuous replication, in which the leading strand of DNA growing 5'–3' is extended continuously, but the lagging strand that grows overall in the opposite 3'–5' direction is made as short Okazaki fragments, each synthesized 5'–3'. The leading strand and each Okazaki fragment of the lagging strand initiate with an RNA primer that is extended by DNA polymerase. Bacteria and eukaryotes each possess more than one DNA polymerase activity. DNA polymerase III synthesizes both lagging and leading strands in *E. coli*. It is part of the replisome, which assembles at the origin before a round of replication.

The replisome contains an asymmetric dimer of DNA polymerase III; each new DNA strand is synthesized by a different core complex containing a catalytic (α) subunit. Processivity of the core complex is maintained by the β clamp, which forms a ring around DNA. The clamp is loaded onto DNA by the clamp loader complex. Clamp/clamp loader pairs with similar structural features are widely found in both prokaryotic and eukaryotic replication systems.

The looping model for the replication fork proposes that, as one half of the dimer advances to synthesize the leading strand, the other half of the dimer pulls DNA through as a single loop that provides the template for the lagging strand. The transition from completion of one Okazaki fragment to the start of the next requires the lagging strand catalytic subunit to dissociate from DNA and then to reattach to a ß clamp at the priming site for the next Okazaki fragment.

DnaB provides the helicase activity at a replication fork; this depends on ATP cleavage. DnaB may function by itself in *oriC* replicons to provide primosome activity by interacting periodically with DnaG, which provides the primase that synthesizes RNA.

The φX priming event also requires DnaB, DnaC, and DnaT. PriA is the component that defines the primosome assembly site (*pas*) for φX replicons; it displaces SSB from DNA in an action that cleaves ATP. PriB and PriC are additional components of the primosome. The importance of the primosome for the bacterial cell is that it is used to restart replication at forks that stall when they encounter damaged DNA.

Origin activation involves an initial limited melting of the double helix, followed by more general unwinding to create single strands. Several proteins act sequentially at the *E. coli* origin. Replication is initiated at *oriC* in *E. coli* when DnaA binds to a series of 9 bp repeats. This is followed by binding to a series of 13 bp repeats, where it uses hydrolysis of ATP to generate the energy to separate the DNA strands. The prepriming complex of DnaC–DnaB displaces DnaA. DnaC is released in a reaction that depends on ATP hydrolysis; DnaB is joined by the replicase enzyme, and replication is initiated by two forks that set out in opposite directions.

Homologous and Site-Specific Recombination

19.1 Introduction

Key Terms

- **Synapsis** describes the association of the two pairs of sister chromatids (representing homologous chromosomes) that occurs at the start of meiosis.

- A **bivalent** is the structure containing all four chromatids (two representing each homologue) that is produced by synapsis.

- The **synaptonemal complex** describes the morphological structure of synapsed chromosomes.

- **Breakage and reunion** describes the mode of genetic recombination, in which two DNA duplex molecules are broken at corresponding points and then rejoined crosswise (involving formation of a length of heteroduplex DNA around the site of joining).

- A **chiasma** (*pl.* **chiasmata**) is a site at which two homologous chromosomes appear to have exchanged material during meiosis.

Recombination occurs during the protracted prophase of meiosis. **Figure 19.1** compares the visible progress of chromosomes through the five stages of meiotic prophase with the molecular interactions that exchange material between duplexes of DNA.

The beginning of meiosis is marked by the point at which individual chromosomes become visible. Each of these chromosomes has replicated previously, and consists of two sister chromatids, each of which contains a duplex DNA. The homologous chromosomes approach one another and begin to pair in one or more regions, forming **bivalents**. Pairing extends until the entire length of each chromosome is apposed with its homologue. The process is called **synapsis** or **chromosome pairing**. When the process is completed, the chromosomes are laterally associated in the form of a **synaptonemal complex**, which has a characteristic structure in each species, although there is wide variation in the details between species.

Recombination between chromosomes is a physical exchange of parts, usually represented as a **breakage and reunion**, in which two nonsister chromatids (each containing a duplex of DNA) have been broken and then linked each with the other. When the chromosomes begin to separate, they can be seen to be held together at discrete sites, the **chiasmata**, which represent the crossing-over events.

What is the molecular basis for these events? Each sister chromatid contains a single DNA duplex, so each bivalent contains four duplex molecules of DNA. Recombination requires a mechanism that allows the duplex DNA of one sister chromatid to interact with the duplex DNA of a sister chromatid from the other chromosome. We know of only one mechanism for nucleic acids to recognize one another on the basis of sequence: complementarity between single strands. Figure 19.1 shows a general model for the involvement of single strands in recombination. The first step in providing single strands is to make a break in each DNA duplex. Then one or both of the strands of that duplex can be released. If (at least) one strand displaces the corresponding strand in the other duplex, the two duplex molecules become connected at corresponding sequences. By exchanging both strands and later cutting them, it is possible to connect the parental duplex molecules by means of a crossover.

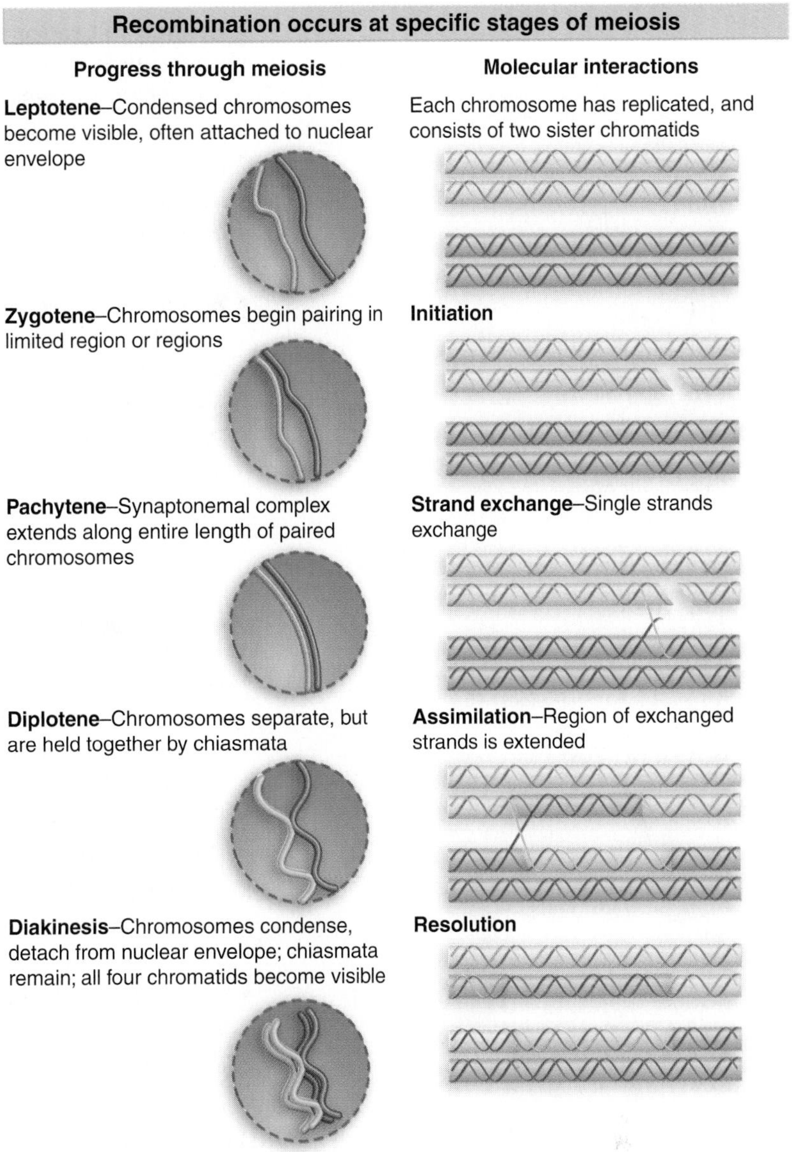

Recombination occurs at specific stages of meiosis

Progress through meiosis

Leptotene—Condensed chromosomes become visible, often attached to nuclear envelope

Zygotene—Chromosomes begin pairing in limited region or regions

Pachytene—Synaptonemal complex extends along entire length of paired chromosomes

Diplotene—Chromosomes separate, but are held together by chiasmata

Diakinesis—Chromosomes condense, detach from nuclear envelope; chiasmata remain; all four chromatids become visible

Molecular interactions

Each chromosome has replicated, and consists of two sister chromatids

Initiation

Strand exchange—Single strands exchange

Assimilation—Region of exchanged strands is extended

Resolution

Figure 19.1 Recombination occurs during the first meiotic prophase. The stages of prophase are defined by the appearance of the chromosomes, each of which consists of two replicas (sister chromatids), although the duplicated state becomes visible only at the end. The molecular interactions of any individual crossing-over event involve two of the four duplex DNAs.

19.2 Breakage and Reunion Involves Heteroduplex DNA

Key Terms

- A **joint molecule** is a pair of DNA duplexes that are connected together through a reciprocal exchange of genetic material.

- A **recombinant joint** is the point at which two recombining molecules of duplex DNA are connected.

- **Heteroduplex DNA (hybrid DNA)** is generated by base pairing between complementary single strands derived from the different parental duplex molecules; it forms during genetic recombination.

- **Branch migration** describes the ability of a DNA strand partially paired with its complement in a duplex to extend its pairing by displacing the resident strand with which it is homologous.

- A **Holliday structure** is an intermediate structure in homologous recombination, where the two duplexes of DNA are connected by the genetic material exchanged between two of the four strands, one from each duplex. A joint molecule is said to be resolved when nicks in the structure restore two separate DNA duplexes.

- **Splice recombinant** DNA results from a Holliday junction being resolved by cutting the nonexchanged strands. Both strands of DNA before the exchange point come from one chromosome; the DNA after the exchange point comes from the homologous chromosome.

- **Patch recombinant** DNA results from a Holliday junction being resolved by cutting the exchanged strands. The duplex is largely unchanged, except for a DNA sequence on one strand that came from the homologous chromosome.

Key Concepts

- The key event in recombination between two duplex DNA molecules is exchange of single strands.
- When a single strand from one duplex displaces its counterpart in the other duplex, it creates a branched structure.
- The exchange generates a stretch of heteroduplex DNA consisting of one strand from each parent.
- Two (reciprocal) exchanges are necessary to generate a joint molecule.
- The joint molecule is resolved into two separate duplex molecules by nicking two of the connecting strands.
- Whether recombinants are formed depends on whether the strands involved in the original exchange or the other pair of strands are nicked during resolution.

The act of connecting two duplex molecules of DNA is at the heart of the recombination process. Our molecular analysis of recombination therefore starts by expanding our view of the use of base pairing between complementary single strands in recombination. It is useful to imagine the recombination reaction in terms of single-strand exchanges (although we shall see that this is not how it is actually initiated), because the properties of the molecules created in this way are central to understanding the processes of recombination.

Figure 19.2 illustrates a process that starts with breakage at the corresponding points of the homologous strands of two paired DNA duplexes. The breakage allows movement of the free ends created by the nicks. Each strand leaves its partner and crosses over to pair with its complement in the other duplex.

The reciprocal exchange creates a connection between the two DNA duplexes. The connected pair of duplexes is called a **joint molecule**. The point at which an individual strand of DNA crosses from one duplex to the other is called the **recombinant joint**.

At the site of recombination, each duplex has a region consisting of one strand from each of the parental DNA molecules. This region is called **heteroduplex DNA** or **hybrid DNA**.

An important feature of a recombinant joint is its ability to move along the duplex. Such mobility is called **branch migration**. **Figure 19.3** illustrates the migration of a single strand in a duplex. The branch point

A recombination reaction has two possible outcomes

DNA duplexes pair

Homologous strands are nicked

Broken strands exchange between duplexes

Crossover point moves by branch migration

Nicks are sealed

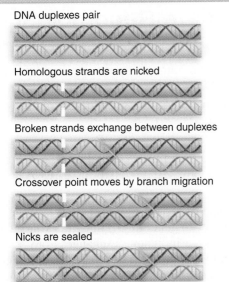

Second nicks made in same strand

Nicks are sealed

Genomes are not recombinant, but contain heteroduplex region

Second nicks made in other strand

Second strands crossover between duplexes

Reciprocal recombinant genomes are generated

Figure 19.2 Recombination between two paired duplex DNAs could involve reciprocal single-strand exchange, branch migration, and nicking.

Branch sites can migrate

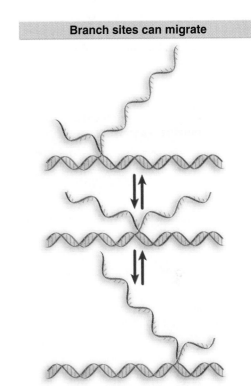

Figure 19.3 Branch migration can occur in either direction when an unpaired single strand displaces a paired strand.

Recombination has alternative resolutions

Rotation shows structure of Holliday junction

Nicking controls outcome

Nicks in same strands (yellow, blue) release patch recombinants

Nicks in other strands (red, green) release splice recombinants

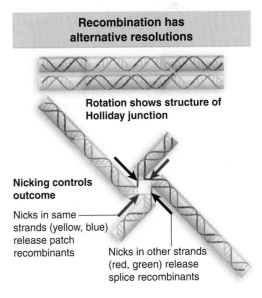

Figure 19.4 Resolution of a Holliday junction can generate parental or recombinant duplexes, depending on which strands are nicked. Both types of product have a region of heteroduplex DNA.

can migrate in either direction as one strand is displaced by the other. This allows the point of crossover in the recombination intermediate to move in either direction.

The joint molecule formed by strand exchange must be *resolved* into two separate duplex molecules. Resolution requires a further pair of nicks. We can most easily visualize the outcome by viewing the joint molecule in one plane as a **Holliday structure**. This is illustrated in **Figure 19.4**, which represents the structure of Figure 19.2 with one duplex rotated relative to the other. The outcome of the reaction depends on which pair of strands is nicked.

If the nicks are made in the pair of strands that were not originally nicked (the pair that did not initiate the strand exchange), all four of the original strands have been nicked. This releases **splice recombinant** DNA molecules. The duplex of one DNA parent is covalently linked to the duplex of the other DNA parent, via a stretch of heteroduplex DNA. There has been a conventional recombination event between markers located on either side of the heteroduplex region.

If the same two strands involved in the original nicking are nicked again, the other two strands remain intact. The nicking releases the original parental duplexes, which remain intact except that each has a residuum of the event in the form of a length of heteroduplex DNA. These are called **patch recombinants**.

These alternative resolutions of the joint molecule establish the principle that *a strand exchange between duplex DNAs always leaves behind a region of heteroduplex DNA, but the exchange may or may not be accompanied by recombination of the flanking regions.*

19.3 Double-Strand Breaks Initiate Recombination

Key Term

- A **double-strand break (DSB)** occurs when both strands of a DNA duplex are cleaved at the same site. Genetic recombination is initiated by double-strand breaks. The cell also has repair systems that act on double-strand breaks created at other times.

Key Concepts

- Recombination is initiated by making a double-strand break in one (recipient) DNA duplex.
- Exonuclease action generates 3′ single-stranded ends that invade the other (donor) duplex.
- New DNA synthesis replaces the material that has been degraded.
- This generates a recombinant joint molecule in which the two DNA duplexes are connected by heteroduplex DNA.

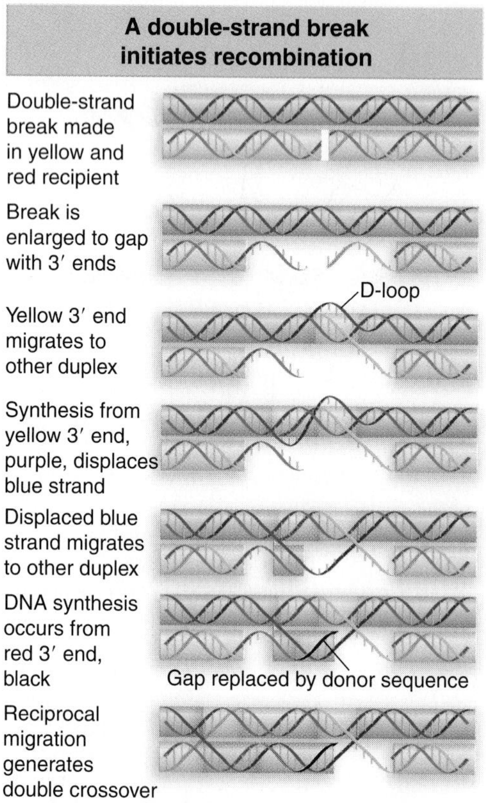

A double-strand break initiates recombination

Double-strand break made in yellow and red recipient

Break is enlarged to gap with 3′ ends

Yellow 3′ end migrates to other duplex

D-loop

Synthesis from yellow 3′ end, purple, displaces blue strand

Displaced blue strand migrates to other duplex

DNA synthesis occurs from red 3′ end, black

Gap replaced by donor sequence

Reciprocal migration generates double crossover

Figure 19.5 Recombination is initiated by a double-strand break, followed by formation of single-stranded 3′ ends, one of which migrates to a homologous duplex.

The general model of Figure 19.1 shows that a break must be made in one duplex in order to generate a point from which single strands can unwind to participate in genetic exchange. Both strands of a duplex must be broken to accomplish a genetic exchange. Figure 19.2 shows a model in which individual breaks in single strands occur successively. However, genetic exchange is actually initiated by a **double-strand break (DSB)**. The model is illustrated in **Figure 19.5**.

Recombination is initiated by an endonuclease that cleaves one of the partner DNA duplexes, the "recipient." The cut is enlarged to a gap by exonuclease action. The exonuclease(s) nibble away one strand on either side of the break, generating 3′ single-stranded termini. One of the free 3′ ends then invades a homologous region in the other, "donor" duplex. This is called single-strand invasion. The formation of heteroduplex DNA generates a *D loop*, in which one strand of the donor duplex is displaced. The D loop is extended by repair DNA synthesis, using the free 3′ end as a primer to generate double-stranded DNA.

Eventually the D loop becomes large enough to correspond to the entire length of the gap on the recipient chromatid. When the extruded single strand reaches the far side of the gap, the complementary single-stranded sequences anneal. Now there is heteroduplex DNA on either side of the gap, and the gap itself is represented by the single-stranded D loop.

The duplex integrity of the gapped region can be restored by repair synthesis using the 3′ end on the left side of the gap as a primer. Overall, the gap has been repaired by two individual rounds of single-strand DNA synthesis.

Branch migration converts this structure into a molecule with two recombinant joints. The joints must be resolved by cutting.

If both joints are resolved in the same way, the original noncrossover molecules are released, each with a region of altered genetic information that is a footprint of the exchange event. If the two joints are resolved in opposite ways a genetic crossover is produced.

The structure of the two-jointed molecule before it is resolved illustrates a critical difference between the double-strand break model and models that invoke only single-strand exchanges.

- Following the double-strand break, heteroduplex DNA has been formed at each end of the region involved in the exchange. Between the two heteroduplex segments is the region corresponding to the gap, which now has the sequence of the donor DNA in both molecules (Figure 19.5). So the arrangement of heteroduplex sequences is asymmetric, and part of one molecule has been converted to the sequence of the other (which is why the initiating chromatid is called the recipient).
- Following reciprocal single-strand exchange, each DNA duplex has heteroduplex material covering the region from the initial site of exchange to the migrating branch (Figure 19.2).

The double-strand break model does not reduce the importance of the formation of heteroduplex DNA, which remains the only plausible means by which two duplex molecules can interact. However, by shifting the responsibility for initiating recombination from single-strand to double-strand breaks, it influences our perspective about the ability of the cell to manipulate DNA.

19.4 Recombining Chromosomes Are Connected by the Synaptonemal Complex

Key Terms

- The **synaptonemal complex** describes the morphological structure of synapsed chromosomes.
- An **axial element** is a proteinaceous structure around which the chromosomes condense at the start of synapsis.
- A **lateral element** forms when a pair of sister chromatids condenses on to an axial element.
- The **central element** is a structure that lies in the middle of the synaptonemal complex, along which the lateral elements of homologous chromosomes align. It is formed from Zip proteins.
- **Cohesin** proteins form the lateral elements that hold sister chromatids together. They include some SMC proteins.

Key Concepts

- During the early part of meiosis, homologous chromosomes are paired in the synaptonemal complex.
- The mass of chromatin of each homologue is separated from the other by a proteinaceous complex.

A basic paradox in recombination is that the parental chromosomes never seem to be in close enough contact for recombination of DNA to occur. The chromosomes enter meiosis in the form of replicated (sister chromatid) pairs, visible as a mass of chromatin. They pair to form the **synaptonemal complex**, and it has been assumed for many years that this represents some stage involved with recombination, possibly a necessary preliminary to exchange of DNA. A more recent view is that the synaptonemal complex is a consequence rather than a cause of recombination. However, we have yet to define how the structure of the synaptonemal complex relates to molecular contacts between DNA molecules.

Synapsis begins when each chromosome (sister chromatid pair) condenses around a proteinaceous structure called the **axial element**. Then the axial elements of corresponding chromosomes become aligned, and the synaptonemal complex forms as a tripartite structure, in which the axial elements, now called **lateral elements**, are separated from each other by a **central element**. **Figure 19.6** shows an example.

Each chromosome at this stage appears as a mass of chromatin bounded by a lateral element. The two lateral elements are separated from each other by a fine but dense central element. The triplet of parallel dense strands lies in a single plane that curves and twists along its axis. The distance between the homologous chromosomes is considerable in molecular terms, more than 200 nm (the diameter of DNA is 2 nm). So a major problem in understanding the role of the complex is

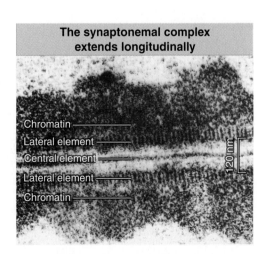

The synaptonemal complex extends longitudinally

Chromatin
Lateral element
Central element
Lateral element
Chromatin

120 nm

Figure 19.6 Chromosomes are closely juxtaposed in the synaptonemal complex. This example of *Neotellia* was kindly provided by M. Westergaard and D. Von Wettstein.

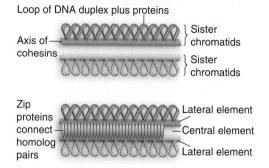

The synaptonemal complex links the homolog pairs

Loop of DNA duplex plus proteins

Axis of cohesins

Sister chromatids

Sister chromatids

Zip proteins connect homolog pairs

Lateral element

Central element

Lateral element

Figure 19.7 Each pair of sister chromatids has an axis made of cohesins. Loops of chromatin project from the axis. The synaptonemal complex is formed by linking together the axes via Zip proteins.

that, although it aligns homologous chromosomes, it is far from bringing homologous DNA molecules into contact.

Two groups of proteins have been identified by mutations in yeast with generating the distinctive features of the synaptonemal complex. **Figure 19.7** shows their roles in its formation.

- The **cohesins** belong to a general group of proteins that organize chromosome structure in both mitosis and meiosis. They bind to specific sites along the chromosomes. At meiosis, cohesins connect the sister chromatids by forming a single linear axis for each pair of sister chromatids that is equivalent to the lateral element in Figure 19.6. The loops of chromatin extend from this axis. The formation of the lateral elements may be necessary for the later stages of recombination, because mutations in the cohesin genes do not prevent the formation of double-strand breaks, but do block formation of recombinants.

- The lateral elements are connected by transverse filaments that are equivalent to the central element in Figure 19.6. These are formed from Zip proteins. A group of three Zip proteins form transverse filaments that connect the lateral elements of the sister chromatid pairs. The N-terminal domain of Zip1 protein is localized in the central element, but the C-terminal domain is localized in the lateral elements. The role of the Zip proteins in bringing the lateral elements together is shown by *zip* mutations, which allow lateral elements to form and to become aligned, but they do not become closely synapsed.

Pairing between chromosomes is the first event in meiosis. After the chromosomes have paired, interactions between homologous sequences lead to the recombination event that triggers synaptonemal complex formation. But what is responsible for initially bringing homologous chromosomes together?

The specificity of association between homologous chromosomes is controlled by the gene *hop2* in *S. cerevisiae*. In *hop2* mutants, normal amounts of synaptonemal complex form at meiosis, but the individual complexes contain nonhomologous chromosomes. This suggests that the formation of synaptonemal complexes as such is independent of homology (and therefore cannot be based on any extensive comparison of DNA sequences). The usual role of Hop2 is to prevent nonhomologous chromosomes from interacting.

19.5 The Synaptonemal Complex Forms After Double-Strand Breaks

Key Concepts

- Double-strand breaks that initiate recombination occur before the synaptonemal complex forms.
- If recombination is blocked, the synaptonemal complex cannot form.
- Mutations can occur in either chromosome pairing or synaptonemal complex formation without affecting the other process.

Double-strand breaks initiate recombination in both homologous and site-specific recombination. In homologous recombination, a DSB creates a hotspot at which recombination is initiated. The frequency of recombination declines in a gradient on one or both sides of the hotspot. The gradient results from the declining probability that a single-stranded region will be generated as distance increases from the site of the double-strand break.

There are few systems in which it is possible to compare molecular and cytological events at recombination, but we can analyze the relative timing of events at meiosis in *S. cerevisiae*, as summarized in **Figure 19.8**.

Double-strand breaks appear and then disappear over a 60-minute period. The first joint molecules, which are putative recombination intermediates, appear soon after the double-strand breaks disappear. The sequence of events suggests that double-strand breaks, individual pairing reactions, and formation of recombinant structures occur in succession at the same chromosomal site.

Double-strand breaks appear during the period when axial elements form. They disappear during the conversion of the paired chromosomes into synaptonemal complexes. This suggests that formation of the synaptonemal complex results from the initiation of recombination via the introduction of double-strand breaks and their conversion into later intermediates of recombination. Recombinant DNA molecules appear only at the end of pachytene, which places the completion of the recombination event after the formation of synaptonemal complexes.

So the synaptonemal complex forms after the double-strand breaks that initiate recombination, and it persists until the formation of recombinant molecules. It does not appear to be necessary for recombination as such, because some mutants that lack a normal synaptonemal complex can generate recombinants. Mutations that abolish recombination, however, also cause failure of the development of a synaptonemal complex. This suggests that the synaptonemal complex forms as a consequence of recombination, following chromosome pairing, and is required for later stages of meiosis.

We may now interpret the role of double-strand breaks in molecular terms. The flush ends created by the double-strand break are rapidly converted on both sides into long 3′ single-stranded ends, as shown in the model of Figure 19.5. A yeast mutation (*rad50*) that blocks the conversion of the flush end into the single-stranded protrusion is defective in recombination. This suggests that double-strand breaks are necessary for recombination.

In *rad50* mutants, the 5′ ends of the double-strand breaks are connected to the protein Spo11, which is homologous to the catalytic subunits of a family of type II topoisomerases. Topoisomerases are enzymes that can change the organization of DNA in space (see *19.9 Topoisomerases Relax or Introduce Supercoils in DNA*). There are close parallels between the actions of enzymes involved in both homologous and site-specific recombination and the mechanisms of topoisomerase action.

Spo11 is a topoisomerase-like enzyme that generates the double-strand breaks. The model for this reaction shown in **Figure 19.9** suggests that Spo11 interacts reversibly with DNA; the break is converted into a permanent structure by an interaction with another protein that dissociates the Spo11 complex. Then removal of Spo11 is followed by nuclease action. At least nine other proteins are required to process the double-strand breaks. One group of proteins is required to convert the double-strand breaks into protruding 3′–OH single-stranded ends. Another group then enables the single-stranded ends to invade homologous duplex DNA.

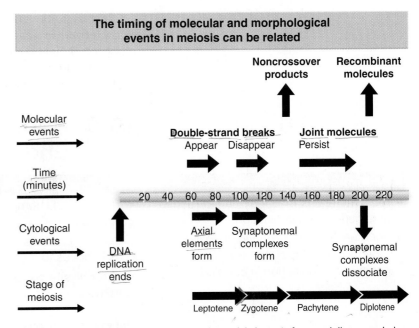

Figure 19.8 Double-strand breaks appear when axial elements form, and disappear during the extension of synaptonemal complexes. Joint molecules appear and persist until DNA recombinants are detected at the end of pachytene.

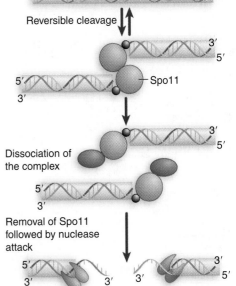

Figure 19.9 Spo11 is covalently joined to the 5′ ends of double-strand breaks.

Although the enzymes involved in recombination have been best characterized in yeast, because of the availability of mutations that block recombination at different stages, the system for generating and processing double-strand breaks is generally conserved. Spo11 homologues have been identified in several higher eukaryotes.

19.6 RecBCD Generates Free Ends for Recombination

Key Terms

- **rec⁻** mutations of *E. coli* cannot undertake general recombination.
- **chi** is an octameric sequence in DNA that provides a hotspot for RecA-mediated genetic recombination in *E. coli*.

Key Concepts

- The RecBCD complex has nuclease and helicase activities.
- The RecBCD complex binds to DNA downstream of a *chi* sequence, unwinds the duplex, and degrades one strand from 3'–5' as it moves to the *chi* site.
- The *chi* site triggers loss of the RecD subunit and nuclease activity.

The events by which sequences are exchanged between DNA molecules were first described in bacterial systems. Here the recognition reaction is part of the recombination mechanism and involves restricted regions of DNA molecules rather than intact chromosomes. But the general order of molecular events is similar: a free single-stranded end is required; it interacts with a partner duplex; the region of pairing is extended; and an endonuclease resolves the interacting DNA molecules.

Bacteria have various ways of initiating recombination, although they do not usually exchange large amounts of duplex DNA. In some cases, DNA may be provided in single-stranded form with free single-stranded 3' ends (as in conjugation; see *16.7 Conjugation Transfers Single-Stranded DNA*). Single-stranded gaps in duplex DNA may be generated by irradiation damage; or single-stranded tails may be generated by phage genomes undergoing replication by a rolling circle. In other cases, when both molecules are duplex, single-stranded regions with 3' ends must be generated.

Bacterial enzymes implicated in recombination have been identified by **rec⁻** mutations in their genes. The phenotype of *rec⁻* mutants is the inability to undertake generalized recombination. Some 10–20 loci have been identified. A key reaction in recombination is the generation of single-stranded 3' ends by the RecBCD enzyme complex, which provides a substrate for the ability of the RecA enzyme to catalyze the interaction of single-stranded DNA with duplex DNA.

RecBCD has several activities. Two of its subunits are helicases, and it has a nuclease activity. Its action is stimulated by the presence of specific short elements in DNA called **chi** sequences. **Figure 19.10** shows how the reactions of RecBCD are coordinated on a substrate DNA that has a *chi* site. RecBCD binds DNA at a double-strand break. It translocates along DNA, using its RecD subunit, which is a helicase with 5'–3' polarity, to unwind the DNA. Its nuclease activity degrades the released single strands. Recognition of one strand of the *chi* sequence in single-stranded form causes the enzyme to stop cleaving the strand with the 3' end. The RecD subunit dissociates or becomes inactivated. However, RecBC continues to translocate along DNA, unwinding it, now using the RecB subunit, which is a helicase with 3'–5' polarity. The net result is to generate single-stranded DNA with a 3' end that terminates in a *chi* sequence.

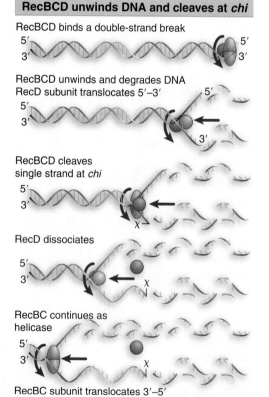

RecBCD unwinds DNA and cleaves at *chi*

RecBCD binds a double-strand break

RecBCD unwinds and degrades DNA
RecD subunit translocates 5'–3'

RecBCD cleaves
single strand at *chi*

RecD dissociates

RecBC continues as
helicase

RecBC subunit translocates 3'–5'

Figure 19.10 RecBCD approaches a *chi* sequence from one side, degrading DNA as it proceeds; at the *chi* site, it stops cleaving from the 3' end, loses RecD and its nuclease activity, and retains only the helicase activity.

19.7 Strand-Transfer Proteins Catalyze Single-Strand Assimilation

Key Concept

- RecA forms filaments with single-stranded or duplex DNA and catalyzes the reaction by which single-stranded DNA with a free 3′ end displaces its counterpart in a DNA duplex.

Key Term

- **Single-strand assimilation (single-strand uptake)** describes the ability of RecA protein to cause a single strand of DNA to displace its homologous strand in a duplex; that is, the single strand is assimilated into the duplex.

The *E. coli* protein RecA was the first example to be discovered of a DNA strand-transfer protein. It is the paradigm for a group that includes several other bacterial and archaeal proteins; the best characterized members are Rad51 in *S. cerevisiae*, and the higher eukaryotic protein Dmc1. Analysis of yeast *rad51* mutants shows that this class of protein plays a central role in recombination. The mutants accumulate double-strand breaks and fail to form normal synaptonemal complexes. This reinforces the idea that exchange of strands between DNA duplexes is involved in formation of the synaptonemal complex.

RecA in bacteria has two quite different types of activity: it can promote base pairing between a single strand of DNA and its complement in a duplex molecule, and it can trigger the *SOS response*, which results in the coordinate activation of a set of repair genes. The ability to trigger the SOS response depends on its ability to inactivate the protein LexA, which is a general repressor of the repair genes. The two activities are independent, but both are activated by single-stranded DNA in the presence of ATP.

The DNA-handling activity of RecA enables a single strand to displace its homologue in a duplex in a reaction that is called **single-strand assimilation** or **single-strand uptake**. The displacement reaction can occur between DNA molecules in several configurations and has three general conditions:

- One of the DNA molecules must have a single-stranded region.
- One of the molecules must have a free 3′ end.
- The single-stranded region and the 3′ end must be located within a region that is complementary between the molecules.

The reaction is illustrated in **Figure 19.11**. When a linear single strand invades a duplex, it displaces the original partner to its complement. The reaction proceeds 5′–3′ along the strand whose partner is being displaced and replaced; that is, the reaction involves an exchange in which (at least) one of the exchanging strands has a free 3′ end.

All of the bacterial and archaeal proteins in the RecA family can aggregate into long filaments with single-stranded or duplex DNA. There are six RecA monomers per turn of the filament, which has a helical structure with a deep groove that contains the DNA. The stoichiometry of binding is 3 nucleotides (or base pairs) per RecA monomer. The DNA is held in a form that is extended 1.5 times relative to duplex B DNA, making a turn every 18.6 nucleotides (or base pairs). When duplex DNA is bound, it contacts RecA via its minor groove, leaving the major groove accessible for possible reaction with a second DNA molecule.

The interaction between two DNA molecules occurs within these filaments. When a single strand is assimilated into a duplex, the first step is for RecA to bind the single strand into a filament. Then the duplex is incorporated, probably forming some sort of triple-stranded structure. In this system, synapsis precedes physical exchange of material, because the pairing reaction can take place even in the absence of free ends, when strand exchange is impossible. A free 3′ end is required

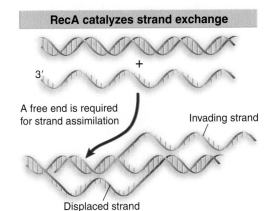

RecA catalyzes strand exchange

3′

A free end is required for strand assimilation

Invading strand

Displaced strand

Figure 19.11 RecA promotes the assimilation of invading single strands into duplex DNA so long as one of the reacting strands has a free end.

for strand exchange. The reaction occurs within the filament, and RecA remains bound to the strand that was originally single, so that at the end of the reaction RecA is bound to the duplex molecule.

Single-strand assimilation is a key stage in the initiation of recombination. All recombination models call for an intermediate in which one or both single strands cross over from one duplex to the other (see Figure 19.2 and Figure 19.5). The functions of strand-transfer proteins are triggered by the availability of a single-stranded 3′ end, and we know several situations in which this occurs:

- In *E. coli*, RecA can take the single strand with the 3′ end that is released when RecBCD cuts at *chi*, and can use it to react with a homologous duplex sequence, thus creating a joint molecule.
- One of the main circumstances in which single-stranded ends are available in bacteria may be created when a replication fork stalls at a site of DNA damage (see *20.8 Recombination Is Important for Correcting Replication Errors*).
- During conjugation, single-stranded DNA is transferred to the target cell, it is converted into duplex form, and then RecA is required to recombine it with the host chromosome.
- In yeast, double-strand breaks may be generated by DNA damage or as part of the normal process of recombination. In either case, processing of the break to generate a 3′ single-stranded end is followed by loading the single strand into a filament with Rad51, followed by a search for matching duplex sequences. This can be used in both repair and recombination reactions.

19.8 The Ruv System Resolves Holliday Junctions

Key Concepts

- The Ruv complex acts on recombinant junctions.
- RuvA recognizes the structure of the junction and RuvB is a helicase that catalyzes branch migration.
- RuvC cleaves junctions to generate recombination intermediates.

One of the most critical steps in recombination is the resolution of the Holliday junction, which determines whether there is a reciprocal recombination or a reversal of the structure that leaves only a short stretch of hybrid DNA (see Figure 19.2 and Figure 19.4). Branch migration from the exchange site (see Figure 19.3) determines the length of the region of hybrid DNA (with or without recombination). The proteins in *E. coli* that stabilize and resolve Holliday junctions are products of the *ruv* genes. Their actions are shown in **Figure 19.12**.

RuvA recognizes the structure of the Holliday junction by binding to all four strands of DNA at the crossover point, where it forms two tetramers that sandwich the DNA. RuvB is a hexameric helicase with an ATPase activity that provides the motor for branch migration. Hexameric rings of RuvB bind around each duplex of DNA upstream of the crossover point. RuvAB displaces RecA from DNA during its action.

RuvC is an endonuclease that cleaves Holliday junctions to resolve recombination intermediates. A common tetranucleotide sequence provides a hotspot for RuvC. The tetranucleotide (ATTG) is asymmetric, and thus may direct resolution with regard to which pair of strands is nicked. This determines whether the outcome is patch recombinant formation (no overall recombination) or splice recombinant formation

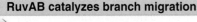

RuvAB catalyzes branch migration

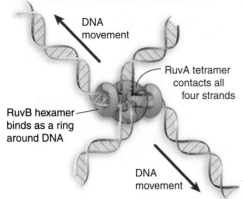

DNA movement

RuvA tetramer contacts all four strands

RuvB hexamer binds as a ring around DNA

DNA movement

Figure 19.12 RuvAB is an asymmetric complex that promotes branch migration of a Holliday junction.

(recombination between flanking markers). Crystal structures of RuvC and other junction-resolving enzymes show that there is a little structural similarity among the group, in spite of their common function.

The Ruv proteins may function as part of a "resolvasome" complex that includes enzymes catalyzing branch migration as well as junction-resolving activity. Mammalian cells contain similar complexes, although their components are not related to the bacterial proteins.

We may now account for the stages of recombination in *E. coli* in terms of individual proteins. **Figure 19.13** shows the events by which recombination repairs a gap in one duplex by retrieving material from the other duplex. The major caveat in applying these conclusions to recombination in eukaryotes is that bacterial recombination generally involves interaction between a fragment of DNA and a whole chromosome. It occurs as a repair reaction that is stimulated by damage to DNA, and this is not entirely equivalent to recombination between genomes at meiosis. Nonetheless, similar molecular activities are involved in manipulating DNA.

19.9 Topoisomerases Relax or Introduce Supercoils in DNA

Key Terms

- A DNA **topoisomerase** is an enzyme that changes the number of times the two strands in a closed DNA molecule cross each other. It does this by cutting the DNA, passing DNA through the break, and resealing the DNA.
- A **type I topoisomerase** is an enzyme that changes the topology of DNA by nicking and resealing one strand of DNA.
- A **type II topoisomerase** is an enzyme that changes the topology of DNA by nicking and resealing both strands of DNA.

Key Concept

- Topoisomerases are required to change the structure of DNA so that replication and transcription can proceed.

Changes in the topology of DNA can be caused in several ways. **Figure 19.14** shows some examples. In order to start replication or transcription, the two strands of DNA must be unwound. For replication, the two strands separate permanently, and each re-forms a duplex with a newly synthesized daughter strand. For transcription, the movement of RNA polymerase creates a region of positive supercoiling in front of the enzyme and a region of negative supercoiling behind it. This must be resolved before the positive supercoils impede the movement of the enzyme (see *11.10 Supercoiling Is an Important Feature of Transcription*). When a circular DNA molecule is replicated, the circular products may be catenated, with one passed through the other. They must be separated in order for the daughter molecules to segregate to separate daughter cells. Yet another situation in which supercoiling is important is the folding of the DNA thread into a chain of nucleosomes in the eukaryotic nucleus (see *29.6 The Nucleosome Absorbs Some Supercoiling*). All of the situations are resolved by the actions of topoisomerases.

Topoisomerases are enzymes that catalyze changes in the topology of DNA by transiently breaking one or both strands of DNA, passing the unbroken strand(s) through the gap, and then resealing the gap. The ends that are generated by the break are never free, but are manipulated

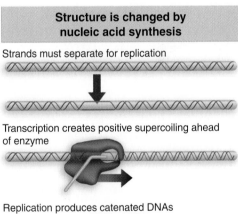

Bacterial enzymes catalyze recombination

Gap generated by replication of damaged DNA

RecA–Strand exchange

Second strand exchange

DNA polymerase–Gap filled by DNA synthesis

RuvA, B–Branch migration

RuvC–Cleave Holliday junction

Figure 19.13 Bacterial enzymes can catalyze all stages of recombination in the repair pathway following the production of suitable substrate DNA molecules.

Structure is changed by nucleic acid synthesis

Strands must separate for replication

Transcription creates positive supercoiling ahead of enzyme

Replication produces catenated DNAs

Figure 19.14 The topological structure of DNA is changed during replication and transcription. Strand separation requires a base turn of DNA to be unwound. Transcription creates positive supercoils ahead of the RNA polymerase. Replication of a circular template produces two catenated daughter templates.

exclusively within the confines of the enzyme—in fact, they are covalently linked to the enzyme. Topoisomerases act on DNA irrespective of its sequence, but some enzymes involved in site-specific recombination function in the same way and also fit the definition of topoisomerases (see *19.11 Site-Specific Recombination Resembles Topoisomerase Activity*).

Topoisomerases are divided into two classes, according to the nature of the mechanisms they employ. **Type I topoisomerases** act by making a transient break in one strand of DNA. **Type II topoisomerases** act by introducing a transient double-strand break. Topoisomerases in general vary with regard to the types of topological change they introduce. Some topoisomerases can relax (remove) only negative supercoils from DNA; others can relax both negative and positive supercoils. Enzymes that can introduce negative supercoils are called gyrases; those that can introduce positive supercoils are called reverse gyrases.

There are four topoisomerase enzymes in *E. coli*, called topoisomerase I, III, IV, and DNA gyrase. DNA topoisomerase I and III are type I topoisomerases. Gyrase and DNA topoisomerase IV are type II. Each of the four enzymes is important in one or more of the situations described in Figure 19.14:

- The overall level of negative supercoiling in the bacterial nucleoid is the result of a balance between the introduction of supercoils by gyrase and their relaxation by topoisomerases I and IV. This is a crucial aspect of nucleoid structure and affects initiation of transcription at certain promoters (see *11.10 Supercoiling Is an Important Feature of Transcription*).
- To resolve the problems created by transcription, gyrase converts the positive supercoils that are generated ahead of RNA polymerase into negative supercoils, and topoisomerases I and IV remove the negative supercoils that are left behind the polymerase. There are similar, but more complicated, effects during replication, and the topoisomerases have similar roles in dealing with them.
- As replication proceeds, the daughter duplexes can become twisted around one another, in a stage known as precatenation. The precatenanes are removed by topoisomerase IV, which also decatenates any catenated genomes that are left at the end of replication. The functions of topoisomerase III partially overlap those of topoisomerase IV.

The corresponding enzymes in eukaryotes follow the same principles, although the detailed division of responsibilities may be different. They do not show sequence or structural similarity with the prokaryotic enzymes. Most eukaryotes contain a single topoisomerase I enzyme that is required both for replication fork movement and for relaxing supercoils generated by transcription. A topoisomerase II enzyme (or more than one) is required to unlink chromosomes following replication. Other individual topoisomerases have been implicated in recombination and repair activities.

19.10 Topoisomerases Break and Reseal Strands

Key Concept

- Type I topoisomerases function by forming a covalent bond to one of the broken ends, moving one strand around the other, and then transferring the bound end to the other broken end. Because bonds are conserved, no input of energy is required.

The common action for all topoisomerases is to link one end of each broken strand to a tyrosine residue in the enzyme. A type I enzyme links to the single broken strand; a type II enzyme links to one end of each broken strand. The topoisomerases are further divided into the A and B groups according to whether the linkage is to a 5′ phosphate or 3′ phosphate. The use of the transient phosphodiester-tyrosine bond suggests a mechanism for the action of the enzyme; it transfers a phosphodiester bond in DNA to the protein, manipulates the structure of one or both DNA strands, and then rejoins the bond in the original strand.

The *E. coli* enzymes are all of type A, using links to 5′ phosphate. This is the general pattern for bacteria, which have almost no type B topoisomerases. All four possible types of topoisomerase (IA, IB, IIA, IIB) are found in eukaryotes.

A model for the action of topoisomerase IA is illustrated in **Figure 19.15**. The enzyme binds to a region in which duplex DNA becomes separated into its single strands; then it breaks one strand, pulls the other strand through the gap, and finally seals the gap. The transfer of bonds from nucleic acid to protein explains how the enzyme can function without requiring any input of energy. There has been no irreversible hydrolysis of bonds; their energy has been conserved through the transfer reactions.

Type II topoisomerases generally relax both negative and positive supercoils. The reaction requires ATP, with one ATP hydrolyzed for each catalytic event. As illustrated in **Figure 19.16**, the reaction is mediated by making a double-stranded break in one DNA duplex. The double-strand is cleaved with a 4-base stagger between the ends, and each subunit of the dimeric enzyme attaches to a protruding broken end. Then another duplex region is passed through the break. The ATP is used in the following religation/release step, when the ends are rejoined and the DNA duplexes are released.

The reaction probably represents a nonspecific recognition of duplex DNA in which the enzyme binds any two double-stranded segments that cross each other. The hydrolysis of ATP may be used to drive the enzyme through conformational changes that provide the force needed to push one DNA duplex through the break made in the other. Because of the topology of supercoiled DNA, the relationship of the crossing segments allows supercoils to be removed from either positively or negatively supercoiled circles.

19.11 Site-Specific Recombination Resembles Topoisomerase Activity

Key Concepts

- Integrases are related to topoisomerases, and the recombination reaction resembles topoisomerase action except that nicked strands from *different* duplexes are sealed together.
- The reaction conserves energy by using a catalytic tyrosine in the enzyme to break a phosphodiester bond and link to the broken 3′ end.
- In a typical site-specific recombination, two enzyme units bind to each recombination site and the two dimers synapse to form a complex in which the transfer reactions occur.

Specialized recombination involves a reaction between two specific sites. The target sites are short, typically 14–50 bp. In some cases, the two sites

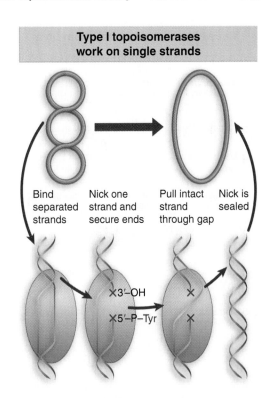

Type I topoisomerases work on single strands

Bind separated strands   Nick one strand and secure ends   Pull intact strand through gap   Nick is sealed

×3′–OH
×5′–P–Tyr

Figure 19.15 Bacterial type I topoisomerases recognize partially unwound segments of DNA and pass one strand through a break made in the other.

Type II topoisomerases handle double strands

DNA duplexes are brought into apposition

Enzyme makes a double-stranded break in one of the duplexes

Unbroken duplex is passed through the ends of the break

Break is sealed and the enzyme releases the DNA

Figure 19.16 Type II topoisomerases can pass a duplex DNA through a double-strand break in another duplex.

Recombinases break and rejoin DNA

—1. Two enzyme subunits bind to each duplex DNA—

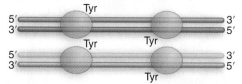

—2. Each duplex is cleaved on one strand to generate a P–Tyr bond and an –OH end

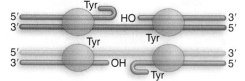

—3. Each hydroxyl attacks the Tyr–phosphate link in the other duplex

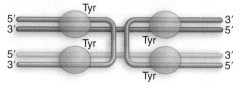

—4. The reactions are repeated by the other subunits to join the other strands

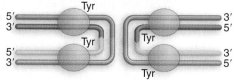

Figure 19.17 Integrases catalyze recombination by a mechanism similar to topoisomerases. Staggered cuts are made in DNA and the 3′-phosphate end is covalently linked to a tyrosine in the enzyme. Then the free hydroxyl group of each strand attacks the P–Tyr link of the other strand. The first exchange shown in the figure generates a Holliday structure. The structure is resolved by repeating the process with the other pair of strands.

Cre–*lox* recombination occurs in a tetramer

Figure 19.18 A synapsed *loxA* recombination complex has a tetramer of Cre recombinases, with one enzyme monomer bound to each half site. Two of the four active sites are in use, acting on complementary strands of the two DNA sites.

have the same sequence, but in other cases they are nonhomologous. The reaction is used to insert a free phage DNA into the bacterial chromosome or to excise an integrated phage DNA from the chromosome, and in this case the two recombining sequences are different from one another. It is also used before division to regenerate monomeric circular chromosomes from a dimer that has been created by a generalized recombination event (see *17.7 Chromosomal Segregation May Require Site-Specific Recombination*). In this case, the recombining sequences are identical.

The enzymes that catalyze site-specific recombination are generally called recombinases, and >100 of them are now known. Those involved in phage integration or related to this group are also known as the integrase family. Prominent members of the integrase family are the prototype Int from phage lambda, Cre from phage P1, and the yeast FLP enzyme (which catalyzes a chromosomal inversion).

Integrases use a mechanism similar to that of type I topoisomerases, in which a break is made in one DNA strand at a time. The corresponding strands on each duplex are cut at the same position, the free 3′ ends exchange between duplexes, the branch migrates along the region of homology, and then the structure is resolved by cutting the other pair of corresponding strands.

The difference between an integrase and a topoisomerase is that a recombinase reconnects the ends crosswise, whereas a topoisomerase makes a break, manipulates the ends, and then rejoins the original ends. The basic principle of the system is that four molecules of the recombinase are required, one to cut each of the four strands of the two duplexes that are recombining.

Figure 19.17 shows the nature of the reaction catalyzed by an integrase. The enzyme is a monomeric protein that has an active site capable of cutting and ligating DNA. The reaction involves an attack by a tyrosine on a phosphodiester bond. The 3′ end of the DNA chain is linked through a phosphodiester bond to a tyrosine in the enzyme. This releases a free 5′–hydroxyl end.

Two enzyme units are bound to each of the recombination sites. At each site, only one of the enzyme active sites attacks the DNA. The symmetry of the system ensures that complementary strands are broken in each recombination site. Then the free 5′–OH end in each site attacks the 3′–phosphotyrosine link in the other site. This generates a Holliday junction.

The structure is resolved when the other two enzyme units (which had not been involved in the first cycle of breakage and reunion) act on the other pair of complementary strands.

The successive interactions accomplish a conservative strand exchange, in which there are no deletions or additions of nucleotides at the exchange site, and there is no need for input of energy. The transient 3′–phosphotyrosine link between protein and DNA conserves the energy of the cleaved phosphodiester bond.

One of the best characterized systems is found in the bacteriophage P1. The Cre recombinase coded by the phage catalyzes a recombination between two identical target sequences. Each consists of a 34 bp-long sequence called *loxP*. The Cre recombinase is sufficient for the reaction; no accessory proteins are required. Because of its simplicity and its efficiency, what is now known as the Cre–*lox* system has been adapted for use in eukaryotic cells, where it has become one of the standard techniques for undertaking site-specific recombination (see *32.7 Gene Targeting Allows Genes to Be Replaced or Knocked Out*).

Figure 19.18 shows the reaction intermediate, based on the crystal structure. The structure of the Cre–*lox* complex shows two Cre molecules, each bound to a 15 bp length of DNA. The DNA is bent by ~100° at the center of symmetry. Two of these complexes assemble in an antiparallel way to form a tetrameric protein structure bound to two synapsed DNA molecules. Strand exchange takes place in a central cavity of the protein structure that contains the central six bases of the crossover region.

The tyrosine that is responsible for cleaving DNA in any particular half site is provided by the enzyme subunit that is bound to that half site. This is called *cis* cleavage. This is true also for the Int integrase and XerD recombinase. However, the FLP recombinase cleaves in *trans*, involving a mechanism in which the enzyme subunit that provides the tyrosine is *not* the subunit bound to that half site, but is one of the other subunits.

19.12 Specialized Recombination in Phage Lambda Involves Specific Sites

Key Concepts

- Specialized recombination involves reaction between specific sites that are not necessarily homologous.
- Phage lambda integrates into the bacterial chromosome by recombination between the *attB* site on the phage and the *attP* site on the *E. coli* chromosome.
- The phage is excised from the chromosome by recombination between the *attL* and *attR* sites at the end of the linear prophage.
- Phage lambda *int* codes for an integrase that catalyzes the integration reaction.
- Lambda integration takes place in a large complex (the intasome) that also includes the host protein IHF.
- The excision reaction requires Int and Xis and recognizes the ends of the prophage DNA as substrates.

Key Terms

- **Prophage** is a phage genome covalently integrated as a linear part of the bacterial chromosome.
- **Integration** of viral or another DNA sequence describes its insertion into a host genome as a region covalently linked on either side to the host sequences.
- The **excision** of phage or episome or other sequence describes its release from the host chromosome as an autonomous DNA molecule.
- ***att*** sites are the loci on a lambda phage and the bacterial chromosome at which recombination integrates the phage into, or excises it from, the bacterial chromosome.
- The **core sequence** is the segment of DNA that is common to the attachment sites on both the phage lambda and bacterial genomes. The recombination event occurs within the core sequence.
- The **arms** of a lambda phage attachment site are the sequences flanking the core region where the recombination event occurs.
- An **intasome** is a protein–DNA complex between the phage lambda integrase (Int) and the phage lambda attachment site (*attP*).

The integration of phage lambda provided one of the first systems in which site-specific recombination was characterized. It is actually one of the more complex systems, with ancillary functions in addition to the enzymes required for the recombination reaction.

The conversion of lambda DNA between its different life forms comprises two types of event. The pattern of gene expression is regulated as described in *Chapter 14 Phage Strategies*. And the physical condition of the DNA is different in the lysogenic and lytic states:

- In the lytic state, lambda DNA exists as an independent, circular molecule in the infected bacterium.
- In the lysogenic state, the phage DNA is an integral part of the bacterial chromosome (called the **prophage**).

Transition between these states requires site-specific recombination:

- To enter the lysogenic state, free lambda DNA must be inserted into the host DNA. This is called **integration**.
- To be released from lysogeny into the lytic cycle, prophage DNA must be released from the chromosome. This is called **excision**.

Integration and excision occur by recombination at specific loci on the bacterial and phage DNAs called attachment (***att***) sites. The bacterial attachment site (att^λ) is called *attB*, consisting of the sequence components *BOB'*. The attachment site on the phage, *attP*, consists of the

Integration and excision require recombination

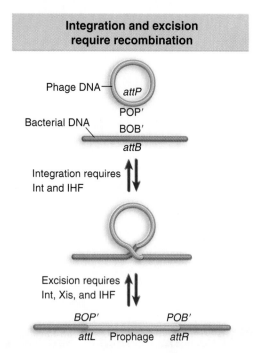

Figure 19.19 Circular phage DNA is converted to an integrated prophage by a reciprocal recombination between *attP* and *attB*; the prophage is excised by reciprocal recombination between *attL* and *attR*.

The *attP* intasome binds *attB*

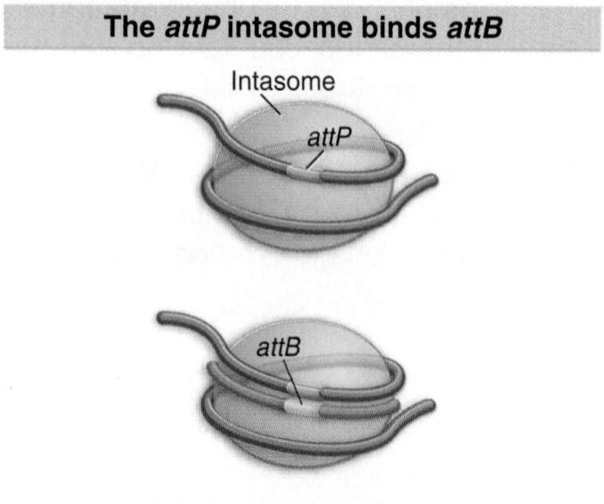

Figure 19.20 Multiple copies of Int protein may organize *attP* into an intasome, which initiates site-specific recombination by recognizing *attB* on free DNA.

components *POP'*. **Figure 19.19** outlines the recombination reaction between these sites. The sequence *O* is common to *attB* and *attP*. It is called the **core sequence**; and the recombination event occurs within it. The flanking regions *B*, *B'* and *P*, *P'* are referred to as the **arms**; each is distinct in sequence. Because the phage DNA is circular, the recombination event inserts it into the bacterial chromosome as a linear sequence. The prophage is bounded by two new *att* sites, the products of the recombination, called *attL* and *attR*, which have the components *BOP'* and *POB'*, respectively.

An important consequence of the constitution of the *att* sites is that the integration and excision reactions do not involve the same pair of reacting sequences. Integration requires recognition between *attP* and *attB*; while excision requires recognition between *attL* and *attR*. The directional character of site-specific recombination is controlled by the identity of the recombining sites.

Although the recombination event is reversible, different conditions prevail for each direction of the reaction. This is an important feature in the life of the phage, since it offers a means to ensure that an integration event is not immediately reversed by an excision, and vice versa.

The difference in the pairs of sites reacting at integration and excision is reflected by a difference in the proteins that mediate the two reactions:

- Integration (*attB* × *attP*) requires the product of the phage gene *int*, which codes for an integrase enzyme, and a bacterial protein called integration host factor (IHF).
- Excision (*attL* × *attR*) requires the product of phage gene *xis*, in addition to Int and IHF.

So Int and IHF are required for both reactions. Xis plays an important role in controlling the direction; it is required for excision, but inhibits integration.

Topoisomerases and many integrases (including the *Cre/lox* system) function as independent proteins. However, the lambda reaction has additional components. IHF is required for both integration and excision. IHF is involved in many reactions in *E. coli* that call for manipulation of DNA structure. It has the ability to bend DNA by wrapping it around the surface of a complex formed by multiple copies of IHF.

Int and IHF bind to *attP* to form a complex in which all the binding sites are pulled together on the surface of a protein. Supercoiling of *attP* is needed for the formation of this **intasome**. The intasome is the intermediate that "captures" *attB*, as indicated schematically in **Figure 19.20**. According to this model, the initial recognition between *attP* and *attB* does not depend directly on DNA homology, but instead is determined by the ability of Int proteins to recognize both *att* sequences. The two *att* sites then are brought together in an orientation predetermined by the structure of the intasome. Sequence homology becomes important at this stage, when it is required for the strand exchange reaction.

The asymmetry of the integration and excision reactions is shown by the fact that Int can form a similar complex with *attR* only if Xis is added. This complex can pair with a condensed complex that Int forms at *attL*. IHF is not needed for this reaction.

Much of the complexity of site-specific recombination may be caused by the need to regulate the reaction so that integration occurs preferentially when the virus is entering the lysogenic state, while excision is preferred when the prophage is entering the lytic cycle. Through control of the amounts of Int and Xis, the appropriate reaction occurs.

19.13 Yeast Mating Type Is Changed by Recombination

The yeast *S. cerevisiae* can propagate in either the haploid or diploid condition. Conversion between these states takes place by mating (fusion of haploid spores to give a diploid) and by sporulation (meiosis of diploids to give haploid spores). The ability to engage in these activities is determined by the **mating type** of the strain. The mating type is determined by the sequence of the *MAT* locus, and can be changed by a recombination event that substitutes a different sequence at this locus. The recombination event is initiated by a double-strand break, like a homologous recombination event, but then the subsequent events ensure a unidirectional replacement of the sequence at the *MAT* locus.

Mating behavior is determined by the genetic information present at the *MAT* locus. Cells that carry the *MATa* allele at this locus are type **a**; likewise, cells that carry the *MATα* allele are type α. Cells of opposite types can mate; cells of the same type cannot. Recognition between cells of opposite mating type is accomplished by the secretion of **pheromones**: α cells secrete the small polypeptide α-factor; **a** cells secrete a-factor. A cell of one mating type carries a surface receptor for the pheromone of the opposite type. When an **a** cell and an α cell encounter one another, their pheromones act on their receptors to arrest the cells in the G1 phase of the cell cycle,

Haploids mate to give diploids

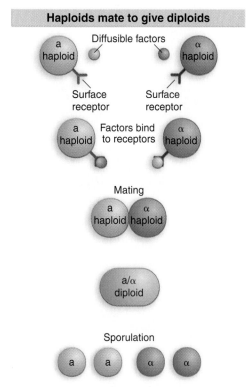

Figure 19.21 The yeast life cycle proceeds through mating of *MATa* and *MATα* haploids to give heterozygous diploids that sporulate to generate haploid spores.

There are silent cassettes for mating type

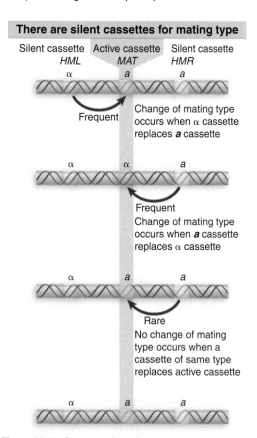

Figure 19.22 Changes of mating type occur when silent cassettes replace active cassettes of opposite genotype; when transpositions occur between cassettes of the same type, the mating type remains unaltered.

and various morphological changes occur. In a successful mating, the cell cycle arrest is followed by cell and nuclear fusion to produce an *a*/α diploid cell.

The *a*/α diploid cell carries both the *MATa* and *MATα* alleles and has the ability to sporulate. **Figure 19.21** demonstrates how this maintains the normal haploid/diploid life cycle. The basic function of the *MAT* locus is to control expression of pheromone and receptor genes; it also has other functions involved in mating. *MATα* codes for two proteins, α1 and α2. *MATa* codes for a single protein, **a1**. The **a** and α proteins directly control transcription of various target genes; they function by both positive and negative regulation. They function independently in haploids, and in conjunction in diploids.

Much of the information about the yeast mating type pathway was deduced from the properties of mutations that eliminate the ability of **a** and/or α cells to mate. The genes identified by such mutations are called *STE* (for sterile). Mutations in the genes for the pheromones or receptors are specific for individual mating types; but mutations in the other *STE* genes eliminate mating in both **a** and α cells. This situation is explained by the fact that the events that follow the interaction of factor with receptor are identical for both types.

Mating is a symmetrical process that is initiated by the interaction of pheromone secreted by one cell type with the receptor carried by the other cell type. The only genes that are uniquely required for the response pathway in a particular mating type are those coding for the receptors. Either the **a** factor–receptor interaction or the α factor–receptor interaction switches on the same response pathway. Mutations that eliminate steps in the common pathway have the same effects in both cell types. The pathway consists of a signal transduction cascade that leads to the synthesis of products that make the necessary changes in cell morphology and gene expression for mating to occur.

Some yeast strains have the remarkable ability to switch their mating types from *MATa* to *MATα* or vice versa. These strains carry a dominant allele *HO* and change their mating type frequently, as often as once every generation.

The existence of switching suggests that all cells contain the potential information needed to be either *MATa* or *MATα*, but express only one type. Where does the information to change mating types come from? Two additional loci are needed for switching. *HMLα* is needed for switching to give a *MATα* type; *HMRa* is needed for switching to give a *MATa* type. These loci lie on the same chromosome that carries *MAT*. *HML* is far to the left; *HMR* far to the right.

The **cassette model** for mating type is illustrated in **Figure 19.22**. The *MAT* locus has an *active cassette* of either type α or type *a*. *HML* and *HMR* have *silent cassettes*. Usually *HML* carries an α cassette, while *HMR* carries an *a* cassette. All cassettes carry information that codes for mating type, but only the active cassette at *MAT* is expressed. Mating-type switching occurs when the active cassette is replaced by information from a silent cassette. The newly installed cassette is then expressed. Switching is nonreciprocal; the copy at *HML* or *HMR* replaces the allele at *MAT*. Switching usually involves replacement of *MATa* by the copy at *HMLα* or replacement of *MATα* by the copy at *HMRa*.

Several groups of genes participate in establishing and switching mating type. As well as the genes that directly determine mating type, they include genes needed to repress the silent cassettes, to switch mating type, or to execute the functions of mating.

19.14 Unidirectional Transposition Is Initiated by the Recipient *MAT* Locus

Key Concept

- Mating type switching is initiated by a double-strand break made at the *MAT* locus by the HO endonuclease.

By comparing the sequences of the two silent cassettes (*HMLα* and *HMRa*) with the sequences of the two types of active cassette (*MATa* and *MATα*), we can delineate the sequences that determine mating type. The organization of the mating type loci is summarized in **Figure 19.23**. Each cassette contains common sequences that flank a central region that differs in the *a* and α types of cassette (called *Ya* or *Yα*). On either side of this region, the flanking sequences are virtually identical, although they are shorter at *HMR*. The active cassette at *MAT* is transcribed from a promoter within the *Y* region.

A switch in mating type is accomplished by a gene conversion in which the recipient site (*MAT*) acquires the sequence of the donor type (*HML* or *HMR*). Switching is initiated when the HO endonuclease makes a double-strand break just on the right of the *Y–Z* boundary. Cleavage generates the single-stranded ends of four bases drawn in **Figure 19.24**. The recognition site is 24 bp, relatively long for a nuclease, and it occurs only at the three mating-type cassettes.

Only the *MAT* locus and not the *HML* or *HMR* loci are targets for the endonuclease. It seems plausible that the same mechanisms that keep the silent cassettes from being transcribed also keep them inaccessible to the HO endonuclease. This inaccessibility ensures that switching is unidirectional.

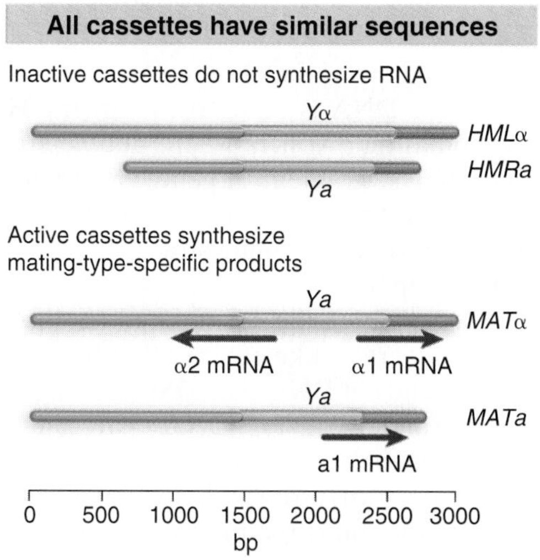

All cassettes have similar sequences

Inactive cassettes do not synthesize RNA

Yα — *HMLα*
Ya — *HMRa*

Active cassettes synthesize mating-type-specific products

Yα — *MATα*
α2 mRNA   α1 mRNA

Ya — *MATa*
a1 mRNA

0   500   1000   1500   2000   2500   3000
bp

Figure 19.23 Silent cassettes have the same sequences as the corresponding active cassettes, except for the absence of the extreme flanking sequences in *HMRa*. Only the *Y* region changes between *a* and α types.

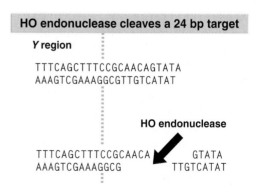

HO endonuclease cleaves a 24 bp target

Y region

TTTCAGCTTTCCGCAACAGTATA
AAAGTCGAAAGGCGTTGTCATAT

HO endonuclease

TTTCAGCTTTCCGCAACA   GTATA
AAAGTCGAAAGGCG   TTGTCATAT

Figure 19.24 HO endonuclease cleaves *MAT* just to the right of the *Y* region, generating sticky ends with a 4–base overhang.

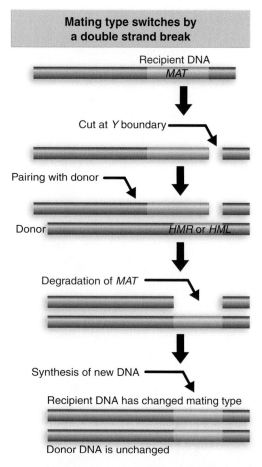

Mating type switches by a double strand break

Recipient DNA
MAT

Cut at *Y* boundary

Pairing with donor

Donor *HMR* or *HML*

Degradation of *MAT*

Synthesis of new DNA

Recipient DNA has changed mating type

Donor DNA is unchanged

Figure 19.25 Cassette substitution is initiated by a double-strand break in the recipient (*MAT*) locus, and may involve pairing on either side of the *Y* region with the donor (*HMR* or *HML*) locus.

The reaction triggered by the cleavage is illustrated schematically in **Figure 19.25** in terms of the general reaction between donor and recipient regions. In terms of the interactions of individual strands of DNA, it follows the scheme for recombination via a double-strand break drawn in Figure 19.5; and the stages following the initial cut require the enzymes that participate in general recombination.

Suppose that the free end of *MAT* invades either the *HML* or *HMR* locus and pairs with the region of homology on the right side. The *Y* region of *MAT* is degraded until a region with homology on the left side is exposed. At this point, *MAT* is paired with *HML* or *HMR* at either the left side and the right side. The *Y* region of *HML* or *HMR* is copied to replace the region lost from *MAT* (which might extend beyond the limits of *Y* itself). The paired loci separate.

Like the double-strand break model for recombination, the process is initiated by *MAT*, the locus that is to be replaced. In this sense, the description of *HML* and *HMR* as donor loci refers to their ultimate role, but not to the mechanism of the process.

19.15 SUMMARY

Recombination involves the physical exchange of parts between corresponding DNA molecules. This results in a duplex DNA in which two regions of opposite parental origins are connected by a stretch of hybrid (heteroduplex) DNA, in which one strand is derived from each parent. Correction events may occur at sites that are mismatched within the hybrid DNA. Hybrid DNA can also be formed without recombination occurring between markers on either side. Gene conversion occurs when an extensive region of hybrid DNA forms during normal recombination (or between nonallelic genes in an aberrant event) and is corrected to the sequence of only one parental strand; then one gene takes on the sequence of the other.

Recombination is initiated by a double-strand break in DNA. The break is enlarged to a gap with a single-stranded end; then the free single-stranded end forms a heteroduplex with the allelic sequence. The DNA in which the break occurs actually incorporates the sequence of the chromosome that it invades, so the initiating DNA is called the recipient. Hotspots for recombination are sites where double-strand breaks are initiated. A gradient of gene conversion is determined by the likelihood that a sequence near the free end will be converted to a single strand; this decreases with distance from the break.

Recombination is initiated in yeast by Spo11, a topoisomerase–like enzyme that becomes linked to the free 5′ ends of DNA. The DSB is then processed to generate single-stranded DNA that can anneal with its complement in the other chromosome. Yeast mutations that block synaptonemal complex formation show that recombination is required for its formation. Formation of the synaptonemal complex may be initiated by double-strand breaks, and it may persist until recombination is completed. Mutations in components of the synaptonemal complex block its formation but do not prevent chromosome pairing, so homologue recognition is independent of recombination and synaptonemal complex formation.

The full set of reactions required for recombination can be undertaken by the Rec and Ruv proteins of *E. coli*. A single-stranded region with a free end is generated by the RecBCD nuclease. The enzyme binds to DNA on one side of a *chi* sequence, and then moves to the *chi* sequence, unwinding DNA as it progresses. A single-strand break is made at the *chi* sequence, which provides a hotspot for recombination. The single strand provides a substrate for RecA, which has the ability to synapse homologous DNA molecules by sponsoring a reaction in which a single strand from one molecule invades a duplex of the other molecule. Heteroduplex DNA is formed by displacing one of the original strands of the duplex. These actions create a recombination junction, which is resolved by the Ruv proteins. RuvA and RuvB act at a heteroduplex, and RuvC cleaves Holliday junctions.

Recombination, like replication and transcription, requires topological manipulation of DNA. Topoisomerases may relax (or introduce) supercoils in DNA, and are required to disentangle DNA molecules that have become catenated by recombination or by replication. Type I topoisomerases introduce a break in one strand of a DNA duplex; type II topoisomerases make double-stranded breaks. The enzyme becomes linked to the DNA by a bond from tyrosine to either 5′ phosphate (type A enzymes) or 3′ phosphate (type B enzymes).

The enzymes for site-specific recombination have actions related to those of topoisomerases. Among this general class of recombinases, those concerned with phage integration form the subclass of integrases. The Cre–*lox* system uses two molecules of Cre to bind to each *lox* site, so that the recombining complex is a tetramer. This is one of the standard systems for inserting DNA into a foreign genome. Phage lambda integration requires the phage Int protein and host IHF protein and involves a precise breakage and reunion in the absence of any synthesis of DNA. The reaction involves wrapping of the *attP* sequence of phage DNA into the nucleoprotein structure of the intasome, which contains several copies of Int and IHF; then the host *attB* sequence is bound, and recombination occurs. Reaction in the reverse direction requires the phage protein Xis. Some integrases function by *cis*-cleavage, where the tyrosine that reacts with DNA in a half site is provided by the enzyme subunit bound to that half site; others function by *trans*-cleavage, in which a different protein subunit provides the tyrosine.

Yeast mating type is determined by whether the *MAT* locus carries the *a* or α sequence. Expression in haploid cells of the sequence at *MAT* leads to expression of genes specific for the mating type and to repression of genes specific for the other mating type. Additional, silent copies of the mating-type sequences are carried at the loci *HML*α and *HMR*a. Cells that carry the HO endonuclease display a unidirectional transfer process in which the sequence at *HML*α replaces an *a* sequence at *MAT*, or the sequence at *HMR*a replaces an α sequence at *MAT*. The endonuclease makes a double-strand break at *MAT*, and a free end invades either *HML*α or *HMR*a. *MAT* initiates the transfer process, but is the recipient of the new sequence.

20

Repair Systems Handle Damage to DNA

20.1 Introduction

Key Concepts

- Repair systems recognize DNA sequences that do not conform to standard base pairs.

- Excision systems remove one strand of DNA at the site of damage and then replace it.

- Recombination-repair systems use recombination to replace the double-stranded region that has been damaged.

- Repair systems may be prone to introducing errors during the repair process.

- Photoreactivation is a nonmutagenic repair system that acts specifically on pyrimidine dimers.

Any event that introduces a deviation from the usual double-helical structure of DNA is a threat to the genetic constitution of the cell. Injury to DNA is minimized by systems that recognize and correct the damage. The repair systems are as complex as the replication apparatus itself, which indicates their importance for the survival of the cell. When a repair system reverses a change to DNA, there is no consequence. But a mutation may result when it fails to do so. The measured rate of mutation reflects a balance between the number of damaging events occurring in DNA and the number that have been corrected (or miscorrected).

Repair systems often can recognize a range of distortions in DNA as signals for action, and a cell is likely to have several systems able to deal with DNA damage. The importance of DNA repair in eukaryotes is indicated by the identification of >130 repair genes in the human genome. We may divide the repair systems into several general types, as summarized in **Figure 20.1**:

- Some enzymes directly reverse specific sorts of damage to DNA.
- There are pathways for base excision repair, nucleotide excision repair, and mismatch repair, all of which function by removing and replacing material.
- There are systems that function by using recombination to retrieve an undamaged copy that is then used to replace a damaged duplex sequence.
- The nonhomologous end-joining pathway rejoins broken double-stranded ends.
- Several different DNA polymerases can resynthesize stretches of replacement DNA.

Direct repair is rare and involves the reversal or simple removal of the damage. *Photoreactivation* of pyrimidine dimers, in which the offending covalent bonds are reversed by a light-dependent enzyme, is a good example. This system is widespread in nature, and appears to be especially important in plants. In *E. coli* it depends on the product of a single gene (*phr*) that codes for an enzyme called photolyase.

Mismatches between the strands of DNA are one of the major targets for repair systems. *Mismatch repair* is accomplished by scrutinizing DNA for apposed bases that do not pair properly. Mismatches that arise during replication are corrected by distinguishing between the "new" and "old" strands and preferentially correcting the sequence of the newly synthesized strand. Other systems deal with mismatches generated by base conversions, such as the result of deamination. The importance of these systems is emphasized by the fact that cancer is caused in human populations by mutation of genes related to those involved in mismatch repair in yeast.

Mismatches are usually corrected by *excision repair*, which is initiated by a recognition enzyme that sees an actual damaged base or a change in the spatial path of DNA. There are two types of excision repair system.

- *Base excision repair* systems directly remove the damaged base and replace it in DNA. A good example is DNA uracil glycolase, which removes uracils that are mispaired with guanines (see *20.4 Base Flipping Is Used by Methylases and Glycosylases*).
- *Nucleotide excision repair* systems excise a sequence that includes the damaged base(s); then a new stretch of DNA is synthesized to replace the excised material. **Figure 20.2** summarizes the main events in the operation of such a system. Such systems are common. Some recognize general damage to DNA. Others act upon specific types of base damage. There are often multiple excision repair systems in a single cell type.

Recombination-repair systems handle situations in which damage remains in a daughter molecule, and replication has been forced to bypass the site, typically creating a gap in the daughter strand. A retrieval system uses recombination to obtain another copy of the sequence from an undamaged source; the copy is then used to repair the gap.

A major feature in recombination and repair is the need to handle double-strand breaks. DSBs initiate crossovers in homologous recombination. They can also be created by problems in replication, when they may trigger the use of recombination-repair systems. When DSBs are created by environmental damage (for example, by radiation damage) or because of the shortening of telomeres, they can cause mutations. One system for handling DSBs can join together nonhomologous DNA ends.

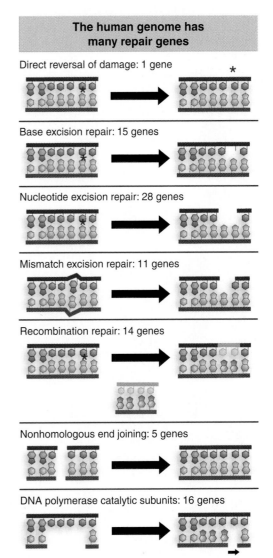

The human genome has many repair genes

Direct reversal of damage: 1 gene

Base excision repair: 15 genes

Nucleotide excision repair: 28 genes

Mismatch excision repair: 11 genes

Recombination repair: 14 genes

Nonhomologous end joining: 5 genes

DNA polymerase catalytic subunits: 16 genes

Figure 20.1 Repair genes can be classified into pathways that use different mechanisms to reverse or bypass damage to DNA.

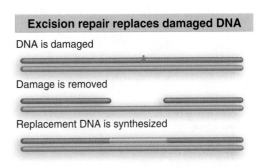

Excision repair replaces damaged DNA

DNA is damaged

Damage is removed

Replacement DNA is synthesized

Figure 20.2 Excision repair directly removes damaged DNA and then resynthesizes a replacement stretch for the damaged strand.

Mutations that affect the ability of *E. coli* cells to engage in DNA repair fall into groups, which correspond to several repair pathways (not necessarily all independent). The major known pathways are the *uvr* excision repair system, the methyl-directed mismatch-repair system, and the *recB* and *recF* recombination and recombination-repair pathways. The enzyme activities associated with these systems are endonucleases and exonucleases (important in removing damaged DNA), resolvases (endonucleases that act specifically on recombinant junctions), helicases to unwind DNA, and DNA polymerases to synthesize new DNA. Some of these enzyme activities are unique to particular repair pathways, but others participate in multiple pathways.

The replication apparatus devotes a lot of attention to quality control. DNA polymerases use proofreading to check the daughter strand sequence and to remove errors. Some of the repair systems are less accurate when they synthesize DNA to replace damaged material. For this reason, these systems have been known historically as *error-prone* systems.

20.2 Mutational Damage Falls into Two General Types

Key Terms

- A **structural distortion** is a change in the conformation of DNA caused by bases or base pairs that do not fit into the normal duplex.
- A **pyrimidine dimer** is formed when ultraviolet irradiation generates a covalent link directly between two adjacent pyrimidine bases in DNA. It blocks DNA replication and transcription.

Key Concepts

- Single base substitutions create a mispair that persists only until the next replication.
- Covalent adducts or nicks block transcription and replication, and cause new errors in every replication cycle.

The types of damage that trigger repair systems can be divided into two general classes: changes that affect only a single base; and structural distortions, such as covalent links between bases or nicks in one strand.

Single base changes affect the sequence but not the overall structure of DNA. They do not affect transcription or replication when the strands of the DNA duplex are separated. So these changes exert their damaging effects on future generations through the consequences of the change in DNA sequence. The cause of this type of effect is the conversion of one base into another that is not properly paired with the partner base. They may be happen as the result of mutation of a base *in situ* or by replication errors. **Figure 20.3** shows that deamination of cytosine to uracil (spontaneously or by a chemical mutagen) creates a mismatched U · G pair. **Figure 20.4** shows that a replication error might insert adenine instead of cytosine to create an A · G pair. These changes may result in very minor structural distortion (as in the case of a U · G pair) or

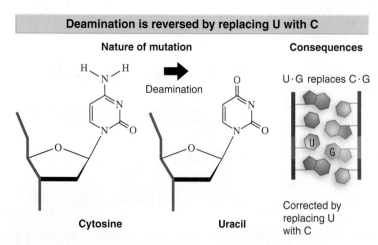

Figure 20.3 Deamination of cytosine creates a U · G base pair. Uracil is preferentially removed from the mismatched pair.

quite significant change (as in the case of an A·G pair), but the common feature is that *the mismatch persists only until the next replication*. If the damage is repaired before then, there are no consequences. But if it is not repaired, one of the products has a mutation that will be perpetuated in its descendants. Base substitutions typically provide substrates for mismatch repair systems that excise one of the strands and replace it with an exact match to the other strand.

Structural distortions may provide a physical impediment to replication or transcription. Introduction of covalent links between bases on one strand of DNA or between bases on opposite strands inhibits replication and transcription. **Figure 20.5** shows the example of ultraviolet irradiation, which introduces covalent bonds between two adjacent thymine bases, giving an intrastrand **pyrimidine dimer**. **Figure 20.6** shows that similar consequences could result from addition of a bulky adduct to a base that distorts the structure of the double helix. A single-strand nick or the removal of a base, as shown in **Figure 20.7**, prevents a strand from serving as a proper template for synthesis of RNA or DNA. The common feature in all these changes is that *the damaged adduct remains in the DNA*, continuing to cause structural problems and/or induce mutations, until it is removed. Some types of structural damage are repaired by removing the nucleotide, or a stretch of DNA containing it, and resynthesizing a replacement stretch. Others, including nicks, form substrates for systems that use recombination to provide a replacement.

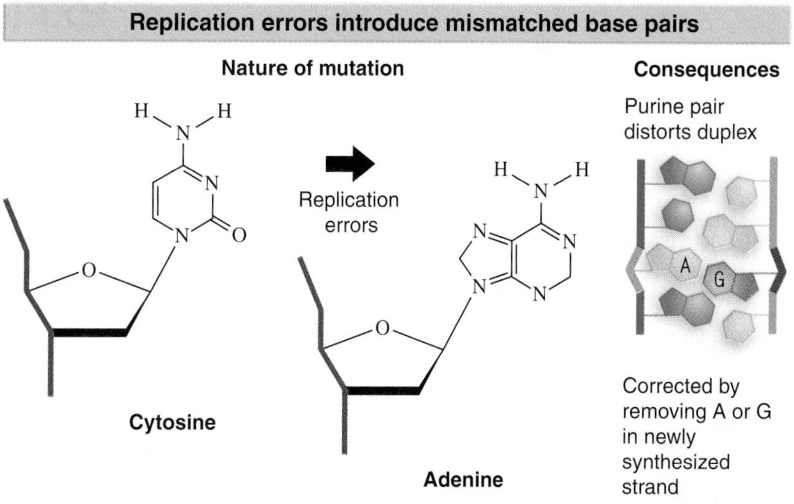

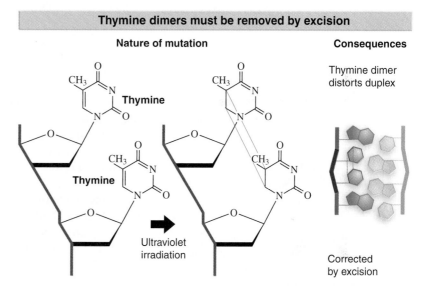

Figure 20.4 A replication error creates a mismatched pair that may be corrected by replacing one base; if uncorrected, a mutation is fixed in one daughter duplex.

Figure 20.5 Ultraviolet irradiation causes dimer formation between adjacent thymines. The dimer blocks replication and transcription.

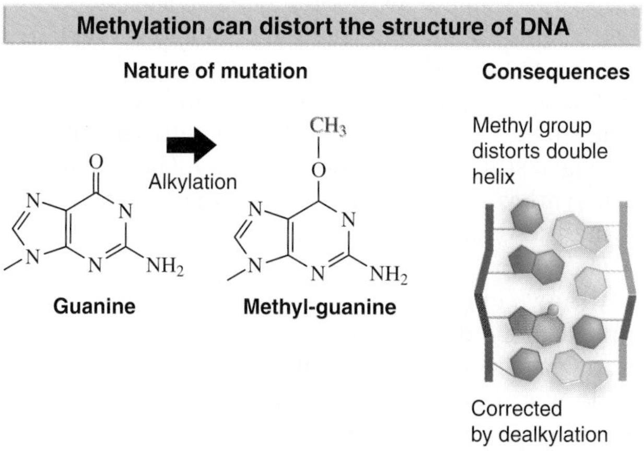

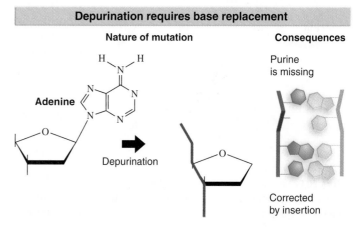

Figure 20.6 Methylation of a base distorts the double helix and causes mispairing at replication.

Figure 20.7 Depurination removes a base from DNA, blocking replication and transcription.

Excision repair replaces a damaged strand

Damage–Mutant base is mismatched and/or distorts structure

Incision–Endonuclease cleaves on both sides of damaged base

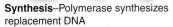

Excision–Endonuclease removes DNA between nicks

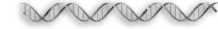

Synthesis–Polymerase synthesizes replacement DNA

Ligase seals nick

Figure 20.8 Excision repair removes and replaces a stretch of DNA that includes the damaged base(s).

The Uvr system removes damaged material

UvrA scans DNA and looks for damage

UvrA recognizes damage and binds with UvrB

UvrB

UvrA is released; UvrC binds

UvrC nicks DNA on both sides of damage

UvrD unwinds region and releases damaged strand

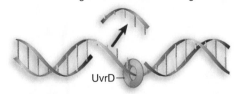

UvrD

Figure 20.9 The Uvr system operates in stages in which UvrAB recognizes damage, UvrBC nicks the DNA, and UvrD unwinds the marked region.

20.3 Excision Repair Systems in *E. coli*

Key Terms

- **Excision repair** describes a type of repair system in which one strand of DNA is directly excised and then replaced by resynthesis using the complementary strand as template.

- **Incision** is a step in a mismatch excision-repair system. An endonuclease recognizes the damaged area in the DNA, and isolates it by cutting the DNA strand on both sides of the damage.

- The **excision** step in an excision-repair system consists of removing a single-stranded stretch of DNA by the action of a 5′–3′ exonuclease.

Key Concept

- The Uvr system makes incisions ~12 bases apart on both sides of damaged DNA, removes the DNA between them, and resynthesizes new DNA.

Excision-repair systems vary in their specificity, but share the same general features. Each system removes mispaired or damaged bases from DNA and then synthesizes a new stretch of DNA to replace them. The main type of pathway for excision repair is illustrated in **Figure 20.8**.

In the **incision** step, the damaged structure is recognized by an endonuclease that cleaves the DNA strand on both sides of the damage.

In the **excision** step, a 5′–3′ exonuclease removes a stretch of the damaged strand.

In the *synthesis* step, the resulting single-stranded region serves as a template for a DNA polymerase to synthesize a replacement for the excised sequence. (Synthesis of the new strand could be associated with removal of the old strand, in one coordinated action.) Finally, DNA ligase covalently links the 3′ end of the new material to the old material.

Virtually all of the excision-repair events in *E. coli* are undertaken by the *uvr* system of excision repair. The genes *uvrA,B,C* code for the components of a repair endonuclease. It functions in the stages indicated in **Figure 20.9**. First, a UvrAB complex recognizes lesions in DNA. Then UvrA dissociates (this requires ATP), and UvrC joins UvrB. The UvrBC complex makes an incision on each side, one 7 nucleotides from the 5′ side of the damaged site, and the other 3–4 nucleotides away from the 3′ side. This also requires ATP. Another Uvr protein, UvrD, is a helicase that helps to unwind the DNA to allow release of the single strand between the two cuts. The enzyme that excises the damaged strand is DNA polymerase I. The enzyme involved in the repair synthesis probably also is DNA polymerase I (although DNA polymerases II and III can substitute for it). The events are basically the same, although their order is different, in the eukaryotic repair pathway shown in Figure 20.20.

The Uvr complex can be directed to damaged sites by other proteins. For example, damage to DNA may prevent transcription, but the situation is handled by a protein called Mfd that displaces the RNA polymerase and recruits the Uvr complex.

20.4 Base Flipping Is Used by Methylases and Glycosylases

Key Concepts

- Uracil and alkylated bases are recognized by glycosylases and removed directly from DNA.
- Pyrimidine dimers are reversed by breaking the covalent bonds between them.
- Methylases add a methyl group to cytosine.
- Methylases and glycosylases act by flipping the base out of the double helix, where, depending on the reaction, it is either removed or is modified and returned to the helix.

Glycosylases and lyases are enzymes that can directly remove bases from a polynucleotide chain. **Figure 20.10** shows that a glycosylase cleaves the bond between the damaged or mismatched base and the deoxyribose. A lyase takes the reaction further by opening the sugar ring.

The interaction of these enzymes with DNA is remarkable. It follows the model first demonstrated for methyltransferases—enzymes that add a methyl group to cytosine in DNA. The methylase flips the target cytosine completely out of the helix. **Figure 20.11** shows that it enters a cavity in the enzyme where it is modified. Then it is returned to its normal position in the helix. All this occurs without input of an external energy source.

One of the most common reactions in which a base is directly removed from DNA is catalyzed by uracil-DNA glycosylase. Uracil occurs in DNA most typically because of a spontaneous deamination of cytosine. It is recognized by the glycosylase and removed. The reaction is similar to that of the methylase: the uracil is flipped out of the helix and into the active site in the glycosylase.

Alkylated bases (typically in which a methyl group has been added to a base) are removed by a similar mechanism. A single human enzyme, alkyladenine DNA glycosylase (AAG) recognizes and removes a variety of alkylated substrates, including 3-methyladenine, 7-methylguanine, and hypoxanthine.

Another enzyme to use base flipping is the photolyase that reverses the bonds between pyrimidine dimers (see Figure 20.5). The pyrimidine dimer is flipped into a cavity in the enzyme. Close to this cavity is an active site that contains an electron donor, which provides the electrons to

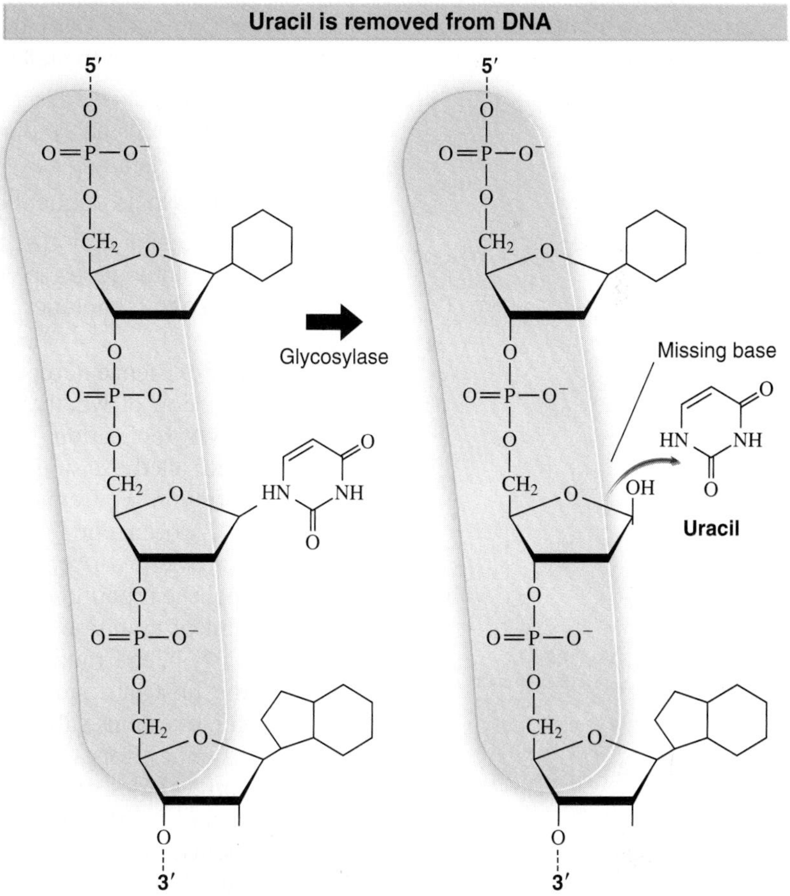

Figure 20.10 A glycosylase removes a base from DNA by cleaving the bond to the deoxyribose.

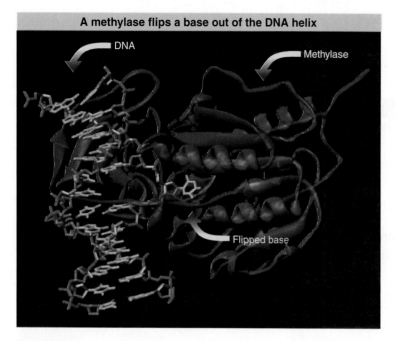

A methylase flips a base out of the DNA helix

DNA

Methylase

Flipped base

Figure 20.11 A methylase flips the target cytosine out of the double helix in order to modify it. Photograph kindly provided by Rich Roberts.

break the bonds. Energy for the reaction is provided by light in the visible wavelength.

The common feature of all of these enzymes is the flipping of the target base into the enzyme structure. A variation on this theme is used by T4 endonuclease V, now renamed T4-pdg (pyrimidine dimer glycosylase) to reflect its mode of action. It flips out the adenine base that is *complementary* to the thymine on the 5′ side of the pyrimidine dimer. So in this case, the target for the catalytic action of the enzyme remains in the DNA duplex, and the enzyme uses flipping as an indirect mechanism to get access to its target.

When a base is removed from DNA, the reaction is followed by excision of the phosphodiester backbone by an endonuclease, DNA synthesis by a DNA polymerase to fill the gap, and ligation by a ligase to restore the integrity of the polynucleotide chain.

20.5 Error-Prone Repair and Mutator Phenotypes

Key Term

- **Error-prone synthesis** occurs when DNA incorporates noncomplementary bases into the daughter strand.

Key Concepts

- Damaged DNA that has not been repaired causes DNA polymerase III to stall during replication.
- DNA polymerase V (coded by *umuCD*), or DNA polymerase IV (coded by *dinB*) can synthesize a complement to the damaged strand.
- The DNA synthesized by the repair DNA polymerase often has errors in its sequence.
- Proteins that affect the fidelity of replication may be identified by mutator genes, in which mutation causes an increased rate of spontaneous mutation.

The existence of repair systems that engage in DNA synthesis raises the question of whether their quality control is comparable with that of DNA replication. So far as we know, most systems, including *uvr*-controlled excision repair, do not differ significantly from DNA replication in the frequency of mistakes. However, **error-prone synthesis** of DNA occurs in *E. coli* under certain circumstances.

The error-prone synthesis was first observed when it was found that the repair of damaged λ phage DNA is accompanied by the induction of mutations if the phage is introduced into cells that had previously been irradiated with UV. This suggests that the UV irradiation of the host has activated functions that generate mutations. The mutagenic response also operates on the bacterial host DNA.

The error-prone activity comes from a DNA polymerase that inserts incorrect bases, which represent mutations, when it passes any site at which it cannot insert complementary base pairs in the daughter strand. The polymerase is coded by the genes *umuD* and *umuC*, whose expression is induced by DNA damage. Their products form a complex UmuD′₂C, consisting of two subunits of a truncated UmuD protein and one subunit of UmuC. UmuD is cleaved by RecA, which is activated by DNA damage.

The UmuD′₂C complex is DNA polymerase V, and is responsible for synthesizing new DNA to replace sequences that have been damaged by UV. This is the only enzyme in *E. coli* that can bypass the classic pyrimidine dimers produced by UV (or other bulky adducts). The polymerase activity is "error prone." Mutations of *umuC* or *umuD* inactivate the enzyme, which makes UV irradiation lethal. Some plasmids carry genes called *mucA* and *mucB* that are homologues of *umuD* and *umuC;* introduction of these genes into a bacterium increases resistance to UV killing and susceptibility to mutagenesis.

How does an alternative DNA polymerase get access to the DNA? When the replicase (DNA polymerase III) encounters a block, such as a thymidine dimer, it stalls. Then it is displaced from the replication fork and replaced by DNA polymerase V. In fact, DNA polymerase V uses some of the same ancillary proteins as DNA polymerase III. The same situation is true for DNA polymerase IV, the product of *dinB,* which is another enzyme that acts on damaged DNA. DNA polymerases IV and V are part of a larger family, including eukaryotic DNA polymerases, that are involved in repairing damaged DNA (see *20.9 Eukaryotic Cells Have Conserved Repair Systems*).

20.6 Controlling the Direction of Mismatch Repair

Key Concepts

- The *mut* genes code for a mismatch repair system that deals with mismatched base pairs.
- There is a bias in the selection of which strand to replace at mismatches.
- The strand lacking methylation at a hemimethylated $\frac{GATC}{CTAG}$, is usually replaced.
- At G·T and C·T mismatches, the T is preferentially removed.

When a structural distortion is removed from DNA, the wild-type sequence is restored. In most cases, the distortion is due to the creation of a base that is not naturally found in DNA, which is therefore recognized and removed by the repair system.

A problem arises if the target for repair is a mispaired partnership of normal bases created when one was mutated. The repair system has no intrinsic means of knowing which is the wild-type base and which is the mutant! All it sees are two improperly paired bases, either of which can provide the target for excision repair.

If the mutated base is excised, the wild-type sequence is restored. But if it happens to be the unmutated base that is excised, the new mutant sequence becomes fixed. Often, however, the direction of excision repair is not random, but is biased in a way that is likely to lead to restoration of the wild-type sequence.

Some precautions are taken to direct repair in the right direction. For example, for cases such as the deamination of 5-methyl-cytosine to thymine, there is a special system to restore the proper sequence (see also *1.17 Many Hotspots Result from Modified Bases*). The deamination generates a G·T pair, and the system that acts on such pairs has a bias to correct them to G·C pairs (rather than to A·T pairs). The system that undertakes this reaction includes the *mutL,S* products that remove T from both G·T and C·T mismatches.

The *mutT,M,Y* system handles the consequences of oxidative damage. A major type of chemical damage is caused by oxidation of G to 8-oxo-G. **Figure 20.12** shows that the system operates at three levels. MutT hydrolyzes

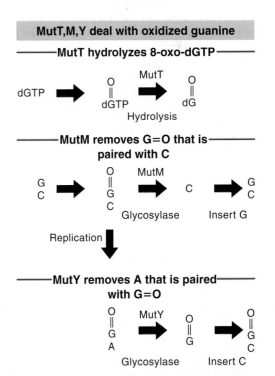

Figure 20.12 Preferential removal of bases in pairs that have oxidized guanine is designed to minimize mutations.

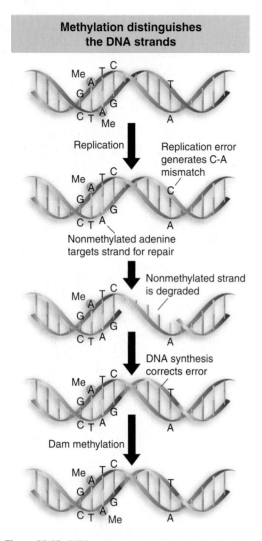

Figure 20.13 GATC sequences are targets for the Dam methylase after replication. During the period before this methylation occurs, the nonmethylated strand is the target for repair of mismatched bases.

the damaged precursor (8-oxo-dGTP), which prevents it from being incorporated into DNA. When guanine is oxidized in DNA, its partner is cytosine; and MutM preferentially removes the 8-oxo-G from 8-oxo-G · C pairs. Oxidized guanine mispairs with A, and so when 8-oxo-G survives and is replicated, it generates an 8-oxo-G · A pair. MutY removes A from these pairs. MutM and MutY are glycosylases that directly remove a base from DNA. This creates an apurinic site that is recognized by an endonuclease whose action triggers the involvement of the excision repair system.

When mismatch errors occur during replication in *E. coli*, it is possible to distinguish the original strand of DNA. Immediately after replication of methylated DNA, only the original parental strand carries the methyl groups. In the period while the newly synthesized strand awaits the introduction of methyl groups, the two strands can be distinguished.

This provides the basis for a system to correct replication errors. The *dam* gene codes for a methylase whose target is the adenine in the sequence $\frac{GATC}{CTAG}$ (see Figure 20.6). The hemimethylated state is used to distinguish replicated origins from nonreplicated origins. The same target sites are used by a replication-related repair system.

Figure 20.13 shows that DNA containing mismatched base pairs is repaired preferentially by excising the strand that lacks the methylation. The excision is quite extensive; mismatches can be repaired preferentially for >1 kb around a GATC site. The result is that the newly synthesized strand is corrected to the sequence of the parental strand.

This repair system reduces the number of mutations caused by errors in replication. It consists of several proteins, coded by the *mut* genes. MutS binds to the mismatch and is joined by MutL. MutS can use two DNA-binding sites, as illustrated in **Figure 20.14**. The first specifically recognizes mismatches. The second is not specific for sequence or structure, and is used to translocate along DNA until a GATC

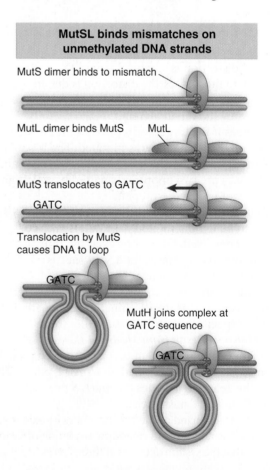

Figure 20.14 MutS recognizes a mismatch and translocates to a GATC site. MutH will cleave the unmethylated strand at the GATC.

sequence is encountered. Hydrolysis of ATP is used to drive the translocation. Because MutS is bound to both the mismatch site and to DNA as it translocates, it creates a loop in the DNA.

Recognition of the GATC sequence causes the MutH endonuclease to bind to MutSL. The endonuclease then cleaves the unmethylated strand. This strand is then excised from the GATC site to the mismatch site. The excision can occur in either the 5′–3′ direction (using RecJ or exonuclease VII) or in the 3′–5′ direction (using exonuclease I), assisted by the helicase UvrD. The new DNA strand is synthesized by DNA polymerase III.

Homologues of the MutSL system are found in higher eukaryotic cells. They are responsible for repairing mismatches that arise as the result of replication slippage. In a region such as a microsatellite where a very short sequence is repeated several times, realignment between the newly synthesized daughter strand and its template can lead to a stuttering in which the DNA polymerase slips backward and synthesizes extra repeating units. These units in the daughter strand are extruded as a single-stranded loop from the double helix (see Figure 6.25). They are repaired by homologues of the MutSL system, as shown in **Figure 20.15**.

The importance of the MutSL system for mismatch repair is indicated by the high rate at which it is found to be defective in human cancers. Loss of this system leads to an increased mutation rate (see *20.11 Defects in Repair Systems Cause Mutations to Accumulate in Tumors*).

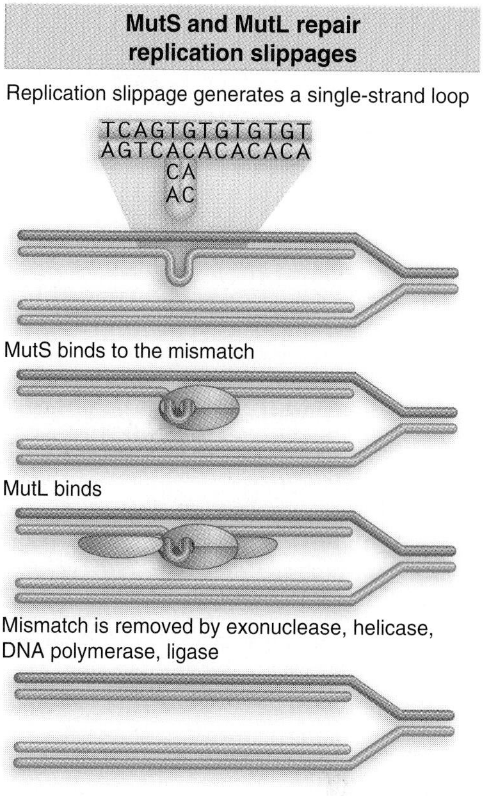

20.7 Recombination-Repair Systems in *E. coli*

Key Terms

- **Recombination repair** is a mode of filling a gap in one strand of duplex DNA by retrieving a homologous single strand from another duplex.
- **Single-strand exchange** is a reaction in which one of the strands of a duplex of DNA leaves its former partner and instead pairs with the complementary strand in another molecule, displacing its homologue in the second duplex.

Key Concepts

- The *rec* genes of *E. coli* code for the principal retrieval system.
- The retrieval system functions when replication leaves a gap in the newly synthesized strand opposite a damaged sequence.
- The single strand of another duplex is used to replace the gap.
- Then the damaged sequence is removed and resynthesized.

Recombination-repair systems use activities that overlap with those of genetic recombination. These systems are sometimes called "post-replication repair," because they function after replication. Such systems are effective in dealing with the defects produced in daughter duplexes by replication of a template that contains damaged bases. An example is illustrated in **Figure 20.16**. Restarting stalled replication forks could be the major role of the recombination-repair systems (see *18.14 The Primosome Is Needed to Restart Replication*).

Consider a structural distortion, such as a pyrimidine dimer, on one strand of a double helix. When the DNA is replicated, the dimer prevents the damaged site from acting as a template. Replication is forced to skip past it.

DNA polymerase probably proceeds up to or close to the pyrimidine dimer. Then the polymerase ceases synthesis of the corresponding

---

Replication slippage generates a single-strand loop

MutS binds to the mismatch

MutL binds

Mismatch is removed by exonuclease, helicase, DNA polymerase, ligase

Figure 20.15 The MutS/MutSL system initiates repair of mismatches produced by replication slippage.

Recombination-repair uses two duplexes

Damage—Bases on one strand of DNA are damaged

Replication—Creates one copy with a gap opposite the damage and one normal copy

Retrieval—The gap is repaired by retrieving the correct sequence from the normal copy

The gap in the normal copy is repaired

Figure 20.16 An *E. coli* retrieval system uses a normal strand of DNA to replace the gap left in a newly synthesized strand opposite a site of unrepaired damage.

daughter strand. Replication restarts some distance farther along. A substantial gap is left in the newly synthesized strand.

The resulting daughter duplexes are different. One has the parental strand containing the damaged adduct, facing a newly synthesized strand with a lengthy gap. The other duplicate has the undamaged parental strand, which has been copied into a normal complementary strand. The retrieval system takes advantage of the normal daughter strand.

The gap opposite the damaged site in the first duplex is filled by stealing the homologous single strand of DNA from the normal duplex. Following this **single-strand exchange**, the recipient duplex has a parental (damaged) strand facing a wild-type strand. The donor duplex has a normal parental strand facing a gap; the gap can be filled by repair synthesis in the usual way, generating a normal duplex. So the damage is confined to the original distortion (although the same recombination-repair events must be repeated after every replication cycle unless and until the damage is removed by an excision-repair system).

Two pathways in *E. coli* undertake recombination repair in different circumstances. The RecBC pathway is involved in restarting stalled replication forks (see next section). The RecF pathway is involved in repairing the gaps in a daughter strand that are left after replication past a pyrimidine dimer. The RecBC and RecF pathways both lead to the association of RecA with a single-stranded DNA. The ability of RecA to exchange single strands allows it to perform the retrieval step in Figure 20.16. Nuclease and polymerase activities then complete the repair action.

The RecF pathway contains a group of three genes: *recF*, *recO*, and *recR*. The proteins form two types of complex, RecOR and RecOF. They promote the formation of RecA filaments on single-stranded DNA. One of their functions is to make it possible for the filaments to assemble in spite of the presence of the SSB (single-strand binding protein), which is inhibitory.

20.8 Recombination Is Important for Correcting Replication Errors

Key Concepts

- A replication fork may stall when it encounters a damaged site or a nick in DNA.
- A stalled fork may reverse by pairing between the two newly synthesized strands.
- A stalled fork may restart after repairing the damage and using a helicase to move the fork forward.
- The structure of the stalled fork is the same as a Holliday junction and may be converted to a duplex and DSB by resolvases.

All cells have many pathways to repair damage in DNA. Which pathway is used depends upon the type of damage and the situation. Excision-repair pathways can in principle be used at any time, but recombination repair can be used only when there is a second duplex with a copy of the damaged sequence, that is, postreplication. A special situation is presented when damaged DNA is replicated, because the replication fork may stall at the site of damage. Recombination-repair pathways allow the fork to be restored after the damage has been repaired or allow it to bypass the damage.

Figure 20.17 shows one possible outcome when a replication fork stalls. The fork stops moving forward when it encounters the damage.

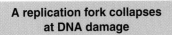

A replication fork collapses at DNA damage

Replication fork stalls at damaged site

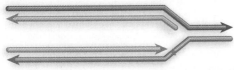

Replication fork reverses and collapses

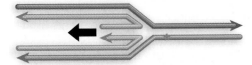

Damage is repaired

Helicase restores replication fork

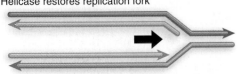

Figure 20.17 A replication fork stalls when it reaches a damaged site in DNA. Reversing the fork allows the two daughter strands to pair. After the damage has been repaired, the fork is restored by forward branch migration catalyzed by a helicase. Arrowheads indicate 3′ ends.

The replication apparatus disassembles, at least partially. This allows branch migration to occur, when the fork effectively moves backward, and the new daughter strands pair to form a duplex structure. After the damage has been repaired, a helicase rolls the fork forward to restore its structure. Then the replication apparatus can reassemble, and replication is restarted (see *18.14 The Primosome Is Needed to Restart Replication*).

The pathway for handling a stalled replication fork requires repair enzymes. In *E. coli*, RecA and the RecBC system have an important role in this reaction (in fact, this may be their major function in the bacterium). One possible pathway is for RecA to bind to single-stranded DNA at the stalled replication fork, stabilizing it, and possibly acting as the sensor that detects the stalling event. RecBC is involved in excision repair of the damage. After the damage has been repaired, replication can resume.

Another pathway may use recombination repair, possibly the strand-exchange reactions of RecA. **Figure 20.18** shows that the structure of the stalled fork is essentially the same as a Holliday junction created by recombination between two duplex DNAs. This makes it a target for resolvases. A double-strand break is generated if a resolvase cleaves either pair of complementary strands. In addition, if the damage is in fact a nick, another double-strand break is created at this site.

Stalled replication forks can be rescued by recombination repair. We don't know the exact sequence of events, but one possible scenario is outlined in **Figure 20.19**. The principle is that a recombination event occurs on either side of the damaged site, allowing an undamaged single strand to pair with the damaged strand. This allows the replication fork to be reconstructed, so that replication can continue, effectively bypassing the damaged site.

In *E. coli*, the RecBC system has an important role in recombination-repair at stalled replication forks. RecBC is involved in generating a single strand end on one daughter duplex, which RecA can then cause to pair with the other daughter duplex.

20.9 Eukaryotic Cells Have Conserved Repair Systems

Key Concepts

- The yeast *RAD* mutations, identified by radiation-sensitive phenotypes, are in genes that code for repair systems.
- Xeroderma pigmentosum is a human disease caused by mutations in any one of several repair genes.
- Transcriptionally active genes are preferentially repaired.

The types of repair functions recognized in *E. coli* are common to a wide range of organisms. The best characterized eukaryotic systems are in yeast, where Rad51 is the counterpart to RecA. In yeast, the main function of the strand-transfer protein is homologous recombination. Many of the repair systems found in yeast have direct counterparts in higher eukaryotic cells, and in several cases these systems are involved with human diseases (see also *20.11 Defects in Repair Systems Cause Mutations to Accumulate in Tumors*).

Genes involved in repair functions have been characterized genetically in yeast by virtue of their sensitivity to radiation. They are called *RAD* genes. There are three general groups of repair genes in the yeast *S. cerevisiae*, identified by the *RAD3* group (involved in excision repair),

Double strand breaks can be generated at stalled forks

Replication fork stalls at damaged site

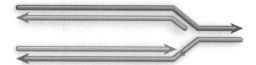

Replication fork reverses and collapses

A resolvase cuts at the junction

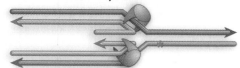

1. A double-strand break has been created

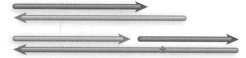

2. Another double-strand break is created if the damage is a nick

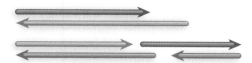

Figure 20.18 The structure of a stalled replication fork resembles a Holliday junction and can be resolved in the same way by resolvases. The results depend on whether the site of damage contains a nick. Result 1 shows that a double-strand break is generated by cutting a pair of strands at the junction. Result 2 shows a second DSB is generated at the site of damage if it contains a nick. Arrowheads indicate 3′ ends.

Recombination restarts replication

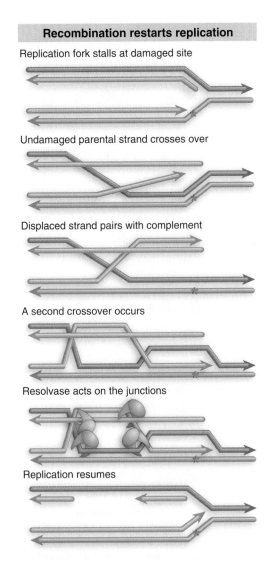

Replication fork stalls at damaged site

Undamaged parental strand crosses over

Displaced strand pairs with complement

A second crossover occurs

Resolvase acts on the junctions

Replication resumes

Figure 20.19 When a replication fork stalls, recombination-repair can place an undamaged strand opposite the damaged site. This allows replication to continue.

A repair system is associated with TFIIH

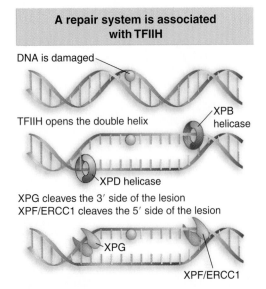

DNA is damaged

TFIIH opens the double helix

XPB helicase

XPD helicase

XPG cleaves the 3′ side of the lesion
XPF/ERCC1 cleaves the 5′ side of the lesion

XPG

XPF/ERCC1

Figure 20.20 Helicases unwind DNA at a damaged site and endonucleases cut on either side of the lesion.

the *RAD6* group (required for post-replication repair), and the *RAD52* group (concerned with recombination–like mechanisms). The RAD52 group is divided into two subgroups by a difference in mutant phenotypes. In one subgroup homologous recombination is affected, as seen by a reduction in mitotic recombination in *RAD52*, *RAD51*, *RAD54*, *RAD55*, and *RAD57*. By contrast, recombination rates are increased in *RAD50*, *MRE11*, and *XRS2* mutants; this subgroup is not deficient in homologous recombination, but is deficient in nonhomologous-DNA-joining reactions.

A superfamily of DNA polymerases involved in synthesizing DNA to replace material at damaged sites is identified by the *dinB* and *umuCD* genes that code for DNA polymerases IV and V in *E. coli*, the *rad30* gene coding for DNA polymerase η of *S. cerevisiae*, and the human homologue *XPV*. Unlike the bacterial enzyme, the eukaryotic enzymes are not error-prone at thymine dimers: they accurately introduce an A–A pair opposite a T–T dimer. When they replicate through other sites of damage, however, they are more prone to introduce errors.

An interesting feature of repair that has been best characterized in yeast is its connection with transcription. Transcriptionally active genes are preferentially repaired on the transcribed strand (removing the impediment to transcription). The cause appears to be a mechanistic connection between the repair apparatus and RNA polymerase. The RAD3 protein, which is a helicase required for the incision step, is a component of a transcription factor associated with RNA polymerase.

An indication of the existence and importance of the mammalian repair systems is given by certain human hereditary disorders. The best investigated of these is xeroderma pigmentosum (XP), a recessive disease resulting in hypersensitivity to sunlight, in particular to ultraviolet. The deficiency results in skin disorders (and sometimes more severe defects).

The disease is caused by a deficiency in excision repair, and can result from mutations in any one of eight genes, called *XP–A* to *XP–G*. They have homologues in the *RAD* genes of yeast, showing that this pathway is widely used in eukaryotes.

A protein complex that includes products of several *XP* genes is responsible for excision of thymine dimers. **Figure 20.20** shows its role in the repair pathway. The complex binds to DNA at a site of damage. Then the strands of DNA are unwound for ~20 bp around the damaged site. This action is undertaken by the helicase activity of the transcription factor TF$_{II}$H, itself a large complex, which includes the products of *XPB* and *XPD* genes, and which is involved with the repair of damaged DNA that is encountered by RNA polymerase during transcription. Then cleavages are made on either side of the lesion by an endonucleases coded by *XPG* and *XPF*. The single-stranded stretch including the damaged bases is then replaced by new synthesis.

In cases where replication encounters a thymine dimer that has not been removed, replication requires the DNA polymerase η activity in order to proceed past the dimer. This is coded by the *XPV* gene. Skin cancers that occur in *XPV* mutants are presumably due to loss of the DNA polymerase.

20.10 A Common System Repairs Double-Strand Breaks

Key Term

- **Nonhomologous end joining (NHEJ)** ligates blunt ends. It is common to many repair pathways and to certain recombination pathways (such as immunoglobulin recombination).

Double-strand breaks occur in cells in various circumstances. They initiate the process of homologous recombination and are an intermediate in the recombination of immunoglobulin genes (see *23.8 The RAG Proteins Catalyze Breakage and Reunion*). Double-strand breaks also occur as the result of damage to DNA, for example, by irradiation. The major mechanism to repair these breaks is called **nonhomologous end joining (NHEJ)**, and consists of ligating the blunt ends together.

The steps in NHEJ are summarized in **Figure 20.21**. The same enzyme complex undertakes the process in both NHEJ and immune recombination. The first stage is recognition of the broken ends by a heterodimer consisting of the proteins Ku70 and Ku80. They form a scaffold that holds the ends together and allows other enzymes to act on them. A key component is the DNA-dependent protein kinase (DNA-PKcs), which is activated by DNA to phosphorylate protein targets. One of these targets is the protein Artemis, which in its activated form has both exonuclease and endonuclease activities, and can both trim overhanging ends and cleave the hairpins generated by recombination of immunoglobulin genes. The DNA polymerase activity that fills in any remaining single-stranded protrusions is not known. The actual joining of the double-stranded ends is undertaken by the DNA ligase IV, which functions in conjunction with the protein XRCC4. Mutations in any of these components may render eukaryotic cells more sensitive to radiation. Some of the genes for these proteins are mutated in patients who have diseases due to deficiencies in DNA repair.

The Ku heterodimer is the sensor that detects DNA damage by binding to the broken ends. The crystal structure in **Figure 20.22** shows why it binds only to ends. The bulk of the protein extends for about two turns along one face of DNA (lower panel), but a narrow bridge between the subunits, located in the center of the structure, completely encircles DNA. This means that the heterodimer needs to slip onto a free end.

Ku can bring broken ends together by binding two DNA molecules. The ability of Ku heterodimers to associate with one another suggests that the reaction might take place as illustrated in **Figure 20.23**. The ligase would act by binding in the region between the bridges on the individual heterodimers. Presumably Ku must change its structure in order to be released from DNA.

Deficiency in DNA repair causes several human diseases. The common feature is that an inability to repair double-strand breaks in DNA leads to chromosomal instability. The instability is revealed by chromosomal aberrations, which are associated with an increased rate of mutation, in turn leading to an increased susceptibility to cancer in patients with the disease. Examples are Ataxia telangiectasia (AT), which is caused by failure of a pathway that activates repair mechanisms, and Nijmegen breakage syndrome (NBS), which is caused by a mutation of a repair enzyme. One of the lessons that we learn from characterizing the repair pathways is that they are conserved in mammals, yeast, and bacteria.

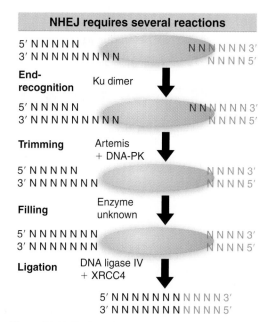

NHEJ requires several reactions

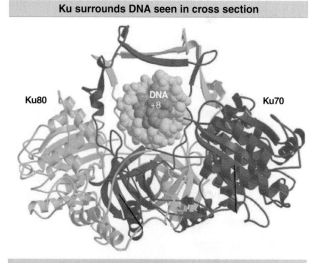

Figure 20.21 Nonhomologous end joining requires recognition of the broken ends, trimming of overhanging ends and/or filling, followed by ligation.

Ku surrounds DNA seen in cross section

Ku extends for 2 helical turns along DNA

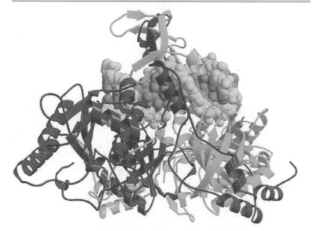

Figure 20.22 The Ku70–Ku80 heterodimer binds along two turns of the DNA double helix and surrounds the helix at the center of the binding site. Photograph kindly provided by Jonathan Goldberg.

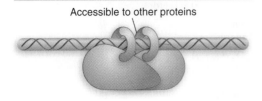

The active form of Ku may be two heterodimers

Accessible to other proteins

Figure 20.23 If two heterodimers of Ku bind to DNA, the distance between the two bridges that encircle DNA is ~12 bp.

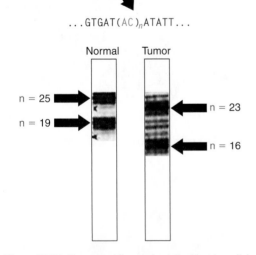

Tumors have mutations

AC microsatellite

...GTGAT(AC)$_n$ATATT...

Normal   Tumor

n = 25

n = 19

n = 23

n = 16

Figure 20.24 The normal tissue of a patient has two alleles for a microsatellite, each with a different number of repeats of the dinucleotide (AC). In tumor cells, both of these alleles have suffered deletions, one reducing the repeat number from 25 to 23, the other reducing it from 19 to 16. The repeat number of each allele in each situation is in fact probably unique, but some additional bands are generated as an artifact during the amplification procedure used to generate the samples. Data kindly provided by Bert Vogelstein, John Hopkins University Medical Center. The bands remaining at the normal position in the tumor samples are due to contamination of the tumor sample with normal tissue.

20.11 Defects in Repair Systems Cause Mutations to Accumulate in Tumors

Key Term

- **Mismatch repair** corrects recently inserted bases that do not pair properly. The process preferentially corrects the sequence of the daughter strand by distinguishing the daughter strand and parental strand, sometimes on the basis of their states of methylation.

Key Concept

- Loss of mismatch-repair systems generates a high mutation rate in HNPCC.

All cells have systems to protect themselves against damage from the environment or errors that may occur during replication. The overall mutation rate is the result of the balance between the introduction of mutations and their removal by these systems. One means by which cancer cells increase the rate of mutation is to inactivate some of their repair systems, so that spontaneous mutations accumulate instead of being removed. In effect, a mutation that occurs in a mutator gene causes mutations to accumulate in other genes. (A mutator gene can be any type of gene—such as a DNA polymerase or a repair enzyme—whose function affects the integrity of DNA sequences.)

The MutSL system is a particularly important target. This **mismatch repair** system is responsible for removing mismatches in newly replicated bacterial DNA. Its homologues perform similar functions in eukaryotic cells. During replication of a microsatellite DNA, DNA polymerase may slip backward by one or more of the short repeating units. The additional units are extruded as a single-stranded region from the duplex. If not removed, they result in an increase in the length of the microsatellite in the next replication cycle (see Figure 6.24). This is averted when homologues of the MutSL system recognize the single-stranded extrusion and replace the newly synthesized material with a nucleotide sequence that properly matches the template (see Figure 20.15).

In the human disease of HNPCC (hereditary nonpolyposis colorectal cancer), new microsatellite sequences are found at a high frequency in tumor cells when their DNA sequences are compared with somatic cells of the same patient. **Figure 20.24** shows an example. This microsatellite has a repeat sequence of AC (reading just one strand of DNA). The length of the repeat varies from 14–27 copies in the population. Any particular individual shows two repeat lengths, one corresponding to each allele in the diploid cell. In the tumor cells of many patients, the repeat length is changed at both alleles, as shown in the example in the figure.

The idea that this type of change might be the result of loss of the mismatch-repair system was confirmed by showing that *mutS* and *mutL* homologues (*hMSH2, hMLH1*) are mutated in the tumors. As expected, the tumor cells are deficient in mismatch-repair. Change in the microsatellite sequences is of course only one of the types of mutation that result from the loss of the mismatch-repair system (it is especially easy to diagnose).

The case of HNPCC illustrates both the role of multiple mutations in malignancy and the contribution that is made by mutator genes.

At least seven independent genetic events are required to form a fully tumorigenic colorectal cancer. More than 90% of cases have mutations in the mismatch-repair system, and the tumor cells have mutation rates that are elevated by 2–3 orders of magnitude from normal somatic cells. The high mutation rate is responsible for creating new variants in the tumor that provide the raw material from which cells with more aggressive growth properties will arise.

20.12 SUMMARY

All cells contain systems that maintain the integrity of their DNA sequences in the face of damage or errors of replication and that distinguish the DNA from sequences of a foreign source.

Repair systems can recognize mispaired, altered, or missing bases in DNA, or other structural distortions of the double helix. Excision repair systems cleave DNA near a site of damage, remove one strand, and synthesize a new sequence to replace the excised material. The Uvr system provides the main excision repair pathway in *E. coli*. The *dam* system is involved in correcting mismatches generated by incorporation of incorrect bases during replication and functions by preferentially removing the base on the strand of DNA that is not methylated at the *dam* target sequence. Eukaryotic homologues of the *E. coli* MutSL system are involved in repairing mismatches that result from replication slippage; mutations in this pathway are common in certain types of cancer.

Recombination-repair systems retrieve information from a DNA duplex and use it to repair a sequence that has been damaged on both strands. The RecBC and RecF pathways both act prior to RecA, whose strand-transfer function is involved in all bacterial recombination. A major use of recombination repair may be to recover from the situation created when a replication fork stalls.

Repair systems can be connected with transcription in both prokaryotes and eukaryotes. Human diseases are caused by mutations in genes coding for repair activities that are associated with the transcription factor TFIIH. They have homologues in the *RAD* genes of yeast, suggesting that this repair system is widespread.

Nonhomologous end joining (NHEJ) is a general reaction for repairing broken ends in eukaryotic DNA. The Ku heterodimer brings the broken ends together so they can be ligated. Several human diseases are caused by mutations in enzymes of this pathway.

Tumors can be caused in man by the failure of repair systems that usually prevent the accumulation of mutations.

Transposons

21.1 Introduction

Key Term

- A **transposon (transposable element)** is a DNA sequence able to insert itself (or a copy of itself) at a new location in the genome, without having any sequence relationship with the target locus.

Transposable elements or **transposons** are discrete sequences in the genome that are *mobile*; they are able to transport themselves to other locations within the genome. The mark of a transposon is that it does not utilize an independent form of the element (like a phage or plasmid DNA), but moves directly from one site in the genome to another. Unlike most other processes that restructure genomes, transposition does not rely on any relationship between the sequences at the donor and recipient sites. Transposition provides a major source of mutations in the genome.

Transposons fall into two general classes. The groups of transposons reviewed in this chapter exist as sequences of DNA coding for proteins that are able directly to manipulate DNA so as to propagate themselves within the genome. The transposons reviewed in *Chapter 22 Retroviruses and Retroposons* are related to retroviruses, and the source of their mobility is the ability to make DNA copies of their RNA transcripts; the DNA copies then become integrated at new sites in the genome.

Transposons that mobilize via DNA are found in both prokaryotes and eukaryotes. Each transposon carries genes that code for the enzyme activities required for its own transposition, although it may also require ancillary functions of the genome in which it resides (such as DNA polymerase or DNA gyrase). A genome may contain both functional and nonfunctional (defective) elements. Often the majority of elements in a eukaryotic genome are defective, and have lost the ability to transpose independently, although they may still be recognized as substrates for transposition by the enzymes produced by functional transposons. A eukaryotic genome contains a large number and variety of transposons. The fly genome has >50 types of transposons, with a total of several hundred individual elements.

Transposable elements can promote rearrangements of the genome, directly or indirectly:

- The transposition event itself may cause deletions or inversions or lead to the movement of a host sequence to a new location.
- Transposons serve as substrates for cellular recombination systems by functioning as "portable regions of homology"; two copies of a transposon at different locations (even on different chromosomes) may provide sites for reciprocal recombination. Such exchanges result in deletions, insertions, inversions, or translocations.

21.2 Insertion Sequences Are Simple Transposition Modules

Key Concepts

- An insertion sequence is a transposon that codes for the enzyme(s) needed for transposition flanked by short inverted terminal repeats.
- The target site at which a transposon is inserted is duplicated during the insertion process to form two repeats in direct orientation at the ends of the transposon.
- The length of each direct repeat is 5–9 bp and is characteristic for any particular transposon.

Key Terms

- An **insertion sequence (IS)** is a small bacterial transposon that carries only the genes needed for its own transposition.
- **Inverted terminal repeats** are the short related or identical sequences present in reverse orientation at the ends of some transposons.
- **Direct repeats** are identical (or closely related) sequences present in two or more copies in the same orientation in the same molecule of DNA.
- A **transposase** provides the enzyme activity for insertion of a transposon at a new site.

Transposable elements were first identified at the molecular level in the form of spontaneous insertions in bacterial operons. Such an insertion prevents transcription and/or translation of the gene in which it is inserted. The simplest transposons are called **insertion sequences** (reflecting the way in which they were detected). Each type is given the prefix IS, followed by a number that identifies the type. The IS elements are normal constituents of bacterial chromosomes and plasmids. A standard strain of *E. coli* is likely to contain several (<10) copies of any one of the more common IS elements. To describe an insertion into a particular site, a double colon is used; so λ::IS1 describes an IS1 element inserted into phage lambda.

The IS elements are autonomous units, each of which codes only for the proteins needed to sponsor its own transposition. Each IS element is different in sequence, but there are some common features in organization. The structure of a generic transposon before and after insertion at a target site is illustrated in **Figure 21.1**, which also summarizes the details of some common IS elements.

An IS element ends in short **inverted terminal repeats**; usually the two copies of the repeat are closely related rather than identical. As shown in the figure, the presence of the inverted terminal repeats means that the same sequence is encountered proceeding into the element from the flanking DNA on either side of it.

When an IS element transposes, a sequence of host DNA at the site of insertion is duplicated. The nature of the duplication is revealed by comparing the sequence of the target site before and after an insertion has occurred. Figure 21.1 shows that at the site of insertion, the IS DNA is always flanked by very short **direct repeats**. (In this context, "direct" indicates that two copies of a sequence are repeated in the same orientation, not that the repeats are adjacent.) But in the original gene (prior to insertion), the target site has the sequence of only one of these repeats. In the figure, the target site consists of the sequence $\frac{ATGCA}{TACGT}$. After transposition,

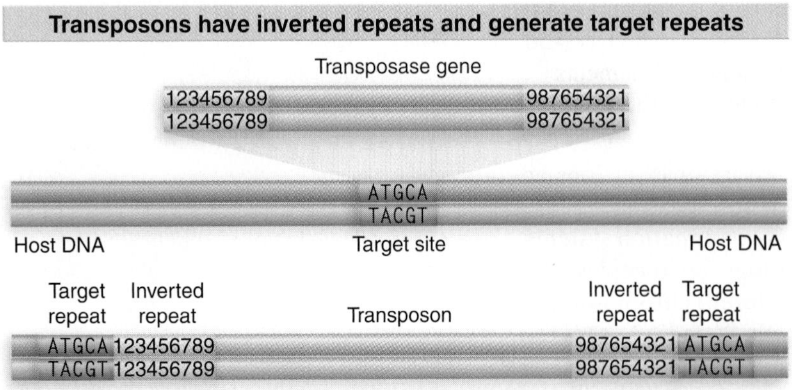

Transposons have inverted repeats and generate target repeats

Transposase gene

| 123456789 | 987654321 |
| 123456789 | 987654321 |

ATGCA
TACGT

Host DNA   Target site   Host DNA

| Target | Inverted | | Inverted | Target |
| repeat | repeat | Transposon | repeat | repeat |

| ATGCA | 123456789 | | 987654321 | ATGCA |
| TACGT | 123456789 | | 987654321 | TACGT |

Transposon	Target repeat (bp)	Inverted repeat (bp)	Overall length	Target selection
IS1	9	23	768	Random
IS2	5	41	1327	Hotspots
IS4	11–13	18	1428	$AAAN_{20}TTT$
IS5	4	16	1195	Hotspots
IS10R	9	22	1329	NGTNAGCN
IS50R	9	9	1531	Hotspots
IS903	9	18	1057	Random

Figure 21.1 Transposons have inverted terminal repeats and generate direct repeats of flanking DNA at the target site. In this example, the target is a 5 bp sequence. The ends of the transposon consist of inverted repeats of 9 bp, where the numbers 1 through 9 indicate a sequence of base pairs.

one copy of this sequence is present on either side of the transposon.

The sequence of the direct repeat varies among individual transposition events undertaken by a transposon, but the length is constant for any particular IS element (a reflection of the mechanism of transposition). The most common length for the direct repeats is 9 bp.

An IS element therefore displays a characteristic structure in which its ends are identified by the inverted terminal repeats, while the adjacent ends of the flanking host DNA are identified by the short direct repeats. When observed in a sequence of DNA, this type of organization is taken to be diagnostic of a transposon, and suggests that the sequence originated in a transposition event.

The inverted repeats define the ends of a transposon. Recognition of the ends is common to transposition events sponsored by all types of transposons. *Cis*-acting mutations that prevent transposition are located in the ends, which are recognized by a protein responsible for transposition. The protein is called a **transposase**. The transposase is responsible both for creating a target site and for recognizing the ends of the transposon. Only the ends are needed for a transposon to serve as a substrate for transposition.

All the IS elements except IS1 contain a single long coding region, starting just inside the inverted repeat at one end, and terminating just before or within the inverted repeat at the other end. This codes for the transposase. IS1 has a more complex organization, with two separate reading frames; the transposase is produced by making a frameshift during translation to allow both reading frames to be used.

The frequency of transposition varies among different elements. The overall rate of transposition is $\sim 10^{-3}$–10^{-4} per element per generation. Insertions in individual targets occur at a level comparable with the spontaneous mutation rate, usually $\sim 10^{-5}$–10^{-7} per generation. Reversion (by precise excision of the IS element) is usually infrequent, with a range of rates of 10^{-6} to 10^{-10} per generation, $\sim 10^{3}$ times less frequent than insertion.

21.3 Composite Transposons Have IS Modules

Key Terms

- Bacterial transposons carrying contain markers that are not related to their function, e.g., drug resistance, are named as **Tn** followed by a number.
- **Composite transposons (composite elements)** have a central region flanked on each side by insertion sequences, either or both of which may enable the entire element to transpose.

Key Concepts

- Transposons can carry other genes in addition to those coding for transposition.
- Composite transposons have a central region flanked by an IS element at each end.
- Either one or both of the IS elements of a composite transposon may be able to undertake transposition.
- A composite transposon may transpose as a unit, but an active IS element at either end may also transpose independently.

Some transposons carry drug resistance (or other) markers in addition to the functions concerned with transposition. These transposons are named **Tn** followed by a number. One class of larger transposons is called **composite elements**, because a central region carrying the drug marker is flanked on either side by "arms" that consist of IS elements.

The arms may be in either the same or (more commonly) inverted orientation. So a composite transposon with arms that are direct repeats has the structure:

If the arms are inverted repeats, the structure is:

ArmL   Central Region   ArmR

The arrows indicate the orientation of the arms, which are identified as L and R according to an arbitrary orientation of the genetic map of the transposon from left to right. Since arms consist of IS modules, and each module has the usual structure ending in inverted repeats, the composite transposon also ends in the same short inverted repeats. A functional IS module can transpose either itself or the entire transposon.

In some cases, the modules of a composite transposon are identical. **Figure 21.2** shows the example of Tn9, where the arms are direct repeats of IS1. Either copy of IS1 can cause transposition of Tn9.

In other cases, the modules are closely related, but not identical. **Figure 21.3** shows the example of Tn10, where the arms are inverted repeats of IS10, but the two copies are different. IS10L is nonfunctional, and only IS10R can cause transposition.

We assume that composite transposons evolved when two originally independent modules associated with the central region. Such a situation could arise when an IS element transposes to a recipient site close to the donor site. The two identical modules may remain identical or diverge. The ability of a single module to transpose the entire composite element explains the lack of selective pressure for both modules to remain active.

A major force supporting the transposition of composite transposons is selection for the marker carried in the central region. An IS10 module is free to move around on its own, and moves an order of magnitude more frequently than Tn10. But Tn10 is held together by selection for tet^R; so that under selective conditions, the relative frequency of intact Tn10 transposition is much increased.

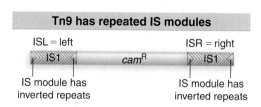

Figure 21.2 The composite transposon Tn9 has a central region with a drug resistance marker, flanked by two copies of IS1 in the same orientation. Both copies of IS1 are functional.

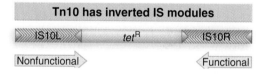

Figure 21.3 Tn10 has a central region with a drug resistance marker, flanked by two copies of IS10 in opposite orientations. Only the IS10R copy is functional.

21.4 Transposition Occurs by Both Replicative and Nonreplicative Mechanisms

Key Terms

- **Replicative transposition** describes the movement of a transposon by a mechanism in which first it is replicated, and then one copy is transferred to a new site.

- **Resolvase** is an enzyme that provides the activity for site-specific recombination between two copies of a transposon that has been duplicated.

- **Nonreplicative transposition** describes the movement of a transposon that leaves a donor site (usually generating a double-strand break) and moves to a new site.

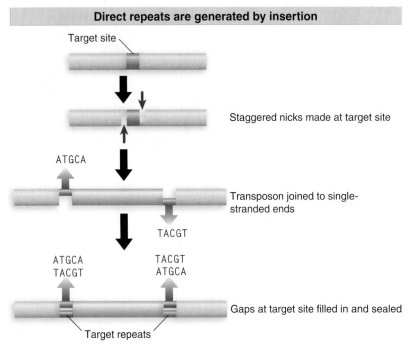

Direct repeats are generated by insertion

Target site

Staggered nicks made at target site

ATGCA

Transposon joined to single-stranded ends

TACGT

ATGCA   TACGT
TACGT   ATGCA

Gaps at target site filled in and sealed

Target repeats

Figure 21.4 The direct repeats of target DNA flanking a transposon are generated by the introduction of staggered cuts whose protruding ends are linked to the transposon.

- All transposons use a common mechanism in which staggered nicks are made in target DNA, the transposon is joined to the protruding ends, and the gaps are filled.

- The order of events and exact nature of the connections between transposon and target DNA determine whether transposition is replicative or nonreplicative.

The insertion of a transposon into a new site is illustrated in **Figure 21.4**. It consists of making staggered breaks in the target DNA, joining the transposon to the protruding single-stranded ends, and filling in the gaps. The generation and filling of the staggered ends explain the occurrence of the direct repeats of target DNA at the site of insertion. The stagger between the cuts on the two strands determines the length of the direct repeats; so the target repeat characteristic of each transposon reflects the geometry of the enzyme that cuts the target DNA.

The use of staggered ends is common to all means of transposition, but we can distinguish two different types of mechanism by which a transposon moves:

- In **replicative transposition**, the element is duplicated during the reaction, so that the transposing entity is a copy of the original element. **Figure 21.5** summarizes the results of such a transposition. The transposon is copied as part of its movement. One copy remains at the original site, while the other inserts at the new site. So transposition is accompanied by an increase in the number of copies of the transposon. Replicative transposition involves two types of enzymatic activity: a transposase acts on the ends of the original transposon; and a **resolvase** acts on the duplicated copies. A group of transposons related to TnA moves only by replicative transposition (see *21.7 Replicative Transposition Proceeds through a Cointegrate*).

- In **nonreplicative transposition**, the transposing element moves as a physical entity directly from one site to another, and is conserved. The insertion sequences and composite transposons Tn10 and Tn5 use the mechanism shown in **Figure 21.6**, which involves the release of the transposon from the flanking donor DNA during transfer. This type of mechanism requires only a transposase. Another mechanism utilizes the connection of donor and target DNA sequences and shares some steps with replicative transposition (see *21.6 Common Intermediates for Transposition*). Both mechanisms of nonreplicative transposition cause the element to be inserted at the target site and lost from the donor site. What happens to the donor molecule after a nonreplicative transposition? Its survival requires that host repair systems recognize the double-strand break and repair it.

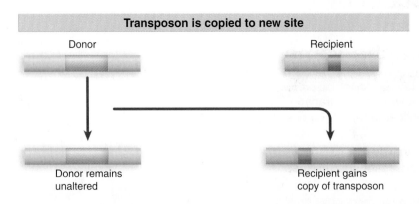

Transposon is copied to new site

Donor   Recipient

Donor remains   Recipient gains
unaltered   copy of transposon

Figure 21.5 Replicative transposition creates a copy of the transposon, which inserts at a recipient site. The donor site remains unchanged, so both donor and recipient have a copy of the transposon.

ments, prokaryotic and eukaryotic transposons, and bacteriophage Mu. Insertion of the DNA copy of retroviral RNA uses a similar mechanism (see *22.2 The Retrovirus Life Cycle Involves Transposition–like Events*). The first stages of immunoglobulin recombination also are similar (see *23.8 The RAF Proteins Catalyze Breakage and Reunion*).

Transposition starts with a common mechanism for joining the transposon to its target. **Figure 21.9** shows that the transposon is nicked at both ends, and the target site is nicked on both strands. The nicked ends are joined crosswise to generate a covalent connection between the transposon and the target. The two ends of the transposon are brought together in this process; for simplicity in following the cleavages, the synapsis stage is shown after cleavage, but it actually occurs previously.

Much of this pathway was first revealed with phage Mu, which uses the process of transposition in two ways. Upon infecting a host cell, Mu integrates into the genome by nonreplicative transposition; during the ensuing lytic cycle, the number of copies is amplified by replicative transposition. Both types of transposition involve the same type of reaction between the transposon and its target, but the subsequent reactions are different.

The initial manipulations of the phage DNA are performed by the MuA transposase. Three MuA-binding sites with a 22 bp consensus are located at each end of Mu DNA. L1, L2, and L3 are at the left end; R1, R2, and R3 are at the right end. A monomer of MuA can bind to each site. MuA also binds to an internal site in the phage genome. Binding of MuA at both the left and right ends and the internal site forms a complex. The role of the internal site is not clear; it appears to be necessary for formation of the complex, but not for strand cleavage and subsequent steps.

Joining the Mu transposon DNA to a target site passes through the three stages illustrated in **Figure 21.10**. This involves only the two sites closest to each end of the transposon. MuA subunits bound to these sites form a tetramer. This achieves synapsis of the two ends of the transposon. The tetramer now functions in a way that ensures a coordinated reaction on both ends of Mu DNA. MuA has two sites for manipulating DNA, and their mode of action compels subunits of the transposase to act in *trans*. The consensus-binding site binds to the 22 bp sequences that constitute the L1, L2, R1, and R2 sites. The active site cleaves the Mu DNA strands at positions adjacent to the MuA-binding sites L1 and R1. But the active site cannot cleave the DNA sequence that is adjacent to the consensus sequence in the consensus-binding site. However, it can cleave the appropriate sequence on a different stretch of DNA.

The ends of the transposon are thus cleaved by MuA subunits acting in *trans*. The *trans* mode of action means that the monomers actually bound to L1 and R1 do not cleave the adjacent sites. One of the monomers bound to the left end nicks the site at the right end, and vice versa. (We do not know which monomer is active at this stage of the reaction.) The strand transfer reaction also occurs in *trans*; the monomer at L1 transfers the strand at R1, and vice versa. It could be the case that different monomers catalyze the cleavage and strand transfer reactions for a given end.

A second protein, MuB, assists the reaction. It has an influence on the choice of target sites. Mu has a preference for transposing to a target site >10–15 kb away from the original insertion. This is called "target immunity." It results from an interaction between MuA and MuB that prevents transposition from occurring close to the donor.

The product of these reactions is a strand transfer complex in which the transposon is connected to the target site through one strand at each

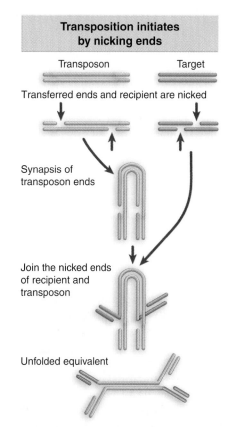

Figure 21.9 Transposition is initiated by nicking the transposon ends and target site and joining the nicked ends into a strand transfer complex.

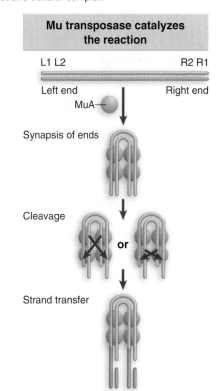

Figure 21.10 Mu transposition passes through three stable stages. MuA transposase forms a tetramer that synapses the ends of phage Mu. Transposase subunits act in *trans* to nick each end of the DNA; then a second *trans* action joins the nicked ends to the target DNA.

end. The next step of the reaction differs and determines the type of transposition. We see in the next two sections how the common structure can be a substrate for replication (leading to replicative transposition) or used directly for breakage and reunion (leading to nonreplicative transposition).

21.7 Replicative Transposition Proceeds Through a Cointegrate

Key Terms

- A **cointegrate structure** is produced by fusion of two replicons, one originally possessing a transposon, the other lacking it; the cointegrate has copies of the transposon present at both junctions of the replicons, oriented as direct repeats.
- **Resolution** occurs by a homologous recombination reaction between the two copies of the transposon in a cointegrate. The reaction generates the donor and target replicons, each with a copy of the transposon.

Key Concepts

- Replication of a strand transfer complex generates a cointegrate, which is a fusion of the donor and target replicons.
- The cointegrate has two copies of the transposon, which lie between the original replicons.
- Recombination between the transposon copies regenerates the original replicons, but the recipient has gained a copy of the transposon.
- The recombination reaction is catalyzed by a resolvase coded by the transposon.

The basic structures involved in replicative transposition are illustrated in **Figure 21.11**. The process starts with the formation of the strand transfer complex (sometimes called a crossover complex). The donor and target strands are ligated so that each end of the transposon sequence is joined to one of the protruding single strands generated at the target site. The strand transfer complex generates a crossover-shaped structure held together at the duplex transposon. The fate of the crossover structure determines the mode of transposition.

The principle of replicative transposition is that replication through the transposon duplicates it, creating copies at both the target and donor sites. The crossover structure contains a single-stranded region at each of the staggered ends. These regions are pseudoreplication forks that provide a template for DNA synthesis. (Use of the ends as primers for replication implies that the strand breakage must occur with a polarity that generates a $3'$—OH terminus at this point.)

If replication continues from both the pseudoreplication forks, it will proceed through the transposon, separating its strands, and terminating at its ends. Replication is probably accomplished by host-coded functions. The product is called a **cointegrate**. It has direct repeats of the transposon at the junctions between the replicons.

Figure 21.12 shows that a homologous recombination between the two copies of the transposon releases two individual replicons, each of which has a copy of the transposon. One of the replicons is the original donor replicon.

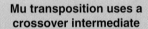

Mu transposition uses a crossover intermediate

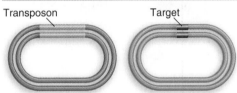

Transposon   Target

Nicking—Single-strand cuts generate staggered ends in both transposon and target

Crossover structure (strand transfer complex)—Nicked ends of transposon are joined to nicked ends of target

Replication from free 3′ ends generates cointegrate—Single molecule has two copies of transposon

Cointegrate drawn as a continuous path shows that transposons are at junctions between replicons

Figure 21.11 Mu transposition generates a crossover structure, which is converted by replication into a cointegrate.

The other is a target replicon that has gained a transposon flanked by short direct repeats of the host target sequence. The recombination reaction is called **resolution**; the enzyme activity responsible is called the resolvase.

21.8 Nonreplicative Transposition Proceeds by Breakage and Reunion

Key Concepts

- Nonreplicative transposition results if a crossover structure is nicked on the unbroken pair of donor strands, and the target strands on either side of the transposon are ligated.
- Two pathways for nonreplicative transposition differ according to whether the first pair of transposon strands are joined to the target before the second pair are cut (Tn5), or whether all four strands are cut before joining to the target (Tn10).

The crossover structure shown in Figure 21.11 can also be used in nonreplicative transposition. The principle of nonreplicative transposition by this mechanism is that a breakage and reunion reaction allows the target to be reconstructed with the insertion of the transposon; the donor remains broken. No cointegrate is formed.

Figure 21.13 shows the cleavage events that generate nonreplicative transposition of phage Mu. Once the unbroken donor strands have been nicked, the target strands on either side of the transposon can be ligated. The single-stranded regions generated by the staggered cuts must be filled in by repair synthesis. The product of this reaction is a target replicon in which the transposon has been inserted between repeats of the sequence created by the original single-strand nicks. The donor replicon has a double-strand break across the site where the transposon was originally located.

Nonreplicative transposition can also occur by an alternative pathway in which nicks are made in target DNA, but a double-strand break is made on either side of the transposon, releasing it entirely from flanking donor sequences (as envisaged in Figure 21.6). This "cut and paste" pathway is used by Tn10, as illustrated in **Figure 21.14**.

The basic difference in Figure 21.14 from the model of Figure 21.13 is that both strands of Tn10 are cleaved before any connection is made

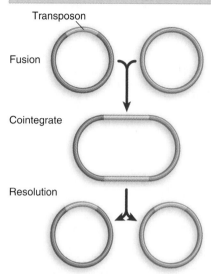

Figure 21.12 Transposition may fuse a donor and recipient replicon into a cointegrate. Resolution releases two replicons, each containing a copy of the transposon.

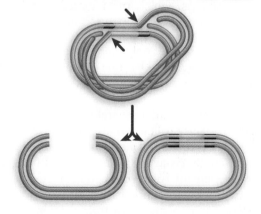

Figure 21.13 Nonreplicative transposition results when a crossover structure is released by nicking. This inserts the transposon into the target DNA, flanked by the direct repeats of the target, and the donor is left with a double-strand break.

Transposition can use cleavage and ligation

Transposase binds to both ends of Tn

Transferred ends are nicked

Other strands ends are nicked, recipient is nicked

Donor is released, Tn joined to target

Figure 21.14 Both strands of Tn10 are cleaved sequentially, and then the transposon is joined to the nicked target site.

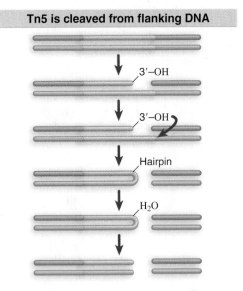

Tn5 is cleaved from flanking DNA

3'—OH

3'—OH

Hairpin

H₂O

Figure 21.15 Cleavage of Tn5 from flanking DNA involves nicking, interstrand reaction, and hairpin cleavage.

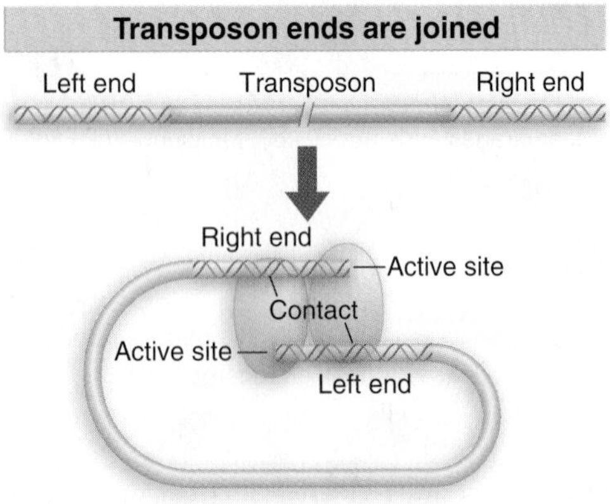

Transposon ends are joined

Left end   Transposon   Right end

Right end
——Active site
Contact
Active site —
Left end

Figure 21.16 Each subunit of the Tn5 transposase has one end of the transposon located in its active site and also makes contact at a different site with the other end of the transposon.

to the target site. The first step in the reaction is recognition of the transposon ends by the transposase, forming a proteinaceous structure within which the reaction occurs. At each end of the transposon, the strands are cleaved in a specific order—first the transferred strand (the one to be connected to the target site) is cleaved, then the other strand.

Tn5 also transposes by nonreplicative transposition, and **Figure 21.15** shows the interesting cleavage reaction that separates the transposon from the flanking sequences. First one DNA strand is nicked. The 3'—OH end that is released then attacks the other strand of DNA. This releases the flanking sequence and joins the two strands of the transposon in a hairpin. Then an activated water molecule attacks the hairpin to generate free ends for both strands of the transposon.

Then the cleaved donor DNA is released, and the transposon is joined to the nicked ends at the target site. The transposon and the target site remain constrained in the proteinaceous structure created by the transposase (and other proteins). The double-strand cleavage at each end of the transposon precludes any replicative-type transposition and forces the reaction to proceed by nonreplicative transposition, thus giving the same outcome as in Figure 21.11, but with the individual cleavage and joining steps occurring in a different order.

The Tn5 and Tn10 transposases both function as dimers. Each subunit in the dimer has an active site that successively catalyzes the double-strand breakage of the two strands at one end of the transposon and then catalyzes staggered cleavage of the target site. **Figure 21.16** illustrates the structure of the Tn5 transposase bound to the cleaved transposon. Each end of the transposon is located in the active site of one subunit. One end of the subunit also contacts the other end of the transposon. This controls the geometry of the transposition reaction. Each of the active sites will cleave one strand of the target DNA. It is the geometry of the complex that determines the distance between these sites on the two target strands (9 base pairs in the case of Tn5).

21.9 TnA Transposition Requires Transposase and Resolvase

Key Concepts

- Replicative transposition of TnA requires a transposase to form the cointegrate structure and a resolvase to release the two replicons.
- The action of the resolvase resembles lambda Int protein and belongs to the general family of topoisomerase–like site-specific recombination reactions, which pass through an intermediate in which the protein is covalently bound to the DNA.

Replicative transposition is the only mode of mobility of the TnA family, which consists of large (~5 kb) transposons. They are not composites relying on IS-type transposition modules, but are independent units carrying genes for transposition as well as for features such as drug resistance. The TnA family includes several related transposons, of which Tn3 and Tn1000 (formerly called γδ) are the best characterized. They have the usual terminal feature of closely related inverted repeats, generally ~38 bp in length. *cis*-acting deletions in either repeat prevent transposition of an element. A 5 bp direct repeat is generated at the target site. They carry resistance markers such as *amp*ʳ.

The two stages of TnA-mediated transposition are accomplished by the transposase and the resolvase, whose genes, *tnpA* and *tnpR*, are identified by recessive mutations. The transposition stage involves the ends of the element, as it does in IS-type elements. Resolution requires a specific internal site. This feature is unique to the TnA family.

Mutants in *tnpA* cannot transpose. The gene product is a transposase that binds to a sequence of ~25 bp located within the 38 bp of the inverted terminal repeat. A binding site for the *E. coli* protein IHF exists adjacent to the transposase binding site, and transposase and IHF bind cooperatively. The transposase recognizes the ends of the element and also makes the staggered 5 bp breaks in target DNA where the transposon is to be inserted. IHF is a DNA-binding protein that is often involved in assembling large structures in *E. coli*; its role in the transposition reaction may not be essential.

The *tnpR* gene product has dual functions. It acts as a repressor of gene expression and it provides the resolvase function.

Mutations in *tnpR* increase the transposition frequency. The reason is that TnpR represses the transcription of both *tnpA* and its own gene. So inactivation of TnpR protein allows increased synthesis of TnpA, which results in an increased frequency of transposition. This implies that the amount of the TnpA transposase must be a limiting factor in transposition.

The *tnpA* and *tnpR* genes are expressed divergently from an A·T-rich intercistronic control region, indicated in the map of Tn3 given in **Figure 21.17**. Both effects of TnpR are mediated by its binding in this region.

In its capacity as the resolvase, TnpR is involved in recombination between the direct repeats of Tn3 in a cointegrate structure. A cointegrate can in principle be resolved by a homologous recombination between any corresponding pair of points in the two copies of the transposon. But the Tn3 resolution reaction occurs only at a specific site.

The site of resolution is called *res*. It is identified by *cis*-acting deletions that block completion of transposition, causing the accumulation of cointegrates. At the molecular level, Tnp resolvase binds to three sites, as summarized in the lower part of Figure 21.17. The sites are 30–40 bp long, and share a consensus sequence that has dyad symmetry. Binding to these sites represses transcription of *tnpR* and *tnpA* as well as catalyzing the resolution reaction.

Resolution occurs by breaking and rejoining bonds without input of energy. The products consist of resolvase covalently attached to both 5′ ends of double-stranded cuts made at the *res* site. The cleavage occurs symmetrically at a short palindromic region to generate two base extensions. Expanding the view of the crossover region located in site I, we can describe the cutting reaction as:

The reaction resembles the action of lambda Int at the *att* sites. Indeed, 15 of the 20 bp of the *res* site are identical to the bases at corresponding positions in *att*. This suggests that the site-specific recombination of lambda and resolution of TnA have evolved from a common type of recombination reaction; and indeed, we see in *23.8 The RAG Proteins Catalyze Breakage and Reunion* that recombination involving immunoglobulin genes has the same basis. The common feature in all these reactions is the transfer of the broken end to the catalytic protein as an intermediate stage before it is rejoined to another broken end (see *15.18 Site-specific Recombination Resembles Topoisomerase Activity*).

The reactions themselves are analogous in terms of manipulation of DNA, although resolution occurs only between intramolecular sites,

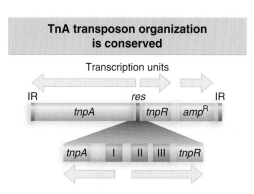

Figure 21.17 Transposons of the TnA family have inverted terminal repeats, an internal *res* site, and three known genes.

whereas the recombination between *att* sites is intermolecular and directional (as seen by the differences in *attB* and *attP* sites). However, the mechanism of protein action is different in each case. Resolvase functions in a manner in which four subunits bind to the recombining *res* sites. Each subunit makes a single-strand cleavage. Then a reorganization of the subunits relative to one another physically moves the DNA strands, placing them in a recombined conformation. This allows the nicks to be sealed, along with the release of resolvase.

21.10 Controlling Elements in Maize Cause Breakage and Rearrangements

Key Terms

- **Controlling elements** of maize are transposable units originally identified solely by their genetic properties. They may be autonomous (able to transpose independently) or nonautonomous (able to transpose only in the presence of an autonomous element).
- A **sector** is a patch of cells made up of a single altered cell and its progeny.
- **Variegation** of phenotype is produced by a change in genotype during somatic development.
- An **acentric fragment** of a chromosome (generated by breakage) lacks a centromere and is lost at cell division.
- A **dicentric chromosome** is the product of fusing two chromosome fragments, each of which has a centromere. It is unstable and may be broken when the two centromeres are pulled to opposite poles in mitosis.
- The **breakage-fusion-bridge** cycle is a type of chromosomal behavior in which a broken chromatid fuses to its sister, forming a "bridge." When the centromeres separate at mitosis, the chromosome breaks again (not necessarily at the bridge), thereby restarting the cycle.

Key Concepts

- Transposition in maize was discovered because of the effects of the chromosome breaks generated by transposition of "controlling elements."
- A chromosome break generates one chromosome that has a centromere and a broken end, and one acentric fragment.
- An acentric fragment is lost during mitosis, and this can be detected by the disappearance of dominant alleles in a heterozygote.
- Fusion between the broken ends of two chromatids generates a dicentric chromosome, which undergoes further cycles of breakage and fusion.
- The fusion-breakage-bridge cycle is responsible for somatic variegation.

One of the most visible consequences of the existence and mobility of transposons occurs during plant development. The actions of transposons (originally called **controlling elements**) result in somatic variation. Two features of maize have helped to follow transposition events. Controlling elements often insert near genes that have visible but nonlethal effects on the phenotype. And because maize displays clonal development, the occurrence and timing of a transposition event can be visualized as depicted diagrammatically in **Figure 21.18**.

Transpositions are clonally inherited

Break in one chromosome causes loss of dominant allele

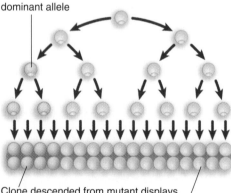

Clone descended from mutant displays recessive phenotype; cells of original genotype display dominant phenotype

Kernel with selector of recessive phenotype

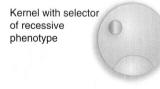

Figure 21.18 Clonal analysis identifies a group of cells descended from a single ancestor in which a transposition-mediated event altered the phenotype. Timing of the event during development is indicated by the number of cells; tissue specificity of the event may be indicated by the location of the cells.

The nature of the event does not matter: it may be a point mutation, insertion, excision, or chromosome break. What is important is that in a heterozygote it alters the expression of one allele. Then the descendants of a cell that has suffered the event display a new phenotype, while the descendants of cells not affected by the event continue to display the original phenotype.

Mitotic descendants of a given cell remain in the same location and give rise to a **sector** of tissue. A change in phenotype during somatic development is called **variegation**; it is revealed by a sector of the new phenotype residing within the tissue of the original phenotype. The size of the sector depends on the number of divisions in the lineage giving rise to it; so the size of the area of the new phenotype is determined by the timing of the change in genotype. The earlier its occurrence in the cell lineage, the greater the number of descendants and thus the size of patch in the mature tissue. This is seen most vividly in the variation in kernel color, when patches of one color appear within another color.

Insertion of a controlling element may affect the activity of adjacent genes. Deletions, duplications, inversions, and translocations all occur at the sites where controlling elements are present. Chromosome breakage is a common consequence of the presence of some elements. A unique feature of the maize system is that the activities of the controlling elements are regulated during development. The elements transpose and promote genetic rearrangements at characteristic times and frequencies during plant development.

The characteristic behavior of controlling elements in maize is typified by the *Ds* element, which was originally identified by its ability to provide a site for chromosome breakage. The consequences are illustrated in **Figure 21.19**. Consider a heterozygote in which *Ds* lies on one homologue between the centromere and a series of dominant markers. The other homologue lacks *Ds* and has recessive markers (*C*, *bz*, *wx*). Breakage at *Ds* generates an **acentric fragment** carrying the dominant markers. Because of its lack of a centromere, this fragment is lost at mitosis. So the descendant cells have only the recessive markers carried by the intact chromosome. This gives the type of situation whose results are depicted in Figure 21.18.

Figure 21.20 shows that breakage at *Ds* leads to the formation of two unusual chromosomes. These are generated by joining the broken ends of the products of replication. One is a U-shaped acentric fragment consisting of the joined sister chromatids for the region distal to *Ds* (on the left as drawn in the figure). The other is a U-shaped **dicentric chromosome** comprising the sister chromatids proximal to *Ds* (on its right in the figure). The latter structure leads to the classic **breakage-fusion-bridge** cycle illustrated in the figure.

Follow the fate of the dicentric chromosome when it attempts to segregate on the mitotic spindle. Each of its two centromeres pulls toward an opposite pole. The tension breaks the chromosome at a random site between the centromeres. In the example of the figure, breakage occurs between loci *A* and *B*, with the result that one daughter chromosome has a duplication of *A*, while the other has a deletion. If *A* is a dominant marker, the cells with the duplication will retain **A** phenotype, but cells with the deletion will display a recessive loss-of-function phenotype.

The breakage-fusion-bridge cycle continues through further cell generations, allowing genetic changes to continue in the descendants. For example, consider the deletion chromosome that has lost *A*. In the next cycle, a break occurs between *B* and *C*, so that the descendants are divided into those with a duplication of *B* and those with a deletion. Successive losses of dominant markers are revealed by subsectors within sectors.

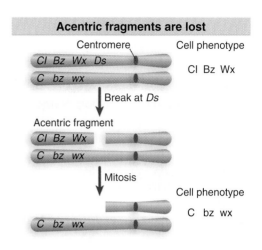

Figure 21.19 A break at a controlling element causes loss of an acentric fragment; if the fragment carries the dominant markers of a heterozygote, its loss changes the phenotype. The effects of the dominant markers, *Cl*, *Bz*, *Wx*, can be visualized by the color of the cells or by appropriate staining.

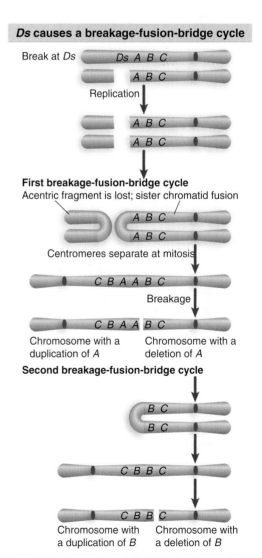

Figure 21.20 *Ds* provides a site to initiate the chromatid breakage-fusion-bridge cycle. The products can be followed by clonal analysis.

21.11 Controlling Elements Form Families of Transposons

Key Terms

- An **autonomous controlling element** in maize is an active transposon with the ability to transpose.
- A **nonautonomous controlling element** is a transposon in maize that encodes a nonfunctional transposase; it can transpose only in the presence of a *trans*-acting autonomous member of the same family.

Key Concepts

- Each family of transposons in maize has both autonomous and nonautonomous controlling elements.
- Autonomous controlling elements code for proteins that enable them to transpose.
- Nonautonomous controlling elements have mutations that eliminate their capacity to catalyze transposition, but they can transpose when an autonomous element provides the necessary proteins.
- Autonomous controlling elements have changes of phase, when their properties alter as a result of changes in methylation.

The maize genome contains several families of controlling elements. The numbers, types, and locations of the elements are characteristic for each individual maize strain. They may occupy a significant part of the genome. The members of each family are divided into two classes:

- **Autonomous controlling elements** have the ability to excise and transpose. Because of the continuing activity of an autonomous element, its insertion at any locus creates an unstable or "mutable" allele. Loss of the autonomous element itself, or of its ability to transpose, converts a mutable allele to a stable allele.

- **Nonautonomous controlling elements** are stable; they do not transpose or suffer other spontaneous changes in condition. They become unstable only when an autonomous member of the same family is present elsewhere in the genome. When complemented in *trans* by an autonomous element, a nonautonomous element displays the usual range of activities associated with autonomous elements, including the ability to transpose to new sites. Nonautonomous elements are derived from autonomous elements by loss of *trans*-acting functions needed for transposition.

Families of controlling elements are defined by the interactions between autonomous and nonautonomous elements. A family consists of a single type of autonomous element accompanied by many varieties of nonautonomous elements. A nonautonomous element is placed in a family by its ability to be activated in *trans* by the autonomous element. The major families of controlling elements in maize are summarized in **Figure 21.21**.

Characterized at the molecular level, the maize transposons share the usual form of organization—inverted repeats at the ends and short direct repeats in the adjacent target DNA—but otherwise vary in size and coding capacity. All families of transposons share the same type of relationship between the autonomous and nonautonomous elements. The autonomous elements have open reading frames between the terminal repeats, whereas the nonautonomous elements do not code for functional proteins. Sometimes the internal sequences are related to those of autonomous elements; sometimes they have diverged completely.

There are typically several members (~10) of each transposon family in a plant genome. By analyzing autonomous and nonautonomous elements of the *Ac/Ds* family, we have

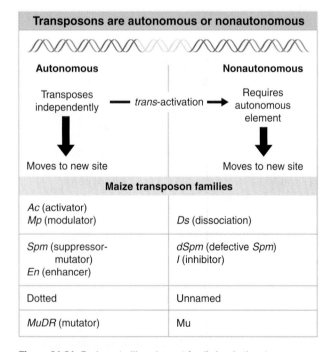

Transposons are autonomous or nonautonomous

Autonomous	Nonautonomous
Transposes independently — *trans*-activation →	Requires autonomous element
↓	↓
Moves to new site	Moves to new site

Maize transposon families

Ac (activator) *Mp* (modulator)	*Ds* (dissociation)
Spm (suppressor-mutator) *En* (enhancer)	*dSpm* (defective *Spm*) *I* (inhibitor)
Dotted	Unnamed
MuDR (mutator)	Mu

Figure 21.21 Each controlling element family has both autonomous and nonautonomous members. Autonomous elements are capable of transposition. Nonautonomous elements are deficient in transposition. Pairs of autonomous and nonautonomous elements can be classified in >4 families.

molecular information about many individual examples of these elements. **Figure 21.22** summarizes their structures.

Most of the length of the autonomous *Ac* element is occupied by a single gene consisting of five exons. The product is the transposase. The element itself ends in inverted repeats of 11 bp; and a target sequence of 8 bp is duplicated at the site of insertion.

Nonautonomous elements lack internal sequences, but possess the terminal inverted repeats (and possibly other sequence features). Nonautonomous elements are derived from autonomous elements by deletions (or other changes) that inactivate the *trans*-acting transposase, but leave intact the sites (including the termini) on which the transposase acts. The nonautonomous elements related to *Ac* define the *Ds* family. *Ds* elements vary in both length and sequence, but are related to *Ac*. They end in the same 11 bp inverted repeats. Their structures range from minor (but inactivating) mutations of *Ac* to sequences that have major deletions or rearrangements, which in an extreme may have only the ends of the element.

Transposition of *Ac/Ds* occurs by a nonreplicative mechanism, and is accompanied by its disappearance from the donor location. Clonal analysis suggests that transposition of *Ac/Ds* almost always occurs soon after the donor element has been replicated. These features resemble transposition of the bacterial element Tn10. The cause is the same: transposition does not occur when the DNA of the transposon is methylated on both strands (the typical state before replication), and is activated when the DNA is hemimethylated (the typical state immediately after replication). The recipient site is frequently on the same chromosome as the donor site, and often quite close to it.

Replication generates two copies of a potential *Ac/Ds* donor, but usually only one copy actually transposes. What happens to the donor site? The rearrangements that are found at sites from which controlling elements have been lost could be explained in terms of the consequences of a chromosome break, as illustrated previously in Figure 21.19.

Autonomous and nonautonomous elements are subject to a variety of changes in their condition. Some of these changes are genetic, others are epigenetic (heritable although not resulting from changes in nucleotide sequence).

The major change is (of course) the conversion of an autonomous element into a nonautonomous element, but further changes may occur in the nonautonomous element. *cis*-acting defects may render a nonautonomous element impervious to autonomous elements. So a nonautonomous element may become permanently stable because it can no longer be activated to transpose.

Autonomous elements are subject to "changes of phase," heritable but relatively unstable alterations in their properties. These take the form of a reversible inactivation in which the element cycles between an active and inactive condition during plant development.

Phase changes in both the *Ac* and *Mu* types of autonomous element result from changes in the methylation of DNA. Comparisons of the susceptibilities of active and inactive elements to restriction enzymes suggest that the inactive form of the element is methylated in the

target sequence $\frac{CAG}{GTC}$. In *MuDR*, demethylation of the terminal repeats increases transposase expression, suggesting that the effect may be mediated through control of the promoter for the transposase gene.

The effect of methylation is common among transposons in plants. The best demonstration of the effect of methylation on activity comes from observations made with the *Arabidopsis* mutant *ddm1*, which causes a loss of methylation in heterochromatin. Among the targets

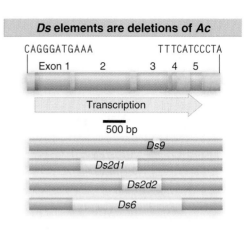

Figure 21.22 The *Ac* element has five exons that code for a transposase; *Ds* elements have internal deletions.

that lose methyl groups is a family of transposons related to *MuDR*. Direct analysis of genome sequences shows that the demethylation causes transposition events to occur. Methylation is probably the major mechanism that prevents transposons from damaging the genome by transposing too frequently.

There may be self-regulating controls of transposition, analogous to the immunity effects displayed by bacterial transposons. An increase in the number of *Ac* elements in the genome decreases the frequency of transposition. The *Ac* element may code for a repressor of transposition; the activity could be carried by the same protein that provides transposase function.

21.12 Transposition of *P* Elements Causes Hybrid Dysgenesis

Key Terms

- **Hybrid dysgenesis** describes the inability of certain strains of *D. melanogaster* to interbreed, because the hybrids are sterile (although otherwise they may be phenotypically normal).

- A ***P* element** is a type of transposon in *D. melanogaster*.

- **Cytotype** is a cytoplasmic condition that affects *P* element activity. The effect of cytotype is due to the presence or absence of repressors of transposition, which are provided by the mother to the egg.

Key Concepts

- *P* elements are transposons that are carried in P strains of *D. melanogaster* but not in M strains.

- *P* elements are activated in the germline of P male × M female crosses because a tissue-specific splicing event removes one intron, generating the coding sequence for the transposase.

- The insertion of *P* elements at new sites in these crosses inactivates many genes and makes the crosses infertile.

- The *P* element also produces a repressor of transposition, which is inherited maternally in the cytoplasm.

- The presence of the repressor explains why M male × P female crosses remain fertile.

Certain strains of *D. melanogaster* encounter difficulties in interbreeding. When flies from two of these strains are crossed, the progeny display "dysgenic traits," a series of defects including mutations, chromosomal aberrations, distorted segregation at meiosis, and sterility. The appearance of these correlated defects is called **hybrid dysgenesis**.

In one of the systems responsible for hybrid dysgenesis in *D. melanogaster*, flies are divided into the two types P (paternal contributing) and M (maternal contributing). **Figure 21.23** illustrates the asymmetry of the system; a cross between a P male and an M female causes dysgenesis, but the reverse cross does not.

Dysgenesis is principally a phenomenon of the germ cells. In crosses within the P–M system, the F1 hybrid flies have normal somatic tissues. However, their gonads do not develop. The morphological defect in gamete development dates from the stage at which rapid cell divisions commence in the germline.

Any one of the chromosomes of a P male can induce dysgenesis in a cross with an M female. The construction of recombinant chromosomes

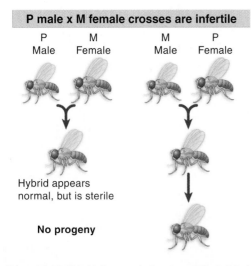

P male x M female crosses are infertile

P Male   M Female   M Male   P Female

Hybrid appears normal, but is sterile

No progeny

Figure 21.23 Hybrid dysgenesis is asymmetrical; it is induced by P male × M female crosses, but not by M male × P female crosses.

shows that several regions within each P chromosome are able to cause dysgenesis. This suggests that a P male has sequences at many different chromosomal locations that can induce dysgenesis. The locations differ between individual P strains. The P-specific sequences are absent from chromosomes of M flies. The inserted sequence is called the ***P* element**.

The *P* element insertions form a classic transposable system. Individual elements vary in length but are homologous in sequence. All *P* elements possess inverted terminal repeats of 31 bp, and generate direct repeats of target DNA of 8 bp upon transposition. The longest *P* elements are ~2.9 kb long and have four open reading frames. The shorter elements arise, apparently rather frequently, by internal deletions of a full-length *P* factor. At least some of the shorter *P* elements have lost the capacity to produce the transposase, but may be activated in *trans* by the enzyme coded by a complete *P* element.

A P strain carries 30–50 copies of the *P* element, about a third of them full length. The elements are absent from M strains. In a P strain, the elements are carried as inert components of the genome. But they become activated to transpose when a P male is crossed with an M female.

Chromosomes from P–M hybrid dysgenic flies have *P* elements inserted at many new sites. The insertions inactivate the genes in which they are located and often cause chromosomal breaks. The result of the transpositions is therefore to inactivate the genome.

Activation of *P* elements is tissue-specific: it occurs only in the germline. But *P* elements are transcribed in both germline and somatic tissues. Tissue-specificity is conferred by a change in the splicing pattern.

Figure 21.24 depicts the organization of the element and its transcripts. The primary transcript extends for 2.5 kb or 3.0 kb, the difference probably reflecting merely the leakiness of the termination site. Two protein products can be produced:

- In somatic tissues, only the first two introns are excised, creating a coding region of ORF0-ORF1-ORF2. Translation of this RNA yields a protein of 66 kD. This protein is a repressor of transposon activity.
- In germline tissues, an additional splicing event occurs to remove intron 3. This connects all four open reading frames into an mRNA that is translated to generate a protein of 87 kD. This protein is the transposase.

So whenever ORF3 is spliced to the preceding reading frame, the *P* element becomes active. This is the crucial regulatory event, and usually it occurs only in the germline. What is responsible for the tissue-specific splicing? Somatic cells contain a protein that binds to sequences in exon 3 to prevent splicing of the last intron (see *26.11 Alternative Splicing Involves Differential Use of Splice Junctions*). The absence of this protein in germline cells allows splicing to generate the mRNA that codes for the transposase.

Transposition of a *P* element requires ~150 bp of terminal DNA. The transposase binds to 10 bp sequences that are adjacent to the 31 bp inverted repeats. Transposition occurs by a nonreplicative "cut and paste" mechanism resembling that of Tn10. (It contributes to hybrid dysgenesis in two ways. Insertion of the transposed element at a new site may cause mutations. And the break that is left at the donor site—see Figure 21.6—has a deleterious effect.)

The dependence of hybrid dysgenesis on the sexual orientation of a cross shows that the cytoplasm is important as well as the *P* elements themselves. The contribution of the cytoplasm is described as the **cytotype**; a line of flies containing *P* elements has P cytotype, while a line of flies lacking *P* elements has M cytotype. Hybrid dysgenesis occurs only when chromosomes containing *P* elements find themselves in M cytotype, that is, when the male parent has *P* elements and the female parent does not.

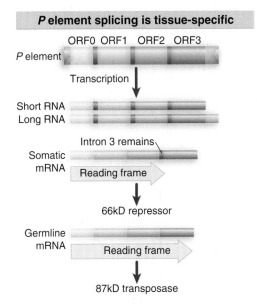

Figure 21.24 The *P* element has four exons. The first three are spliced together in somatic expression; all four are spliced together in germline expression.

Cytotype shows a heritable cytoplasmic effect; when a cross occurs through P cytotype (the female parent has *P* elements), hybrid dysgenesis is suppressed for several generations of crosses with M female parents. So something in P cytotype, which can be diluted out over some generations, suppresses hybrid dysgenesis.

The effect of cytotype is explained in molecular terms by the model of **Figure 21.25**. It depends on the ability of the 66 kD protein to repress transposition. The protein is provided as a maternal factor in the egg. In a P line, there must be sufficient protein to prevent transposition from occurring, even though the *P* elements are present. In any cross involving a P female, its presence prevents either synthesis or activity of the transposase. But when the female parent is M type, there is no repressor in the egg, and the introduction of a *P* element from the male parent results in activity of transposase in the germline. The ability of P cytotype to exert an effect through more than one generation suggests that there must be enough repressor protein in the egg, and it must be stable enough, to be passed on through the adult to be present in the eggs of the next generation.

Because hybrid dysgenesis reduces interbreeding, it is a step on the path to speciation. Suppose that a dysgenic system is created by a transposable element in some geographic location. Another element may create a different system in some other location. Flies in the two areas will be dysgenic for two (or possibly more) systems. If this renders them intersterile and the populations become genetically isolated, further separation may occur. Multiple dysgenic systems therefore lead to inability to mate—and to speciation.

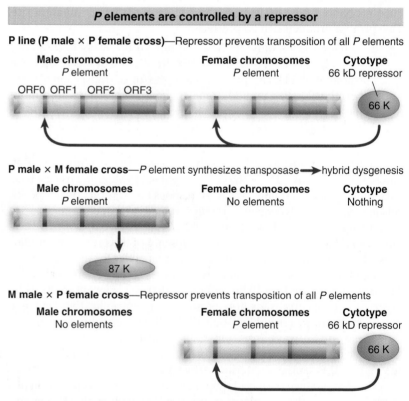

Figure 21.25 Hybrid dysgenesis is determined by the interactions between *P* elements in the genome and 66 kD repressor in the cytotype.

21.13 SUMMARY

Prokaryotic and eukaryotic cells contain a variety of transposons that mobilize by moving or copying DNA sequences. The transposon can be identified only as an entity within the genome; its mobility does not involve an independent form. All transposons have systems to limit the extent of transposition, but the molecular mechanisms are different in each case.

The archetypal transposon has inverted repeats at its termini and generates direct repeats of a short sequence at the site of insertion. The simplest types are the bacterial insertion sequences (IS), which consist essentially of the inverted terminal repeats flanking a coding frame whose product provides transposition activity. Composite transposons have terminal modules that consist of IS elements; one or both of the IS modules provide transposase activity, and the sequences between them (often carrying antibiotic resistance), are treated as passengers.

The generation of target repeats flanking a transposon reflects a common feature of transposition. The target site is cleaved at points that are staggered on each DNA strand by a fixed distance (often 5 or 9 base pairs). The transposon is in effect inserted between protruding single-stranded ends generated by the staggered cuts. Target repeats are generated by filling in the single-stranded regions.

IS elements, composite transposons, and *P* elements mobilize by nonreplicative transposition, in which the element moves directly from a donor site to a recipient site. A single transposase enzyme undertakes the reaction. It occurs by a "cut and paste" mechanism in which the transposon is separated from flanking DNA. Cleavage of the transposon ends, nicking of the target site, and connection of the transposon ends to the staggered nicks, all occur in a nucleoprotein complex containing the transposase. Loss of the transposon from the donor creates a double-strand break. In the case of Tn10, transposition becomes possible immediately after DNA replication, when sites recognized by the *dam* methylation system are transiently hemimethylated. This imposes a demand for the existence of two copies of the donor site, which may enhance the cell's chances for survival.

The TnA family of transposons mobilizes by replicative transposition. After the transposon at the donor site becomes connected to the target site, replication generates a cointegrate molecule that has two copies of the transposon. A resolution reaction, involving recombination between two particular sites, then frees the two copies of the transposon, so that one remains at the donor site and one appears at the target site. Two enzymes coded by the transposon are required: transposase recognizes the ends of the transposon and connects them

to the target site; and resolvase provides a site-specific recombination function.

Phage Mu undergoes replicative transposition by the same mechanism as TnA. It also can use its cointegrate intermediate to transpose by a nonreplicative mechanism. The difference between this reaction and the nonreplicative transposition of IS elements is that the cleavage events occur in a different order.

The best characterized transposons in plants are the controlling elements of maize, which fall into several families. Each family contains a single type of autonomous element, analogous to bacterial transposons in its ability to mobilize. A family also contains many different nonautonomous elements, derived by mutations (usually deletions) of the autonomous element. The nonautonomous elements lack the ability to transpose, but display transposition activity and other abilities of the autonomous element, when an autonomous element is present to provide the necessary *trans*-acting functions.

In addition to the direct consequences of insertion and excision, the maize elements may also control the activities of genes at or near the sites where they are inserted; this control may be subject to developmental regulation. Maize elements inserted into genes may be excised from the transcripts, which explains why they do not simply impede gene activity. Control of target gene expression involves a variety of molecular effects, including activation by provision of an enhancer and suppression by interference with posttranscriptional events.

Transposition of maize elements (in particular *Ac*) is nonreplicative, probably requiring only a single transposase enzyme coded by the element. Transposition occurs preferentially after replication of the element. There are probably mechanisms to limit the frequency of transposition. Advantageous rearrangements of the maize genome may have been effected by the elements.

P elements in *D. melanogaster* are responsible for hybrid dysgenesis, which could be a forerunner of speciation. A cross between a male carrying *P* elements and a female lacking them generates hybrids that are sterile. A *P* element has four open reading frames, separated by introns. Splicing of the first three ORFs generates a 66 kD repressor, and occurs in all cells. Splicing of all four ORFs to generate the 87 kD transposase occurs only in the germline, by a tissue-specific splicing event. *P* elements mobilize when exposed to cytoplasm lacking the repressor. The burst of transposition events inactivates the genome by random insertions. Only a complete *P* element can generate transposase, but defective elements can be mobilized in *trans* by the enzyme.

Retroviruses and Retroposons

22.1 Introduction

Key Terms

- A **retrovirus** is an RNA virus with the ability to convert its sequence into DNA by reverse transcription.
- A **retroposon (retrotransposon)** is a transposon that mobilizes via an RNA form; the DNA element is transcribed into RNA, and then reverse-transcribed into DNA, which is inserted at a new site in the genome. The difference from retroviruses is that the retroposon does not have an infective (viral) form.

Transposition that involves an obligatory intermediate of RNA is unique to eukaryotes, and was discovered by the ability of **retroviruses** to insert DNA copies (proviruses) of an RNA viral genome into the chromosomes of a host cell. A similar mechanism is used by eukaryotic transposons that transpose through RNA intermediates. These elements are called **retroposons** (or sometimes **retrotransposons**). The unifying feature of retroviruses and retroposons is the use of a reverse transcriptase enzyme activity to generate a DNA copy of the element from an RNA. Retroviruses and retroposons share with all transposons the diagnostic feature of generating short direct repeats of target DNA at the site of an insertion. The major distinction between them is that retroviruses are packaged into infectious protein coats, but retroposons are solely intracellular.

Even in genomes where active transposons have not been detected, footprints of ancient transposition events are found in the form of direct target repeats flanking dispersed repetitive sequences. The features of these sequences sometimes implicate an RNA sequence as the progenitor of the genomic (DNA) sequence. This suggests that the RNA must have been a substrate that was converted into a duplex

DNA by the reverse transcriptase of a retrovirus or retroposon, and then inserted into the genome by a transposition–like event.

Like any other reproductive cycle, the cycle of a retrovirus or retroposon is continuous; it is arbitrary at which point we interrupt it to consider a "beginning." But our perspectives of these elements are biased by the forms in which we usually observe them, as indicated in **Figure 22.1**. Retroviruses were first observed as infectious virus particles, capable of transmission between cells, and so the intracellular cycle (involving duplex DNA) is thought of as the means of reproducing the RNA virus. Retroposons were discovered as components of the genome, and the RNA forms have been mostly characterized for their functions as mRNAs. So we think of retroposons as genomic (duplex DNA) sequences that may transpose within a genome; they do not migrate between cells.

22.2 The Retrovirus Life Cycle Involves Transposition–like Events

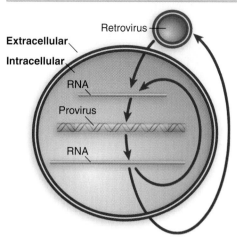

Figure 22.1 The reproductive cycles of retroviruses and retroposons alternate reverse transcription from RNA to DNA with transcription from DNA to RNA. Only retroviruses can generate infectious particles. Retroposons are confined to an intracellular cycle.

Key Terms

- **Provirus** is a duplex sequence of DNA integrated into a eukaryotic genome; it represents the sequence of the RNA genome of a retrovirus.

- **Reverse transcriptase** is an enzyme that uses a template of single-stranded RNA to generate a double-stranded DNA copy.

- An **integrase** is an enzyme that is responsible for a site-specific recombination that inserts one molecule of DNA into another.

Key Concepts

- A retrovirus has two copies of its genome of single-stranded RNA.

- An integrated provirus is a double-stranded DNA sequence.

- A retrovirus generates a provirus by reverse transcription of the retroviral genome.

Retroviruses have genomes of single-stranded RNA that are replicated through a double-stranded DNA intermediate. The life cycle of the virus has an obligatory stage in which the double-stranded DNA is inserted into the host genome by a transposition-like event that generates short direct repeats of target DNA.

The significance of this reaction extends beyond the perpetuation of the virus. Some of its consequences are that:

- A retroviral sequence that is integrated in the germline remains in the cellular genome as an endogenous **provirus**. Like a lysogenic bacteriophage, a provirus behaves as part of the genetic material of the organism.
- Cellular sequences occasionally recombine with the retroviral sequence and then are transposed with it; these sequences may be inserted into the genome as duplex sequences in new locations.
- Cellular sequences that are transposed by a retrovirus may change the properties of a cell that becomes infected with the virus.

The particulars of the retroviral life cycle are expanded in **Figure 22.2**. The crucial steps are that the viral RNA is converted into DNA, the DNA becomes integrated into the host genome, and then the DNA provirus is transcribed into RNA.

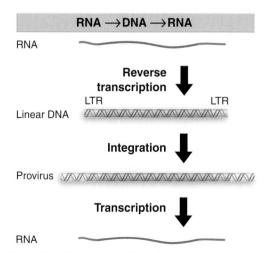

Figure 22.2 The retroviral life cycle proceeds by reverse transcribing the RNA genome into duplex DNA, which is inserted into the host genome, in order to be transcribed into RNA. LTRs are long terminal repeats.

The enzyme responsible for generating the initial DNA copy of the RNA is **reverse transcriptase**. The enzyme converts the RNA into a linear duplex of DNA in the cytoplasm of the infected cell. The DNA also is converted into circular forms, but these do not appear to be involved in reproduction.

The linear DNA makes its way to the nucleus. One or more DNA copies become integrated into the host genome. A single enzyme, called **integrase**, is responsible for integration. The provirus is transcribed by the host machinery to produce viral RNAs, which serve both as mRNAs and as genomes for packaging into virions. Integration is a normal part of the life cycle and is necessary for transcription.

Two copies of the RNA genome are packaged into each virion, making the individual virus particle effectively diploid. When a cell is simultaneously infected by two different but related viruses, it is possible to generate heterozygous virus particles carrying one genome of each type. The diploidy may be important in allowing the virus to acquire cellular sequences. The enzymes reverse transcriptase and integrase are carried with the genome in the viral particle.

22.3 Retroviral Genes Code for Polyproteins

Key Concepts

- A typical retrovirus has three genes: *gag, pol, env*.
- Gag and Pol proteins are translated from a full-length transcript of the genome.
- Translation of Pol requires a frameshift by the ribosome.
- Env is translated from a separate mRNA that is generated by splicing.
- Each of the three protein products is processed by proteases to give multiple proteins.

A typical retroviral sequence contains three or four "genes," the term here identifying coding regions each of which actually gives rise to multiple proteins by processing reactions. A typical retrovirus genome with three genes is organized in the sequence *gag–pol–env* as indicated in **Figure 22.3**.

The *gag* gene gives rise to the protein components of the nucleo-protein core of the virion. The *pol* gene codes for functions concerned with nucleic acid synthesis and recombination. The *env* gene codes for components of the envelope of the particle, which also sequesters components from the cellular cytoplasmic membrane.

Both the Gag or Gag–Pol and the Env products are polyproteins that are cleaved by a protease to release the individual proteins found in mature virions. The protease activity is coded by the virus in various forms: it may be part of Gag or Pol, or sometimes takes the form of an additional independent reading frame.

The Gag and Pol polyproteins are translated from an mRNA corresponding to the full length of the virus. The Gag product is translated by reading from the initiation codon to the first termination codon. The termination codon at the end of *gag* must be bypassed to express *pol*. Different mechanisms are used in different viruses to proceed beyond the *gag* termination codon, depending on the relationship between the *gag* and *pol* reading frames. When *gag* and *pol* follow continuously, suppression by a glutamyl-tRNA that recognizes the termination codon allows a single protein to be generated. When *gag* and *pol* are in different reading frames, a ribosomal frameshift occurs to generate a single protein.

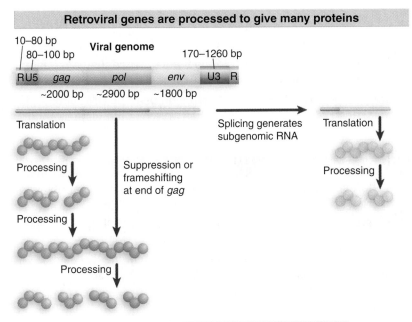

Figure 22.3 The genes of the retrovirus are expressed as polyproteins that are processed into individual products.

Each gene generates several protein products
Gag MA = matrix (between nucleocapsid and viral envelope) CA = capsid (major structural component) NC = nucleocapsid (packaging the dimer of RNA)
Pol PR = protease (cleaves Gag–Pol and Env) RT = reverse transcriptase (synthesizes DNA) IN = integrase (integrates provirus DNA into genome)
Env SU = surface protein (spikes on virion interact with host) TM = transmembrane (mediates virus–host fusion)

Usually the readthrough is ~5% efficient, so Gag protein outnumbers Gag–Pol protein about 20-fold.

The Env polyprotein is expressed by another means: splicing of the full-length transcript generates a shorter *subgenomic* messenger that is translated into the Env product.

A retroviral particle is produced by packaging the RNA into a core, surrounding it with capsid proteins, and pinching off a segment of membrane from the host cell. The release of infective particles by such means is shown in **Figure 22.4**. The process is reversed during infection; a virus infects a new host cell by fusing with the plasma membrane and then releasing the contents of the virion into the cell.

22.4 Viral DNA Is Generated by Reverse Transcription

Key Terms

- A **plus strand virus** has a single-stranded nucleic acid genome whose sequence directly codes for the protein products.

- **Minus strand DNA** is the single-stranded DNA sequence that is complementary to the viral RNA genome of a plus strand virus.

- **Plus strand DNA** is the strand of the duplex sequence representing a retrovirus that has the same sequence as the RNA.

- The **R segments** are the sequences that are repeated at the ends of a retroviral RNA. They are called R-U5 and U3-R.

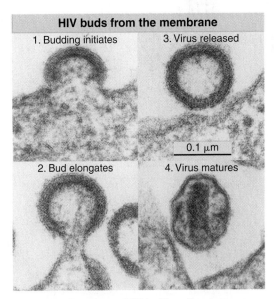

Figure 22.4 Retroviruses (HIV) bud from the plasma membrane of an infected cell. Photograph kindly provided by Dr. Matthew A. Gonda.

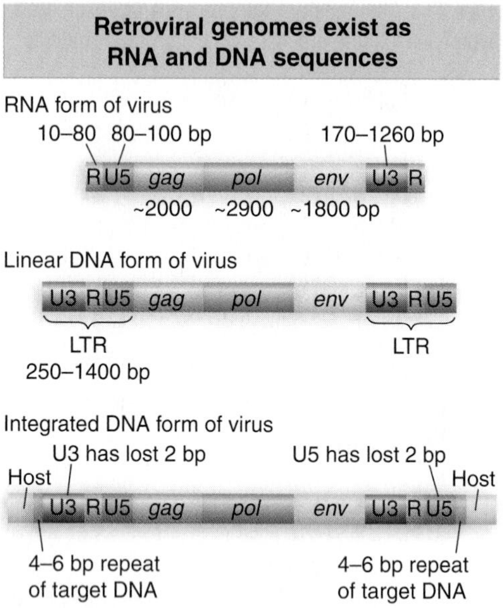

Retroviral genomes exist as RNA and DNA sequences

RNA form of virus
10–80 80–100 bp   170–1260 bp

R U5 | gag | pol | env | U3 R
~2000 ~2900 ~1800 bp

Linear DNA form of virus

U3 R U5 | gag | pol | env | U3 R U5

LTR   LTR
250–1400 bp

Integrated DNA form of virus

U3 has lost 2 bp   U5 has lost 2 bp
Host   Host

U3 R U5 | gag | pol | env | U3 R U5

4–6 bp repeat   4–6 bp repeat
of target DNA   of target DNA

Figure 22.5 Retroviral RNA ends in direct repeats (R), the free linear DNA ends in LTRs (long terminal repeats), and the provirus ends in LTRs that are shortened by two bases each.

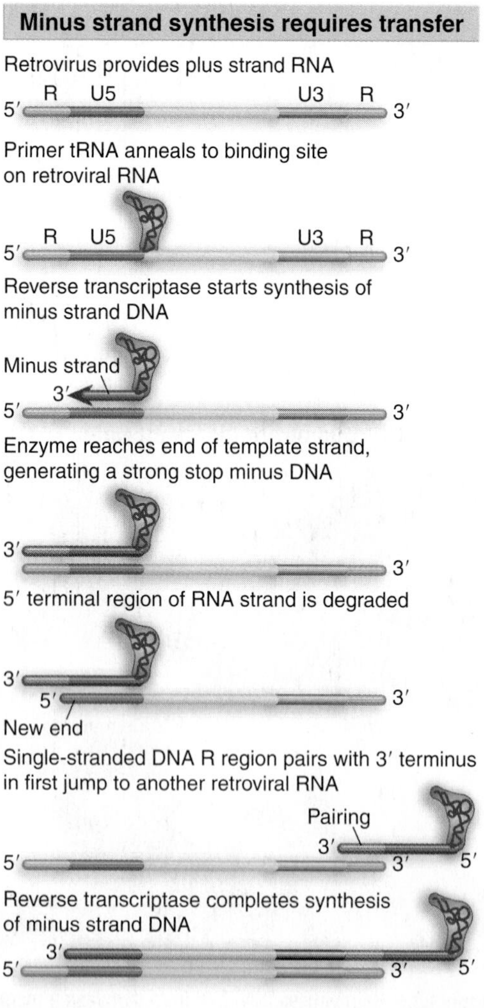

Minus strand synthesis requires transfer

Retrovirus provides plus strand RNA

5′   R   U5   U3   R   3′

Primer tRNA anneals to binding site on retroviral RNA

5′   R   U5   U3   R   3′

Reverse transcriptase starts synthesis of minus strand DNA

Minus strand
5′   3′   3′

Enzyme reaches end of template strand, generating a strong stop minus DNA

3′   3′

5′ terminal region of RNA strand is degraded

3′   3′
5′
New end

Single-stranded DNA R region pairs with 3′ terminus in first jump to another retroviral RNA

Pairing
3′
5′   3′   5′

Reverse transcriptase completes synthesis of minus strand DNA

5′   3′   3′   5′

Figure 22.6 Minus strand DNA is generated by switching templates during reverse transcription.

- **U5** is the repeated sequence at the 5′ end of a retroviral RNA.
- **U3** is the repeated sequence at the 3′ end of a retroviral RNA.
- The **long terminal repeat (LTR)** is the sequence that is repeated at each end of the integrated retroviral genome.
- **Copy choice** is a type of recombination used by RNA viruses, in which the RNA polymerase switches from one template to another during synthesis.

Key Concepts

- A short sequence (R) is repeated at each end of the viral RNA, so the 5′ and 3′ ends respectively are R-U5 and U3-R.
- Reverse transcriptase starts synthesis when a tRNA primer binds to a site 100–200 bases from the 5′ end.
- When the enzyme reaches the end, the 5′–terminal bases of RNA are degraded, exposing the 3′ end of the DNA product.
- The exposed 3′ end base pairs with the 3′ terminus of another RNA genome.
- Synthesis continues, generating a product in which the 5′ and 3′ regions are repeated, giving each end the structure U3–R–U5.
- Similar strand switching events occur when reverse transcriptase uses the DNA product to generate a complementary strand.
- Strand switching is an example of the copy choice mechanism of recombination.

Retroviruses are called **plus strand viruses**, because the viral RNA itself codes for the protein products. The enzyme reverse transcriptase is responsible for converting the genome (plus strand RNA) into a complementary DNA strand, which is called the **minus strand DNA**. Reverse transcriptase also catalyzes subsequent stages in the production of duplex DNA. It has a DNA polymerase activity, which enables it to synthesize a duplex DNA from the single-stranded reverse transcript of the RNA. The second DNA strand in this duplex is called **plus strand DNA**. And as a necessary adjunct to this activity, the reverse transcriptase has an RNAase H activity, which can degrade the RNA part of the RNA–DNA hybrid. All retroviral reverse transcriptases share considerable similarities of amino acid sequence, and homologous sequences can be recognized in some other retroposons.

The structures of the DNA forms of the virus are compared with the RNA in **Figure 22.5**. The viral RNA has direct repeats at its ends. These **R segments** vary in different strains of virus from 10–80 nucleotides. The sequence at the 5′ end of the virus is R-**U5**, and the sequence at the 3′ end is **U3**-R. The R segments are used during the conversion from RNA to DNA to generate the more extensive direct repeats that are found in linear DNA (see **Figure 22.6** and **Figure 22.7**). The shortening of 2 bp at each end in the integrated form is a consequence of the mechanism of integration (see Figure 22.9).

Like other DNA polymerases, reverse transcriptase requires a primer. The native primer is tRNA. An uncharged host tRNA is present in the virion. A sequence of 18 bases at the 3′ end of the tRNA is base paired to a site 100–200 bases from the 5′ end of one of the viral RNA molecules. The tRNA may also be base paired to another site near the 5′ end of the other viral RNA, thus assisting in dimer formation between the viral RNAs.

Here is a dilemma. Reverse transcriptase starts to synthesize DNA at a site only 100–200 bases downstream from the 5′ end. How can

DNA be generated to represent the intact RNA genome? (This is an extreme variant of the general problem in replicating the ends of any linear nucleic acid; see *16.2 The Ends of Linear DNA Are a Problem for Replication*.)

Synthesis *in vitro* proceeds to the end, generating a short DNA sequence called minus strong-stop DNA. This molecule is not found *in vivo* because it is immediately used in the continuing reaction illustrated in Figure 22.6. Reverse transcriptase switches templates, carrying the nascent DNA with it to the new template. This is the first of two jumps between templates.

In this reaction, the R region at the 5′ terminus of the RNA template is degraded by the RNAase H activity of reverse transcriptase. Its removal allows the R region at a 3′ end to base pair with the newly synthesized DNA. The source of the R region that pairs with the strong-stop minus DNA is usually the 3′ end of a different RNA molecule (intermolecular pairing). Then reverse transcription continues through the U3 region into the body of the RNA.

The result of the switch and extension is to add a U3 segment to the 5′ end. The stretch of sequence U3–R–U5 is called the **long terminal repeat** (**LTR**) because a similar series of events adds a U5 segment to the 3′ end, giving it the same structure of U5–R–U3. Its length varies from 250–1400 bp (see Figure 22.5).

We now need to generate the plus strand of DNA and to generate the LTR at the other end. The reaction is shown in Figure 22.7. Reverse transcriptase primes synthesis of plus strand DNA from a fragment of RNA that is left after degrading the original RNA molecule. A strong-stop plus strand DNA is generated when the enzyme reaches the end of the template. This DNA is then transferred to the other end of a minus strand. Probably it is released by a displacement reaction when a second round of DNA synthesis occurs from a primer fragment farther upstream (to its left in the figure). It uses the R region to pair with the 3′ end of a minus strand DNA. This double-stranded DNA then requires completion of both strands to generate a duplex LTR at each end.

Each retroviral particle carries two RNA genomes. This makes it possible for recombination to occur during a viral life cycle. In principle this could occur during minus strand synthesis and/or during plus strand synthesis:

- The intermolecular pairing shown in Figure 22.6 allows recombination to occur between sequences of the two successive RNA templates when minus strand DNA is synthesized. Retroviral recombination is mostly due to strand transfer at this stage, when the nascent DNA strand is transferred from one RNA template to another during reverse transcription.
- Plus strand DNA may be synthesized discontinuously, in a reaction with several internal initiations. Strand transfer during this reaction can also occur, but is less common.

The common feature of both events is that recombination results from a change in the template during the act of DNA synthesis. This is a general example of a mechanism for recombination called **copy choice**. For many years this was regarded as a possible mechanism for general recombination. It is unlikely to be employed by cellular systems, but is a common basis for recombination during infection by RNA viruses, including those that replicate exclusively through RNA forms, such as poliovirus.

Strand switching occurs with a certain frequency during each cycle of reverse transcription, that is, in addition to the transfer reaction that is forced at the end of the template strand. The principle is illustrated in **Figure 22.8**, although we do not know much about the mechanism.

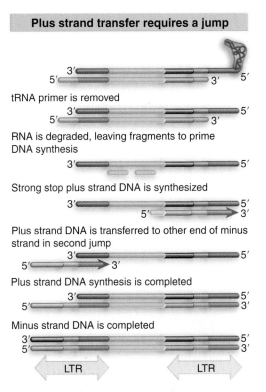

Plus strand transfer requires a jump

tRNA primer is removed

RNA is degraded, leaving fragments to prime DNA synthesis

Strong stop plus strand DNA is synthesized

Plus strand DNA is transferred to other end of minus strand in second jump

Plus strand DNA synthesis is completed

Minus strand DNA is completed

LTR   LTR

Figure 22.7 Synthesis of plus strand DNA requires a second jump.

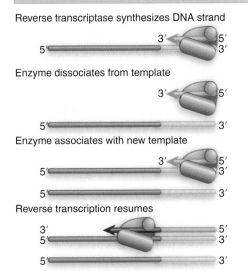

Strand transfer causes recombination

Reverse transcriptase synthesizes DNA strand

Enzyme dissociates from template

Enzyme associates with new template

Reverse transcription resumes

Figure 22.8 Copy choice recombination occurs when reverse transcriptase releases its template and resumes DNA synthesis using a new template. Transfer between template strands probably occurs directly, but is shown here in separate steps to illustrate the process.

22.5 Viral DNA Integrates into the Chromosome

Key Concepts

- The organization of proviral DNA in a chromosome is the same as a transposon, with the provirus flanked by short direct repeats of a sequence at the target site.
- Linear DNA is inserted directly into the host chromosome by the retroviral integrase enzyme.
- Two base pairs of DNA are lost from each end of the retroviral sequence during the integration reaction.

The organization of the integrated provirus resembles that of the linear DNA. The LTRs at each end of the provirus are identical. The 3′ end of U5 consists of a short inverted repeat relative to the 5′ end of U3, so the LTR itself ends in short inverted repeats. The integrated proviral DNA is like a transposon: the proviral sequence ends in inverted repeats and is flanked by short direct repeats of target DNA.

The provirus is generated by directly inserting a linear DNA into a target site. Integration of linear DNA is catalyzed by a single viral product, the integrase. Integrase acts on both the retroviral linear DNA and the target DNA. The reaction is illustrated in **Figure 22.9**.

The ends of the viral DNA are important; as is the case with transposons; mutations in the ends prevent integration. The most conserved feature is the presence of the dinucleotide sequence CA close to the end of each inverted repeat. The integrase brings the ends of the linear DNA together in a ribonucleoprotein complex, and converts the blunt ends into recessed ends by removing the bases beyond the conserved CA; usually this involves loss of two bases.

Target sites are chosen at random with respect to sequence. The integrase makes staggered cuts at a target site. In the example of Figure 22.9, the cuts are separated by 4 bp. The length of the target repeat depends on the particular virus; it may be 4, 5, or 6 bp. Presumably it is determined by the geometry of the reaction of integrase with target DNA.

The 5′ ends generated by the cleavage of target DNA are covalently joined to the 3′ recessed ends of the viral DNA. At this point, both termini of the viral DNA are joined by one strand to the target DNA. The single-stranded region is repaired by enzymes of the host cell, and in the course of this reaction the protruding two bases at each 5′ end of the viral DNA are removed. The result is that the integrated viral DNA has lost 2 bp at each LTR; this corresponds to the loss of 2 bp from the left end of the 5′ terminal U3 and loss of 2 bp from the right end of the 3′ terminal U5. There is a characteristic short direct repeat of target DNA at each end of the integrated retroviral genome.

The viral DNA integrates into the host genome at randomly selected sites. A successfully infected cell gains 1–10 copies of the provirus. (An infectious virus enters the cytoplasm, of course, but the DNA form becomes integrated into the genome in the nucleus. Retroviruses can replicate only in proliferating cells, because entry into the nucleus requires the cell to pass through mitosis, when the viral genome gains access to the nuclear material.)

The U3 region of each LTR carries a promoter. The promoter in the left LTR is responsible for initiating transcription of the provirus. Recall that the generation of proviral DNA is required to place the U3 sequence at the left LTR; so we see that the promoter is in fact generated by the conversion of the RNA into duplex DNA.

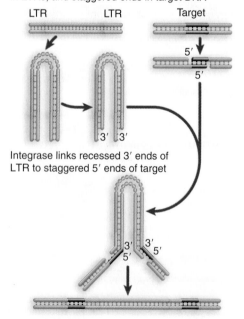

Integrase catalyzes all the stages of integration

Integrase generates two base recessed 3′ ends in LTRs, and staggered ends in target DNA

LTR   LTR   Target

Integrase links recessed 3′ ends of LTR to staggered 5′ ends of target

Figure 22.9 Integrase is the only viral protein required for the integration reaction, in which each LTR loses 2 bp and is inserted between 4 bp repeats of target DNA.

Sometimes (probably rather rarely), the promoter in the right LTR sponsors transcription of the host sequences that are adjacent to the site of integration. The LTR also carries an enhancer (a sequence that activates promoters in the vicinity) that can act on cellular as well as viral sequences. Integration of a retrovirus can be responsible for converting a host cell into a tumorigenic state when certain types of genes are activated in this way.

Can integrated proviruses be excised from the genome? Homologous recombination could take place between the LTRs of a provirus; solitary LTRs that could be relics of an excision event are present in some cellular genomes.

We have dealt so far with retroviruses in terms of the infective cycle, in which integration is necessary for the production of further copies of the RNA. However, when a viral DNA integrates in a germline cell, it becomes an inherited "endogenous provirus" of the organism. Endogenous viruses usually are not expressed, but sometimes they are activated by external events, such as infection with another virus.

22.6 Retroviruses May Transduce Cellular Sequences

Key Terms

- A **transducing virus** carries part of the host genome in place of part of its own sequence. The best known examples are retroviruses in eukaryotes and DNA phages in *E. coli*.
- A **replication-defective virus** cannot perpetuate an infective cycle because some of the necessary genes are absent (replaced by host DNA in a transducing virus) or mutated.
- A **helper virus** provides functions absent from a defective virus, enabling the latter to complete the infective cycle during a mixed infection.

Key Concept

- Transforming retroviruses are generated by a recombination event in which a cellular RNA sequence replaces part of the retroviral RNA.

An interesting light on the viral life cycle is cast by the occurrence of **transducing viruses**, variants that have acquired cellular sequences in the form illustrated in **Figure 22.10**. Part of the viral sequence has been replaced by the *v–onc* gene. Protein synthesis generates a Gag-v-Onc protein instead of the usual Gag, Pol, and Env proteins. The resulting virus is **replication defective**; it cannot sustain an infective cycle by itself. However, it can be perpetuated in the company of a **helper virus** that provides the missing viral functions.

A *v–onc* gene confers upon a virus the ability to transform a certain type of host cell. Every *v–onc* gene has homologous sequences in the host genome; these are called *c–onc* genes. A revealing discrepancy in the structures of *c–onc* and *v–onc* genes is that the *c–onc* genes usually are interrupted by introns, but the *v–onc* genes are uninterrupted. This suggests that the *v–onc* genes originate from spliced RNA copies of the *c–onc* genes.

A model for the formation of transforming viruses is illustrated in **Figure 22.11**. A retrovirus has integrated near a *c–onc* gene. A deletion

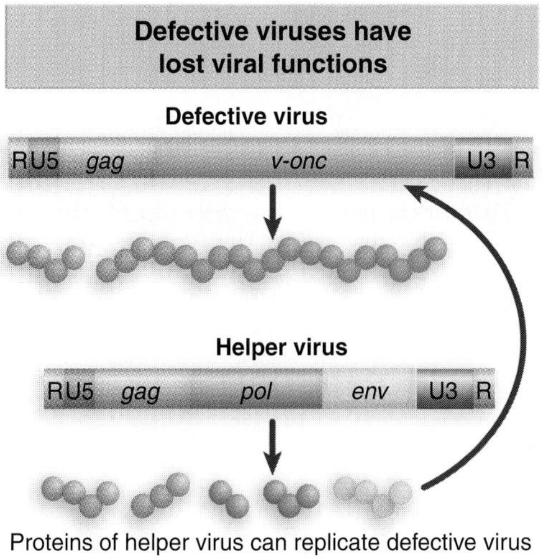

Defective viruses have lost viral functions

Defective virus

RU5   *gag*   *v-onc*   U3 R

Helper virus

RU5   *gag*   *pol*   *env*   U3 R

Proteins of helper virus can replicate defective virus

Figure 22.10 Replication-defective transforming viruses have a cellular sequence substituted for part of the viral sequence. The defective virus may replicate with the assistance of a helper virus that carries the wild-type functions.

Replication-defective viruses are generated by deletion

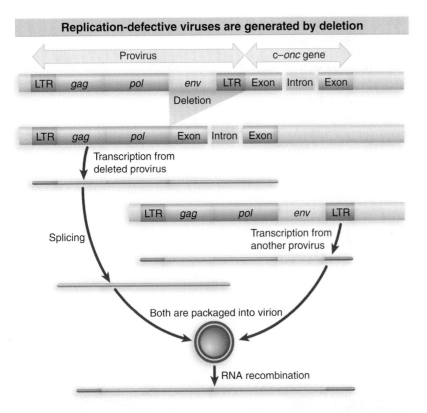

Figure 22.11 Replication-defective viruses may be generated through integration and deletion of a viral genome to generate a fused viral–cellular transcript that is packaged with a normal RNA genome. Nonhomologous recombination is necessary to generate the replication-defective transforming genome.

occurs to fuse the provirus to the *c–onc* gene; then transcription generates a joint RNA, containing viral sequences at one end and cellular *onc* sequences at the other end. Splicing removes the introns in both the viral and cellular parts of the RNA. The RNA has the appropriate signals for packaging into the virion; virions will be generated if the cell also contains another, intact copy of the provirus. Then some of the diploid virus particles may contain one fused RNA and one viral RNA.

A recombination between these sequences could generate the transforming genome, in which the viral repeats are present at both ends. (Recombination occurs at a high frequency during the retroviral infective cycle, by various means. We do not know anything about its demands for homology in the substrates, but we assume that the nonhomologous reaction between a viral genome and the cellular part of the fused RNA proceeds by the same mechanisms responsible for viral recombination.)

The common features of the entire retroviral class suggest that it may be derived from a single ancestor. Primordial IS elements could have surrounded a host gene for a nucleic acid polymerase; the resulting unit would have the form LTR–pol–LTR. It might evolve into an infectious virus by acquiring more sophisticated abilities to manipulate both DNA and RNA substrates, including the incorporation of genes whose products allowed packaging of the RNA. Other functions, such as transforming genes, might be incorporated later. (There is no reason to suppose that the mechanism involved in acquisition of cellular functions is unique for *onc* genes; but viruses carrying these genes may have a selective advantage because of their stimulatory effect on cell growth.)

22.7 Yeast *Ty* Elements Resemble Retroviruses

Key Term

- *Ty* stands for transposon yeast, the first transposable element to be identified in yeast.

Key Concepts

- *Ty* transposons have a similar organization to endogenous retroviruses.
- They are retroposons, with a reverse transcriptase activity, that transpose via an RNA intermediate.

Ty elements constitute a family of dispersed repetitive DNA sequences that are found at different sites in different strains of yeast. *Ty* is an abbreviation for "transposon yeast." A transposition event creates a characteristic footprint: 5 bp of target DNA are repeated on either side of the inserted *Ty* element. *Ty* elements are **retroposons** that transpose by the same mechanism as retroviruses. The frequency of *Ty* transposition is lower than that of most bacterial transposons, $\sim 10^{-7}$–10^{-8}.

There is considerable divergence between individual *Ty* elements. Most elements fall into one of two major classes, called *Ty1* and *Ty917*.

They have the same general organization illustrated in **Figure 22.12**. Each element is 6.3 kb long; the last 330 bp at each end constitute direct repeats, called δ. There are ~30 copies of the *Ty1* type and ~6 of the *Ty917* type in a typical yeast genome. In addition, there are ~100 independent *delta* elements, called solo δs.

The sequence of the *Ty* element has two open reading frames, *TyA* and *TyB*, expressed in the same direction, but read in different phases and overlapping by 13 amino acids. The TyA DNA-binding protein represents the *TyA* reading frame, and terminates at its end. The *TyB* reading frame is expressed only as part of a joint protein, in which the *TyA* region is fused to the *TyB* region by a specific frameshift event that allows the termination codon to be bypassed (analogous to *gag–pol* translation in retroviruses). The sequence of *TyB* contains regions that have homologies with reverse transcriptase, protease, and integrase sequences of retroviruses.

Ty elements can excise by homologous recombination between the directly repeated *delta* sequences. The large number of solo *delta* elements may be footprints of such events. An excision of this nature may be associated with reversion of a mutation caused by the insertion of *Ty*; the level of reversion may depend on the exact *delta* sequences left behind.

Although the *Ty* element does not give rise to infectious particles, viruslike particles (VLPs) accumulate within the cells in which transposition has been induced. The particles can be seen in **Figure 22.13**. They contain full-length RNA, double-stranded DNA, reverse transcriptase activity, and a TyB product with integrase activity. The TyA product is cleaved like a *gag* precursor to produce the mature core proteins of the VLP. This takes the analogy between the *Ty* transposon and the retrovirus even further. The *Ty* element behaves in short like a retrovirus that has lost its *env* gene and therefore cannot properly package its genome.

Only some of the *Ty* elements in any yeast genome are active: most have lost the ability to transpose (and are analogous to inert endogenous proviruses). Since these "dead" elements retain the δ repeats, however, they provide targets for transposition in response to the proteins synthesized by an active element.

22.8 Many Transposable Elements Reside in *D. melanogaster*

Key Concept

- *copia* is a retroposon that is abundant in *D. melanogaster*.

The *Drosophila* genome contains several types of transposable sequences, as illustrated in **Figure 22.14**. They include the *copia* retroposon, the *FB* family, and the *P* elements discussed previously in *21.12 Transposition of P Elements Causes Hybrid Dysgenesis*.

The best-characterized family of retroposons is *copia*. The *copia* family is taken as a paradigm for several other types of elements whose sequences are unrelated, but whose structure and general behavior appear to be similar. The number of copies of the *copia* element depends on the strain of fly; usually it is 20–60. The locations of *copia* elements show a different (although overlapping) spectrum in each strain of *D. melanogaster*.

The *copia* element is ~5000 bp long, with identical direct terminal repeats of 276 bp. Each of the direct repeats itself ends in related inverted repeats. A direct repeat of 5 bp of target DNA is generated at the site of insertion. The divergence between individual members of the *copia*

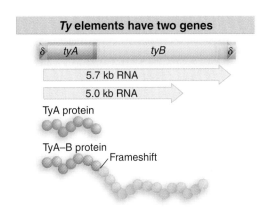

Figure 22.12 *Ty* elements terminate in short direct repeats and are transcribed into two overlapping RNAs. They have two reading frames, with sequences related to the retroviral *gag* and *pol* genes.

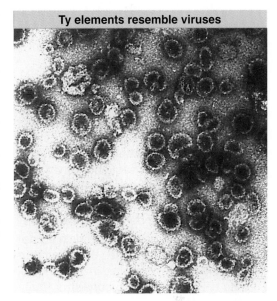

Figure 22.13 *Ty* elements generate viruslike particles. Photograph kindly provided by Alan Kingsman, Oxford Bio-Medica plc.

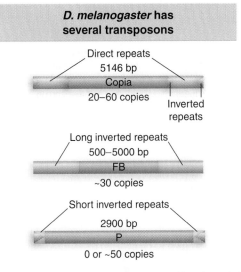

Figure 22.14 Three types of transposable element in *D. melanogaster* have different structures.

family is slight, <5%; variants often contain small deletions. All of these features are common to the other *copia*-like families, although their individual members display greater divergence.

The identity of the two direct repeats of each *copia* element implies either that they interact to permit correction events, or that both are generated from one of the direct repeats of a progenitor element during transposition. As in the similar case of *Ty* elements, this is suggestive of a relationship with retroviruses.

The *copia* sequence contains a single long reading frame of 4227 bp. There are homologies between parts of the *copia* open reading frame and the *gag* and *pol* sequences of retroviruses. A notable absence from the homologies is any relationship with retroviral *env* sequences required for the envelope of the virus, which means that *copia* is unlikely to be able to generate viruslike particles.

Transcripts of *copia* are found as abundant poly(A)$^+$ mRNAs, representing both full-length and part-length transcripts. The mRNAs have a common 5′ terminus, resulting from initiation in the middle of one of the terminal repeats. Several proteins are produced, probably involving events such as splicing of RNA and cleavage of polyproteins.

22.9 Retroposons Fall into Three Classes

Key Terms

- The **viral superfamily** comprises transposons that are related to retroviruses. They are defined by sequences that code for reverse transcriptase or integrase.

- The **nonviral superfamily** of transposons originated independently of retroviruses.

- **Interspersed repeats** were originally defined as short sequences that are common and widely distributed in the genome. They are now known to consist of transposable elements.

Key Concepts

- Retroposons of the viral superfamily are transposons that mobilize via an RNA that does not form an infectious particle.

- Some directly resemble retroviruses in their use of LTRs, but others do not have LTRs.

- Other elements can be found that were generated by an RNA-mediated transposition event, but do not themselves code for enzymes that can catalyze transposition.

- Transposons and retroposons constitute almost half of the human genome.

Retroposons are defined by their use of mechanisms for transposition that involve reverse transcription of RNA into DNA. Three classes of retroposons are distinguished in **Figure 22.15**:

- Members of the **viral superfamily** code for reverse transcriptase and/or integrase activities. Like other retroposons, they reproduce like retroviruses but differ from them in not passing through an

Figure 22.15 Retroposons can be divided into the viral superfamilies that are either retroviruslike or LINES and the nonviral superfamilies that do not have coding functions.

Eukaryotic genomes have three types of retroposons			
	Viral superfamily	LINES	Nonviral superfamily
Common types	*Ty (S. cerevisiae)* *copia (D. melanogaster)*	L1 (human) B1, B2 ID, B4 (mouse)	SINES (mammals) pseudogenes of pol III transcripts
Termini	Long terminal repeats	No repeats	No repeats
Target repeats	4–6 bp	7–21 bp	7–21 bp
Enzyme activities	Reverse transcriptase and/or integrase	Reverse transcriptase/ endonuclease	None (or none coding for transposon products)
Organization	May contain introns (removed in subgenomic mRNA)	1 or 2 uninterrupted ORFs	No introns

independent infectious form. They are best characterized in the *Ty* and *copia* elements of yeast and flies.

- LINES also have reverse transcriptase activity (and may therefore be considered to comprise more distant members of the viral superfamily), but they lack LTRs and use a different mechanism from retroviruses to prime the reverse transcription reaction. They are derived from RNA polymerase II transcripts. A minority of the elements in the LINES genome are fully functional and can transpose autonomously; others have mutations and so can only transpose as the result of the action of a *trans*-acting autonomous element.

- Members of the **nonviral superfamily** are identified by external and internal features that suggest they originated in RNA sequences, although in these cases we can only speculate on how a DNA copy was generated. We assume that they were targets for a transposition event by an enzyme system coded elsewhere; that is, they are always nonautonomous. They originated in cellular transcripts. They do not code for proteins that have transposition functions. The most prominent component of this family is called SINES. They are derived from RNA polymerase III transcripts.

Figure 22.16 shows the organization and sequence relationships of elements that code for reverse transcriptase. Like retroviruses, the LTR-containing retroposons can be classified into groups according to the number of independent reading frames for *gag*, *pol*, and *int*, and the order of the genes. In spite of these superficial differences of organization, the common feature is the presence of reverse transcriptase and integrase activities. Typical mammalian LINES elements have two reading frames, one coding for a nucleic-acid-binding protein, the other for reverse transcriptase and endonuclease activity.

LTR-containing elements can vary from integrated retroviruses to retroposons that have lost the capacity to generate infectious particles. Yeast and fly genomes have the *Ty* and *copia* elements that cannot generate infectious particles. Mammalian genomes have endogenous retroviruses that, when active, can generate infectious particles. The mouse genome has several active endogenous retroviruses which are able to generate particles that propagate horizontal infections. By contrast, almost all endogenous retroviruses in the human lineage lost their activity some 50 million years ago, and the human genome now has mostly inactive remnants of the endogenous retroviruses.

LINES and SINES are prominent components of the animal genome. They were defined originally by the existence of a large number of relatively short sequences that are related to one another (consisting of the moderately repetitive DNA described in *4.6 Eukaryotic Genomes Contain Both Nonrepetitive and Repetitive DNA Sequences*). LINES comprise long interspersed sequences, and SINES comprise short interspersed sequences. (They are described as interspersed sequences or **interspersed repeats** because of their common occurrence and widespread distribution.)

LINES and SINES make up a significant part of the repetitive DNA of animal genomes. In many higher eukaryotic genomes, they are ~50% of the total DNA. **Figure 22.17** summarizes the distribution of the different types of transposons of the human genome. Except for the SINES, which are always nonfunctional, the other

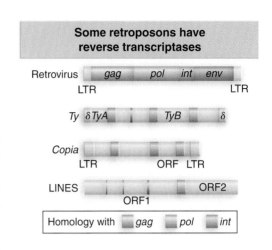

Figure 22.16 Retroposons that are closely related to retroviruses have a similar organization, but LINES share only the reverse transcriptase activity.

		Length (kb)	Human genome Number	Fraction
Retroviruses and transposons constitute half the human genome				
Element	Organization			
Retrovirus/retroposon	LTR *gag pol (env)* LTR	1–11	450,000	8%
LINES (autonomous) e.g., L1	ORF1 *(pol)* (A)n	6–8	850,000	17%
SINES (nonautonomous) e.g., Alu	(A)n	<0.3	1,500,000	15%
DNA transposon	Transposase	2–3	300,000	3%

Figure 22.17 Four types of transposable elements constitute almost half of the human genome.

types of elements all consist of both functional elements and elements that have suffered deletions eliminating parts of the reading frames that code for proteins needed for transposition. The relative proportions of these types of transposons are generally similar in mouse and human genomes.

A common LINES in mammalian genomes is called L1. The typical member is ~6,500 bp long, terminating in an A-rich tract. The two open reading frames of a full-length element are called ORF1 and ORF2. The number of full-length elements is usually small (~50), and the remainder of the copies are truncated. Transcripts can be found. As implied by its presence in repetitive DNA, the LINES family shows sequence variation among individual members. However, the members of the family within a species are relatively homogeneous compared to the variation shown between species. L1 is the only member of the LINES family that has been active in either the mouse or human lineages, and it seems to have remained highly active in the mouse, but has declined in the human lineage.

22.10 The Alu Family Has Many Widely Dispersed Members

Key Term

- The **Alu family** is a set of dispersed, related sequences, each ~300 bp long, in the human genome. The individual members have *Alu* cleavage sites at each end (hence the name).

Key Concept

- A major part of repetitive DNA in mammalian genomes consists of repeats of a single family organized like transposons and derived from RNA polymerase III transcripts.

The most prominent SINES comprises members of a single family. Its short length and high degree of repetition make it comparable to simple sequence (satellite) DNA, except that the individual members of the family are dispersed around the genome instead of being confined to tandem clusters. Again there is significant similarity between the members within a species compared with variation between species.

In the human genome, a large part of the moderately repetitive DNA exists as sequences of ~300 bp that are interspersed with nonrepetitive DNA. At least half of the renatured duplex material is cleaved by the restriction enzyme AluI at a single site, located 170 bp along the sequence. The cleaved sequences all are members of a single family, known as the **Alu family** after the means of its identification. There are ~300,000 members in the haploid genome (equivalent to one member per 6 kb of DNA). The individual *Alu* sequences are widely dispersed. A related sequence family is present in the mouse (where the 50,000 members are called the B1 family), in the Chinese hamster (where it is called the Alu-equivalent family), and in other mammals.

The members of the Alu family resemble transposons in being flanked by short direct repeats. The human family seems to have originated by a 130 bp tandem duplication, with an unrelated sequence of 31 bp inserted in the right half of the dimer. The two repeats are sometimes called the "left half" and "right half" of the *Alu* sequence. The individual members of the Alu family are related rather than identical, and have an average identity with the consensus sequence of 87%. The mouse B1 repeating unit is 130 bp long, corresponding to a monomer of the human unit. It has 70–80% homology with the human sequence.

The *Alu* sequence is related to 7SL RNA, a component of the signal recognition particle (see *10.4 The SRP Interacts with the SRP Receptor*). The 7SL RNA corresponds to the left half of an *Alu* sequence with an insertion in the middle. So the 90 5′ terminal bases of 7SL RNA are homologous to the left end of *Alu*, the central 160 bases of 7SL RNA have no homology to *Alu*, and the 40 3′ terminal bases of 7SL RNA are homologous to the right end of *Alu*. The 7SL RNA is coded by genes that are actively transcribed by RNA polymerase III. It is possible that these genes (or genes related to them) gave rise to the inactive *Alu* sequences.

Members of the Alu family may be included within structural gene transcription units, as seen by their presence in long nuclear RNA. The presence of multiple copies of the *Alu* sequence in a single nuclear molecule can generate secondary structure. In fact, the presence of Alu family members in the form of inverted repeats is responsible for most of the secondary structure found in mammalian nuclear RNA.

The *Alu* element is the only SINES that has been active in the human lineage. Its counterpart *B1* has been active in the mouse genome, which also has other active SINES (*B2, ID, B4*). The other mouse SINES appear to have originated from reverse transcripts of tRNAs. The transposition of the SINES probably results from their recognition as substrates by an active *L1* element.

22.11 Processed Pseudogenes Originated as Substrates for Transposition

Key Concept

- A processed pseudogene is an inactive gene copy that lacks introns, contrasted with the interrupted structure of the active gene. Such genes originate by reverse transcription of mRNA.

When a sequence generated by reverse transcription of an mRNA is inserted into the genome, we can recognize its relationship to the gene from which the mRNA was transcribed. Such a sequence is called a **processed pseudogene** to reflect the fact that it was processed from RNA and is not active. The characteristic features of a processed pseudogene are compared in **Figure 22.18** with the features of the original gene and the mRNA. The figure shows all the relevant diagnostic features, only some of which are found in any individual example. Any transcript of RNA polymerase II could in principle give rise to such a pseudogene, and there are many examples, including the processed globin pseudogenes that were the first to be discovered (*3.11 Pseudogenes Are Dead Ends of Evolution*).

The pseudogene may start at the point equivalent to the 5′ terminus of the RNA, which would be expected only if the DNA had originated from the RNA. Several pseudogenes consist of precisely joined exon sequences; we know of no mechanism to recognize introns in DNA, so this feature argues for an RNA-mediated stage. The pseudogene may end in a short stretch of A · T base pairs, presumably derived from the poly(A) tail of the RNA. On either side of the pseudogene is a short direct repeat, presumed to have been generated by a transposition–like event. Processed pseudogenes reside at locations unrelated to their presumed sites of origin.

The processed pseudogenes do not carry any information that might be used to sponsor a transposition event (or to carry out the

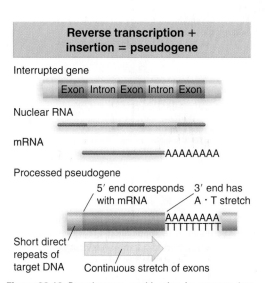

Figure 22.18 Pseudogenes could arise by reverse transcription of RNA to give duplex DNAs that become integrated into the genome.

preceding reverse transcription of the RNA). This suggests that the RNA was a substrate for another system, coded by a retroposon. In fact, it seems likely that the active LINES elements provide most of the reverse transcriptase activity, and they are responsible not only for their own transposition, but also for acting on the SINES and for generating processed pseudogenes.

22.12 LINES Use an Endonuclease to Generate a Priming End

Key Concept

- LINES do not have LTRs and require the retroposon to code for an endonuclease that generates a nick to prime reverse transcription.

LINES and some others elements, do not terminate in the LTRs that are typical of retroviral elements. This poses the question: how is reverse transcription primed? It does not involve the typical reaction in which a tRNA primer pairs with the LTR (see Figure 22.6). The open reading frames in these elements lack many of the retroviral functions, such as protease or integrase domains, but typically have reverse transcriptaselike sequences and code for an endonuclease activity. In the human LINES *L1*, ORF1 is a DNA-binding protein and ORF2 has both reverse transcriptase and endonuclease activities; both products are required for transposition.

Figure 22.19 shows how these activities support transposition. A nick is made in the DNA target site by an endonuclease activity coded by the retroposon. The RNA product of the element associates with the protein bound at the nick. The nick provides a 3′–OH end that primes synthesis of cDNA on the RNA template. A second cleavage event is required to open the other strand of DNA, and the RNA/DNA hybrid is linked to the other end of the gap either at this stage or after it has been converted into a DNA duplex. A similar mechanism is used by some mobile introns (see Figure 27.10).

One of the reasons why LINES are so effective lies with their method of propagation. When a LINES mRNA is translated, the protein products show a *cis*-preference for binding to the mRNA from which they were translated. **Figure 22.20** shows that the ribonucleoprotein complex then moves to the nucleus, where the proteins insert a DNA copy into the genome. Often reverse transcription does not proceed fully to the end, so the copy is inactive. However, there is the potential for insertion of an active copy, because the proteins are acting on a transcript of the original active element.

By contrast, the proteins produced by the DNA transposons of Chapter 21 must be imported into the nucleus after being synthesized in the cytoplasm, but they have no means of distinguishing full-length transposons from inactive deleted transposons. **Figure 22.21** shows that instead, they will indiscriminately recognize any element by virtue of the repeats that mark the ends, much reducing their chance of acting on a full-length element rather than one with a deletion. The consequence is that inactive elements accumulate and eventually the family dies out because a transposase has such a small chance of finding a target that is a fully functional transposon.

Are transposition events currently occurring in these genomes or are we seeing only the footprints of ancient systems? This varies with

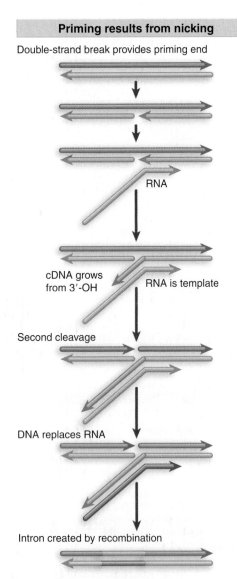

Priming results from nicking

Double-strand break provides priming end

RNA

cDNA grows from 3′-OH | RNA is template

Second cleavage

DNA replaces RNA

Intron created by recombination

Figure 22.19 Retrotransposition of non-LTR elements occurs by nicking the target to provide a primer for cDNA synthesis on an RNA template. The arrowheads indicate 3′ ends.

the species. There are few currently active transposons in the human genome, but by contrast several active transposons are known in the mouse genome. This explains the fact that spontaneous mutations caused by LINES insertions occur at a rate of ~3% in mouse, but at only 0.1% in man. There appear to be ~10–50 active LINES in the human genome. Some human diseases can be pinpointed as the result of transposition of *L1* into genes, and others result from unequal crossing-over events involving repeated copies of *L1*. A model system in which LINES transposition occurs in tissue culture cells suggests that a transposition event can introduce several types of collateral damage as well as inserting into a new site; the damage includes chromosomal rearrangements and deletions. Such events may be viewed as agents of genetic change. Neither DNA transposons nor retroviral–like retroposons seem to have been active in the human genome for 40–50 million years, but several active examples of both are found in the mouse.

Note that for transpositions to survive, they must occur in the germline. Presumably, similar events occur in somatic cells, but do not survive beyond one generation.

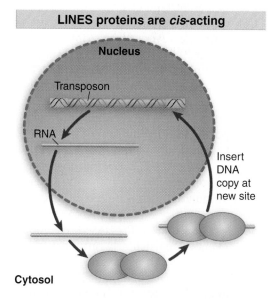

LINES proteins are *cis*-acting

Nucleus

Transposon

RNA

Insert DNA copy at new site

Cytosol

Figure 22.20 A LINES is transcribed into an RNA that is translated into proteins that assemble into a complex with the RNA. The complex translocates to the nucleus, where it inserts a DNA copy into the genome.

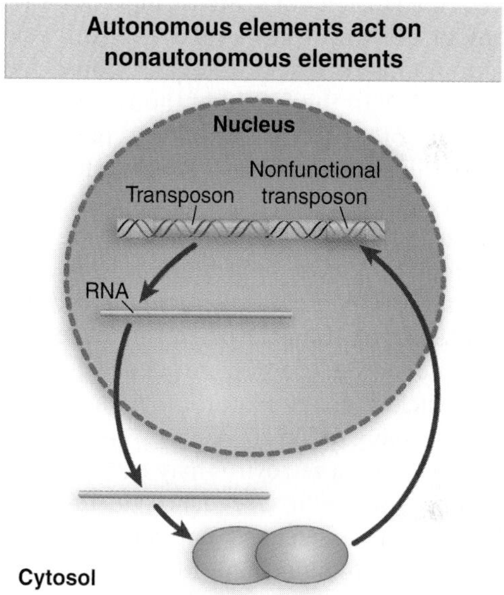

Autonomous elements act on nonautonomous elements

Nucleus

Transposon   Nonfunctional transposon

RNA

Cytosol

Figure 22.21 A transposon is transcribed into an RNA that is translated into proteins that move independently to the nucleus, where they act on any pair of inverted repeats with the same sequence as the original transposon.

22.13 SUMMARY

Reverse transcription is the unifying mechanism for reproduction of retroviruses and perpetuation of retroposons. The cycle of each type of element is in principle similar, although retroviruses are usually regarded from the perspective of the free viral (RNA) form, while retroposons are regarded from the stance of the genomic (duplex DNA) form.

Retroviruses have genomes of single-stranded RNA that are replicated through a double-stranded DNA intermediate. An individual retrovirus contains two copies of its genome. The genome contains the *gag*, *pol*, and *env* genes, which are translated into polyproteins, each of which is cleaved into smaller functional proteins. The Gag and Env components are concerned with packing RNA and generating the virion; the Pol components are concerned with nucleic acid synthesis.

Reverse transcriptase is the major component of Pol, and is responsible for synthesizing a DNA (minus strand) copy of the viral (plus strand) RNA. By switching template strands, reverse transcriptase copies the 3′ sequence of the RNA to the 5′ end of the DNA, and copies the 5′ sequence of the RNA to the 3′ end of the DNA. This generates the characteristic LTRs (long terminal repeats) of the DNA. A similar switch of templates occurs when the plus strand of DNA is synthesized using the minus strand as template. Linear duplex DNA is inserted into a host genome by the integrase enzyme. Transcription of the integrated DNA from a promoter in the left LTR generates further copies of the RNA sequence.

Switches in template during nucleic acid synthesis allow recombination to occur by copy choice. During an infective cycle, a retrovirus may exchange part of its usual sequence for a cellular sequence; the resulting virus is usually replication-defective, but can be perpetuated in the course of a joint infection with a helper virus. Many of the defective viruses have gained an RNA version (*v–onc*) of a cellular gene (*c–onc*). The *onc* sequence may be any one of a number of genes whose expression in *v–onc* form causes the cell to be transformed into a tumorigenic phenotype.

The integration event generates direct target repeats (like transposons that mobilize via DNA). An inserted provirus therefore has direct terminal repeats of the LTRs, flanked by short repeats of target DNA. Mammalian and avian genomes have endogenous (inactive) proviruses with such structures. Other elements with this organization have been found in a variety of genomes, most notably in *S. cerevisiae* and *D. melanogaster*. *Ty* elements of yeast and *copia* elements of flies have coding sequences with homology to reverse transcriptase, and mobilize via an RNA form. They may generate particles that resemble viruses but that do not have infectious capability. The LINES sequences of mammalian genomes are further removed from the retroviruses, but retain enough similarities to suggest a common origin. They use a different type of priming event to initiate reverse transcription, in which an endonuclease activity associated with the reverse transcriptase makes a nick that provides a 3′—OH end for priming synthesis on an RNA template. The frequency of LINES transposition is increased because its protein products are *cis*-acting; they associate with the mRNA from which they were translated to form a ribonucleoprotein complex that is transported into the nucleus.

Another class of retroposons has the hallmarks of transposition via RNA, but has no coding sequences (or at least none resembling retroviral functions). They may have originated as passengers in a retrovirallike transposition event, in which an RNA was a target for a reverse transcriptase. Processed pseudogenes arise by such events. A particularly prominent family apparently originating from a processing event is the mammalian SINES, including the human Alu family. Some snRNAs, including 7SL snRNA (a component of the SRP), are related to this family.

23

Recombination in the Immune System

23.1 Introduction

Key Terms

- An **immune response** is an organism's reaction, mediated by components of the immune system, to an antigen.

- An **antigen** is any foreign substance whose entry into an organism provokes an immune response by stimulating the synthesis of an antibody (an immunoglobulin protein that can bind to the antigen).

- A **B cell** is a lymphocyte that produces antibodies. B cells develop primarily in bone marrow.

- **T cells** are lymphocytes of the T (thymic) lineage; they are subdivided into several functional types. They carry T-cell receptor (TCR) and participate in the cell-mediated immune response.

- An **immunoglobulin** (antibody) is a class of protein that is produced by B cells in response to antigen.

- An **antibody** is a protein (immunoglobulin), produced by a B lymphocyte cell, that recognizes a particular "foreign antigen" and thus triggers the immune response.

- The **T-cell receptor (TCR)** is the antigen receptor on T lymphocytes. It is clonally expressed and binds to a complex of MHC class I or class II protein and antigen-derived peptide.

- **Clonal expansion** is the proliferation of mature lymphocytes stimulated by antigen binding. This proliferation stage is necessary for adaptive immune responses because it substantially increases the number of antigen-specific lymphocytes. After proliferation, lymphocytes differentiate into effector cells.

- A **superfamily** is a set of genes related by descent from a common ancestor but now showing considerable variation.

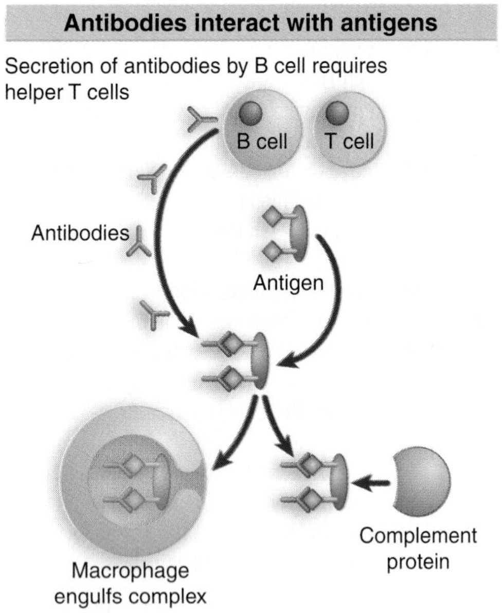

Antibodies interact with antigens

Secretion of antibodies by B cell requires helper T cells

Figure 23.1 Humoral immunity is conferred by the binding of free antibodies to antigens to form antigen-antibody complexes that are removed from the bloodstream by macrophages or that are attacked directly by the complement proteins.

T-cell receptor binds antigen fragments

Infected target cell degrades antigen into fragments

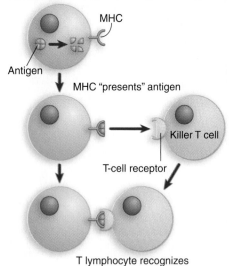

Figure 23.2 In cell-mediated immunity, killer T cells use the T-cell receptor to recognize a fragment of the foreign antigen that is presented on the surface of the target cell by the MHC protein.

It is an axiom of genetics that the genetic constitution created in the zygote by the combination of sperm and egg is inherited by all somatic cells of the organism. We look to differential control of gene expression, rather than to changes in DNA content, to explain the different phenotypes of particular somatic cells. Yet there are exceptional situations in which the reorganization of certain DNA sequences regulates gene expression or creates new genes. The immune system provides a striking and extensive case in which the content of the genome changes when recombination creates active genes in lymphocytes.

The **immune response** of vertebrates provides a protective system that distinguishes foreign proteins from the proteins of the organism itself. Foreign material (or part of the foreign material) is recognized as an **antigen**. Usually the antigen is a protein (or protein-attached moiety) that has entered the bloodstream of the animal—for example, the coat protein of an infecting virus. Exposure to an antigen initiates production of an immune response that *specifically recognizes the antigen and destroys it.*

Immune reactions are the responsibility of white blood cells—the B and T lymphocytes and macrophages. The lymphocytes are named after the tissues that produce them. In mammals, **B cells** mature in the bone marrow, while **T cells** mature in the thymus. *Each class of lymphocyte uses the rearrangement of DNA as a mechanism for producing the proteins that enable it to participate in the immune response.* B cells trigger the *humoral response*, and T cells trigger the *cell-mediated response.*

B cells secrete antibodies, which are **immunoglobulin** (Ig) proteins. *Production of an **antibody** specific for a foreign molecule is the primary event responsible for recognition of an antigen.* Recognition requires the antibody to bind to a small region or structure on the antigen.

The function of antibodies is represented in **Figure 23.1**. Foreign material circulating in the bloodstream, for example, a toxin or pathogenic bacterium, has a surface that presents antigens. The antigens are recognized by the antibodies, which form an antigen-antibody complex. This complex then attracts the attention of other components of the immune system, which destroy the antigen.

The T cells that recognize foreign antigens are called *cytotoxic* or *killer* T cells. Their basic function is indicated in **Figure 23.2**. They are usually activated by an intracellular parasite, such as a virus that infects the body's own cells. As a result of the viral infection, fragments of viral antigens are displayed on the surface of the cell in association with a host cell protein that belongs to the MHC class (major histocompatibility antigen). These fragments are recognized by the **T-cell receptor (TCR)**, which is the T cells' equivalent of the antibody produced by a B cell.

When an antigen is recognized by an antibody or T-cell receptor, the recognition triggers a change in the B- or T-lymphocyte that causes it to divide. By dividing many times, a large number of lymphocytes are produced, all carrying the same immunoglobulin or T-cell receptor. This **clonal expansion** gives the organism the cells required to fight the infection and confers immunity to new infections by the same antigen.

Immunoglobulins and T-cell receptors are direct counterparts, each produced by its own type of lymphocyte. The proteins are related in structure, and their genes are related in organization. The sources of variability are similar. The MHC proteins also share some common features with the antibodies, as do other lymphocyte-specific proteins. In dealing with the genetic organization of the immune system, we are therefore concerned with a series of related gene families—indeed, a **superfamily**—that may have evolved from some common ancestor representing a primitive immune response.

23.2   Immunoglobulin Genes Are Assembled from Their Parts in Lymphocytes

Key Terms

- The immunoglobulin **light chain** is one of two types of subunits in an antibody tetramer. Each antibody contains two light chains. The N-terminus of the light chain forms part of the antigen recognition site.
- The immunoglobulin **heavy chain** is one of two types of subunits in an antibody tetramer. Each antibody contains two heavy chains. The N-terminus of the heavy chain forms part of the antigen recognition site, whereas the C-terminus determines the subclass (isotype).
- The **variable region (V region)** of an immunoglobulin chain is coded by the V gene and varies extensively when different chains are compared, as the result of multiple (different) genomic copies and changes introduced during construction of an active immunoglobulin.
- **Constant regions (C regions)** of immunoglobulins are coded by C genes and are the parts of the chain that vary least. Those of heavy chains identify the type of immunoglobulin.
- A **V gene** is a sequence coding for the major part of the variable (N-terminal) region of an immunoglobulin chain.
- **C genes** code for the constant regions of immunoglobulin protein chains.
- **Somatic recombination** describes the process of joining a V gene to a C gene in a lymphocyte to generate an immunoglobulin or T-cell receptor.

Key Concepts

- An immunoglobulin is a tetramer of two light chains and two heavy chains.
- Light chains fall into the lambda and kappa families; heavy chains form a single family.
- Each chain has an N-terminal variable region (V) and a C-terminal constant region (C).
- The V domain recognizes antigen and the C domain provides the effector response.
- V domains and C domains are separately coded by V genes and C genes.
- A gene coding for an intact immunoglobulin chain is generated by somatic recombination to join a V gene with a C gene.

A remarkable feature of the immune response is an animal's ability to produce an appropriate antibody whenever it is exposed to a new antigen. How can the organism be prepared to produce antibody proteins each designed specifically to recognize an antigen with an unpredictable structure?

Each antibody is an immunoglobulin tetramer consisting of two identical **light chains** (L) and two identical **heavy chains** (H). The structure of the immunoglobulin tetramer is illustrated in **Figure 23.3**. Light chains and heavy chains share the same general type of organization in which each protein chain consists of two principal regions: the N-terminal **variable region (V region)**; and the C-terminal **constant region (C region)**. They were defined originally by comparing the amino acid sequences of different immunoglobulin chains. As the names suggest, the

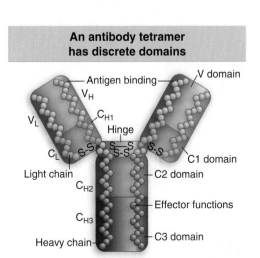

An antibody tetramer has discrete domains

Figure 23.3 Heavy and light chains combine to generate an immunoglobulin with several discrete domains.

variable regions show considerable changes in sequence from one protein to the next, while the constant regions show substantial homology. There are two types of light chain and ~10 types of heavy chain. Different classes of immunoglobulins have different effector functions. The class is determined by the heavy chain constant region, which exercises the effector function (see Figure 23.15).

Corresponding regions of the light and heavy chains associate to generate distinct domains in the immunoglobulin protein.

The variable (V) domain is generated by association between the variable regions of the light chain and heavy chain. *The V domain is responsible for recognizing the antigen.* An immunoglobulin has a Y-shaped structure in which the arms of the Y are identical, and each arm has a pair of V_L-V_H domains. Production of V domains of different specificities creates the ability to respond to diverse antigens. The total number of variable regions for either light- or heavy-chain proteins is measured in hundreds. *So the protein displays the maximum versatility in the region responsible for binding the antigen.*

The number of constant regions is vastly smaller than the number of variable regions—typically there are only 1–10 types of C region for any particular type of chain. The constant regions in the subunits of the immunoglobulin tetramer associate to generate several individual C domains. The first domain results from association of the single constant region of the light chain (C_L) with the C_{H1} part of the heavy-chain constant region. The two copies of this domain complete the arms of the Y-shaped molecule. Association between the C regions of the heavy chains generates the remaining C domains, which vary in number depending on the type of heavy chain.

Comparing the characteristics of the variable and constant regions, we see the central dilemma in immunoglobulin gene structure. How does the genome code for a set of proteins in which any individual polypeptide chain must have one of <10 possible C regions, but can have any one of several hundred possible V regions? It turns out that the number of coding sequences for each type of region reflects its variability. There are many genes coding for V regions, but only a few genes coding for C regions.

In this context, *"gene" means a sequence of DNA coding for a discrete part of the final immunoglobulin polypeptide* (heavy or light chain). So **V genes** code for variable regions and **C genes** code for constant regions, although *neither type of gene is expressed as an independent unit.* To construct a unit that can be expressed in the form of an authentic light or heavy chain, a V gene must be joined physically to a C gene. In this system, two "genes" code for one polypeptide. To avoid confusion, we will refer to these units as "gene segments" rather than "genes."

The sequences coding for light chains and heavy chains are assembled in the same way: *any one of many V gene segments may be joined to any one of a few C gene segments.* This **somatic recombination** occurs *in the B lymphocyte in which the antibody is expressed.* The large number of available V gene segments is responsible for a major part of the diversity of immunoglobulins. However, not all diversity is coded in the genome; some is generated by changes that occur during the process of constructing a functional gene.

The crucial fact about the synthesis of immunoglobulins, therefore, is that *the arrangement of V gene segments and C gene segments is different in the cells producing the immunoglobulins (or T-cell receptors) from all other somatic cells or germ cells.* The entire process occurs in somatic cells and does not affect the germline; so the response to an antigen is not inherited by an organism's progeny.

Figure 23.4 summarizes the overall process of creating an active immunoglobulin. There are two families of immunoglobulin light chains, κ and λ, and one family containing all the types of heavy chain (H). Each family resides on a different chromosome, and consists of its own set of both V gene segments and C gene segments. This is called the *germline pattern*, and is found in the germline and in somatic cells of all lineages other than the immune system.

But in a cell expressing an antibody, each of its chains—one light type (either κ or λ) and one heavy type—is coded by a single intact gene. The recombination event that brings a V gene segment to partner a C gene segment creates an active gene consisting of exons that correspond precisely with the functional domains of the protein. The introns are removed in the usual way by RNA splicing.

The principles by which functional genes are assembled are the same in each family, but there are differences in the details of the organization of the V and C gene segments, and correspondingly of the recombination reaction between them. In addition to the V and C gene segments, other short DNA sequences (including J segments and D segments) are included in the functional somatic loci.

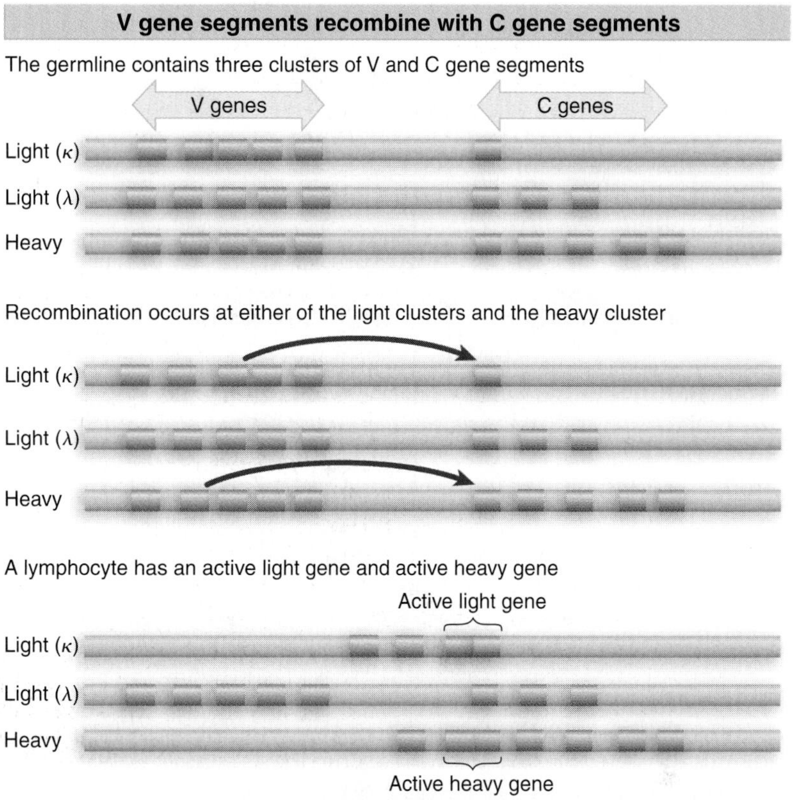

V gene segments recombine with C gene segments

Figure 23.4 The germline pattern has three separate clusters of V gene segments separated from C gene segments. Recombination can occur in each of the clusters to create an active gene by linking a V gene segment with a C gene segment. To produce an immunoglobulin protein, a lymphocyte must have a recombination at one of the two light clusters and also at the heavy cluster.

23.3 Light Chains Are Assembled by a Single Recombination

Key Concepts

- A lambda light chain is assembled by a single recombination between a V gene segment and a J-C gene segment.
- The lambda V gene has a leader exon, intron, and variable-coding region.
- The lambda J-C gene has a short J-coding exon, intron, and C-coding region.
- A kappa light chain is assembled by a single recombination between a V gene segment and one of five J segments preceding the C gene segment.

Key Term

- **J segments (joining segments)** are coding sequences in the immunoglobulin and T-cell receptor loci. The J segments are between the variable (V) and constant (C) gene segments.

A λ light chain is assembled from two parts, as illustrated in **Figure 23.5**. The V gene segment consists of the leader exon (L) separated by a single intron from the variable (V) segment. The C gene segment consists of the J segment separated by a single intron from the constant (C) exon.

The name of the **J segment** is an abbreviation for joining, since it identifies the region to which the V segment becomes connected. So the joining reaction does not directly involve V and C gene segments, but occurs via the J segment; when we discuss the joining of "V and C gene

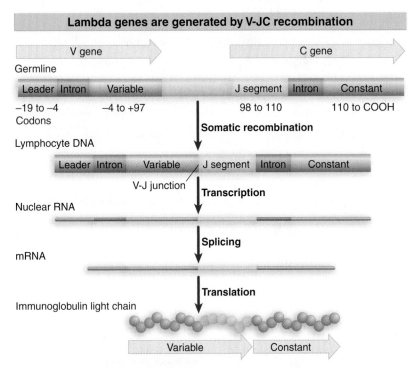

Figure 23.5 The lambda C gene segment is preceded by a J segment, so that V-J recombination generates a functional lambda light-chain gene. (The first amino acid of the mature protein is +1; negative numbers indicate a leader sequence that is cleaved)

segments" for light chains, we really mean V-JC joining.

The J segment is short and codes for the last few (13) amino acids of the variable region, as defined by amino acid sequences. In the intact gene generated by recombination, the V-J segment constitutes a single exon coding for the entire variable region.

The consequences of the κ joining reaction are illustrated in **Figure 23.6**. A κ light chain also is assembled from two parts, but there is a difference in the organization of the C gene segment. A group of five J segments is spread over a region of 500–700 bp, separated by an intron of 2–3 kb from the C_κ exon. In the mouse, the central J segment is nonfunctional (ψJ3). A V_κ segment may be joined to any one of the J segments.

Whichever J segment is used becomes the terminal part of the intact variable exon. Any J segments on the left of the recombining J segment are lost (J1 has been lost in the figure). Any J segment on the right of the recombining J segment is treated as part of the intron between the variable and constant exons (J3–5 are included in the intron that is spliced out in the figure).

All functional J segments possess a signal at the left boundary that makes it possible to recombine with the V segment, and they possess a signal at the right boundary that can be used for splicing to the C exon. Whichever J segment is recognized in V–J DNA joining uses its splicing signal in RNA processing.

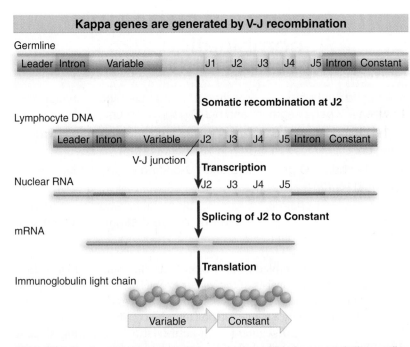

Figure 23.6 The kappa C gene segment is preceded by multiple J segments in the germline. V-J joining may recognize any one of the J segments, which is then spliced to the C gene segment during RNA processing.

23.4 Heavy Chains Are Assembled by Two Recombinations

Key Concepts

- The units for heavy-chain recombination are a V gene segment, D segment, and J-C gene segment.
- The first recombination joins D to J-C.
- The second recombination joins V to D-J-C.
- The C segment consists of several exons.

Key Term

- The **D segment** is an additional sequence that is found between the V and J regions of an immunoglobulin heavy chain.

Heavy-chain construction involves an additional segment. The **D segment** (for diversity) was discovered by the presence in the protein of an extra 2–13 amino acids between the sequences coded by the V segment and the J segment. An array of >10 D segments lies on the chromosome between the V_H segments and the four J_H segments.

V-D-J joining takes place in two stages, as illustrated in **Figure 23.7**. First one of the D segments recombines with a J_H segment; then a V_H segment recombines with the DJ_H combined segment. The reconstruction leads to expression of the adjacent C_H segment (which consists of several exons).

The D segments are organized in a tandem array. The mouse heavy-chain locus contains 12 D segments of variable length; the human locus has ~30 D segments (not all necessarily active). (When we discuss joining of V and C gene segments for heavy chains, we assume the process has been completed by V-D and D-J joining reactions.)

The V gene segments of all three immunoglobulin families are similar in organization. The first exon codes for the signal sequence (involved in membrane attachment), and the second exon codes for the major part of the variable region itself (<100 codons long). The remainder of the variable region is provided by the D segment (in the H family only) and by a J segment (in all three families).

The structure of the constant region depends on the type of chain. For both κ and λ light chains, the constant region is coded by a single exon (which becomes the third exon of the reconstructed, active gene). For H chains, the constant region is coded by several exons; corresponding with

Figure 23.7 Heavy genes are assembled by sequential joining reactions. First a D segment is joined to a J segment; then a V gene segment is joined to the D segment.

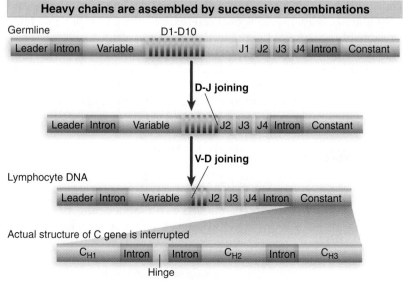

Heavy chains are assembled by successive recombinations

Germline — Leader | Intron | Variable | D1-D10 | J1 J2 J3 J4 | Intron | Constant

D-J joining

Leader | Intron | Variable | J2 J3 J4 | Intron | Constant

V-D joining

Lymphocyte DNA — Leader | Intron | Variable | J2 J3 J4 | Intron | Constant

Actual structure of C gene is interrupted

C_{H1} | Intron | Intron | C_{H2} | Intron | C_{H3}

Hinge

the protein chain shown in Figure 23.3, separate exons code for the regions C_{H1}, hinge, C_{H2}, and C_{H3}. Each C_H exon is ~100 codons long; the hinge is shorter. The introns usually are relatively small (~300 bp).

23.5 Recombination Generates Extensive Diversity

Key Concepts

- A light-chain locus can produce >1000 chains by combining 300 V genes with 4–5 C genes.
- An H locus can produce >4000 chains by combining 300 V genes, 20 D segments, and 4 J segments.

Now we must examine the different types of V and C gene segments to see how much diversity can be accommodated by the variety of the coding regions carried in the germline. In each light Ig gene family, many V gene segments are linked to a much smaller number of C gene segments.

Figure 23.8 shows that the human λ locus has ~300 V gene segments, followed by several C gene segments, each preceded by its own J segment. The λ locus in mouse is much less diverse, because at some time in the past, it suffered a catastrophic deletion of most of its germline V_λ gene segments, leaving only two.

Figure 23.9 shows that the κ locus has only one C gene segment, although it is preceded by five J segments (one of them inactive). The V_κ gene segments occupy a large cluster on the chromosome, upstream of the constant region. The human cluster has two regions. Just preceding the C_κ gene segment, a region of 600 kb contains the 5 J_κ segments and 40 V_κ gene segments. A gap of 800 kb separates this region from another group of 36 V_κ gene segments.

The V_κ gene segments can be subdivided into families, defined by the criterion that members of a family have >80% amino acid identity. The mouse family is unusually large, ~1000 genes, and there are ~18 V_κ families, varying in size from 2–100 members. Like other families of related genes, related V gene segments form subclusters, generated by duplication and divergence of individual ancestral members. However, many of the V segments are inactive pseudogenes, and <50 are likely to be used to generate immunoglobulins.

A given lymphocyte generates *either* a κ *or* a λ light chain to associate with the heavy chain. In man, ~60% of the light chains are κ and ~40% are λ. In mouse, 95% of B cells express the κ type of light chain, presumably because of the reduced number of λ gene segments.

The single locus for heavy-chain production in human consists of several discrete sections, as summarized in **Figure 23.10**. It is similar in the mouse, where there are more V_H gene segments, fewer D and J segments, and a slight difference in the number and organization of C gene segments. The 3' member of the V_H cluster is separated by only 20 kb from the first D segment. The D segments are spread over ~50 kb, and then comes the cluster of J segments. Over the next 220 kb lie all the C_H gene segments. There are nine functional C_H gene segments and two pseudogenes. The organization suggests that a γ gene segment must have been duplicated to give the subcluster of γ-γ-ε-α, after which the entire group was then duplicated.

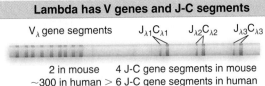

Lambda has V genes and J-C segments

V_λ gene segments $J_{\lambda 1}C_{\lambda 1}$ $J_{\lambda 2}C_{\lambda 2}$ $J_{\lambda 3}C_{\lambda 3}$

2 in mouse   4 J-C gene segments in mouse
~300 in human > 6 J-C gene segments in human

Figure 23.8 The lambda family consists of V gene segments linked to a small number of J-C gene segments.

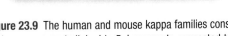

The kappa family has only one C gene

36V_κ 40V_κ J_{1-5} C_κ

Figure 23.9 The human and mouse kappa families consist of V gene segments linked to 5 J segments connected to a single C gene segment.

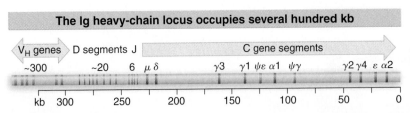

The Ig heavy-chain locus occupies several hundred kb

V_H genes D segments J   C gene segments

~300   ~20   6   μ δ   γ3   γ1 ψε α1   ψγ   γ2 γ4 ε α2

kb   300   250   200   150   100   50   0

Figure 23.10 A single gene cluster in human contains all the information for heavy-chain gene assembly.

How far is the diversity of germline information responsible for V region diversity in immunoglobulin proteins? By combining any one of ~50 V gene segments with any one of 4–5 J segments, a typical light-chain locus has the potential to produce some 250 chains. There is even greater diversity in the H chain locus; by combining any one of ~50 V_H gene segments, 20 D segments, and 4 J segments, the genome potentially can produce 4000 variable regions to accompany any C_H gene segment. In mammals, this is the starting point for diversity, but additional mechanisms introduce further changes. *When closely related variants of immunoglobulins are examined, there often are more proteins than can be accounted for by the number of corresponding V gene segments.* The new members are created by somatic changes in individual genes during or after the recombination process (see *23.10 Somatic Mutation Is Induced by Cytidine Deaminase and Uracil Glycosylase*).

23.6 Immune Recombination Uses Two Types of Consensus Sequence

Key Concepts

- The consensus sequence used for recombination is a heptamer separated by either 12 or 23 base pairs from a nonamer.
- Recombination occurs between two consensus sequences that have different spacings.

Assembly of light- and heavy-chain genes proceeds by the same mechanism (although the number of parts is different). The same consensus sequences are found at the boundaries of all germline segments that participate in joining reactions. Each consensus sequence consists of a heptamer separated by either 12 or 23 bp from a nonamer.

Figure 23.11 illustrates the relationship between the consensus sequences at the mouse Ig loci. At the κ locus, each $V_κ$ gene segment is followed by a consensus sequence with a 12 bp spacing. Each $J_κ$ segment is preceded by a consensus sequence with a 23 bp spacing. The V and J consensus sequences are inverted in orientation. At the λ locus, each $V_λ$ gene segment is followed by a consensus sequence with 23 bp spacing, while each $J_λ$ gene segment is preceded by a consensus of the 12 bp spacer type.

The rule that governs the joining reaction is that *a consensus sequence with one type of spacing can be joined only to a consensus sequence with the other type of spacing.* Since the consensus sequences at V and J segments can lie in either order, the different spacings do not impart any directional information, but serve to prevent one V gene segment from recombining with another, or one J segment from recombining with another.

This concept is borne out by the structure of the components of the heavy gene segments. Each V_H gene segment is followed by a consensus sequence of the 23 bp spacer type. The D segments are flanked on either side by consensus sequences of the 12 bp spacer type. The J_H segments are preceded by consensus sequences of the 23 bp spacer type. So the V gene segment must be

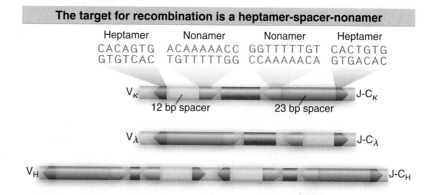

The target for recombination is a heptamer-spacer-nonamer

Heptamer	Nonamer	Nonamer	Heptamer
CACAGTG	ACAAAAACC	GGTTTTTGT	CACTGTG
GTGTCAC	TGTTTTTGG	CCAAAAACA	GTGACAC

$V_κ$ — 12 bp spacer — 23 bp spacer — J-$C_κ$

$V_λ$ — J-$C_λ$

V_H — J-C_H

Figure 23.11 Consensus sequences are present in inverted orientation at each pair of recombining sites. One member of each pair has a spacing of 12 bp between its components; the other has 23 bp spacing.

joined to a D segment; and the D segment must be joined to a J segment. A V gene segment cannot be joined directly to a J segment, because both possess the same type of consensus sequence.

The spacing between the components of the consensus sequences corresponds almost to one or two turns of the double helix. This may reflect a geometric relationship in the recombination reaction. For example, the recombination protein(s) may approach the DNA from one side, in the same way that RNA polymerase and repressors approach recognition elements such as promoters and operators.

23.7 Recombination Generates Deletions or Inversions

Key Terms

- A **signal end** is produced during recombination of immunoglobulin and T-cell receptor genes. The signal ends are at the termini of the cleaved fragment containing the recombination signal sequences. The subsequent joining of the signal ends yields a signal joint.
- A **coding end** is produced during recombination of immunoglobulin and T-cell receptor genes. Coding ends are at the termini of the cleaved V and (D)J coding regions. The subsequent joining of the coding ends yields a coding joint.
- **Allelic exclusion** describes the expression in any particular lymphocyte of only one allele coding for the expressed immunoglobulin. This is caused by feedback from the first immunoglobulin allele to be expressed that prevents activation of a copy on the other chromosome.

Key Concepts

- Recombination occurs by double-strand breaks at the heptamers of two consensus sequences.
- The signal ends of the fragment excised between the breaks usually join to generate a circular molecule.
- The coding ends are covalently linked to join V to J-C (L chain) or D to J-C and V to D-J-C (H chain).
- If the recombining genes are in an inverted rather than a direct orientation, there is an inversion instead of the deletion of an excised circle.
- Recombination to generate an intact immunoglobulin gene is productive if it leads to expression of an active protein.
- A productive rearrangement prevents any further rearrangement from occurring, but a nonproductive rearrangement does not.
- Allelic exclusion applies separately to light chains (only one kappa *or* lambda may be productively rearranged) and to heavy chains (one heavy chain is productively rearranged).

Recombination of the components of immunoglobulin genes is accomplished by a physical rearrangement of sequences, involving breakage and reunion, but the mechanism is different from homologous recombination. The general nature of the reaction is illustrated in **Figure 23.12** for the example of a κ light chain. (The reaction is similar at a heavy-chain locus, except that there are two recombination events: first D-J, then V-DJ.)

Breakage and reunion occur as separate reactions. A double-strand break is made at the heptamers that lie at the ends of the coding units.

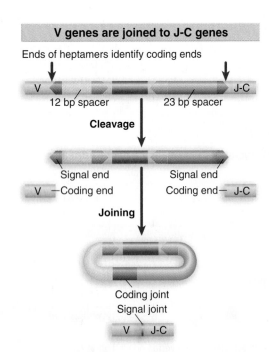

Figure 23.12 Breakage and reunion at consensus sequences generates immunoglobulin genes.

This releases the entire fragment between the V gene segment and J-C gene segment; the cleaved termini of this fragment are called **signal ends**. The cleaved termini of the V and J-C loci are called **coding ends**. The two coding ends are covalently linked to form a coding joint; this is the connection that links the V and J segments. If the two signal ends are also connected, the excised fragment forms a circular molecule.

The V and J-C loci are usually organized in the same orientation. As a result, the cleavage at each consensus sequence releases the region between them as a linear fragment. If the signal ends are joined, the fragment is converted into a circular molecule, as indicated in Figure 23.12. Deletion to release an excised circle is the predominant mode of recombination at the immunoglobulin and TCR loci.

Each B cell expresses a single type of light chain and a single type of heavy chain, because only a single productive rearrangement of each type occurs in a given lymphocyte, to produce one light- and one heavy-chain gene. Because each event involves the genes of only *one* of the homologous chromosomes, *the alleles on the other chromosome are not expressed in the same cell*. This phenomenon is called **allelic exclusion**.

The crux of the model is that the cell keeps trying to recombine V and C gene segments until a productive rearrangement is achieved. Allelic exclusion is caused by the suppression of further rearrangement as soon as an active chain is produced. The use of this mechanism *in vivo* is demonstrated by the creation of transgenic mice whose germline has a rearranged immunoglobulin gene. Expression of the transgene in B cells suppresses the rearrangement of endogenous genes.

Allelic exclusion is independent for the heavy- and light-chain loci. Heavy-chain genes usually rearrange first. Allelic exclusion for light chains must apply equally to both families (cells may have *either* active κ or λ light chains). It is likely that the cell rearranges its κ genes first, and tries to rearrange λ only if both κ attempts are unsuccessful.

23.8 The RAG Proteins Catalyze Breakage and Reunion

Key Concepts

- The RAG proteins are necessary and sufficient for the cleavage reaction.
- RAG1 recognizes the nonamer consensus sequences for recombination. RAG2 binds to RAG1 and cleaves at the heptamer.
- The reaction resembles the topoisomerase-like resolution reaction that occurs in transposition.
- It proceeds through a hairpin intermediate at the coding end; opening of the hairpin is responsible for insertion of extra bases (P nucleotides) in the recombined gene.
- Deoxynucleoside transferase inserts additional N nucleotides at the coding end.
- The codon at the site of the V-(D)J joining reaction has an extremely variable sequence and codes for amino acid 96 in the antigen-binding site.
- The double-strand breaks at the coding joints are repaired by the same system involved in nonhomologous end joining of damaged DNA.
- An enhancer in the C gene activates the promoter of the V gene after recombination has generated the intact immunoglobulin gene.

Key Terms

- A **P nucleotide** sequence is a short palindromic (inverted repeat) sequence that is generated during rearrangement of immunoglobulin and T-cell receptor gene segments. P nucleotides are generated at coding joints when RAG proteins cleave the hairpin ends generated during rearrangement.
- An **N nucleotide** sequence is a short non-templated sequence that is added randomly by the enzyme deoxynucleoside transferase at coding joints during rearrangement of immunoglobulin and T-cell receptor genes. N nucleotides augment the diversity of antigen receptors.

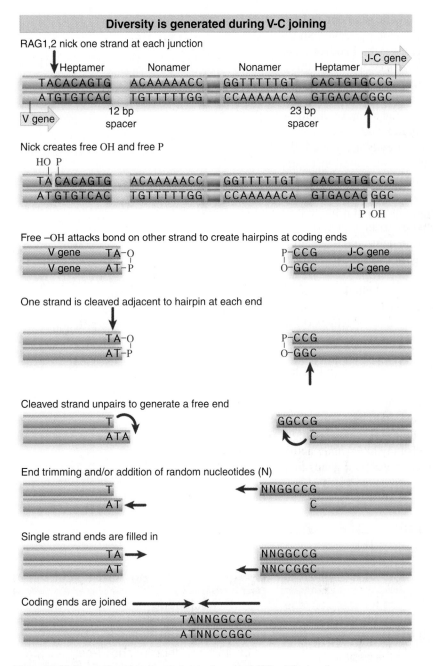

Figure 23.13 Processing of coding ends introduces variability at the junction.

Two proteins, RAG1 and RAG2, together undertake the catalytic reactions of cleaving and rejoining DNA, and also provide a structural framework within which the reactions occur. RAG1 recognizes the heptamer/nonamer signals with the appropriate 12/23 spacing and recruits RAG2 to the complex. The nonamer provides the site for initial recognition, and the heptamer directs the site of cleavage.

The reactions involved in recombination are shown in **Figure 23.13**. At each junction, one strand is nicked to generate $3' — OH$ and $5' — P$ ends. The free $3' — OH$ end then attacks the phosphate bond at the corresponding position *in the other strand of the duplex*. This creates a hairpin at the coding end, in which the $3'$ end of one strand is covalently linked to the $5'$ end of the other strand; it leaves a blunt double-strand break at the signal end.

This second cleavage is a transesterification reaction in which bond energies are conserved. It resembles the topoisomerase-like reactions catalyzed by the resolvase proteins of bacterial transposons and the reactions catalyzed by bacterial integrases (see *21.9 TnA Transposition Requires Transposase and Resolvase*). This suggests that somatic recombination of immune genes evolved from an ancestral transposon.

The hairpins at the coding ends provide the substrate for the next stage of reaction. If a single-strand break is introduced into one strand close to the hairpin, an unpairing reaction at the end generates a single-stranded protrusion. Synthesis of a complement to the exposed single strand then converts the coding end to an extended duplex. This reaction explains the introduction of **P nucleotides** at coding ends; they consist of a few extra base pairs, related to, but reversed in orientation from, the original coding end.

Some extra bases also may be inserted, apparently with random sequences, between the coding ends. They are called **N nucleotides**. Their insertion occurs via the activity of the enzyme deoxynucleoside transferase (known to be an active component of lymphocytes) at a free $3'$ coding end generated during the joining process.

Changes in sequence during recombination are therefore a consequence of the enzymatic mechanisms involved in breaking and rejoining the DNA. In heavy-chain recombination, base pairs are lost or inserted at the V_H-D or D-J junctions or both. Deletion also occurs in V_λ-J_λ joining, but insertion at these joints is unusual. The changes in sequence affect the amino acid coded at V-D and D-J junctions in heavy chains or at the V-J junction in light chains.

These various mechanisms together ensure that a coding joint may have a sequence that is different from what would be predicted by a direct joining of the coding ends of the V, D, and J regions.

Changes in the sequence at the junction make it possible for a great variety of amino acids to be coded at this site. It is interesting that the amino acid at position 96 is created by the V-J joining reaction. It forms part of the antigen-binding site and also is involved in making contacts between the light and heavy chains. So the maximum diversity is generated at the site that contacts the target antigen.

Changes in the number of base pairs at the coding joint affect the reading frame. The joining process appears to be random with regard to reading frame, so that probably only one third of the joined sequences retain the proper frame of reading through the junctions. If the V-J region is joined so that the J segment is out of phase, translation is terminated prematurely by a nonsense codon in the incorrect frame. We may think of the formation of aberrant genes as the price the cell must pay for the increased diversity that it gains by being able to adjust the sequence at the joining site.

Similar although even greater diversity is generated in the joining reactions for the D segment of the heavy chain. The same result is seen with regard to reading frame; nonproductive genes are generated by joining events that place J and C out of phase with the preceding V gene segment.

The joining reaction that works on the coding end uses the same pathway of nonhomologous end joining (NHEJ) that repairs double-strand breaks in cells (See *20.9 Eukaryotic Cells Have Conserved Repair Systems*). The initial stages of the reaction were identified by the mouse *SCID* mutation, which greatly reduces activity in immunoglobulin and TCR recombination. *SCID* mice accumulate broken molecules that terminate in double-strand breaks at the coding ends, and are thus deficient in completing some aspect of the joining reaction.

The *SCID* mutation inactivates a DNA-dependent protein kinase (DNA-PK). The kinase is recruited to DNA by the Ku70 and Ku80 proteins, which bind to the DNA ends. DNA-PK phosphorylates and thereby activates the protein Artemis, which nicks the hairpin ends (it also has exonuclease and endonuclease activities that function in the NHEJ pathway). The actual ligation is undertaken by DNA ligase IV and also requires the protein XRCC4.

What is the connection between joining of V and C gene segments and their activation? Unrearranged V gene segments are not actively represented in RNA. But when a V gene segment is joined productively to a C_κ gene segment, the resulting unit is transcribed. However, since the sequence upstream of a V gene segment is not altered by the joining reaction, *the promoter must be the same in unrearranged, nonproductively rearranged, and productively rearranged genes.*

A promoter lies upstream of every V gene segment, but is inactive. It is activated by its relocation to the C region. The effect must depend on sequences downstream. What role might they play? An enhancer located within or downstream of the C gene segment activates the promoter at the V gene segment. The enhancer is tissue specific; it is active only in B cells. Its existence suggests the model illustrated in **Figure 23.14**, in which the V gene segment promoter is activated when it is brought within the range of the enhancer.

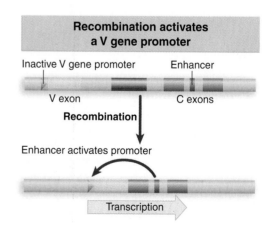

Figure 23.14 A V gene promoter is inactive until recombination brings it into the proximity of an enhancer in the C gene segment. The enhancer is active only in B lymphocytes.

23.9 Class Switching Is Caused by a Novel Type of DNA Recombination

Key Terms

- **Class switching** describes a change in Ig gene organization in which the C region of the heavy chain is changed but the V region remains the same.
- An **S region** is a sequence involved in immunoglobulin class switching. S regions consist of repetitive sequences at the 5′ ends of gene segments encoding the heavy-chain constant regions.

Key Concepts

- Immunoglobulins are divided into five classes according to the type of constant region in the heavy chain.
- Class switching to change the C_H region occurs by a recombination between S regions that deletes the region between the old C_H region and the new C_H region.
- Multiple successive switch recombinations can occur.
- Switching occurs by a double-strand break followed by the non-homologous end-joining reaction.
- The important feature of a switch region is the presence of inverted repeats.
- Switching requires activation of promoters that are upstream of the switch sites.
- A cytidine deaminase is required for class switching.

The *class* of an immunoglobulin is defined by the type of C_H region. **Figure 23.15** summarizes the five Ig classes. IgM (the first immunoglobulin to be produced by any B cell) and IgG (the most common immunoglobulin) possess the central ability to activate complement, which leads to destruction of invading cells. IgA is found in secretions (such as saliva), and IgE is associated with the allergic response and defense against parasites.

All lymphocytes start productive life as immature cells engaged in synthesis of IgM. Cells expressing IgM have the germline arrangement of the C_H gene segment cluster shown in Figure 23.10. The V-D-J joining reaction triggers expression of the C_μ gene segment. A lymphocyte generally produces only a single class of immunoglobulin at any one time, but the class may change during the cell lineage. A change in expression is called **class switching**. It is accomplished by a substitution in the type of C_H region that is expressed. Switching can be stimulated by environmental effects; for example, the growth factor TGFβ causes switching from C_μ to C_α.

Switching involves only the C_H gene segment; the same V_H gene segment continues to be expressed. So a given V_H gene segment may be expressed successively in combination with more than one C_H gene segment. The same light chain continues to be expressed throughout the lineage of the cell. Class switching therefore allows the type of effector response (mediated by the C_H region) to change, while maintaining the same capacity to recognize antigen (mediated by the V regions).

Switching is accomplished by a new type of DNA recombination event, which deletes a region including the expressed C_H gene segment, and extending up to a new C_H gene segment that will be expressed instead. The sites of switching are called **S regions**, and they lie upstream of the C_H gene segments themselves. **Figure 23.16** depicts two successive switches.

There are five types of heavy chain					
Type	**IgM**	**IgD**	**IgG**	**IgA**	**IgE**
Heavy chain	μ	δ	γ	α	ε
Structure	$(\mu_2 L_2)_5 J$	$\delta_2 L_2$	$\gamma_2 L_2$	$(\alpha_2 L_2)_2 J$	$\varepsilon_2 L_2$
Proportion	5%	1%	80%	14%	<1%
Effector function	Activates complement	Development of tolerance (?)	Activates complement	Found in secretions	Allergic response

Figure 23.15 Immunoglobulin type and function are determined by the heavy chain. J is a joining protein in IgM; all other Ig types exist as tetramers.

In the first switch, expression of C_μ is succeeded by expression of $C_{\gamma 1}$. The $C_{\gamma 1}$ gene segment is brought into the expressed position by recombination between the sites S_μ and $S_{\gamma 1}$. The S_μ site lies between V-D-J and the C_μ gene segment. The $S_{\gamma 1}$ site lies upstream of the $C_{\gamma 1}$ gene segment. The DNA sequence between the two switch sites is excised as a circular molecule. The second switch, which must be to a C_H gene segment *downstream* of the expressed gene, is accomplished in this example by recombination between $S_{\alpha 1}$ and the switch region $S_{\mu,\gamma 1}$ that was generated by the original switch.

We know that switch sites are not uniquely defined, because different cells expressing the same C_H gene segment prove to have recombined at different points. Switch regions vary in length (as defined by the limits of the sites involved in recombination) from 1–10 kb. They contain groups of short inverted repeats, with repeating units that vary from 20–80 nucleotides in length. The primary sequence of the switch region does not seem to be important; what matters is the presence of the inverted repeats.

An S region typically is located ~2 kb upstream of a C_H gene segment. Two of the proteins required for the joining phase of VDJ recombination (and also for the general nonhomologous end-joining pathway, NHEJ), Ku and DNA-PKcs, are required, suggesting that the joining reaction may use the NHEJ pathway. Basically this implies that the reaction occurs by a double-strand break followed by rejoining of the cleaved ends.

We can put together the features of the reaction to propose a model for the generation of the double-strand break. The critical points are:

- transcription through the S region is required;
- the inverted repeats are crucial;
- and the break can occur at many different places within the S region.

Figure 23.17 shows the stages of the class switching reaction. A promoter (I) lies immediately upstream of each switch region. Switching requires transcription from this promoter. The promoter may respond to activators that respond to environmental conditions, such as stimulation by cytokines, thus creating a mechanism to regulate switching.

The key insight into the mechanism of switching was the discovery of the requirement for the enzyme AID (activation-induced cytidine deaminase). In the absence of AID, class switching is blocked before the nicking stage. Somatic mutation is also blocked, showing an interesting connection between two important processes in immune diversification (see next section).

AID is a member of a class of enzymes that act on RNA to change a cytidine to a uridine (see *27.9 RNA Editing Occurs at Individual Bases*). However, AID has a different specificity, and acts on single-stranded DNA. Its sites of action then become targets for the enzyme UNG, a uracil DNA glycosylase, which

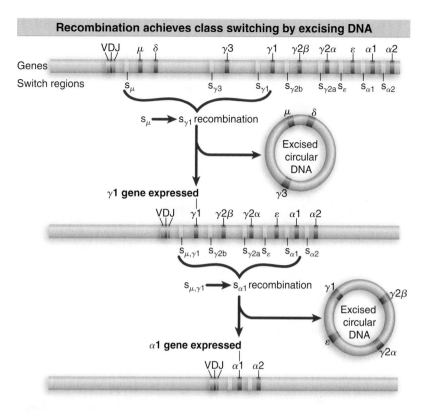

Recombination achieves class switching by excising DNA

Figure 23.16 Class switching of heavy genes may occur by recombination between switch regions (S), deleting the material between the recombining S sites. Successive switches may occur.

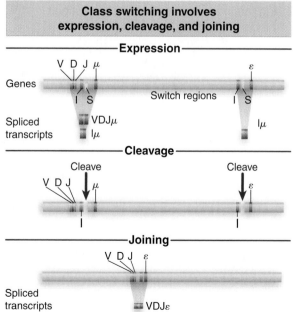

Class switching involves expression, cleavage, and joining

Figure 23.17 Class switching passes through discrete stages. The I promoters initiate transcription of sterile transcripts. The switch regions are cleaved. Joining occurs at the cleaved regions.

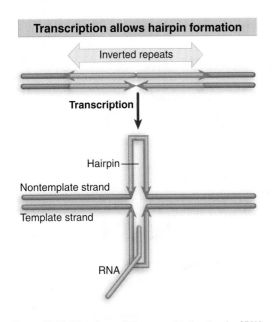

Transcription allows hairpin formation

Inverted repeats

Transcription

Hairpin

Nontemplate strand

Template strand

RNA

Figure 23.18 When transcription separates the strands of DNA, one strand may form a hairpin if the sequence is palindromic.

removes the uracil that AID generates by deaminating cytidine. This suggests a model in which the successive actions of AID and UNG create sites from which a base has been removed in DNA.

The source of the single-stranded DNA target for AID is generated by the process of sterile transcription, most probably by exposing the nontemplate strand of DNA that is displaced when the other strand is used as template for RNA synthesis. To cause class switching, these sites are converted into breaks in the nucleotide chain that provide the cleavage events shown in Figure 23.17. The broken ends are joined by the NHEJ pathway, which is a repair system that acts on double-strand breaks in DNA (see *20.10 A Common System Repairs Double-Strand Breaks*). We do not know yet how the abasic sites are converted into double-strand breaks. The MSH system that is involved in repairing mismatches in DNA may be required, because mutations in the gene *MSH2* reduce class switching.

One unexplained feature is the involvement of inverted repeats. One possibility is that hairpins are formed by an interaction between the inverted repeats on the displaced nontemplate strand, as shown in **Figure 23.18**. In conjunction with the generation of abasic sites on this strand, this might lead to breakage.

Another critical question that remains to be answered is how the system is targeted to the appropriate regions in the heavy-chain locus, and what controls the use of switching sites.

23.10 Somatic Mutation Is Induced by Cytidine Deaminase and Uracil Glycosylase

Key Term

- **Hypermutation** describes the introduction of somatic mutations in a rearranged immunoglobulin gene. The mutations can change the sequence of the corresponding antibody, especially in its antigen-binding site.

Key Concepts

- Active immunoglobulin genes have V regions with sequences that are changed from the germline because of somatic mutation.
- The mutations occur as substitutions of individual bases.
- The sites of mutation are concentrated in the antigen-binding site.
- The process depends on the enhancer that activates transcription at the Ig locus.
- A cytidine deaminase is required for somatic mutation as well as for class switching.
- Uracil-DNA glycosylase activity influences the pattern of somatic mutations.
- Hypermutation may be initiated by the sequential action of these enzymes.

Comparisons between the sequences of expressed immunoglobulin genes and the corresponding V gene segments of the germline show that new sequences appear in the expressed population. Some of this additional diversity results from sequence changes at the V-J or V-D-J junctions that occur during the recombination process. However, other changes occur upstream at locations within the variable domain.

Two types of mechanism can generate changes in V gene sequences after rearrangement has generated a functional immunoglobulin gene. In mouse and man, mutations are induced at individual locations within the gene specifically in the active lymphocyte. The process is sometimes called **hypermutation**. In chicken, rabbit, and pig, a different mechanism uses gene conversion to change a segment of the expressed V gene into the corresponding sequence from a different V gene (see next section).

Figure 23.19 shows that sequence changes extend around the V gene segment. They take the form of substitutions of individual nucleotide pairs. Usually there are ~3–15 substitutions, corresponding to <10 amino acid changes in the protein. They are concentrated in the antigen-binding site (thus generating the maximum diversity for recognizing new antigens). Only some of the mutations affect the amino acid sequence, since others lie in third-base coding positions as well as in nontranslated regions.

The large proportion of ineffectual mutations suggests that somatic mutation occurs more or less at random in a region including the V gene segment and extending beyond it. There is a tendency for some mutations to recur on multiple occasions. These may represent hotspots as a result of some intrinsic preference in the system.

In many cases, a single family of V gene segments is used consistently to respond to a particular antigen. Upon exposure to an antigen, presumably the V region with highest intrinsic affinity provides a starting point. Then somatic mutation increases the repertoire. Random mutations have unpredictable effects on protein function; some inactivate the protein, others confer high specificity for a particular antigen. The proportion and effectiveness of the lymphocytes that respond are increased by selection among the lymphocyte population for those cells bearing antibodies in which mutation has increased the affinity for the antigen.

Somatic mutation has many of the same requirements as class switching (discussed in previous section):

- transcription must occur in the target region;
- it requires the enzymes AID (activation induced cytidine deaminase) and UNG (uracil-DNA-glycosylase);
- the MSH mismatch-repair system is involved.

When AID deaminates cytosine, it generates uracil, which then is removed from the DNA by UNG. Then the MSH repair system is recruited to the damaged site. The simplest possibility is that the stretch of DNA containing the damage is replaced by an error-prone DNA polymerase, introducing mutations. Another possibility is that so many abasic sites are created that the repair systems are overwhelmed. When replication occurs, this could lead to the random insertion of bases opposite the abasic sites. We don't know yet what restricts the action of this system to the target region for hypermutation.

The difference in the systems is at the end of the process, when double-strand breaks are introduced in class switching, but individual point mutations are created during somatic mutation. We do not yet know exactly where the systems diverge. One possibility is that breaks are introduced at abasic sites in class switching, but are repaired in an error-prone manner in somatic mutation.

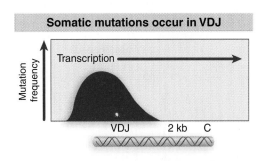

Figure 23.19 Somatic mutation occurs in the region surrounding the V segment and extends over the joined VDJ segments.

23.11 Avian Immunoglobulins Are Assembled from Pseudogenes

Key Concept

- An immunoglobulin gene in chicken is generated by copying a sequence from one of 25 pseudogenes into the V gene at a single active locus.

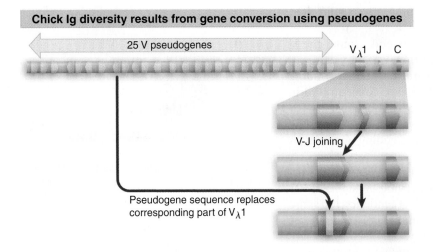

Chick Ig diversity results from gene conversion using pseudogenes

25 V pseudogenes

$V_\lambda 1$ J C

V-J joining

Pseudogene sequence replaces corresponding part of $V_\lambda 1$

Figure 23.20 The chicken lambda light locus has 25 V pseudogenes upstream of the single functional V-J-C region. But sequences derived from the pseudogenes are found in active rearranged V-J-C genes.

The chick immune system is the paradigm for rabbits, cows, and pigs, which rely upon using the diversity that is coded in the genome. A similar mechanism is used by both the single light-chain locus (of the λ type) and the H-chain locus. The organization of the λ locus is drawn in **Figure 23.20**. It has only one functional V gene segment, one J segment, and one C gene segment. Upstream of the functional $V_{\lambda 1}$ gene segment lie 25 V_λ pseudogenes, organized in either orientation. They are classified as pseudogenes because either the coding segment is deleted at one or both ends, or proper signals for recombination are missing (or both). This assignment is confirmed by the fact that only the $V_{\lambda 1}$ gene segment recombines with the J-C_λ gene segment.

But sequences of active rearranged V_λ-J-C_λ gene segments show considerable diversity! A rearranged gene has one or more positions at which a cluster of changes has occurred in the sequence. A sequence identical to the new sequence can almost always be found in one of the pseudogenes (which themselves remain unchanged). The exceptional sequences that are not found in a pseudogene always represent changes at the junction between the original sequence and the altered sequence.

So a novel mechanism is employed to generate diversity. Sequences from the pseudogenes, between 10 and 120 bp in length, are substituted into the active $V_{\lambda 1}$ region by gene conversion. The unmodified $V_{\lambda 1}$ sequence is not expressed, even at early times during the immune response. A successful conversion event probably occurs every 10–20 cell divisions to every rearranged $V_{\lambda 1}$ sequence. At the end of the immune maturation period, a rearranged $V_{\lambda 1}$ sequence has 4–6 converted segments spanning its entire length, derived from different donor pseudogenes. If all pseudogenes participate, this allows 2.5×10^8 possible combinations!

The enzymatic basis for copying pseudogene sequences into the expressed locus depends on enzymes involved in recombination and is related to the mechanism for somatic hypermutation that introduces diversity in mouse and man. Some of the genes involved in recombination are required for the gene conversion process; for example, it is prevented by deletion of *RAD54*. Deletion of other recombination genes (*XRCC2*, *XRCC3*, and *RAD51B*) has another, very interesting effect: somatic mutation occurs at the V gene in the expressed locus. The frequency of the somatic mutation is ~10× greater than the usual rate of gene conversion.

These results show that the absence of somatic mutation in chick is not due to a deficiency in the enzymatic systems that are responsible in mouse and man. The most likely explanation for a connection between (lack of) recombination and somatic mutation is that unrepaired breaks at the locus trigger the induction of mutations. The reason why somatic mutation occurs in mouse and human but not in chick may therefore lie with the details of the operation of the repair system that operates on breaks at the locus. It is more efficient in chick, so that the gene is repaired by gene conversion before mutations can be induced.

23.12 T-Cell Receptors Are Related to Immunoglobulins

Key Concepts

- T cells use a similar mechanism of V(D)J-C joining to B cells to produce either of two types of T-cell receptor.
- TCR $\alpha\beta$ is found on >95% of T lymphocytes, and TCR $\gamma\delta$ is found on <5%.

The lymphocyte lineage presents an example of evolutionary opportunism: a similar procedure is used in both B cells and T cells to generate proteins that have a variable region able to provide significant diversity, while constant regions are more limited and account for a small range of effector functions. T cells produce either of two types of T-cell receptor.

The $\gamma\delta$ receptor is found on <5% of T lymphocytes. It is synthesized only at an early stage of T-cell development. In mice, it is the only receptor detectable at <15 days of gestation, but has virtually been lost by birth at day 20.

TCR $\alpha\beta$ is found on >95% of lymphocytes. It is synthesized later in T-cell development than $\gamma\delta$. In mice, it first becomes apparent at 15–17 days after gestation. By birth it is the predominant receptor. It is synthesized by a separate lineage of cells from those involved in TCR $\gamma\delta$ synthesis, and involves independent rearrangement events.

Like immunoglobulins, a TCR must recognize a foreign antigen of unpredictable structure. The problem of antigen recognition by B cells and T cells is resolved in the same way, and the organization of the T-cell receptor genes resembles the immunoglobulin genes in the use of variable and constant regions. *Each locus is organized in the same way as the immunoglobulin genes, with separate segments that are brought together by a recombination reaction specific to the lymphocyte.* The components are the same as those found in the three Ig families.

The organization of the TCR proteins resembles that of the immunoglobulins. The V sequences have the same general internal organization in both Ig and TCR proteins. The TCR C region is related to the constant Ig regions and has a single constant domain followed by transmembrane and cytoplasmic portions. Exon-intron structure is related to protein function. TCRα resembles an Ig light chain, and TCRβ resembles an Ig heavy chain.

As summarized in **Figure 23.21**, the organization of TCRα resembles that of Ig κ, with V gene segments separated from a cluster of J segments that precedes a single C gene segment. The organization of the locus is similar in both man and mouse, with some differences only in the number of V_α gene segments and J_α segments.

The components of TCRβ resemble those of IgH. **Figure 23.22** shows that the organization is different, with V gene segments separated from two clusters each containing a D segment, several J segments, and a C gene segment. Again the only differences between human and mouse are in the numbers of the V_β and J_β units.

Diversity is generated by the same mechanisms as in immunoglobulins. Intrinsic diversity results from the combination of a variety of V, D, J, and C segments; some additional diversity results from the introduction of new sequences at the junctions between these components (in the form of P and N nucleotides; see Figure 23.13). Some TCR β chains incorporate two D segments, generated by D-D joins (directed by an appropriate organization of the nonamer and heptamer sequences). A difference between TCR and Ig is that somatic mutation does not occur at the TCR loci. Measurements of the extent of diversity show that the

The TCRα locus has α and δ genes

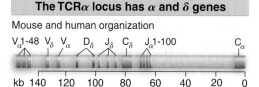

Mouse and human organization

kb 140 120 100 80 60 40 20 0

Human α summary: 42 V   61 J

Figure 23.21 The human TCRα locus has interspersed α and δ segments. A V_δ segment is located within the V_α cluster. The D-J-C$_\delta$ segments lie between the V gene segments and the J-C$_\alpha$ segments. The mouse locus is similar, but has more V_δ segments.

TCRβ is similar in mouse and man

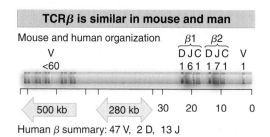

Mouse and human organization

Human β summary: 47 V, 2 D, 13 J

Figure 23.22 The TCRβ locus contains many V gene segments spread over ~500 kb, and lying ~280 kb upstream of the two D-J-C clusters.

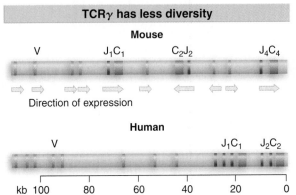

Figure 23.23 The TCRγ locus contains a small number of functional V gene segments (and also some pseudogenes; not shown), lying upstream of the J-C loci.

10^{12} T cells in man contain 2.5×10^7 different α chains associated with 10^6 different β chains.

The same mechanisms are likely to be involved in the reactions that recombine Ig genes in B cells and TCR genes in T cells. The recombining TCR segments are surrounded by nonamer and heptamer consensus sequences identical to those used by the Ig genes. This argues strongly that the same enzymes are involved. Most rearrangements probably occur by the deletion model (see Figure 23.12). We do not know how the process is controlled so that Ig loci are rearranged in B cells, while T-cell receptors are rearranged in T cells.

The organization of the γ locus resembles that of Igλ, with V gene segments separated from a series of J-C segments. **Figure 23.23** shows that this locus has relatively little diversity, with ~8 functional V segments. The organization is different in human and mouse. Mouse has three functional J-C loci, but some segments are inverted in orientation. Man has multiple J segments for each C gene segment.

The δ subunit is coded by segments that lie at the TCRα locus, as illustrated previously in Figure 23.21. The segments D_δ-D_δ-J_δ-C_δ lie between the V gene segments and the J_α-C_α segments. Both of the D segments may be incorporated into the δ chain to give the structure VDDJ.

Rearrangements at the TCR loci, like those of immunoglobulin genes, may be productive or nonproductive. The β locus shows allelic exclusion in much the same way as immunoglobulin loci; rearrangement is suppressed once a productive allele has been generated. The α locus may be different; several cases of continued rearrangement suggest the possibility that substitution of V_α sequences may continue after a productive allele has been generated.

23.13 SUMMARY

Immunoglobulins and T-cell receptors are proteins that play analogous functions in the roles of B cells and T cells in the immune system. An Ig or TCR protein is generated by rearrangement of DNA in a single lymphocyte; exposure to an antigen recognized by the Ig or TCR leads to clonal expansion to generate many cells which have the same specificity as the original cell. Many different rearrangements occur early in the development of the immune system, creating a large repertoire of cells of different specificities.

Each immunoglobulin protein is a tetramer containing two identical light chains and two identical heavy chains. A TCR is a dimer containing two different chains. Each polypeptide chain is expressed from a gene created by linking one of many V segments via D and J segments to one of a few C segments. Ig L chains (either κ or λ) have the general structure V-J-C, Ig H chains have the structure V-D-J-C, TCR α and γ have components like Ig L chains, and TCR δ and β are like Ig H chains.

Each type of chain is coded by a large cluster of V genes separated from the cluster of D, J, and C segments. The numbers of each type of segment, and their organization, are different for each type of chain, but the principle and mechanism of recombination appear to be the same. The same nonamer and heptamer consensus sequences are involved in each recombination; the reaction always involves joining of a consensus with 23 bp spacing to a consensus with 12 bp spacing. The cleavage reaction is catalyzed by the RAG1 and RAG2 proteins, and the joining reaction is catalyzed by the same NHEJ pathway that repairs double-strand breaks in cells. The mechanism of action of the RAG proteins is related to the action of site-specific recombination catalyzed by resolvases.

Although considerable diversity is generated by joining different V, D, J segments to a C segment, additional variations are introduced in the form of changes at the junctions between segments during the recombination process. Changes are also induced in immunoglobulin genes by somatic mutation, which requires the actions of cytidine deaminase and uracil glycosylase. Mutations induced by cytidine deaminase probably lead to removal of uracil by uracil glycosylase, followed by the induction of mutations at the sites where bases are missing.

Allelic exclusion ensures that a given lymphocyte synthesizes only a single Ig or TCR. A productive rearrangement inhibits the occurrence of further rearrangements. Although the use of the V region is fixed by the first productive rearrangement, B cells switch use of C_H genes from the initial μ chain to one of the H chains coded farther downstream. This process involves a different type of recombination in which the sequences between the VDJ region and the new C_H gene are deleted. More than one switch occurs in C_H gene usage. Class switching requires the same cytidine deaminase and uracil glycosylase enzymes that are required for somatic mutation.

24

Promoters and Enhancers

24.1 Introduction

Initiation of transcription requires the enzyme RNA polymerase and transcription factors. Any protein that is needed for the initiation of transcription, but which is not itself part of RNA polymerase, is defined as a transcription factor. Many transcription factors act by recognizing *cis*-acting sites on DNA. However, binding to DNA is not the only means of action for a transcription factor. A factor may recognize another factor, or may recognize RNA polymerase, or may be incorporated into an initiation complex only in the presence of several other proteins. The ultimate test for membership of the transcription apparatus is functional: a protein must be needed for transcription to occur at a specific promoter or set of promoters.

A major difference between transcription in prokaryotes and eukaryotes is the relative importance of the RNA polymerase and ancillary transcription factors. In bacteria, RNA polymerase binds to the promoter by directly recognizing short consensus sequences in the vicinity of the startpoint (see *Chapter 11 Transcription*); in some cases an ancillary factor is required to assist or to stabilize the binding of RNA polymerase. In eukaryotes, the promoter is defined by a series of short consensus sequences that lie in the region upstream of the startpoint, but they are recognized by transcription factors, not by RNA polymerase. The transcription factors are needed for initiation, but are not required subsequently. RNA polymerase itself binds to the startpoint because it is able to interact with the factors bound there, and it does not directly contact the extended upstream region of the promoter.

Transcription in eukaryotic cells is divided into three classes. Each class of genes is transcribed by a different RNA polymerase:

- RNA polymerase I transcribes rRNA.
- RNA polymerase II transcribes mRNA.
- RNA polymerase III transcribes tRNA and other small RNAs.

Each eukaryotic RNA polymerase works in conjunction with its own set of transcription factors. For RNA polymerases I and III, these factors are relatively simple, but for RNA polymerase II they form a large complex known as the basal apparatus.

The promoters for RNA polymerases I and II are (mostly) upstream of the startpoint, but some promoters for RNA polymerase III lie downstream of the startpoint. Each promoter contains characteristic sets of short conserved sequences that are recognized by the appropriate class of factors. RNA polymerases I and III each recognize a relatively restricted set of promoters, and rely upon a small number of accessory factors.

Promoters utilized by RNA polymerase II show more variation in sequence, and have a modular organization. Short sequence elements that are recognized by transcription factors lie upstream of the startpoint. These *cis*-acting sites usually are spread out over a region of >200 bp. Some of these elements and the factors that recognize them are common: they are found in a variety of promoters and are used constitutively. Others are specific: they identify particular classes of genes and their use is regulated. The elements occur in different combinations in individual promoters.

All RNA polymerase II promoters have sequence elements close to the startpoint that are bound by the basal apparatus and that establish the site of initiation. The sequences farther upstream determine whether the promoter is expressed in all cell types or is specifically regulated. Promoters that are constitutively expressed (their genes are sometimes called housekeeping genes) have upstream sequence elements that are recognized by ubiquitous activators. No single element–factor combination is an essential component of the promoter, which suggests that initiation by RNA polymerase II may be sponsored in many different ways. Promoters that are expressed only in certain times or places have sequence elements that require activators that are available only at those times or places.

Sequence components of the promoter are defined operationally by the demand that they must be located in the general vicinity of the startpoint and are required for initiation. The *enhancer* is another type of site involved in initiation. It is identified by sequences that stimulate initiation, but that are located a considerable distance from the startpoint. Enhancer elements are often targets for tissue-specific or temporal regulation. **Figure 24.1** illustrates the general properties of promoters and enhancers.

The components of an enhancer resemble those of the promoter; they consist of a variety of modular elements. However, the elements are organized in a closely packed array. The elements in an enhancer function like those in the promoter, but the enhancer does not need to be near the startpoint. However, proteins bound at enhancer elements interact with proteins bound at promoter elements.

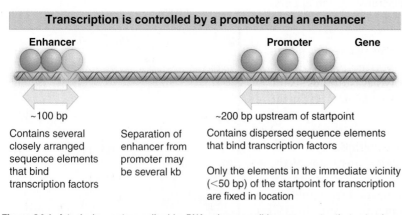

Transcription is controlled by a promoter and an enhancer

Enhancer   Promoter   Gene

~100 bp

~200 bp upstream of startpoint

Contains several closely arranged sequence elements that bind transcription factors

Separation of enhancer from promoter may be several kb

Contains dispersed sequence elements that bind transcription factors

Only the elements in the immediate vicinity (<50 bp) of the startpoint for transcription are fixed in location

Figure 24.1 A typical gene transcribed by RNA polymerase II has a promoter that extends upstream from the site where transcription is initiated. The promoter contains several short (<10 bp) sequence elements that bind transcription factors, dispersed over >200 bp. An enhancer containing a more closely packed array of elements that also bind transcription factors may be located several kb distant. (DNA may be coiled or otherwise rearranged so that transcription factors at the promoter and at the enhancer interact to form a large protein complex.)

The distinction between promoters and enhancers is operational, rather than implying a fundamental difference in mechanism. This view is strengthened by the fact that some types of elements are found in both promoters and enhancers.

Eukaryotic transcription is most often under positive regulation: a transcription factor is provided under tissue-specific control to activate a promoter or set of promoters that contains a common target sequence. Regulation by specific repression of a target promoter is less common.

A eukaryotic transcription unit generally contains a single gene, and termination occurs beyond the end of the coding region. Termination lacks the regulatory importance that applies in prokaryotic systems. RNA polymerases I and III terminate at discrete sequences in defined reactions, but the mode of termination by RNA polymerase II is not clear. However, the significant event in generating the 3′ end of an mRNA is not the termination event itself, but instead results from a cleavage reaction in the primary transcript (see *26.14 The 3′ Ends of mRNAs Are Generated by Cleavage and Polyadenylation*).

24.2 Eukaryotic RNA Polymerases Consist of Many Subunits

Key Terms

- The **carboxy-terminal domain (CTD)** of eukaryotic RNA polymerase II is phosphorylated at initiation and is involved in coordinating several activities with transcription.
- **Amanitin** (more fully α-amanitin) is a bicyclic octapeptide derived from the poisonous mushroom *Amanita phalloides*; it inhibits transcription by certain eukaryotic RNA polymerases, especially RNA polymerase II.

Key Concepts

- RNA polymerase I synthesizes rRNA in the nucleolus.
- RNA polymerase II synthesizes mRNA in the nucleoplasm.
- RNA polymerase III synthesizes small RNAs in the nucleoplasm.
- All eukaryotic RNA polymerases have ~12 subunits and are aggregates of >500 kD.
- Some subunits are common to all three RNA polymerases.
- The largest subunit in RNA polymerase II has a CTD (carboxy-terminal domain) consisting of multiple repeats of a heptamer.

The three eukaryotic RNA polymerases have different locations in the nucleus, corresponding with the genes that they transcribe.

The most prominent activity is the enzyme RNA polymerase I, which resides in the nucleolus and is responsible for transcribing the genes coding for rRNA. It accounts for most cellular RNA synthesis (in terms of quantity).

The other major enzyme is RNA polymerase II, located in the nucleoplasm (the part of the nucleus excluding the nucleolus). It represents most of the remaining cellular activity and is responsible for synthesizing heterogeneous nuclear RNA (hnRNA), the precursor for mRNA.

RNA polymerase III is a minor enzyme activity. This nucleoplasmic enzyme synthesizes tRNAs and other small RNAs.

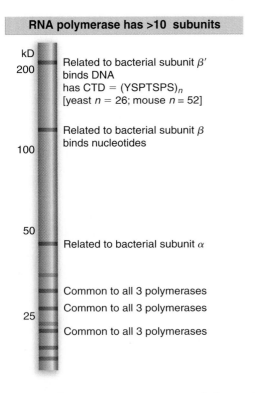

Figure 24.2 Some subunits are common to all classes of eukaryotic RNA polymerases and some are related to bacterial RNA polymerase.

All eukaryotic RNA polymerases are large proteins, appearing as aggregates of >500 kD. They typically have ~12 subunits. The purified enzyme can undertake template-dependent transcription of RNA, but is not able to initiate selectively at promoters. The general constitution of a eukaryotic RNA polymerase II as typified in *S. cerevisiae* is illustrated in **Figure 24.2**. The two largest subunits are homologous to the β and β′ subunits of bacterial RNA polymerase. Three of the remaining subunits are common to all the RNA polymerases; that is, they are also components of RNA polymerases I and III.

The largest subunit in RNA polymerase II has a **carboxy-terminal domain (CTD)**, which consists of multiple repeats of a consensus sequence of seven amino acids (YSPTSPS). The sequence is unique to RNA polymerase II. There are ~26 repeats in yeast and ~50 in mammals. The number of repeats is important, because deletions that remove (typically) more than half of the repeats are lethal (in yeast). The CTD can be highly phosphorylated on serine or threonine residues; this is involved in the initiation reaction (see *24.8 Initiation Is Followed by Promoter Clearance*).

Mitochondria and chloroplasts have their own RNA polymerases. These are smaller, and resemble bacterial RNA polymerase rather than any of the nuclear enzymes. Of course, the organelle genomes are much smaller, the resident polymerase needs to transcribe relatively few genes, and the control of transcription is likely to be very much simpler (if existing at all).

A major practical distinction between the eukaryotic enzymes is drawn from their response to the bicyclic octapeptide α-**amanitin**. In basically all eukaryotic cells the activity of RNA polymerase II is rapidly inhibited by low concentrations of α-amanitin. RNA polymerase I is not inhibited. The response of RNA polymerase III to α-amanitin is less well conserved; in animal cells it is inhibited by high levels, but in yeast and insects it is not inhibited.

24.3 RNA Polymerase I Has a Bipartite Promoter

Key Terms

- A **spacer** is a sequence in a gene cluster that separates the repeated copies of the transcription unit.
- A **core promoter** is the shortest sequence at which an RNA polymerase can initiate transcription (typically at much lower level than that displayed by a promoter containing additional elements).

Key Concepts

- The RNA polymerase I promoter consists of a core promoter and an upstream control element.
- The factor UBF1 binds to both regions and enables the factor SL1 to bind.
- SL1 includes the factor TBP that is involved in initiation by all three RNA polymerases.
- RNA polymerase binds to the UBF1–SL1 complex at the core promoter.

RNA polymerase I transcribes only the genes for ribosomal RNA, from a single type of promoter. The transcript includes the sequences of both large and small rRNAs, which are later released by cleavages and processing. There are many copies of the transcription unit, alternating with nontranscribed **spacers** and organized in a cluster, as discussed in *6.7 Genes for rRNA Form Tandem Repeats Including an Invariant Transcription Unit*. The organization of the promoter, and the events involved in initiation, are illustrated in **Figure 24.3**. RNA polymerase I exists as a holoenzyme that contains additional factors required for initiation, and is recruited directly as a giant complex to the promoter.

The promoter consists of two separate regions. The **core promoter** surrounds the startpoint, extending from -45 to $+20$, and is sufficient for transcription to initiate. It is generally $G \cdot C$-rich (unusual for a promoter) except for the only conserved sequence element, a short $A \cdot T$-rich sequence around the startpoint called the Inr. However, its efficiency is very much increased by the upstream promoter element (UPE), another $G \cdot C$-rich sequence, related to the core promoter sequence, which extends from -180 to -107. This type of organization is common to pol I promoters in many species, although the actual sequences vary widely.

RNA polymerase I requires two ancillary factors. For high-frequency initiation, the factor UBF is required. This is a single polypeptide that binds to the narrow groove of a $G \cdot C$-rich element in the upstream promoter element. It wraps the DNA into a $360°$ turn on the protein surface, with the result that the core promoter and UPE come into close proximity. This enables UBF to stimulate binding of the second factor, SL1 (also known as TIF-IB, Rib1 in different species), to the core promoter.

The core-binding factor SL1 has primary responsibility for ensuring that the RNA polymerase is properly localized at the startpoint. It consists of four proteins, and one of its components, called TBP, is a component of "positioning factors" that are required for initiation by RNA polymerases II and III (see *24.6 TBP Is a Component of TF$_{II}$D and Binds the TATA Box*). TBP probably interacts with a common subunit or feature that has been conserved among RNA polymerases. Its exact mode of action is different in each of the positioning factors; at the promoter for RNA polymerase I, it does not bind to DNA, whereas at the promoter for RNA polymerase II it is the principal means of locating the factor on DNA.

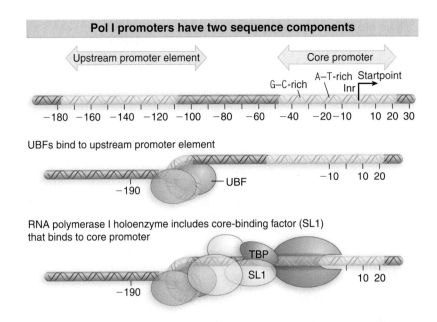

Figure 24.3 Transcription units for RNA polymerase I have a core promoter separated by ~70 bp from the upstream promoter element. UBF binding to the UPE increases the ability of core-binding factor (SL1) to bind to the core promoter.

24.4 RNA Polymerase III Uses Both Downstream and Upstream Promoters

Key Concepts

- RNA polymerase III has two types of promoters.
- Internal promoters have short consensus sequences located within the transcription unit and cause initiation to occur a fixed distance upstream.
- Upstream promoters contain three short consensus sequences upstream of the startpoint that are bound by transcription factors.
- TF$_{III}$A and TF$_{III}$C bind to the consensus sequences and enable TF$_{III}$B to bind at the startpoint.
- TF$_{III}$B has TBP as one subunit and enables RNA polymerase to bind.

Key Terms

- An **assembly factor** is a protein that is required for formation of a macromolecular structure but is not itself part of that structure.
- **Preinitiation complex** in eukaryotic transcription describes the assembly of transcription factors at the promoter before RNA polymerase binds.

Recognition of promoters by RNA polymerase III strikingly illustrates the relative roles of transcription factors and the polymerase enzyme. The promoters fall into two general classes that are recognized

There are three types of pol III promoters

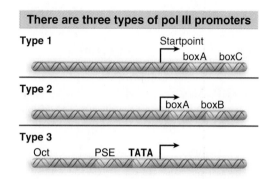

Figure 24.4 Promoters for RNA polymerase III may consist of bipartite sequences downstream of the startpoint, with boxA separated from either boxC or boxB. Or they may consist of separated sequences upstream of the startpoint (Oct, PSE, TATA).

Type 2 internal promoters use TF$_{III}$C

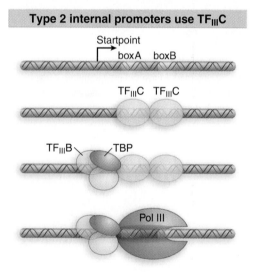

Figure 24.5 Internal type 2 pol III promoters use binding of TF$_{III}$C to boxA and boxB sequences to recruit the positioning factor TF$_{III}$B, which recruits RNA polymerase III.

Type 1 pol III promoters use TF$_{III}$A/C

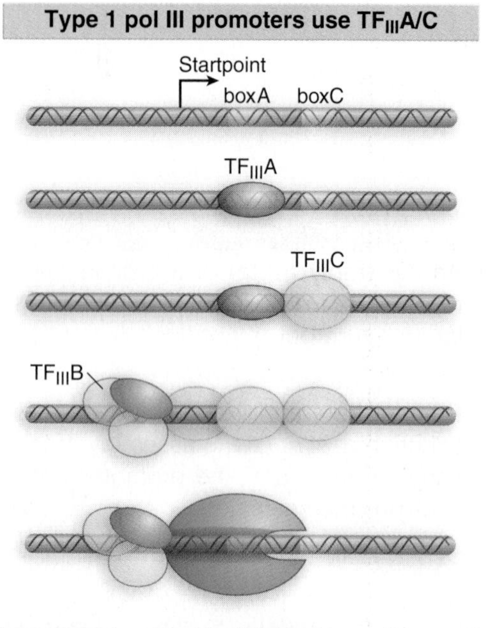

Figure 24.6 Internal type 1 pol III promoters use the assembly factors TF$_{III}$A and TF$_{III}$C, at boxA and boxC, to recruit the positioning factor TF$_{III}$B, which recruits RNA polymerase III.

in different ways by different groups of factors. The promoters for 5S and tRNA genes are *internal*; they lie downstream of the startpoint. The promoters for snRNA (small nuclear RNA) genes lie upstream of the startpoint in the more conventional manner of other promoters. In both cases, the individual elements that are necessary for promoter function consist exclusively of sequences recognized by transcription factors, which in turn direct the binding of RNA polymerase.

The structures of three types of promoters for RNA polymerase III are summarized in **Figure 24.4**. There are two types of internal promoters. Each contains a bipartite structure, in which two short sequence elements are separated by a variable sequence. Type 1 consists of a boxA sequence separated from a boxC sequence, and type 2 consists of a boxA sequence separated from a boxB sequence. Type 3 promoters have three sequence elements all located upstream of the startpoint.

The detailed interactions at the two types of internal promoters differ, but the principle is the same. TF$_{III}$C binds downstream of the startpoint, either independently (type 2 promoters) or in conjunction with TF$_{III}$A (type 1 promoters). The presence of TF$_{III}$C enables the positioning factor TF$_{III}$B to bind at the startpoint. Then RNA polymerase is recruited.

Figure 24.5 summarizes the stages of reaction at type 2 internal promoters. The initial event is the binding of TF$_{III}$C to both boxA and boxB. The difference at type 1 internal promoters is that TF$_{III}$A must bind at boxA to enable TF$_{III}$C to bind at boxC, as shown in **Figure 24.6.**

Once TF$_{III}$C has bound, events follow the same course as at type 2 promoters, with TF$_{III}$B binding at the startpoint, and RNA polymerase III joining the complex. Type 1 promoters are found only in the genes for 5S rRNA.

TF$_{III}$A and TF$_{III}$C are **assembly factors**, whose sole role is to assist the binding of TF$_{III}$B at the right location. Once TF$_{III}$B has bound, TF$_{III}$A and TF$_{III}$C can be removed from the promoter without affecting the initiation reaction. *TF$_{III}$B remains bound in the vicinity of the startpoint and its presence is sufficient to allow RNA polymerase III to bind at the startpoint.* So TF$_{III}$B is the only true initiation factor required by RNA polymerase III. This sequence of events explains how the promoter boxes downstream can cause RNA polymerase to bind at the startpoint, farther upstream. Although the ability to transcribe these genes is conferred by the internal promoter, changes in the region immediately upstream of the startpoint can alter the efficiency of transcription.

The upstream region has a conventional role in the third class of polymerase III promoters. In the example shown in Figure 24.4, there are three upstream elements. These elements are also found in promoters for snRNA genes that are transcribed by RNA polymerase II. (Genes for some snRNAs are transcribed by RNA polymerase II, while others are transcribed by RNA polymerase III.) The upstream elements function in a similar manner in promoters for both polymerases II and III.

Initiation at an upstream promoter for RNA polymerase III can occur on a short region that immediately precedes the startpoint and contains only the TATA element. However, efficiency of transcription is much increased by the presence of the PSE and OCT elements. The factors that bind at these elements interact cooperatively.

The TATA element confers specificity for the type of polymerase (II or III) that is recognized by an snRNA promoter. It is bound by a factor that includes the TBP, which actually recognizes the sequence in DNA. (The factor is TF$_{III}$B for RNA polymerase III, and is TF$_{II}$D for RNA polymerase II). The function of TBP and its associated proteins is to position the RNA polymerase correctly at the startpoint. We discuss this in more detail for RNA polymerase II (see *24.6 TBP Is a Component of TF$_{II}$D and Binds the TATA Box*).

As the positioning factor that enables RNA polymerase III to bind at the promoter, $TF_{III}B$ works in the same way for all the types of promoters. *It binds at the promoter to form a* **preinitiation complex** *that directs binding of the RNA polymerase.* The basic difference between the types of promoters lies with whether assembly factors are required for $TF_{III}B$ to bind (as in type 1 and type 2) or whether it can bind directly (as in type 3). Irrespective of the location of the promoter sequences, $TF_{III}B$ binds close to the startpoint in order to direct binding of RNA polymerase III.

24.5 The Startpoint for RNA Polymerase II

Key Terms

- A **basal factor** is a transcription factor required by RNA polymerase II to form the initiation complex at all promoters. Factors are identified as $TF_{II}X$, where X is a letter.

- The **core promoter** for RNA polymerase II is the minimal sequence at which the basal transcription apparatus can assemble. It is typically ~40 bp long, and includes the InR and either a TATA box or a DPE.

- The **initiator (Inr)** is the sequence of a pol II promoter between -3 and $+5$ and has the general sequence Py_2CAPy_5. It is the simplest possible pol II promoter.

- **TATA box** is a conserved A·T-rich octamer found about 25 bp before the startpoint of each eukaryotic RNA polymerase II transcription unit; it is involved in positioning the enzyme for correct initiation.

- A **TATA-less promoter** does not have a TATA box in the sequence upstream of its startpoint.

- The **DPE** is a common component of RNA polymerase II promoters that do not contain a TATA box.

The basic organization of the apparatus for transcribing protein-coding genes was revealed by the discovery that purified RNA polymerase II can catalyze synthesis of mRNA, but cannot initiate transcription unless an additional extract is added. The purification of this extract led to the definition of the general transcription factors, or **basal factors**—a group of proteins that are needed for initiation by RNA polymerase II at all promoters. RNA polymerase II in conjunction with these factors constitutes the basal transcription apparatus that is needed to transcribe any promoter. The general factors are described as $TF_{II}X$, where X is a letter that identifies the individual factor. The subunits of RNA polymerase II and the general transcription factors are conserved among eukaryotes.

Our starting point for considering promoter organization is to define the **core promoter** as the shortest sequence at which RNA polymerase II can initiate transcription. A core promoter can in principle be expressed in any cell. It is the minimum sequence that enables the general transcription factors to assemble at the startpoint. These factors are involved in the mechanics of binding to DNA, and enable RNA polymerase II to initiate transcription. A core promoter functions at only a low efficiency. Other proteins, called activators, are required for a proper level of function (see *24.9 Short Sequence Elements Bind Activators*). The activators are not described systematically, but have casual names reflecting their histories of identification.

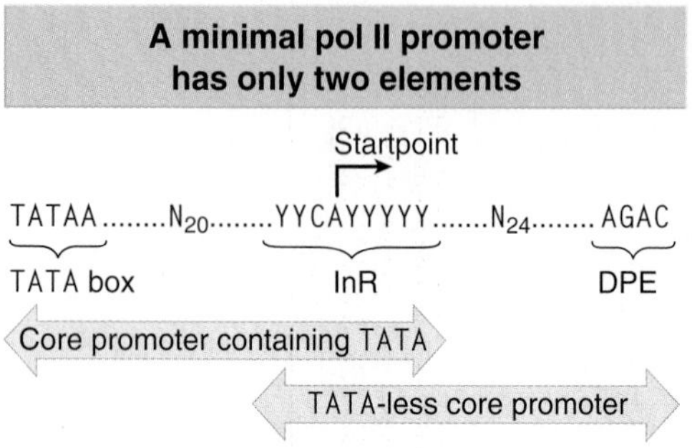

A minimal pol II promoter has only two elements

Startpoint

TATAA........N$_{20}$........YYCAYYYYY.......N$_{24}$........ AGAC

TATA box   InR   DPE

Core promoter containing TATA

TATA-less core promoter

Figure 24.7 The minimal pol II promoter has a TATA box ~25 bp upstream of the InR. The TATA box has the consensus sequence of TATAA. The Inr has pyrimidines (Y) surrounding the CA at the startpoint. The DPE is downstream of the startpoint. The sequence shows the coding strand.

We may expect any sequence components involved in the binding of RNA polymerase and general transcription factors to be conserved at most or all promoters. As with bacterial promoters, when promoters for RNA polymerase II are compared, homologies in the regions near the startpoint are restricted to rather short sequences. These elements correspond with the sequences implicated in promoter function by mutation. **Figure 24.7** shows the construction of a typical pol II core promoter.

At the startpoint, there is no extensive homology of sequence, but there is a tendency for the first base of mRNA to be A, flanked on either side by pyrimidines. (This description is also valid for the CAT start sequence of bacterial promoters.) This region is called the **initiator (Inr)**, and may be described in the general form Py$_2$CAPy$_5$. The InR is contained between positions −3 and +5.

Many promoters have a sequence called the **TATA box**, usually located ~25 bp upstream of the startpoint. It constitutes the only upstream promoter element that has a relatively fixed location with respect to the startpoint. The core sequence is TATAA, usually followed by three more A · T base pairs. The TATA box tends to be surrounded by G · C-rich sequences, which could be a factor in its function. It is almost identical with the −10 sequence found in bacterial promoters.

Promoters that do not contain a TATA element are called **TATA-less promoters**. Surveys of promoter sequences suggest that 50% or more of promoters may be TATA-less. When a promoter does not contain a TATA box, it usually contains another element, the **DPE** (downstream promoter element), which is located at +28 to +32.

A core promoter can consist either of a TATA box plus InR or of an InR plus DPE.

24.6 TBP Is a Component of TF$_{II}$D and Binds the TATA Box

Key Terms

- **TF$_{II}$D** is the transcription factor that binds to the TATA sequence upstream of the startpoint of promoters for RNA polymerase II. It consists of TBP (TATA-binding protein) and the TAF subunits that bind to TBP.

- **TAFs** are the subunits of TF$_{II}$D that assist TBP in binding to DNA. They also provide points of contact for other components of the transcription apparatus.

Key Concepts

- TBP is a component of the positioning factor that is required for each type of RNA polymerase to bind its promoter.

- The factor for RNA polymerase II is TF$_{II}$D, which consists of TBP and 11 TAFs, with a total mass ~800 kD.

- TBP binds to the TATA box in the minor groove of DNA.

- TBP forms a saddle around the DNA and bends it by ~80°.

- Some of the TAFs resemble histones and may form a structure resembling a histone octamer.

The first stage in initiation at all classes of eukaryotic promoters is for a positioning factor to bind. Each class of RNA polymerase is assisted by a positioning factor that consists of the small protein TBP associated with other components. The positioning factor for RNA polymerase I is

SL1 (see *24.3 RNA Polymerase I Has a Bipartite Promoter*), and for RNA polymerase III is TF$_{III}$B (see *24.4 RNA Polymerase III Uses Both Downstream and Upstream Promoters*). For RNA polymerase II, the position factor is **TF$_{II}$D**, which consists of TBP associated with 11 other subunits, which are called **TAFs** (for TBP-associated factors). The total mass of TF$_{II}$D typically is ~800 kD. The TAFs in TF$_{II}$D are named in the form TAF$_{II}$00, where "00" gives the molecular mass of the subunit; an alternative nomenclature labels them as TAF$_1$ to TAF$_{13}$, where the numbers increase with declining size.

Figure 24.8 shows that the positioning factor recognizes the promoter in a different way in each case. At promoters for RNA polymerase III, TF$_{III}$B binds adjacent to TF$_{III}$C. At promoters for RNA polymerase I, SL1 binds in conjunction with UBF. By contrast, TF$_{II}$D is solely responsible for recognizing promoters for RNA polymerase II. At a promoter that has a TATA element, TBP binds specifically to the DNA, so that TF$_{II}$D occupies a region extending upstream from the TATA sequence. The key feature in identifying the startpoint for RNA polymerase is its fixed distance from the TATA box. At TATA-less promoters, TF$_{II}$D may be incorporated by association with other proteins that bind to DNA. Whatever its means of entry into the initiation complex, it has the common purpose of interaction with the RNA polymerase.

The crystal structure of TBP suggests a detailed model for its binding to DNA. **Figure 24.9** shows that it surrounds one face of DNA, forming a "saddle" around the double helix. In effect, the inner surface of TBP binds to DNA, and the larger outer surface is available to extend contacts to other proteins. The DNA-binding site consists of a C-terminal domain that is conserved between species, while the variable N-terminal tail is exposed to interact with other proteins. It is a measure of the conservation mechanism in transcriptional initiation that the DNA-binding sequence of TBP is 80% conserved between yeast and man.

TBP binds to the minor groove and bends the DNA by ~80°, as illustrated in **Figure 24.10**. The TATA box bends towards the major groove, widening the minor groove. The distortion is restricted to the 8 bp of the TATA box; at each end of the sequence, the minor groove has its usual width of ~5 Å, but at the center of the sequence the minor groove is >9 Å. This is a deformation of the structure, but does not actually separate the strands of DNA, because base pairing is maintained. By changing the spatial organization of DNA on either side of the TATA box, this structure allows the transcription factors and RNA polymerase to form a closer association than would be possible on linear DNA. The angle of the bend depends on the exact sequence of the TATA box and is correlated with promoter efficiency, suggesting that it influences interactions between proteins binding to sequences at the ends of the bend.

Within TF$_{II}$D as a free protein complex, the factor TAF$_{II}$230 binds to TBP, where it occupies the concave DNA-binding surface. In fact, the structure of the binding site, which lies in the N-terminal domain of TAF$_{II}$230, mimics the surface of the minor groove in DNA. This molecular mimicry allows TAF$_{II}$230 to control the ability of TBP to bind to DNA; the N-terminal domain of TAF$_{II}$230 must be displaced from the DNA-binding surface of TBP in order for TF$_{II}$D to bind to DNA.

What happens at TATA-less promoters? The same general transcription factors, including TF$_{II}$D, are needed. The Inr provides the positioning element; TF$_{II}$D binds to it via an ability of one or more of the TAFs to recognize the Inr directly. Other TAFs in TF$_{II}$D also recognize

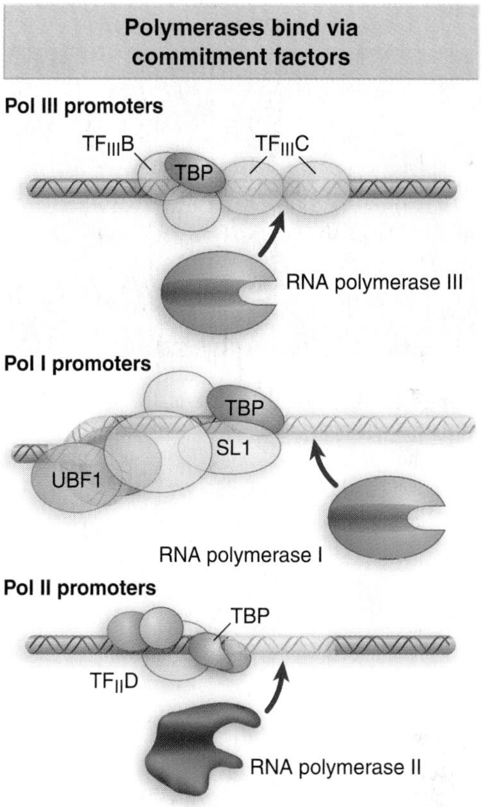

Polymerases bind via commitment factors

Pol III promoters

Pol I promoters

Pol II promoters

Figure 24.8 RNA polymerases are positioned at all promoters by a factor that contains TBP.

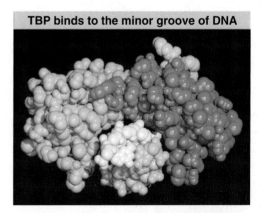

TBP binds to the minor groove of DNA

Figure 24.9 A view in cross section shows that TBP surrounds DNA from the side of the minor groove. TBP consists of two related (40% identical) conserved domains, which are shown in light and dark blue. The N-terminal region varies extensively and is shown in green. The two strands of the DNA double helix are in light and dark grey. Photograph kindly provided by Stephen K. Burley.

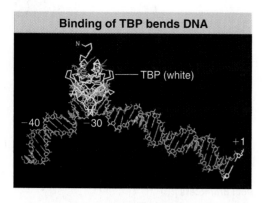

Binding of TBP bends DNA

TBP (white)

−40   −30   +1

Figure 24.10 The crystal structure of TBP with DNA from −40 to the startpoint shows a bend at the TATA box that widens the minor groove where TBP binds. Photograph kindly provided by Stephen K. Burley.

Factors assemble into an initiation complex

TATA   Startpoint

−60 −40 −20 −10   10 20 30 40 50

TBP binds minor groove, bends DNA

TAF

TF$_{II}$230

TBP

TF$_{II}$A

TF$_{II}$B

TF$_{II}$F

−60

40

TF$_{II}$E

Polymerase

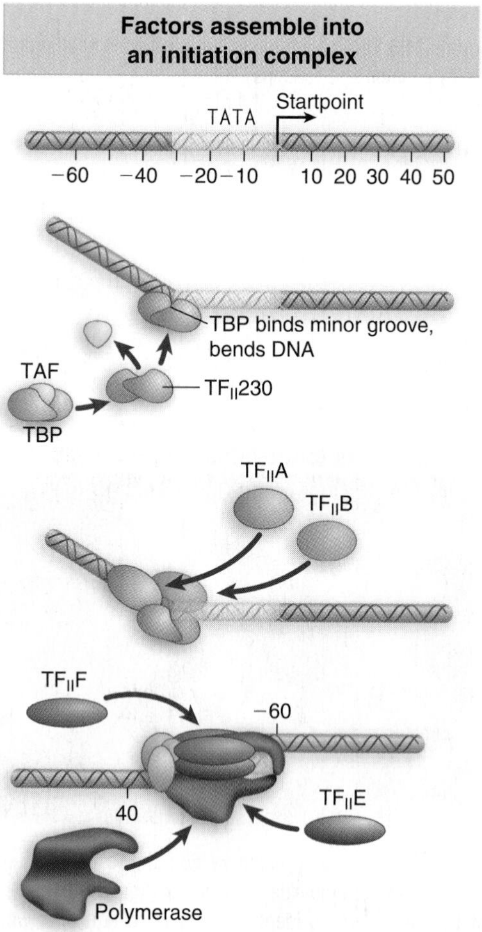

Figure 24.11 An initiation complex assembles at promoters for RNA polymerase II by an ordered sequence of association with transcription factors.

the DPE element downstream from the startpoint. The function of TBP at these promoters is more like that at promoters for RNA polymerase I and at internal promoters for RNA polymerase III.

When a TATA box is present, it determines the location of the startpoint. Its deletion causes the site of initiation to become erratic, although any overall reduction in transcription is relatively small. Indeed, some TATA-less promoters lack unique startpoints; initiation occurs instead at any one of a cluster of startpoints. The TATA box aligns the RNA polymerase (via the interaction with TF$_{II}$D and other factors) so that it initiates at the proper site. This explains why its location is fixed with respect to the startpoint. Binding of TBP to TATA is the predominant feature in recognition of the promoter, but two large TAFs (TAF$_{II}$250 and TAF$_{II}$150) also contact DNA in the vicinity of the startpoint and influence the efficiency of the reaction.

24.7 The Basal Apparatus Assembles at the Promoter

Key Concepts

- Binding of TF$_{II}$D to the promoter is the first step in initiation.

- Other transcription factors bind to the complex in a defined order, extending the length of the protected region on DNA.

- When RNA polymerase II binds to the complex, it initiates transcription.

Initiation requires the transcription factors to act in a defined order to build a complex that is joined by RNA polymerase. The sequence of events is summarized in **Figure 24.11**.

Commitment to a promoter is initiated when TF$_{II}$D binds the TATA box. (TF$_{II}$D also recognizes the InR sequence at the startpoint.) TF$_{II}$A may activate TBP by relieving the repression that is caused by the TAF$_{II}$230.

TF$_{II}$B binds downstream of the TATA box, adjacent to TBP, extending contacts along one face of DNA as shown in **Figure 24.12**. It makes contacts in the minor groove downstream of the TATA box, and contacts the major groove upstream of the TATA box, in a region called the BRE. In archaea, the homologue of TF$_{II}$B actually makes sequence-specific contacts with the promoter in the BRE region.

The schematic of **Figure 24.13** shows the relationship between TF$_{II}$D, TF$_{II}$B, and RNA polymerase II. TF$_{II}$B binds adjacent to TF$_{II}$D, and its N-terminal region contacts RNA polymerase near the RNA exit site. Its C-terminal region extends across the enzyme, with a protrusion into the active site. TF$_{II}$B determines the path of the DNA where it contacts the factors TF$_{II}$E, TF$_{II}$F, and TF$_{II}$H, which may align them in the basal factor complex, and determine the startpoint.

The factor TF$_{II}$F is a heterotetramer consisting of two types of subunit. The larger subunit (RAP74) has an ATP-dependent DNA helicase activity that may help melt the DNA at initiation. The smaller subunit (RAP38) has some homology to the regions of bacterial sigma factor that contact the core polymerase; it binds tightly to RNA polymerase II. TF$_{II}$F may bring RNA polymerase II to the assembling transcription complex and provide the means by which it binds. The complex of TBP and TAFs may interact with the CTD tail of RNA polymerase, and interaction with TF$_{II}$B may also be important when TF$_{II}$F/polymerase joins the complex.

Assembly of the RNA polymerase II initiation complex provides an interesting contrast with prokaryotic transcription. Bacterial RNA polymerase is essentially a coherent aggregate with intrinsic ability to bind DNA; the sigma factor, needed for initiation but not for elongation, becomes part of the enzyme before DNA is bound, although it is later released. But RNA polymerase II can bind to the promoter only after separate transcription factors have bound. The factors play a role analogous to that of bacterial sigma factor—to allow the basic polymerase to recognize DNA specifically at promoter sequences—but have evolved to have more independence. Indeed, the factors are primarily responsible for the specificity of promoter recognition. Only some of the factors participate in protein–DNA contacts (and only TBP makes sequence-specific contacts); thus protein–protein interactions are important in the assembly of the complex.

Although assembly can take place just at the core promoter *in vitro*, this reaction is not sufficient for transcription *in vivo*, where interactions with activators that recognize the more upstream elements are required. The activators interact with the basal apparatus at various stages during its assembly (see *25.4 Activators Interact with the Basal Apparatus*).

24.8 Initiation Is Followed by Promoter Clearance

Key Concepts

- TF$_{II}$E and TF$_{II}$H are required to melt DNA to allow polymerase movement.
- Phosphorylation of the CTD may be required for elongation to begin.
- Further phosphorylation of the CTD is required at some promoters to end abortive initiation.
- The CTD may coordinate processing of RNA with transcription.
- TF$_{II}$H provides the link to a complex of repair enzymes.
- Mutations in the XPD component of TF$_{II}$H cause three types of human diseases.

Some final steps are needed to release RNA polymerase from the promoter once the first nucleotide bonds have been formed. The last factors to join the initiation complex are TF$_{II}$E, TF$_{II}$H, and TF$_{II}$J. They act at the later stages of initiation.

TF$_{II}$H is a general transcription factor that has several enzymatic activities. Its several activities include an ATPase, helicases of both polarities, and a kinase activity that can phosphorylate the CTD tail of RNA polymerase II. TF$_{II}$H is an exceptional factor that may play a role also in elongation. Its interaction with DNA downstream of the startpoint is required for RNA polymerase to escape from the promoter. TF$_{II}$H also participates in repairing damage to DNA.

Figure 24.14 proposes a model in which phosphorylation of the tail is needed to release RNA polymerase II from the transcription factors so that it can make the transition to the elongating form. Most of the transcription factors are released from the promoter at this stage.

The CTD may also be involved, directly or indirectly, in processing RNA after it has been synthesized by RNA polymerase II. **Figure 24.15** summarizes processing reactions in which the CTD may participate. The capping enzyme (guanylyl transferase), which adds the guanine cap to the 5′ end of newly synthesized mRNA, binds to the phosphorylated CTD: this may be important in enabling it to modify the 5′ end as soon as it is

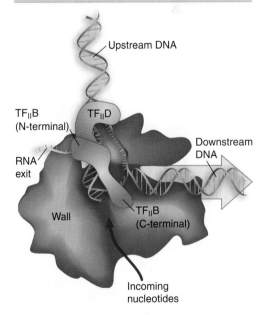

TF$_{II}$B binds to bent DNA downstream from TBP

TBP

TF$_{II}$B

DNA

TBP

DNA

TF$_{II}$B

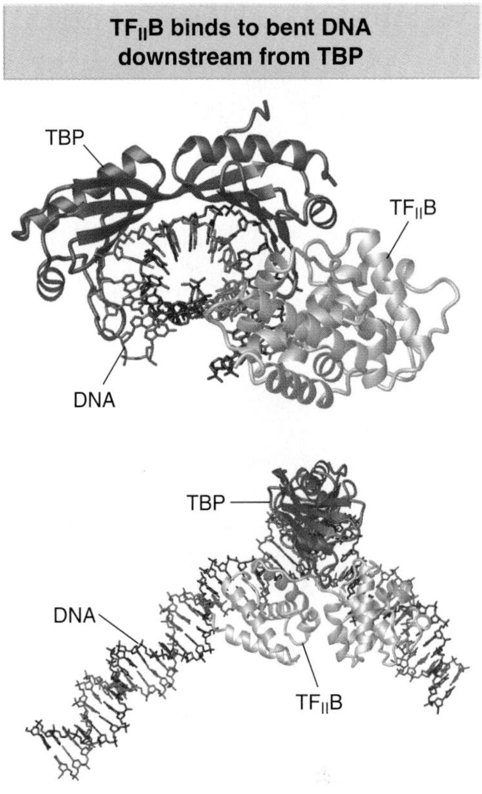

Figure 24.12 Two views of the ternary complex of TF$_{II}$B-TBP-DNA show that TF$_{II}$B binds along the bent face of DNA. Photograph kindly provided by Stephen K. Burley.

TF$_{II}$B helps position RNA polymerase II

Upstream DNA

TF$_{II}$B (N-terminal)

TF$_{II}$D

Downstream DNA

RNA exit

Wall

TF$_{II}$B (C-terminal)

Incoming nucleotides

Figure 24.13 TF$_{II}$B binds to DNA and contacts RNA polymerase near the RNA exit site and at the active center, and orients it on DNA. Compare with Figure 11.10, which shows the polymerase structure engaged in transcription.

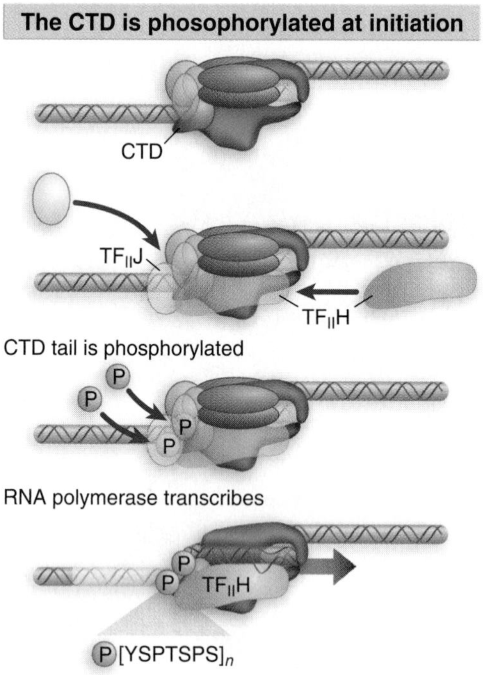

The CTD is phosophorylated at initiation

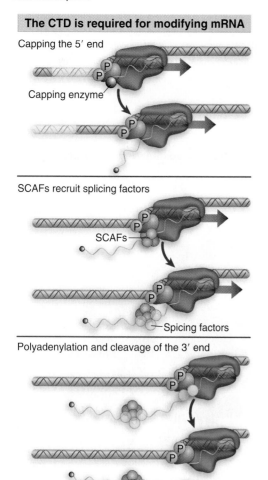

CTD

TF_{II}J

TF_{II}H

CTD tail is phosphorylated

RNA polymerase transcribes

TF_{II}H

P[YSPTSPS]$_n$

Figure 24.14 Phosphorylation of the CTD by the kinase activity of TF$_{II}$H may be needed to release RNA polymerase to start transcription.

The CTD is required for modifying mRNA

Capping the 5′ end

Capping enzyme

SCAFs recruit splicing factors

SCAFs

Spicing factors

Polyadenylation and cleavage of the 3′ end

AAAAAA

Figure 24.15 The CTD is important in recruiting enzymes that modify RNA.

synthesized. A set of proteins called SCAFs binds to the CTD, and they may in turn bind to splicing factors. This may be a means of coordinating transcription and splicing. Some components of the cleavage/polyadenylation apparatus also bind to the CTD. Oddly enough, they do so at the time of initiation, so that RNA polymerase is all ready for the 3′ end processing reactions as soon as it sets out! All of this suggests that the CTD may be a general focus for connecting other processes with transcription. For capping and splicing, the CTD functions indirectly to promote formation of the protein complexes that undertake the reactions. For 3′ end generation, it may participate directly in the reaction.

The general process of initiation is similar to that catalyzed by bacterial RNA polymerase. Binding of RNA polymerase generates a closed complex, which is converted at a later stage to an open complex in which the DNA strands have been separated. In the bacterial reaction, formation of the open complex completes the necessary structural change to DNA; a difference in the eukaryotic reaction is that further unwinding of the template is needed after this stage.

TF$_{II}$H has a common function in both initiating transcription and repairing damage. The same helicase subunit (XPD) creates the initial transcription bubble and melts DNA at a damaged site. The other functions of TF$_{II}$H differ between transcription and repair, which require different forms of the complex.

Figure 24.16 shows that the TF$_{II}$H used in transcription is a core (of five subunits) associated with other subunits that have a kinase activity.

The alternative complex consists of the core associated with a large group of proteins that are coded by repair genes. (The basic model for repair is shown in Figure 20.20.) The repair proteins include a subunit (XPC) that recognizes damaged DNA, which provides the coupling function that enables a template strand to be preferentially repaired when RNA polymerase becomes stalled at damaged DNA. Other proteins associated with the complex include endonucleases (XPG, XPF, ERCC1). Homologous proteins are found in the complexes in yeast (where they are often identified by *rad* mutations that are defective in repair) and in man, where they are identified by mutations that cause diseases resulting from deficiencies in repairing damaged DNA. Subunits with the name XP are coded by genes in which mutations cause the disease xeroderma pigmentosum (see *20.9 Eukaryotic Cells Have Conserved Repair Systems*).

The kinase complex and the repair complex can associate and dissociate reversibly from the core TF$_{II}$H. This suggests a model in which the first form of TF$_{II}$H is required for initiation, but may be replaced by the other form (perhaps in response to encountering DNA damage). TF$_{II}$H dissociates from RNA polymerase at an early stage of elongation (after transcription of ~50 bp); its reassociation at a site of damaged DNA may require additional coupling components. The importance of the TF$_{II}$H factor for repair is shown by the fact that mutations in the genes coding for some of its subunits cause human diseases resulting from the inability to repair irradiation or other damage to DNA.

24.9 Short Sequence Elements Bind Activators

Key Terms

- An **activator** is a protein that stimulates the expression of a gene, typically by acting at a promoter to stimulate RNA polymerase. In eukaryotes, the sequence to which it binds in the promoter is called a response element.

- A **CAAT box** is part of a conserved sequence located upstream of the startpoints of eukaryotic transcription units; it is recognized by a large group of transcription factors.
- The **GC box** is a common pol II promoter element consisting of the sequence GGGCGG.

Key Concepts

- Short conserved sequence elements are dispersed in the region preceding the startpoint.
- The upstream response elements increase the frequency of initiation.
- No individual upstream element is essential for promoter function, although one or more elements must be present for efficient initiation.

A promoter for RNA polymerase II consists of two types of regions. The startpoint itself is identified by the Inr and/or by the TATA box close by. In conjunction with the general transcription factors, RNA polymerase II forms an initiation complex surrounding the startpoint, as we have just seen. The efficiency and specificity with which a promoter is recognized, however, depend upon short sequences, farther upstream, which are recognized by a different group of factors, usually called **activators**. Usually the target sequences are ~100 bp upstream of the startpoint, but sometimes they are more distant. Binding of activators at these sites may influence the formation of the initiation complex at any one of several stages.

An analysis of a typical promoter is summarized in **Figure 24.17**. Individual base substitutions were introduced at almost every position in the 100 bp upstream of the β-globin startpoint. The striking result is that *most mutations do not affect the ability of the promoter to initiate transcription*. Down mutations occur in three locations, corresponding to three short discrete elements. The two upstream elements have a greater effect on the level of transcription than the element closest to the startpoint. Up mutations occur in only one of the elements. We conclude that the three short sequences centered at

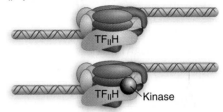

TF$_{II}$H plays multiple roles

TF$_{II}$H provides a kinase at initiation

TF$_{II}$H provides a repair complex for damaged DNA at elongation

Figure 24.16 The TF$_{II}$H core may associate with a kinase at initiation and associate with a repair complex when damaged DNA is encountered.

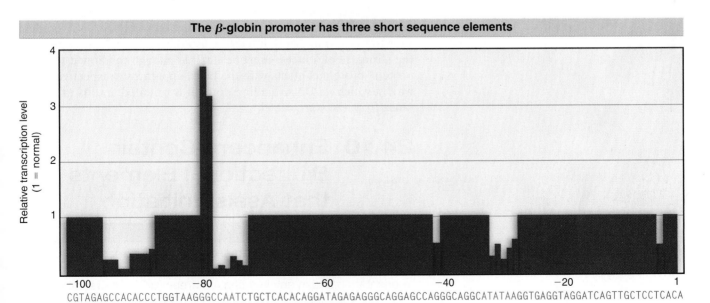

The β-globin promoter has three short sequence elements

CGTAGAGCCACACCCTGGTAAGGGCCAATCTGCTCACACAGGATAGAGAGGGCAGGAGCCAGGGCAGGCATATAAGGTGAGGTAGGATCAGTTGCTCCTCACA

Figure 24.17 Saturation mutagenesis of the upstream region of the β-globin promoter identifies three short regions (centered at −30, −75, and −90) that are needed to initiate transcription. These correspond to the TATA, CAAT, and GC boxes.

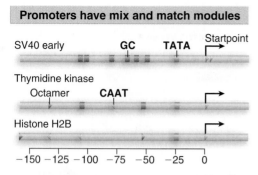

Figure 24.18 Promoters contain different combinations of TATA boxes, CAAT boxes, GC boxes, and other elements.

−30, −75, and −90 constitute the promoter. Each of them corresponds to the consensus sequence for a common type of promoter element.

The TATA box (centered at −30) is the least effective component of the promoter as measured by the reduction in transcription that is caused by mutations. But although initiation is not prevented when a TATA box is mutated, the startpoint varies from its usual precise location. This confirms the role of the TATA box as a crucial positioning component of the core promoter.

The basal elements and the elements upstream of them have different types of functions. The basal elements (the TATA box and Inr) primarily determine the location of the startpoint, but can sponsor initiation only at a rather low level. They identify the *location* at which the general transcription factors assemble to form the basal complex. The sequence elements farther upstream influence the *frequency* of initiation; they are the binding sites for activators that interact with the general transcription factors to enhance the efficiency of assembly into an initiation complex (see *25.4 Activators Interact with the Basal Apparatus*).

The sequence at −75 is the **CAAT box**. Named for its consensus sequence, it was one of the first common elements to be described. It is often located close to −80, but it can function at distances that vary considerably from the startpoint. It functions in either orientation. Susceptibility to mutations suggests that the CAAT box plays a strong role in determining the efficiency of the promoter, but does not influence its specificity.

The **GC box** at −90 contains the sequence GGGCGG. Often multiple copies are present in the promoter, and they occur in either orientation. It too is a relatively common promoter component.

Promoters are organized on a principle of "mix and match." A variety of elements can contribute to promoter function, but none is essential for all promoters. Some examples are summarized in **Figure 24.18**. Four types of elements are found altogether in these promoters: TATA, GC boxes, CAAT boxes, and the octamer (an 8 bp element). The elements found in any individual promoter differ in number, location, and orientation. No element is common to all of the promoters. Although the promoter conveys directional information (transcription proceeds only in the downstream direction), the GC and CAAT boxes seem to be able to function in either orientation. This implies that the elements function solely as DNA-binding sites to bring transcription factors into the vicinity of the startpoint; the structure of a factor must be flexible enough to allow it to make protein–protein contacts with the basal apparatus irrespective of the way in which its DNA-binding domain is oriented and its exact distance from the startpoint.

24.10 Enhancers Contain Bidirectional Elements that Assist Initiation

Key Terms

- An **enhancer** is a *cis*-acting sequence that increases the utilization of some eukaryotic promoters, and that can function in either orientation and in any location (upstream or downstream) relative to the promoter.

- An **upstream activator sequence (UAS)** is the equivalent in yeast of the enhancer in higher eukaryotes.

Key Concepts

- An enhancer activates the nearest promoter to it.
- A UAS (upstream activator sequence) in yeast behaves like an enhancer but works only upstream of the promoter.
- Similar sequence elements are found in enhancers and promoters.
- Enhancers form complexes of activators that interact directly or indirectly with the promoter.

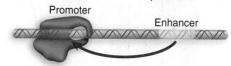

Figure 24.19 An enhancer can activate a promoter from upstream or downstream.

We have considered the promoter so far as an isolated region responsible for binding RNA polymerase. But eukaryotic promoters do not necessarily function alone. In at least some cases, the activity of a promoter is enormously increased by the presence of an **enhancer**, which consists of another group of elements, but is located at a variable distance from those regarded as being part of the promoter itself.

The concept that the enhancer is distinct from the promoter reflects two characteristics. The position of the enhancer relative to the promoter need not be fixed, but can vary substantially. **Figure 24.19** shows that it can be either upstream or downstream. And it can function in either orientation (that is, it can be inverted) relative to the promoter. Manipulations of DNA show that an enhancer can stimulate any promoter placed in its vicinity.

For operational purposes, it is sometimes useful to define the promoter *as a sequence or sequences of DNA that must be in a (relatively) fixed location with regard to the startpoint*. By this definition, the TATA box and other upstream elements are included, but the enhancer is excluded. This is, however, a working definition rather than a rigid classification.

Elements analogous to enhancers, called **upstream activator sequences (UAS)**, are found in yeast. They can function in either orientation, at variable distances upstream of the promoter, but cannot function when located downstream. They have a regulatory role: in several cases the UAS is bound by the regulatory protein(s) that activates the genes downstream.

Reconstruction experiments in which the enhancer sequence is removed from the DNA and then is inserted elsewhere show that normal transcription can be sustained so long as it is present *anywhere* on the DNA molecule. If a β-globin gene is placed on a DNA molecule that contains an enhancer, its transcription is increased *in vivo* more than 200-fold, even when the enhancer is several kb upstream or downstream of the startpoint, in either orientation. We have yet to discover at what distance the enhancer fails to work.

24.11 Enhancers Contain the Same Elements that Are Found at Promoters

Key Term

- An **enhanceosome** is a complex of transcription factors that assembles cooperatively at an enhancer.

Key Concepts

- Enhancers are made of the same short sequence elements that are found in promoters.
- The density of sequence components is greater in the enhancer than in the promoter.

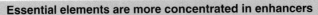

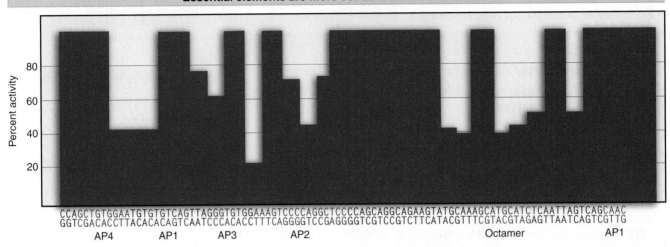

Figure 24.20 An enhancer contains several structural motifs. The histogram plots the effect of all mutations that reduce enhancer function to <75% of wild type. Binding sites for proteins are indicated below the histogram.

A difference between the enhancer and a typical promoter is presented by the density of regulatory elements. **Figure 24.20** summarizes the susceptibility of the SV40 enhancer to damage by mutation; and we see that a much greater proportion of its sites directly influences its function than is the case with the promoter analyzed in the same way in Figure 24.17. There is a corresponding increase in the density of protein-binding sites. Many of these sites are common elements in promoters, for example, AP1 and the octamer.

The specificity of transcription may be controlled by either a promoter or an enhancer. A promoter may be specifically regulated, and a nearby enhancer used to increase the efficiency of initiation; or a promoter may lack specific regulation, but become active only when a nearby enhancer is specifically activated. An example is provided by immunoglobulin genes, which carry enhancers *within* the transcription unit. The immunoglobulin enhancers appear to be active only in the B lymphocytes in which the immunoglobulin genes are expressed. Such enhancers provide part of the regulatory network by which gene expression is controlled.

A difference between enhancers and promoters may be that an enhancer shows greater cooperativity between the binding of factors. A complex that assembles at the enhancer that responds to IFNγ (interferon) assembles cooperatively to form a functional structure called the **enhanceosome**. Binding of the nonhistone protein, HMGI(Y) bends the DNA into a structure that then binds several activators (NF–κB, IRF, ATF–Jun). In contrast with the "mix and match" construction of promoters, all of these components are required to create an active structure at the enhancer. These components do not themselves directly bind to RNA polymerase, but they create a surface that binds a *coactivating complex*. The complex helps the preinitiation complex of basal transcription factors that is assembling at the promoter to recruit RNA polymerase. We discuss the function of coactivators in more detail in *25.4 Activators Interact with the Basal Apparatus*.

24.12 Enhancers Work by Increasing the Concentration of Activators Near the Promoter

Key Concepts

- Enhancers usually work only in *cis* configuration with a target promoter.
- Enhancers can be made to work in *trans* configuration by linking the DNA that contains the target promoter to the DNA that contains the enhancer via a protein bridge or by catenating the two molecules.
- The principle is that an enhancer works in any situation in which it is constrained to be in proximity with the promoter.

Enhancers work much like promoters, except for their ability to function in either orientation and at variable distances from the startpoint. They have the same sort of interaction with the basal apparatus as the interactions sponsored by upstream promoter elements. Enhancers are modular, like promoters. Some elements are found in both enhancers and promoters. Some individual elements found in promoters share with enhancers the ability to function at variable distance and in either orientation. So the distinction between enhancers and promoters is blurred: enhancers might be viewed as containing promoter elements that are grouped closely together, with the ability to function at increased distances from the startpoint.

The essential role of the enhancer is to increase the concentration of activators in the vicinity of the promoter (vicinity in this sense being a relative term). Two types of experiment illustrated in **Figure 24.21** suggest that this is the case.

A fragment of DNA that contains an enhancer at one end and a promoter at the other is not effectively transcribed, but the enhancer can stimulate transcription from the promoter when they are connected by a protein bridge. Since structural effects, such as changes in supercoiling, could not be transmitted across such a bridge, this suggests that the critical feature is bringing the enhancer and promoter into close proximity.

A bacterial enhancer provides a binding site for the regulator NtrC, which acts upon RNA polymerase using promoters recognized by σ^{54}. When the enhancer is placed upon a circle of DNA that is catenated (interlocked) with a circle that contains the promoter, initiation is almost as effective as when the enhancer and promoter are on the same circular molecule. But there is no initiation when the enhancer and promoter are on separated circles. Again this suggests that the critical

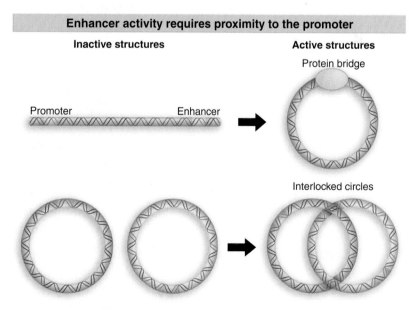

Enhancer activity requires proximity to the promoter

Inactive structures   Active structures

Protein bridge

Promoter   Enhancer

Interlocked circles

Figure 24.21 An enhancer may function by bringing proteins into the vicinity of the promoter. An enhancer does not act on a promoter at the opposite end of a long linear DNA, but becomes effective when the DNA is joined into a circle by a protein bridge. An enhancer and promoter on separate circular DNAs do not interact, but can interact when the two molecules are catenated.

feature is localization of the protein bound at the enhancer, to increase its chance of contacting a protein bound at the promoter.

If proteins bound at an enhancer several kb distant from a promoter interact directly with proteins bound in the vicinity of the start-point, the organization of DNA must be flexible enough to allow the enhancer and promoter to be closely located. This requires the intervening DNA to be extruded as a large "loop." Such loops have been directly observed in the case of the bacterial enhancer.

What limits the activity of an enhancer? Typically it works upon the nearest promoter. There are situations in which an enhancer is located between two promoters, but activates only one of them on the basis of specific protein–protein contacts between the complexes bound at the two elements. The action of an enhancer may be limited by an insulator—an element in DNA that prevents it from acting on promoters beyond (see *29.14 Insulators Block the Actions of Enhancers and Heterochromatin*).

24.13 CpG Islands Are Regulatory Targets

Key Term

- A **CpG island** is a stretch of 1–2 kb in a mammalian genome that is rich in unmethylated CpG doublets.

Key Concepts

- Demethylation at the 5′ end of the gene is necessary for transcription.
- CpG islands surround the promoters of constitutively expressed genes where they are unmethylated.
- They are also found at the promoters of some tissue-regulated genes.
- There are ~29,000 CpG islands in the human genome.
- Methylation of a CpG island prevents activation of a promoter within it.
- Repression is caused by proteins that bind to methylated CpG doublets.

Methylation of DNA is one of the parameters that controls transcription. The typical relationship is that methylation in the vicinity of the promoter prevents transcription, and demethylation is required for gene expression. The methylation events usually occur in regions called **CpG islands** that are found in the 5′ region of the gene. These islands are detected by the presence of an increased density of the dinucleotide sequence, CpG.

The CpG doublet occurs in vertebrate DNA at only ~20% of the frequency that would be expected from the proportion of $G \cdot C$ base pairs. (This may be because CpG doublets are often methylated on C, and spontaneous deamination of methyl-C converts it to T.) In certain regions, however, the density of CpG doublets reaches the predicted value; in fact, it is increased by relative $10\times$ to the rest of the genome. The CpG doublets in these regions are unmethylated.

These CpG-rich islands have an average G·C content of ~60%, compared with the 40% average in bulk DNA. They take the form of stretches of DNA typically 1–2 kb long. There are ~45,000 such islands altogether in the human genome. Some of the islands are present in repeated Alu elements, and may just be the consequence of their high G·C-content. The human genome sequence confirms that, excluding these examples, there are ~29,000 islands. There are fewer in the mouse genome, ~15,500. About 10,000 of the predicted islands in both species appear to reside in a context of sequences that are conserved between the species, suggesting that these may be the islands with regulatory significance. The structure of chromatin in these regions has changes associated with gene expression (see *30.10 Promoter Activation Involves an Ordered Series of Events*).

In several cases, CpG-rich islands begin just upstream of a promoter and extend downstream into the transcribed region before petering out. **Figure 24.22** compares the density of CpG doublets in a "general" region of the genome with a CpG island identified from the DNA sequence. The CpG island surrounds the 5' region of the APRT gene, which is constitutively expressed.

All of the "housekeeping" genes that are constitutively expressed have CpG islands; this accounts for about half of the islands altogether. The other half of the islands occur at the promoters of tissue-regulated genes; only a minority (<40%) of these genes have islands. In these cases, the islands are unmethylated irrespective of the state of expression of the gene. The presence of unmethylated CpG-rich islands may be necessary, but therefore is not sufficient, for transcription. So the presence of unmethylated CpG islands may be taken as an indication that a gene is potentially active, rather than inevitably transcribed. Many islands that are nonmethylated in the animal become methylated in cell lines in tissue culture, and this could be connected with the inability of these lines to express all of the functions typical of the tissue from which they were derived.

Methylation of a CpG island can affect transcription. Two mechanisms can be involved:

- Methylation of a binding site for some factor may prevent it from binding. This happens in a case of binding to a regulatory site other than the promoter (see *31.6 DNA Methylation Is Responsible for Imprinting*).
- Or methylation may cause specific repressors to bind to the DNA.

Repression is caused by either of two types of protein that bind to methylated CpG sequences. The protein MeCP1 requires the presence of several methyl groups to bind to DNA, while MeCP2 and a family of related proteins can bind to a single methylated CpG base pair. This explains why a methylation-free zone is required for initiation of transcription.

The absence of methyl groups is associated with gene expression. However, there are some difficulties in supposing that the state of methylation provides a general means for controlling gene expression. In *D. melanogaster* (and other Dipteran insects), there is very little methylation of DNA (although there is a gene potentially coding a methyltransferase), and in the nematode *C. elegans* there is no methylation of DNA. The other differences between inactive and active chromatin appear to be the same as in species that display methylation. So in these organisms, any role that methylation has in vertebrates is replaced by some other mechanism.

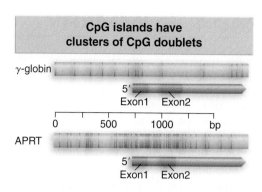

Figure 24.22 The typical density of CpG doublets in mammalian DNA is ~1/100 bp, as seen for a γ-globin gene. In a CpG-rich island, the density is increased to >10 doublets/100 bp. The island in the APRT gene starts ~100 bp upstream of the promoter and extends ~400 bp into the gene. Each vertical line represents a CpG doublet.

24.14 SUMMARY

Of the three eukaryotic RNA polymerases, RNA polymerase I transcribes rDNA and accounts for the majority of activity, RNA polymerase II transcribes structural genes for mRNA and has the greatest diversity of products, and RNA polymerase III transcribes small RNAs. The enzymes have similar structures, with two large subunits and many smaller subunits; there are some common subunits among the enzymes.

None of the three RNA polymerases recognizes its promoters directly. A unifying principle is that transcription factors have primary responsibility for recognizing the characteristic sequence elements of any particular promoter, and they serve in turn to bind the RNA polymerase and to position it correctly at the startpoint. At each type of promoter, the initiation complex is assembled by a series of reactions in which individual factors join (or leave) the complex. The factor TBP is required for initiation by all three RNA polymerases. In each case it provides one subunit of a "positioning" factor that binds in the vicinity of the startpoint.

A promoter for RNA polymerase II consists of a number of short sequence elements in the region upstream of the startpoint. Each element is bound by a transcription factor. The basal apparatus, which consists of the TF_{II} factors, assembles at the startpoint and enables RNA polymerase to bind. The TATA box (if there is one) near the startpoint, and the initiator region immediately at the startpoint, are responsible for selection of the exact startpoint at promoters for RNA polymerase II. TBP binds directly to the TATA box when there is one; in TATA-less promoters it is located near the startpoint by binding to the DPE downstream. After binding of $TF_{II}D$, the other general transcription factors for RNA polymerase II assemble the basal transcription apparatus at the promoter. Other elements in the promoter, located upstream of the TATA box, bind activators that interact with the basal apparatus. The activators and basal factors are released when RNA polymerase begins elongation.

The CTD of RNA polymerase II is phosphorylated during the initiation reaction. $TF_{II}D$ and SRB proteins both may interact with the CTD. It may also provide a point of contact for proteins that modify the RNA transcript, including the 5' capping enzyme, splicing factors, and the 3' processing complex.

Promoters may be stimulated by enhancers, sequences that can act at great distances and in either orientation on either side of a gene. Enhancers also consist of sets of elements, although they are more compactly organized. Some elements are found in both promoters and enhancers. Enhancers probably function by assembling a protein complex that interacts with the proteins bound at the promoter, requiring that DNA between is "looped out."

CpG islands contain concentrations of CpG doublets and often surround the promoters of constitutively expressed genes, although they are also found at the promoters of regulated genes. The island including a promoter must be unmethylated for that promoter to be able to initiate transcription. A specific protein binds to the methylated CpG doublets and prevents initiation of transcription.

25

Regulating Eukaryotic Transcription

25.1 Introduction

Key Concept

- Eukaryotic gene expression is usually controlled at the level of initiation of transcription.

The phenotypic differences that distinguish the various kinds of cells in a higher eukaryote are largely due to differences in the expression of genes that code for proteins, that is, those transcribed by RNA polymerase II. In principle, the expression of these genes might be regulated at any one of several stages. We can distinguish (at least) five potential control points, forming the series:

Activation of gene structure
↓
Initiation of transcription
↓
Processing the transcript
↓
Transport to cytoplasm
↓
Translation of mRNA

As we see in **Figure 25.1**, gene expression in eukaryotes is largely controlled at the initiation of transcription. For most eukaryotic genes, this is the major control point in their expression. Control at the initiation of transcription involves two types of event.

- A change in the structure of the gene at the promoter makes it accessible to transcription factors (see *30.10 Promoter Activation Involves an Ordered Series of Events*).
- Initiation requires binding to the promoter by the general transcription factors and RNA polymerase II, and this binding is activated for each promoter by other transcription factors.

449

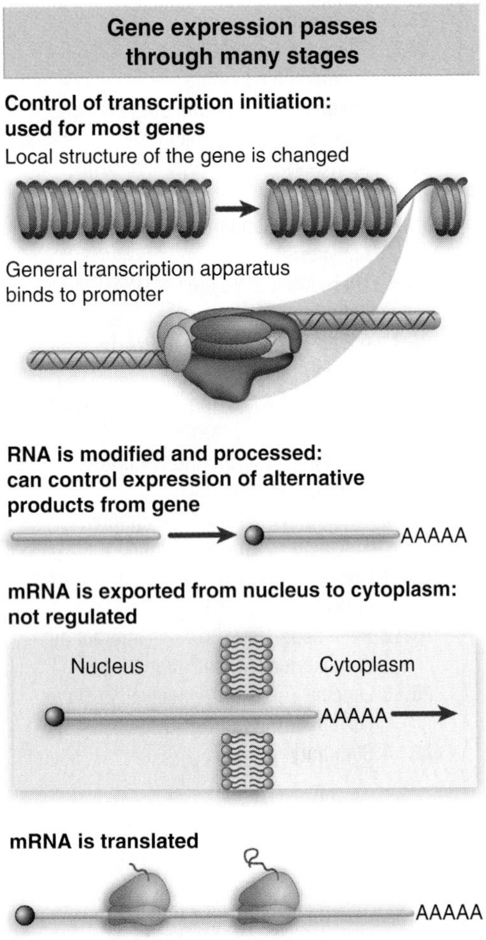

Gene expression passes through many stages

Control of transcription initiation: used for most genes

Local structure of the gene is changed

General transcription apparatus binds to promoter

RNA is modified and processed: can control expression of alternative products from gene

AAAAA

mRNA is exported from nucleus to cytoplasm: not regulated

Nucleus   Cytoplasm

AAAAA →

mRNA is translated

AAAAA

Figure 25.1 Gene expression is controlled principally at the initiation of transcription, and it is rare for the subsequent stages to be used to determine whether a gene is expressed, although control of processing may be used to determine which form of a gene is represented in mRNA.

A regulatory transcription factor controls a large number of target genes, and we seek to answer two questions about this control: how does the transcription factor identify its group of target genes; and how is the activity of the transcription factor itself regulated in response to intrinsic or extrinsic signals?

When the RNA is synthesized, the primary transcript is modified by capping at the 5′ end, and by polyadenylation at the 3′ end. Introns must be excised from the transcripts of interrupted genes. The mature RNA must be exported from the nucleus to the cytoplasm. Regulation of gene expression by selection of sequences of nuclear RNA might involve any or all of these stages, but the one for which we have most evidence concerns changes in splicing; some genes are expressed by means of alternative splicing patterns that can change the coding sequence in mRNA (see *26.11 Alternative Splicing Involves Differential Use of Splice Junctions*).

25.2 There Are Several Types of Transcription Factors

Key Terms

- A **basal factor** is a transcription factor required by RNA polymerase II to form the initiation complex at all promoters. Factors are identified as $TF_{II}X$, where X is a letter.
- An **activator** is a protein that stimulates the expression of a gene, typically by acting at a promoter to stimulate RNA polymerase binding. In eukaryotes, the sequence to which it binds in the promoter is called a response element.
- A **response element** is a sequence in a eukaryotic promoter or enhancer that is recognized by a specific transcription factor.
- **Coactivators** are factors required for transcription that do not bind DNA but are required for (DNA-binding) activators to interact with the basal transcription factors.

Key Concepts

- The basal apparatus determines the startpoint for transcription.
- Activators determine the frequency of transcription.
- Activators work by making protein–protein contacts with the basal factors.
- Activators may work via coactivators.

Initiation of transcription involves many protein–protein interactions among transcription factors bound at the promoter or at an enhancer as well as interactions with RNA polymerase. We can divide the factors required for transcription into several classes. **Figure 25.2** summarizes their properties:

- **Basal factors**, together with RNA polymerase, bind at the startpoint and TATA box (see *24.7 The Basal Apparatus Assembles at the Promoter*).
- **Activators** are transcription factors that recognize specific short consensus elements. They bind to sites in the promoter or in enhancers (see *24.9 Short Sequence Elements Bind Activators*). They act by increasing the efficiency with which the basal apparatus binds to the promoter. They therefore increase the frequency of transcription, and are required for a promoter to function at an adequate level.

Some activators act constitutively (they are ubiquitous), but others have a regulatory role, and are synthesized or activated at specific times or in specific tissues. These factors are therefore responsible for the control of transcription patterns in time and space. The sequences that activators bind are called **response elements**.

- Another group of factors necessary for efficient transcription do not themselves bind DNA. **Coactivators** provide a connection between activators and the basal apparatus (see *25.4 Activators Interact with the Basal Apparatus*). They work by protein–protein interactions, forming bridges between activators and the basal transcription apparatus.

- Some regulators make changes in the structure of the gene that are needed to assist transcription (see *30.4 Nucleosome Organization May Be Changed at the Promoter*).

The diversity of elements from which a functional promoter may be constructed, and the variations in their locations relative to the startpoint, argues that the activators have multiple ways to interact with one another by protein–protein interactions. The modular nature of the promoter is illustrated by experiments in which equivalent regions of different promoters have been exchanged. Such hybrid promoters work well. This suggests that the main purpose of the elements is to bring the activators they bind into the vicinity of the initiation complex, where protein–protein interactions determine the efficiency of the initiation reaction.

The organization of RNA polymerase II promoters contrasts with that of bacterial promoters where all transcription factors interact directly with RNA polymerase. In the eukaryotic system, only the basal factors interact directly with the enzyme. Activators may interact with the basal factors, or may interact with coactivators that in turn interact with the basal factors. The construction of the apparatus through layers of interactions explains the flexibility with which elements may be arranged, and the distance over which they can be dispersed.

25.3 Independent Domains Bind DNA and Activate Transcription

Key Concepts

- DNA-binding activity and transcription activation are carried out by independent domains of an activator.
- The DNA-binding domain determines specificity for the target promoter or enhancer.
- The role of the DNA-binding domain is to bring the transcription-activation domain into the vicinity of the promoter.

Activators and other regulatory proteins require two types of ability:

- They recognize specific target sequences in enhancers, promoters, or other regulatory elements that affect a particular target gene.
- Having bound to DNA, an activator exercises its function by binding to other components of the transcription apparatus.

Can we characterize domains in the activator that are responsible for these activities? Often an activator has separate domains that bind DNA and activate transcription. Each domain behaves as a separate module that functions independently when it is linked to a domain of the other type. The geometry of the overall transcription complex must

Several types of factors affect transcription

RNA polymerase and basal factors bind at promoter

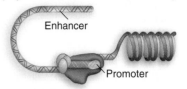

Activators bind at promoter

Activators bind to distal sites in promoter or to enhancers

Coactivators connect activators to basal factors

Regulators act on local structure of gene

Figure 25.2 Factors in gene expression include RNA polymerase and the basal apparatus, activators that bind directly to DNA at the promoter or at enhancers, coactivators that bind to both activators and the basal apparatus, and regulators that act on local chromatin structure.

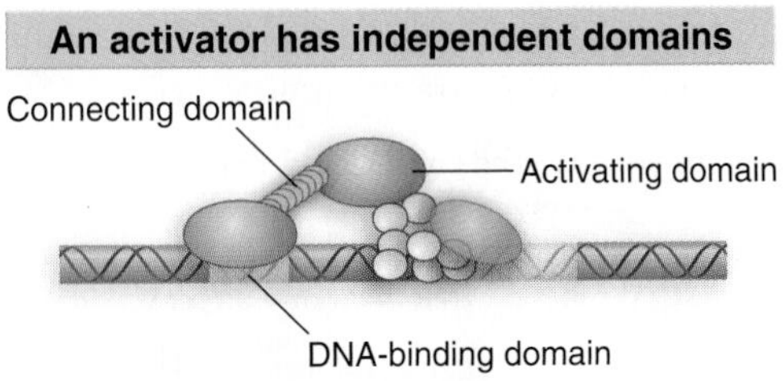

An activator has independent domains

Figure 25.3 DNA-binding and activating functions in a transcription factor may be in independent domains of the protein.

allow the activating domain to contact the basal apparatus irrespective of the exact location and orientation of the DNA-binding domain.

Upstream promoter elements may be an appreciable distance from the startpoint, and in many cases may be oriented in either direction. Enhancers may be even farther away and always show orientation independence. This organization has implications for both the DNA and proteins. The DNA may be looped or condensed in some way to allow the formation of the transcription complex. And the domains of the activator may be connected in a flexible way, as illustrated diagrammatically in **Figure 25.3**. The main point here is that the DNA-binding and activating domains are independent and connected in a way that allows the activating domain to interact with the basal apparatus irrespective of the orientation and exact location of the DNA-binding domain.

Binding to DNA is necessary for activating transcription. But does activation depend on the *particular* DNA-binding domain? This question has been answered by making hybrid proteins that consist of the DNA-binding domain of one activator linked to the transcription-activating domain of another activator. The hybrid functions in transcription at sites dictated by its DNA-binding domain, but in a way determined by its transcription-activating domain.

This result fits the modular view of transcription activators. The function of the DNA-binding domain is *to bring the activating domain into the vicinity of the startpoint*. Precisely how or where it is bound to DNA is irrelevant, but, once it is there, the transcription-activating domain can play its role. This explains why the exact locations of DNA-binding sites can vary within the promoter. The ability of the two types of module to function in hybrid proteins suggests that each domain of the protein folds independently into an active structure that is not influenced by the rest of the protein.

25.4 Activators Interact with the Basal Apparatus

Key Term

- **Mediator** is a large protein complex associated with yeast bacterial RNA polymerase II. It contains factors that are necessary for transcription from many or most promoters.

Key Concepts

- An activator that works directly has a DNA-binding domain and an activating domain.
- An activator that does not have an activating domain may work by binding a coactivator that has an activating domain.
- Several factors in the basal apparatus are targets with which activators or coactivators interact.
- RNA polymerase may be associated with various alternative sets of transcription factors in a holoenzyme complex.
- Repression is usually achieved by affecting chromatin structure, but there are repressors that act by binding to specific promoters.

An activator may work directly when it consists of a DNA-binding domain linked to a transcription-activating domain, as illustrated in Figure 25.3. In other cases, the activator does not itself have a transcription-activating domain, but binds another protein—a coactivator—that has the transcription-activating domain. **Figure 25.4** shows the action of such an activator. We may regard coactivators as transcription factors whose specificity is conferred by the ability to bind to DNA-binding transcription factors instead of directly to DNA. A particular activator may require a specific coactivator.

But although the protein components are organized differently, the mechanism is the same. An activator that contacts the basal apparatus directly has an activation domain covalently connected to the DNA-binding domain. When an activator works through a coactivator, the connections are noncovalent binding between protein subunits (compare Figure 25.3 and Figure 25.4). The same interactions are responsible for activation, irrespective of whether the various domains are present in the same protein subunit or divided into multiple protein subunits.

A transcription-activating domain works by making protein–protein contacts with general transcription factors that promote assembly of the basal apparatus. Contact with the basal apparatus may be made with any one of several basal factors, typically TF$_{II}$D, TF$_{II}$B, or TF$_{II}$A. All of these factors participate in early stages of assembly of the basal apparatus (see Figure 24.11). **Figure 25.5** illustrates the situation when such a contact is made. The major effect of the activators is to influence the assembly of the basal apparatus.

TF$_{II}$D is the most common target for activators, which may contact any one of several TAFs. In fact, a major role of the TAFs is to provide the connection from the basal apparatus to activators. This explains why TBP alone can support basal-level transcription, but the TAFs of TF$_{II}$D are required for the higher levels of transcription that are stimulated by activators. Different TAFs in TF$_{II}$D may provide surfaces that interact with different activators. Some activators interact only with individual TAFs; others interact with multiple TAFs. We assume that the interaction either assists binding of TF$_{II}$D to the TATA box or assists the binding of other activators around the TF$_{II}$D-TATA box complex. In either case, the interaction stabilizes the basal transcription complex; this speeds the process of initiation, and thereby increases use of the promoter.

How does an activator stimulate transcription? We can imagine two general models:

- The recruitment model argues that the sole effect of an activator is to increase the binding of RNA polymerase to the promoter.
- An alternative model argues that an activator induces some change in the transcriptional complex, for example, in the conformation of the enzyme, that increases its efficiency.

A test of these models in yeast showed that recruitment can account for activation. When the concentration of RNA polymerase was increased sufficiently, the activator failed to produce any increase in transcription, suggesting that its sole effect is to increase the effective concentration of RNA polymerase at the promoter. However, although some transcription factors influence transcription directly by interacting with RNA polymerase or the basal apparatus, others work by manipulating the structure of chromatin (see *30.2 Chromatin Remodeling Is an Active Process*).

Adding up all the components required for efficient transcription—basal factors, RNA polymerase, activators, coactivators—we get a very large apparatus, consisting of >40 proteins. Is it feasible for this apparatus to assemble step by step at the promoter? Some activators, coactivators, and basal factors may assemble stepwise at the promoter, but then may be joined by a very large complex consisting of RNA polymerase preassembled with further activators and coactivators, as illustrated in **Figure 25.6**.

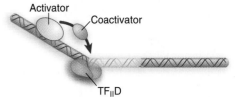

An activator may use a coactivator

Figure 25.4 An activator may bind a coactivator that contacts the basal apparatus.

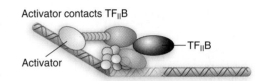

Activators contact the basal apparatus

Figure 25.5 Activators may work at different stages of initiation, by contacting the TAFs of TFIID or contacting TFIIB.

RNA polymerase exists as a holoenzyme

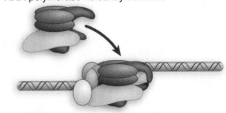

Figure 25.6 RNA polymerase exists as a holoenzyme containing many activators.

Several forms of RNA polymerase have been found in which the enzyme is associated with various transcription factors. The most prominent "holoenzyme complex" in yeast (defined as being capable of initiating transcription without additional components) consists of RNA polymerase associated with a 20-subunit complex called **mediator**. Mediator is necessary for transcription of most yeast genes, and homologous complexes are required for the transcription of most higher eukaryotic genes. Mediator undergoes a conformational change when it interacts with the CTD domain of RNA polymerase. It can transmit either activating or repressing effects from upstream components to the RNA polymerase. It is probably released when a polymerase starts elongation.

Repression of transcription in eukaryotes is generally accomplished at the level of influencing chromatin structure; regulator proteins that function like *trans*-acting bacterial repressors to block transcription are relatively rare, but some examples are known. One case is the global repressor NC2/Dr1/DRAP1, a heterodimer that binds to TBP to prevent it from interacting with other components of the basal apparatus. Repressors that work in this way have an active role in inhibiting basal apparatus function.

25.5 Response Elements Are Recognized by Activators

Key Terms

- The **heat shock response element (HSE)** is a sequence in a promoter or enhancer that is used to activate a gene by an activator induced by heat shock.
- The **glucocorticoid response element (GRE)** is a sequence in a promoter or enhancer that is recognized by the glucocorticoid receptor, which is activated by glucocorticoid steroids.
- The **serum response element (SRE)** is a sequence in a promoter or enhancer that is activated by transcription factor(s) induced by treatment with serum. This activates genes that stimulate cell growth.
- **Heat shock genes** are a set of loci activated in response to an increase in temperature (and other abuses to the cell). All organisms have heat shock genes. Their products usually include chaperones that act on denatured proteins.

Key Concepts

- Response elements may be located in promoters or enhancers.
- Each response element is recognized by a specific activator.
- A promoter may have many response elements, which may activate transcription independently or in certain combinations.

The principle that emerges from characterizing groups of genes under common control is that *they share a promoter (or enhancer) element that is recognized by an activator*. A sequence that causes a gene to respond to such a factor is called a **response element**; examples are **HSE (heat shock response element)**, **GRE (glucocorticoid response element)**, **SRE (serum response element)**. Response elements contain short consensus sequences; copies of the response elements found in different genes are closely related, but not necessarily identical.

The region bound by the factor extends for a short distance on either side of the consensus sequence. In promoters, the response elements are not present at fixed distances from the startpoint, but are usually <200 bp upstream of it. The presence of a single response element usually is sufficient to confer the regulatory response, but sometimes there are multiple copies.

Response elements may be located in promoters or in enhancers. Some types are typically found in one rather than the other: usually an HSE is found in a promoter, while a GRE is found in an enhancer. All response elements function by the same general principle: the element binds an activator that interacts with the general transcription factors that are required for binding of RNA polymerase II. The availability or activity of the activator may control expression of the gene.

An example of a situation in which many genes are controlled by a single factor is provided by the heat shock response. This is common to a wide range of prokaryotes and eukaryotes and involves multiple controls of gene expression: an increase in temperature turns off transcription of some genes, turns on transcription of the **heat shock genes**, and causes changes in the translation of mRNAs. The control of the heat shock genes illustrates the differences between prokaryotic and eukaryotic modes of control. In bacteria, a new sigma factor is synthesized that directs RNA polymerase holoenzyme to recognize an alternative −10 sequence common to the promoters of heat shock genes (see *11.11 Substitution of Sigma Factors May Control Initiation*). In eukaryotes, the heat shock genes also possess a common consensus sequence (HSE), but it is located at various positions relative to the startpoint, and is recognized by an independent activator, HSTF. The activation of this factor therefore provides a means to initiate transcription at the specific group of ~20 genes that contains the appropriate target sequence at its promoter. For example, all the heat shock genes of *D. melanogaster* contain multiple copies of the HSE. The HSTF binds cooperatively to adjacent response elements. Both the HSE and HSTF have been conserved in evolution, and it is striking that a heat shock gene from *D. melanogaster* can be activated in species as distant as mammals or sea urchins.

The metallothionein (MT) gene provides an example of how a single gene may be regulated by many different circuits. The metallothionein protein protects the cell against excess concentrations of heavy metals by binding the metal and removing it from the cell. The gene is expressed at a basal level, but is induced to greater levels of expression by heavy metal ions (such as cadmium) or by glucocorticoids. The control region combines several different kinds of regulatory elements.

The organization of the promoter for an MT gene is summarized in **Figure 25.7**. A major feature of this map is the high density of elements that can activate transcription. The TATA and GC boxes are located at their usual positions fairly close to the startpoint. Also needed for the basal level of expression are the two basal level elements (BLE), which fit the formal description of enhancers. Although located near the startpoint, they can be moved elsewhere without loss of effect. They contain sequences related to those found in other enhancers, and are bound by proteins that also bind the SV40 enhancer.

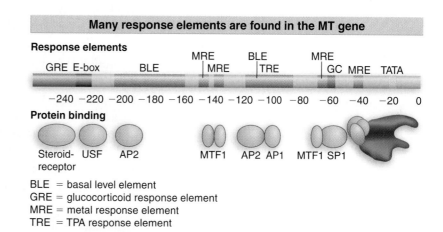

Many response elements are found in the MT gene

BLE = basal level element
GRE = glucocorticoid response element
MRE = metal response element
TRE = TPA response element

Figure 25.7 The regulatory region of a human metallothionein gene contains regulator elements in both its promoter and enhancer. The promoter has elements for metal induction; an enhancer has an element for response to glucocorticoid. Promoter elements are shown above the map, and proteins that bind them are indicated below.

The TRE is a consensus sequence that is present in several enhancers, including one BLE of metallothionein and the 72 bp repeats of the virus SV40. The TRE has a binding site for factor AP1; this interaction is part of the mechanism for constitutive expression, for which AP1 is an activator. However, AP1 binding also has a second function. The TRE confers a response to phorbol esters such as TPA (an agent that promotes tumors), and this response is mediated by the interaction of AP1 with the TRE. This binding reaction is one (not necessarily the sole) means by which phorbol esters trigger a series of transcriptional changes.

The inductive response to metals is conferred by the multiple MRE sequences, which function as promoter elements. The presence of one MRE confers the ability to respond to heavy metal; a greater level of induction is achieved by the inclusion of multiple elements. The factor MTF1 binds to the MRE in response to the presence of metal ions.

The response to steroid hormones is governed by a GRE, located 250 bp upstream of the startpoint, which behaves as an enhancer. Deletion of this region does not affect the basal level of expression or the level induced by metal ions. But it is absolutely needed for the response to steroids.

The regulation of metallothionein illustrates the general principle that *any one of several different elements, located in either an enhancer or promoter, can independently activate the gene.* The variety of elements, their independence of action, and the apparently unlimited flexibility of their relative arrangements, suggest that a factor binding to any one element is able independently to increase the efficiency of initiation by the basal transcription apparatus, probably by virtue of protein–protein interactions that stabilize or otherwise assist formation of the initiation complex.

25.6 There Are Many Types of DNA-Binding Domains

Key Terms

- The **zinc finger** is a DNA-binding motif that typifies a class of transcription factor.
- **Steroid receptors** are transcription factors that are activated by binding of a steroid ligand.
- The **helix-turn-helix** motif describes an arrangement of two α-helices that form a site that binds to DNA, one fitting into the major groove of DNA and other lying across it.
- The **homeodomain** is a DNA-binding motif that typifies a class of transcription factors. The DNA sequence that codes for it is called the homeobox.
- The **helix-loop-helix (HLH)** motif is responsible for dimerization of a class of transcription factors called HLH proteins. A bHLH protein has a basic DNA-binding sequence close to the dimerization motif.
- The **leucine zipper** is a dimerization motif adjacent to a basic DNA-binding region that is found in a class of transcription factors.

Key Concepts

- Activators are classified according to the type of DNA-binding domain.
- Members of the same group have sequence variations of a specific motif that confer specificity for individual target sites.

It is common for an activator to have a modular structure in which different domains are responsible for binding to DNA and for activating transcription. Factors are often classified according to the type of DNA-binding domain. Typically a relatively short motif in this domain is responsible for binding to DNA:

- The **zinc finger** motif is a DNA-binding domain. It was originally recognized in factor $TF_{III}A$, which is required for RNA polymerase III to transcribe 5S rRNA genes. It has since been identified in several other transcription factors (and presumed transcription factors). A distinct form of the motif is found also in the steroid receptors.

- The **steroid receptors** are defined as a group by a functional relationship: each receptor is activated by binding a particular steroid. The glucocorticoid receptor is the most fully analyzed. Together with other receptors, such as the thyroid hormone receptor or the retinoic acid receptor, the steroid receptors are members of the superfamily of ligand-activated activators with the same general *modus operandi: the protein factor is inactive until it binds a small ligand.*

- The **helix-turn-helix** motif was originally identified as the DNA-binding domain of phage repressors. One α-helix lies in the major groove of DNA; the other lies at an angle across DNA. A related form of the motif is present in the **homeodomain**, a sequence first characterized in several proteins coded by genes concerned with developmental regulation in *Drosophila*. It is also present in genes for mammalian transcription factors.

- The amphipathic **helix-loop-helix (HLH)** motif has been identified in some developmental regulators and in genes coding for eukaryotic DNA-binding proteins. Each amphipathic helix presents a face of hydrophobic residues on one side and charged residues on the other side. The length of the connecting loop varies from 12–28 amino acids. The motif enables proteins to dimerize, and a basic region near this motif contacts DNA.

- **Leucine zippers** consist of a stretch of amino acids with a leucine residue in every seventh position. A leucine zipper in one polypeptide interacts with a zipper in another polypeptide to form a dimer. Adjacent to each zipper is a stretch of positively charged residues that is involved in binding to DNA.

The activity of an inducible activator may be regulated in any one of several ways, as illustrated schematically in **Figure 25.8**:

- A factor is tissue-specific because it is synthesized only in a particular type of cell. This is typical of factors that regulate development, such as homeodomain proteins.

- The activity of a factor may be directly controlled by modification. HSTF is converted to the active form by phosphorylation. AP1 (a heterodimer between the subunits Jun and Fos) is converted to the active form by phosphorylating the Jun subunit.

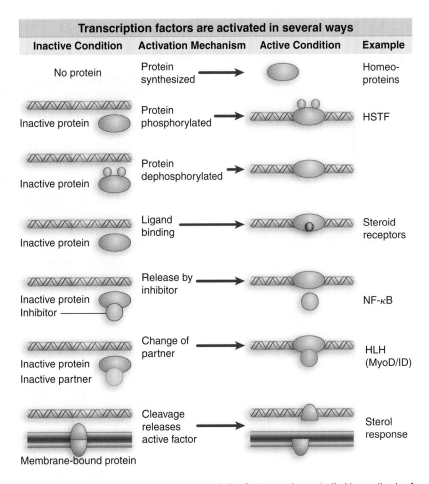

Transcription factors are activated in several ways			
Inactive Condition	**Activation Mechanism**	**Active Condition**	**Example**
No protein	Protein synthesized		Homeo-proteins
Inactive protein	Protein phosphorylated		HSTF
Inactive protein	Protein dephosphorylated		
Inactive protein	Ligand binding		Steroid receptors
Inactive protein Inhibitor	Release by inhibitor		NF-κB
Inactive protein Inactive partner	Change of partner		HLH (MyoD/ID)
Membrane-bound protein	Cleavage releases active factor		Sterol response

Figure 25.8 The activity of a regulatory transcription factor may be controlled by synthesis of protein, covalent modification of protein, ligand binding, or binding of inhibitors that sequester the protein or affect its ability to bind to DNA.

- A factor is activated or inactivated by binding a ligand. The steroid receptors are prime examples. Ligand binding may influence the localization of the protein (causing transport from cytoplasm to nucleus), as well as determining its ability to bind to DNA.
- Availability of a factor may vary; for example, the factor NF-κB (which activates immunoglobulin κ genes in B lymphocytes) is present in many cell types. But it is sequestered in the cytoplasm by the inhibitory protein I-κB. In B lymphocytes, NF-κB is released from I-κB and moves to the nucleus, where it activates transcription.
- An extreme example of control of availability is found when a factor is actually part of a cytoplasmic complex, and is released from that complex to translocate to the nucleus.
- A dimeric factor may have alternative partners. One partner may cause it to be inactive; synthesis of the active partner may displace the inactive partner. Such situations may be amplified into networks in which various alternative partners pair with one another, especially among the HLH proteins.
- The factor may be cleaved from an inactive precursor. One activator is produced as a protein bound to the nuclear envelope and endoplasmic reticulum. The absence of sterols (such as cholesterol) causes the cytosolic domain to be cleaved; it then translocates to the nucleus and provides the active form of the activator.

We now discuss in more detail the DNA-binding and activation reactions that are sponsored by some of these classes of proteins.

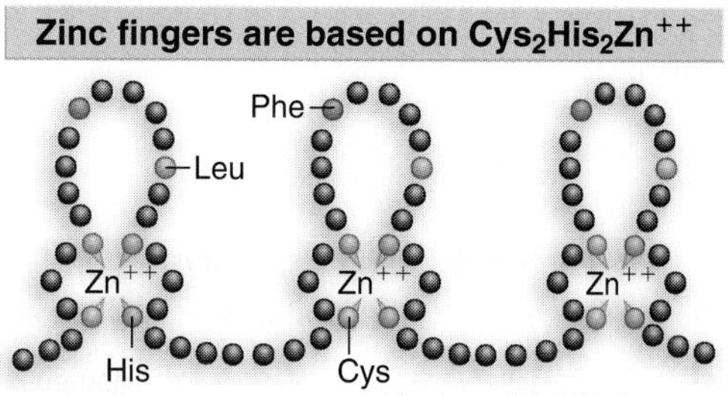

25.7 A Zinc Finger Motif Is a DNA-Binding Domain

Key Concepts

- A zinc finger is a loop of ~23 amino acids that protrudes from a zinc-binding site formed by His and Cys amino acids.
- A zinc finger protein usually has multiple zinc fingers.
- The C-terminal part of each finger forms an α-helix that binds one turn of the major groove of DNA.
- Some zinc finger proteins bind RNA instead of or as well as DNA.

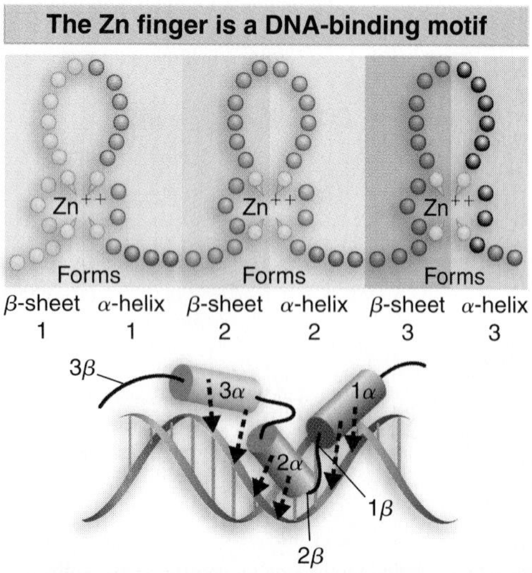

Figure 25.9 Transcription factor SP1 has a series of three zinc fingers, each with a characteristic pattern of cysteine and histidine residues that constitute the zinc-binding site.

Zinc fingers take their name from the structure illustrated in **Figure 25.9**, in which a small group of conserved amino acids binds a zinc ion to form an independent domain in the protein. Two types of DNA-binding proteins have structures of this type: the classic "zinc finger" proteins, and the steroid receptors. Zinc fingers can also be involved in binding RNA, however.

A "finger protein" typically has a series of zinc fingers, as depicted in the figure. The consensus sequence of a single finger is:

$$\text{Cys-X}_{2-4}\text{-Cys-X}_3\text{-Phe-X}_5\text{-Leu-X}_2\text{-His-X}_3\text{-His}$$

The motif takes its name from the loop of amino acids that protrudes from the zinc-binding site and is described as the $\text{Cys}_2/\text{His}_2$ finger. The zinc is held in a tetrahedral structure formed by the conserved Cys and His residues. The finger itself comprises ~23 amino acids, and the linker between fingers is usually 7–8 amino acids.

Zinc fingers are a common motif in DNA-binding proteins. They are usually organized as a single series of tandem repeats. The crystal structure of DNA bound by a protein with three fingers suggests the structure illustrated schematically in **Figure 25.10**. The C-terminal part of each finger forms an α-helix that binds DNA; the N-terminal part forms

Figure 25.10 Zinc fingers form α-helices that insert into the major groove, associated with β-sheets on the other side.

a β-sheet. The three α-helical stretches fit into one turn of the major groove; each α-helix (and thus each finger) makes two sequence-specific contacts with DNA (indicated by the arrows). We expect that the nonconserved amino acids in the C-terminal side of each finger are responsible for recognizing specific target sites.

25.8 Some Steroid Hormone Receptors Are Transcription Factors

Key Concepts

- Steroid receptors are examples of ligand-responsive activators that are activated by binding a steroid (or other related molecules).
- There are separate DNA-binding and ligand-binding domains.

Steroid hormones are synthesized in response to a variety of neuroendocrine activities, and exert major effects on growth, tissue development, and body homeostasis in the animal world. The major groups of steroids and some other compounds with related molecular activities are classified in **Figure 25.11**.

The adrenal gland secretes >30 steroids, the two major groups being the glucocorticoids and mineralocorticoids. Steroids provide the reproductive hormones (androgen male sex hormones and estrogen female sex hormones). Vitamin D is required for bone development.

Other hormones, with unrelated structures and physiological purposes, function at the molecular level in a similar way to the steroid hormones. Thyroid hormones, based on iodinated forms of tyrosine, control basal metabolic rate in animals. Steroid and thyroid hormones also may be important in metamorphosis (ecdysteroids in insects, and thyroid hormones in frogs).

Retinoic acid (vitamin A) is a morphogen responsible for development of the anterior–posterior axis in the developing chick limb bud. Its metabolite, 9-*cis* retinoic acid, is found in tissues that are major sites for storage and metabolism of vitamin A.

We may account for these various actions in terms of pathways for regulating gene expression. These diverse compounds share a common mode of action: *each is a small molecule that binds to a specific receptor that activates gene transcription.* Receptors for the diverse groups of steroid hormones, thyroid hormones, and retinoic acid represent a new "superfamily" of gene regulators, the ligand-responsive activators. All the receptors have independent domains for DNA binding and

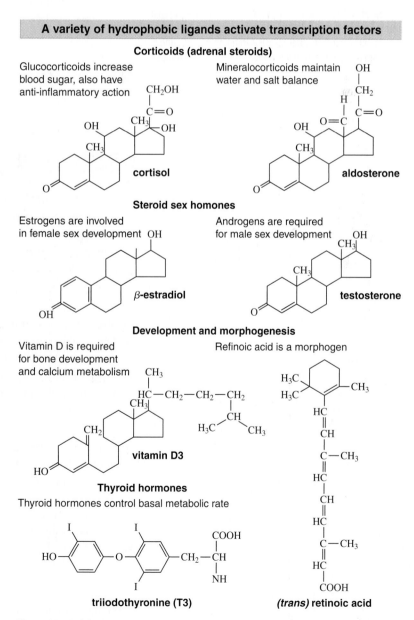

A variety of hydrophobic ligands activate transcription factors

Corticoids (adrenal steroids)

Glucocorticoids increase blood sugar, also have anti-inflammatory action

cortisol

Mineralocorticoids maintain water and salt balance

aldosterone

Steroid sex homones

Estrogens are involved in female sex development

β-estradiol

Androgens are required for male sex development

testosterone

Development and morphogenesis

Vitamin D is required for bone development and calcium metabolism

vitamin D3

Refinoic acid is a morphogen

Thyroid hormones

Thyroid hormones control basal metabolic rate

triiodothyronine (T3)

(trans) retinoic acid

Figure 25.11 Several types of hydrophobic small molecules activate transcription factors.

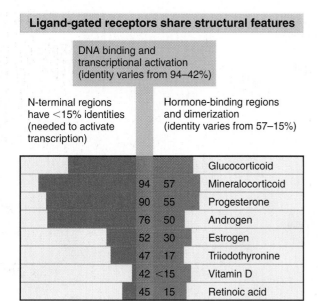

Ligand-gated receptors share structural features

DNA binding and transcriptional activation (identity varies from 94–42%)

N-terminal regions have <15% identities (needed to activate transcription)

Hormone-binding regions and dimerization (identity varies from 57–15%)

			Glucocorticoid
	94	57	Mineralocorticoid
	90	55	Progesterone
	76	50	Androgen
	52	30	Estrogen
	47	17	Triiodothyronine
	42	<15	Vitamin D
	45	15	Retinoic acid

Figure 25.12 Receptors for many steroid and thyroid hormones have a similar organization, with an individual N-terminal region, conserved DNA-binding region, and a C-terminal hormone-binding region. Identities are relative to GR.

hormone binding, in the same relative locations. Their general organization is summarized in **Figure 25.12**.

The central part of the protein is the DNA-binding domain. These regions are closely related for the various steroid receptors (from the most closely related pair with 94% sequence identity to the least well-related pair at 42% identity). The act of binding DNA cannot be disconnected from the ability to activate transcription, because mutations in this domain affect both activities.

The N-terminal regions of the receptors show the least conservation of sequence. They include other regions that are needed to activate transcription.

The C-terminal domains bind the hormones. Those in the steroid receptor family show identities ranging from 30–57%, reflecting specificity for individual hormones. Their relationships with the other receptors are minimal, reflecting specificity for a variety of compounds—thyroid hormones, vitamin D, retinoic acid, etc. This domain also has the motifs responsible for dimerization and a region involved in transcriptional activation.

Some ligands have multiple receptors that are closely related, such as the three retinoic acid receptors (RARα, β, γ) and the three receptors for 9-cis-retinoic acid (RXRα, β, γ).

25.9 Zinc Fingers of Steroid Receptors Use a Combinatorial Code

Key Concepts

- The DNA-binding domain of a steroid receptor is a type of zinc finger that has Cys but not His residues.
- Glucocorticoid and estrogen receptors each have two zinc fingers.
- A steroid response element consists of two short half sites that may be palindromic or directly repeated.
- A receptor is a dimer, and each subunit binds to a half site in DNA.
- A receptor recognizes its response element by the orientation and spacing of the half sites.
- The sequence of the half site is recognized by the first zinc finger.
- The second zinc finger is responsible for dimerization, which determines the distance between the subunits.
- Subunit separation in the receptor determines the recognition of spacing in the response element.
- Some steroid receptors function as homodimers but others form heterodimers.
- Homodimers recognize palindromic response elements; heterodimers recognize response elements with directly repeated half sites.
- There are only two types of half sites.

Steroid receptors (and some other proteins) have another type of zinc finger that is different from Cys_2/His_2 fingers. The structure is based on a sequence with the zinc-binding consensus:

$$Cys-X_2-Cys-X_{13}-Cys-X_2-Cys$$

These are called Cys_2/Cys_2 fingers. Proteins with Cys_2/Cys_2 fingers often have nonrepetitive fingers, in contrast with the tandem repetition of the Cys_2/His_2 type. Their binding sites in DNA are short and palindromic.

The glucocorticoid and estrogen receptors each have two fingers, each with a zinc atom at the center of a tetrahedron of cysteines. The two fingers form α-helices that fold together to form a large globular domain. The aromatic sides of the α-helices form a hydrophobic center together with a β-sheet that connects the two helices. One side of the N-terminal helix makes contacts in the major groove of DNA. Two glucocorticoid receptors dimerize upon binding to DNA, and each engages a successive turn of the major groove. This fits with the palindromic nature of the response element.

Each finger controls one important property of the receptor. **Figure 25.13** identifies the relevant amino acids. Those on the right side of the first finger determine the sequence of the target in DNA; those on the left side of the second finger control the spacing between the target sites recognized by each subunit in the dimer.

Direct evidence that the first finger binds DNA was obtained by a "specificity swap" experiment. The finger of the estrogen receptor was deleted and replaced by the sequence of the glucocorticoid receptor. The new protein recognized the GRE sequence (the usual target of the glucocorticoid receptor) instead of the ERE (the usual target of the estrogen receptor). This region therefore establishes the specificity with which DNA is recognized. In reverse, substitution at the two positions shown in **Figure 25.14** allows the glucocorticoid receptor to bind at an ERE instead of a GRE.

Each receptor recognizes a response element that consists of two short repeats (or half sites). This immediately suggests that the receptor binds as a dimer, so that each half of the consensus is contacted by one subunit (reminiscent of the λ operator-repressor interaction described in *14.10 Repressor Uses a Helix-Turn-Helix Motif to Bind DNA*).

The half sites may be arranged either as palindromes or as repeats in the same orientation. They are separated by 0–4 base pairs whose sequence is irrelevant. Only two types of half site are used by the various receptors. Their orientation and spacing determine which receptor recognizes the response element. This behavior allows response elements that have restricted consensus sequences to be recognized specifically by a variety of receptors. The rules that govern recognition are not absolute, but may be modified by context, and there are also cases in which palindromic response elements are recognized permissively by more than one receptor.

The receptors fall into two groups:

- Glucocorticoid (GR), mineralocorticoid (MR), androgen (AR), and progesterone (PR) receptors all form homodimers. They recognize response elements whose half sites have the consensus sequence TGTTCT. **Figure 25.15** shows that the half sites are arranged as palindromes, and the spacing between the sites determines the type of element. The estrogen (ER) receptor functions in the same way, but has the half site sequence TGACCT.
- The 9-*cis*-retinoic acid (RXR) receptor forms homodimers and also forms heterodimers with ~15 other receptors, including thyroid (T3R), vitamin D (VDR), and retinoic acid (RAR). **Figure 25.16** shows that the dimers recognize half elements with the sequence TGACCT. The half sites are arranged as direct repeats, and recognition is controlled by spacing between them. Some of the heterodimeric receptors are activated when the ligand binds to the partner for RXR; others can be activated by ligand binding either to this subunit or to the RXR subunit. These receptors can also form homodimers, which recognize palindromic sequences.

Now we are in a position to understand the basis for specificity of recognition. Recall that Figure 25.13 shows how recognition of the sequence of the half site is conferred by the amino acid sequence in the first finger. Specificity for the spacing between half sites is carried by

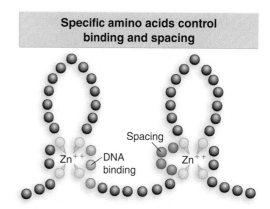

Specific amino acids control binding and spacing

Figure 25.13 The first finger of a steroid receptor controls which DNA sequence is bound (positions shown in red); the second finger controls spacing between the sequences (positions shown in blue).

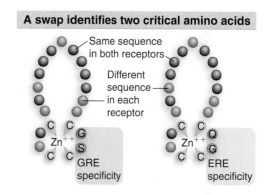

A swap identifies two critical amino acids

Figure 25.14 Discrimination between GRE and ERE target sequences is determined by two amino acids at the base of the first zinc finger in the receptor.

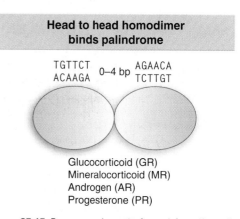

Head to head homodimer binds palindrome

TGTTCT   0–4 bp   AGAACA
ACAAGA   TCTTGT

Glucocorticoid (GR)
Mineralocorticoid (MR)
Androgen (AR)
Progesterone (PR)

Figure 25.15 Response elements formed from the palindromic half site TGTTCT are recognized by several different receptors depending on the spacing between the half sites.

Heterodimer binds direct repeats

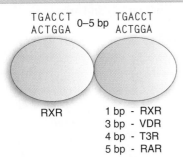

TGACCT 0–5 bp TGACCT
ACTGGA   ACTGGA

RXR

1 bp - RXR
3 bp - VDR
4 bp - T3R
5 bp - RAR

Figure 25.16 Response elements with the direct repeat TGACCT are recognized by heterodimers of which one member is RXR.

Repression prevails in absence of ligand

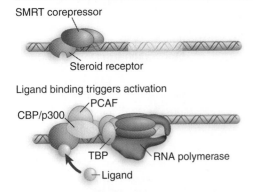

SMRT corepressor

Steroid receptor

Ligand binding triggers activation

PCAF

CBP/p300

TBP   RNA polymerase

Ligand

Figure 25.17 The steroid receptors TR and RAR bind the SMRT corepressor in the absence of ligand. The promoter is not expressed. When SMRT is displaced by binding of ligand, the receptor binds a coactivator complex. This leads to activation of transcription by the basal apparatus.

amino acids in the second finger. The structure of the dimer determines the distance between the subunits that sit in successive turns of the major groove, and thus controls the response to the spacing of half sites. The exact positions of the residues responsible for dimerization differ in individual pairwise combinations.

25.10 Binding to the Response Element Is Activated by Ligand Binding

Key Concept

- Binding of ligand to the C-terminal domain of an activator increases the affinity of the DNA-binding domain for its specific target site in DNA.

How do the steroid receptors activate transcription? They do not act directly on the basal apparatus, but function via a coactivating complex. The coactivator includes various activities, including the common component CBP/p300, one of whose functions is to modify the structure of chromatin by acetylating histones (see Figure 30.13).

All receptors in the superfamily are ligand-dependent activators of transcription. However, some are also able to repress transcription. The TR and RAR receptors, in the form of heterodimers with RXR, bind to certain loci in the *absence* of ligand and repress transcription by means of their ability to interact with a corepressor protein. The corepressor functions by the reverse of the mechanism used by coactivators: it inhibits the function of the basal transcription apparatus (see also *30.8 Deacetylases Are Associated with Repressors*).

The effect of ligand binding on the receptor is to convert it from a repressing complex to an activating complex, as shown in **Figure 25.17**. In the absence of ligand, the receptor is bound to a corepressor complex. The component of the corepressor that binds to the receptor is SMRT. Binding of ligand causes a conformational change that displaces SMRT. This allows the coactivator to bind.

The activated receptor recognizes the sequence of the GRE, the glucocorticoid response element.

The C-terminal (steroid-binding) region regulates the activity of the receptor in a way that varies for the individual receptor. With the glucocorticoid receptor, it prevents the receptor from recognizing the GRE; the addition of steroid inactivates the inhibition. With the estrogen receptor, the C-terminal region directly activates the ability to bind DNA.

25.11 Homeodomains Bind Related Targets in DNA

Key Concepts

- The homeodomain is a DNA-binding domain of 60 amino acids that has three α-helices.
- The C-terminal α-helix-3 is 17 amino acids and binds in the major groove of DNA.
- The N-terminal arm of the homeodomain projects into the minor groove of DNA.
- Proteins containing homeodomains may be either activators or repressors of transcription.

The homeobox is a sequence that codes for a domain of 60 amino acids present in proteins of many or even all eukaryotes. Its name derives from its original identification in *Drosophila* homeotic loci (whose genes determine the identity of body structures). It is present in many of the genes that regulate early development in *Drosophila*, and a related motif is found in genes in a wide range of higher eukaryotes. The homeodomain is found in many genes concerned with developmental regulation. Sequences related to the homeodomain are found in several types of animal transcription factors.

In *Drosophila* homeotic genes, the homeodomain often (but not always) is close to the C-terminal end. Some examples of genes containing homeoboxes are summarized in **Figure 25.18**. Often the genes have little conservation of sequence except in the homeobox. The conservation of the homeobox sequence varies. A major group of homeobox-containing genes in *Drosophila* has a well conserved sequence, with 80–90% similarity in pairwise comparisons. Other genes have less closely related homeoboxes.

The homeodomain is responsible for binding to DNA, and experiments to swap homeodomains between proteins suggest that the specificity of DNA recognition lies within the homeodomain, but (as with phage repressors) no simple code relating protein and DNA sequences can be deduced. The C-terminal region of the homeodomain shows homology with the helix-turn-helix motif of prokaryotic repressors. We recall from *14.10 Repressor Uses a Helix-Turn-Helix Motif to Bind DNA* that the λ repressor has a "recognition helix" (α-helix-3) that makes contacts in the major groove of DNA, while the other helix (α-helix-2) lies at an angle across the DNA. The homeodomain can be organized into three potential helical regions; the sequences of three examples are compared in **Figure 25.19**. The best-conserved part of the sequence lies in the third helix. The difference between these structures and the prokaryotic repressor structures lies in the length of the helix that recognizes DNA, helix-3, which is 17 amino acids long in the homeodomain, compared to 9 residues long in the λ repressor.

The structure of the homeodomain of the *D. melanogaster* engrailed protein is represented schematically in **Figure 25.20**. Helix 3 binds in the major groove of DNA and makes the majority of the contacts between protein and nucleic acid. Many of the contacts that orient the helix in the major groove are made with the phosphate backbone, so they are not specific for DNA sequence. They lie largely on one face of the double helix, and flank the bases with which specific contacts are made. The remaining contacts are made by the N-terminal arm of the homeodomain, the sequence that just precedes the first helix. It projects into the minor groove. So the N-terminal and C-terminal regions of the homeodomain are primarily responsible for contacting DNA.

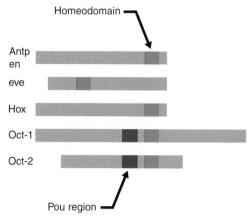

The homeodomain is a discrete module

Figure 25.18 The homeodomain may be the sole DNA-binding motif in a transcriptional regulator or may be combined with other motifs. It represents a discrete (60 residue) part of the protein.

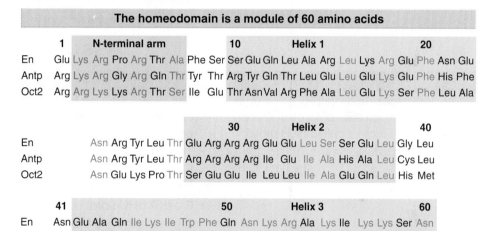

The homeodomain is a module of 60 amino acids

	1	N-terminal arm						10	Helix 1								20		
En	Glu	Lys	Arg	Pro	Arg	Thr	Ala	Phe	Ser	Ser	Glu	Gln	Leu	Ala	Arg	Leu	Lys	Arg	Glu	Phe	Asn	Glu
Antp	Arg	Lys	Arg	Gly	Arg	Gln	Thr	Tyr	Thr	Arg	Tyr	Gln	Thr	Leu	Glu	Leu	Glu	Lys	Glu	Phe	His	Phe
Oct2	Arg	Arg	Lys	Lys	Arg	Thr	Ser	Ile	Glu	Thr	Asn	Val	Arg	Phe	Ala	Leu	Glu	Lys	Ser	Phe	Leu	Ala

	30	Helix 2							40	
En	Asn	Arg	Tyr	Leu	Thr	Glu	Arg	Arg	Arg	Glu	Glu	Leu	Ser	Ser	Glu	Leu	Gly	Leu
Antp	Asn	Arg	Tyr	Leu	Thr	Arg	Arg	Arg	Arg	Ile	Glu	Ile	Ala	His	Ala	Leu	Cys	Leu
Oct2	Asn	Glu	Lys	Pro	Thr	Ser	Glu	Glu	Ile	Leu	Leu	Ile	Ala	Glu	Gln	Leu	His	Met

	41			50	Helix 3				60	
En	Asn	Glu	Ala	Gln	Ile	Lys	Ile	Trp	Phe	Gln	Asn	Lys	Arg	Ala	Lys	Ile	Lys	Lys	Ser	Asn
Antp	Thr	Glu	Arg	Gln	Ile	Lys	Ile	Trp	Phe	Gln	Asn	Arg	Arg	Met	Lys	Trp	Lys	Lys	Glu	Asn
Oct2	Glu	Lys	Glu	Val	Ile	Arg	Val	Trp	Phe	Cys	Asn	Arg	Arg	Gln	Lys	Glu	Lys	Arg	Ile	Asn

Figure 25.19 The homeodomain of the *Antennapedia* gene represents the major group of genes containing homeoboxes in *Drosophila*; *engrailed* (*en*) represents another type of homeotic gene; and the mammalian factor Oct2 represents a distantly related group of transcription factors. The homeodomain is conventionally numbered from 1 to 60. It starts with the N-terminal arm, and the three helical regions occupy residues 10–22, 28–38, and 42–58. Amino acids in red are conserved in all three examples.

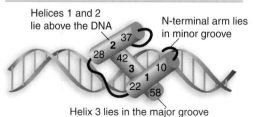

The homeodomain has 3 α-helices

Figure 25.20 Helix 3 of the homeodomain binds in the major groove of DNA, with helices 1 and 2 lying outside the double helix. Helix 3 contacts both the phosphate backbone and specific bases. The N-terminal arm lies in the minor groove, and makes additional contacts.

A striking demonstration of the generality of this model derives from a comparison of the crystal structure of the homeodomain of engrailed with that of the α2 mating protein of yeast. The DNA-binding domain of this protein resembles a homeodomain, and can form three similar helices: its structure in the DNA groove can be superimposed almost exactly on that of the engrailed homeodomain. These similarities suggest that all homeodomains bind to DNA in the same manner. This means that a relatively small number of residues in helix-3 and in the N-terminal arm are responsible for specificity of contacts with DNA.

Homeodomain proteins can be either transcriptional activators or repressors. The function of the factor depends on the other domain(s)—the homeodomain is responsible solely for binding to DNA. The activator or repressor domains both act by influencing the basal apparatus. Activator domains may interact with coactivators that in turn bind to components of the basal apparatus. Repressor domains also interact with the transcription apparatus (that is, they do not act by blocking access to DNA as such).

25.12 Helix-Loop-Helix Proteins Interact by Combinatorial Association

Key Terms
• The **helix-loop-helix (HLH)** motif is responsible for dimerization of a class of transcription factors called HLH proteins.
• A **bHLH protein** has a basic DNA-binding region adjacent to the helix-loop-helix motif.

Key Concepts
• Helix-loop-helix proteins have a motif of 40–50 amino acids that comprises two amphipathic α-helices of 15–16 residues separated by a loop.
• The helices are responsible for dimer formation.
• bHLH proteins have a basic sequence adjacent to the HLH motif that is responsible for binding to DNA.
• Class A bHLH proteins are ubiquitously expressed. Class B bHLH proteins are tissue specific.
• A class B protein usually forms a heterodimer with a class A protein.
• HLH proteins that lack the basic region prevent a bHLH partner in a heterodimer from binding to DNA.
• HLH proteins form combinatorial associations that may be changed during development by the addition or removal of specific proteins.

Two common features in DNA-binding proteins are the presence of helical regions that bind DNA, and the ability of the protein to dimerize. Both features are represented in the group of **helix-loop-helix** proteins that share a common type of sequence motif: a stretch of 40–50 amino acids contains two amphipathic α-helices separated by a linker region (the loop) of varying length. (An amphipathic helix forms two faces, one presenting hydrophobic amino acids, the other presenting charged amino acids.) The proteins in this group form both homodimers and heterodimers by means of interactions between the hydrophobic residues on the corresponding faces of the two helices. The helical regions are 15–16 amino acids long, and each contains several conserved residues. Two examples are compared in **Figure 25.21**. The ability to form dimers resides with these amphipathic helices, and is common to all HLH proteins. The loop is probably important only for allowing the freedom for the two helical regions to interact independently of one another.

Most HLH proteins contain a region adjacent to the HLH motif itself that is highly basic, and which is needed for binding to DNA. There are ~6 conserved residues in a stretch of 15 amino acids (see Figure 25.21). Members of the group with such a region are called **bHLH proteins**. A dimer in which both subunits have the basic region can bind to DNA. The HLH domains probably correctly orient the two basic regions contributed by the individual subunits.

HLH proteins have two helical regions

		Basic region 6 conserved residues are absent from Id
MyoD	Ala Asp Arg Arg Lys Ala Ala Thr Met Arg Gln Arg Arg Arg	
Id	Arg Leu Pro Ala Leu Leu Asp Gln Glu Glu Val Asn Val Leu	

		Helix 1 Conserved residues are found in both MyoD and Id
MyoD	Leu Ser Lys Val Asn Gln Ala Phe Gln Thr Leu Lys Arg Cys Thr	
Id	Leu Tyr Asp Met Asn Gly Cys Tyr Ser Arg Leu Lys Gln Leu Val	

		Helix 2
MyoD	Lys Val Gln Ile Leu Arg Asn Ala Ile Arg Tyr Ile Gln Gly Leu Glu	
Id	Lys Val Gln Ile Leu Glu His Val Ile Asp Tyr Ile Arg Asp Leu Glu	

Figure 25.21 All HLH proteins have regions corresponding to helix 1 and helix 2, separated by a loop of 10–24 residues. Basic HLH proteins have a region with conserved positive charges immediately adjacent to helix 1.

The bHLH proteins fall into two general groups. Class A consists of proteins that are ubiquitously expressed. Class B consists of proteins that are expressed in a tissue-specific manner. A common *modus operandi* for a tissue-specific bHLH protein is to form a heterodimer with a ubiquitous partner. Dimers formed from bHLH proteins differ in their abilities to bind to DNA. So both dimer formation and DNA binding may represent important regulatory points.

Differences in DNA binding result from properties of the region in or close to the HLH motif. Some HLH proteins lack the basic region and/or contain proline residues that appear to disrupt its function. The example of the protein Id is shown in Figure 25.21. Proteins of this type have the same capacity to dimerize as bHLH proteins, but a dimer that contains one subunit of this type can no longer bind to DNA specifically. This is a forceful demonstration of the importance of doubling the DNA-binding motif in DNA-binding proteins.

A model for the functions of bHLH and nonbasic HLH proteins in forming a regulatory network is illustrated in **Figure 25.22**. Two bHLH proteins form a dimer that activates a gene. But a nonbasic HLH protein can prevent the action of a bHLH protein by binding to it to form a dimer that cannot activate transcription. A set of HLH proteins can form combinatorial associations where each pair of dimers has a particular ability to activate or to suppress gene expression; and gene expression can therefore be regulated by controlling the availability of particular members of the family.

The behavior of the HLH proteins therefore illustrates two general principles of transcriptional regulation. Proteins can form combinatorial associations. Particular combinations have different functions with regard to DNA binding and transcriptional regulation. Differentiation may depend either on the presence or on the removal of particular partners.

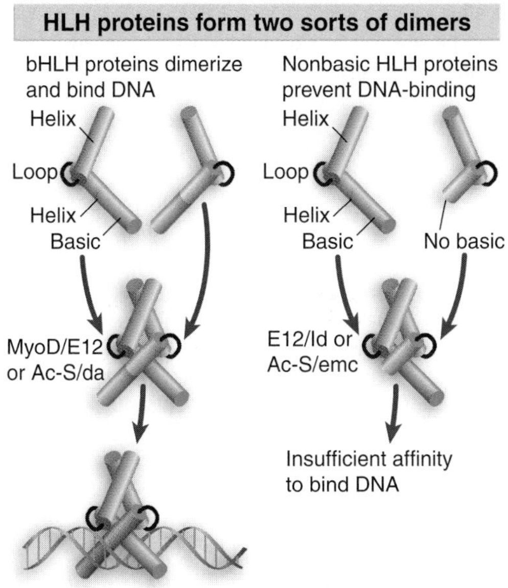

Figure 25.22 An HLH dimer in which both subunits are of the bHLH type can bind DNA, but a dimer in which one subunit lacks the basic region cannot bind DNA.

25.13 Leucine Zippers Are Involved in Dimer Formation

Key Concept

• Dimerization between proteins with leucine zippers forms the bZIP motif in which the two basic regions symmetrically bind inverted repeats in DNA.

Key Terms

• The **leucine zipper** is a dimerization motif that is found in a class of transcription factors.

• A **bZIP** protein has a basic DNA-binding region adjacent to a leucine zipper dimerization motif.

Interactions between proteins are a common theme in building a transcription complex, and a motif found in several activators (and other proteins) is involved in both homo- and heteromeric interactions.

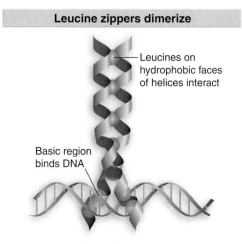

Leucine zippers dimerize

Leucines on hydrophobic faces of helices interact

Basic region binds DNA

Figure 25.23 The basic regions of the bZIP motif are held together by the dimerization at the adjacent zipper region when the hydrophobic faces of two leucine zippers interact in parallel orientation.

The **leucine zipper** is a stretch of amino acids rich in leucine residues that provide a dimerization motif. Dimer formation itself has emerged as a common principle in the action of proteins that recognize specific DNA sequences, and in the leucine zipper, the relationship of dimer formation to DNA binding is especially clear, because we can see how dimerization juxtaposes the DNA-binding regions of each subunit. The reaction is depicted diagrammatically in **Figure 25.23**.

An amphipathic α-helix has a structure in which the hydrophobic groups (including leucine) face one side, while charged groups face the other side. A leucine zipper forms an amphipathic helix in which the leucines of the zipper on one protein could protrude from the α-helix and interdigitate with the leucines of the zipper of another protein in parallel. The two right-handed helices wind around each other, with 3.5 residues per turn, so the pattern repeats integrally every 7 residues.

How is this structure related to DNA binding? The region adjacent to the leucine repeats is highly basic in each of the zipper proteins, and could be a DNA-binding site. The two leucine zippers in effect form a Y-shaped structure, in which the zippers compose the stem, and the two basic regions stick out to form the arms that bind to DNA. This is known as the **bZIP** structural motif. It explains why the target sequences for such proteins are inverted repeats with no separation.

Zippers may be used to sponsor formation of homodimers or heterodimers. They are lengthy motifs. Leucine (or another hydrophobic amino acid) occupies every seventh residue in the potential zipper. There are four repeats of the zipper (Leu-X_6) in the protein C/EBP (a factor that binds as a dimer to both the CAAT box and the SV40 core enhancer), and five repeats in the factors Jun and Fos (which form the heterodimeric activator, AP1).

25.14 SUMMARY

Transcription factors include basal factors, activators, and coactivators. Basal factors interact with RNA polymerase at the startpoint. Activators bind specific short response elements located in promoters or enhancers. Activators function by making protein–protein interactions with the basal apparatus. Some activators interact directly with the basal apparatus; others require coactivators to mediate the interaction. Activators often have a modular construction, in which there are independent domains responsible for binding to DNA and for activating transcription. The main function of the DNA-binding domain may be to tether the activating domain in the vicinity of the initiation complex. Some response elements are present in many genes and are recognized by ubiquitous factors; others are present in a few genes and are recognized by tissue-specific factors.

Promoters for RNA polymerase II contain a variety of short *cis*-acting elements, each of which is recognized by a *trans*-acting factor. The *cis*-acting elements are located upstream of the TATA box and may be present in either orientation and at a variety of distances with regard to the startpoint. The upstream elements are recognized by activators that interact with the basal transcription complex to determine the efficiency with which the promoter is used. Some activators interact directly with components of the basal apparatus; others interact via intermediaries called coactivators. The targets in the basal apparatus are the TAFs of TF$_{II}$D, or TF$_{II}$B or TF$_{II}$A. The interaction stimulates assembly of the basal apparatus.

Several groups of transcription factors have been identified by sequence homologies. The homeodomain is a 60 residue sequence found in genes that regulate development in insects and worms and in mammalian transcription factors. It is related to the prokaryotic helix-turn-helix motif and provides the motif by which the factors bind to DNA.

Another motif involved in DNA binding is the zinc finger, which is found in proteins that bind DNA or RNA (or sometimes both). A finger has cysteine residues that bind zinc. One type of finger is found in multiple repeats in some transcription factors; another is found in single or double repeats in others.

Steroid receptors were the first members identified of a group of transcription factors in which the protein is activated by binding a small hydrophobic hormone. The activated factor becomes localized in the nucleus, and binds to its specific response element, where it activates transcription. The DNA-binding domain has zinc fingers. The receptors are homodimers or heterodimers. The homodimers all recognize palindromic response elements with the same consensus sequence; the difference between the response elements is the spacing between the inverted repeats. The heterodimers recognize direct repeats, again being distinguished by the spacing between the repeats. The DNA-binding motif of these receptors includes two zinc fingers; the first determines which consensus sequence is recognized, and the second responds to the spacing between the repeats.

HLH (helix-loop-helix) proteins have amphipathic helices that are responsible for dimerization, adjacent to basic regions that bind to DNA. bHLH proteins have a basic region that binds to DNA, and fall into two groups: ubiquitously expressed and tissue-specific. An active protein is usually a heterodimer between two subunits, one from each group. When a dimer has one subunit that does not have the basic region, it fails to bind DNA, so such subunits can prevent gene expression. Combinatorial associations of subunits form regulatory networks.

The leucine zipper contains a stretch of amino acids rich in leucine that dimerizes in transcription factors. An adjacent basic region is responsible for binding to DNA.

26

RNA Splicing and Processing

26.1 Introduction

Key Terms

- **RNA splicing** is the process of excising introns from RNA and connecting the exons into a continuous mRNA.
- An **hnRNP** is the ribonucleoprotein form of hnRNA (heterogeneous nuclear RNA), in which the hnRNA is complexed with proteins. This is the form of pre-mRNAs while they are being processed in the nucleus.

Interrupted genes are found in all classes of organisms. They represent a minor proportion of the genes of the very lowest eukaryotes, but the vast majority of genes in higher eukaryotic genomes. Removal of introns is a major part of the production of RNA in all eukaryotes. The process by which the introns are removed is called **RNA splicing**, and it occurs in the nucleus, together with the other modifications that are made to newly synthesized RNAs.

We can identify several types of splicing systems:

- Introns are removed from the nuclear pre-mRNAs of higher eukaryotes by a system that recognizes only short consensus sequences conserved at exon–intron boundaries and within the intron. This reaction requires a large splicing apparatus, which takes the form of an array of proteins and ribonucleoproteins that functions as a large particulate complex (the spliceosome). The mechanism of splicing involves transesterifications, and the catalytic center includes RNA as well as proteins.
- Certain RNAs have the ability to excise their introns autonomously. Introns of this type fall into two groups, as distinguished by secondary/tertiary structure. Both groups use transesterification reactions in which the RNA is the catalytic agent (see *Chapter 27 Catalytic RNA*).

• The removal of introns from yeast nuclear tRNA precursors involves enzymatic activities that handle the substrate in a way resembling the tRNA processing enzymes, in which a critical feature is the conformation of the tRNA precursor. These splicing reactions are accomplished by enzymes that use cleavage and ligation.

The process of expressing an interrupted protein-coding gene is reviewed in **Figure 26.1**. The transcript is capped at the 5′ end (see *7.9 The 5′ End of Eukaryotic mRNA Is Capped*), has the introns removed, and is polyadenylated at the 3′ end (see *7.10 The 3′ Eukaryotic Terminus Is Polyadenylated*). The RNA is then transported through nuclear pores to the cytoplasm, where it is available to be translated.

When the pre-RNA is synthesized, it becomes bound by proteins to form a ribonucleoprotein particle. Taking its name from its broad size distribution, the RNA was originally called *heterogeneous nuclear RNA (hnRNA)*, and the particle is called **hnRNP**. Some of the proteins may have a structural role in packaging the hnRNA; several are known to shuttle between the nucleus and cytoplasm, and play roles in exporting the RNA or otherwise controlling its activity.

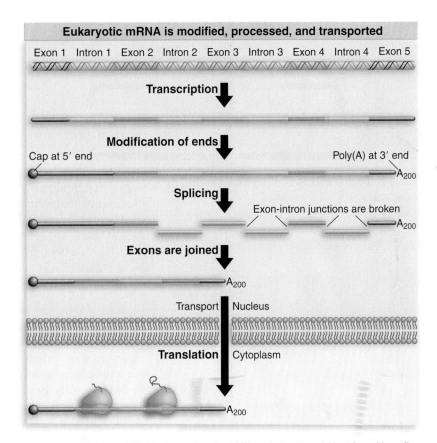

Figure 26.1 RNA is modified in the nucleus by additions to the 5′ and 3′ ends and by splicing to remove the introns. The splicing event requires breakage of the exon–intron junctions and joining of the ends of the exons. Mature mRNA is transported through nuclear pores to the cytoplasm, where it is translated.

26.2   Nuclear Splice Junctions Are Short Sequences

Key Terms

• **Splice sites** are the sequences immediately surrounding the exon–intron boundaries.
• The **GT–AG rule** describes the presence of these constant dinucleotides at the first two and last two positions of introns of nuclear genes.

Key Concepts

• The 5′ splice site at the 5′ (left) end of the intron includes the consensus sequence GU.
• The 3′ splice site at the 3′ (right) end of the intron includes the consensus sequence AG.
• Most introns follow the GU–AG rule (originally called the GT–AG rule in terms of DNA sequence), but there are minor classes of introns with the ends GC–AG or AU–AC.

To focus on the molecular events involved in nuclear intron splicing, we must consider the nature of the **splice sites**, the boundaries at each end

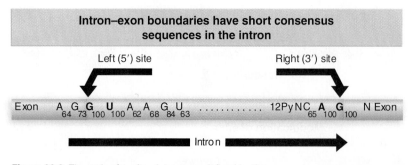

Figure 26.2 The ends of nuclear introns are defined by the GU–AG rule.

of an intron that include the sites of breakage and reunion.

By comparing the nucleotide sequence of mRNA with that of the structural gene, the junctions between exons and introns can be assigned. There is no extensive homology or complementarity between the two ends of an intron. However, the junctions have well conserved, though rather short, consensus sequences. A specific end can be assigned to every intron by aligning the exon–intron junctions to conform with the consensus sequence given in **Figure 26.2**.

The subscripts indicate the percent occurrence of the specified base at each consensus position. High conservation is found only *immediately within the intron* at the presumed junctions. This identifies the sequence of a generic intron as:

$$GU \ldots AG$$

Because the intron defined in this way starts with the dinucleotide GU and ends with the dinucleotide AG, the junctions are often described as conforming to the **GT–AG rule**. (This reflects the fact that the sequences were originally analyzed in terms of DNA, but of course the GT in the coding strand sequence of DNA becomes a GU in the RNA.)

Note that the two sites have different sequences and so they define the ends of the intron *directionally*. They are named proceeding from left to right along the intron as the 5′ splice site (sometimes called the left or donor site) and the 3′ splice site (also called the right or acceptor site). The consensus sequences are implicated as the sites recognized in splicing by point mutations that prevent splicing *in vivo* and *in vitro*.

GU–AG introns constitute the vast majority (>98% of splicing junctions in the human genome). A small number of introns (<1%) use the related junctions GC–AG. In addition, there is a minor class of introns marked by the ends AU–AC (0.1%).

26.3 Splice Junctions Are Read in Pairs

Key Concepts

- Splicing depends only on recognition of pairs of splice junctions.
- All 5′ splice sites are functionally equivalent, and all 3′ splice sites are functionally equivalent.

A typical mammalian mRNA has many introns. The basic problem of pre-mRNA splicing results from the simplicity of the splice sites, and is illustrated in **Figure 26.3**: what ensures that the correct pairs of sites are spliced together? The corresponding GU–AG pairs must be connected across great distances (some introns are >10 kb long). Experiments using hybrid RNA precursors show that any 5′ splice site can in principle be connected to any 3′ splice site. Such experiments make two general points:

- *Splice sites are generic:* they do not have specificity for individual RNA precursors, and individual precursors do not have specific information (such as secondary structure) that is needed for splicing.
- *The apparatus for splicing is not tissue specific:* an RNA can usually be properly spliced by any cell, whether or not it is usually synthesized in that cell. (We discuss exceptions in which there are tissue-specific alternative splicing patterns in *26.11 Alternative Splicing Involves Differential Use of Splice Junctions*.)

Here is a paradox. Although in principle any 5′ splice site may be able to react with any 3′ splice site, splicing usually occurs only between the 5′ and 3′ sites of the same intron. What rules ensure that recognition of splice sites is restricted so that only the 5′ and 3′ sites of the same intron are spliced?

Are introns removed in a specific *order* from a particular RNA? Using RNA blotting, we can identify nuclear RNAs that represent intermediates from which some introns have been removed. **Figure 26.4** shows a blot of the precursors to ovomucoid mRNA. There is a discrete series of bands, which suggests that splicing occurs via definite pathways. (If the seven introns were removed in an entirely random order, there would be more than 300 precursors with different combinations of introns, and we should not see discrete bands.)

There does not seem to be a *unique* pathway, since intermediates can be found in which different combinations of introns have been removed. However, there is evidence for a *preferred* pathway in which introns are removed in the order 5/6, 7/4, 2/1, 3. Probably the conformation of the RNA influences the accessibility of the splice sites. As particular introns are removed, the conformation changes, and new pairs of splice sites become available. But the ability of the precursor to remove its introns in more than one order suggests that alternative conformations are available at each stage. One important conclusion of this analysis is that *the reaction does not proceed sequentially along the precursor.*

A simple model to control recognition of splice sites would be for the splicing apparatus to act in a processive manner. Having recognized a 5′ site, the apparatus might scan the RNA in the appropriate direction until it meets the next 3′ site. This would restrict splicing to adjacent sites. But this model is excluded by experiments that show that splicing can occur in *trans* as an intermolecular reaction under special circumstances (see *26.12 Trans-Splicing Reactions Use Small RNAs*) or in RNA molecules in which part of the nucleotide chain is replaced by a chemical linker. This means that there cannot be a requirement for strict scanning along the RNA from the 5′ splice site to the 3′ splice site. Another problem with the scanning model is that it cannot explain the existence of alternative splicing patterns, where (for example) a common 5′ site is spliced to more than one 3′ site.

26.4 pre-mRNA Splicing Proceeds Through a Lariat

Key Terms

- The **lariat** is an intermediate in RNA splicing in which a circular structure with a tail is created by a 5′–2′ bond.
- The **branch site** is a short sequence just before the end of an intron at which the lariat intermediate is formed by joining the 5′ nucleotide of the intron to the 2′ position of an adenosine.
- A **transesterification** reaction breaks and makes chemical bonds in a coordinated transfer so that no energy is required.

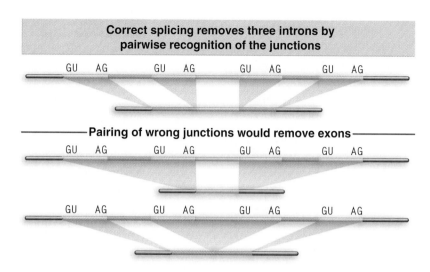

Figure 26.3 Splicing junctions are recognized only in the correct pairwise combinations.

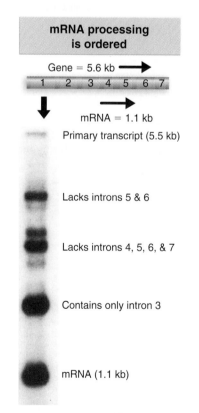

Figure 26.4 Northern blotting of nuclear RNA with an ovomucoid probe identifies discrete precursors to mRNA. The contents of the more prominent bands are indicated. Photograph kindly provided by Bert W. O'Malley, M.D., Baylor College of Medicine.

Splicing proceeds through a lariat

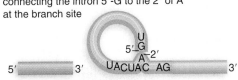

Py₈₀ N Py₈₀ Py₈₇ Pu₇₅ A Py₉₅
Animal consensus

Cut at 5′ site and form lariat by 5′–2′ bond connecting the intron 5′-G to the 2′ of A at the branch site

Cut at 3′ site and join exons; intron released as lariat

Debranch intron

Figure 26.5 Splicing occurs in two stages. First the 5′ exon is cleaved off; then it is joined to the 3′ exon.

Splicing uses transesterification

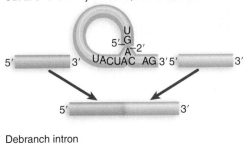

Figure 26.6 Nuclear splicing occurs by two transesterification reactions in which an OH group attacks a phosphodiester bond.

Key Concepts

- Splicing requires the 5′ and 3′ splice sites and a branch site just upstream of the 3′ splice site.
- The branch sequence is conserved in yeast but less well conserved in higher eukaryotes.
- A lariat is formed when the intron is cleaved at the 5′ splice site, and the 5′ end is joined to a 2′ position at an A at the branch site in the intron.
- The intron is released as a lariat when it is cleaved at the 3′ splice site, and the left and right exons are then ligated together.
- The reactions occur by transesterifications in which a bond is transferred from one location to another.

The stages of splicing are illustrated in the pathway of **Figure 26.5**. We discuss the reaction in terms of the individual RNA species that can be identified, but remember that *in vivo* the species containing exons are not released as free molecules, but remain held together by the splicing apparatus.

The first step is to make a cut at the 5′ splice site, separating the left exon and the right intron–exon molecule. The left exon takes the form of a linear molecule. The right intron–exon molecule forms a **lariat**, in which the 5′ terminus generated at the end of the intron becomes linked by a 5′–2′ bond to a base within the intron. The target base is an A in a sequence that is called the **branch site**.

Cutting at the 3′ splice site releases the free intron in lariat form, while the right exon is ligated (spliced) to the left exon. The cleavage and ligation reactions are shown separately in the figure for illustrative purposes, but actually occur as one coordinated transfer.

The lariat is then "debranched" to give a linear excised intron, which is rapidly degraded.

The sequences needed for splicing are the short consensus sequences at the 5′ and 3′ splice sites and at the branch site.

The branch site plays an important role in identifying the 3′ splice site. The branch site in yeast is highly conserved, and has the consensus sequence UACUAAC. The branch site in higher eukaryotes is not well conserved, but has a preference for purines or pyrimidines at each position and retains the target A nucleotide (see Figure 26.5). The branch site lies 18–40 nucleotides upstream of the 3′ splice site. *Its role is to identify the nearest 3′ splice site as the target for connection to the 5′ splice site*; it is always the 3′ consensus sequence nearest to the 3′ side of the branch that becomes the target for splicing. This can be explained by the fact that an interaction occurs between protein complexes that bind to these two sites.

The bond that forms the lariat goes from the 5′ position of the invariant G that was at the 5′ end of the intron to the 2′ position of the invariant A in the branch site. This corresponds to the third A residue in the yeast UACUAAC box.

The chemical reactions proceed by **transesterification**: a bond is in effect *transferred* from one location to another. **Figure 26.6** shows that the first step is a nucleophilic attack by the 2′ — OH of the invariant A of the UACUAAC sequence on the 5′ splice site. In the second step, the free 3′ — OH of the exon that was released by the first reaction now attacks the bond at the 3′ splice site. Note that the number of phosphodiester bonds is conserved. There were originally two 5′–3′ bonds at the exon–intron splice sites; one has been replaced by the 5′–3′ bond between the exons, and the other has been replaced by the 5′–2′ bond that forms the lariat.

26.5 snRNAs Are Required for Splicing

Key Concepts

- The five snRNPs involved in splicing are U1, U2, U5, U4, and U6.
- Together with some additional proteins, the snRNPs form the spliceosome.
- All the snRNPs except U6 contain a conserved sequence that binds the Sm proteins that are recognized by antibodies generated in autoimmune disease.

The 5′ and 3′ splice sites and the branch sequence are recognized by components of the splicing apparatus that assemble to form a large complex. This complex brings together the 5′ and 3′ splice sites before any reaction occurs, explaining why a deficiency in any one of the sites may prevent the reaction from initiating. The complex assembles sequentially on the pre-mRNA, and passes through several "presplicing complexes" before forming the final, active complex, which is called the **spliceosome**. Splicing occurs only after all the components have assembled.

The splicing apparatus contains both proteins and RNAs (in addition to the pre-mRNA). The RNAs take the form of small molecules that exist as ribonucleoprotein particles. Both the nucleus and cytoplasm of eukaryotic cells contain many discrete small RNA species (typically 100–300 bases long in higher eukaryotes). Those restricted to the nucleus are called **small nuclear RNAs (snRNA)**; those found in the cytoplasm are called **small cytoplasmic RNAs (scRNA)**. In their natural state, they exist as the ribonucleoprotein particles **snRNP** and **scRNP**. Colloquially, they are sometimes known as **snurps** and **scyrps**. Small RNAs found in the nucleolus that are involved in processing ribosomal RNA are called snoRNAs (see *26.15 Small RNAs Are Required for rRNA Processing*).

The spliceosome is a large body, greater in mass than the ribosome, and contains five snRNPs as well as many additional proteins. The snRNPs involved in splicing are U1, U2, U5, U4, and U6. They are named according to the snRNAs that are present. Each snRNP contains a single snRNA and several (<20) proteins. The U4 and U6 snRNPs are usually found as a single (U4/U6) particle. A common structural core for each snRNP consists of a group of eight proteins, all of which are recognized by an autoimmune antiserum called **anti-Sm**; conserved sequences in the proteins form the target for the antibodies. The other proteins in each snRNP are unique to it. The Sm proteins bind to the conserved sequence $PuAU_{3-6}GPu$, which is present in all snRNAs except U6. The U6 snRNP contains instead a set of Sm-like (Lsm) proteins. The Sm proteins must be involved in the autoimmune reaction, although their relationship to the phenotype of the autoimmune disease is not clear.

Figure 26.7 summarizes the components of the spliceosome. Its five snRNAs account for more than a quarter of the mass; together with their 41 associated proteins, they account for almost half of the mass. Some 70 other proteins found in the spliceosome are described as **splicing factors**. They include proteins required for assembly of the spliceosome, proteins required for it to bind to the RNA substrate, and proteins involved in the catalytic process. In addition to these proteins, another ~30 proteins associated with the spliceosome have been implicated in acting at other stages of gene expression, suggesting that the spliceosome may serve as a coordinating apparatus.

Key Terms

- The **spliceosome** is a complex formed by the snRNPs that are required for splicing together with additional protein factors.
- A **small nuclear RNA (snRNA)** is one of many small RNA species confined to the nucleus; several of the snRNAs are involved in splicing or other RNA processing reactions.
- **Small cytoplasmic RNAs (scRNA)** are present in the cytoplasm and sometimes are also found in the nucleus.
- **snRNPs (snurps)** are small nuclear ribonucleoproteins (snRNAs associated with proteins).
- **scRNPs (scyrps)** are small cytoplasmic ribonucleoproteins (scRNAs associated with proteins).
- **Anti-Sm** is an autoimmune antiserum that defines the Sm epitope that is common to a group of proteins found in snRNPs that are involved in RNA splicing.
- A **splicing factor** is a protein component of the spliceosome that is not part of one of the snRNPs.

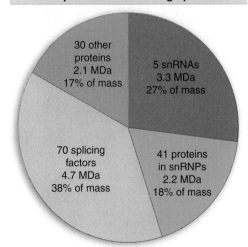

The spliceosome is a large particle

30 other proteins 2.1 MDa 17% of mass

5 snRNAs 3.3 MDa 27% of mass

70 splicing factors 4.7 MDa 38% of mass

41 proteins in snRNPs 2.2 MDa 18% of mass

Figure 26.7 The spliceosome is ~12 MDa. Five snRNPs account for almost half of the mass. The remaining proteins include known splicing factors and also proteins involved in other stages of gene expression.

Some of the proteins in the snRNPs may be involved directly in splicing; others may be required in structural roles or just for assembly or interactions between the snRNP particles. About one third of the proteins involved in splicing are components of the snRNPs. Increasing evidence for a direct role of RNA in the splicing reaction suggests that relatively few of the splicing factors play a direct role in catalysis; most are involved in structural or assembly roles.

26.6 U1 snRNP Initiates Splicing

Key Concept

- U1 snRNP initiates splicing by binding to the 5′ splice site by means of an RNA–RNA pairing reaction.

Splicing can be broadly divided into two stages:

- First the consensus sequences at the 5′ splice site, branch sequence, and adjacent pyrimidine tract are recognized. A complex assembles that contains all of the splicing components.
- Then the cleavage and ligation reactions change the structure of the substrate RNA. Components of the complex are released or reorganized as it proceeds through the splicing reactions.

The important point is that all of the splicing components are assembled and have assured that the splice sites are available before any irreversible change is made to the RNA.

Recognition of the consensus sequences involves both RNAs and proteins. Certain snRNAs have sequences that are complementary to the consensus sequences or to one another, and base pairing between snRNA and pre-mRNA, or between snRNAs, plays an important role in splicing.

Binding of U1 snRNP to the 5′ splice site is the first step in splicing. The human U1 snRNP contains eight proteins as well as the RNA. The secondary structure of the U1 snRNA is drawn in **Figure 26.8**. It contains several domains. The Sm-binding site is required for interaction with the common snRNP proteins. Domains identified by the individual stem-loop structures provide binding sites for proteins that are unique to U1 snRNP.

U1 snRNA base pairs with the 5′ site by means of a single-stranded region at its 5′ terminus, which usually includes a stretch of 4–6 bases that is complementary with the splice site. **Figure 26.9** describes an experiment that directly demonstrated the need for this base pairing. The wild-type sequence of the splice site of the 12S adenovirus pre-mRNA pairs at five out of six positions with U1 snRNA. A mutant in the 12S RNA that

U1 snRNA base pairs with donor splicing junction

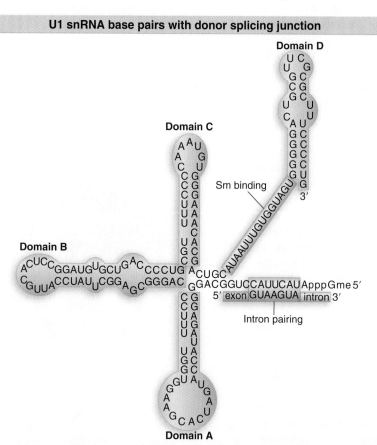

Figure 26.8 U1 snRNA has a base-paired structure that creates several domains. The 5′ end remains single stranded and can base pair with the 5′ splicing site.

cannot be spliced has two sequence changes; the GG residues at positions 5–6 in the intron are changed to AU. Splicing can be restored by a compensating change in the U1 snRNA that restores pairing.

26.7 The E Complex Commits an RNA to Splicing

Key Terms

- The **E complex** is the first complex to form at a splice site. It consists of U1 snRNP bound at the 5′ splice site together with factor ASF/SF2, U2AF bound at the branch site, and the bridging protein SF1/BBP.
- An **SR protein** has a variable length of an Arg–Ser-rich region and is involved in splicing.
- **Intron definition** is the recognition of a pair of splicing sites by interactions involving only the 5′ site and the branchpoint/3′ site.
- **Exon definition** is the recognition of a pair of splicing sites by interactions involving the 5′ site of the intron and also the 5′ site of the next intron downstream.
- The E presplicing complex is converted to the **A complex** by the binding of U2 snRNP to the branch site.

Key Concepts

- The direct way of forming an E complex is for U1 snRNP to bind at the 5′ splice site and U2AF to bind at a pyrimidine tract between the branch site and the 3′ splice site. This is intron definition.
- Another possibility is for the complex to form between U2AF at the pyrimidine tract and U1 snRNP at a downstream 5′ splice site. This is exon definition.

Figure 26.10 shows the early stages of splicing. The first complex formed during splicing is the **E complex** (early presplicing complex), which contains U1 snRNP, a factor called ASF/SF2 which binds with it to the 5′ splice site, the splicing factor U2AF, and members of a family called **SR proteins**, which are an important group of splicing factors and regulators. They take their name from the presence of an Arg–Ser-rich region that is variable in length. SR proteins interact with one another via their Arg–Ser-rich regions. They also bind to RNA. They are an essential component of the spliceosome, forming a framework on the RNA substrate. The E complex is sometimes called the commitment complex, because its formation identifies a pre-mRNA as a substrate for formation of the splicing complex.

In the E complex, the factor U2AF is bound to the region between the branch site and the 3′ splice site. The name of U2AF reflects its original isolation as the U2 auxiliary factor. In most organisms, it has a large subunit (U2AF65) that contacts a pyrimidine tract downstream of the branch site, while a small subunit (U2AF35) directly contacts the dinucleotide AG at the 3′ splice site.

The E complex can form in either of the ways summarized in **Figure 26.11**. The most direct reaction is for both splice sites to be recognized across the intron. The presence of U1 snRNP at the 5′ splice site enables U2AF to bind at the pyrimidine tract near the branch site. A splicing factor, an SR protein called SF1 in mammals (the equivalent protein is called BBP in yeast, for branch point binding protein), connects U2AF to

U1 snRNA selects the donor splicing junction

Wild-type U1 RNA and 12S pre-mRNA
Normal splicing

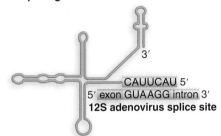

Wild-type U1 snRNA and mutant 12S pre-mRNA
No splicing

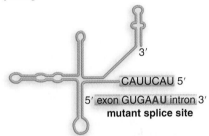

Mutant U1 snRNA and mutant 12S RNA
Splicing restored

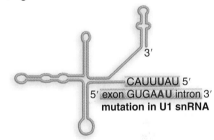

Figure 26.9 Mutations that abolish function of the 5′ splicing site can be suppressed by compensating mutations in U1 snRNA that restore base pairing.

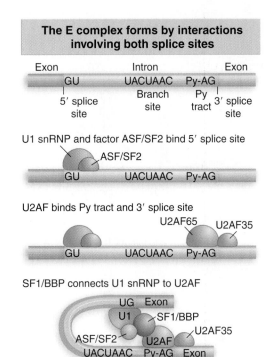

Figure 26.10 The commitment (E) complex forms by the successive addition of U1 snRNP to the 5′ splice site, U2AF to the pyrimidine tract/3′ splice site, and the bridging protein SF1/BBP.

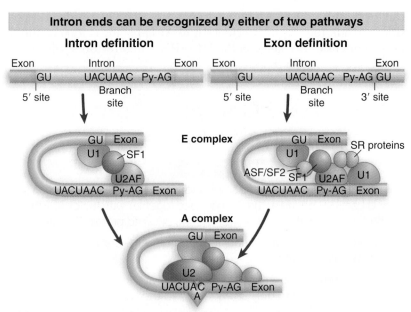

Figure 26.11 There may be multiple routes for initial recognition of 5′ and 3′ splice sites.

the U1 snRNP bound at the 5′ splice site. This interaction is probably responsible for making the first connection between the two splice sites across the intron. *The basic feature of this route for splicing is that the two splice sites are recognized without requiring any sequences outside of the intron.* This process is called **intron definition**.

The E complex is converted to the **A complex** when U2 snRNP binds to the branch site. Both U1 snRNP and U2AF are needed for U2 binding. The U2 snRNA includes sequences complementary to the branch site. A sequence near the 5′ end of the snRNA base pairs with the branch sequence in the intron. In yeast this typically involves formation of a duplex with the UACUAAC box (see Figure 26.13). Several proteins of the U2 snRNP are bound to the substrate RNA just upstream of the branch site. The binding of U2 snRNP requires ATP hydrolysis, and commits a pre-mRNA to the splicing pathway.

An alternative route to form the spliceosome may be followed when the introns are long and the splice sites are weak, as shown on the right of Figure 26.11. The 5′ splice site is recognized by U1 snRNA in the usual way. However, the 3′ splice site is recognized as part of a complex that forms across the *next exon*, in which the next 5′ splice site is also bound by U1 snRNA. This U1 snRNA is connected by SR proteins to the U2AF at the pyrimidine tract. When U2 snRNP joins to generate the A complex, there is a rearrangement, in which the correct (leftmost) 5′ splice site displaces the downstream 5′ splice site in the complex. The important feature of this route for splicing is that sequences downstream of the intron itself are required. Usually these sequences include the next 5′ splice site. This process is called **exon definition**. This mechanism is not universal: neither SR proteins nor exon definition are found in *S. cerevisiae*.

26.8 Five snRNPs Form the Spliceosome

Key Concepts

- Binding of U5 and U4/U6 snRNPs converts the A complex to the B1 spliceosome, which contains all the components necessary for splicing.

- The spliceosome passes through a series of further complexes as splicing proceeds.
- Release of U1 snRNP allows U6 snRNA to interact with the 5′ splice site, and converts the B1 spliceosome to the B2 spliceosome.
- When U4 dissociates from U6 snRNP, U6 snRNA can pair with U2 snRNA to form the catalytic active site.
- An alternative splicing pathway uses another set of snRNPs that constitute the U12 spliceosome.
- The target introns for U12 spliceosome have longer consensus sequences at the splice junctions, but usually include the same GU–AG junctions.
- Some introns have the splice junctions AU–AC, including some that are U1-dependent and some that are U12-dependent.

snRNPs and factors involved in splicing associate with the assembling complex in a defined order. **Figure 26.12** shows the components of the complexes that can be identified as the reaction proceeds.

The B1 complex is formed when a trimer containing the U5 and U4/U6 snRNPs binds to the A complex containing U1 and U2 snRNPs. It is converted to the B2 complex after U1 is released. The dissociation of U1 is necessary to allow other components to come into juxtaposition with the 5′ splice site, most notably U6 snRNA. At this point U5 snRNA changes its position; initially it is close to exon sequences at the 5′ splice site, but it must shift its position to bind to the 3′ splice site for the catalytic reaction to occur.

The catalytic reaction is triggered by the release of U4; this requires hydrolysis of ATP. The role of U4 snRNA may be to sequester U6 snRNA until it is needed. **Figure 26.13** shows the changes that occur in the base-pairing interactions between snRNAs during splicing. In the U6/U4 snRNP, a continuous length of 26 bases of U6 is paired with two separated regions of U4. When U4 dissociates, the region in U6 that is released becomes free

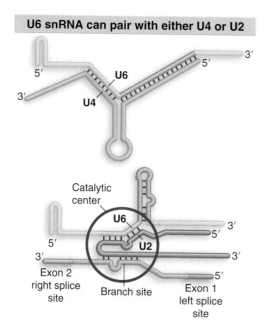

Figure 26.13 U6–U4 pairing is incompatible with U6–U2 pairing. When U6 joins the spliceosome it is paired with U4. Release of U4 allows a conformational change in U6; one part of the released sequence forms a hairpin (purple), and the other part (pink) pairs with U2. Because an adjacent region of U2 is already paired with the branch site, this brings U6 into juxtaposition with the branch. Note that the substrate RNA is reversed from the usual orientation and is shown 3′–5′.

A spliceosome forms through several discrete complexes

E complex

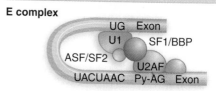

A complex–U2 binds branch site

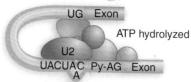

B1 complex–U5/U4/U6 trimer binds, U5 binds exon at 5′ site, U6 binds U2

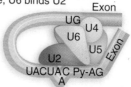

B2 complex–U1 is released, U5 shifts from exon to intron, U6 binds at 5′ splice site

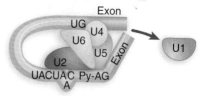

C1 complex–U4 is released, U6/U2 catalyzes transesterification, U5 binds exon at 3′ splice site, 5′ site cleaved and lariat is formed

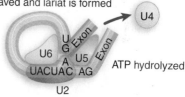

C2 complex–U2/U5/U6 remain bound to lariat, 3′ site cleaved and exons ligated

Spliced RNA is released

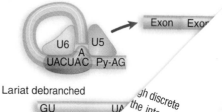

Lariat debranched

GU   UA

Figure 26.12 The splic[...] stages in which splice[...] tion of components that [...]

snRNA pairing is important in splicing

U1 pairs with the 5′ splice site

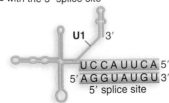

U2 pairs with the branch site

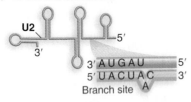

U6 pairs with the 5′ splice site

U5 is close to both exons

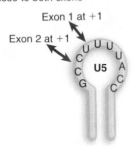

Figure 26.14 Splicing utilizes a series of base-pairing reactions between snRNAs and splice sites.

to take up another structure. The first part of it pairs with U2; the second part forms an intramolecular hairpin. The interaction between U4 and U6 is mutually incompatible with the interaction between U2 and U6, so the release of U4 controls the ability of the spliceosome to proceed.

Although for clarity the figure shows the RNA substrate in extended form, the 5′ splice site is actually close to the U6 sequence immediately on the 5′ side of the stretch bound to U2. This sequence in U6 snRNA pairs with sequences in the intron just downstream of the conserved GU at the 5′ splice site (mutations that enhance such pairing improve the efficiency of splicing).

So several pairing reactions between snRNAs and the substrate RNA occur in the course of splicing. They are summarized in **Figure 26.14**. The snRNPs have sequences that pair with the substrate and with one another. They also have single-stranded regions in loops that are in close proximity to sequences in the substrate, and which play an important role, as judged by the ability of mutations in the loops to block splicing.

The base pairing between U2 and the branch point, and between U2 and U6, creates a structure that resembles the active center of group II self-splicing introns (see Figure 26.19). This suggests the possibility that the catalytic component could be an RNA structure generated by the U2–U6 interaction. U6 is paired with the 5′ splice site, and crosslinking experiments show that a loop in U5 snRNA is immediately adjacent to the first-base positions in both exons. But although we can define the proximities of the substrate (5′ splice site and branch site) and snurps (U2 and U6) at the catalytic center (as shown in Figure 26.13), the components that undertake the transesterifications have not been directly identified.

The important conclusion suggested by these results is that *the snRNA components of the splicing apparatus interact both among themselves and with the substrate RNA by means of base-pairing interactions, and these interactions allow for changes in structure that may bring reacting groups into apposition and may even create catalytic centers.* Furthermore, the conformational changes in the snRNAs are reversible; for example, U6 snRNA is not used up in a splicing reaction, and at completion must be released from U2, so that it can reform the duplex structure with U4 to undertake another cycle of splicing.

A small proportion of introns is spliced by an alternative apparatus, called the U12 spliceosome, consisting of U11 and U12 (related to U1 and U2, respectively), a U5 variant, and the $U4_{atac}$ and $U6_{atac}$ snRNAs. The splicing reaction is essentially similar to that at U2-dependent introns, and the snRNAs play analogous roles. Whether there are differences in the protein components of this apparatus is not known.

The dependence on the type of spliceosome is influenced by sequences in the intron. A strong consensus sequence at the left end defines the U12-dependent type of intron: $5'^G_AUAUCCUUU\ldots PyA^G_C3'$. In addition, U12-dependent introns have a highly conserved branch point, UCCUUPuAPy, which pairs with U12. Both U12-dependent and U2-dependent introns may have either GU–AG or AU–AC termini.

26.9 Splicing Is Connected to Export of mRNA

Key Concepts

- The REF proteins bind to splicing junctions by associating with the spliceosome.
- After splicing, they remain attached to the RNA at the exon–exon junction.
- They interact with the transport protein TAP/Mex that exports the RNA through the nuclear pore.

After it has been synthesized and processed, mRNA is exported from the nucleus to the cytoplasm in the form of a ribonucleoprotein complex. One means for ensuring that transport occurs only after the completion of splicing may be that introns can prevent export of mRNA because they are associated with the splicing apparatus. The spliceosome also may provide the initial point of contact for the export apparatus. **Figure 26.15** shows a model in which a protein complex binds to the RNA via the splicing apparatus. The complex consists of >9 proteins and is called the EJC (exon junction complex).

The EJC is involved in several functions of spliced mRNAs. Some of the proteins of the EJC are directly involved in these functions, and others recruit additional proteins for particular functions. The first contact in assembling the EJC is made with one of the splicing factors. Then after splicing, the EJC remains attached to the mRNA just upstream of the exon–exon junction. The EJC is not associated with RNAs transcribed from genes that lack introns, so its involvement in the process is unique for spliced products.

If introns are deleted from a gene, its RNA product is exported much more slowly to the cytoplasm. This suggests that the intron may provide a signal for attachment of the export apparatus. We can now account for this phenomenon in terms of a series of protein interactions, as shown in **Figure 26.16**. The EJC includes a group of proteins called the REF family (the best characterized member is called Aly). The REF proteins in turn interact with a transport protein (variously called TAP and Mex) which has direct responsibility for interaction with the nuclear pore.

A similar system may be used to identify a spliced RNA so that nonsense mutations prior to the last exon trigger its degradation in the cytoplasm (see *7.13 Nonsense Mutations Trigger a Surveillance System*).

Splicing is required for mRNA export

Exon   Intron   Exon

Splicing

Protein binds splicing complex

Protein remains at exon–exon junction

Complex (EJC) assembles at exon–exon junction

EJC binds proteins involved in
RNA export, localization, decay

Figure 26.15 The EJC (exon junction complex) binds to RNA by recognizing the splicing complex.

26.10 Group II Introns Autosplice via Lariat Formation

Key Term

- **Autosplicing (self-splicing)** describes the ability of an intron to excise itself from an RNA by a catalytic action that depends only on the sequence of RNA in the intron.

Key Concepts

- Group II introns excise themselves from RNA by an autocatalytic splicing event.
- The splice junctions and mechanism of splicing of group II introns are similar to splicing of nuclear introns.
- A group II intron folds into a secondary structure that generates a catalytic site resembling the structure of U6–U2 bound to a nuclear intron.

Two groups of introns that are quite separate from the introns in nuclear protein-coding genes are found in organelles and in bacteria. Group I and group II introns are classified according to their internal organization. Each can be folded into a typical type of secondary structure. Group I introns are found also in the nucleus in lower eukaryotes.

The group I and group II introns have the remarkable ability to excise themselves from an RNA. This is called **autosplicing**. Group I introns are more common than group II introns. There is little relationship between the two classes, but in each case the RNA can perform the splicing

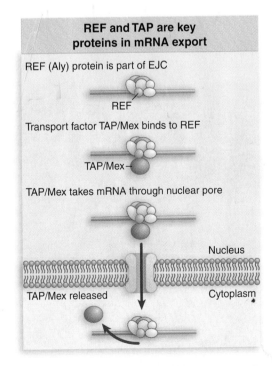

**REF and TAP are key
proteins in mRNA export**

REF (Aly) protein is part of EJC

REF

Transport factor TAP/Mex binds to REF

TAP/Mex

TAP/Mex takes mRNA through nuclear pore

Nucleus

TAP/Mex released   Cytoplasm

Figure 26.16 A REF protein binds to a splicing factor and remains with the spliced RNA product. REF binds to a transport factor that binds to the nuclear pore.

Splicing uses transesterification

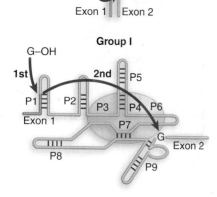

Figure 26.17 Three classes of splicing reactions proceed by two transesterifications. First, a free OH group attacks the exon 1–intron junction. Second, the OH created at the end of exon 1 attacks the intron–exon 2 junction.

reaction *in vitro* by itself, without requiring enzymatic activities provided by proteins; however, proteins are almost certainly required *in vivo* to assist with folding (see *Chapter 27 Catalytic RNA*).

Figure 26.17 shows that three classes of introns are excised by two successive transesterifications (shown previously for nuclear introns in Figure 26.5). In the first reaction, the 5' exon–intron junction is attacked by a free hydroxyl group (provided by an internal 2'—OH position in nuclear and group II introns, and by a free guanine nucleotide in group I introns). In the second reaction, the free 3'—OH at the end of the released exon in turn attacks the 3' intron–exon junction.

There are parallels between group II introns and pre-mRNA splicing. Group II mitochondrial introns are excised by the same mechanism as nuclear pre-mRNAs, via a lariat that is held together by a 5'–2' bond. An example of a lariat produced by splicing a group II intron is shown in **Figure 26.18**. When an isolated group II RNA is incubated *in vitro* in the absence of additional components, it is able to perform the splicing reaction. This means that the two transesterification reactions shown in Figure 26.17 can be performed by the group II intron RNA sequence itself. Because the number of phosphodiester bonds is conserved in the reaction, an external supply of energy is not required; this could have been an important feature in the evolution of splicing.

A group II intron forms into a secondary structure that contains several domains formed by base-paired stems and single-stranded loops. Domain 5 is separated by two bases from domain 6, which contains an A residue that donates the 2'—OH group for the first transesterification. This constitutes a catalytic domain in the RNA. **Figure 26.19** compares this secondary structure with the structure formed by the combination of U6 with U2 and of U2 with the branch site. The similarity suggests that U6 may have a catalytic role.

Group II introns form lariats

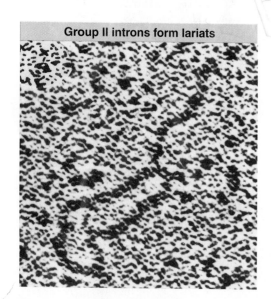

Figure 26.18 Splicing releases a mitochondrial group II intron in the form of a stable lariat. Photograph kindly provided by Leslie Grivell and Annika Arnberg.

Nuclear and group II splicing are similar

Nuclear splicing constructs an active site from pairing between U6–U2 and U2–intron

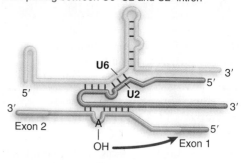

Group II splicing constructs an active center from the base paired regions of domains 5 and 6

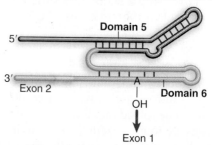

Figure 26.19 Nuclear splicing and group II splicing have similar secondary structures. The sequences are more specific in nuclear splicing; group II splicing uses positions that may be occupied by either purine (R) or either pyrimidine (Y).

The features of group II splicing suggest that splicing evolved from an autocatalytic reaction undertaken by an individual RNA molecule, in which it accomplished a controlled deletion of an internal sequence. Probably such a reaction required the RNA to fold into a specific conformation, or series of conformations, and would occur exclusively in *cis* conformation.

The ability of group II introns to remove themselves by an autocatalytic splicing event stands in great contrast to the requirement of nuclear introns for a complex splicing apparatus. We may regard the snRNAs of the spliceosome as compensating for the lack of sequence information in the intron, and providing the information required to form particular structures in RNA. The functions of the snRNAs may have evolved from the original autocatalytic system. These snRNAs act in *trans* upon the substrate pre-mRNA; we might imagine that the ability of U1 to pair with the 5′ splice site, or of U2 to pair with the branch sequence, replaced a similar reaction that required the relevant sequence to be carried by the intron. So the snRNAs may undergo reactions with the pre-mRNA substrate and with one another that have substituted for the series of conformational changes that occur in RNAs that splice by group II mechanisms. In effect, these changes have relieved the substrate pre-mRNA of the obligation to carry the sequences needed to sponsor the reaction. As the splicing apparatus has become more complex (and as the number of potential substrates has increased), proteins have played a more important role.

26.11 Alternative Splicing Involves Differential Use of Splice Junctions

Key Term

• **Alternative splicing** describes the production of different RNA products from a single product by changes in the usage of splicing junctions.

Key Concepts

• Specific exons may be excluded or included in the RNA product by using or failing to use a pair of splicing junctions.
• Exons may be extended by changing one of the splice junctions to use an alternative junction.
• Sex determination in *Drosophila* involves a series of alternative splicing events in genes coding for successive products of a pathway.

When an interrupted gene is transcribed into an RNA that gives rise to a single type of spliced mRNA, there is no ambiguity in assignment of exons and introns. But the RNAs of some genes follow patterns of **alternative splicing**, when a single gene gives rise to more than one mRNA sequence. In some cases, the ultimate pattern of expression is dictated by the primary transcript, because the use of different startpoints or the generation of alternative 3′ ends alters the pattern of splicing. In other cases, a single primary transcript is spliced in more than one way, and internal exons are substituted, added, or deleted. In some cases, the multiple products all are made in the same cell, but in others the process is regulated so that particular splicing patterns occur only under particular conditions.

Alternative splicing generates multiple RNAs

SV40 T/t antigens splice two 5′ sites to a common 3′ site

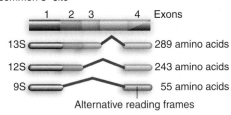

Adenovirus E1A splices variable 5′ sites to a common 3′ site

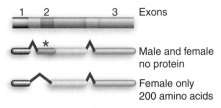

D. melanogaster tra splices 5′ site to alternative 3′ sites

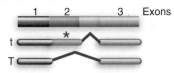

Figure 26.20 Alternative forms of splicing may generate a variety of protein products from an individual gene. Changing the splice sites may introduce termination codons (shown by asterisks) or change reading frames.

Alternative splicing may substitute exons

D. melanogaster dsx skips an exon

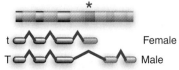

α-tropomyosin splices alternative exons

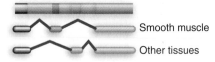

P elements splice out an extra intron

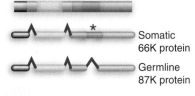

Figure 26.21 Alternative splicing events may cause exons to be added or substituted.

Figure 26.20 shows examples of alternative splicing in which one splice site remains constant, but the other varies. The large T/ small t antigens of SV40 and the products of the adenovirus E1A region are generated by connecting a varying 5′ site to a constant 3′ site. In the T/t antigens, the 5′ site used for T antigen removes a termination codon that is present in the t antigen mRNA, so that T antigen is larger than t antigen. In the E1A transcripts, one of the 5′ sites connects to the last exon in a different reading frame, again making a significant change in the C-terminal part of the protein. In these examples, all the relevant splicing events take place in every cell in which the gene is expressed, so all the protein products are made.

There are differences in the ratios of T/t antigens in different cell types. The relative usage of the alternative splice sites is determined by the splicing factor ASF/SF2 (a component of the E complex). When a pre-mRNA has more than one 5′ splice site preceding a single 3′ splice site, increased concentrations of ASF/SF2 promote use of the 5′ site nearest to the 3′ site at the expense of the other site. This effect of ASF/SF2 can be counteracted by another splicing factor, SF5. As a general rule, alternative splicing involving different 5′ sites may be influenced by proteins in spliceosome assembly that either stimulate or repress the usage of one of the possible sites.

Figure 26.21 shows examples of cases in which splice sites are used to add or to substitute exons or introns, again with the consequence that different protein products are generated. In the *Drosophila doublesex (dsx)* gene, females splice the 5′ site of intron 3 to the 3′ site of that intron; as a result translation terminates at the end of exon 4. Males splice the 5′ site of intron 3 directly to the 3′ site of intron 4, thus omitting exon 4 from the mRNA, and allowing translation to continue through exon 6. The result of the alternative splicing is that different proteins are produced in each sex: the male product blocks female sexual differentiation, while the female product represses expression of male-specific genes. Alternative splicing of *dsx* RNA is controlled by competition between 3′ splice sites, which rests on the binding of splicing factors to those sites.

26.12 *trans*-Splicing Reactions Use Small RNAs

Key Term

- **SL RNA (spliced leader RNA)** is a small RNA that donates an exon in the *trans*-splicing reaction of trypanosomes and nematodes.

Key Concepts

- Splicing reactions usually occur only in *cis* between splice junctions on the same molecule of RNA.
- *trans*-splicing occurs in trypanosomes and worms where a short sequence (SL RNA) is spliced to the 5′ ends of many precursor mRNAs.
- SL RNA has a structure resembling the Sm-binding site of U snRNAs and may play an analogous role in the reaction.

In both mechanistic and evolutionary terms, splicing has been viewed as an *intramolecular* reaction, amounting essentially to a controlled deletion of the intron sequences from RNA. In genetic terms, splicing occurs only in *cis*. This means that *only sequences on the same molecule of RNA can be spliced together*. The upper part of **Figure 26.22** shows the normal situation. The introns can be removed from each RNA molecule, allowing the exons of that

RNA molecule to be spliced together, but there is no *intermolecular* splicing of exons between different RNA molecules. However, *trans*-splicing can occur in the artificial situation in which introns are created with complementary sequences.

Although *trans*-splicing is rare, it occurs *in vivo* in some special situations. One is revealed by the presence of a common 35 base leader sequence at the end of numerous mRNAs in the trypanosome. But the leader sequence is not coded upstream of the individual transcription units. Instead it is transcribed into an independent RNA, carrying additional sequences at its 3′ end, from a repetitive unit located elsewhere in the genome. **Figure 26.23** shows that this RNA carries the 35 base leader sequence followed by a 5′ splice site sequence. The sequences coding for the mRNAs carry a 3′ splice site just preceding the sequence found in the mature mRNA. A similar situation is found in *C. elegans*, which contains two types of **SL RNA** (spliced leader RNA).

When the leader and the mRNA are connected by a *trans*-splicing reaction, the 3′ region of the leader RNA and the 5′ region of the mRNA in effect constitute the 5′ and 3′ halves of an intron. When splicing occurs, a 5′–2′ link forms by the usual reaction between the GU of the 5′ intron and the branch sequence near the AG of the 3′ intron. Because the two parts of the intron are not covalently linked, this generates a Y-shaped molecule instead of a lariat.

The SL RNAs found in several species of trypanosomes and also in the nematode (*C. elegans*) have some common features. They fold into a secondary structure that has three stem-loops and a single-stranded region that resembles the Sm-binding site. The SL RNAs therefore exist as snRNPs that count as members of the Sm snRNP class. Trypanosomes possess the U2, U4, and U6 snRNAs, but do not have U1 or U5 snRNAs. The absence of U1 snRNA can be explained by the properties of the SL RNA, which can carry out the functions that U1 snRNA usually performs at the 5′ splice site; thus SL RNA in effect consists of an snRNA sequence possessing U1 function, linked to the exon–intron site that it recognizes.

The *trans*-splicing reaction of the SL RNA may represent a step towards the evolution of the pre-mRNA splicing apparatus. The SL RNA provides in *cis* the ability to recognize the 5′ splice site, and this probably depends upon the specific conformation of the RNA. The remaining functions required for splicing are provided by independent snRNPs.

26.13 Yeast tRNA Splicing Involves Cutting and Rejoining

Key Term

- An **RNA ligase** is an enzyme that functions in tRNA splicing to make a phosphodiester bond between the two exon sequences that are generated by cleavage of the intron.

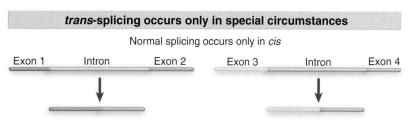
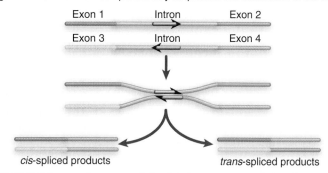

Figure 26.22 Splicing usually occurs only in *cis* between exons carried on the same physical RNA molecule, but *trans* splicing can occur when special constructs are made that support base pairing between introns.

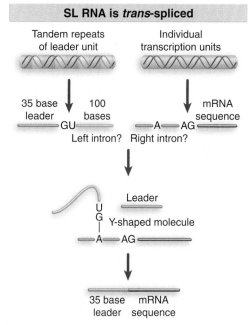

Figure 26.23 The SL RNA provides an exon that is connected to the first exon of an mRNA by *trans*-splicing. The reaction involves the same interactions as nuclear *cis*-splicing, but generates a Y-shaped RNA instead of a lariat.

Key Concepts

- tRNA splicing occurs by successive cleavage and ligation reactions.
- An endonuclease cleaves the tRNA precursors at both ends of the intron.
- Release of the intron generates two half-tRNAs that pair to form the mature structure.
- The halves have the unusual ends 5′ hydroxyl and 2′–3′ cyclic phosphate.
- The 5′—OH end is phosphorylated by a polynucleotide kinase, the cyclic phosphate group is opened by phosphodiesterase to generate a 2′–phosphate terminus and 3′—OH group, exon ends are joined by an RNA ligase, and the 2′–phosphate is removed by a phosphatase.
- The yeast endonuclease is a heterotetramer, with two (related) catalytic subunits.
- It uses a measuring mechanism to determine the sites of cleavage by their positions relative to a point in the tRNA structure.
- The archaeal nuclease has a simpler structure and recognizes a bulge–helix–bulge structural motif in the substrate.

Most splicing reactions depend on short consensus sequences and occur by transesterification reactions in which breaking and making of bonds is coordinated. The splicing of tRNA genes is achieved by a different mechanism that relies upon separate cleavage and ligation reactions.

Some 59 of the 272 nuclear tRNA genes in the yeast *S. cerevisiae* are interrupted. Each has a single intron, located just one nucleotide beyond the 3′ side of the anticodon. The introns vary in length from 14–60 bp. Those in related tRNA genes are related in sequence, but the introns in tRNA genes representing different amino acids are unrelated. *There is no consensus sequence that could be recognized by the splicing enzymes.* This is also true of interrupted nuclear tRNA genes of plants, amphibians, and mammals.

All the introns include a sequence that is complementary to the anticodon of the tRNA. This creates an alternative conformation for the anticodon arm in which the anticodon is base paired to form an extension of the usual arm. An example is drawn in **Figure 26.24**. Only the anticodon arm is affected—the rest of the molecule retains its usual structure.

The exact sequence and size of the intron is not important. Most mutations in the intron do not prevent splicing. *Splicing of tRNA depends principally on recognition of a common secondary structure in tRNA rather than a common sequence of the intron.* Regions in various parts of the molecule are important, including the stretch between the acceptor arm and D arm, in the TψC arm, and especially the anticodon arm. This is reminiscent of the structural demands placed on tRNA for protein synthesis (see *Chapter 8 Protein Synthesis*).

The intron is not entirely irrelevant, however. Pairing between a base in the intron loop and an unpaired base in the stem is required for splicing. Mutations at other positions that influence this pairing (for example, to generate alternative patterns for pairing) influence splicing. The rules that govern availability of tRNA precursors for splicing resemble the rules that govern recognition by aminoacyl-tRNA synthetases (see *9.8 tRNAs Are Charged with Amino Acids by Synthetases*).

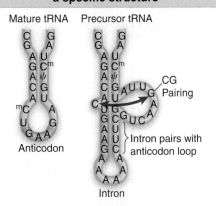

tRNA splicing recognizes a specific structure

Mature tRNA   Precursor tRNA

Anticodon

CG Pairing

Intron pairs with anticodon loop

Intron

Figure 26.24 The intron in yeast tRNA^Phe base pairs with the anticodon to change the structure of the anticodon arm. Pairing between an excluded base in the stem and the intron loop in the precursor may be required for splicing.

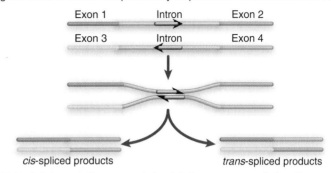

The reaction occurs in two stages that are catalyzed by different enzymes:

- The first step does not require ATP. It cleaves a phosphodiester bond by an atypical nuclease reaction. It is catalyzed by an endonuclease.
- The second step requires ATP and forms a bond; it is a ligation reaction, and the responsible enzyme activity is an **RNA ligase**.

The overall tRNA splicing reaction is summarized in **Figure 26.25**. The products of cleavage are a linear intron and two half-tRNA molecules. These intermediates have unique ends. Each 5′ terminus ends in a hydroxyl group; each 3′ terminus ends in a 2′,3′–cyclic phosphate group. (All other known RNA splicing enzymes cleave on the other side of the phosphate bond.)

The two half-tRNAs base pair to form a tRNA-like structure. When ATP is added, the second reaction occurs. Both of the unusual ends generated by the endonuclease must be altered.

The cyclic phosphate group is opened to generate a 2′–phosphate terminus. This reaction requires cyclic phosphodiesterase activity. The product has a 2′–phosphate group and a 3′—OH group.

The 5′—OH group generated by the nuclease must be phosphorylated to give a 5′–phosphate. This generates a site in which the 3′—OH is next to the 5′–phosphate. Covalent integrity of the polynucleotide chain is then restored by ligase activity.

All three activities—phosphodiesterase, polynucleotide kinase, and adenylate synthetase (which provides the ligase function)—are arranged in different functional domains on a single protein. They act sequentially to join the two tRNA halves.

The spliced molecule is now uninterrupted, with a 5′–3′ phosphate linkage at the site of splicing, but it also has a 2′–phosphate group marking the event. The surplus group must be removed by a phosphatase.

The endonuclease is responsible for the specificity of intron recognition. It cleaves the precursor at both ends of the intron. The yeast endonuclease is a heterotetrameric protein that functions as shown in **Figure 26.26**. The related subunits Sen34 and Sen2 cleave the 3′ and 5′ splice sites, respectively. Subunit Sen54 may determine the sites of cleavage by "measuring" distance from a point in the tRNA structure. This point is in the elbow of the (mature) L-shaped structure.

An interesting insight into the evolution of tRNA splicing is provided by the endonucleases of archaea. These are homodimers or homotetramers, in which each subunit has an active site (although only two of the sites function in the tetramer) that cleaves one of the splice sites. The subunit has sequences related to the sequences of the active sites in the Sen34 and Sen2 subunits of the yeast enzyme. However, the archaeal enzymes recognize their substrates in a different way. Instead of measuring distance from particular sequences, they recognize a structural

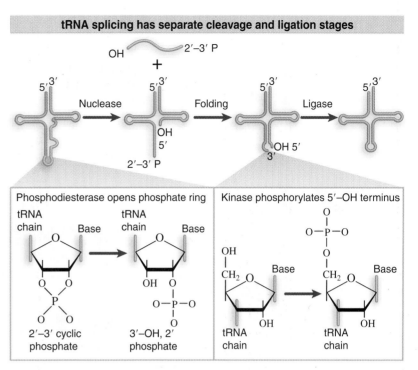

Figure 26.25 Splicing of tRNA requires separate nuclease and ligase activities. The exon–intron boundaries are cleaved by the nuclease to generate 2′–3′ cyclic phosphate and 5′ OH termini. The cyclic phosphate is opened to generate 3′—OH and 2′ phosphate groups. The 5′—OH is phosphorylated. After releasing the intron, the tRNA half molecules fold into a tRNA-like structure that now has a 3′—OH, 5′—P break. This is sealed by a ligase.

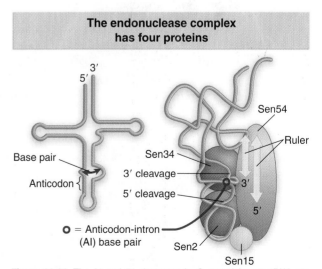

Figure 26.26 The 3′ and 5′ cleavages in *S. cerevisiae* pre-tRNA are catalyzed by different subunits of the endonuclease. Another subunit may determine location of the cleavage sites by measuring distance from the mature structure. The AI base pair is also important.

Pre-tRNA is cleaved at bulges

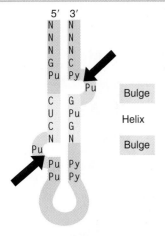

Figure 26.27 Archaeal tRNA splicing endonuclease cleaves each strand at a bulge in a bulge–helix–bulge motif.

feature, called the bulge–helix–bulge. **Figure 26.27** shows that cleavage occurs in the two bulges.

The existence of tRNA splicing in both archaea and eukaryotes indicates that its development must have preceded their evolutionary separation. If interrupted tRNA genes originated by insertion of the intron into tRNAs, this must have therefore been a very ancient event.

26.14 The 3′ Ends of mRNAs Are Generated by Cleavage and Polyadenylation

Key Term

- **Poly(A) polymerase** is the enzyme that adds the stretch of polyadenylic acid to the 3′ end of eukaryotic mRNA. It does not use a template.

Key Concepts

- The sequence AAUAAA is a signal for cleavage to generate a 3′ end of mRNA that is polyadenylated.
- The reaction requires a protein complex that contains a specificity factor, an endonuclease, and poly(A) polymerase.
- The specificity factor and endonuclease cleave RNA downstream of AAUAAA.
- The specificity factor and poly(A) polymerase add ~200 A residues processively to the 3′ end.

It is not clear whether RNA polymerase II actually engages in a termination event at a specific site. It is possible that its termination is only loosely specified. In some transcription units, termination occurs >1000 bp downstream of the site corresponding to the mature 3′ end of the mRNA (which is generated by cleavage at a specific sequence). Instead of using specific terminator sequences, the enzyme ceases RNA synthesis within multiple sites located in rather long "terminator regions." The nature of the individual termination sites is not known.

The 3′ ends of mRNAs are generated by cleavage followed by polyadenylation. The reactions are coupled by enzymes that are present in a single multi-protein complex.

Generation of the 3′ end is illustrated in **Figure 26.28**. RNA polymerase transcribes past the site corresponding to the 3′ end, and sequences in the RNA are recognized as targets for an endonucleolytic cut followed by polyadenylation. A single processing complex undertakes both the cutting and polyadenylation. The polyadenylation stabilizes the mRNA against degradation from the 3′ end. Its 5′ end is already stabilized by the cap. RNA polymerase continues transcription after the cleavage, but the 5′ end that is generated by the cleavage is unprotected. As a result, the rest of the transcript is rapidly degraded. This makes it difficult to determine what is happening beyond the point of cleavage.

A common feature of mRNAs in higher eukaryotes (but not in yeast) is the presence of the highly conserved sequence AAUAAA in the region from 11–30 nucleotides upstream of the site of poly(A) addition. Deletion or mutation of the AAUAAA hexamer prevents generation of the polyadenylated 3′ end. The signal is needed for both cleavage and polyadenylation.

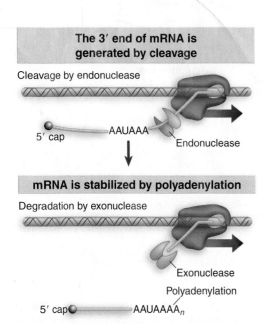

Figure 26.28 The sequence AAUAAA is necessary for cleavage to generate a 3′ end for polyadenylation.

The formation and functions of the complex that undertakes 3′ processing are illustrated in **Figure 26.29**. Generation of the proper 3′ terminal structure requires an *endonuclease* (consisting of the components CFI and CFII) to cleave the RNA, a **poly(A) polymerase** (PAP) to synthesize the poly(A) tail, and a *specificity component* (CPSF) that recognizes the AAUAAA sequence and directs the other activities. A stimulatory factor, CstF, binds to a G–U-rich sequence that is downstream from the cleavage site itself.

The specificity factor contains four subunits, which together bind specifically to RNA containing the sequence AAUAAA. The individual subunits are proteins that have common RNA-binding motifs, but which by themselves bind nonspecifically to RNA. Protein–protein interactions between the subunits may be needed to generate the specific AAUAAA-binding site. CPSF binds strongly to AAUAAA only when CstF is also present to bind to the G–U-rich site.

The specificity factor is needed for both the cleavage and polyadenylation reactions. It exists in a complex with the endonuclease and poly(A) polymerase, and this complex usually undertakes cleavage followed by polyadenylation in a tightly coupled manner.

The two components CFI and CFII (cleavage factors I and II), together with specificity factor, are necessary and sufficient for the endonucleolytic cleavage.

The poly(A) polymerase has a nonspecific catalytic activity. When it is combined with the other components, the synthetic reaction becomes specific for RNA containing the sequence AAUAAA. The polyadenylation reaction passes through two stages. First, a rather short oligo(A) sequence (~10 residues) is added to the 3′ end. This reaction is absolutely dependent on the AAUAAA sequence, and poly(A) polymerase performs it under the direction of the specificity factor. In the second phase, the oligo(A) tail is extended to the full ~200 residue length. This reaction requires another stimulatory factor that recognizes the oligo(A) tail and directs poly(A) polymerase specifically to extend the 3′ end of a poly(A) sequence.

The poly(A) polymerase by itself adds A residues individually to the 3′ position. Its intrinsic mode of action is distributive; it dissociates after each nucleotide has been added. However, in the presence of CPSF and PABP (poly(A)-binding protein), it functions processively to extend an individual poly(A) chain. The PABP is a 33 kD protein that binds stoichiometrically to the poly(A) stretch. The length of poly(A) is controlled by the PABP, which in some way limits the action of poly(A) polymerase to ~200 additions of A residues. The limit may represent the accumulation of a critical mass of PABP on the poly(A) chain. PABP binds to the translation initiation factor eIF4G, thus generating a closed loop in which a protein complex contains both the 5′ and 3′ ends of the mRNA.

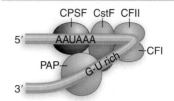

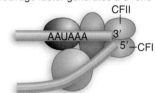
Cleavage factor generates a 3′ end

Poly(A) polymerase (PAP) adds A residues

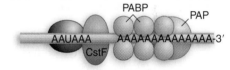
Poly(A)-binding protein (PABP) binds to poly(A)

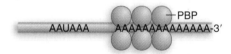
Complex dissociates after adding ~200 A residues

Figure 26.29 The 3′ processing complex consists of several activities. CPSF and CstF each consist of several subunits; the other components are monomeric. The total mass is >900 kD.

26.15 Small RNAs Are Required for rRNA Processing

Key Concepts
- The C/D group of snoRNAs is required for modifying the 2′ position of ribose with a methyl group.
- The H/ACA group of snoRNAs is required for converting uridine to pseudouridine.
- In each case the snoRNA base pairs with a sequence of rRNA that contains the target base to generate a structure that is the substrate for modification.

Key Term
- A **snoRNA** is a small nuclear RNA that is localized in the nucleolus.

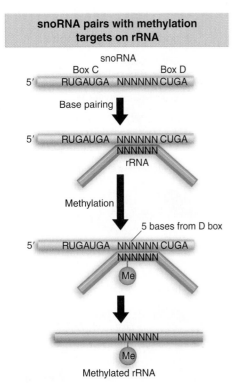

snoRNA pairs with methylation targets on rRNA

Figure 26.30 A snoRNA base pairs with a region of rRNA that is to be methylated.

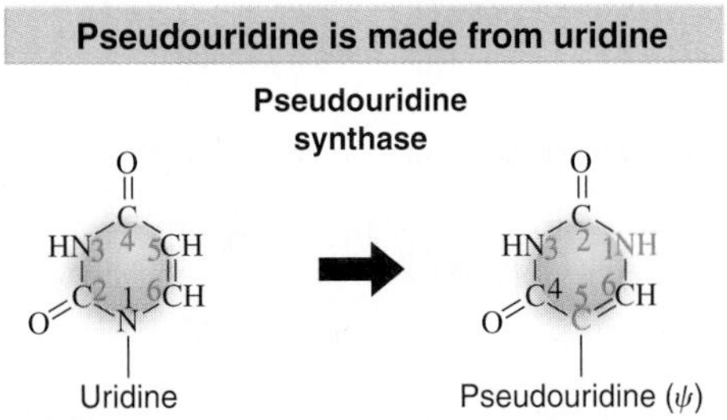

Pseudouridine is made from uridine

Figure 26.31 Uridine is converted to pseudouridine by replacing the N1–sugar bond with a C5–sugar bond and rotating the base relative to the sugar.

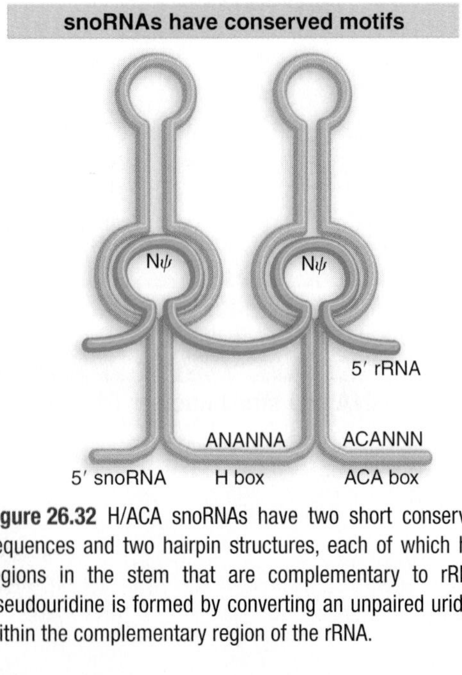

snoRNAs have conserved motifs

Figure 26.32 H/ACA snoRNAs have two short conserved sequences and two hairpin structures, each of which has regions in the stem that are complementary to rRNA. Pseudouridine is formed by converting an unpaired uridine within the complementary region of the rRNA.

The major rRNAs are synthesized as part of a single primary transcript that is processed by cleavage and trimming events to generate the mature products. The precursor contains the sequences of the 18S, 5.8S, and 28S rRNAs. In higher eukaryotes, the precursor is named for its sedimentation rate as 45S RNA. In bacteria, the precursor contains the sequences of 16S and 23S rRNAs (and sometimes also the 5S RNA). The ribosomal precursor RNA is modified by the addition of many methyl groups.

Processing and modification of rRNA in eukaryotes requires a class of small RNAs called **snoRNAs** (small nucleolar RNAs). There are 71 snoRNAs in the yeast (*S. cerevisiae*) genome. They are associated with the protein fibrillarin, which is an abundant component of the nucleolus (the region of the nucleus where the rRNA genes are transcribed). Some snoRNAs are required for cleavage of the precursor to rRNA; one example is U3 snoRNA, which is required for the first cleavage event in both yeast and *Xenopus*. We do not know what role the snoRNA plays in cleavage. It could be required to pair with the rRNA sequence to form a secondary structure that is recognized by an endonuclease.

Two groups of snoRNAs are required for the modifications that are made to bases in the rRNA. The members of each group are identified by very short conserved sequences and common features of secondary structure.

The C/D group of snoRNAs is required for adding a methyl group to the 2′ position of ribose. There are >100 2′–O–methyl groups at conserved locations in vertebrate rRNAs. This group takes its name from two short conserved sequences motifs called boxes C and D. Each snoRNA contains a sequence near the D box that is complementary to a region of the 18S or 28S rRNA that is methylated. Loss of a particular snoRNA prevents methylation in the rRNA region to which it is complementary.

Figure 26.30 suggests that the snoRNA base pairs with the rRNA to create the duplex region that is recognized as a substrate for methylation. Methylation occurs within the region of complementarity, at a position that is fixed 5 bases on the 5′ side of the D box. Probably each methylation event is specified by a different snoRNA; ~40 snoRNAs have been characterized so far. The methylases have not been characterized; one possibility is that the snoRNA itself provides part of the methylase activity.

Another group of snoRNAs is involved in the synthesis of pseudouridine (ψ). There are 43 ψ residues in yeast rRNAs and ~100 in vertebrate rRNAs. Pseudouridine is synthesized by the reaction shown in **Figure 26.31** in which the N1 bond from uridylic acid to ribose is broken, the base is rotated, and C5 is rejoined to the sugar.

Pseudouridine formation in rRNA requires the H/ACA group of ~20 snoRNAs. They are named for the presence of an ACA triplet located three nucleotides from the 3′ end and a partially conserved sequence (the H box) that lies between two stem-loop hairpin structures. Each of these snoRNAs has a sequence complementary to rRNA within the stem of each hairpin. **Figure 26.32** shows the structure that would be produced by pairing with the rRNA. Within each pairing region, there are two unpaired bases, one of which is a uridine that is converted to pseudouridine. The enzymatic activity that catalyzes the reaction has not yet been identified.

26.16 SUMMARY

Splicing accomplishes the removal of introns and the joining of exons into the mature sequence of RNA. There are at least four types of reactions, including those for eukaryotic nuclear introns, group I introns, group II introns, and tRNA introns. Each reaction changes the organization within an individual RNA molecule, and is therefore a *cis*-acting event.

pre-mRNA splicing follows preferred but not obligatory pathways. Only very short consensus sequences are necessary; the rest of the intron appears irrelevant. All 5′ splice sites are probably equivalent, as are all 3′ splice sites. The required sequences are given by the GT–AG rule, which describes the ends of the intron. The UACUAAC branch site of yeast, or a less well conserved consensus in mammalian introns, is also required. The reaction with the 5′ splice site forms a lariat that joins the GU end of the intron via a 5′–2′ linkage to the A at position 6 of the branch site. Then the 3′—OH end of the exon attacks the 3′ splice site, so that the exons are ligated and the intron is released as a lariat. Both reactions are transesterifications in which bonds are conserved. Several stages of the reaction require hydrolysis of ATP, probably to drive conformational changes in the RNA and/or protein components. Lariat formation is responsible for choice of the 3′ splice site. Alternative splicing patterns are caused by protein factors that either stimulate use of a new site or that block use of the default site.

pre-mRNA splicing requires formation of a spliceosome, a large particle that assembles the consensus sequences into a reactive conformation. The spliceosome most often forms by the process of intron definition, which includes recognition of the 5′ splice site, branch site, and 3′ splice site. An alternative pathway is exon definition, which includes initial recognition of the 5′ splice sites of both the substrate intron and the next intron. The formation of the spliceosome passes through a series of stages from the E (commitment) complex that contains U1 snRNP and splicing factors, through the A and B complexes as additional components are added.

The spliceosome contains the U1, U2, U4/U6, and U5 snRNPs and some additional splicing factors. The U1, U2, and U5 snRNPs each contain a single snRNA and several proteins; the U4/U6 snRNP contains two

snRNAs and several proteins. Some proteins are common to all snRNP particles. The snRNPs recognize consensus sequences. U1 snRNA base pairs with the 5′ splice site, U2 snRNA base pairs with the branch sequence, U5 snRNP acts at the 5′ splice site. When U4 releases U6, the U6 snRNA base pairs with U2, and this may create the catalytic center for splicing. An alternative set of snRNPs provides analogous functions for splicing the U12-dependent subclass of introns. The snRNA molecules may have catalytic–like roles in splicing and other processing reactions.

In the nucleolus, two groups of snoRNAs are responsible for pairing with rRNAs at sites that are modified; group C/D snoRNAs indicate target sites for methylation, and group ACA snoRNAs identify sites where uridine is converted to pseudouridine.

Splicing is usually intramolecular, but *trans*-(intermolecular) splicing occurs in trypanosomes and nematodes. It proceeds by a reaction between a small SL RNA and the pre-mRNA. The SL RNA resembles U1 snRNA and may combine the role of providing the exon and the functions of U1. In worms there are two types of SL RNA, one used for splicing to the 5′ end of an mRNA, the other for splicing to an internal site.

Group II introns share with nuclear introns the use of a lariat as intermediate, but are able to perform the reaction as a self-catalyzed property of the RNA. These introns follow the GT–AG rule, but form a characteristic secondary structure that holds the reacting splice sites in the appropriate apposition.

Yeast tRNA splicing includes separate endonuclease and ligase reactions. The endonuclease recognizes the secondary (or tertiary) structure of the precursor and cleaves both ends of the intron. The two half-tRNAs released by loss of the intron are ligated in the presence of ATP.

The termination capacity of RNA polymerase II has not been characterized, and 3′ ends of its transcripts are generated by cleavage. The sequence AAUAAA, located 11–30 bases upstream of the cleavage site, provides the signal for both cleavage and polyadenylation. An endonuclease and the poly(A) polymerase are associated in a complex with other factors that confer specificity for the AAUAAA signal.

Catalytic RNA

27.1 Introduction

Key Terms

- A **ribozyme** is an RNA that has catalytic activity.
- **RNA editing** changes the sequence of RNA following transcription.

Several types of catalytic reactions are now known to reside in RNA. **Ribozyme** has become a general term used to describe an RNA with catalytic activity, and it is possible to characterize the enzymatic activity in the same way as a more conventional (proteinaceous) enzyme. Some RNA catalytic activities are directed against separate substrates, while others are intramolecular (which limits the catalytic action to a single cycle).

Introns of the group I and group II classes possess the ability to splice themselves out of the pre-mRNA that contains them. Engineering of group I introns has generated RNA molecules that have several other catalytic activities related to the original activity.

The enzyme ribonuclease P is a ribonucleoprotein that contains a single RNA molecule bound to a protein. The RNA possesses the ability to catalyze cleavage in a tRNA substrate, while the protein component plays an indirect role, probably to maintain the structure of the catalytic RNA.

The common theme of these reactions is that the RNA can perform an intramolecular or intermolecular reaction that involves cleavage or joining of phosphodiester bonds *in vitro*. Although the specificity of the reaction and the basic catalytic activity is provided by RNA, proteins associated with the RNA may be needed for the reaction to occur efficiently *in vivo*.

RNA splicing is not the only means by which changes can be introduced in the informational content of RNA. In the process of **RNA**

editing, individual bases are changed or added at particular positions within an mRNA. The insertion of bases (most commonly uridine residues) occurs for several genes in the mitochondria of certain lower eukaryotes; like splicing, it involves the breakage and reunion of bonds between nucleotides, but also requires a template for coding the information of the new sequence.

27.2 Group I Introns Undertake Self-Splicing by Transesterification

Key Term

- **Autosplicing (self-splicing)** describes the ability of an intron to excise itself from an RNA by a catalytic action that depends only on the sequence of RNA in the intron.

Key Concepts

- The only factors required for autosplicing *in vitro* by group I introns are a monovalent cation, a divalent cation, and a guanine nucleotide.
- Splicing occurs by two transesterifications, without requiring input of energy.
- The 3′—OH end of the guanine cofactor attacks the 5′ end of the intron in the first transesterification.
- The 3′—OH end generated at the end of the first exon attacks the junction between the intron and second exon in the second transesterification.
- The intron is released as a linear molecule that circularizes when its 3′—OH terminus attacks a bond at one of two internal positions.
- The G^{414}–A^{16} internal bond of the *Tetrahymena* intron can also be attacked by other nucleotides in a *trans*-splicing reaction.

Group I introns are found in diverse locations. They occur in the genes coding for rRNA in the nuclei of the lower eukaryotes *Tetrahymena thermophila* (a ciliate) and *Physarum polycephalum* (a slime mold). They are common in the genes of fungal mitochondria. They are present in three genes of phage T4 and also are found in bacteria. Group I introns have an intrinsic ability to splice themselves. This is called **self-splicing** or **autosplicing**.

Self-splicing was discovered as a property of the transcripts of the rRNA genes in *T. thermophila*. The genes for the two major rRNAs follow the usual organization, in which both are expressed as part of a common transcription unit. The product is a 35S precursor RNA with the sequence of the small rRNA in the 5′ part, and the sequence of the larger (26S) rRNA toward the 3′ end.

In some strains of *T. thermophila*, the sequence coding for 26S rRNA is interrupted by a single, short intron. When the 35S precursor RNA is incubated *in vitro*, splicing occurs as an autonomous reaction. The intron is excised from the precursor and accumulates as a linear fragment of 400 bases, which is subsequently converted to a circular RNA. These events are summarized in **Figure 27.1**.

The reaction requires only a monovalent cation, a divalent cation, and a guanine nucleotide cofactor. No other base can be substituted for G; but a triphosphate is not needed; GTP, GDP, GMP, and guanosine

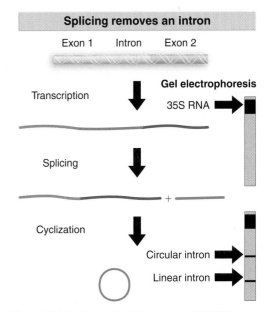

Figure 27.1 Splicing of the *Tetrahymena* 35S rRNA precursor can be followed by gel electrophoresis. The removal of the intron is revealed by the appearance of a rapidly moving small band. When the intron becomes circular, it electrophoreses more slowly, as seen by a higher band.

Self-splicing occurs by successive transesterifications

First transfer–3'—OH end of G attacks 5' end of intron

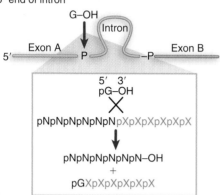

Second transfer–3'—OH of exon A attacks 5' end of exon B

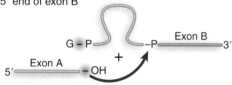

Third transfer–3'—OH of intron attacks bond 15 bases from 5' end

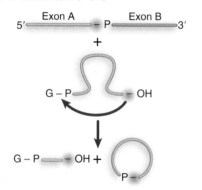

Figure 27.2 Self-splicing occurs by transesterification reactions in which bonds are exchanged directly. The bonds that have been generated at each stage are indicated by the shaded boxes.

itself all can be used, so there is no net energy requirement. The guanine nucleotide must have a 3'—OH group.

Figure 27.2 shows that three transfer reactions occur. In the first transfer, the guanine nucleotide behaves as a cofactor to provide a free 3'—OH group that attacks the 5' end of the intron. This reaction creates the G–intron link and generates a 3'—OH group at the end of the exon. The second transfer involves a similar chemical reaction, in which this 3'—OH then attacks the second exon. The two transfers are connected, so there is no release of free exons. The intron is released as a linear molecule, but the third transfer reaction converts it to a circle.

Each stage of the self-splicing reaction occurs by a transesterification, in which one phosphate ester is converted directly into another, without any intermediary hydrolysis. Bonds are exchanged directly, and energy is conserved, so the reaction does not require input of energy from hydrolysis of ATP or GTP.

If each of the consecutive transesterification reactions involves no net change of energy, why does the splicing reaction proceed to completion instead of coming to equilibrium between spliced product and nonspliced precursor? The concentration of GTP is high relative to that of RNA, and therefore drives the reaction forward; and a change in secondary structure in the RNA prevents the reverse reaction.

The in vitro system includes no protein so the ability to splice is intrinsic to the RNA. The RNA forms a specific secondary/tertiary structure in which the relevant groups are brought into juxtaposition so that a guanine nucleotide can be bound to a specific site and then the bond breakage and reunion reactions shown in Figure 27.2 can occur. Although a property of the RNA itself, the reaction is assisted *in vivo* by proteins, which stabilize the RNA structure.

The ability to engage in these transfer reactions resides with the sequence of the intron, which continues to be reactive after its excision as a linear molecule. **Figure 27.3** summarizes its activities.

The intron can circularize when the 3' terminal G attacks either of two positions near the 5' end. The internal bond is broken and the new 5' end is transferred to the 3'—OH end of the intron. The *primary cyclization* usually involves reaction between the terminal G^{414} and the A^{16}. This is the most common reaction (shown as the third transfer in Figure 27.2). Less frequently, the G^{414} reacts with U^{20}. Each reaction generates a circular intron and a linear fragment that represents the original 5' region (15 bases long for attack on A^{16}, 19 bases long for attack on U^{20}). The released 5' fragment contains the original added guanine nucleotide.

Either type of circle can regenerate a linear molecule *in vitro* by specifically hydrolyzing the bond (G^{414}–A^{16} or G^{414}–U^{20}) that had closed the circle. This is called a *reverse cyclization*. The linear molecule generated by reversing the primary cyclization at A^{16} remains reactive, and can perform a secondary cyclization by attacking U^{20}.

The final product of the spontaneous reactions following release of the intron is the L-19 RNA, a linear molecule generated by reversing the shorter circular form. This molecule has an enzymatic activity that allows it to catalyze the extension of short oligonucleotides (not shown in the figure, but see Figure 27.7).

The reactivity of the released intron extends beyond merely reversing the cyclization reaction. Addition of the oligonucleotide UUU reopens the primary circle by reacting with the G^{414}–A^{16} bond. The UUU (which resembles the 3' end of the 15-mer released by the primary cyclization) becomes the 5' end of the linear molecule that is formed. This is an *intermolecular* reaction, and thus demonstrates the ability to connect together two different RNA molecules.

27.3 Group I Introns Form a Characteristic Secondary Structure

Key Concepts

- Group I introns form a secondary structure with nine duplex regions.
- The core of regions P3, P4, P6, and P7 has catalytic activity.
- Regions P4 and P7 are both formed by pairing between conserved consensus sequences.
- A sequence adjacent to P7 base pairs with the sequence that contains the reactive G.

All group I introns can be organized into a characteristic secondary structure, with nine helices (P1–P9). **Figure 27.4** shows a model for the secondary structure of the *Tetrahymena* intron.

The group I splicing reaction depends on the formation of secondary structure between pairs of consensus sequences within the intron. The principle established by this work is that *sequences distant from the splice junctions themselves are required to form the active site that makes self-splicing possible.*

Two of the base-paired regions are generated by pairing between conserved sequence elements that are common to group I introns. P4 is constructed from the sequences *P* and *Q*; P7 is formed from sequences *R* and *S*. The other base-paired regions vary in sequence in individual introns. Mutational analysis identifies an intron "core," containing P3, P4, P6, and P7, which provides the minimal region that can undertake a catalytic reaction.

Some of the pairing reactions are directly involved in bringing the splice junctions into a conformation that supports the enzymatic reaction. P1 includes the 3′ end of the left exon. The sequence within the intron that pairs with the exon is called the IGS, or internal guide sequence. A very short sequence, sometimes as short as two bases, between P7 and P9, base pairs with the sequence that immediately precedes the reactive G (position 414 in *Tetrahymena*) at the 3′ end of the intron.

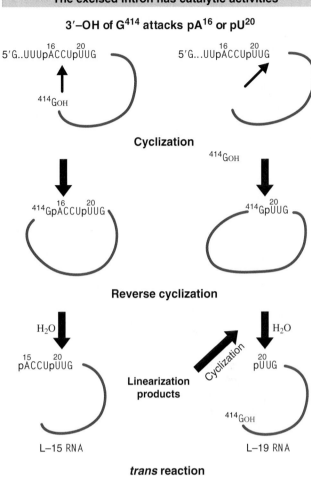

The excised intron has catalytic activities

3′–OH of G^{414} attacks pA^{16} or pU^{20}

Figure 27.3 The excised intron can form circles by using either of two internal sites for reaction with the 5′ end, and can reopen the circles by reaction with water or oligonucleotides.

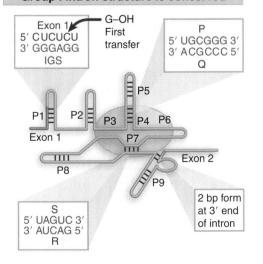

Group I intron structure is conserved

Exon 1	G–OH
5′ CUCUCU	First
3′ GGGAGG	transfer
IGS	

| P |
| 5′ UGCGGG 3′ |
| 3′ ACGCCC 5′ |
| Q |

| S |
| 5′ UAGUC 3′ |
| 3′ AUCAG 5′ |
| R |

| 2 bp form at 3′ end of intron |

Figure 27.4 Group I introns have a common secondary structure that is formed by nine base-paired regions. The sequences of regions P4 and P7 are conserved, and identify the individual sequence elements *P*, *Q*, *R*, and *S*. P1 is created by pairing between the end of the left exon and the IGS of the intron; a region between P7 and P9 pairs with the 3′ end of the intron.

RNA catalysis uses two sites

Catalytic RNA has a guanosine-binding site and substrate-binding site

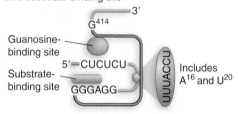

First transfer—G–OH occupies G-binding site; 5′ exon occupies substrate-binding site

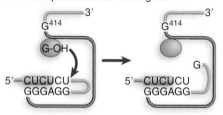

Second transfer—G^{414} is in G-binding site; 5′ exon is in substrate-binding site

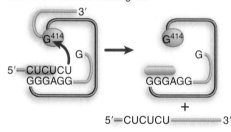

Third transfer—G^{414} is in G-binding site; 5′ end of intron is in substrate-binding site

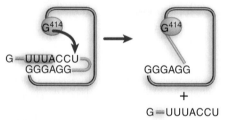

Figure 27.5 Excision of the group I intron in *Tetrahymena* rRNA occurs by successive reactions between the occupants of the guanosine-binding site and substrate-binding site. The left exon is red, and the right exon is purple.

27.4 Ribozymes Have Various Catalytic Activities

Key Term

- A **riboswitch** is a catalytic RNA whose activity responds to a small ligand.

Key Concepts

- By changing the substrate-binding site of a group I intron, it is possible to introduce alternative sequences that interact with the reactive G.
- The reactions follow classical enzyme kinetics with a low catalytic rate.
- Reactions using 2′—OH bonds could have been the basis for evolving the original catalytic activities in RNA.

The catalytic activity of group I introns was discovered by virtue of their ability to autosplice, but they are able to undertake other catalytic reactions *in vitro*. All of these reactions are based on transesterifications. We analyze these reactions in terms of their relationship to the splicing reaction itself.

The catalytic activity of a group I intron is conferred by its ability to generate a particular secondary and tertiary structure that creates active sites, equivalent to the active sites of a conventional (proteinaceous) enzyme. **Figure 27.5** illustrates the splicing reaction in terms of these sites (this is the same series of reactions shown previously in Figure 27.2).

The substrate-binding site is formed from the P1 helix, in which the 3′ end of the first intron base pairs with the IGS in an intermolecular reaction. A guanosine-binding site is formed by sequences in P7. This site may be occupied either by a free guanosine nucleotide or by the G residue in position 414. In the first transfer reaction, it is used by free guanosine nucleotide, but it is subsequently occupied by G^{414}. The second transfer releases the joined exons. The third transfer creates the circular intron.

Binding to the substrate involves a change of conformation; before substrate binding, the 5′ end of the IGS is close to P2 and P8, but after binding, when it forms the P1 helix, it is close to conserved bases that lie between P4 and P5. The reaction is visualized by contacts that are detected in the secondary structure in **Figure 27.6**. In the tertiary structure, the

The structure changes on substrate binding

Contacts found before substrate binding

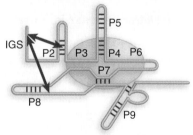

Contact found after substrate binding

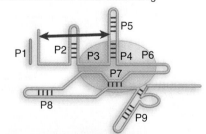

Figure 27.6 The position of the IGS in the tertiary structure changes when P1 is formed by substrate binding.

two sites alternately contacted by P1 are 37 Å apart, which implies a substantial movement in the position of P1.

The L-19 RNA is generated by opening the circular intron (shown as the last stage of the intramolecular rearrangements in Figure 27.3). It still retains enzymatic abilities. These resemble the activities of the original splicing reaction, and we may consider ribozyme function in terms of the ability to bind an intramolecular sequence complementary to the IGS in the substrate-binding site, while binding either the terminal G^{414} or a free G-nucleotide in the G-binding site.

Figure 27.7 illustrates the mechanism by which the oligonucleotide C_5 is extended to generate a C_6 chain. The C_5 oligonucleotide binds in the substrate-binding site, while G^{414} occupies the G-binding site. By transesterification reactions, a C is transferred from C_5 to the 3′–terminal G, and then back to a new C_5 molecule. Further transfer reactions lead to the accumulation of longer cytosine oligonucleotides. The reaction is a true catalysis, because the L-19 RNA remains unchanged, and is available to catalyze multiple cycles. The ribozyme is behaving as a nucleotidyl transferase.

The reactions catalyzed by RNA can be characterized in the same way as classical enzymatic reactions in terms of Michaelis-Menten kinetics. **Figure 27.8** analyzes the reactions catalyzed by RNA. The K_M values for RNA-catalyzed reactions are low, and therefore imply that the RNA can bind its substrate with high specificity. The turnover numbers are low, which reflects a low catalytic rate. In effect, the RNA molecules behave in the same general manner as traditionally defined for enzymes, although they are relatively slow compared to protein catalysts (where a typical range of turnover numbers is 10^3–10^6).

How does RNA provide a catalytic center? Its ability seems reasonable if we think of an active center as a surface that exposes a series of active groups in a fixed relationship. In a protein, the active groups are provided by the side chains of the amino acids, which have appreciable variety, including positive and negative ionic groups and hydrophobic groups. In an RNA, the available moieties are more restricted, consisting primarily of the exposed groups of bases. Short regions are held in a particular structure by the secondary/tertiary conformation of the molecule, providing a surface of active groups able to maintain an environment in which bonds can be broken and made in another molecule. It seems inevitable that the interaction between the RNA catalyst and the RNA substrate will rely on base pairing to create the environment. Divalent cations (typically Mg^{2+}) play an important role in structure, typically being present at the active site where they coordinate the positions of the various groups. They play a direct role in the endonucleolytic activity of virusoid ribozymes (see *27.8 Viroids Have Catalytic Activity*).

A powerful extension of the activities of ribozymes has been made with the discovery that they can be regulated by ligands (see *13.7 Small*

RNA catalysis is enzymatic			
Enzyme	Substrate	K_M (mM)	Turnover (/min)
19 base virusoid	24 base RNA	0.0006	0.5
L-19 intron	CCCCCC	0.04	1.7
RNAase P RNA	pre-tRNA	0.00003	0.4
RNAase P complete	pre-tRNA	0.00003	29.0
RNAase T1	GpA	0.05	5,700.0
β-galactosidase	Lactose	4.0	12,500.0

A reactive G–OH catalyzes successive transfers

C_5 pairs with IGS site near 5′ end of RNA

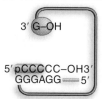

G–OH attacks CpC bond

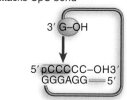

C is transferred to 3′–G; C_4 is released

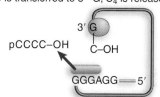

Another C_5 binds; transfer reaction is reversed

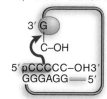

C_6 is released, regenerating L–19 RNA

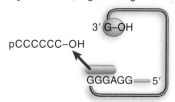

Process is repeated

Figure 27.7 The L-19 linear RNA can bind C in the substrate-binding site; the reactive G–OH 3′ end is located in the G-binding site, and catalyzes transfer reactions that convert two C_5 oligonucleotides into a C_4 and a C_6 oligonucleotide.

Figure 27.8 Reactions catalyzed by RNA have the same features as those catalyzed by proteins, although the rate is slower. The K_M gives the concentration of substrate required for half-maximum velocity; this is an inverse measure of the affinity of the enzyme for substrate. The turnover number gives the number of substrate molecules transformed in unit time by a single catalytic site.

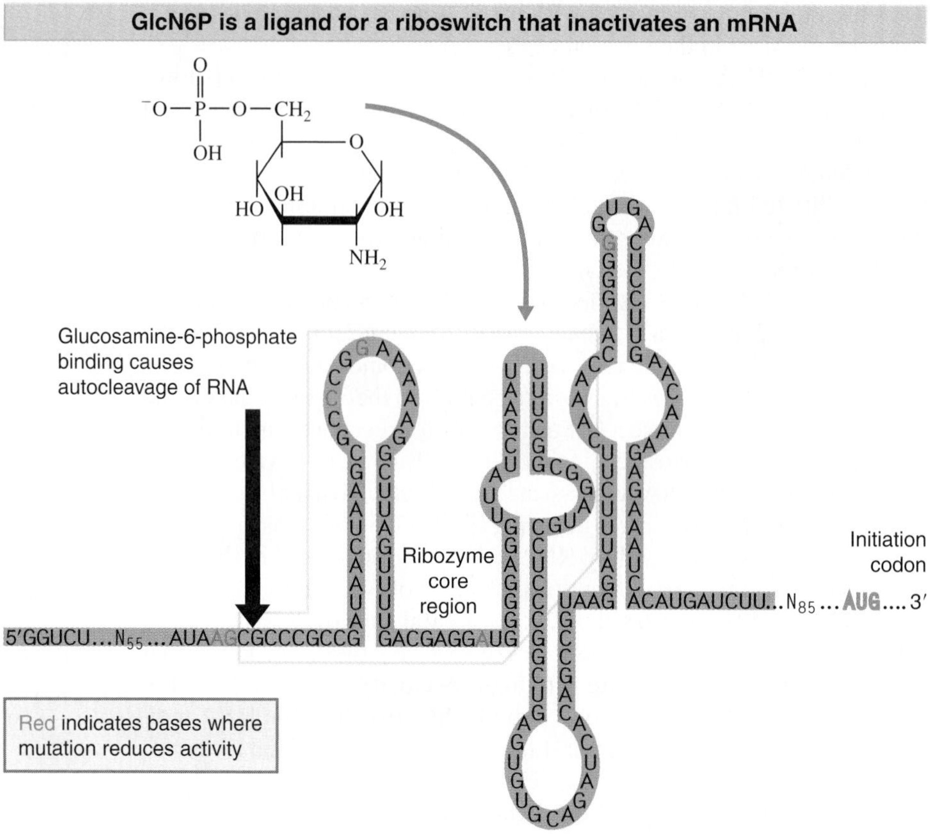

Figure 27.9 A ribozyme is contained within the 5′ untranslated region of the mRNA coding for the enzyme that produces glucosamine-6-phosphate. When Glc-6-P binds to the ribozyme, it cleaves off the 5′ end of the mRNA, inactivating it, and preventing further production of the enzyme.

GlcN6P is a ligand for a riboswitch that inactivates an mRNA

Glucosamine-6-phosphate binding causes autocleavage of RNA

Ribozyme core region

Initiation codon

5′GGUCU...N₅₅...AUAAGCGCCCGCCG GACGAGGAUG UAAG ACAUGAUCUU...N₈₅...AUG....3′

Red indicates bases where mutation reduces activity

RNA Molecules Can Regulate Translation). **Figure 27.9** summarizes the regulation of a **riboswitch**. The small metabolite GlcN6P binds to a ribozyme and activates its ability to cleave the RNA in an intramolecular reaction. The purpose of the system is to regulate production of GlcN6P; the ribozyme is located in the 5′ untranslated region of the mRNA that codes for the enzyme active in producing GlcN6P, and the cleavage prevents translation.

27.5 Some Group I Introns Code for Endonucleases that Sponsor Mobility

Key Term

- **Intron homing** describes the ability of certain introns to insert themselves into a target DNA. The reaction is specific for a single target sequence.

Key Concepts

- Mobile introns are able to insert themselves into new sites.
- Mobile group I introns code for an endonuclease that makes a double-strand break at a target site.
- The intron transposes into the site of the double-strand break by a DNA-mediated replicative mechanism.

Certain introns of both the group I and group II classes contain open reading frames that are translated into proteins. Expression of the proteins allows the intron (either in its original DNA form or as a DNA copy of the

RNA) to be *mobile*: it is able to insert itself into a new genomic site. Introns of both groups I and II are extremely widespread, being found in both prokaryotes and eukaryotes. Group I introns migrate by DNA-mediated mechanisms, whereas group II introns migrate by RNA-mediated mechanisms.

Intron mobility was first detected by crosses in which the alleles for the relevant gene differ with regard to their possession of the intron. Polymorphisms for the presence or absence of introns are common in fungal mitochondria. This is consistent with the view that these introns originated by insertion into the gene. The first example resulted from an analysis of recombination in crosses involving the large rRNA gene of the yeast mitochondrion.

This gene has a group I intron that contains a coding sequence. The intron is present in some strains of yeast (called ω^+) but absent in others (ω^-). Genetic crosses between ω^+ and ω^- are *polar:* the progeny are usually ω^+.

If we think of the ω^+ strain as a donor and the ω^- strain as a recipient, we form the view that in $\omega^+ \times \omega^-$ crosses a new copy of the intron is generated in the ω^- genome. As a result, the progeny are all ω^+.

Mutations can occur in either parent to abolish the polarity. Mutants show normal segregation, with equal numbers of ω^+ and ω^- progeny. The mutations indicate the nature of the process. Mutations in the ω^- strain occur close to the site where the intron would be inserted. Mutations in the ω^+ strain lie in the reading frame of the intron and prevent production of the protein. This suggests the model of **Figure 27.10**, in which the protein coded by the intron in an ω^+ strain recognizes the site where the intron should be inserted in an ω^- strain and causes it to be preferentially inherited.

What is the action of the protein? The product of the ω intron is an endonuclease *that recognizes the ω^- gene as a target for a double-strand break.* The endonuclease recognizes an 18 bp target sequence that contains the site where the intron is inserted. The target sequence is cleaved on each strand of DNA 2 bases to the 3′ side of the insertion site. So the cleavage sites are 4 bp apart, and generate overhanging single strands.

This type of cleavage is related to the cleavage characteristic of transposons when they migrate to new sites (see *Chapter 21 Transposons*). The double-strand break probably initiates a gene conversion process in which the sequence of the ω^+ gene is copied to replace the sequence of the ω^- gene. The reaction involves transposition by a duplicative mechanism, and occurs solely at the level of DNA. Insertion of the intron interrupts the sequence recognized by the endonuclease, thus ensuring stability.

Many group I introns code for endonucleases that make them mobile. Similar introns often carry quite different endonucleases. The dissociation between the intron sequence and the endonuclease sequence is emphasized by the fact that the same endonuclease sequences are found in inteins (sequences that code for self-splicing proteins; see *27.11 Protein Splicing Is Autocatalytic*).

The variation in the endonucleases means that there is no homology between the sequences of their target sites. The target

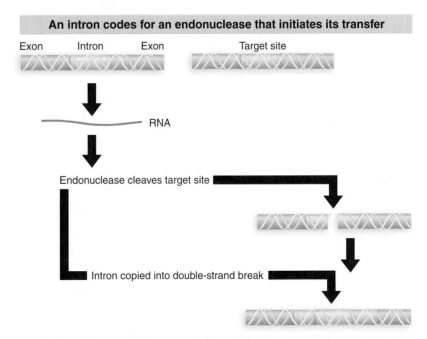

Figure 27.10 An intron codes for an endonuclease that makes a double-strand break in DNA. The sequence of the intron is duplicated and then inserted at the break.

sites are among the longest and therefore the most specific known for any endonucleases (with a range of 14–40 bp). The specificity ensures that the intron perpetuates itself only by insertion into a single target site and not elsewhere in the genome. This is called **intron homing**.

Introns carrying sequences that code for endonucleases are found in a variety of bacteria and lower eukaryotes. These results strengthen the view that introns carrying coding sequences originated as independent elements.

27.6 Some Group II Introns Code for Reverse Transcriptases

Key Concept

- Some group II introns code for a reverse transcriptase that generates a DNA copy of the RNA sequence that transposes by a retroposon-like mechanism.

Most of the open reading frames contained in group II introns have regions that are related to reverse transcriptases. Introns of this type are found in organelles of lower eukaryotes and also in some bacteria. The reverse transcriptase activity is specific for the intron, and is involved in homing. The reverse transcriptase generates a DNA copy of the intron from the pre-mRNA, and thus allows the intron to become mobile by a mechanism resembling that of retroviruses (see *22.2 The Retrovirus Life Cycle Involves Transposition-Like Events*). The type of retrotransposition involved in this case resembles that of a group of retroposons that lack LTRs, and which generate the 3′—OH needed for priming by making a nick in the target (see Figure 22.19). The best-characterized mobile group II introns code for a single protein in a region of the intron beyond its catalytic core. The typical protein contains an N-terminal reverse transcriptase activity, a central domain associated with maturase activity (see next section), and a C-terminal endonuclease domain. The endonuclease initiates the transposition reaction, and thus plays the same role in homing as its counterpart in a group I intron. The reverse transcriptase generates a DNA copy of the intron that is inserted at the homing site. The endonuclease also cleaves target sites that resemble, but are not identical to, the homing site, at much lower frequency, leading to insertion of the intron at new locations.

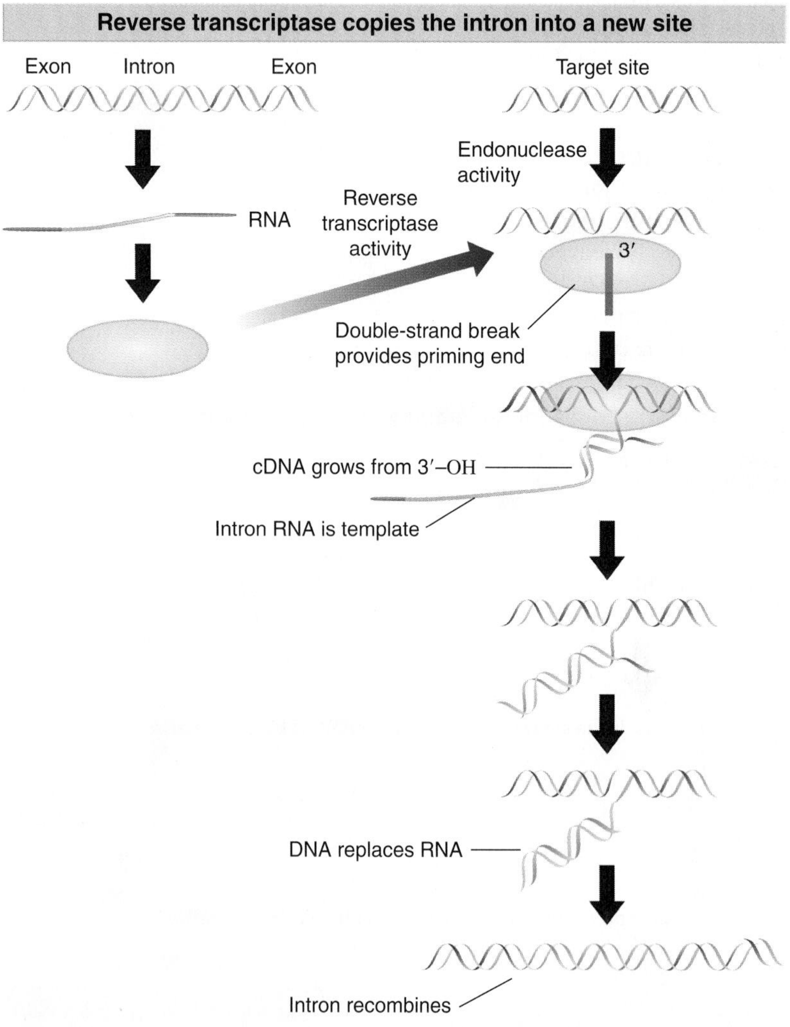

Figure 27.11 illustrates the transposition reaction for a typical group II intron. The endonuclease makes a double-strand break at the target site. A 3′ end is generated at the site of the break, and provides a primer for the reverse transcriptase. The intron RNA provides the template for the synthesis of cDNA.

Reverse transcriptase copies the intron into a new site

Exon Intron Exon   Target site

RNA

Reverse transcriptase activity

Endonuclease activity

3′

Double-strand break provides priming end

cDNA grows from 3′–OH

Intron RNA is template

DNA replaces RNA

Intron recombines

Figure 27.11 Reverse transcriptase coded by an intron allows a copy of the RNA to be inserted at a target site generated by a double-strand break.

Because the RNA includes exon sequences on either side of the intron, the cDNA product is longer than the region of the intron itself, so that it can span the double-strand break, allowing the cDNA to repair the break. The result is the insertion of the intron.

27.7 Some Autosplicing Introns Require Maturases

Key Concept

- Autosplicing introns may require maturase activities encoded within the intron to assist folding into the active catalytic structure.

Both types of autosplicing intron may code for proteins involved in their perpetuation and homing, such as endonucleases in group I and reverse transcriptases in group II. In addition, both types of intron may code for **maturase** activities that are required to assist the splicing reaction.

The maturase activity is part of the single open reading frame coded by the intron. In the example of introns that code for homing endonucleases, the single protein product has both endonuclease and maturase activity. The activities are provided by different active sites in the protein, each located in a separate domain. The endonuclease site binds to DNA, but the maturase site binds to the intron RNA. **Figure 27.12** shows the structure of one such protein bound to DNA. A characteristic feature of the endonuclease is the presence of parallel α helices, containing hallmark LAGLIDADG sequences, leading to the two catalytic amino acids. The maturase activity is located some distance away on the surface of the protein.

Introns that code for maturases may be unable to splice themselves effectively in the absence of the protein activity. The maturase is in effect a splicing factor that is required specifically for splicing of the sequence that codes for it. It functions to assist the folding of the catalytic core to form an active site.

Some group II introns that do not code for maturase activities may use comparable proteins that are coded by sequences in the host genome. This suggests a possible route for the evolution of general splicing factors. The factor may have originated as a maturase that specifically assisted the splicing of a particular intron. The coding sequence became isolated from the intron in the host genome, and then it evolved to function with a wider range of substrates than the original intron sequence. The catalytic core of the intron could have evolved into an snRNA.

27.8 Viroids Have Catalytic Activity

Key Concepts

- Viroids and virusoids form a hammerhead structure that has a self-cleaving activity.
- Similar structures can be generated by pairing a substrate strand with an enzyme strand.
- When an enzyme strand is introduced into a cell, it can pair with a substrate strand target that is then cleaved.

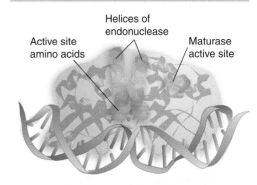

Endonuclease and maturase active sites are separate

Figure 27.12 A homing intron codes for an endonuclease of the LAGLIDADG family that also has maturase activity. The LAGLIDADG sequences are part of the two α helices that terminate in the catalytic amino acids close to the DNA duplex. The maturase active site is identified by an arginine residue elsewhere on the surface of the protein.

Key Terms

- A **viroid** is a small infectious nucleic acid that does not have a protein coat.
- A **virusoid (satellite RNA)** is a small infectious nucleic acid that is encapsidated by a plant virus together with its own genome.

Another example of the ability of RNA to function as an endonuclease is provided by some small plant RNAs (~350 bases) that undertake a self-cleavage reaction. As with the case of the *Tetrahymena* group I intron, however, it is possible to engineer constructs that can function on external substrates.

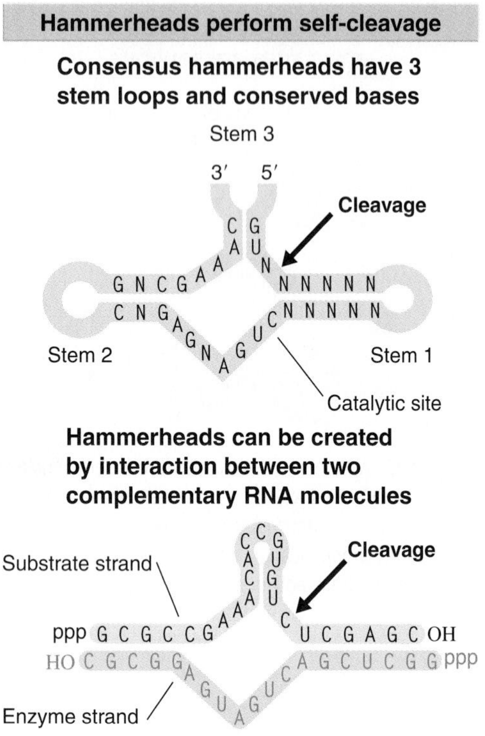

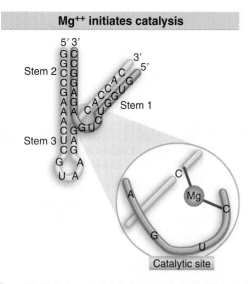

Hammerheads perform self-cleavage

Consensus hammerheads have 3 stem loops and conserved bases

Hammerheads can be created by interaction between two complementary RNA molecules

Figure 27.13 Self-cleavage sites of viroids and virusoids have a consensus sequence and form a hammerhead secondary structure by intramolecular pairing. Hammerheads can also be generated by pairing between a substrate strand and an "enzyme" strand.

Mg⁺⁺ initiates catalysis

Figure 27.14 A hammerhead ribozyme forms a V-shaped tertiary structure in which stem 2 is stacked upon stem 3. The catalytic center lies between stem 2/3 and stem 1. It contains a magnesium ion that initiates the hydrolytic reaction.

These small plant RNAs fall into two general groups: viroids and virusoids. The **viroids** are infectious RNA molecules that function independently, without encapsidation by any protein coat. The **virusoids** are similar in organization, but are encapsidated by plant viruses, being packaged together with a viral genome. The virusoids cannot replicate independently, but require assistance from the virus. The virusoids are sometimes called satellite RNAs.

Viroids and virusoids both replicate via rolling circles (see Figure 16.6). The strand of RNA that is packaged into the virus is called the plus strand. The complementary strand, generated during replication of the RNA, is called the minus strand. Multimers of both plus and minus strands are found. Both types of monomer are generated by cleaving the tail of a rolling circle; circular plus-strand monomers are generated by ligating the ends of the linear monomer.

Both plus and minus strands of viroids and virusoids undergo self-cleavage *in vitro*. The cleavage reaction is promoted by divalent metal cations; it generates 5′—OH and 2′–3′–cyclic phosphodiester termini. Some of the RNAs cleave *in vitro* under physiological conditions. Others do so only after a cycle of heating and cooling; this suggests that the isolated RNA has an inappropriate conformation, but can generate an active conformation when it is denatured and renatured.

The viroids and virusoids that undergo self-cleavage form a "hammerhead" secondary structure at the cleavage site, as drawn in the upper part of **Figure 27.13**. The sequence of this structure is sufficient for cleavage. When the surrounding sequences are deleted, the need for a heating–cooling cycle is obviated, and the small RNA self-cleaves spontaneously. This suggests that the sequences beyond the hammerhead usually interfere with its formation.

The active site is a sequence of only 58 nucleotides. The hammerhead contains three stem-loop regions whose position and size are constant, and 13 conserved nucleotides, mostly in the regions connecting the center of the structure. The conserved bases and duplex stems generate an RNA with the intrinsic ability to cleave.

An active hammerhead can also be generated by pairing an RNA representing one side of the structure with an RNA representing the other side. The lower part of Figure 27.13 shows an example of a hammerhead generated by hybridizing a 19 base molecule with a 24 base molecule. The hybrid mimics the hammerhead structure, with the omission of loops I and III. When the 19 base RNA is added to the 24 base RNA, cleavage occurs at the appropriate position in the hammerhead.

We may regard the top (24 base) strand of this hybrid as the "substrate," and the bottom (19 base) strand as the "enzyme." When the 19 base RNA is mixed with an excess of the 24 base RNA, multiple copies of the 24 base RNA are cleaved. This suggests that there is a cycle of 19 base–24 base pairing, cleavage, dissociation of the cleaved fragments from the 19 base RNA, and pairing of the 19 base RNA with a new 24 base substrate. The 19 base RNA is therefore a ribozyme with endonuclease activity. The parameters of the reaction are similar to those of other RNA-catalyzed reactions.

The crystal structure of a hammerhead shows that it forms a compact V-shape, in which the catalytic center lies in a turn, as indicated diagrammatically in **Figure 27.14**. An Mg²⁺ ion located in the catalytic site plays a crucial role in the reaction. It is positioned by the target cytidine and by the cytidine at the base of stem 1; it may also be connected to the adjacent uridine. It extracts a proton from the 2′—OH of the target cytidine, and then directly attacks the labile phosphodiester bond. Mutations in the hammerhead sequence that affect the transition state of the cleavage reaction occur in both the active site and other locations, suggesting that there may be a substantial rearrangement of structure prior to cleavage.

It is possible to design many enzyme-substrate combinations that can form hammerhead structures, and these have been used to demonstrate that introduction of the appropriate RNA molecules into a cell can allow the enzymatic reaction to occur *in vivo*. A ribozyme designed in this way essentially provides a highly specific restriction-like activity directed against an RNA target. By placing the ribozyme under control of a regulated promoter, it can be used in the same way as, for example, antisense RNAs, to specifically turn off expression of a target gene under defined circumstances.

27.9 RNA Editing Occurs at Individual Bases

Key Concept

- Apolipoprotein-B and glutamate receptors have site-specific deaminations catalyzed by cytidine and adenosine deaminases that change the coding sequence.

Key Term

- **RNA editing** describes a change of sequence at the level of RNA following transcription.

A prime axiom of molecular biology is that the sequence of an mRNA can only represent what is coded in the DNA. The central dogma envisaged a linear relationship in which a continuous sequence of DNA is transcribed into a sequence of mRNA that is in turn directly translated into protein. The occurrence of interrupted genes and the removal of introns by RNA splicing introduce an additional step into the process of gene expression: the coding sequences (exons) in DNA must be reconnected in RNA. But the process remains one of information transfer, in which the actual coding sequence in DNA remains unchanged.

Changes in the information coded by DNA occur in some exceptional circumstances, most notably in the generation of new sequences coding for immunoglobulins in mammals and birds. These changes occur specifically in the somatic cells (B lymphocytes) in which immunoglobulins are synthesized (see *Chapter 23 Recombination in the Immune System*). New information is generated in the DNA of an individual during the process of reconstructing an immunoglobulin gene, and information coded in the DNA is changed by somatic mutation. The information in DNA continues to be faithfully transcribed into RNA.

RNA editing is a process in which *information changes at the level of mRNA*. It is revealed by situations in which the coding sequence in an RNA differs from the sequence of DNA from which it was transcribed. RNA editing occurs in two different situations, with different causes. In mammalian cells there are cases in which a substitution occurs in an individual base in mRNA, causing a change in the sequence of the protein that is coded. In trypanosome mitochondria, more widespread changes occur in transcripts of several genes, when bases are systematically added or deleted.

Figure 27.15 summarizes the sequences of the apolipoprotein-B gene and mRNA in mammalian intestine and liver. The genome contains a single interrupted gene whose sequence is identical in all tissues, with a coding region of 4563 codons. This gene is transcribed into an mRNA that is translated into a protein of 512 kD representing the full coding sequence in the liver.

A shorter form of the protein, ~250 kD, is synthesized in intestine. This protein consists of the N-terminal half of the full-length protein. It is translated from an mRNA whose sequence is identical with that of liver except for a change from C to U at codon 2153. This substitution changes the codon CAA for glutamine into the ochre codon UAA for termination.

What is responsible for this substitution? No alternative gene or exon is available in the genome to code for the new sequence, and no change in

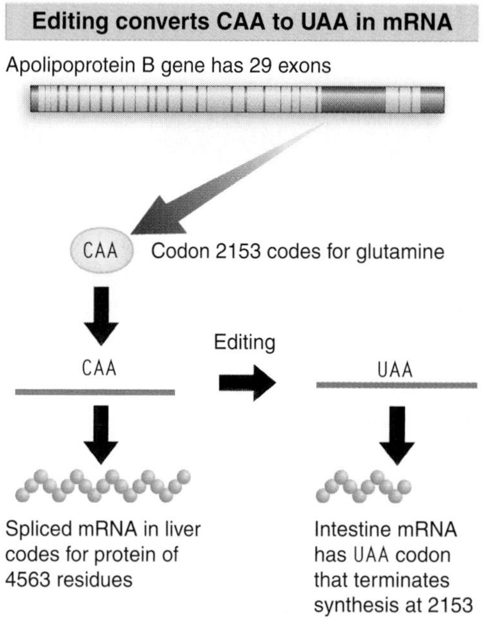

Editing converts CAA to UAA in mRNA

Apolipoprotein B gene has 29 exons

CAA — Codon 2153 codes for glutamine

Editing

CAA → UAA

Spliced mRNA in liver codes for protein of 4563 residues

Intestine mRNA has UAA codon that terminates synthesis at 2153

Figure 27.15 The sequence of the apo-B gene is the same in intestine and liver, but the sequence of the mRNA is modified by a base change that creates a termination codon in intestine.

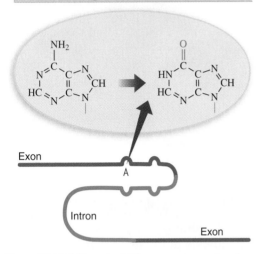

The editing enzyme is a deaminase

Figure 27.16 Editing of mRNA occurs when a deaminase acts on an adenine in an imperfectly paired RNA duplex region.

the pattern of splicing can be discovered. We are forced to conclude that a change has been made directly in the sequence of the transcript.

The editing event in apo-B causes C2153 to be changed to U. Similar events in rat brain glutamate receptors change an A to I (inosine). These events are deaminations in which the amino group on the nucleotide ring is removed. Such events are catalyzed by enzymes called cytidine and adenosine deaminases, respectively. This type of editing appears to occur largely in the nervous system. There are 16 (potential) targets for cytosine deaminase in *D. melanogaster*, and all are genes involved in neurotransmission. In many cases, the editing event changes an amino acid at a functionally important position in the protein.

What controls the specificity of an editing reaction? Enzymes that undertake deamination as such often have broad specificity—for example, the best-characterized adenosine deaminase acts on any A residue in a duplex RNA region. Editing enzymes are related to the general deaminases, but have other regions or additional subunits that control their specificity. In the case of apoB editing, the catalytic subunit of an editing complex is related to bacterial cytidine deaminase, but has an additional RNA-binding region that helps to recognize the specific target site for editing. A special adenosine deaminase enzyme recognizes the target sites in the glutamate receptor RNA, and similar events occur in a serotonin receptor RNA.

The complex may recognize a particular region of secondary structure in a manner analogous to tRNA-modifying enzymes or could directly recognize a nucleotide sequence. The development of an *in vitro* system for the apo-B editing event suggests that a relatively small sequence (~26 bases) surrounding the editing site provides a sufficient target. **Figure 27.16** shows that in GluR-B RNA, a base-paired region that is necessary for recognition of the target site is formed between the edited region in the exon and a complementary sequence in the downstream intron. A pattern of mispairing within the duplex region is necessary for specific recognition. So different editing systems may have different types of requirement for sequence specificity in their substrates.

27.10 RNA Editing Can Be Directed by Guide RNAs

Key Term

• A **guide RNA** is a small RNA whose sequence is complementary to the sequence of an RNA that has been edited. It is used as a template for changing the sequence of the pre-edited RNA by inserting or deleting nucleotides.

Key Concepts

• Extensive RNA editing in trypanosome mitochondria occurs by insertions or deletions of uridine.

• The substrate RNA base pairs with a guide RNA on both sides of the region to be edited.

• The guide RNA provides the template for addition (or less often deletion) of uridines.

• Editing is catalyzed by a complex of endonuclease, terminal uridyltransferase activity, and RNA ligase.

Another type of editing is revealed by dramatic changes in sequence in the products of several genes of trypanosome mitochondria. In the first case to be discovered, the sequence of the cytochrome oxidase subunit II protein has a frameshift relative to the sequence of the *coxII* gene. The sequences of the gene and protein given in **Figure 27.17** are conserved in several trypanosome species. How does this gene function?

The *coxII* mRNA has an insert of an additional four nucleotides (all uridines) around the site of the frameshift. The insertion changes the reading frame; it inserts an extra amino acid and changes the amino acids downstream. No second gene with this sequence can be discovered, and we are forced to conclude that the extra bases are inserted during or after transcription. A similar discrepancy between mRNA and genomic sequences is found in genes of the SV5 and measles paramyxoviruses, in these cases from adding G residues in the mRNA.

Similar editing of RNA sequences occurs for other genes, and includes deletions as well as additions of uridine. The extraordinary case of the *coxIII* gene of *T. brucei* is summarized in **Figure 27.18**.

More than half of the residues in the mRNA consist of uridines that are not coded in the gene. Comparison between the genomic DNA and the mRNA shows that no stretch longer than seven nucleotides is represented in the mRNA without alteration; and runs of uridine up to seven bases long are inserted.

What provides the information for the specific insertion of uridines? A **guide RNA** contains a sequence that is complementary to the correctly edited mRNA. **Figure 27.19** shows a model for its action in the cytochrome *b* gene of *Leishmania*.

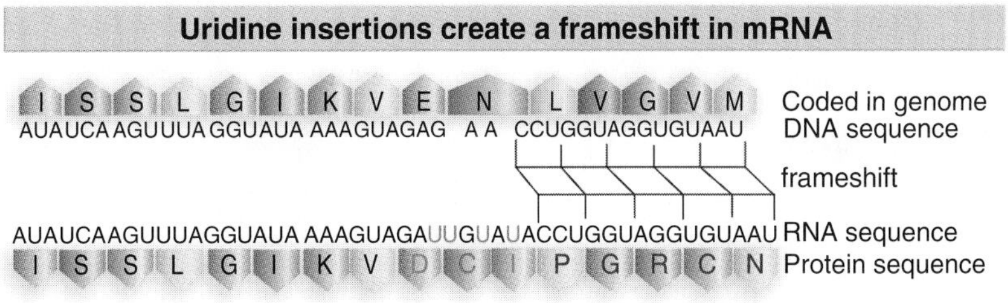

Uridine insertions create a frameshift in mRNA

I	S	S	L	G	I	K	V	E	N		L	V	G	V	M

AUAUCA AGUUUA GGUAUA AAAGUAGAG A A CCUGGUAGGUGUAAU Coded in genome DNA sequence

frameshift

AUAUCAAGUUUAGGUAUA AAAGUAGAUUGUAUACCUGGUAGGUGUAAU RNA sequence

| I | S | S | L | G | I | K | V | D | C | I | P | G | R | C | N | Protein sequence |

Figure 27.17 The mRNA for the trypanosome *coxII* gene has a frameshift relative to the DNA; the correct reading frame is created by the insertion of four uridines.

CoxIII mRNA has extensive editing by both insertion and deletion of uridine

UAUAUGUUUUGUUGUUUAUUAUGUGAUUAUGGUUUUGUUUUUUAUUGGUAUUUUUUAGAUUUAUUUAAUUUGUUGAUA

AAUACAUUUUAUUUGUUUGUUAAUUUUUUUGUUUUGUGUUUUGGUUUAGGUUUUUUUGUUGUUGUUGUUUUUGUAUUAU

Figure 27.18 Part of the mRNA sequence of *T. brucei coxIII* shows many uridines that are not coded in the DNA (shown in red) or that are removed from the RNA (shown as T).

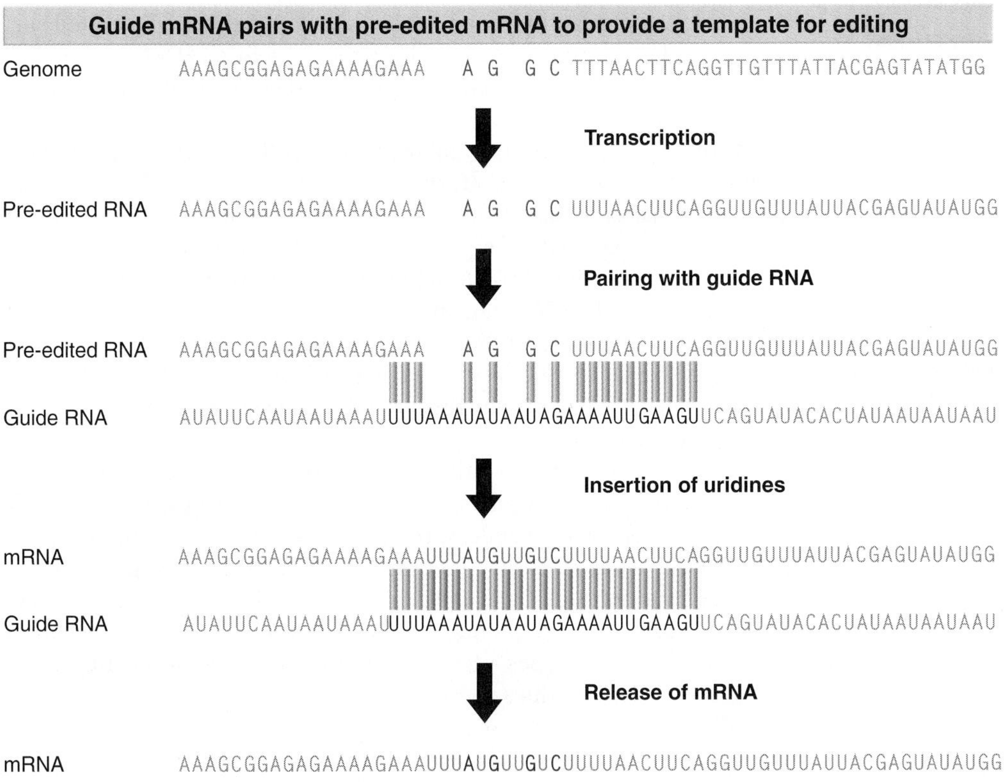

Guide mRNA pairs with pre-edited mRNA to provide a template for editing

Genome   AAAGCGGAGAGAAAAGAAA   A G   G C TTTAACTTCAGGTTGTTTATTACGAGTATATGG

Transcription

Pre-edited RNA   AAAGCGGAGAGAAAAGAAA   A G   G C UUUAACUUCAGGUUGUUUAUUACGAGUAUAUGG

Pairing with guide RNA

Pre-edited RNA   AAAGCGGAGAGAAAAGAAA   A G   G C UUUAACUUCAGGUUGUUUAUUACGAGUAUAUGG

Guide RNA   AUAUUCAAUAAUAAAUUUUAAAUAUAUAAUAGAAAAUUGAAGUUCAGUAUACACUAUAAUAAUAAU

Insertion of uridines

mRNA   AAAGCGGAGAGAAAAGAAAUUUAUGUUGUCUUUUAACUUCAGGUUGUUUAUUACGAGUAUAUGG

Guide RNA   AUAUUCAAUAAUAAAUUUUAAAUAUAUAAUAGAAAAUUGAAGUUCAGUAUACACUAUAAUAAUAAU

Release of mRNA

mRNA   AAAGCGGAGAGAAAAGAAAUUUAUGUUGUCUUUUAACUUCAGGUUGUUUAUUACGAGUAUAUGG

Figure 27.19 Pre-edited RNA base pairs with a guide RNA on both sides of the region to be edited. The guide RNA provides a template for the insertion of uridines. The mRNA produced by the insertions is complementary to the guide RNA.

Genes for pre-edited mRNAs are interspersed with guide RNA genes

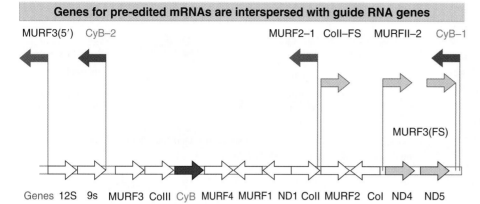

Figure 27.20 The *Leishmania* genome contains genes coding for pre-edited RNAs interspersed with units that code for the guide RNAs required to generate the correct mRNA sequences. Some genes have multiple guide RNAs. CyB is the gene for pre-edited cytocrome b, and CyB–1 and CyB–2 are genes for guide RNAs involved in its editing.

Editing occurs by cleavage and ligation

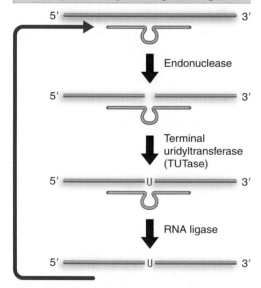

Figure 27.21 Addition or deletion of U residues occurs by cleavage of the RNA, removal or addition of the U, and ligation of the ends. The reactions are catalyzed by a complex of enzymes under the direction of guide RNA.

The sequence at the top shows the original transcript, or pre-edited RNA. Gaps show where bases will be inserted in the editing process. Eight uridines must be inserted into this region to create the valid mRNA sequence.

The guide RNA is complementary to the mRNA for a significant distance including and surrounding the edited region. Typically the complementarity is more extensive on the 3′ side of the edited region and is rather short on the 5′ side. Pairing between the guide RNA and the pre-edited RNA leaves gaps where unpaired A residues in the guide RNA do not find complements in the pre-edited RNA. The guide RNA provides a template that allows the missing U residues to be inserted at these positions. When the reaction is completed, the guide RNA separates from the mRNA, which becomes available for translation.

Specification of the final edited sequence can be quite complex; in this example, a lengthy stretch of the transcript is edited by the insertion altogether of 39 U residues, and this appears to require two guide RNAs that act at adjacent sites. The first guide RNA pairs at the 3′ most site, and the edited sequence then becomes a substrate for further editing by the next guide RNA.

The guide RNAs are encoded as independent transcription units. **Figure 27.20** shows a map of the relevant region of the *Leishmania* mitochondrial DNA. It includes the "gene" for cytochrome *b*, which codes for the pre-edited sequence, and two regions that specify guide RNAs. Genes for the major coding regions and for their guide RNAs are interspersed.

The characterization of intermediates that are partially edited suggests that the reaction proceeds along the pre-edited RNA in the 3′–5′ direction. The guide RNA determines the specificity of uridine insertions by its pairing with the pre-edited RNA.

Editing of uridines is catalyzed by a 20S enzyme complex that contains an endonuclease, a terminal uridyltransferase (TUTase), and an RNA ligase, as illustrated in **Figure 27.21**. It binds the guide RNA and uses it to pair with the pre-edited mRNA. The substrate RNA is cleaved at a site that is presumably identified by the absence of pairing with the guide RNA, a uridine is inserted or deleted to base pair with the guide RNA, and then the substrate RNA is ligated. UTP provides the source for the uridyl residue. It is added by the TUTase activity; it is not clear whether this activity, or a separate exonuclease, is responsible for deletion.

The structures of partially edited molecules suggest that the U residues are added one at a time, and not in groups. It is possible that the reaction proceeds through successive cycles in which U residues are added, tested for complementarity with the guide RNA, retained if acceptable and removed if not, so that the construction of the correct edited sequence occurs gradually. We do not know whether the same types of reactions are involved in editing reactions that add C residues.

27.11 Protein Splicing Is Autocatalytic

Key Terms

- **Protein splicing** is the autocatalytic process by which an intein is removed from a protein and the exteins on either side become connected by a standard peptide bond.
- **Extein** sequences remain in the mature protein that is produced by processing a precursor via protein splicing.
- An **intein** is the part that is removed from a protein that is processed by protein splicing.

Key Concepts

- An intein has the ability to catalyze its own removal from a protein in such a way that the flanking exteins are connected.
- Most inteins have two independent activities: protein splicing and a homing endonuclease.

Protein splicing has the same effect as RNA splicing: a sequence that is represented within the gene fails to be represented in the protein. The parts of the protein are named by analogy with RNA splicing: **exteins** are the sequences that are represented in the mature protein, and **inteins** are the sequences that are removed. The mechanism of removing the intein is completely different from RNA splicing. **Figure 27.22** shows that the gene is translated into a protein precursor that contains the intein, and then the intein is excised from the protein. About 100 examples of protein splicing are known, spread through all classes of organisms. The typical gene whose product undergoes protein splicing has a single intein.

The first intein was discovered in an archaeal DNA polymerase gene in the form of an intervening sequence in the gene that does not conform to the rules for introns. Then it was demonstrated that the purified protein can splice this sequence out of itself in an autocatalytic reaction. The reaction does not require input of energy and occurs through the series of bond rearrangements shown in **Figure 27.23**. It is a function of the intein, although its efficiency can be influenced by the exteins.

The first reaction is an attack by an –OH or –SH side chain of the first amino acid in the intein on the peptide bond that connects it to the first extein. This transfers the extein from the amino-terminal group of the intein to an N–O or N–S acyl connection. Then this bond is attacked by the –OH or –SH side chain of the first amino acid in the second extein. The result is to transfer extein–1 to the side chain of the amino-terminal acid of extein–2. Finally, the C-terminal asparagine of the intein cyclizes, and the terminal NH of extein–2 attacks the acyl bond to replace it with a conventional peptide bond. Each of these reactions can occur spontaneously at very low rates, but their occurrence in a coordinate manner rapidly enough to achieve protein splicing requires catalysis by the intein.

Inteins have characteristic features. They are found as in-frame insertions into coding sequences. They can be recognized as such because

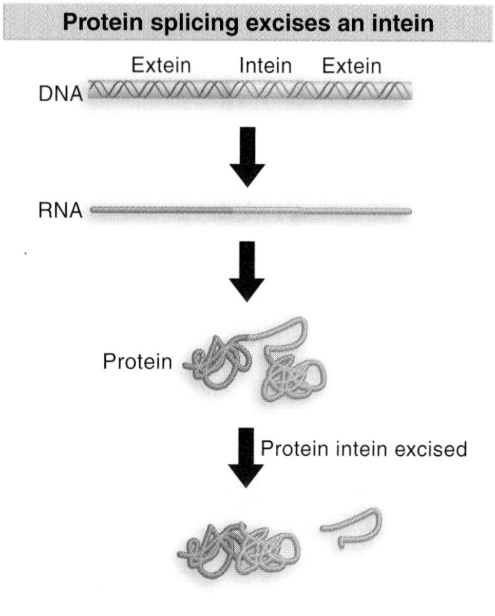

Figure 27.23 Bonds are rearranged through a series of transesterifications involving the –OH groups of serine or threonine or the –SH group of cysteine until finally the exteins are connected by a peptide bond and the intein is released with a circularized C-terminus.

Protein splicing excises an intein

Extein   Intein   Extein

DNA

RNA

Protein

Protein intein excised

Figure 27.22 In protein splicing the exteins are connected by removing the intein from the protein.

Protein splicing uses transesterifications

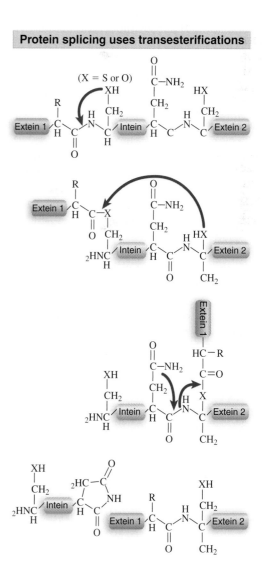

of the existence of homologous genes that lack the insertion. They have an N-terminal serine or cysteine (to provide the –XH side chain) and a C-terminal asparagine. A typical intein has a sequence of ~150 amino acids at the N-terminal end and ~50 amino acids at the C-terminal end that are involved in catalyzing the protein splicing reaction. The sequence in the center of the intein can have other functions.

An extraordinary feature of many inteins is that they have homing endonuclease activity. A homing endonuclease cleaves a target DNA to create a site into which the DNA sequence coding for the intein can be inserted (see Figure 27.10). The protein splicing and homing endonuclease activities of an intein are independent.

27.12 SUMMARY

Self-splicing is a property of two groups of introns, which are widely dispersed in lower eukaryotes, prokaryotic systems, and mitochondria. The information necessary for the reaction resides in the intron sequence (although the reaction is actually assisted by proteins *in vivo*). For both group I and group II introns, the reaction requires formation of a specific secondary/tertiary structure involving short consensus sequences. Group I intron RNA creates a structure in which the substrate sequence is held by the IGS region of the intron, and other conserved sequences generate a guanine nucleotide binding site. It occurs by a transesterification involving a guanosine residue as cofactor. No input of energy is required. The guanosine breaks the bond at the exon–intron junction and becomes linked to the intron; the hydroxyl at the free end of the exon then attacks the 3' exon–intron junction. The intron cyclizes and loses the guanosine and the terminal 15 bases. A series of related reactions can be catalyzed via attacks by the terminal G—OH residue of the intron on internal phosphodiester bonds. By providing appropriate substrates, it has been possible to engineer ribozymes that perform a variety of catalytic reactions, including nucleotidyl transferase activities.

Some group I and some group II mitochondrial introns have open reading frames. The proteins coded by group I introns are endonucleases that make double-stranded cleavages in target sites in DNA; the cleavage initiates a gene conversion process in which the sequence of the intron itself is copied into the target site. The proteins coded by group II introns include an endonuclease that initiates the transposition process, and a reverse transcriptase that enables an RNA copy of the intron to be copied into the target site. These types of introns probably originated by insertion events. The proteins coded by both groups of introns may include maturase activities that assist splicing of the intron by stabilizing the formation of the secondary/tertiary structure of the active site.

Virusoid RNAs can undertake self-cleavage at a "hammerhead" structure. Hammerhead structures can form between a substrate RNA and a ribozyme RNA, allowing cleavage to be directed at highly specific sequences. These reactions support the view that RNA can form specific active sites that have catalytic activity.

RNA editing changes the sequence of an RNA after or during its transcription. The changes are required to create a meaningful coding sequence. Substitutions of individual bases occur in mammalian systems; they take the form of deaminations in which C is converted to U, or A is converted to I. A catalytic subunit related to cytidine or adenosine deaminase functions as part of a larger complex that has specificity for a particular target sequence.

Additions and deletions (most usually of uridine) occur in trypanosome mitochondria and in paramyxoviruses. Extensive editing reactions occur in trypanosomes in which as many as half of the bases in an mRNA are derived from editing. The editing reaction uses a template consisting of a guide RNA that is complementary to the mRNA sequence. The reaction is catalyzed by an enzyme complex that includes an endonuclease, terminal uridyltransferase, and RNA ligase, using free nucleotides as the source for additions, or releasing cleaved nucleotides following deletion.

Protein splicing is an autocatalytic reaction that occurs by bond transfer reactions and input of energy is not required. The intein catalyzes its own splicing out of the flanking exteins. Many inteins have a homing endonuclease activity that is independent of the protein splicing activity.

28

Chromosomes

28.1 Introduction

Key Terms

- The **nucleoid** is the region in a prokaryotic cell that contains the genome. The DNA is bound to proteins and is not enclosed by a membrane.
- **Chromatin** describes the state of nuclear DNA and its associated proteins during the interphase (between mitoses) of the eukaryotic cell cycle.
- A **chromosome** is a discrete unit of the genome carrying many genes. Each chromosome consists of a very long molecule of duplex DNA and an approximately equal mass of proteins. It is visible as a morphological entity only during mitosis.

A common problem is presented by the packaging of all genomes. The length of the DNA as an extended molecule would vastly exceed the dimensions of the compartment that contains it. The DNA (or in the case of some viruses, the RNA) must be compressed exceedingly tightly to fit into the space available. *So in contrast with the customary picture of DNA as an extended double helix, structural deformation of DNA to bend or fold it into a more compact form is the rule rather than exception.*

The magnitude of the discrepancy between the length of the nucleic acid and the size of its compartment is evident from the examples summarized in **Figure 28.1**. For all types of viruses, the nucleic acid genome effectively fills the container (which can be rodlike or spherical).

For bacteria or for eukaryotic cell compartments, the discrepancy is hard to calculate exactly, because the DNA is contained in a compact area that occupies only part of the compartment. The genetic material is seen in the form of the **nucleoid** in bacteria and as the mass of **chromatin** in eukaryotic nuclei at interphase (between divisions).

Figure 28.1 The length of nucleic acid is much greater than the dimensions of the surrounding compartment.

DNA is highly compressed in all types of genomes				
Compartment	Shape	Dimensions	Type of nucleic acid	Length
TMV	Filament	$0.008 \times 0.3\ \mu m$	1 single-stranded RNA	$2\ \mu m = 6.4$ kb
Phage fd	Filament	$0.006 \times 0.85\ \mu m$	1 single-stranded DNA	$2\ \mu m = 6.0$ kb
Adenovirus	Icosahedron	$0.07\ \mu m$ diameter	1 double-stranded DNA	$11\ \mu m = 35.0$ kb
Phage T4	Icosahedron	$0.065 \times 0.10\ \mu m$	1 double-stranded DNA	$55\ \mu m = 170.0$ kb
E. coli	Cylinder	$1.7 \times 0.65\ \mu m$	1 double-stranded DNA	$1.3\ \mu m = 4.2 \times 10^3$ kb
Mitochondrion (human)	Oblate spheroid	$3.0 \times 0.5\ \mu m$	~10 identical double-stranded DNAs	$50\ \mu m = 16.0$ kb
Nucleus (human)	Spheroid	$6\ \mu m$ diameter	46 chromosomes of double-stranded DNA	1.8 m $= 6 \times 10^6$ kb

The condensed state of nucleic acid results from its binding to basic proteins. The positive charges of these proteins neutralize the negative charges of the nucleic acid. The natural state of the DNA therefore takes the form of a nucleoprotein complex. The density of DNA in the complex is high, reaching a concentration that is equivalent to a gel of great viscosity. We do not entirely understand the physiological implications, such as what effect this has upon the ability of proteins to find their binding sites on DNA.

The confinement of the cellular genetic material to a limited space raises important questions as to how its various activities, such as replication and transcription, are accomplished within this space. Furthermore, the organization of the material must accommodate transitions between inactive and active states. This is seen most dramatically in the transitions that are visible during the eukaryotic cell cycle. At the time of division (mitosis or meiosis), the genetic material becomes even more tightly packaged, and individual **chromosomes** become recognizable. The usual reckoning is that mitotic chromosomes are likely to be 5–10× more tightly packaged than interphase chromatin.

28.2 Viral Genomes Are Packaged into Their Coats

Key Terms

- A **capsid** is the external protein coat of a virus particle.
- The **nucleation center** of tobacco mosaic virus (TMV) is a duplex hairpin where assembly of coat protein with RNA is initiated.
- A **terminase** enzyme cleaves multimers of a viral genome and then uses hydrolysis of ATP to provide the energy to translocate the DNA into an empty viral capsid starting with the cleaved end.

Key Concepts

- The length of DNA that can be incorporated into a virus is limited by the structure of the head shell.
- Nucleic acid within the head shell is extremely condensed.
- Filamentous RNA viruses condense the RNA genome as they assemble the head shell around it.
- Spherical DNA viruses insert the DNA into a preassembled protein shell.

From the perspective of packaging the *individual* sequence, there is an important difference between a cellular genome and a virus. The cellular genome is essentially indefinite in size; the number and location of individual sequences can be changed by duplication, deletion, and rearrangement. So it requires a *generalized* method for packaging its DNA, insensitive to the total content or distribution of sequences. By contrast, two restrictions define the needs of a virus. The amount of nucleic acid to be packaged is *predetermined* by the size of the genome. And it must all fit within a coat assembled from a protein or proteins coded by the viral genes.

A virus particle is deceptively simple in its superficial appearance. The nucleic acid genome is contained within a **capsid**, a symmetrical or quasi-symmetrical structure assembled from one or only a few proteins. Capsids fall into two general types and are either filamentous (rod-like) or have icosahedral symmetry (pseudospherical). Attached to the capsid, or incorporated into it, are other structures, assembled from distinct proteins, and necessary for infection of the host cell.

The virus particle is tightly constructed. The internal volume of the capsid is rarely much greater than the volume of the nucleic acid it must hold. The difference is usually less than twofold, and often the internal volume is barely larger than the nucleic acid. There are two types of solution to the problem of how to construct a capsid that contains nucleic acid:

- The protein shell can be assembled around the nucleic acid, condensing the DNA or RNA by protein–nucleic acid interactions during the process of assembly.
- The capsid can be constructed from its components in the form of an empty shell, into which the nucleic acid must be inserted, being condensed as it enters.

The capsid is assembled around the genome for single-stranded RNA viruses. The principle of assembly is that *the position of the RNA within the capsid is determined directly by its binding to the proteins of the shell.* The best characterized example is tobacco mosaic virus (TMV). Assembly starts at a duplex hairpin that lies within the RNA sequence. From this **nucleation center**, it proceeds bidirectionally along the RNA, until reaching the ends. The unit of the capsid is a two-layer disk, each layer containing 17 identical protein subunits. The disk is a circular structure, which forms a helix as it interacts with the RNA. At the nucleation center, the RNA hairpin inserts into the central hole in the disk, and the disk changes conformation into a helical structure that surrounds the RNA. Then further disks are added, each disk pulling a new stretch of RNA into its central hole. The RNA becomes coiled in a helical array on the inside of the protein shell, as illustrated in **Figure 28.2**.

The spherical capsids of DNA viruses are assembled in a different way, as best characterized for the phages lambda and T4. In each case, an empty head shell is assembled from a small set of proteins. *Then the duplex genome is inserted into the head,* accompanied by a structural change in the capsid.

Figure 28.3 summarizes the assembly of lambda. It starts with a small head shell that contains a protein "core." This is converted to an empty head shell of more distinct shape. Then DNA packaging begins, the head shell expands in size though remaining the same shape, and finally the full head is sealed by addition of the tail.

Inserting DNA into a phage head involves two types of reaction: translocation and condensation. Both are energetically unfavorable.

Translocation is an active process in which the DNA is driven into the head by an ATP-dependent mechanism. A common mechanism is used for many viruses that replicate by a rolling circle mechanism to generate long tails that contain multimers of the viral genome. The best-characterized example is phage lambda. The genome is packaged into the empty capsid by the **terminase** enzyme. **Figure 28.4** summarizes the process.

The terminase was first recognized for its role in generating the ends of the linear phage DNA by cleaving at *cos* sites. (The name *cos* reflects the fact that it generates cohesive ends that have complementary

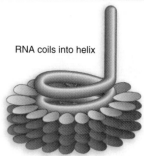

The headshell condenses around TMV RNA

RNA coils into helix

Figure 28.2 A helical path for TMV RNA is created by the stacking of protein subunits in the virion.

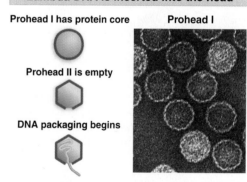

Lambda DNA is inserted into the head

Prohead I has protein core   Prohead I

Prohead II is empty

DNA packaging begins

Grizzled particle is part full with expanded head shell

Mature phage particle

Black particle has full head

Tail is attached

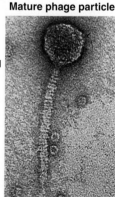

Figure 28.3 Maturation of phage lambda passes through several stages. The empty head changes shape and expands when it becomes filled with DNA. The electron micrographs show the particles at the start and end of the maturation pathway. Photographs kindly provided by A. F. Howatson.

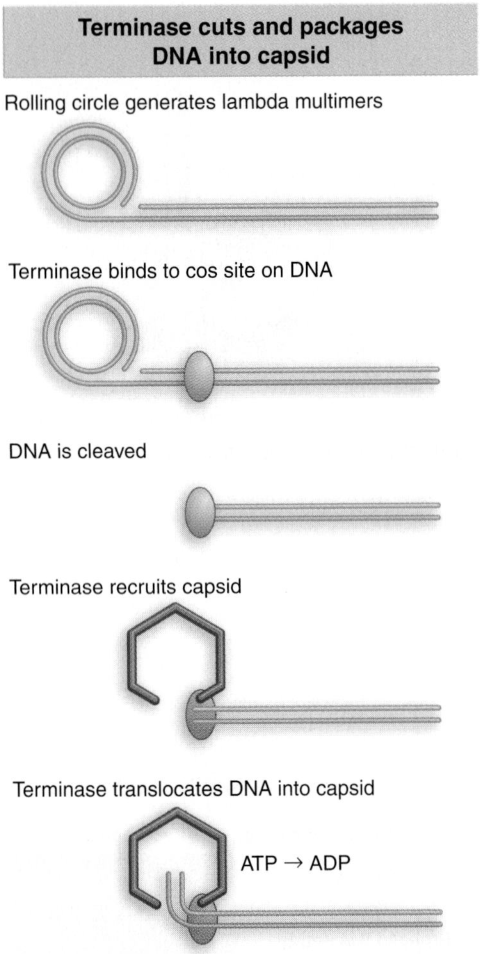

Terminase cuts and packages DNA into capsid

Rolling circle generates lambda multimers

Terminase binds to cos site on DNA

DNA is cleaved

Terminase recruits capsid

Terminase translocates DNA into capsid

ATP → ADP

Figure 28.4 Terminase protein binds to specific sites on a multimer of virus genomes generated by rolling circle replication. Terminase cuts the DNA and binds to an empty virus capsid. It uses energy from hydrolysis of ATP to insert the DNA into the capsid.

single-stranded tails.) The phage genome codes two subunits that make up the terminase. One subunit binds to a *cos* site; then it is joined by the other subunit, which cuts the DNA. The terminase assembles into a heterooligomer in a complex that also includes IHF (the integration host factor coded by the bacterial genome). It then binds to an empty capsid and uses ATP hydrolysis to power translocation along the DNA. The translocation drives the DNA into the empty capsid.

Another method of packaging uses a structural component of the phage. In the *B. subtilis* phage φ29, the motor that inserts the DNA into the phage head is the structure that connects the head to the tail. It functions as a rotary motor, where the motor action effects the linear translocation of the DNA into the phage head. The same motor is used to eject the DNA from the phage head when it infects a bacterium.

Little is known about the mechanism of condensation into an empty capsid, except that the capsid contains "internal proteins" as well as DNA. One possibility is that they provide some sort of "scaffolding" onto which the DNA condenses. (This would be a counterpart to the use of the proteins of the shell in the plant RNA viruses.)

How specific is the packaging? It cannot depend on particular sequences, because deletions, insertions, and substitutions all fail to interfere with the assembly process. The relationship between DNA and the head shell has been investigated directly by determining which regions of the DNA can be chemically crosslinked to the proteins of the capsid. The surprising answer is that all regions of the DNA are more or less equally susceptible. This probably means that when DNA is inserted into the head, it follows a general rule for condensing, but the pattern is not determined by particular sequences.

28.3 The Bacterial Genome Is a Supercoiled Nucleoid

Key Terms

- The **nucleoid** is the region in a prokaryotic cell that contains the genome. The DNA is bound to proteins and is not enclosed by a membrane.
- A chromosomal **domain** is a region within which supercoiling is independent of other domains.

Key Concepts

- The bacterial nucleoid is ~80% DNA by mass and can be unfolded by agents that act on RNA or protein.
- The proteins that are responsible for condensing the DNA have not been identified.
- The nucleoid has ~100 independently negatively supercoiled domains.
- The average density of supercoiling is ~1 turn/100 bp.

Although bacteria do not display structures with the distinct morphological features of eukaryotic chromosomes, their genomes nonetheless are organized into definite bodies. The genetic material can be seen as a fairly compact clump or series of clumps that occupies about a third of the volume of the cell. **Figure 28.5** displays a thin section through a bacterium in which this **nucleoid** is evident.

When *E. coli* cells are lysed, fibers are released in the form of loops attached to the broken envelope of the cell. As can be seen

Bacterial DNA is a compact nucleoid

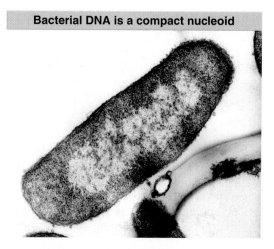

Figure 28.5 A thin section shows the bacterial nucleoid as a compact mass in the center of the cell. Photograph kindly provided by Jack Griffith, University of North Carolina School of Medicine.

Bacterial DNA is a tightly coiled thread

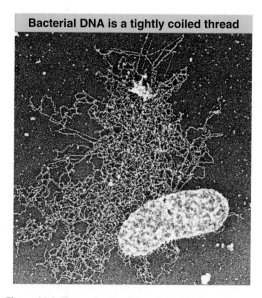

Figure 28.6 The nucleoid spills out of a lysed *E. coli* cell in the form of loops of a fiber. Photograph kindly provided by Jack Griffith, University of North Carolina School of Medicine.

from **Figure 28.6**, the DNA of these loops is not found in the extended form of a free duplex, but is compacted by association with proteins. We don't yet know exactly which proteins are bound to DNA and how they interact with it.

The nucleoid can be isolated directly in the form of a very rapidly sedimenting complex, consisting of ~80% DNA by mass. It can be unfolded by treatment with reagents that act on RNA or protein. The DNA of the bacterial nucleoid behaves as a closed duplex structure, as judged by its response to ethidium bromide. This small molecule intercalates between base pairs to generate *positive* superhelical turns in "closed" circular DNA molecules—that is, molecules in which both strands have covalent integrity. (In "open" circular molecules, which contain a nick in one strand, or with linear molecules, the DNA can rotate freely in response to the intercalation, thus relieving the tension.)

In a natural closed DNA that is *negatively* supercoiled, the intercalation of ethidium bromide first removes the negative supercoils and then introduces positive supercoils. The amount of ethidium bromide needed to achieve zero supercoiling is a measure of the original density of negative supercoils.

Some nicks occur in the compact nucleoid during its isolation; they can also be generated by limited treatment with DNAase. But this does not abolish the ability of ethidium bromide to introduce positive supercoils. This capacity of the genome to retain its response to ethidium bromide in the face of nicking means that it must have many independent chromosomal **domains**; *the supercoiling in each domain is not affected by events in the other domains.*

This autonomy suggests that the structure of the bacterial chromosome has the general organization depicted diagrammatically in **Figure 28.7**. Each domain consists of a loop of DNA, the ends of which are secured in some (unknown) way that does not allow rotational events to propagate from one domain to another. There are ~100 such domains per genome; each consists of ~40 kb (13 μm) of DNA, organized into some more compact fiber whose structure has yet to be characterized.

The existence of separate domains could permit different degrees of supercoiling to be maintained in different regions of the genome. This could be relevant in considering the different susceptibilities of particular bacterial promoters to supercoiling (see *11.10 Supercoiling Is an Important Feature of Transcription*).

Bacterial DNA has independently coiled domains

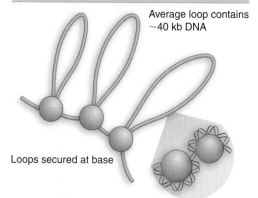

Average loop contains ~40 kb DNA

Loops secured at base

Loop consists of duplex DNA condensed by basic proteins

Figure 28.7 The bacterial genome consists of a large number of loops of duplex DNA (in the form of a fiber), each secured at the base to form an independent structural domain.

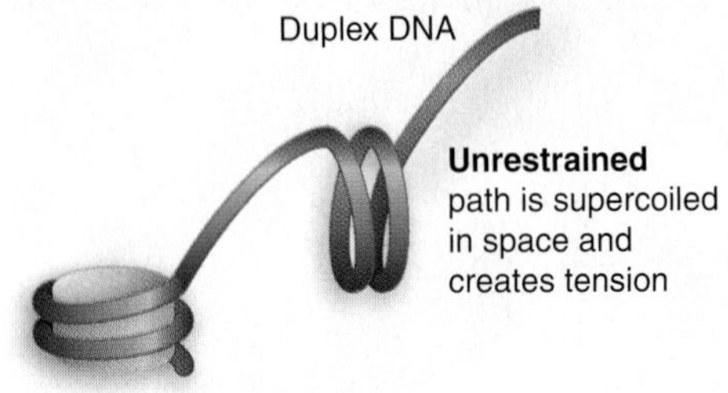

Protein binding restrains supercoils

Duplex DNA

Unrestrained
path is supercoiled
in space and
creates tension

Restrained
path is supercoiled around
protein but creates no tension

Figure 28.8 An unrestrained supercoil in the DNA path creates tension, but no tension is transmitted along DNA when a supercoil is restrained by protein binding.

Supercoiling in the genome can in principle take either of two forms, as summarized in **Figure 28.8**:

• If a supercoiled DNA is free, its path is *unrestrained*, and negative supercoils generate a state of torsional tension that is transmitted freely along the DNA within a domain. It can be relieved by unwinding the double helix, as described in *1.7 Supercoiling Affects the Structure of DNA*. The DNA is in a dynamic equilibrium between the states of tension and unwinding.

• Supercoiling can be *restrained* if proteins are bound to the DNA to hold it in a particular three-dimensional configuration. In this case, the supercoils are represented by the path the DNA follows in its fixed association with the proteins. The energy of interaction between the proteins and the supercoiled DNA stabilizes the nucleic acid, so that no tension is transmitted along the molecule.

Measurements of how supercoiling is released by making nicks in DNA, or assays of the ability of agents to bind DNA where the reaction is influenced by supercoiling, both suggest that about half of the total supercoils in the DNA path are free and can transmit tension along the DNA. The overall supercoiling density is one negative superhelical turn / 100 bp. There may be variation in the density between different regions, but we can see that the level can be sufficient to exert significant effects on DNA structure, for example, in assisting melting in particular regions such as origins or promoters.

28.4 Eukaryotic DNA Has Loops and Domains Attached to a Scaffold

Key Terms

• A chromosome **scaffold** is a proteinaceous structure in the shape of a sister chromatid pair, generated when chromosomes are depleted of histones.
• A **matrix attachment region (MAR)** is a region of DNA that attaches to the nuclear matrix. It is also known as a scaffold attachment site (SAR).

Key Concepts

• DNA of interphase chromatin is negatively supercoiled into independent domains of ~85 kb.
• Metaphase chromosomes have a protein scaffold to which the loops of supercoiled DNA are attached.
• DNA is attached to a proteinaceous matrix in interphase cells at specific sequences called MARs or SARs.
• The MARs are A·T-rich but do not have any specific consensus sequence.

Interphase chromatin is a tangled mass occupying a large part of the nuclear volume, in contrast with the highly organized and reproducible ultrastructure of mitotic chromosomes. What controls the distribution of interphase chromatin within the nucleus?

Some indirect evidence on its nature is provided by the isolation of the genome as a single, compact body. Like a bacterial nucleoid, the genome of *D. melanogaster* can be visualized as a compactly folded fiber (10 nm in diameter), consisting of DNA bound to proteins. Supercoiling measured by the response to ethidium bromide corresponds to about one negative supercoil / 200 bp. These supercoils can be removed by nicking with DNAase, although the DNA remains in the form of the 10 nm fiber. This suggests that the supercoiling is caused by the arrangement of the fiber in space, and represents the existing torsion.

Full relaxation of the supercoils requires one nick / 85 kb, identifying the average length of "closed" DNA. This region may be a loop or domain similar to those identified in the bacterial genome. Loops can be seen directly when the majority of proteins are extracted from mitotic chromosomes. The resulting complex consists of the DNA associated with ~8% of the original protein content. As seen in **Figure 28.9**, the protein-depleted chromosomes take the form of a central **scaffold** surrounded by a halo of DNA.

The metaphase scaffold consists of a dense network of fibers. Threads of DNA emanate from the scaffold, apparently as loops of average length 10–30 μm (30–90 kb). The DNA can be digested without affecting the integrity of the scaffold, which consists of a set of specific proteins. This suggests a form of organization in which loops of DNA of ~60 kb are anchored in a central proteinaceous scaffold.

Is DNA attached to the scaffold via specific sequences? DNA sites attached to proteinaceous structures in interphase nuclei are called **MAR (matrix attachment regions)**; they are sometimes also called *SAR* (scaffold attachment regions). Chromatin often appears to be attached to a matrix, and there have been many suggestions that this attachment is necessary for transcription or replication. When nuclei are depleted of proteins, the DNA extrudes as loops from a residual proteinaceous structure.

The MAR fragments are usually ~70% A · T-rich, but otherwise lack any consensus sequences. However, other interesting sequences often are in the DNA stretch containing the MAR. *cis*-acting sites that regulate transcription are common. And a recognition site for topoisomerase II is usually present in the MAR. It is therefore possible that a MAR serves more than one function, providing a site for attachment to the matrix, but also containing other sites at which topological changes in DNA are effected.

The interphase matrix and chromosome scaffold consist of different proteins, although there are some common components. Topoisomerase II is a prominent component of the chromosome scaffold, and is a constituent of the interphase matrix, suggesting that the control of topology is important in both cases.

Loops of DNA are attached to a protein scaffold

Figure 28.9 Histone-depleted chromosomes consist of a protein scaffold to which loops of DNA are anchored. Photograph kindly provided by Professor Ulrich K. Laemmli, Dept. of Biochemistry & Molecular Biology, University of Geneva, Switzerland.

28.5 Chromatin Is Divided into Euchromatin and Heterochromatin

Key Terms

- **Euchromatin** composes all of the genome in the interphase nucleus except for the heterochromatin. The euchromatin is less tightly coiled than heterochromatin, and contains the active or potentially active genes.

- **Heterochromatin** describes regions of the genome that are highly condensed, are not transcribed, and are late-replicating. Heterochromatin is divided into two types, which are called constitutive and facultative.

- The **chromocenter** is an aggregate of heterochromatin from different chromosomes.

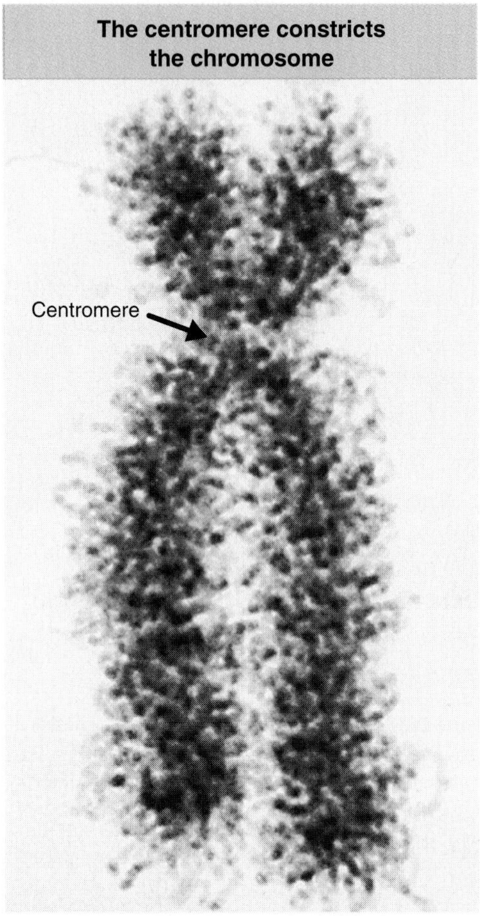

The centromere constricts the chromosome

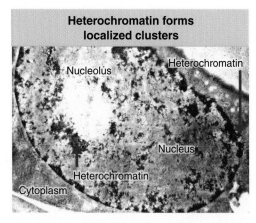

Centromere

Figure 28.10 The sister chromatids of a mitotic pair each consist of a fiber (~30 nm in diameter) compactly folded into the chromosome. Photograph kindly provided by E. J. DuPraw.

Heterochromatin forms localized clusters

Nucleolus

Heterochromatin

Nucleus

Heterochromatin

Cytoplasm

Figure 28.11 A thin section through a nucleus stained with Feulgen shows heterochromatin as compact regions clustered near the nucleolus and nuclear membrane. Photograph kindly provided by Edmund Puvion, Institut Puvion de Lille.

Key Concepts

- Individual chromosomes can be seen only during mitosis.

- During interphase, the general mass of chromatin is in the form of euchromatin, which is less tightly packed than mitotic chromosomes.

- Regions of heterochromatin remain densely packed throughout interphase.

Each chromosome contains a single, very long duplex of DNA, folded into a fiber that runs continuously throughout the chromosome. *So in accounting for interphase chromatin and mitotic chromosome structure, we have to explain the packaging of a single, exceedingly long molecule of DNA into a form in which it can be transcribed and replicated, and can become cyclically more and less compressed.*

Individual eukaryotic chromosomes come into the limelight for a brief period, during the act of cell division. Only then can each be seen as a compact unit. **Figure 28.10** is an electron micrograph of a sister chromatid pair, captured at metaphase. (The sister chromatids are daughter chromosomes produced by the previous replication event, still joined together at this stage of mitosis.) Each consists of a fiber with a diameter of ~30 nm and a nubbly appearance.

During most of the life cycle of the eukaryotic cell, however, its genetic material occupies an area of the nucleus in which individual chromosomes cannot be distinguished. The structure of the interphase chromatin does not change visibly between divisions. No disruption is evident during the period of replication, when the amount of chromatin doubles. Chromatin consists of a fiber that is similar or identical to that of the mitotic chromosomes.

Chromatin can be divided into two types of material, which can be seen in the nuclear section of **Figure 28.11**:

- In most regions, the fibers are much less densely packed than in the mitotic chromosome. This material is called **euchromatin**. It has a relatively dispersed appearance in the nucleus, and occupies most of the nuclear region in Figure 28.11.

- Some regions of chromatin are very densely packed with fibers, displaying a condition comparable to that of the chromosome at mitosis. This material is called **heterochromatin**. It is typically found at centromeres, but occurs at other locations also. It passes through the cell cycle with relatively little change in its degree of condensation. It forms a series of discrete clumps in Figure 28.11, but often the various heterochromatic regions aggregate into a densely staining **chromocenter**. (This description applies to regions that are always heterochromatic, called constitutive heterochromatin; in addition, there is another sort of heterochromatin, called facultative heterochromatin, in which regions of euchromatin are converted to a heterochromatic state.)

The same fibers run continuously between euchromatin and heterochromatin, which implies that these states represent different degrees of condensation of the genetic material. In the same way, euchromatic regions exist in different states of condensation during interphase and during mitosis. So the genetic material is organized in a manner that permits alternative states to be maintained side by side in chromatin, and allows cyclical changes to occur in the packaging of euchromatin between interphase and division. We discuss the molecular basis for these states in *Chapter 30 Chromatin Structure Is a Focus for Regulation.*

The characteristic features of heterochromatin correlate with a lack of genetic activity:

- It is permanently condensed.
- Constitutive heterochromatin often consists of multiple repeats of a few sequences of DNA that are not transcribed.
- The density of genes in constitutive heterochromatin is very much reduced compared with euchromatin, and genes that are translocated into or near it are often inactivated.
- Probably resulting from the condensed state, it replicates late in S phase and has a reduced frequency of genetic recombination.

We have some molecular markers for changes in the properties of the DNA and protein components (see *30.13 Heterochromatin Depends on Interactions with Histones*). They include reduced acetylation of histone proteins, increased methylation of one histone protein, and hypermethylation of cytidine bases in DNA. These molecular changes cause the condensation of the material, which is responsible for its inactivity.

Although active genes are contained within euchromatin, only a small minority of the sequences in euchromatin are transcribed at any time. So location in euchromatin is *necessary* for gene expression, but is not *sufficient* for it.

28.6 Chromosomes Have Banding Patterns

Key Term

- **G-bands** are generated on eukaryotic chromosomes by staining techniques and appear as a series of lateral striations. They are used for karyotyping (identifying chromosomal regions by the banding pattern).

Key Concepts

- Certain staining techniques cause the chromosomes to have the appearance of a series of striations called G-bands.
- The bands are lower in G · C content than the interbands.
- Genes are concentrated in the G · C-rich interbands.

Because of the diffuse state of chromatin, we cannot directly determine the specificity of its organization. But we can ask whether the structure of the mitotic chromosome is ordered. Do particular sequences always lie at particular sites, or is the folding of the fiber into the overall structure a more random event?

At the level of the chromosome, each member of the complement has a different and reproducible ultrastructure. When subjected to certain treatments and then stained with the chemical dye Giemsa, chromosomes generate a series of **G-bands**. **Figure 28.12** presents an example of the human set. The bands are large structures, each containing $\sim 10^7$ bp of DNA.

G-banding allows each chromosome to be identified by its characteristic banding pattern. This pattern allows translocations from one chromosome to another to be identified by comparison with the original diploid set. **Figure 28.13** shows a diagram of the bands of the human X chromosome.

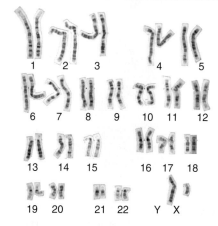

Every chromosome has a distinct G-banding pattern

Figure 28.12 G-banding generates a characteristic lateral series of bands in each member of the chromosome set. Photograph kindly provided by Dr. Lisa Shaffer, Baylor College of Medicine.

The X chromosome has many G-bands

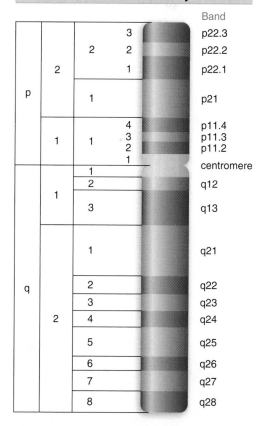

Figure 28.13 The human X chromosome can be divided into distinct regions by its banding pattern. The short arm is p and the long arm is q; each arm is divided into larger regions that are further subdivided. This map shows a low-resolution structure; at higher resolution, some bands are further subdivided into smaller bands and interbands, e.g., p21 is divided into p21.1, p21.2, and p21.3.

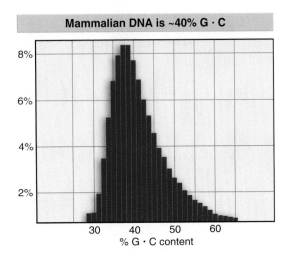

Mammalian DNA is ~40% G · C

Figure 28.14 There are large fluctuations in G · C content over short distances. Each bar shows the percent of 20 kb fragments with the given G · C content.

Bands have a lower G · C content than the interbands. **Figure 28.14** shows that there are distinct fluctuations in G · C content when the genome is divided into small tranches. The average of 41% G · C is common to mammalian genomes. There are regions as low as 30% or as high as 65%. When longer tranches are examined, there is less variation. The average length of regions with >43% G · C is 200–250 kb. This makes it clear that the band/interband structure does not represent homogeneous segments that alternate in G · C content, although the bands do contain a higher content of low G · C segments. Genes are concentrated in regions of higher G · C content, that is, in the interbands. All of this argues for some long-range sequence-dependent organization, but we do not yet know the details or how the G · C content affects chromosome structure.

28.7 Lampbrush Chromosomes Are Extended

Key Terms

- **Chromomeres** are densely staining granules visible in chromosomes under certain conditions, especially early in meiosis, when a chromosome may appear to consist of a series of chromomeres.
- **Lampbrush chromosomes** are the extremely extended meiotic bivalents of certain amphibian oocytes.

Key Concept

- Sites of gene expression on lampbrush chromosomes show loops that are extended from the chromosomal axis.

It would be extremely useful to visualize gene expression in its natural state, to see what structural changes are associated with transcription. Gene expression can be visualized directly in certain unusual situations, in which the chromosomes are found in a highly extended form that allows individual loci (or groups of loci) to be distinguished. Lateral differentiation of structure is evident in many chromosomes when they first appear for meiosis. At this stage, the chromosomes resemble a series of beads on a string. The beads are densely staining granules, properly known as **chromomeres**. However, usually there is little gene expression at meiosis, and it is not practical to use this material to identify the activities of individual genes. But an exceptional situation that allows the material to be examined is presented by **lampbrush chromosomes**, which have been best characterized in certain amphibians.

Lampbrush chromosomes are formed during an unusually extended meiosis, which can last up to several months! During this period, the chromosomes are held in a stretched-out form in which they can be visualized in the light microscope. Later during meiosis, the chromosomes revert to their usual compact size. So the extended state essentially proffers an unfolded version of the normal condition of the chromosome.

The lampbrush chromosomes are meiotic bivalents, each consisting of two pairs of sister chromatids. **Figure 28.15** shows an example in which the sister chromatid pairs have mostly separated so that they are held together only by chiasmata. Each sister chromatid pair forms a series of ellipsoidal chromomeres, ~1–2 μm in diameter, which are connected by a very fine thread. This thread contains the two sister duplexes of DNA, and runs continuously along the chromosome, through the chromomeres.

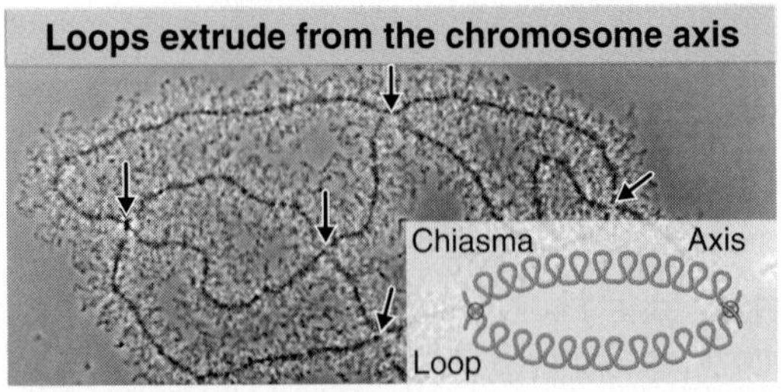

The lengths of the individual lampbrush chromosomes in the newt *Notophthalmus viridescens* range from 400–800 μm, compared with the

Loops extrude from the chromosome axis

Chiasma   Axis

Loop

Figure 28.15 A lampbrush chromosome is a meiotic bivalent in which the two pairs of sister chromatids are held together at chiasmata (indicated by arrows). Photograph kindly provided by Joseph G. Gall, Dept. of Embryology, Carnegie Institution of Washington.

range of 15–20 μm seen later in meiosis. So the lampbrush chromosomes are ~30 times less tightly packed. The total length of the entire lamp-brush chromosome set is 5–6 mm, organized into ~5000 chromomeres.

Expressed genes form lateral loops that extend from the chromomeres. The loops extend in pairs, one from each sister chromatid. The loops are continuous with the axial thread, which suggests that they represent chromosomal material extruded from its more compact organization in the chromomere. The loops are surrounded by a matrix of ribonucleoproteins. These contain nascent RNA chains. Often a transcription unit can be defined by the increase in the length of the RNP moving around the loop. **Figure 28.16** shows an example.

So the loop is an extruded segment of DNA that is being actively transcribed. In some cases, loops corresponding to particular genes have been identified. Then the structure of the transcribed gene, and the nature of the product, can be scrutinized *in situ.*

28.8 Polytene Chromosomes Form Bands that Puff at Sites of Gene Expression

Key Terms

- **Polytene** chromosomes are generated by successive replications of a chromosome set without separation of the replicas.
- **Bands** of polytene chromosomes are visible as dense regions that contain the majority of DNA. They include active genes.
- **Interbands** are the relatively dispersed regions of polytene chromosomes that lie between the bands.
- A **puff** is an expansion of a band of a polytene chromosome associated with the synthesis of RNA at some locus in the band.
- *In situ* **hybridization (cytological hybridization)** is performed by denaturing the DNA of cells squashed on a microscope slide so that reaction is possible with an added single-stranded RNA or DNA; the added preparation is radioactively labeled and its hybridization is followed by autoradiography.

Key Concepts

- Polytene chromosomes of Dipterans have a series of bands that can be used as a cytological map.
- The puffs allow sites of gene expression on polytene chromosomes to be examined directly.

The interphase nuclei of some tissues of the larvae of Dipteran flies contain chromosomes that are greatly enlarged relative to their usual condition. They possess both increased diameter and greater length. **Figure 28.17** shows an example of a chromosome set from the salivary gland of *D. melanogaster.* They are called **polytene** chromosomes.

Each member of the polytene set consists of a visible series of **bands** (more properly, but rarely, described as chromomeres). The bands range in size from the largest with a breadth of ~0.5 μm to the smallest of ~0.05 μm. The bands contain most of the mass of DNA and stain intensely with appropriate reagents. The regions between them stain more lightly and are called **interbands**. There are ~5000 bands in the *D. melanogaster* set.

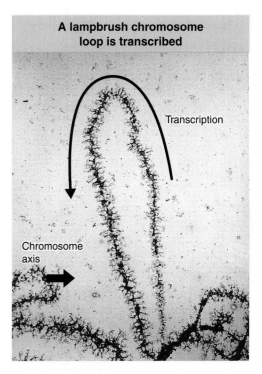

A lampbrush chromosome loop is transcribed

Transcription

Chromosome axis

Figure 28.16 A lampbrush chromosome loop is surrounded by a matrix of ribonucleoprotein. Photograph kindly provided by Oscar L. Miller, Department of Biology, University of Virginia.

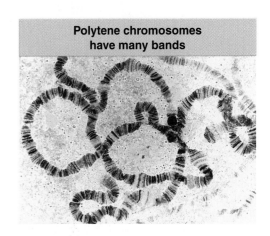

Polytene chromosomes have many bands

Figure 28.17 The polytene chromosomes of *D. melanogaster* form an alternating series of bands and interbands. Photograph kindly provided by Jose Bonner, Dept. of Biology, Indiana University.

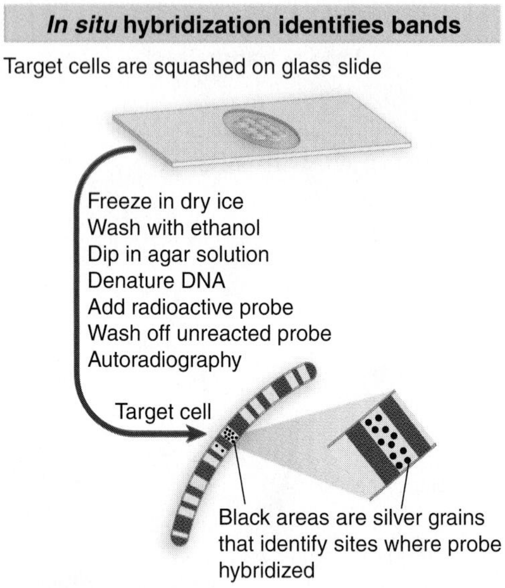

In situ hybridization identifies bands

Target cells are squashed on glass slide

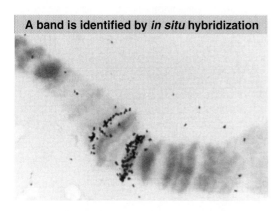

Freeze in dry ice
Wash with ethanol
Dip in agar solution
Denature DNA
Add radioactive probe
Wash off unreacted probe
Autoradiography

Target cell

Black areas are silver grains
that identify sites where probe
hybridized

Figure 28.18 Individual bands containing particular genes can be identified by *in situ* hybridization.

A band is identified by *in situ* hybridization

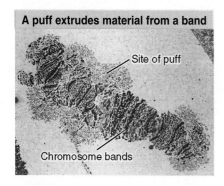

Figure 28.19 A magnified view of bands 87A and 87C shows their hybridization *in situ* with labeled RNA extracted from heat-shocked cells. Photograph kindly provided by Jose Bonner, Dept. of Biology, Indiana University.

A puff extrudes material from a band

Site of puff

Chromosome bands

Figure 28.20 Chromosome IV of the insect *C. tentans* has three Balbiani rings in the salivary gland. Photograph kindly provided by Bertil Daneholt, Dept. of Cell and Molecular Biology, Medical Nobel Institute.

The centromeres of all four chromosomes of *D. melanogaster* aggregate to form a chromocenter that consists largely of heterochromatin (in the male it includes the entire Y chromosome). Allowing for this, ~75% of the haploid DNA set is organized into alternating bands and interbands. The length of the chromosome set is ~2000 μm. The DNA in extended form would stretch for ~40,000 μm, so the packing ratio is ~20. This demonstrates vividly the extension of the genetic material relative to the usual states of interphase chromatin or mitotic chromosomes.

What is the structure of these giant chromosomes? Each is produced by the successive replications of a synapsed diploid pair. The replicas do not separate, but remain attached to each other in their extended state. At the start of the process, each synapsed pair has a DNA content of 2C (where C represents the DNA content of the individual chromosome). Then this doubles up to nine times, at its maximum giving a content of 1024C. The number of doublings is different in the various tissues of the *D. melanogaster* larva.

Each chromosome can be visualized as a large number of parallel fibers running longitudinally, tightly condensed in the bands, less condensed in the interbands. Probably each fiber represents a single (C) haploid chromosome. This gives rise to the name *polytene*. The degree of polyteny is the number of haploid chromosomes contained in the giant chromosome.

The banding pattern is characteristic for each strain of *Drosophila*. The constant number and linear arrangement of the bands were first noted in the 1930s, when it was realized that they form a *cytological map* of the chromosomes. Rearrangements—such as deletions, inversions, or duplications—result in alterations of the order of bands.

The linear array of bands can be equated with the linear array of genes. So genetic rearrangements, as seen in a linkage map, can be correlated with structural rearrangements of the cytological map. Ultimately, a particular mutation can be located in a particular band. Since the total number of genes in *D. melanogaster* exceeds the number of bands, there are probably multiple genes in most or all bands.

The positions of particular genes on the cytological map can be determined directly by the technique of *in situ* hybridization. The protocol is summarized in **Figure 28.18**. A radioactive probe representing a gene (most often a labeled cDNA clone derived from the mRNA) is hybridized with the denatured DNA of the polytene chromosomes *in situ*. Autoradiography identifies the position or positions of the corresponding genes by the superimposition of grains at a particular band or bands. An example is shown in **Figure 28.19**. With this type of technique at hand, it is possible to determine directly the band within which a particular sequence lies.

One of the intriguing features of the polytene chromosomes is that active sites can be visualized. Some of the bands pass transiently through an expanded state in which they appear like a **puff** on the chromosome, when chromosomal material is extruded from the axis. An example of some very large puffs (called Balbiani rings) is shown in **Figure 28.20**.

The puffs are *sites where RNA is being synthesized*. The accepted view of puffing has been that expansion of the band is a consequence of the need to relax its structure in order to synthesize RNA. Puffing has therefore been viewed as a consequence of transcription. A puff can be generated by a single active gene. The sites of puffing differ from ordinary bands in accumulating additional proteins, which include RNA polymerase II and other proteins associated with transcription.

The features displayed by lampbrush and polytene chromosomes suggest a general conclusion. In order to be transcribed, the genetic material is dispersed from its usual more tightly packed state. The question to keep in mind is whether this dispersion at the gross level of the chromosome mimics the events that occur at the molecular level within the mass of ordinary interphase euchromatin.

28.9 Centromeres Often Have Extensive Repetitive DNA

Key Concepts

- A eukaryotic chromosome is held on the mitotic spindle by the attachment of microtubules to the kinetochore that forms in its centromeric region.
- Centromeres in higher eukaryotic chromosomes contain large amounts of repetitious DNA.
- The function of the repetitious DNA is not known.

Key Terms

- The **spindle** guides the movement of chromosomes during cell division. The structure is made up of microtubules.
- A **microtubule organizing center (MTOC)** is a region from which microtubules emanate. The major MTOCs in a mitotic cell are the centrosomes.
- The **centromere** is a constricted region of a chromosome that includes the site of attachment (the kinetochore) to the mitotic or meiotic spindle.
- An **acentric fragment** of a chromosome (generated by breakage) lacks a centromere and is lost at cell division.
- The **kinetochore** is the structural feature of the chromosome to which microtubules of the mitotic spindle attach. Its location determines the centromeric region.
- **C-bands** are generated by staining techniques that react with centromeres. The centromere appears as a darkly staining dot.

During mitosis, the sister chromatids move to opposite poles of the cell. Their movement depends on the attachment of the chromosome to microtubules, which are connected at their other end to the poles. (The microtubules constitute a cellular filamentous system, reorganized at mitosis into a structure called the **spindle**, so that they connect the chromosomes to the poles of the cell.) The sites in the two regions where microtubule ends are organized—in the vicinity of the centrioles at the poles and at the chromosomes—are called **MTOCs (microtubule organizing centers)**.

Figure 28.21 illustrates the separation of sister chromatids as mitosis proceeds from metaphase to telophase. The region of the chromosome that is responsible for its segregation at mitosis and meiosis is called the **centromere**. The centromeric region on each sister chromatid is pulled by microtubules to the opposite pole. Opposing this motive force, "glue" proteins called cohesins hold the sister chromatids together. Initially the sister chromatids separate at their centromeres, and then they are released completely from one another during anaphase when the cohesins are degraded.

The centromere is pulled toward the pole during mitosis. Because it holds the sister chromatids together prior to the separation of the individual chromosomes, it shows as a constricted region connecting all four chromosome arms, as in the photograph of Figure 28.10.

The centromere is essential for segregation, as shown by the behavior of chromosomes that have been broken. A single break generates one piece that retains the centromere, and another, an **acentric fragment**, that lacks it. The acentric fragment does not become attached to the mitotic spindle, and as a result it fails to be included in either of the daughter nuclei.

The region of the chromosome at which the centromere forms is defined by specific DNA sequences. The centromeric DNA binds specific proteins that are responsible for establishing the structure that attaches the chromosome

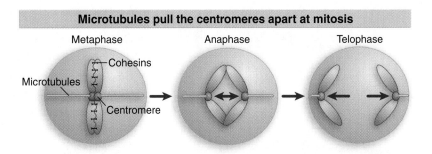

Figure 28.21 Chromosomes are pulled to the poles via microtubules that attach at the centromeres. The sister chromatids are held together until anaphase by glue proteins (cohesins). The centromere is shown here in the middle of the chromosome (metacentric), but can be located anywhere along its length, including close to the end (acrocentric) and at the end (telocentric).

Centromeres are bound to microtubules

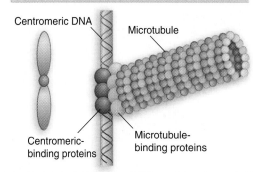

Figure 28.22 The centromere is identified by a DNA sequence that binds specific proteins. These proteins do not themselves bind to microtubules, but establish the site at which the microtubule-binding proteins in turn bind.

to the microtubules. This structure is called the **kinetochore**. It is a darkly staining fibrous object of diameter or length ~400 nm. The kinetochore provides the MTOC on a chromosome. **Figure 28.22** shows the hierarchy of organization that connects centromeric DNA to the microtubules. Proteins bound to the centromeric DNA bind other proteins that bind to microtubules.

In those cases where we can equate specific DNA sequences with the centromeric region, they usually include repetitive sequences. As result, the centromere is often embedded in heterochromatin. Because the entire mitotic chromosome is condensed, centromeric heterochromatin is not immediately evident. However, it can be visualized by a technique that generates **C-bands**. In the example of **Figure 28.23**, all the centromeres show as darkly staining regions. Although it is common, heterochromatin cannot be identified around *every* known centromere, which suggests that it is unlikely to be essential for the division mechanism.

A DNA molecule lacking a centromere does not segregate properly at cell division. This enables centromeric DNA to be identified by its ability to confer proper segregation onto any molecule containing it. Using this criterion, centromeric DNA has been identified in several organisms. In all cases except one (*S. cerevisiae*, discussed in the next section), the length of DNA required for centromeric function is quite long. In the yeast *S. pombe*, all three centromeric regions consist of stretches 40–100 kb long that consist largely or entirely of repetitious DNA. Attempts to localize centromeric functions in *Drosophila* chromosomes suggest that they are dispersed in a large region, consisting of 200–600 kb. The large size of this type of centromere suggests that it is likely to contain several separate specialized functions, including sequences required for kinetochore assembly, sister chromatid pairing, etc.

The primary motif of the heterochromatin of primate centromeres is the α satellite DNA, which consists of tandem arrays of a 170 bp repeating unit. There is significant variation between individual repeats, although those at any centromere tend to be better related to one another than to members of the family in other locations. It is clear that the sequences required for centromeric function reside within the blocks of α satellite DNA, but it is not clear whether the α satellite sequences themselves provide this function, or whether other sequences are embedded within the α satellite arrays.

C-bands identify centromeres

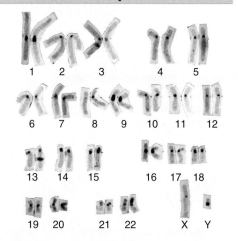

Figure 28.23 C-banding generates intense staining at the centromeres of all chromosomes. Photograph kindly provided by Dr. Lisa Shaffer, Baylor College of Medicine.

28.10 *S. cerevisiae* Centromeres Have Short Protein-Binding DNA Sequences

Key Concepts

- *CEN* elements are identified in *S. cerevisiae* by the ability to allow a plasmid to segregate accurately at mitosis.
- *CEN* elements consists of short conserved sequences CDE-I and CDE-III that flank the A·T-rich region CDE-II.
- A specialized protein complex that is an alternative to the usual chromatin structure is formed at CDE-II.
- The CBF3 protein complex that binds to CDE-III is essential for centromeric function.
- The complexes bound to *CEN* bind proteins that may provide the connection to microtubules.

The *S. cerevisiae* centromere has short conserved sequences and a long A · T stretch

```
TCACATGATGATATTTGATTTTATTATATTTTTAAAAAAAGTAAAAAAATAAAAAGTAGTTTATTTTTAAAAAATAAAATTTAAAATATTTCACAAAATGATTTCCGAA
AGTGTACTACTATAAACTAAAATAATATAAAAATTTTTTTCATTTTTTATTTTTCATCAAATAAAAATTTTTTATTTTAAATTTTATAAAGTGTTTTACTAAAGGCTT
```

CDE-I   *CDE-II*   80–90 bp, >90% A + T   *CDE-III*

Figure 28.24 Three conserved regions can be identified by the sequence homologies between yeast *CEN* elements.

Yeast chromosomes do not display visible kinetochores comparable to those of higher eukaryotes, but otherwise divide at mitosis and segregate at meiosis by the same mechanisms. A *CEN* fragment is identified as the minimal sequence that can allow a plasmid to segregate accurately at mitosis. Every chromosome has such a region. A *CEN* fragment derived from one chromosome can replace the centromere of another chromosome with no apparent consequence. This result suggests that centromeres are interchangeable. *They are used simply to attach the chromosome to the spindle, and play no role in distinguishing one chromosome from another.*

The sequences required for centromeric function fall within a stretch of ~120 bp. The centromeric region is packaged into a nuclease-resistant structure, and it binds a single microtubule. We may therefore look to the *S. cerevisiae* centromeric region to identify proteins that bind centromeric DNA and proteins that connect the chromosome to the spindle.

Three types of sequence element may be distinguished in the *CEN* region, as summarized in **Figure 28.24**:

- CDE-I is a sequence of 9 bp that is conserved with minor variations at the left boundary of all centromeres.
- CDE-II is a >90% A · T-rich sequence of 80–90 bp found in all centromeres; its function could depend on its length rather than exact sequence. Its constitution is reminiscent of some short tandemly repeated (satellite) DNAs (see *6.10 Arthropod Satellites Have Very Short Identical Repeats*). Its base composition may cause some characteristic distortions of the DNA double helical structure.
- CDE-III is an 11 bp sequence highly conserved at the right boundary of all centromeres. Sequences on either side of the element are less well conserved, and may also be needed for centromeric function. Point mutations in the central CCG of CDE-III completely inactivate the centromere.

A large protein complex assembles at the CDE sequences, and connects the chromosome to microtubules. The structure is summarized in **Figure 28.25**.

CDE-II binds a protein called Cse4p, which resembles one of the histone proteins that comprise the basic subunits of chromatin. A protein called Mif2p may also be part of this complex or connected to it. Cse4p and Mif2p have counterparts that are localized at higher eukaryotic centromeres, called CENP-A and CENP-C, which suggests that this interaction may be a universal aspect of centromere construction. The basic interaction consists of bending the DNA of the CDE-II region around a protein aggregate; the reaction is probably assisted by the occurrence of intrinsic bending in the CDE-II sequence.

CDE-I is bound by the homodimer CBF1; this interaction is not essential for centromere function, but in its absence the fidelity of chromosome segregation is reduced ~10×. A 240 kD complex of four proteins, called CBF3, binds to CDE-III. This interaction is essential for centromeric function.

The proteins bound at CDE-I and CDE-III are connected to each other and also to the protein structure bound at CDE-II by another group of proteins (Ctf19, Mcm21, Okp1). The connection to the microtubule may be made by this complex.

The overall model suggests that the complex is localized at the centromere by a protein structure that resembles the normal building

Proteins bind to the yeast CDE elements

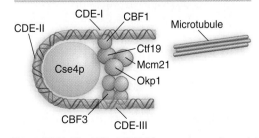

Figure 28.25 The DNA at CDE-II is wound around a protein aggregate including Cse4p, CDE-III is bound to CBF3, and CDE-I is bound to CBF1. These proteins are connected by the group of Ctf19, Mcm21, and Okp1.

block of chromatin (the nucleosome). The bending of DNA at this structure allows proteins bound to the flanking elements to become part of a single complex. Some components of the complex (possibly not those that bind directly to DNA) link the centromere to the microtubule. The construction of kinetochores probably follows a similar pattern, and uses related components, in a wide variety of organisms.

28.11 Telomeres Have Simple Repeating Sequences

Key Term

- A **telomere** is the natural end of a chromosome; the DNA sequence consists of a simple repeating unit with a protruding single-stranded end.

Key Concepts

- The telomere is required for the stability of the chromosome end.
- A telomere consists of a simple repeat where a C + A-rich strand has the sequence $C_{>1}(A/T)_{1-4}$.
- The protein TRF2 catalyzes a reaction in which the 3′ repeating unit of the G + T-rich strand forms a loop by displacing its homologue in an upstream region of the telomere.

Another essential feature in all chromosomes is the **telomere**, which "seals" the end. We know that the telomere must be a special structure, because chromosome ends generated by breakage are "sticky" and tend to react with other chromosomes, whereas natural ends are stable.

We can apply two criteria in identifying a telomeric sequence:

- It must lie at the end of a chromosome (or, at least, at the end of an authentic linear DNA molecule).
- It must confer stability on a linear molecule.

The problem of finding a system that offers an assay for function again has been brought to the molecular level by using yeast. All the plasmids that survive in yeast (by virtue of possessing *ARS* and *CEN* elements) are circular DNA molecules. Linear plasmids are unstable (because they are degraded). Could an authentic telomeric DNA sequence confer stability on a linear plasmid? Fragments from yeast DNA that prove to be located at chromosome ends can be identified by such an assay. And a region from the end of a known natural linear DNA molecule—the extrachromosomal rDNA of *Tetrahymena*—is able to render a yeast plasmid stable in linear form.

Telomeric sequences have been characterized from a wide range of lower and higher eukaryotes. The same type of sequence is found in plants and man, so the construction of the telomere seems to follow a universal principle. Each telomere consists of a long series of short, tandemly repeated sequences. There may be 100–1000 repeats, depending on the organism.

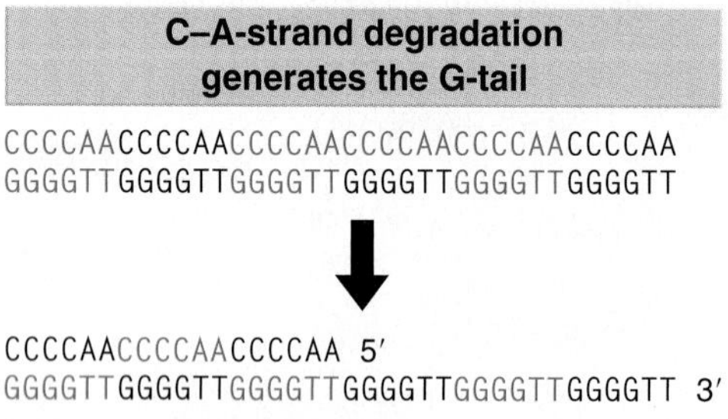

All telomeric sequences can be written in the general form $C_n(A/T)_m$, where $n > 1$ and m is 1–4. **Figure 28.26** shows a generic example. One unusual property of the telomeric sequence is the extension of the G–T-rich strand, usually for 14–16 bases as a single strand. The G-tail is probably generated because there is a specific limited degradation of the C–A-rich strand.

Some indications about how a telomere functions are given by some unusual properties of the ends of linear DNA molecules. In a trypanosome population, the ends are variable in length. When an individ-

C–A-strand degradation generates the G-tail

```
CCCCAACCCCAACCCCAACCCCAACCCCAACCCCAA
GGGGTTGGGGTTGGGGTTGGGGTTGGGGTTGGGGTT
```

```
CCCCAACCCCAACCCCAA 5′
GGGGTTGGGGTTGGGGTTGGGGTTGGGGTTGGGGTT 3′
```

Figure 28.26 A typical telomere has a simple repeating structure with a G–T-rich strand that extends beyond the C–A-rich strand. The G-tail is generated by a limited degradation of the C–A-rich strand.

ual cell clone is followed, the telomere grows longer by 7–10 bp (1–2 repeats) per generation. Even more revealing is the fate of ciliate telomeres introduced into yeast. After replication in yeast, *yeast telomeric repeats are added onto the ends of the* Tetrahymena *repeats*.

Addition of telomeric repeats to the end of the chromosome in every replication cycle could solve the difficulty of replicating linear DNA molecules discussed in *16.2 The Ends of Linear DNA Are a Problem for Replication*. The addition of repeats by *de novo* synthesis would counteract the loss of repeats resulting from failure to replicate up to the end of the chromosome. Extension and shortening would be in dynamic equilibrium.

If telomeres are continually being lengthened (and shortened), their exact sequence may be irrelevant. All that is required is for the end to be recognized as a suitable substrate for addition. This explains how the ciliate telomere functions in yeast.

Isolated telomeric fragments do not behave as though they contain single-stranded DNA; instead, they show aberrant electrophoretic mobility and other properties.

What feature of the telomere is responsible for the stability of the chromosome end? **Figure 28.27** shows that a loop of DNA forms at the telomere. The absence of any free end may be the crucial feature that stabilizes the end of the chromosome. The average length of the loop in animal cells is 5–10 kb.

Figure 28.28 shows that the loop is formed when the 3′ single-stranded end of the telomere (TTAGGG)$_n$ displaces the same sequence in an upstream region of the telomere. This converts the duplex region into a structure like a D-loop, where a series of TTAGGG repeats are displaced to form a single-stranded region, and the tail of the telomere is paired with the homologous strand.

The reaction is catalyzed by the telomere-binding protein TRF2, which together with other proteins forms a complex that stabilizes the chromosome ends. Its importance in protecting the ends is indicated by the fact the deletion of TRF2 causes chromosome rearrangements to occur.

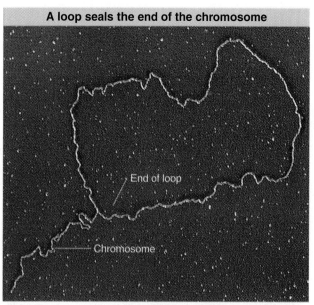

A loop seals the end of the chromosome

End of loop

Chromosome

Figure 28.27 A loop forms at the end of chromosomal DNA. Photograph kindly provided by Jack Griffith, University of North Carolina School of Medicine.

28.12 Telomeres Are Synthesized by a Ribonucleoprotein Enzyme

Key Term

- **Telomerase** is the ribonucleoprotein enzyme that creates repeating units of one strand at the telomere, by adding individual bases to the DNA 3′ end, as directed by an RNA sequence in the RNA component of the enzyme.

Key Concepts

- Telomerase uses the 3′–OH of the G + T telomeric strand to prime synthesis of tandem TTGGGG repeats.
- The RNA component of telomerase has a sequence that pairs with the C + A-rich repeats.
- One of the protein subunits is a reverse transcriptase that uses the RNA as template to synthesize the G + T-rich sequence.

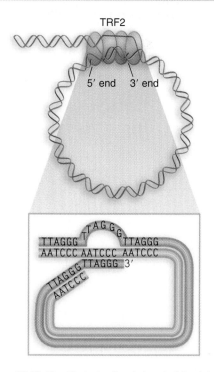

Telomeric DNA forms a T-loop

TRF2

5′ end   3′ end

TTAGGG   T A G G G   TTAGGG
AATCCC AATCCC AATCCC
TTAGGG   3′
TTAGGG
AATCCC

Figure 28.28 The 3′ single-stranded end of the telomere (TTAGGG)n displaces the homologous repeats from duplex DNA to form a t-loop. The reaction is catalyzed by TRF2.

Telomerase is a reverse transcriptase

Binding: RNA template pairs with DNA primer

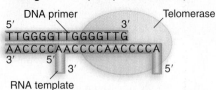

Polymerization: RNA template directs addition of nucleotides to 3′ end of DNA

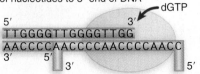

Polymerization continues to end of template region

Translocation: Enzyme moves to template 3′ end

Process repeats

Figure 28.29 Telomerase positions itself by base pairing between the RNA template and the protruding single-stranded DNA primer. It adds G and T bases one at a time to the primer, as directed by the template. The cycle starts again when one repeating unit has been added.

The telomere has two functions:

- One is to protect the chromosome end. Any other DNA end—for example, the end generated by a double-strand break—becomes a target for repair systems. The cell has to be able to distinguish the telomere.
- The second is to allow the telomere to be extended. Otherwise it would become shorter with each replication cycle (because replication cannot start at the very end).

Proteins that bind to the telomere provide the solution for both problems. In yeast, different sets of proteins solve each problem, but both are bound to the telomere via the same protein, Cdc13:

- The Stn1 protein protects against degradation (specifically against any extension of the degradation of the C–A-strand that generates the G-tail).
- A **telomerase** enzyme extends the C–A-rich strand. Its activity is influenced by two proteins that have ancillary roles, such as controlling the length of the extension.

The telomerase uses the 3′—OH of the G + T telomeric strand as a primer for synthesis of tandem TTGGGG repeats. Only dGTP and dTTP are needed for the activity. Telomerase is a large ribonucleoprotein that consists of a templating RNA (coded by *TLC1*) and a protein with catalytic activity (*EST2*). The short RNA component (159 bases long in *Tetrahymena*, 192 bases long in *Euplotes*) includes a sequence of 15–22 bases that is identical to two repeats of the C-rich repeating sequence. This RNA provides the template for synthesizing the G-rich repeating sequence. The protein component of the telomerase is a catalytic subunit that can act only upon the RNA template provided by the nucleic acid component.

Figure 28.29 shows the action of telomerase. The enzyme progresses discontinuously: the template RNA is positioned on the DNA primer, several nucleotides are added to the primer, and then the enzyme translocates to begin again. The telomerase is a specialized example of a reverse transcriptase, an enzyme that synthesizes a DNA sequence using an RNA template (see *22.4 Viral DNA Is Generated by Reverse Transcription*). We do not know how the complementary (C–A-rich) strand of the telomere is assembled, but we may speculate that it could be synthesized by using the 3′—OH of a terminal G–T hairpin as a primer for DNA synthesis.

Telomerase synthesizes the individual repeats that are added to the chromosome ends, but does not itself control the number of repeats. Other proteins are involved in determining the length of the telomere. They can be identified by the *EST1* and *EST3* mutants in yeast that have altered telomere lengths. These proteins may bind telomerase, and influence the length of the telomere by controlling the access of telomerase to its substrate. Proteins that bind telomeres in mammalian cells have been found similarly, but less is known about their functions.

The minimum features required for existence as a chromosome are:

- Telomeres to ensure survival.
- A centromere to support segregation.
- An origin to initiate replication.

All of these elements have been put together to construct a yeast artificial chromosome (YAC). This is a useful method for perpetuating foreign sequences. It turns out that the synthetic chromosome is stable only if it is longer than 20–50 kb. We do not know the basis for this effect, but the ability to construct a synthetic chromosome allows us to investigate the nature of the segregation device in a controlled environment.

28.13 SUMMARY

The genetic material of all organisms and viruses takes the form of tightly packaged nucleoprotein. Some virus genomes are inserted into preformed virions, while others assemble a protein coat around the nucleic acid. The bacterial genome forms a dense nucleoid, with ~20% protein by mass, but details of the interaction of the proteins with DNA are not known. The DNA is organized into ~100 domains that maintain independent supercoiling, with a density of unrestrained supercoils corresponding to ~1/100–200 bp. Interphase chromatin and metaphase chromosomes both appear to be organized into large loops. Each loop may be an independently supercoiled domain. The bases of the loops are connected to a metaphase scaffold or to the nuclear matrix by specific DNA sites.

Transcriptionally active sequences reside within the euchromatin that constitutes the majority of interphase chromatin. The regions of heterochromatin are packaged ~5–10× more compactly, and are transcriptionally inert. All chromatin becomes densely packaged during cell division, when the individual chromosomes can be distinguished. The existence of a reproducible ultrastructure in chromosomes is indicated by the production of G-bands by treatment with Giemsa stain. The bands are very large regions, ~10^7 bp, that can be used to map chromosomal translocations or other large changes in structure.

Lampbrush chromosomes of amphibians and polytene chromosomes of insects have unusually extended structures, with packing ratios <100. Polytene chromosomes of *D. melanogaster* are divided into ~5000 bands, varying in size by an order of magnitude, with an average of ~25 kb. Transcriptionally active regions can be visualized in even more unfolded ("puffed") structures, in which material is extruded from the axis of the chromosome. This may resemble the changes that occur on a smaller scale when a sequence in euchromatin is transcribed.

The centromeric region contains the kinetochore, which is responsible for attaching a chromosome to the mitotic spindle. The centromere often is surrounded by heterochromatin. Centromeric sequences have been identified only in the yeast *S. cerevisiae*, where they consist of short conserved elements, CDE-I and CDE-III that binds CBF1 and the CBF3 complex, respectively, and a long A·T-rich region called CDE-II that binds Cse4p to form a specialized structure in chromatin. Another group of proteins that binds to this assembly provides the connection to microtubules.

Telomeres make the ends of chromosomes stable. Almost all known telomeres consist of multiple repeats in which one strand has the general sequence $C_n(A/T)_m$, where $n > 1$ and $m = 1$–4. The other strand, $G_n(T/A)_m$, has a single protruding end that provides a template for addition of individual bases in defined order. The enzyme telomere transferase is a ribonucleoprotein, whose RNA component provides the template for synthesizing the G-rich strand. This overcomes the problem of the inability to replicate at the very end of a duplex. The telomere stabilizes the chromosome end because the overhanging single strand $G_n(T/A)_m$ displaces its homologue in earlier repeating units in the telomere to form a loop, so there are no free ends.

29

Nucleosomes

29.1 Introduction

Key Terms

- The **nucleosome** is the basic structural subunit of chromatin, consisting of ~200 bp of DNA and an octamer of histone proteins.
- **Histones** are conserved DNA-binding proteins that form the basic subunit of chromatin in eukaryotes. Histones H2A, H2B, H3, H4 form an octameric core around which DNA coils to form a nucleosome. Histone H1 is external to the nucleosome.
- A **nonhistone** is any structural protein found in a chromosome except one of the histones.

The fundamental subunit of chromatin has the same type of structure in all eukaryotes. The **nucleosome** contains ~200 bp of DNA, organized by an octamer of small, basic proteins into a bead-like structure. The proteins are called **histones**. They form an interior core; the DNA lies on the surface of the particle. The nucleosome provides the first level of organization in both euchromatin and heterochromatin.

The second level of organization is the coiling of the series of nucleosomes into a helical array to constitute the fiber of diameter ~30 nm that is found in both interphase chromatin and mitotic chromosomes (see Figure 28.10). The structure of this fiber requires additional proteins.

The 30 nm fiber is then itself coiled or otherwise folded into interphase chromatin, depending on other proteins to form the characteristic densities of euchromatin or heterochromatin in interphase cells. At mitosis, the fiber is densely packed into the structures of mitotic chromosomes.

We need to work through these levels of organization to characterize the events of cyclical packaging, replication, and transcription.

Association with additional proteins, or modifications of existing chromosomal proteins, change the structure of chromatin. Both replication and transcription require unwinding of DNA, and thus must involve an unfolding of the structure that allows the relevant enzymes to manipulate the DNA. This is likely to involve changes in all levels of organization.

The mass of chromatin contains up to twice as much protein as DNA. Approximately half of the protein mass is accounted for by the nucleosomes. The mass of RNA is <10% of the mass of DNA. Much of the RNA consists of nascent transcripts still associated with the template DNA.

The **nonhistones** include all the proteins of chromatin except the histones. They are more variable between tissues and species, and they constitute a smaller proportion of the mass than the histones. The nonhistones comprise a much larger number of proteins, so that any individual nonhistone protein is present in amounts much smaller than any histone.

29.2 The Nucleosome Is the Subunit of All Chromatin

Key Terms

- **Microccocal nuclease** is an endonuclease that cleaves DNA; in chromatin, DNA is cleaved preferentially between nucleosomes.
- A **core histone** is one of the four types (H2A, H2B, H3, H4) found in the core particle derived from the nucleosome (this excludes histone H1).

Key Concepts

- Micrococcal nuclease releases individual nucleosomes from chromatin as 11S particles.
- A nucleosome contains ~200 bp of DNA, two copies of each core histone (H2A, H2B, H3, H4), and one copy of H1.
- DNA is wrapped around the outside surface of the protein octamer.

When interphase nuclei are suspended in a solution of low ionic strength, they swell and rupture to release fibers of chromatin. **Figure 29.1** shows a lysed nucleus in which fibers are streaming out. In some regions, the fibers consist of tightly packed material, but in regions that have become stretched, they can be seen to consist of discrete particles. These are the nucleosomes. In especially extended regions, individual nucleosomes are connected by a fine thread, a free duplex of DNA. *A continuous duplex thread of DNA runs through the series of particles.*

Individual nucleosomes can be obtained by treating chromatin with the endonuclease **micrococcal nuclease**. It cuts the DNA thread at the junction between nucleosomes. First, it releases groups of particles; finally, it releases single nucleosomes. Individual nucleosomes can be seen in **Figure 29.2** as compact particles. They sediment at ~11S.

The nucleosome contains ~200 bp of DNA associated with a histone octamer that consists of two copies each of H2A, H2B, H3, and H4. These are known as the **core histones**. Their association is illustrated diagrammatically in **Figure 29.3**. This model explains the stoichiometry of the core histones in chromatin: H2A, H2B, H3, and H4 are present in equimolar amounts, with two molecules of each per ~200 bp of DNA.

Histones H3 and H4 are among the most conserved proteins known. This suggests that their functions are identical in all eukaryotes.

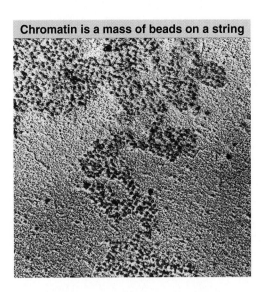

Figure 29.1 Chromatin spilling out of lysed nuclei consists of a compactly organized series of particles. Photograph kindly provided by Pierre Chambon.

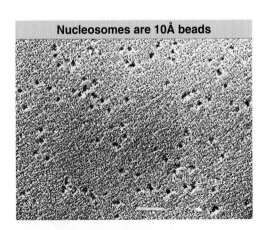

Figure 29.2 Individual nucleosomes are released by digestion of chromatin with micrococcal nuclease. The bar is 100 nm. Photograph kindly provided by Pierre Chambon.

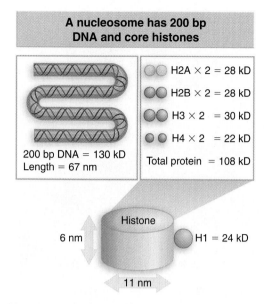

A nucleosome has 200 bp DNA and core histones

H2A × 2 = 28 kD	
H2B × 2 = 28 kD	
H3 × 2 = 30 kD	
H4 × 2 = 22 kD	

200 bp DNA = 130 kD
Length = 67 nm

Total protein = 108 kD

Histone

6 nm

H1 = 24 kD

11 nm

Figure 29.3 The nucleosome consists of approximately equal masses of DNA and histones (including H1). The predicted mass of the nucleosome is 262 kD.

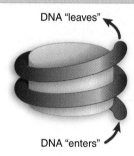

The nucleosome has two turns of DNA

DNA "leaves"

DNA "enters"

Figure 29.4 The nucleosome may be a cylinder with DNA organized into two turns around the surface.

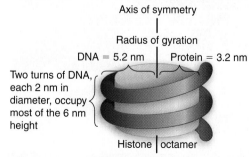

The nucleosome is a flat cylinder

Axis of symmetry

Radius of gyration

DNA = 5.2 nm   Protein = 3.2 nm

Two turns of DNA, each 2 nm in diameter, occupy most of the 6 nm height

Histone octamer

Figure 29.5 The two turns of DNA on the nucleosome lie close together.

The types of H2A and H2B can be recognized in all eukaryotes, but show appreciable species-specific variation in sequence.

Histone H1 comprises a set of closely related proteins that show appreciable variation between tissues and between species. The role of H1 is different from the core histones. It is present in half the amount of a core histone and can be extracted more readily from chromatin (typically with dilute salt [0.5 M] solution). *The H1 can be removed without affecting the structure of the nucleosome, which suggests that its location is external to the particle.*

The shape of the nucleosome corresponds to a flat disk or cylinder, of diameter 11 nm and height 6 nm. The length of the DNA is roughly twice the ~34 nm circumference of the particle. The DNA follows a symmetrical path around the octamer. **Figure 29.4** shows the DNA path diagrammatically as a helical coil that makes two turns around the cylindrical octamer. Note that the DNA "enters" and "leaves" the nucleosome at points close to one another. Histone H1 may be located in this region (see *29.4 Nucleosomes Have a Common Structure*).

Considering this model in terms of a cross section through the nucleosome, in **Figure 29.5** we see that the two circumferences made by the DNA lie close to one another. The height of the cylinder is 6 nm, of which 4 nm is occupied by the two turns of DNA (each of diameter 2 nm).

29.3 DNA Is Coiled in Arrays of Nucleosomes

Key Concepts

- >95% of the DNA is recovered in nucleosomes or multimers when micrococcal nuclease cleaves chromatin.
- The length of DNA per nucleosome varies for individual tissues in a range from 154–260 bp.

When chromatin is digested with the enzyme micrococcal nuclease, the DNA is cleaved into integral multiples of a unit length. Fractionation by gel electrophoresis reveals the "ladder" presented in **Figure 29.6**. Such ladders extend for ~10 steps, and the unit length, determined by the increments between successive steps, is ~200 bp.

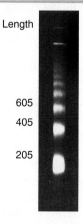

The unit length of DNA is ~200 bp

Length

605

405

205

Figure 29.6 Micrococcal nuclease digests chromatin in nuclei into a multimeric series of DNA bands that can be separated by gel electrophoresis. Photograph kindly provided by Markus Noll, Institute of Molecular Biology, University of Zurich.

Figure 29.7 shows that the ladder is generated by groups of nucleosomes. When nucleosomes are fractionated on a sucrose gradient, they give a series of discrete peaks that correspond to monomers, dimers, trimers, etc. When the DNA is extracted from the individual fractions and electrophoresed, each fraction yields a band of DNA whose size corresponds with a step on the micrococcal nuclease ladder. The monomeric nucleosome contains DNA of the unit length, the nucleosome dimer contains DNA of twice the unit length, and so on.

So each step on the ladder represents the DNA derived from a discrete number of nucleosomes. *We therefore take the existence of the 200 bp ladder in any chromatin to indicate that the DNA is organized into nucleosomes.* The micrococcal ladder is generated when only ~2% of the DNA in the nucleus is rendered acid soluble (degraded to small fragments) by the enzyme. *So a small proportion of the DNA is specifically attacked; it must represent especially susceptible regions.* Virtually all of the DNA is organized into nucleosomes, as shown by the fact that >95% *of the DNA of chromatin can be recovered in the form of the 200 bp ladder.* In their natural state, nucleosomes are likely to be closely packed, with DNA passing directly from one to the next.

The length of DNA present in the nucleosome varies somewhat from the "typical" value of 200 bp. The chromatin of any particular cell type has a characteristic average value ($\pm$5 bp). The average most often is between 180 and 200, but there are extremes as low as 154 bp (in a fungus) or as high as 260 bp (in a sea urchin sperm). The average value may be different in individual tissues of the adult organism. And there can be differences between different parts of the genome in a single cell type.

29.4 Nucleosomes Have a Common Structure

Key Terms

- The **core particle** is a digestion product of the nucleosome that retains the histone octamer and has 146 bp of DNA; its structure is similar to that of the nucleosome itself.

- **Core DNA** is the 146 bp of DNA contained on a core particle that is generated by cleaving the DNA on a nucleosome with micrococcal nuclease.

- **Linker DNA** is all DNA contained on a nucleosome in excess of the 146 bp core DNA. Its length varies from 8–114 bp, and it is cleaved by micrococcal nuclease to leave the core DNA.

Key Concepts

- Nucleosomal DNA is divided into the core DNA and linker DNA depending on its susceptibility to micrococcal nuclease.

- H1 is associated with linker DNA and may lie at the point where DNA enters and leaves the nucleosome.

A common structure underlies the varying amount of DNA that is contained in nucleosomes of different sources. The association of DNA with the histone octamer forms a **core particle** containing 146 bp of DNA, irrespective of the total length of DNA in the nucleosome. The variation in total length of DNA per nucleosome is superimposed on this basic core structure.

The core particle is defined by the effects of micrococcal nuclease on the nucleosome monomer. The initial reaction of the enzyme is to

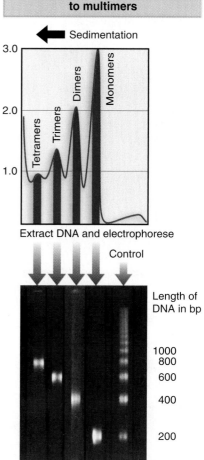

Figure 29.7 Each multimer of nucleosomes contains the appropriate number of unit lengths of DNA. Photograph kindly provided by John Finch.

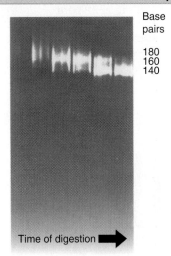

The nucleosome core is 146 bp

Base pairs

180
160
140

Time of digestion ➡

Figure 29.8 Micrococcal nuclease reduces the length of nucleosome monomers in discrete steps. Photograph kindly provided by Roger Kornberg, Dept. of Structural Biology, Stanford University School of Medicine.

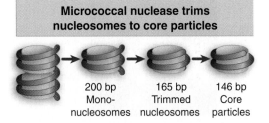

Micrococcal nuclease trims nucleosomes to core particles

200 bp Mono-nucleosomes

165 bp Trimmed nucleosomes

146 bp Core particles

Figure 29.9 Micrococcal nuclease initially cleaves between nucleosomes. Mononucleosomes typically have ~200 bp DNA. End-trimming reduces the length of DNA first to ~165 bp, and then generates core particles with 146 bp.

cut between nucleosomes, but if it is allowed to continue after monomers have been generated, then it proceeds to digest some of the DNA of the individual nucleosome. This occurs by a reaction in which DNA is "trimmed" from the ends of the nucleosome.

The length of the DNA is reduced in discrete steps, as shown in **Figure 29.8**. With rat liver nuclei, the nucleosome monomers initially have 205 bp of DNA. Then some monomers are found in which the length of DNA has been reduced to ~165 bp. Finally this is reduced to the length of the DNA of the core particle, 146 bp. (The core is reasonably stable, but continued digestion generates a "limit digest," in which the longest fragments are the 146 bp DNA of the core, while the shortest are as small as 20 bp.)

This analysis suggests that the nucleosomal DNA can be divided into two regions:

- **Core DNA** has an invariant length of 146 bp, and is relatively resistant to digestion by nucleases.
- **Linker DNA** comprises the rest of the repeating unit. Its length varies from as little as 8 bp to as much as 114 bp per nucleosome.

The sharp size of the band of DNA generated by the initial cleavage with micrococcal nuclease suggests that the region immediately available to the enzyme is restricted. It represents only part of each linker. (If the entire linker DNA were susceptible, the band would range from 146 bp to >200 bp.) But once a cut has been made in the linker DNA, the rest of this region becomes susceptible, and it can be removed relatively rapidly by further enzyme action. The connection between nucleosomes is represented in **Figure 29.9**.

Core particles have properties similar to those of the nucleosomes themselves, although they are smaller. Their shape and size are similar to nucleosomes, which suggests that the essential geometry of the particle is established by the interactions between DNA and the protein octamer in the core particle. Because core particles are more readily obtained as a homogeneous population, they are often used for structural studies in preference to nucleosome preparations. (Nucleosomes tend to vary because it is difficult to obtain a preparation in which there has been no end-trimming of the DNA.)

Where is histone H1 located? The H1 is lost during the degradation of nucleosome monomers. It can be retained on monomers that still have 165 bp of DNA, but is always lost with the final reduction to the 146 bp core particle. This suggests that H1 could be located in the region of the linker DNA immediately adjacent to the core DNA.

If H1 is located at the linker, it could "seal" the DNA in the nucleosome by binding at the point where the nucleic acid enters and leaves (see Figure 29.4). The idea that H1 lies in the region joining adjacent nucleosomes is consistent with old results that H1 is removed the most readily from chromatin, and that H1-depleted chromatin is more readily "solubilized."

29.5 DNA Structure Varies on the Nucleosomal Surface

Key Terms

- The **cutting periodicity** is the spacing between cleavages on each strand when a duplex DNA immobilized on a flat surface is attacked by a DNAase that makes single-strand cuts.
- The **structural periodicity** is the number of base pairs per turn of the double helix of DNA.

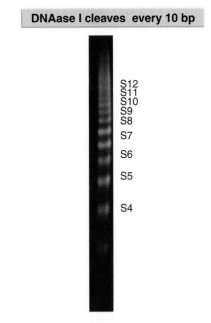

Figure 29.10 Sites for nicking lie at regular intervals along core DNA, as seen in a DNAase I digest of nuclei. Photograph kindly provided by Leonard Lutter, Molecular Biology Research Program, Henry Ford Hospital, Detroit, MI.

Key Concepts

- 1.65 turns of DNA are wound round the histone octamer.
- The structure of the DNA is altered so that it has an increased number of base pairs per turn in the middle, but a decreased number at the ends.

The exposure of DNA on the surface of the nucleosome explains why it is accessible to cleavage by certain nucleases. The reaction with nucleases that attack single strands has been especially informative. The enzymes DNAase I and DNAase II make single-strand nicks in DNA; they cleave a bond in one strand, but the other strand remains intact at this point.

When DNA is free in solution, it is nicked (relatively) at random. The DNA on nucleosomes also can be nicked by the enzymes, *but only at regular intervals*. When the points of cutting are determined by using radioactively end-labeled DNA and then DNA is denatured and electrophoresed, a ladder of the sort displayed in **Figure 29.10** is obtained.

The interval between successive steps on the ladder is 10–11 bases. The ladder extends for the full distance of core DNA. The cleavage sites are numbered as S1 through S13 (where S1 is ~10 bases from the labeled 5′ end, S2 is ~20 bases from it, and so on). Not all sites are cut with equal frequency: some are cut rather efficiently, others are cut scarcely at all.

Because there are two strands of DNA in the core particle, in an end-labeling experiment both 5′ (or both 3′) ends are labeled, one on each strand. This means that each labeled band in fact represents two fragments, generated by cutting the *same* distance from *either* of the labeled ends. The implication is that the path of DNA on the particle is symmetrical (about a horizontal axis through the nucleosome drawn in Figure 29.4). So if, for example, no 80-base fragment is generated by DNAase I, this must mean that the position at 80 bases from the 5′ end of *either* strand is not susceptible to the enzyme.

When DNA is immobilized on a flat surface, sites are cut with a regular separation. **Figure 29.11** suggests that this reflects the recurrence of the exposed site with the helical periodicity of B-form DNA. The **cutting periodicity** (the spacing between cleavage points) coincides with, indeed, is a reflection of, the **structural periodicity** (the number of base pairs per turn of the double helix). So the distance between the sites corresponds to the number of base pairs per turn. Measurements of this type suggest that the average value for double-helical B-type DNA is 10.5 bp/turn.

What is the nature of the target sites on the nucleosome? **Figure 29.12** shows that each site has 3–4 positions at which cutting occurs; that is, the cutting site is defined ±2 bp. So a cutting site represents a short stretch of bonds on both strands, exposed to nuclease action over 3–4 base pairs. The relative intensities indicate that some sites are preferred to others.

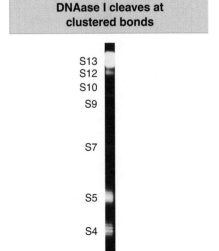

Figure 29.12 High-resolution analysis shows that each site for DNAase I consists of several adjacent susceptible phosphodiester bonds as seen in this example of sites S4 and S5 analyzed in end-labeled core particles. Photograph kindly provided by Leonard Lutter, Molecular Biology Research Program, Henry Ford Hospital, Detroit, MI.

One face of DNA is exposed to DNAase I

Sites exposed to DNAase I

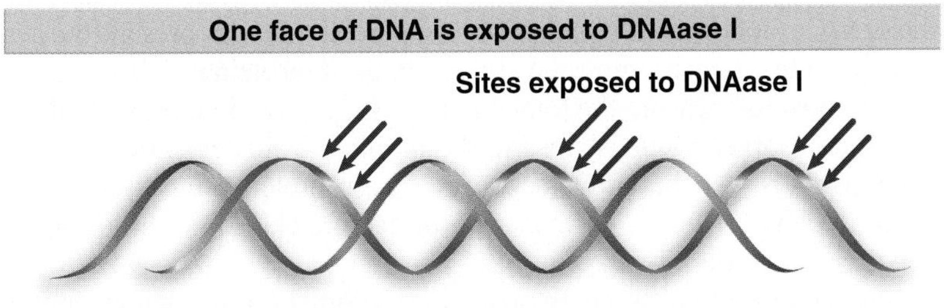

Figure 29.11 The most exposed positions on DNA recur with a periodicity that reflects the structure of the double helix. (For clarity, sites are shown for only one strand.)

From this pattern, we can calculate the "average" point that is cut. At the ends of the DNA, pairs of sites from S1 to S4 or from S10 to S13 lie apart a distance of 10.0 bases each. In the center of the particle, the separation from sites S4 to S10 is 10.7 bases. (Because this analysis deals with *average* positions, sites need not lie an integral number of bases apart.)

The variation in cutting periodicity along the core DNA (10.0 at the ends, 10.7 in the middle) means that there is variation in the structural periodicity of core DNA. The DNA has more base pairs per turn than its solution value in the middle, but has fewer base pairs per turn at the ends. The average periodicity over the nucleosome is only 10.17 bp/turn, which is less than the 10.5 bp/turn of DNA in solution.

29.6 The Nucleosome Absorbs Some Supercoiling

Key Terms

- The **minichromosome** of SV40 or polyoma is the nucleosomal form of the viral circular DNA.
- The **linking number paradox** describes the discrepancy between the existence of −2 supercoils in the path of DNA on the nucleosome compared with the measurement of −1 supercoil released when histones are removed.

Key Concept

- ~0.6 negative turns of DNA are absorbed by the change in base pairs per turn from 10.5 in solution to an average of 10.2 on the nucleosomal surface, explaining the linking number paradox.

Some insights into the structure of nucleosomal DNA emerge when we compare predictions for supercoiling in the path that DNA follows with actual measurements of supercoiling of nucleosomal DNA. Much work on the structure of sets of nucleosomes has been carried out with the virus SV40. The DNA of SV40 is a circular molecule of 5200 bp, with a contour length ~1500 nm. In both the virion and infected nucleus, it is packaged into a series of nucleosomes, called a **minichromosome**.

As usually isolated, the contour length of the minichromosome is ~210 nm, corresponding to a packing ratio of ~7 (essentially the same as the ~6 of the nucleosome itself). Changes in the salt concentration can convert it to a flexible string of beads with a much lower overall packing ratio. This emphasizes the point that nucleosome arrays can take more than one form *in vitro*, depending on the conditions.

The degree of supercoiling on the individual nucleosomes of the minichromosome can be measured as illustrated in **Figure 29.13**. First, the free supercoils of the minichromosome itself are relaxed by treatment with topoisomerase, so that the nucleosomes form a circular string with a superhelical density of 0. Then the histone octamers are extracted. This releases the DNA to follow a free path. Every supercoil that was present but restrained in the minichromosome appears in the deproteinized DNA as −1 turn. So now we measure the total number of supercoils in the SV40 DNA.

The observed value is close to the number of nucleosomes. The reverse result is seen when nucleosomes are assembled *in vitro* onto a supercoiled SV40 DNA: the formation of each nucleosome removes ~1 negative supercoil.

So the DNA follows a path on the nucleosomal surface that generates ~1 negative supercoiled turn when the restraining protein is removed. But the path that DNA follows on the nucleosome corresponds to

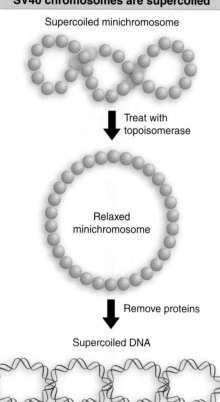

SV40 chromosomes are supercoiled

Supercoiled minichromosome

Treat with topoisomerase

Relaxed minichromosome

Remove proteins

Supercoiled DNA

Figure 29.13 The supercoils of the SV40 minichromosome can be relaxed to generate a circular structure, whose loss of histones then generates supercoils in the free DNA.

−1.65 superhelical turns (see Figure 29.4). This discrepancy is some-times called the **linking number paradox**.

The discrepancy is explained by the difference between the average 10.17 bp/turn of nucleosomal DNA and the 10.5 bp/turn of free DNA. In a nucleosome of 200 bp, there are $200/10.17 = 19.67$ turns. When DNA is released from the nucleosome, it now has $200/10.5 = 19.0$ turns. The path of the less tightly wound DNA on the nucleosome absorbs −0.67 turns, and this explains the discrepancy between the physical path of −1.65 and the measurement of −1.0 superhelical turns. In effect, some of the torsion-al strain in nucleosomal DNA goes into increasing the number of base pairs per turn; only the rest is left to be measured as a supercoil.

29.7 Organization of the Core Particle

Key Term

- The **histone fold** is a motif found in all four core histones in which three α-helices are connected by two loops.

Key Concepts

- The histone octamer has a kernel of a $H3_2 \cdot H4_2$ tetramer associ-ated with two $H2A \cdot H2B$ dimers.

- Each histone is extensively interdigitated with its partner.

- All core histones have the structural motif of the histone fold. N-terminal tails extend out of the nucleosome.

The crystal structure of the core particle suggests that DNA is orga-nized as a flat superhelix, with 1.65 turns wound around the histone oc-tamer. The pitch of the superhelix varies, with a discontinuity in the middle. Regions of high curvature are arranged symmetrically, and occur at positions ±1 and ±4. These correspond to S6 and S8, and to S3 and S11, which are the sites least sensitive to DNAase I.

A high-resolution structure of the nucleosome core shows that most of the supercoiling occurs in the central 129 bp, which are coiled into 1.59 left-handed superhelical turns with a diameter of 80 Å (only 4× the diameter of the DNA duplex itself). The terminal sequences on either end make only a very small contribution to the overall curvature.

The central 129 bp are in the form of B-DNA, but with a substantial curvature that is needed to form the superhelix. The major groove is smoothly bent, but the minor groove has abrupt kinks. These conforma-tional changes may explain why the central part of nucleosomal DNA is not usually a target for binding by regulatory proteins, which typically bind to the terminal parts of the core DNA or to the linker sequences.

The basic framework for the structure of the histone octamer is provided by the abilities of the histones to interact with one another. The core histones form two types of complexes. H3 and H4 form a tetramer ($H3_2 \cdot H4_2$). H2A and H2B form a dimer ($H2A \cdot H2B$). The octamer has a central "kernel" consisting of the $H3_2 \cdot H4_2$ tetramer. This can organize DNA *in vitro* into particles that display some of the properties of the core particle. The full octamer is made by the addition of two $H2A \cdot H2B$ dimers. **Figure 29.14** shows a model for the organiza-tion of the nucleosome.

The crystal structure suggests the model for the histone oc-tamer shown in **Figure 29.15**. Tracing the paths of the individual polypeptide backbones in the crystal structure suggests that the his-

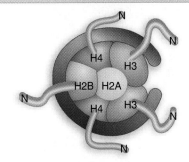

The structures of histone tails are not defined

Figure 29.14 In a symmetrical model for the nucleosome, the $H3_2$–$H4_2$ tetramer provides a kernel for the shape. One H2A–H2B dimer can be seen in this top view; the other is underneath.

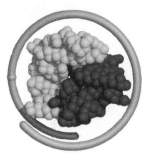

DNA surrounds the histone octamer

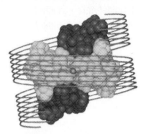

DNA turns twice around the nucleosome

Figure 29.15 The crystal structure of the histone core oc-tamer is represented in a space-filling model with the $H3_2$–$H4_2$ tetramer shown in light blue and the H2A–H2B dimers shown in dark blue. Only one of the H2A–H2B dimers is visible in the top view, because the other is hidden under-neath. The potential path of the DNA is shown in the top view as a narrow tube (one quarter the diameter of DNA), and in the side view by the parallel lines in a 20 Å wide bundle. Pho-tographs kindly provided by Evangelos Moudrianakis, Dept. of Biological Sciences, John Hopkins University.

Histones contact DNA by the histone fold

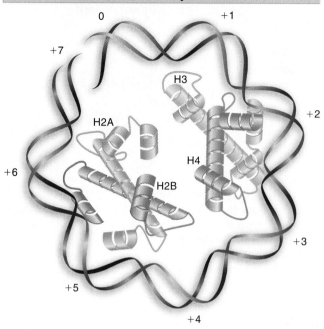

Figure 29.16 All four core histones contact DNA. The structure of a "half nucleosome" shows the contacts with one turn of DNA.

The nucleosome has histone pairs

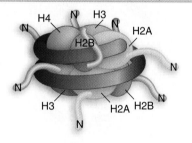

Figure 29.17 The globular bodies of the histones are localized in the histone octamer of the core particle, but the locations of the N-terminal tails, which carry the sites for modification, are not known, and could be more flexible.

Histone tails emerge between DNA turns

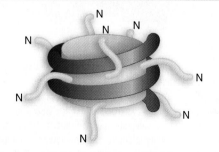

Figure 29.18 The N-terminal histone tails are disordered and exit from the nucleosome between turns of the DNA.

tones are not organized as individual globular proteins, but that each is interdigitated with its partner, H3 with H4, and H2A with H2B. So the model distinguishes the $H3_2 \cdot H4_2$ tetramer (light blue) from the $H2A \cdot H2B$ dimers (dark blue), but does not show individual histones.

The top view shows that the $H3_2 \cdot H4_2$ tetramer accounts for the diameter of the octamer. It forms the shape of a horseshoe. The $H2A \cdot H2B$ pairs fit in as two dimers, but only one can be seen in this view. The side view represents the same perspective that was illustrated in Figure 29.4. Here the responsibilities of the $H3_2 \cdot H4_2$ tetramer and of the separate $H2A \cdot H2B$ dimers can be distinguished. The protein forms a sort of spool, with a superhelical path that could correspond to the binding site for DNA, which would be wound in almost two full turns in a nucleosome. The model displays twofold symmetry about an axis that would run perpendicular through the side view.

All four core histones contact DNA by using a similar type of structure in which three α-helices are connected by two loops. This is called the **histone fold**. These regions interact to form crescent-shaped heterodimers; each heterodimer binds 2.5 turns of the DNA double helix as shown in the illustration of the contacts with one turn of DNA in **Figure 29.16**. H2A–H2B binds at +3.5–+6; H3–H4 binds at +0.5–+3 for the circumference that is illustrated. Binding is mostly to the phosphodiester backbones (consistent with the need to package any DNA irrespective of sequence), involving contacts with a group of positively charged amino acids on the surface of the octamer. In the nucleosome, the $H3_2 \cdot H4_2$ tetramer is formed by contacts between the two H3 histones.

Each of the core histones has a globular domain that contributes to the central protein mass of the nucleosome. Each histone also has a flexible N-terminal tail, which has sites for modification that are important in chromatin function. The positions of the tails, which account for about one quarter of the protein mass, are not so well defined, as indicated in **Figure 29.17**. However, the tails of both H3 and H2B can be seen to pass between the turns of the DNA superhelix and extend out of the nucleosome, as seen in **Figure 29.18**. When histone tails are crosslinked to DNA by UV irradiation, more products are obtained with nucleosomes compared to core particles, which could mean that the tails contact the linker DNA. The tail of H4 appears to contact an H2A–H2B dimer in an adjacent nucleosome; this could be an important feature in the overall structure of folded chromatin.

29.8 The Path of Nucleosomes in the Chromatin Fiber

Key Terms

- The **10 nm fiber** is a linear array of nucleosomes, generated by unfolding from the natural condition of chromatin.
- The **30 nm fiber** is a coiled coil of nucleosomes. It is the basic level of organization of nucleosomes in chromatin.

Key Concepts

- 10 nm chromatin fibers are unfolded from 30 nm fibers and consist of a string of nucleosomes.
- 30 nm fibers have six nucleosomes/turn, organized into a solenoid.
- Histone H1 is required for formation of the 30 nm fiber.

When chromatin is examined in the electron microscope, two types of fibers are seen: the 10 nm fiber and 30 nm fiber. They are described by the approximate diameter of the thread.

The **10 nm fiber** is essentially a continuous string of nucleosomes. Sometimes, indeed, it runs continuously into a more stretched-out region in which nucleosomes are seen as a string of beads, as indicated in the example of **Figure 29.19**. The 10 nm fibril structure is obtained under conditions of low ionic strength and does not require the presence of histone H1. This means that it is a function strictly of the nucleosomes themselves. It may be visualized essentially as a continuous series of nucleosomes, as in **Figure 29.20**. It is most likely a consequence of unfolding during extraction *in vitro*.

When chromatin is visualized in conditions of greater ionic strength the **30 nm fiber** is obtained. An example is given in **Figure 29.21**. The fiber can be seen to have an underlying coiled structure. It has ~6 nucleosomes for every turn, which corresponds to a packing ratio of 40 (that is, each μm along the axis of the fiber contains 40 μm of DNA). The presence of H1 is required. This fiber is the basic constituent of both interphase chromatin and mitotic chromosomes.

The most likely arrangement for packing nucleosomes into the fiber is a solenoid, in which the nucleosomes turn in a helical array, coiled around a central cavity. Recent crosslink data suggest a "two-start" model, in in which the solenoid is formed from a double row of nucleosomes, as shown in **Figure 29.22**.

The 30 nm and 10 nm fibers can be reversibly converted by changing the ionic strength. This suggests that the linear array of nucleosomes in the 10 nm fiber is coiled into the 30 nm structure at higher ionic strength and in the presence of H1.

29.9 Reproduction of Chromatin Requires Assembly of Nucleosomes

Key Concepts

- Histone octamers are not conserved during replication, but H2A · H2B dimers and $H3_2 \cdot H4_2$ tetramers are conserved.
- There are different pathways for the assembly of nucleosomes during replication and independently of replication.
- Accessory proteins are required to assist the assembly of nucleosomes.
- CAF-1 is an assembly protein that is linked to the PCNA subunit of the replisome; it is required for deposition of $H3_2 \cdot H4_2$ tetramers following replication.
- A different assembly protein and a variant of histone H3 may be used for replication-independent assembly.

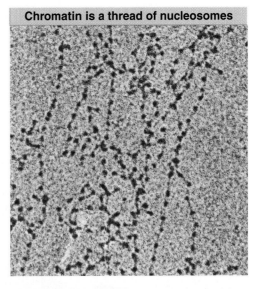

Chromatin is a thread of nucleosomes

Figure 29.19 The 10 nm fiber in partially unwound state can be seen to consist of a string of nucleosomes. Photograph kindly provided by Barbara Hamkalo, Dept. of Molecular Biology and Biochemistry, University of California at Irvine.

10 nm fiber consists of nucleosomes

Figure 29.20 The 10 nm fiber is a continuous string of nucleosomes.

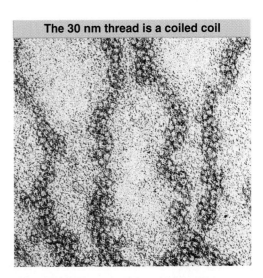

The 30 nm thread is a coiled coil

Figure 29.21 The 30 nm fiber has a coiled structure. Photograph kindly provided by Barbara Hamkalo, Dept. of Molecular Biology and Biochemistry, University of California at Irvine.

The 30 nm fiber is a solenoid

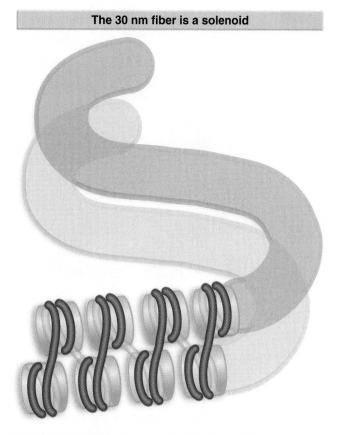

Figure 29.22 The 30 nm fiber is a helical ribbon consisting of two parallel rows of nucleosomes coiled into a solenoid.

Nucleosomes form immediately after replication

Nonreplicated

Replicated

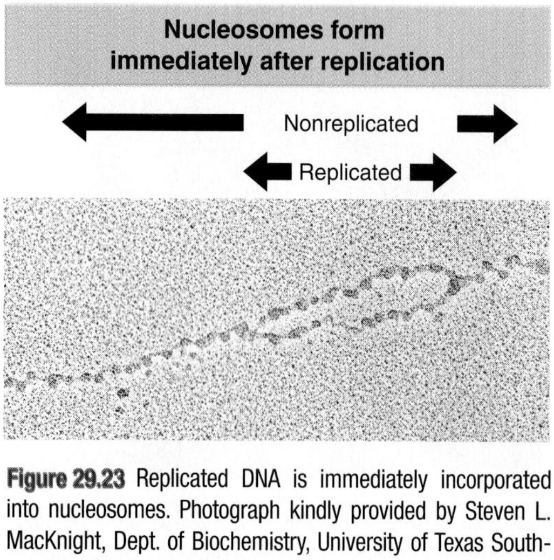

Figure 29.23 Replicated DNA is immediately incorporated into nucleosomes. Photograph kindly provided by Steven L. MacKnight, Dept. of Biochemistry, University of Texas Southwestern Medical Center.

Reproduction of chromatin does not involve any protracted period during which the DNA is free of histones. Once DNA has been replicated, nucleosomes are quickly generated on both the duplicates. This point is illustrated by the electron micrograph of **Figure 29.23**, which shows a recently replicated stretch of DNA, already packaged into nucleosomes on both daughter duplex segments.

Accessory proteins assist histones to associate with DNA. Candidates for this role can be identified by using extracts that assemble histones and exogenous DNA into nucleosomes. Accessory proteins may act as "molecular chaperones" that bind to the histones in order to release either individual histones or complexes ($H3_2 \cdot H4_2$ or $H2A \cdot H2B$) to the DNA in a controlled manner. This could be necessary because the histones, as basic proteins, have a general high affinity for DNA. *Such interactions allow histones to form nucleosomes without becoming trapped in other kinetic intermediates (that is, other complexes resulting from indiscreet binding of histones to DNA).*

Formation of nucleosomes requires an ancillary factor, CAF-1, that consists of >5 subunits, with a total mass of 238 kD. CAF-1 is recruited to the replication fork by PCNA, the processivity factor for DNA polymerase. This provides the link between replication and nucleosome assembly, ensuring that nucleosomes are assembled as soon as DNA has been replicated.

Our working model for the formation of nucleosomes is illustrated in **Figure 29.24**. The replication fork displaces histone octamers, which then dissociate into $H3_2 \cdot H4_2$ tetramers and $H2A \cdot H2B$ dimers. These "old" tetramers and dimers enter a pool that also includes "new" tetramers and dimers, assembled from newly synthesized histones. Nucleosomes assemble ~600 bp behind the replication fork. Assembly is initiated when $H3_2 \cdot H4_2$ tetramers bind to each of the daughter duplexes, assisted by CAF-1. Then two $H2A \cdot H2B$ dimers bind to each $H3_2 \cdot H4_2$ tetramer to complete the histone octamer. The assembly of tetramers and dimers is random with respect to "old" and "new" subunits.

During S phase (the period of DNA replication) in a eukaryotic cell, the duplication of chromatin requires synthesis of sufficient histone proteins to package an entire genome—basically the same quantity of histones must be synthesized that are already contained in nucleosomes. The synthesis of histone mRNAs is controlled as part of the cell cycle, and increases enormously in S phase. The pathway for assembling chromatin from this equal mix of old and new histones during S phase is called the replication-coupled (RC) pathway.

Another pathway, called the replication-independent (RI) pathway, exists for assembling nucleosomes during other phases of cell cycle, when DNA is not being synthesized. This may become necessary as the result of damage to DNA or because nucleosomes are displaced during transcription. The assembly process must necessarily have some differences from the replication-coupled pathway, because it cannot be linked to the replication apparatus. One of the most interesting features of the replication-independent pathway is

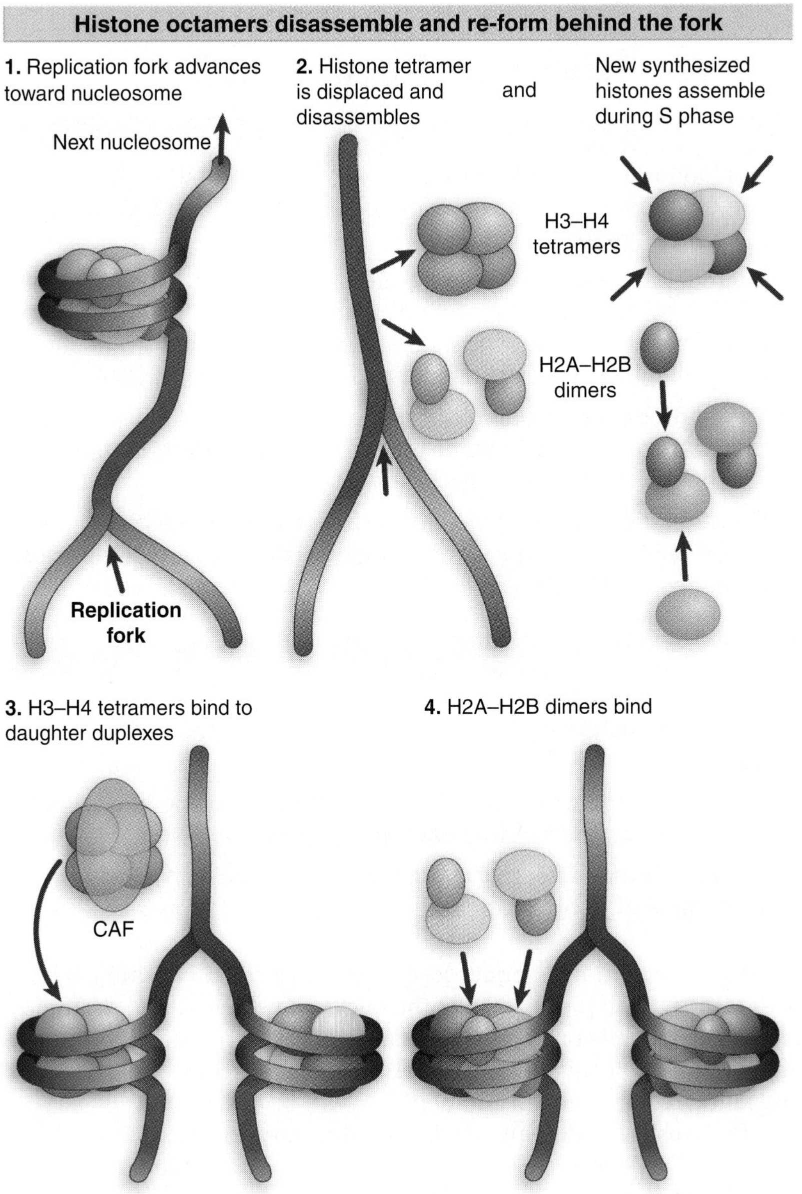

Histone octamers disassemble and re-form behind the fork

1. Replication fork advances toward nucleosome

Next nucleosome

Replication fork

2. Histone tetramer is displaced and disassembles   and

New synthesized histones assemble during S phase

H3–H4 tetramers

H2A–H2B dimers

3. H3–H4 tetramers bind to daughter duplexes

CAF

4. H2A–H2B dimers bind

Figure 29.24 Replication fork passage displaces histone octamers from DNA. They disassemble into H3–H4 tetramers and H2A–H2B dimers. Newly synthesized histones are assembled into H3–H4 tetramers and H2A–H2B dimers. The old and new tetramers and dimers are assembled with the aid of CAF–1 at random into new nucleosomes immediately behind the replication fork.

that it uses different variants of some of the histones from those used during replication.

The histone H3.3 variant differs from the highly conserved H3 histone at four amino acid positions. H3.3 slowly replaces H3 in differentiating cells that do not have replication cycles. This happens as the result of assembly of new histone octamers to replace those that have been displaced from DNA for whatever reason. Different pathways are used in different systems to ensure that H3.3 is used in replacement synthesis.

29.10 Do Nucleosomes Lie at Specific Positions?

Key Terms

- **Nucleosome positioning** describes the placement of nucleosomes at defined sequences of DNA instead of at random locations with regards to sequence.
- **Indirect end labeling** is a technique for examining the organization of DNA by making a cut at a specific site and identifying all fragments containing the sequence adjacent to one side of the cut; it reveals the distance from the cut to the next break(s) in DNA.
- **Translational positioning** describes the location of a histone octamer at successive turns of the double helix, which determines which sequences are located in linker regions.
- **Rotational positioning** describes the location of the histone octamer relative to turns of the double helix, which determines which face of DNA is exposed on the nucleosome surface.

Key Concepts

- Nucleosomes may form at specific positions as the result either of the local structure of DNA or of proteins that interact with specific sequences.
- The most common cause of nucleosome positioning is the binding of proteins to DNA to establish a boundary.
- Positioning may affect which regions of DNA are in the linker and which face of DNA is exposed on the nucleosome surface.

Does a particular DNA sequence always lie in a certain position *in vivo* with regard to the topography of the nucleosome? Or are nucleosomes arranged randomly on DNA, so that a particular sequence may occur at any location, for example, in the core region in one copy of the genome and in the linker region in another?

To investigate this question, it is necessary to use a defined sequence of DNA; more precisely, we need to determine the position relative to the nucleosome of a defined point in the DNA. **Figure 29.25** illustrates the principle of a procedure used to achieve this.

Suppose that the DNA sequence is organized into nucleosomes in only one particular configuration, so that each site on the DNA always is located at a particular position on the nucleosome. This type of organization is called **nucleosome positioning** (or sometimes nucleosome phasing). In a series of positioned nucleosomes, the linker regions of DNA comprise unique sites.

Consider the consequences for just a single nucleosome. Cleavage with micrococcal nuclease generates a monomeric fragment that constitutes a *specific sequence*. If the DNA is isolated and cleaved with a restriction enzyme that has only one target site in this fragment, it should be cut at a unique point. This produces two fragments, each of unique size.

The products of the micrococcal/restriction double digest are separated by gel electrophoresis. A probe representing the sequence on one side of the restriction site is used to identify the corresponding fragment in the double digest. This technique is called **indirect end labeling**.

Reversing the argument, the identification of a single sharp band demonstrates that the position of the restriction site is uniquely defined with respect to the end of the nucleosomal DNA (as defined by the micrococcal nuclease cut). So the nucleosome has a unique sequence of DNA.

Precise fragments identify positioning

Positioning places target sequence (red) at unique position

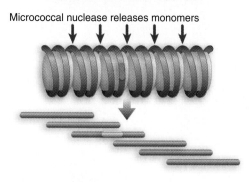

Microccocal nuclease releases monomers

Restriction enzyme cleaves at target sequence

Fragment has restriction cut at one end, micrococcal cut at other end; electrophoresis gives unique band

Figure 29.25 Nucleosome positioning places restriction sites at unique positions relative to the linker sites cleaved by micrococcal nuclease.

What happens if the nucleosomes do *not* lie at a single position? Now the linkers consist of *different* DNA sequences in each copy of the genome. So the restriction site lies at a different position each time; in fact, it lies at all possible locations relative to the ends of the monomeric nucleosomal DNA. **Figure 29.26** shows that the double cleavage then generates a broad smear, ranging from the smallest detectable fragment (~20 bases) to the length of the monomeric DNA.

Nucleosome positioning might be accomplished in either of two ways:

- It is intrinsic: *every nucleosome is deposited specifically at a particular DNA sequence.* This would require a change in our view of the nucleosome as a subunit able to form between any sequence of DNA and a histone octamer.
- It is extrinsic: *the first nucleosome in a region is preferentially assembled at a particular site.* A preferential starting point for nucleosome positioning results from the presence of a region from which nucleosomes are excluded. The excluded region provides a *boundary* that restricts the positions available to the adjacent nucleosome. Then a series of nucleosomes may be assembled sequentially, with a defined repeat length. Positioning of nucleosomes near boundaries is common.

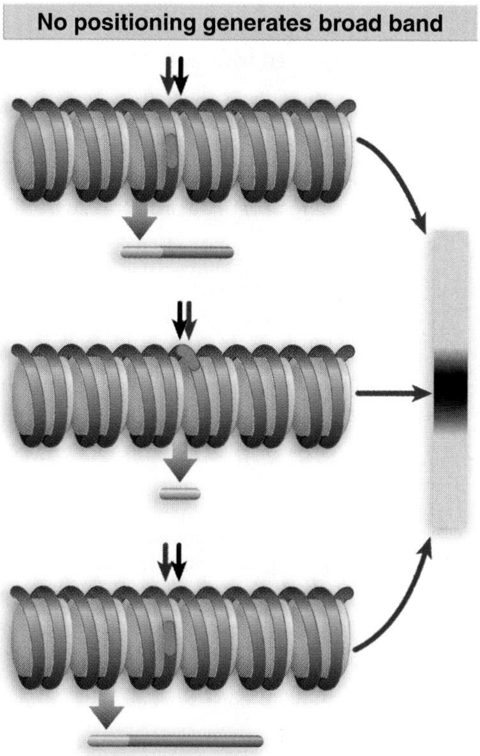

It is now clear that the deposition of histone octamers on DNA is not random with regard to sequence. The pattern is intrinsic in some cases, in which it is determined by structural features in DNA. It is extrinsic in other cases, in which it results from the interactions of other proteins with the DNA and/or histones.

Certain structural features of DNA affect placement of histone octamers. DNA has intrinsic tendencies to bend in one direction rather than another; thus A·T-rich regions locate so that the minor groove faces in towards the octamer, whereas G·C-rich regions are arranged so that the minor groove points out. Long runs of dA·dT (>8 bp) avoid positioning in the central superhelical turn of the core. It is not yet possible to sum all of the relevant structural effects and thus entirely to predict the location of a particular DNA sequence with regard to the nucleosome. Sequences that cause DNA to take up more extreme structures may have effects such as the exclusion of nucleosomes, and thus could cause boundary effects.

The location of DNA on nucleosomes can be described in two ways. **Figure 29.27** shows that **translational positioning** describes the position of DNA with regard to the boundaries of the nucleosome. In particular, it determines which sequences are found in the linker regions. Shifting the DNA by 10 bp brings the next turn into a linker region. So translational positioning determines which regions are more accessible (at least as judged by sensitivity to micrococcal nuclease).

Because DNA lies on the outside of the histone octamer, one face of any particular sequence is obscured by the histones, but the other face is accessible. Depending upon its positioning with regard to the nucleosome, a site in DNA that must be recognized by a regulator protein could be inaccessible or available. The exact position of the histone octamer with respect to DNA sequence may therefore be important. **Figure 29.28** shows the effect of **rotational positioning** of the double helix with regard to the octamer surface. If the DNA is moved by a partial number of turns (imagine the DNA as rotating relative to the protein surface), there is a change in the exposure of sequence to the outside.

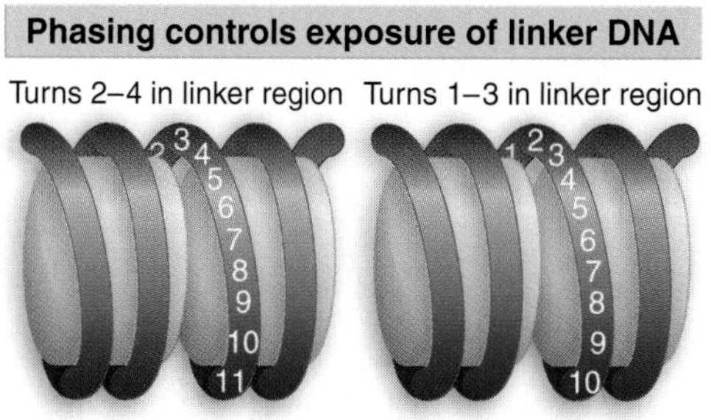

No positioning generates broad band

Figure 29.26 In the absence of nucleosome positioning, a restriction site lies at all possible locations in different copies of the genome. Fragments of all possible sizes are produced when a restriction enzyme cuts at a target site (red) and micrococcal nuclease cuts at the junctions between nucleosomes (black).

Phasing controls exposure of linker DNA

Turns 2–4 in linker region   Turns 1–3 in linker region

Figure 29.27 Translational positioning describes the linear position of DNA relative to the histone octamer. Displacement of the DNA by 10 bp changes the sequences that are in the more exposed linker regions, but does not alter which face of DNA is protected by the histone surface and which is exposed to the exterior.

Bases 1–5 are on the outside

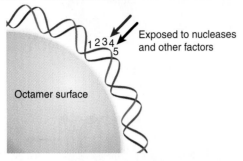

Exposed to nucleases
and other factors

Octamer surface

Bases 1–5 are on the inside

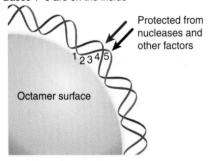

Protected from
nucleases and
other factors

Octamer surface

Figure 29.28 Rotational positioning describes the exposure of DNA on the surface of the nucleosome. Any movement that differs from the helical repeat (~10.2 bp/turn) displaces DNA with reference to the histone surface. Nucleotides on the inside are more protected against nucleases than nucleotides on the outside.

Both translational and rotational positioning can be important in controlling access to DNA. The best-characterized cases of positioning are for the specific placement of nucleosomes at promoters. Translational positioning and/or the exclusion of nucleosomes from a particular sequence may be necessary to allow a transcription complex to form. Some regulatory factors can bind to DNA only if a nucleosome is excluded to make the DNA freely accessible, and this creates a boundary for translational positioning. In other cases, regulatory factors can bind to DNA on the surface of the nucleosome, but rotational positioning is important to ensure that the face of DNA with the appropriate contact points is exposed. We discuss the connection between nucleosomal organization and transcription in *30.4 Nucleosome Organization May Be Changed at the Promoter*.

29.11 Histone Octamers Are Displaced by Transcription

Key Concepts

- Nucleosomes are found at the same frequency when transcribed genes or nontranscribed genes are digested with micrococcal nuclease.
- Some heavily transcribed genes appear to be exceptional cases that are devoid of nucleosomes.
- RNA polymerase displaces histone octamers during transcription in a model system, but octamers reassociate with DNA as soon as the polymerase has passed.
- Nucleosomes are reorganized when transcription passes through a gene.

The first question to ask about the structure of active genes is whether DNA being transcribed remains organized in nucleosomes. If the histone octamers are displaced, do they remain attached in some way to the transcribed DNA?

One experimental approach is to digest chromatin with micrococcal nuclease, and then to use a probe to some specific gene or genes to determine whether the corresponding fragments are present in the usual 200 bp ladder at the expected concentration. The conclusions that we can draw from these experiments are limited but important. *Genes that are being transcribed contain nucleosomes at the same frequency as nontranscribed sequences.* So genes do not necessarily enter an alternative form of organization in order to be transcribed.

But since the average transcribed gene probably only has a single RNA polymerase at any given moment, this does not reveal what is happening at sites actually engaged by the enzyme. Perhaps they retain their nucleosomes; more likely the nucleosomes are temporarily displaced as RNA polymerase passes through, but reform immediately afterward.

Experiments to test whether an RNA polymerase can transcribe directly through a nucleosome suggest that the histone octamer is displaced by the act of transcription. **Figure 29.29** shows what happens in a model system when the phage T7 RNA polymerase transcribes a short piece of DNA containing a single octamer core *in vitro*. The core remains associated with the DNA, but is found in a different location. The core is most likely to rebind to the same DNA molecule from which it was displaced.

Figure 29.30 shows a model for polymerase progression. DNA is displaced as the polymerase enters the nucleosome, but the polymerase

Transcription displaces a histone octamer

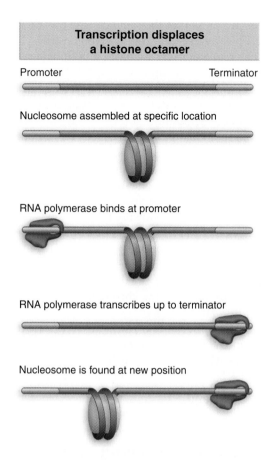

Promoter   Terminator

Nucleosome assembled at specific location

RNA polymerase binds at promoter

RNA polymerase transcribes up to terminator

Nucleosome is found at new position

Figure 29.29 A protocol to test the effect of transcription on nucleosomes shows that the histone octamer is displaced from DNA and rebinds at a new position.

The displaced octamer never leaves DNA

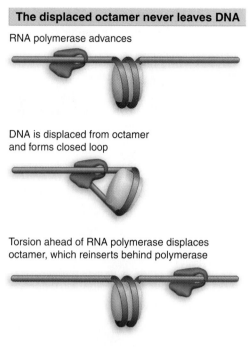

RNA polymerase advances

DNA is displaced from octamer and forms closed loop

Torsion ahead of RNA polymerase displaces octamer, which reinserts behind polymerase

Figure 29.30 RNA polymerase displaces DNA from the histone octamer as it advances. The DNA loops back and attaches (to polymerase or to the octamer) to form a closed loop. As the polymerase proceeds, it generates positive supercoiling ahead. This displaces the octamer, which keeps contact with DNA and/or polymerase, and is inserted behind the RNA polymerase.

reaches a point at which the DNA loops back and reattaches, forming a closed region. As polymerase advances further, unwinding the DNA, it creates positive supercoils in this loop; the effect could be dramatic, because the closed loop is only ~80 bp, so each base pair through which the polymerase advances makes a significant addition to the supercoiling. In fact, the polymerase progresses easily for the first 30 bp into the nucleosome. Then it proceeds more slowly, as though encountering increasing difficulty in progressing. Pauses occur every 10 bp, suggesting that the structure of the loop imposes a constraint related to rotation around each turn of DNA. When the polymerase reaches the midpoint of the nucleosome (the next bases to be added are essentially at the axis of dyad symmetry), pausing ceases, and the polymerase advances rapidly. This suggests that the midpoint of the nucleosome marks the point at which the octamer is displaced (possibly because positive supercoiling has reached some critical level that expels the octamer from DNA). This releases tension ahead of the polymerase and allows it to proceed. The octamer then binds to the DNA behind the polymerase and no longer presents an obstacle to progress. Probably the octamer changes position without ever completely losing contact with the DNA.

The organization of nucleosomes may be changed by transcription. **Figure 29.31** shows what happens to the yeast *URA3* gene when it transcribed under control of an inducible promoter. Positioning is examined by using micrococcal nuclease to examine cleavage sites relative to a restriction site at the 5′ end of the gene. Initially the gene displays a pattern of nucleosomes that are organized from the promoter for a significant distance across the gene; positioning is lost in the 3′ regions. When the gene is expressed, a general smear replaces the positioned pattern of nucleosomes.

Expression changes nucleosome organization

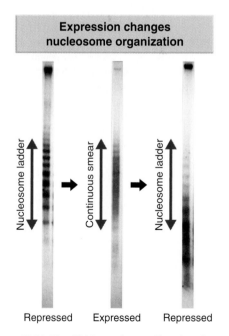

Nucleosome ladder   →   Continuous smear   →   Nucleosome ladder

Repressed   Expressed   Repressed

Figure 29.31 The *URA3* gene has positioned nucleosomes before transcription. When transcription is induced, nucleosome positions are randomized. When transcription is repressed, the nucleosomes resume their particular positions. Photograph kindly provided by Fritz Thoma, Institute of Cell Biology, Zurich, Switzerland.

Factors disassemble and reassemble nucleosomes

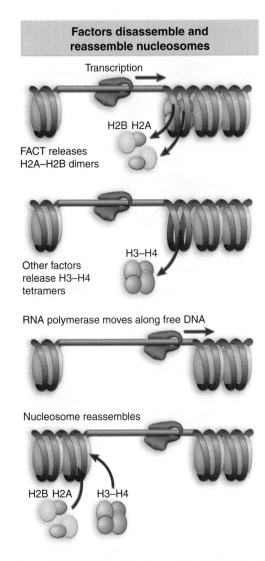

Transcription

FACT releases
H2A–H2B dimers

H2B H2A

Other factors
release H3–H4
tetramers

H3–H4

RNA polymerase moves along free DNA

Nucleosome reassembles

H2B H2A   H3–H4

Figure 29.32 Histone octamers are disassembled ahead of transcription to remove nucleosomes. They reform following transcription. Release of H2A · H2B dimers probably initiates the disassembly process.

So, nucleosomes are present at the same density but are no longer organized in phase. This suggests that transcription destroys the nucleosomal positioning. When repression is reestablished, positioning is restored.

The unifying model is to suppose that RNA polymerase displaces histone octamers as it progresses. In fact, displacing nucleosomes from DNA is a key requirement for all stages of transcription. (Before transcription is initiated, nucleosomes are removed from the promoter by special chromatin remodeling complexes; see *30.2 Chromatin Remodeling Is an Active Process.*) This means that RNA polymerase starts DNA synthesis on a short stretch of DNA unimpeded by nucleosomes. For it to continue advancing during elongation, the histone octamers ahead of it must be displaced. And then, to avoid leaving naked DNA behind it, the octamers must re-form following transcription.

The displacement and re-formation of histone octamers is assisted by ancillary proteins. One is a protein called FACT that behaves like a transcription elongation factor (it is not part of RNA polymerase, but associates with it specifically during the elongation phase of transcription). FACT is associated with the chromatin of active genes. It causes nucleosomes to lose H2A · H2B dimers. It may also be involved in the reassembly of nucleosomes after transcription, because it assists formation of nucleosomes from core histones. This suggests the model shown in **Figure 29.32**, in which FACT detaches H2A · H2B from a nucleosome in front of RNA polymerase, and then helps to add it to a nucleosome that is reassembling behind the enzyme. Other factors are required to complete the process.

29.12 DNAase Hypersensitive Sites Change Chromatin Structure

Key Term

- A **hypersensitive site** is a short region of chromatin detected by its extreme sensitivity to cleavage by DNAase I and other nucleases; it is an area from which nucleosomes are excluded.

Key Concepts

- Hypersensitive sites are found at the promoters of expressed genes.
- They are generated by the binding of transcription factors that displace histone octamers.

In addition to the general changes that occur in active or potentially active regions, structural changes occur at specific sites associated with initiation of transcription or with certain structural features in DNA. These changes were first detected by the effects of digestion with very low concentrations of the enzyme DNAase I.

When chromatin is digested with DNAase I, the first effect is the introduction of breaks in the duplex at specific **hypersensitive sites**. These sites represent chromatin regions in which the DNA is particularly exposed because it is not organized in the usual nucleosomal structure. A typical hypersensitive site is 100× more sensitive to enzyme attack than bulk chromatin.

Many of the hypersensitive sites are related to gene expression. Every active gene has a hypersensitive site, or sometimes more than one site, in the region of the promoter. *Most hypersensitive sites are found only in chromatin of cells in which the associated gene is being expressed;*

they do not occur when the gene is inactive. The 5′ hypersensitive sites appear before transcription begins, and the DNA sequences contained within the hypersensitive sites are required for gene expression.

A particularly well-characterized nuclease-sensitive region lies on the SV40 minichromosome. A short segment near the origin of replication, just upstream of the promoter for the late transcription unit, is cleaved preferentially by DNAase I, micrococcal nuclease, and other nucleases (including restriction enzymes).

Figure 29.33 visualizes a "gap" in the nucleosomal organization of the SV40 minichromosome by electron microscopy. The gap is a region of ~120 nm in length (about 350 bp), surrounded on either side by nucleosomes. The visible gap corresponds with the nuclease-sensitive region. This shows directly that increased sensitivity to nucleases is associated with the exclusion of nucleosomes.

A hypersensitive site is not necessarily uniformly sensitive to nucleases. **Figure 29.34** shows the maps of two hypersensitive sites.

Within the SV40 gap of ~300 bp, there are two hypersensitive DNAase I sites and a "protected" region. The protected region presumably reflects the association of nonhistone protein with the DNA. The gap is associated with the DNA sequence elements that are necessary for promoter function.

The hypersensitive site at the β-globin promoter is preferentially digested by several enzymes, including DNAase I, DNAase II, and micrococcal nuclease. The enzymes have preferred cleavage sites that lie at slightly different points in the same general region. So a region extending from about −70 to −270 is preferentially accessible to nucleases when the gene is transcribable.

What is the structure of the hypersensitive site? Its preferential accessibility to nucleases indicates that it is not protected by histone octamers, but this does not necessarily imply that it is free of protein. A region of free DNA might be vulnerable to damage; and in any case, how would it be able to exclude nucleosomes? *A hypersensitive site results from the binding of specific regulatory proteins that exclude nucleosomes.* The presence of such proteins is the basis for the existence of the protected region within the hypersensitive site. The mechanism for displacing histone octamers depends on chromatin remodeling complexes, which use ATP to provide the energy to make the change. They can be recruited to act at specific sites on chromatin by a variety of factors (see *30.4 Nucleosome Organization May Be Changed at the Promoter*).

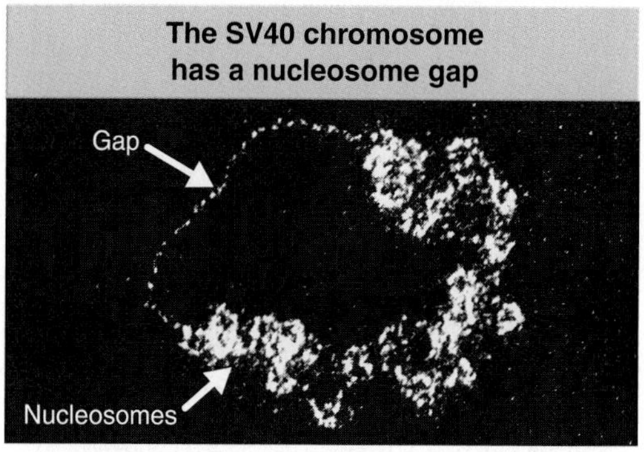

The SV40 chromosome has a nucleosome gap

Figure 29.33 The SV40 minichromosome has a nucleosome gap. Photograph kindly provided by Moshe Yaniv, Unit of Gene Expression and Diseases, Institut Pasteur.

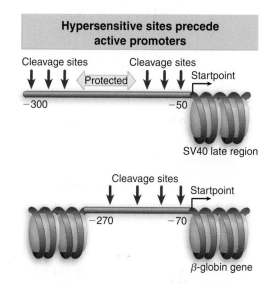

Hypersensitive sites precede active promoters

Figure 29.34 The SV40 gap includes hypersensitive sites, sensitive regions, and a protected region of DNA. The hypersensitive site of a chicken β-globin gene comprises a region that is susceptible to several nucleases.

29.13 Domains Define Regions that Contain Active Genes

Key Concept

- A domain containing a transcribed gene is defined by increased sensitivity to degradation by DNAase I.

Key Term

- A **domain** of a chromosome may refer *either* to a discrete structural entity defined as a region within which supercoiling is independent of other domains *or* to an extensive region including an expressed gene that has heightened sensitivity to degradation by the enzyme DNAase I.

A region of the genome that contains an active gene may have an altered structure. The change in structure precedes, and is different from, the disruption of nucleosome structure that may be caused by the actual passage of RNA polymerase.

One indication of the change in structure of transcribed chromatin is provided by its increased susceptibility to degradation by DNAase I. DNAase I sensitivity defines a chromosomal **domain**, a region of altered structure including at least one active transcription unit, and sometimes extending farther. (Note that use of the term "domain" does

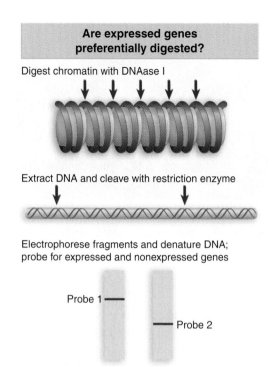

Are expressed genes preferentially digested?

Digest chromatin with DNAase I

Extract DNA and cleave with restriction enzyme

Electrophorese fragments and denature DNA; probe for expressed and nonexpressed genes

Probe 1

Probe 2

Compare intensities of bands in preparations in which chromatin was digested with increasing concentrations of DNAase

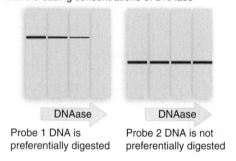

DNAase — DNAase

Probe 1 DNA is preferentially digested — Probe 2 DNA is not preferentially digested

Figure 29.35 Sensitivity to DNAase I can be measured by determining the rate of disappearance of the material hybridizing with a particular probe.

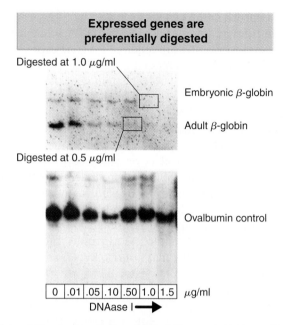

Expressed genes are preferentially digested

Digested at 1.0 μg/ml

Embryonic β-globin

Adult β-globin

Digested at 0.5 μg/ml

Ovalbumin control

| 0 | .01 | .05 | .10 | .50 | 1.0 | 1.5 | μg/ml |

DNAase I ➞

Figure 29.36 In adult erythroid cells, the adult β-globin gene is highly sensitive to DNAase I digestion, the embryonic β-globin gene is partially sensitive (probably due to spreading effects), but ovalbumin is not sensitive. Data kindly provided by Harold Weintraub.

not imply any necessary connection with the structural domains identified by the loops of chromatin or chromosomes.)

When chromatin is digested with DNAase I, it is eventually degraded into acid-soluble material (very small fragments of DNA). The progress of the overall reaction can be followed in terms of the proportion of DNA that is rendered acid soluble. *When only 10% of the total DNA has become acid soluble, more than 50% of the DNA of an active gene has been lost.* This suggests that active genes are preferentially degraded.

The fate of individual genes can be followed by quantitating the amount of DNA that survives to react with a specific probe. The protocol is outlined in **Figure 29.35**. The principle is that the loss of a particular band indicates that the corresponding region of DNA has been degraded by the enzyme.

Figure 29.36 shows what happens to β-globin genes and an ovalbumin gene in chromatin extracted from chicken red blood cells (in which globin genes are expressed and the ovalbumin gene is inactive). The restriction fragments representing the β-globin genes are rapidly lost, while those representing the ovalbumin gene show little degradation. (The ovalbumin gene in fact is digested at the same rate as the bulk of DNA.)

So the bulk of chromatin is relatively resistant to DNAase I and contains nonexpressed genes (as well as other sequences). *A gene becomes relatively susceptible to the enzyme specifically in the tissues in which it is expressed.*

Is preferential susceptibility a characteristic only of rather actively expressed genes, such as globin, or of all active genes? Experiments using probes representing the entire cellular mRNA population suggest that all active genes, whether coding for abundant or for rare mRNAs, are preferentially susceptible to DNAase I. (However, there are variations in the degree of susceptibility.) Because the rarely expressed genes are likely to have very few RNA polymerase molecules actually engaged in transcription at any moment, this implies that the sensitivity to DNAase I does not result from the act of transcription, but is a feature of *genes that are able to be transcribed.*

The critical concept implicit in the description of the domain is that a region of high sensitivity to DNAase I extends over a considerable distance. Often we think of regulation as residing in events that occur at a discrete site in DNA—for example, in the ability to initiate transcription at the promoter. Even if this is true, such regulation must determine, or must be accompanied by, a more wide-ranging change in structure. This is a difference between eukaryotes and prokaryotes.

29.14 Insulators Block the Actions of Enhancers and Heterochromatin

Key Term

- An **insulator** is a sequence that prevents an activating or inactivating effect passing from one side to the other.

Key Concepts

- Insulators are able to block passage of any activating or inactivating effects from enhancers or silencers.
- Insulators may provide barriers against the spread of heterochromatin.
- Two insulators can protect the region between them from all external effects.

Chromatin contains specific structures embedded within the coil of nucleosomes that may have activating or inactivating effects on gene expression. They are often identified as hypersensitive sites because of the change in structure. Hypersensitive sites associated with activation are located in promoters or enhancers. Elements that have the opposite effect and that prevent the passage of activating or inactivating effects are called **insulators**. They have either or both of two key properties:

- When an insulator is placed between an enhancer and a promoter, *it prevents the enhancer from activating the promoter*. The blocking effect is shown in **Figure 29.37**. This may explain how the action of an enhancer is limited to a particular promoter.
- When an insulator is placed between an active gene and heterochromatin, *it provides a barrier that protects the gene against the inactivating effect that spreads from the heterochromatin*. (Heterochromatin is a region of chromatin that is inactive as the result of its higher-order structure; see *31.2 Heterochromatin propagates from a nucleation event*.) The barrier effect is shown in **Figure 29.38**.

Some insulators possess both these properties, but others have only one, or the blocking and barrier functions can be separated. Although both actions are likely to be mediated by changing chromatin structure, they may involve different effects. In either case, however, the insulator defines a limit for long-range effects.

What is the purpose of an insulator? A major function may be to counteract the indiscriminate actions of enhancers on promoters. Most enhancers will work with any promoter in the vicinity. An insulator can restrict an enhancer by blocking the effects from passing beyond a certain point, so that it can act only on a specific promoter. Similarly, when a gene is located near heterochromatin, an insulator can prevent it from being inadvertently inactivated by the spread of the heterochromatin. Insulators therefore function as elements for increasing the precision of gene regulation.

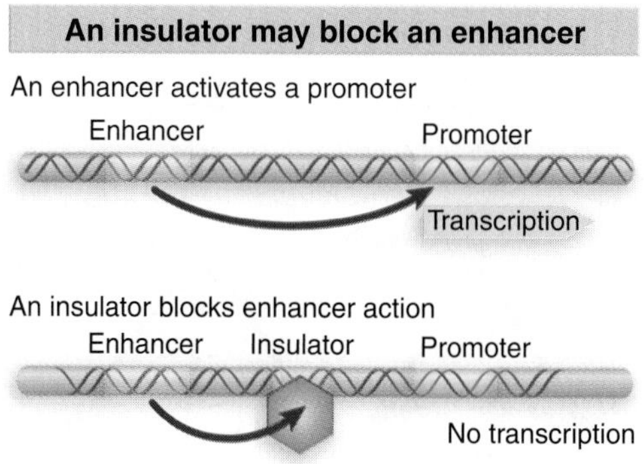

An insulator may block an enhancer

An enhancer activates a promoter

Enhancer   Promoter

Transcription

An insulator blocks enhancer action

Enhancer   Insulator   Promoter

No transcription

Figure 29.37 An enhancer activates a promoter in its vicinity, but may be blocked from doing so by an insulator located between them.

An insulator may block heterochromatin

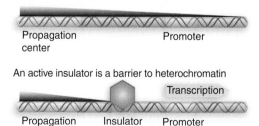

Propagation   Promoter
center

An active insulator is a barrier to heterochromatin

Transcription

Propagation   Insulator   Promoter
center

Figure 29.38 Heterochromatin may spread from a center and then block any promoters that it covers. An insulator may be a barrier to propagation of heterochromatin that allows the promoter to remain active.

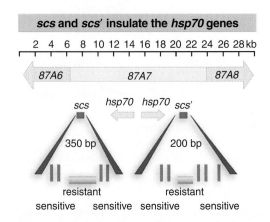

Figure 29.39 Specialized chromatin structures that include hypersensitive sites mark the ends of a domain in the *D. melanogaster* genome and insulate genes between them from the effects of surrounding sequences.

By blocking activating and inactivating effects, two insulators can define the region between them as a domain. This effect was characterized with the first insulators that were discovered, in the region of the *D. melanogaster* genome summarized in **Figure 29.39**. Two genes for the protein Hsp70 lie within an 18 kb region that constitutes band 87A7. Special structures, called *scs* and *scs'* (specialized chromatin structures), are found at the ends of the band. Each consists of a region that is highly resistant to degradation by DNAase I, flanked on either side by sites that are hypersensitive to the enzyme, spaced at about 100 bp.

The *scs* elements insulate the *hsp70* genes from the effects of surrounding regions. If we take *scs* units and place them on either side of a *white* gene, the gene can function anywhere it is placed in the genome, even in sites where it would normally be repressed by context, for example, in heterochromatic regions.

The *scs* and *scs'* units do not seem to play either positive or negative roles in controlling gene expression, but just restrict effects from passing from one region to the next. If adjacent regions have repressive effects, however, the *scs* elements might be needed to block the spread of such effects, and therefore could be essential for gene expression. In this case, deletion of such elements could eliminate the expression of the adjacent gene(s).

The *scs* and *scs'* units have different structures and each appears to have a different basis for its insulator activity. The key sequence in the *scs* element is a stretch of 24 bp that binds the product of the *zw5* gene. The insulator property of *scs'* resides in a series of CGATA repeats. The repeats bind a group of related proteins called BEAF-32. The protein shows discrete localization within the nucleus, but the most remarkable data derive from its localization on polytene chromosomes. **Figure 29.40** shows that an anti-BEAF-32 antibody stains ~50% of the interbands of the polytene chromosomes. This suggests that there are many insulators in the genome, and that BEAF-32 is a common part of the insulating apparatus.

A gene that is surrounded by insulators is usually protected against the propagation of inactivating effects from the surrounding regions. The test is to insert DNA into a genome at random locations by transfection. The expression of a gene in the inserted sequence is often erratic; in some instances it is properly expressed, but in others it is extinguished (see *32.5 Genes Can Be Injected into Animal Eggs*). However, when insulators that have a barrier function are placed on either side of the gene in the inserted DNA, its expression typically is uniform in every case.

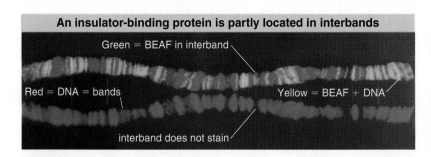

Figure 29.40 A protein that binds to the insulator *scs'* is localized at interbands in *Drosophila* polytene chromosomes. Red staining identifies the DNA (the bands) on both the upper and lower samples; green staining identifies BEAF-32 (often at interbands) on the upper sample. Yellow shows coincidence of the two labels (meaning that BEAF-32 is in a band). Photograph kindly provided by Ulrich K. Laemmli.

29.15 An LCR May Control a Domain

Key Term

- The **locus control region (LCR)** is required for the expression of several genes in a domain.

Key Concept

- An LCR is located at the 5′ end of the domain and consists of several hypersensitive sites.

Every gene is controlled by its promoter, and some genes also respond to enhancers (containing similar control elements but located farther away), as discussed in *Chapter 24 Promoters and Enhancers*. However, these local controls are not sufficient for all genes. In some cases, a gene lies within a domain of several genes all of which are influenced by regulatory elements that act on the whole domain. The existence of these elements was identified by the inability of a region of DNA including a gene and all its known regulatory elements to be properly expressed when introduced into an animal as a transgene.

The best-characterized example of a regulated gene cluster is provided by the mouse β-globin genes. Recall from Figure 6.3 that the α-globin and β-globin genes in mammals each exist as clusters of related genes, expressed at different times during embryonic and adult development. These genes are provided with a large number of regulatory elements, which have been analyzed in detail. In the adult human β-globin gene, regulatory sequences are located both 5′ and 3′ to the gene and include both positive and negative elements in the promoter region, and additional positive elements within and downstream of the gene.

But a human β-globin gene containing all of these control regions is never expressed in a transgenic mouse within an order of magnitude of wild-type levels. Some further regulatory sequence is required. Regions that provide the additional regulatory function are identified by DNAase I hypersensitive sites that are found at the ends of the cluster. The map in **Figure 29.41** shows that the 20 kb upstream of the ε-gene contains a group of four sites, and there is a single site 30 kb downstream of the β-gene. Transfecting various constructs into mouse erythroleukemia cells shows that sequences between the individual hypersensitive sites in the 5′ region can be removed without much effect, but that removal of any of the sites reduces the overall level of expression.

The 5′ regulatory sites are the primary regulators, and the cluster of hypersensitive sites is called the **LCR (locus control region)**. We do not know whether the 3′ site has any function. The LCR is absolutely required for expression of each of the globin genes in the cluster. Each gene is then further regulated by its own specific controls. Some of these controls are autonomous: expression of the ε- and γ-genes appears intrinsic to those loci in conjunction with the LCR. Other controls appear to rely upon position in the cluster, which provides a suggestion that *gene order* in a cluster is important for regulation.

The entire region containing the globin genes, and extending well beyond them, constitutes a chromosomal *domain*. It shows increased sensitivity to digestion by DNAase I (see Figure 29.35). Deletion of the 5′ LCR restores normal resistance to DNAase over the whole region. The leftmost hypersensitive site in this cluster is an insulator that marks the end of

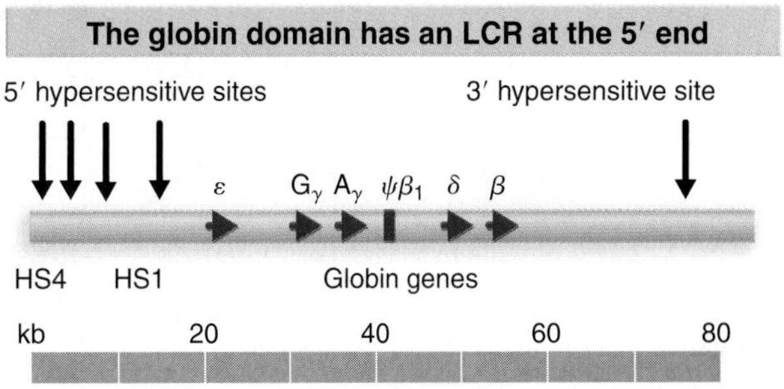

Figure 29.41 A globin domain is marked by hypersensitive sites at either end. The group of sites at the 5′ side constitutes the LCR and is essential for the function of all genes in the cluster.

the domain, and which restricts the LCR to acting only on the globin genes within the domain.

Two models for how an LCR works propose that its action is required in order to activate the promoter, or alternatively, to increase the rate of transcription from the promoter. The exact nature of the interactions between the LCR and the individual promoters has not yet been fully defined.

Does this model apply to other gene clusters? The α-globin locus has a similar organization of genes that are expressed at different times, with a group of hypersensitive sites at one end of the cluster, and increased sensitivity to DNAase I throughout the region. Only a small number of other cases are known in which an LCR controls a group of genes.

29.16 What Constitutes a Regulatory Domain?

Key Concept

- A domain has an insulator, an LCR, a matrix attachment site, and transcription unit(s).

If we now put together the various types of structures that have been found in different systems, we can think about the possible nature of a chromosomal domain. The basic feature of a regulatory domain is that regulatory elements can act only on transcription units within the same domain. A domain might contain more than one transcription unit and/or enhancer. **Figure 29.42** summarizes the structures that might be involved in defining a domain.

An *insulator* stops activating or repressing effects from passing. In its simplest form, an insulator blocks either type of effect from passing across it, but there can be more complex relationships in which the insulator blocks only one type of effect and/or acts directionally. We assume that insulators act by affecting higher order chromatin structure, but we do not know the details and varieties of such effects.

A *matrix attachment region* (MAR) may be responsible for attaching chromatin to a site on the nuclear periphery (see *28.4 Eukaryotic DNA Has Loops and Domains Attached to a Scaffold*). These are likely to be responsible for creating physical domains of DNA that take the form of loops extending out from the attachment sites. This looks very much like one model for insulator action. In fact, some MAR elements behave as insulators in assays *in vitro*, but it seems that the their ability to attach DNA to the matrix can be separated from the insulator function, so there is not a simple cause and effect. It would not be surprising if insulator and MAR elements were associated to maintain a relationship between regulatory effects and physical structure.

An *LCR* functions at a distance and may be required for any and all genes in a domain to be expressed (see previous section). When a domain has an LCR, its function is essential for all genes in the domain, but LCRs do not seem to be common. Several types of *cis*-acting structures could be required for function. As defined originally, the property of the LCR rests with an enhancer–like hypersensitive site that is needed for the full activity of promoter(s) within the domain.

The organization of domains may help to explain the large size of the genome. A certain amount of space could be required for such a structure to operate, for example, to allow chromatin to become decondensed and to become accessible. Although the exact sequences of much of the unit might be irrelevant, there might be selection for the overall amount of DNA within it, or at least selection might prevent the various transcription units from becoming too closely spaced.

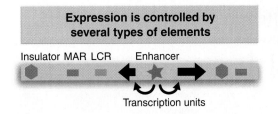

Figure 29.42 Domains may possess three types of sites: insulators to prevent effects from spreading between domains; MARs to attach the domain to the nuclear matrix; and LCRs that are required for initiation of transcription. An enhancer may act on more than one promoter within the domain.

29.17 SUMMARY

All eukaryotic chromatin consists of nucleosomes. A nucleosome contains a characteristic length of DNA, usually ~200 bp, wrapped around an octamer containing two copies each of histones H2A, H2B, H3, and H4. A single H1 protein is associated with each nucleosome. Virtually all genomic DNA is organized into nucleosomes. Treatment with micrococcal nuclease shows that the DNA packaged into each nucleosome can be divided operationally into two regions. The linker region is digested rapidly by the nuclease; the core region of 146 bp is resistant to digestion. Histones H3 and H4 are the most highly conserved and an $H3_2 \cdot H4_2$ tetramer accounts for the diameter of the particle. The H2A and H2B histones are organized as two $H2A \cdot H2B$ dimers. Octamers are assembled by the successive addition of two $H2A \cdot H2B$ dimers to the $H3_2 \cdot H4_2$ kernel.

The path of DNA around the histone octamer creates -1.65 supercoils. The DNA "enters" and "leaves" the nucleosome in the same vicinity, and could be "sealed" by histone H1. Removal of the core histones releases -1.0 supercoils. The difference can be largely explained by a change in the helical pitch of DNA, from an average of 10.5 bp/turn when free in solution to an average of 10.17 bp/turn on the nucleosomal surface. There is variation in the structure of DNA from a periodicity of 10.0 bp/turn at the nucleosome ends to 10.7 bp/turn in the center. There are kinks in the path of DNA on the nucleosome.

Nucleosomes are organized into a fiber of 30 nm diameter which has six nucleosomes per turn and a packing ratio of 40. Removal of H1 allows this fiber to unfold into a 10 nm fiber that consists of a linear string of nucleosomes. The 30 nm fiber is a solenoid made from a double bond of nucleosomes. The 30 nm fiber is the basic constituent of both euchromatin and heterochromatin; nonhistone proteins are responsible for further organization of the fiber into chromatin or chromosome ultrastructure.

There are two pathways for nucleosome assembly. In the replication-coupled pathway, the PCNA processivity subunit of the replisome recruits CAF-1, which is a nucleosome assembly factor. CAF-1 assists the deposition of $H3_2 \cdot H4_2$ tetramers onto the daughter duplexes resulting from replication. The tetramers may be produced either by disruption of existing nucleosomes by the replication fork or as the result of assembly from newly synthesized histones. Similar sources provide the $H2A \cdot H2B$ dimers that then assemble with the $H3_2 \cdot H4_2$ tetramer to complete the nucleosome. Because the $H3_2 \cdot H4_2$ tetramer and the $H2A \cdot H2B$ dimers assemble at random, the new nucleosomes may include both pre-existing and newly synthesized histones.

RNA polymerase displaces histone octamers during transcription. Nucleosomes re-form on DNA after the polymerase has passed, unless transcription is very intensive (such as in rDNA) when they may be displaced completely. The replication-independent pathway for nucleosome assembly is responsible for replacing histone octamers that have been displaced by transcription. It uses the histone variant H3.3 instead of H3. A similar pathway, with another alternative to H3, is used for assembling nucleosomes at centromeric DNA sequences following replication.

Two types of changes in sensitivity to nucleases are associated with gene activity. Chromatin capable of being transcribed has a generally increased sensitivity to DNAase I, reflecting a change in structure over an extensive region that can be defined as a domain containing active or potentially active genes. Hypersensitive sites in DNA occur at discrete locations, and are identified by greatly increased sensitivity to DNAase I. A hypersensitive site forms a boundary that may cause adjacent nucleosomes to be restricted in position. Nucleosome positioning may be important in controlling access of regulatory proteins to DNA.

The variety of situations in which hypersensitive sites occur suggests that their existence reflects a general principle. *Sites at which the double helix initiates an activity are kept free of nucleosomes.* A transcription factor, or some other nonhistone protein concerned with the particular function of the site, modifies the properties of a short region of DNA so that nucleosomes are excluded. The structures formed in each situation need not necessarily be similar (except that each, by definition, creates a site hypersensitive to DNAase I). Hypersensitive sites occur at several types of regulators. Those that regulate transcription include promoters, enhancers, and LCRs. Other sites include origins for replication and centromeres. A promoter or enhancer acts on a single gene, but an LCR contains a group of hypersensitive sites and may regulate a domain containing several genes.

An insulator blocks the transmission of activating or inactivating effects in chromatin. An insulator that is located between an enhancer and a promoter prevents the enhancer from activating the promoter. Most insulators block regulatory effects from passing in either direction, but some are directional. Insulators usually can block both activating effects (enhancer-promoter interactions) and inactivating effects (mediated by spread of heterochromatin), but some are limited to one or the other. Two insulators define the region between them as a regulatory domain; regulatory interactions within the domain are limited to it, and the domain is insulated from outside effects.

30

Chromatin Structure Is a Focus for Regulation

30.1 Introduction

Key Terms

- **Silencing** describes the repression of gene expression in a localized region, usually as the result of a structural change in chromatin.

- **Heterochromatin** describes regions of the genome that are highly condensed, are not transcribed, and are late-replicating. Heterochromatin is divided into two types, which are called constitutive and facultative.

Whether a gene is expressed depends on the structure of chromatin both locally (at the promoter) and in the surrounding domain. Chromatin structure correspondingly can be regulated by individual activation events or by changes that affect a wide chromosomal region. The most localized events concern an individual target gene, where changes in nucleosomal structure and organization occur in the immediate vicinity of the promoter. More general changes may affect regions as large as a whole chromosome.

Local chromatin structure is an integral part of controlling gene expression. Changes at an individual promoter control whether transcription is initiated for a particular gene. These changes may be either activating or repressing.

Activation of a gene requires changes in the state of chromatin: the essential issue is how the transcription factors gain access to the promoter DNA. Genes may exist in either of two structural conditions. Genes are found in an "active" state only in the cells in which they are expressed. The change of structure precedes the act of transcription, and indicates that the gene is "transcribable." This suggests that acquisition of the "active" structure must be the first step in gene expression.

Markers for the change in overall structure are given by the effects of nucleases. Active genes are found in domains of euchromatin with a preferential susceptibility to nucleases (see *29.13 Domains Define Regions That Contain Active Genes*). Hypersensitive sites are created at promoters before a gene is activated (see *29.12 DNAase Hypersensitive Sites Change Cromatin Structure*).

At the molecular level, modifications of histones and of DNA are important. Some activators of gene transcription directly modify histones; in particular, acetylation of histones is associated with gene activation. Conversely, some repressors of transcription function by deacetylating histones. So a reversible change in histone structure in the vicinity of the promoter is involved in the control of gene expression. These changes influence the association of histone octamers with DNA, and are responsible for controlling the presence of nucleosomes at specific sites. This is an important aspect of the mechanism by which a gene is maintained in an active or inactive state.

Changes that affect large regions control the potential of a gene to be expressed. The term **silencing** is used to refer to repression of gene activity in a local chromosomal region. The term **heterochromatin** is used to describe chromosomal regions that are large enough to be seen to have a physically more compact structure in the microscope. The basis for both types of change is the same: additional proteins bind to chromatin and either directly or indirectly prevent transcription factors and RNA polymerase from activating promoters in the region.

30.2 Chromatin Remodeling Is an Active Process

Key Concepts

- Chromatin structure is stable and cannot be changed by altering the equilibrium of transcription factors and histones.
- Chromatin remodeling uses energy provided by hydrolysis of ATP to change the organization of nucleosomes.

Key Term

- **Chromatin remodeling** describes the energy-dependent displacement or reorganization of nucleosomes that occurs in conjunction with activation of genes for transcription.

Two types of models could explain how the state of expression of DNA is changed: equilibrium and discontinuous change-of-state.

Figure 30.1 shows the equilibrium model. Here the only pertinent factor is the concentration of the repressor or activator protein, which drives an equilibrium between free form and DNA-bound form. When the concentration of the protein is high enough, its DNA-binding site is occupied, and the state of expression of the DNA is affected. (Binding might either repress or activate any particular target sequence.) This type of model explains the regulation of transcription in bacterial cells, where gene expression is determined exclusively by the actions of individual repressor and activator proteins (see *Chapter 12 The Operon*). Whether a bacterial gene is transcribed can be predicted from the sum of the concentrations of the various factors that either activate or repress the individual gene. Changes in these concentrations *at any time* will change the state of expression accordingly. In most cases, the protein binding is cooperative, so that once the concentration becomes high enough, there is a rapid association with DNA, resulting in a switch in gene expression.

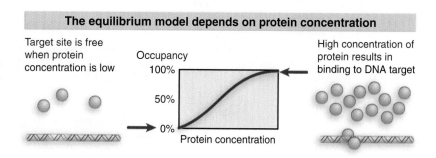

Figure 30.1 In an equilibrium model, the state of a binding site on DNA depends on the concentration of the protein that binds to it.

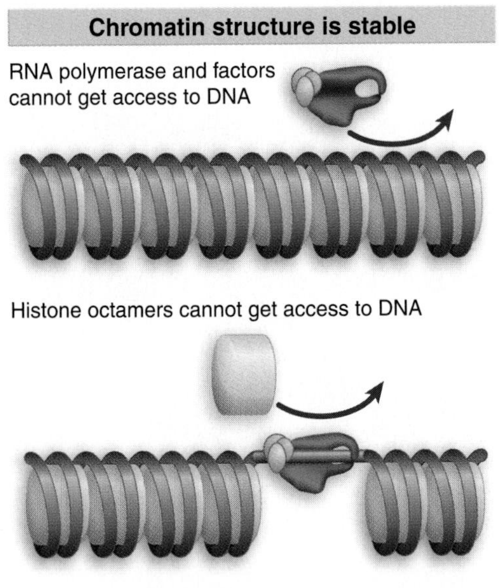

Chromatin structure is stable

RNA polymerase and factors
cannot get access to DNA

Histone octamers cannot get access to DNA

Figure 30.2 If nucleosomes form at a promoter, transcription factors (and RNA polymerase) cannot bind. If transcription factors (and RNA polymerase) bind to the promoter to establish a stable complex for initiation, histones are excluded.

Nucleosome displacement is an active process

Remodeling complex

Octamer is displaced

ATP → ADP + P

Factors and RNA polymerase bind

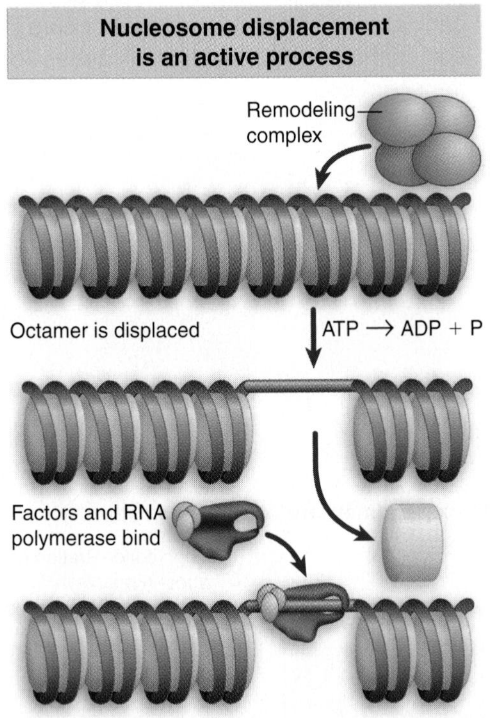

Figure 30.3 The dynamic model for transcription of chromatin relies upon factors that can use energy provided by hydrolysis of ATP to displace nucleosomes from specific DNA sequences.

A different situation applies with eukaryotic chromatin. Early *in vitro* experiments showed that either an active or inactive state can be established, but this is not affected by the subsequent addition of other components. A gene that is wrapped into nucleosomes cannot be activated simply by addition of transcription factors and RNA polymerase. **Figure 30.2** illustrates the two types of conditions that can exist at a eukaryotic promoter. In the inactive state, nucleosomes are present, and they prevent basal factors and RNA polymerase from binding. In the active state, the basal apparatus occupies the promoter, and histone octamers cannot bind to it. Each type of state is stable. The principle that *transcription factors or nucleosomes may form stable structures that cannot be changed merely by changing the equilibrium with free components* now poses the question: how does the state of chromatin change from active to inactive and vice versa?

The general process of inducing changes in chromatin structure is called **chromatin remodeling**. This consists of mechanisms for displacing histones that depend on the input of energy. Many protein–protein and protein–DNA contacts need to be disrupted to release histones from chromatin. There is no free ride: the energy must be provided to disrupt these contacts. **Figure 30.3** illustrates the principle of a *dynamic model* by a factor that hydrolyzes ATP. When the histone octamer is released from DNA, other proteins (transcription factors and RNA polymerase) can bind.

Figure 30.4 summarizes the types of remodeling changes in chromatin:

- Histone octamers may slide along DNA, changing the relationship between the nucleic acid and protein. This alters the position of a particular sequence on the nucleosomal surface.
- The spacing between histone octamers may be changed, again with the result that the positions of individual sequences are altered relative to protein.
- And the most extensive change is that an octamer, or more than one, may be displaced entirely from DNA to generate a nucleosome-free gap.

The most common use of chromatin remodeling is to change the organization of nucleosomes at the promoter of a gene that is to be transcribed. This is required to allow the transcription apparatus to gain access to the promoter. The remodeling most often takes the form of displacing one or more histone octamers. It often results in the creation of a site that is hypersensitive to cleavage with DNAase I (see *29.12 DNAase Hypersensitive Sites Change Chromatin Structure*). Sometimes there are less dramatic changes, for example, in the positioning of nucleosomes.

30.3 There Are Several Chromatin Remodeling Complexes

Key Term

- **SWI/SNF** is a chromatin remodeling complex; it uses hydrolysis of ATP to change the organization of nucleosomes.

Key Concept

- The SWI/SNF, RSC, and NURF remodeling complexes all are very large; they are classified into groups according to the ATPase subunits.

Chromatin remodeling is undertaken by large complexes that use ATP hydrolysis to provide the energy for remodeling. The heart of the remodeling complex is its ATPase subunit. Remodeling complexes are usually classified according to the type of ATPase subunit—those with related ATPase subunits are considered to belong to the same family (usually some other subunits are common also). **Figure 30.5** keeps the names straight. The two major types of complexes are SWI/SNF and ISWI (ISWI stands for imitation SWI). Yeast has at least two complexes of each type. Complexes of both types are also found in fly and in man. Each type of complex may undertake a different range of remodeling activities.

SWI/SNF was the first remodeling complex to be identified. Its name reflects the fact that many of its subunits are coded by genes originally identified by *SWI* or *SNF* mutations in *S. cerevisiae*. Mutations in these loci are pleiotropic, and the range of defects is similar to those shown by mutants that have lost the CTD tail of RNA polymerase II. These mutations also show genetic interactions with mutations in genes that code for components of chromatin, in particular *SIN1*, which codes for a nonhistone protein, and *SIN2*, which codes for histone H3. The *SWI* and *SNF* genes are required for expression of a variety of individual loci (~120 or 2% of *S. cerevisiae* genes are affected). Expression of these loci may require the SWI/SNF complex to remodel chromatin at their promoters.

SWI/SNF acts catalytically *in vitro*, and there are only ~150 complexes per yeast cell. All of the genes encoding the SWI/SNF subunits are nonessential, which implies that yeast must also have other ways of remodeling chromatin. The RSC complex is more abundant and also is essential. It acts at ~700 target loci.

SWI/SNF complexes can either displace histone octamers from DNA or change their positions. Both types of reactions may pass through the same intermediate in which the structure of the target nucleosome is altered, leading either to reformation of a (remodeled) nucleosome on the original DNA or to displacement of the histone octamer to a different DNA molecule. The SWI/SNF complex alters nucleosomal sensitivity to DNAase I at the target site, and induces changes in protein–DNA contacts that persist after it has been released from the nucleosomes. The SWI2 subunit is the ATPase that provides the energy for remodeling by SWI/SNF.

There are many contacts between DNA and a histone octamer—14 are identified in the crystal structure. All of these contacts must be broken for an octamer to be released or for it to move to a new position. How is this achieved? Present thinking is that remodeling complexes in the SWI and ISW classes use the hydrolysis of ATP to twist DNA on the nucleosomal surface. Indirect evidence suggests that this creates a mechanical force that allows a small region of DNA to be released from the surface and then repositioned.

One puzzle about the action of the SWI/SNF complex is its sheer size. It has 11 subunits with a combined molecular weight of $\sim 1.6 \times 10^6$. It dwarfs RNA polymerase and the nucleosome, making it difficult to understand how all of these components could interact with DNA retained on the nucleosomal surface. However, a transcription complex with full activity, called RNA polymerase II holoenzyme, can be found that contains the RNA polymerase itself, all the TF_{II} factors except TBP and $TF_{II}A$, and the SWI/SNF complex, which is associated with the CTD tail of the polymerase. In fact, virtually all of the SWI/SNF complex may be present in holoenzyme preparations. This suggests that the remodeling of chromatin and recognition of promoters are undertaken in a coordinated manner by a single complex.

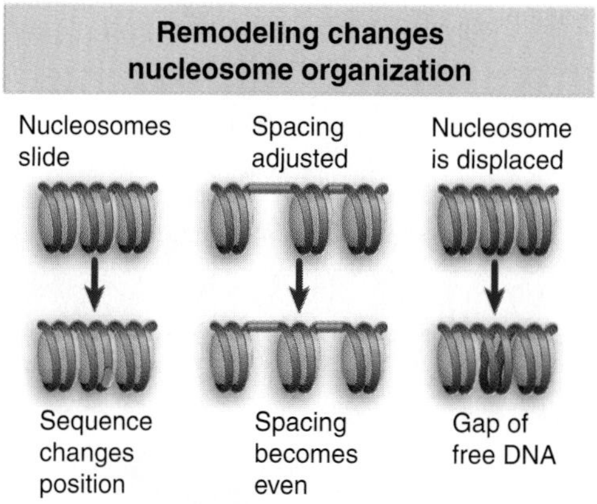

Remodeling changes nucleosome organization

Figure 30.4 Remodeling complexes can cause nucleosomes to slide along DNA, can displace nucleosomes from DNA, or can reorganize the spacing between nucleosomes.

There are several types of remodeling complexes

Type of complex	SWI/SNF	ISWI	Other
Yeast	SWI/SNF RSC	ISW1 ISW2	
Fly	dSWI/SNF (Brahma)	NURF CHRAC ACF	
Human	hSWI/SNF	RSF hACF/WCFR hCHRAC	NuRD
Frog			Mi–2

Figure 30.5 Remodeling complexes can be classified by their ATPase subunits.

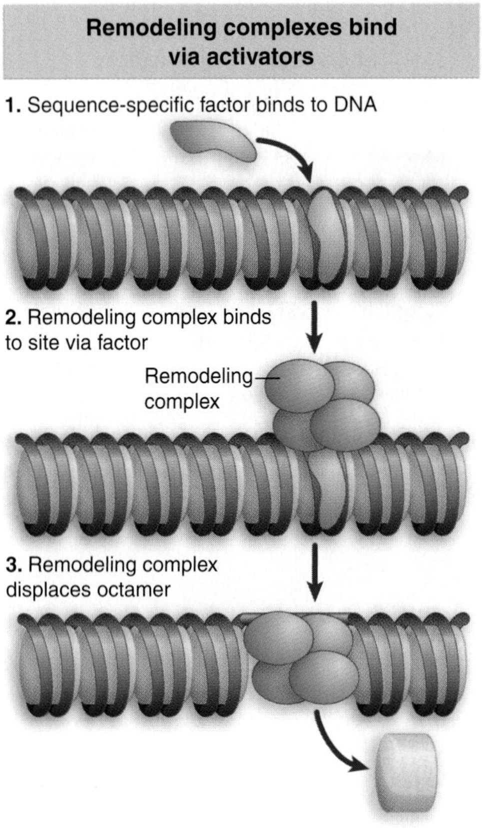

Remodeling complexes bind via activators

1. Sequence-specific factor binds to DNA

2. Remodeling complex binds to site via factor

Remodeling complex

3. Remodeling complex displaces octamer

Figure 30.6 A remodeling complex binds to chromatin via an activator (or repressor).

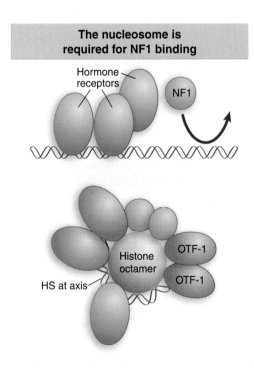

The nucleosome is required for NF1 binding

Hormone receptors

NF1

HS at axis

Histone octamer

OTF-1

OTF-1

Figure 30.7 Hormone receptor and NF1 cannot bind simultaneously to the MMTV promoter in the form of linear DNA, but can bind when the DNA is presented on a nucleosomal surface.

30.4 Nucleosome Organization May Be Changed at the Promoter

Key Concepts

- A remodeling complex does not itself have specificity for any particular target site, but must be recruited by a component of the transcription apparatus.
- The factor may be released once the remodeling complex has bound.
- The MMTV promoter requires a change in rotational positioning of a nucleosome to allow an activator to bind to DNA on the nucleosome.

How are remodeling complexes targeted to specific sites on chromatin? They do not themselves contain subunits that bind specific DNA sequences. This suggests the model shown in **Figure 30.6** in which they are recruited by activators or (sometimes) by repressors.

The interaction between transcription factors and remodeling complexes gives a key insight into their modus operandi. The transcription factor Swi5p activates the *HO* locus in yeast. (Note that Swi5p is not a member of the SWI/SNF complex.) Swi5p enters nuclei toward the end of mitosis and binds to the *HO* promoter. It then recruits SWI/SNF to the promoter. Then Swi5p is released, leaving SWI/SNF at the promoter. This means that a transcription factor can activate a promoter by a "hit and run" mechanism, in which its function is fulfilled once the remodeling complex has bound.

The involvement of remodeling complexes in gene activation was discovered because the complexes are necessary for the ability of certain transcription factors to activate their target genes. One of the first examples was the GAGA factor, which activates the *Drosophila hsp70* promoter *in vitro*. Binding of GAGA to four $(CT)_n$-rich sites on the promoter disrupts the nucleosomes, creates a hypersensitive region, and causes the adjacent nucleosomes to be rearranged so that they occupy preferential instead of random positions. Disruption is an energy-dependent process that requires the NURF remodeling complex. The organization of nucleosomes is altered so as to create a boundary that determines the positions of the adjacent nucleosomes. During this process, GAGA binds to its target sites and DNA, and its presence fixes the remodeled state.

It is not always the case, however, that nucleosomes must be excluded in order to permit initiation of transcription. Some activators can bind to DNA on a nucleosomal surface. Nucleosomes appear to be precisely positioned at some steroid hormone response elements in such a way that receptors can bind. Receptor binding may alter the interaction of DNA with histones, and even lead to exposure of new binding sites. The exact positioning of nucleosomes could be required either because the nucleosome "presents" DNA in a particular rotational phase or because there are protein–protein interactions between the activators and histones or other components of chromatin. So we have now moved some way from viewing chromatin exclusively as a repressive structure to considering which interactions between activators and chromatin can be required for activation.

The MMTV promoter presents an example of the need for specific nucleosomal organization. It contains an array of six partly palindromic sites, each bound by one dimer of hormone receptor (HR), which constitute the HRE. It also has a single binding site for the factor NF1, and two adjacent sites for the factor OTF. HR and NF1 cannot bind simultaneously to their sites in free DNA. **Figure 30.7** shows how the nucleosomal structure controls binding of the factors.

The HR protects its binding sites at the promoter when hormone is added, but does not affect the micrococcal nuclease-sensitive sites that mark either side of the nucleosome. This suggests that HR is binding to the DNA on the nucleosomal surface. However, the rotational positioning of DNA on the nucleosome prior to hormone addition allows access to only two of the four sites. Binding to the other two sites requires a change in rotational positioning on the nucleosome. NF1 can be footprinted on the nucleosome after hormone induction, so these structural changes may be necessary to allow NF1 to bind, perhaps because they expose DNA and abolish the steric hindrance by which HR blocks NF1 binding to free DNA.

30.5 Histone Modification Is a Key Event

Key Concepts

- Histones are modified by acetylation of lysine, and by methylation of lysine and arginine.
- The target amino acids are located in the N-terminal tails of the histones.
- Phosphorylation occurs on serine.

All of the histones are modified by covalently linking extra moieties to the free groups of certain amino acids. Acetylation and methylation occur on the free (ϵ) amino group of lysine. As seen in **Figure 30.8**, acetylation removes the positive charge that resides on the NH_3^+ form of the group. Methylation also occurs on arginine. Phosphorylation occurs on the hydroxyl group of serine and also on threonine. This introduces a negative charge in the form of the phosphate group.

The sites that are modified are concentrated in the N-terminal tails of the histones. The histone tails consist of the N-terminal 20 amino acids, and extend from the nucleosome between the turns of DNA (see Figure 29.18). **Figure 30.9** summarizes the principal sites of modification in histones H3 and H4.

The modifications are transient. Because they change the charge of the protein molecule, they are potentially able to change the functional properties of the octamers. The histone modifications may directly affect nucleosome structure or may create binding sites for the attachment of nonhistone proteins that change the properties of chromatin. Modification of histones is associated with structural changes in chromatin at replication and transcription.

The range of nucleosomes that is targeted for modification can vary. Modification can be a local event, for example, restricted to nucleosomes at the promoter. Or it can be a general event, extending for example to an entire chromosome. **Figure 30.10** shows that there is a general correlation in which acetylation is associated with active chromatin while methylation is associated with inactive chromatin. However, this is not a simple rule, and the particular sites that are modified, as well as combinations of specific modifications, may be important, and there are exceptions in which acetylation of an amino acid in a histone tail is associated with inactive chromatin.

30.6 Histone Acetylation Occurs in Two Circumstances

Key Concepts

- Histones are acetylated transiently at replication.
- Histone acetylation is associated with activation of gene expression.

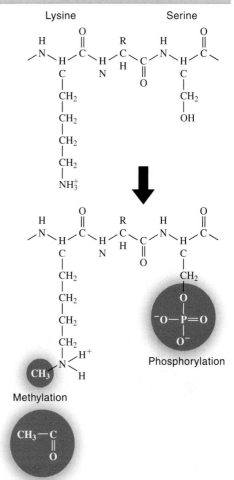

Lysine and serine are targets for modification

Figure 30.8 Acetylation of lysine or phosphorylation of serine reduces the overall positive charge of a protein.

Histone N-terminal tails have many sites of modification

Sites of modification in H3

Ala Arg Thr Lys Gln Thr Ala Arg Lys Ser Thr Glu Glu Lys
 1   2   3   4   5   6   7   8   9 10 11 12 13 14

Sites of modification in H4

Ser Glu Arg Glu Lys Glu Glu Lys Glu Leu Glu Lys Glu Glu
 1   2   3   4   5   6   7   8   9 10 11 12 13 14

Figure 30.9 The N-terminal tails of histones H3 and H4 can be acetylated, methylated, or phosphorylated at several positions.

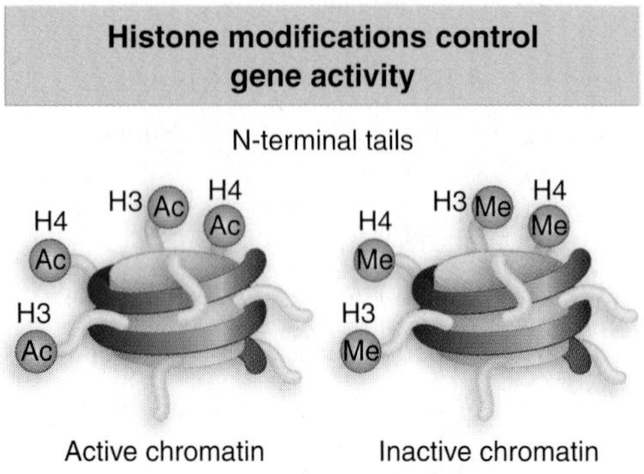

Histone modifications control gene activity

N-terminal tails

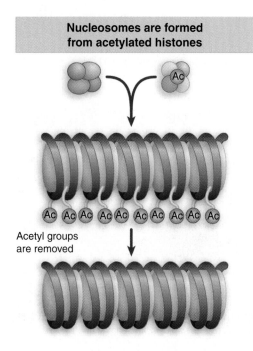

Active chromatin   Inactive chromatin

Figure 30.10 Acetylation of H3 and H4 is associated with active chromatin, while methylation is associated with inactive chromatin.

Nucleosomes are formed from acetylated histones

Acetyl groups are removed

Figure 30.11 Acetylation at replication occurs on histones before they are incorporated into nucleosomes.

Nucleosomal histones are acetylated

Inactive gene

Active gene

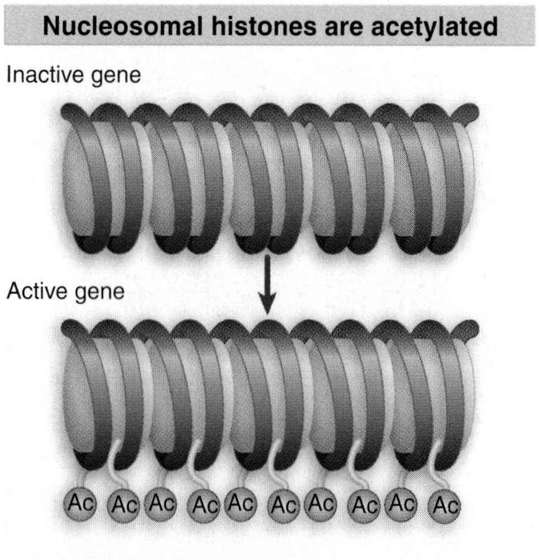

Figure 30.12 Acetylation associated with gene activation occurs by directly modifying histones in nucleosomes.

All the core histones can be acetylated. The major targets for acetylation are lysines in the N-terminal tails of histones H3 and H4. Acetylation occurs in two different circumstances:

- during DNA replication;
- and when genes are activated.

When chromosomes are replicated, during the S phase of the cell cycle, histones are transiently acetylated. **Figure 30.11** shows that this acetylation occurs before the histones are incorporated into nucleosomes. Histones H3 and H4 are acetylated at the stage when they are associated with one another in the $H3_2 \cdot H4_2$ tetramer. The tetramer is then incorporated into nucleosomes. Quite soon after, the acetyl groups are removed.

The importance of the acetylation is indicated by the fact that preventing acetylation of both histones H3 and H4 during replication causes loss of viability in yeast. The two histones are redundant as substrates because yeast can manage perfectly well so long as they can acetylate either one of these histones during S phase. There are two possible roles for the acetylation: it could be needed for the histones to be recognized by factors that incorporate them into nucleosomes; or it could be required for the assembly and/or structure of the new nucleosome.

The factors that are known to be involved in chromatin assembly do not distinguish between acetylated and nonacetylated histones, suggesting that the modification is more likely to be required for subsequent interactions. It has been thought for a long time that acetylation might be needed to help control protein–protein interactions that occur as histones are incorporated into nucleosomes.

Outside of S phase, acetylation of histones in chromatin is generally correlated with the state of gene expression. The correlation was first noticed because histone acetylation is increased in a domain containing active genes, and acetylated chromatin is more sensitive to DNAase I and (possibly) to micrococcal nuclease. **Figure 30.12** shows that this involves the acetylation of histone tails in nucleosomes. We now know that this occurs largely because of acetylation of the nucleosomes in the vicinity of the promoter when a gene is activated.

In addition to events at individual promoters, wide-scale changes in acetylation occur on sex chromosomes. This is part of the mechanism by which the activities of genes on the X chromosome are altered to compensate for the presence of two X chromosomes in one sex but only one X chromosome (in addition to the Y chromosome) in the other sex (see *31.4 X Chromosomes Undergo Global Changes*). The inactive X chromosome in female mammals has underacetylated H4. The superactive X chromosome in *Drosophila* males is acetylated specifically at ^{16}Lys of histone H4. This suggests that the presence of acetyl groups may be a prerequisite for a less condensed, active structure.

30.7 Acetylases Are Associated with Activators

Key Terms

- **Histone acetyltransferase (HAT)** enzymes modify histones by addition of acetyl groups; some transcriptional coactivators have HAT activity.
- **Histone deacetylase (HDAC)** enzymes remove acetyl groups from histones; they may be associated with repressors of transcription.

Key Concepts

- Deacetylated chromatin may have a more condensed structure.
- Transcription activators are associated with histone acetylase activities in large complexes.
- Histone acetylases vary in their target specificity.
- Acetylation could affect transcription in a quantitative or qualitative way.

Acetylation is reversible. Each direction of the reaction is catalyzed by a specific type of enzyme. Enzymes that can acetylate histones are called **histone acetyltransferases**, or **HATs** (also called histone acetylases); the acetyl groups are removed by histone **deacetylases** or **HDACs**. There are two groups of HAT enzymes: group A is involved with transcription; group B is involved with nucleosome assembly.

The breakthrough in analyzing the role of histone acetylation was provided by the characterization of the acetylating and deacetylating enzymes, and their association with other proteins involved in specific events of activation and repression. A basic change in our view of histone acetylation was caused by the discovery that HATs are not necessarily dedicated enzymes associated with chromatin: rather it turns out that known coactivators of transcription have HAT activity, and the activity is required for their ability to activate transcription.

This enables us to redraw our picture for the action of coactivators as shown in **Figure 30.13**, where RNA polymerase is bound at a hypersensitive site and coactivators are acetylating histones on the nucleosomes in the vicinity. Many examples are now known of interactions of this type.

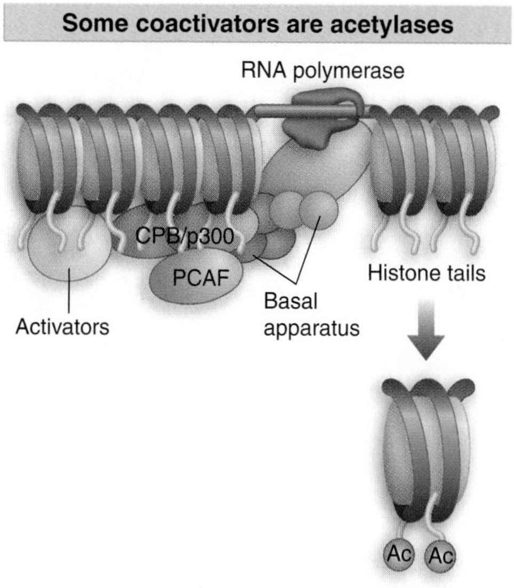

The first transcriptional coactivator to be discovered with HAT activity was the yeast protein Gcn5, which is part of an adaptor complex that is needed for the interaction between certain enhancers and their target promoters. Gcn5 leads us into one of the most important acetylase complexes. In yeast, Gcn5 is part of the 1.8 MDa SAGA complex, which contains several proteins that are involved in transcription. Among these proteins are several TAF$_{II}$s. Also, the TAF$_{II}$145 subunit of TF$_{II}$D is an acetylase. There are some functional overlaps between TF$_{II}$D and SAGA, most notably that yeast can manage with either TAF$_{II}$145 or Gcn5, but is damaged by the deletion of both. This suggests that an acetylase activity is essential for gene expression, but can be provided by either TF$_{II}$D or SAGA. As might be expected from the size of the SAGA complex, acetylation is only one of its functions, although its other functions in gene activation are less well characterized.

A general feature of acetylation is that a HAT is part of a large complex. **Figure 30.14** shows a simplified model for the behavior of such a complex. Typically the complex contains a targeting subunit(s) that determines the binding sites on DNA. This determines the target for the HAT. The complex also contains effector subunits that affect chromatin structure or act directly on transcription. Probably at least some of the effectors require the acetylation event in order to act. Deacetylation, catalyzed by an HDAC, may work in a similar way.

Is the effect of acetylation quantitative or qualitative? One possibility is that a certain number of acetyl groups are required to have an effect, and the exact positions at which they occur are largely irrelevant. An alternative is that individual acetylation events have specific effects. We might interpret the existence of complexes containing multiple HAT activities in either way—if individual enzymes have different specificities, we may need multiple activities either to acetylate a sufficient number of different positions or because the individual events are

Figure 30.13 Coactivators (PCAF and CPB/p300) have HAT activities that acetylate the tails of nucleosomal histones.

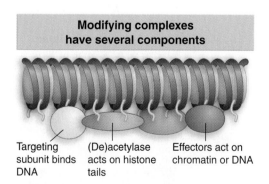

Figure 30.14 Complexes that modify chromatin structure or activity have targeting subunits that determine their sites of action, HAT or HDAC enzymes that acetylate or deacetylate histones, and effector subunits that have other actions on chromatin or DNA.

necessary for different effects upon transcription. At replication, it appears, at least with respect to histone H4, that acetylation at any two of three available positions is adequate, favoring a quantitative model in this case. However, where chromatin structure is changed to affect transcription, acetylation at specific positions is important (see *30.13 Heterochromatin Depends on Interactions with Histones*).

30.8 Deacetylases Are Associated with Repressors

Key Concepts

- Deacetylation is associated with repression of gene activity.
- Deacetylases are present in complexes with repressor activity.

If acetylation is required for gene activation, it seems appropriate that deacetylation will be required to turn off activated genes. The first connection of this sort was established in yeast, via mutations in *SIN3* and *RPD3* that behave as though these loci repress a variety of genes. The proteins form a complex with the DNA-binding protein Ume6, which binds to the *URS1* element. The complex represses transcription at the promoters containing *URS1*, as illustrated in **Figure 30.15**. Rpd3 is a histone deacetylase.

A similar system for repression is found in mammalian cells. The bHLH family of transcription regulators includes activators that function as heterodimers, including MyoD (see *25.12 Helix-Loop-Helix Proteins Interact by Combinatorial Association*). It also includes repressors, in particular the heterodimer Mad:Max, where Mad can be any one of a group of closely related proteins. The Mad:Max heterodimer (which binds to specific DNA sites) interacts with a homologue of Sin3 (called mSin3 in mouse and hSin3 in man). mSin3 is part of a repressive complex that includes histone binding proteins and the histone deacetylases HDAC1 and HDAC2. Deacetylase activity is required for repression. The modular nature of this system is emphasized by other means of employment: a corepressor (SMRT), which enables retinoid hormone receptors to repress certain target genes, functions by binding mSin3, which in turns brings the HDAC activities to the site. Another means of bringing HDAC activities to the site may be a connection with MeCP2, a protein that binds to methylated cytosines (see *24.13 CpG Islands Are Regulatory Targets*).

Absence of histone acetylation is also a feature of heterochromatin. This is true of both constitutive heterochromatin (typically involving regions of centromeres or telomeres) and facultative heterochromatin (regions that are inactivated in one cell although they may be active in another). Typically the N-terminal tails of histones H3 and H4 are not acetylated in heterochromatic regions.

30.9 Methylation of Histones and DNA Is Connected

Key Concepts

- Methylation of both DNA and histones is a feature of inactive chromatin.
- The two types of methylation event may be connected.

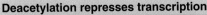

Deacetylation represses transcription

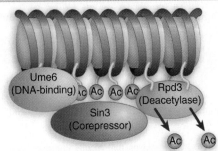

Figure 30.15 A repressor complex contains three components: a DNA-binding subunit, a corepressor, and a histone deacetylase.

Methylation of both histones and DNA is associated with inactivity. Sites that are methylated in histones include two lysines in the tail of H3 and an arginine in the tail of H4.

Methylation of H3 ^{9}Lys is a feature of condensed regions of chromatin, including heterochromatin as seen in bulk and also smaller regions that are known not to be expressed. The histone methyltransferase enzyme that targets this lysine is called SUV39H1. (We see the origin of this peculiar name in *30.12 Some Common Motifs Are Found in Proteins That Modify Chromatin*). Its catalytic site has a region called the SET domain. Other histone methyltransferases act on arginine. In addition, methylation may occur on ^{79}Lys in the globular core region of H3; this may be necessary for the formation of heterochromatin at telomeres.

Most of the methylation sites in DNA are CpG islands (see *24.13 CpG Islands Are Regulatory Targets*). CpG sequences in heterochromatin are usually methylated. Conversely, it is necessary for the CpG islands located in promoter regions to be unmethylated in order for a gene to be expressed.

Methylation of DNA and methylation of histones may be connected. Some histone methyltransferase enzymes contain potential binding sites for the methylated CpG doublet, raising the possibility that a methylated DNA sequence may cause a histone methyltransferase to bind. An alternative type of connection is indicated by the fact that in the fungus *Neurospora*, the methylation of DNA is prevented by a mutation in a gene coding for a histone methylase that acts on ^{9}Lys of histone H3. This suggests that methylation of the histone is a signal that recruits the DNA methylase to chromatin. The important point is not the detailed order of events—which remains to be worked out—but the fact that one type of modification can be the trigger for another.

30.10 Promoter Activation Is an Ordered Series of Events

Key Concepts

- The remodeling complex may recruit the acetylating complex.
- Acetylation of histones may be the event that maintains the complex in the activated state.

Figure 30.16 summarizes three types of differences that are found between active chromatin and inactive chromatin:

- Active chromatin is acetylated on the tails of histones H3 and H4.
- Inactive chromatin is methylated on ^{9}Lys of histone H3.
- Inactive chromatin is methylated on cytosines of CpG doublets.

Acetylases and deacetylases may trigger the initiating events in setting the activity of chromatin. Deacetylation allows methylation to occur, which causes formation of a heterochromatic complex. Acetylation marks a region as active (see *30.13 Heterochromatin Depends on Interactions with Histones*).

How are acetylases (or deacetylases) recruited to their specific targets? As we have seen with remodeling complexes, the process is likely to be indirect. A sequence-specific activator (or repressor) may interact with a component of the acetylase (or deacetylase) complex to recruit it to a promoter.

There may also be direct interactions between remodeling complexes and histone-modifying complexes. Binding by the SWI/SNF remodeling complex may lead in turn to binding by the SAGA acetylase complex. Acetylation of histones may then in fact stabilize the association

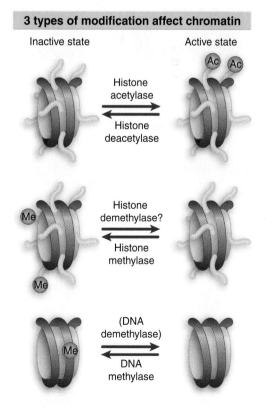

Figure 30.16 Acetylation of histones activates chromatin, and methylation of DNA and histones inactivates chromatin.

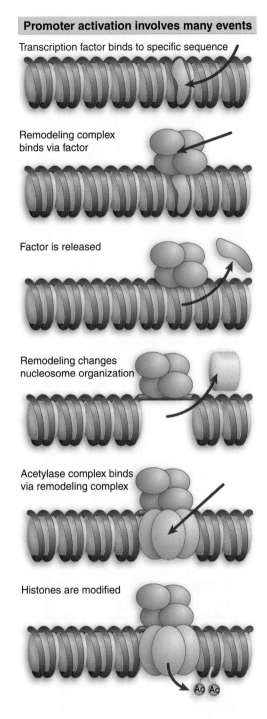

Promoter activation involves many events

Transcription factor binds to specific sequence

Remodeling complex binds via factor

Factor is released

Remodeling changes nucleosome organization

Acetylase complex binds via remodeling complex

Histones are modified

Figure 30.17 Promoter activation involves binding of a sequence-specific activator, recruitment and action of a remodeling complex, and recruitment and action of an acetylating complex.

with the SWI/SNF complex, making a mutual reinforcement of the changes in the components at the promoter.

We can connect all of the events at the promoter into the series summarized in **Figure 30.17**. The initiating event is binding of a sequence-specific component (which is able to find its target DNA sequence in the context of chromatin). This recruits a remodeling complex. Changes occur in nucleosome structure. An acetylating complex binds, and the acetylation of target histones provides a covalent mark that the locus has been activated.

Modification of DNA also occurs at the promoter. Methylation of cytosine at CpG doublets is associated with gene inactivity. The basis for recognition of DNA as a target for methylation is not very well established.

It is clear that chromatin remodeling at the promoter requires a variety of changes that affect nucleosomes, including acetylation, but what changes are required within the gene to allow an RNA polymerase to traverse it? We know that RNA polymerase can transcribe DNA *in vitro* at rates comparable to the *in vivo* rate (~25 nucleotides per second) only with a template of free DNA. Several proteins have been characterized for their abilities to improve the speed with which RNA polymerase transcribes chromatin *in vivo*. The common feature is that they act on chromatin. A current model for their action is that they associate with RNA polymerase and travel with it along the template, modifying nucleosome structure by acting on histones. Among these factors are histone acetylases. One possibility is that the first RNA polymerase to transcribe a gene is a pioneer polymerase carrying factors that change the structure of the transcription unit so as to make it easier for subsequent polymerases.

30.11 Histone Phosphorylation Affects Chromatin Structure

Key Concept

- Most histones are targets for phosphorylation, but the consequences are less well characterized than for acetylation or methylation.

Histones are phosphorylated in two circumstances:

- cyclically during the cell cycle;
- and in association with chromatin remodeling.

It is has been known for a very long time that histone H1 is phosphorylated at mitosis, and more recently it was discovered that H1 is an extremely good substrate for the Cdc2 kinase that controls cell division. This led to speculations that the phosphorylation might be connected with the condensation of chromatin, but so far no direct effect of this phosphorylation event has been demonstrated, and we do not know whether it plays a role in cell division.

Loss of a kinase that phosphorylates histone H3 on [10]Ser has devastating effects on chromatin structure. **Figure 30.18** compares the usual extended structure of the polytene chromosome set of *D. melanogaster* (upper photograph) with the structure that is found in a null mutant that has no JIL-1 kinase (lower photograph).

The cause of the disruption of structure is most likely the failure to phosphorylate histone H3 (of course, JIL-1 may also have other targets). This suggests that H3 phosphorylation is required to generate the more extended chromosome structure of euchromatic regions. Evidence supporting the idea that JIL-1 acts directly on chromatin is that it associates with the complex of proteins that binds to the X chromosome to increase its gene expression in males (see *31.4 X Chromosomes Undergo Global Changes*).

30.12 Some Common Motifs Are Found in Proteins that Modify Chromatin

Key Terms

- The **chromo domain** is found in nonhistone chromatin proteins and is involved in protein–protein interactions.
- The **SET domain** is part of the active site of histone methyltransferase enzymes.
- The **bromo domain** is found in a variety of proteins that interact with chromatin and is used to recognize acetylated sites on histones.

Key Concept

- Several motifs are characteristic of nonhistone proteins, and identify their functions.

Our insights into the molecular mechanisms for controlling the structure of chromatin start with mutants that affect chromatin activity. They were discovered through their effects on position effect variegation, a phenomenon in which a region of constitutive heterochromatin inactivates genes in the vicinity (see *31.2 Heterochromatin Propagates from a Nucleation Event*). Some 30 genes have been identified in *Drosophila*. They are named systematically as *Su(var)* for genes in which mutations suppress variegation (increase gene activity) and *E(var)* for genes in which mutations enhance variegation (suppress gene activity).

- *Su(var)* mutations identify genes whose products are needed for the formation of heterochromatin. They include enzymes that act on chromatin, such as histone deacetylases, and proteins that are localized to heterochromatin.
- *E(var)* mutations identify genes whose products are needed to activate gene expression. They include members of the SWI/SNF complex.

We see immediately from these properties that modification of chromatin structure is important for controlling the formation of heterochromatin. Many of these functions are conserved in evolution, and homologous genes can be recognized in mammals, flies, and yeasts.

Many of the Su(var) and E(var) proteins have a common protein motif of 60 amino acids called the **chromo domain**. The fact that this domain is found in proteins of both groups suggests that it represents a motif that participates in protein–protein interactions with targets in chromatin. The best-characterized action of the chromo domain is in targeting repressive proteins to chromatin. It recognizes methylated lysine in the H3 histone tail. Different chromo domains have different specificities: the two best-characterized examples, in proteins HP1 and Pc, recognize Me-^{9}Lys and Me-^{27}Lys in H3, respectively (see *30.13 Heterochromatin Depends on Interactions with Histones* and *31.3 Polycomb and Trithorax Are Antagonistic Repressors and Activators*).

Su(var)3-9 has a chromo domain and also a **SET domain**, a motif that is found in several Su(var) proteins. Its mammalian homologues localize to centromeric heterochromatin. It is the histone methyltransferase that acts

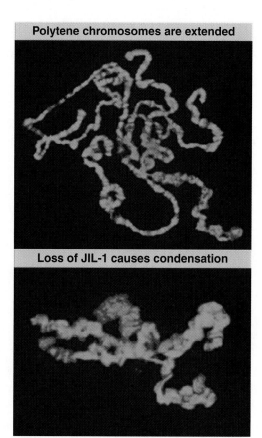

Polytene chromosomes are extended

Loss of JIL-1 causes condensation

Figure 30.18 Polytene chromosomes of flies that have no JIL-1 kinase have abnormal polytene chromosomes that are condensed instead of extended. Photograph kindly provided by Kristen M. Johansen.

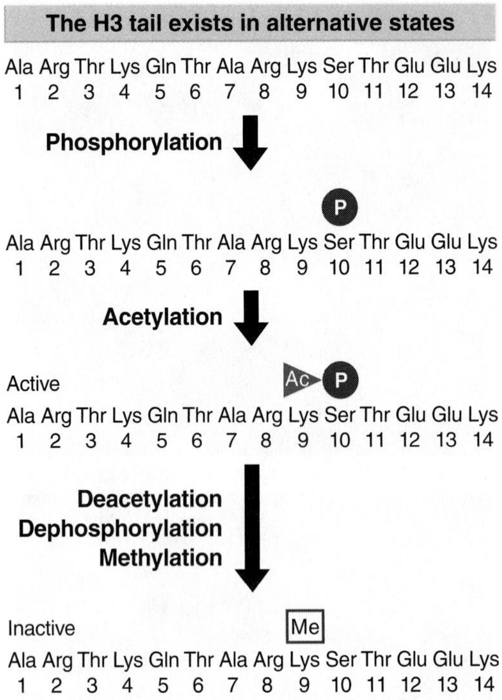

The H3 tail exists in alternative states
Ala Arg Thr Lys Gln Thr Ala Arg Lys Ser Thr Glu Glu Lys
1 2 3 4 5 6 7 8 9 10 11 12 13 14
Phosphorylation
P
Ala Arg Thr Lys Gln Thr Ala Arg Lys Ser Thr Glu Glu Lys
1 2 3 4 5 6 7 8 9 10 11 12 13 14
Acetylation
Active
Ac
P
Ala Arg Thr Lys Gln Thr Ala Arg Lys Ser Thr Glu Glu Lys
1 2 3 4 5 6 7 8 9 10 11 12 13 14
Deacetylation
Dephosphorylation
Methylation
Inactive
Me
Ala Arg Thr Lys Gln Thr Ala Arg Lys Ser Thr Glu Glu Lys
1 2 3 4 5 6 7 8 9 10 11 12 13 14

Figure 30.19 Multiple modifications in the H3 tail affect chromatin activity.

on ^{9}Lys of histone H3 (see *30.9 Methylation of Histones and DNA Is Connected*). The SET domain is part of the active site, and in fact is a marker for the methylase activity.

The **bromo domain** has a binding site for acetylated lysine. The bromo domain itself recognizes only a very short sequence of four amino acids including the acetylated lysine, so specificity for target recognition must depend on interactions in other regions. Besides the acetylases, the bromo domain is found in a range of proteins that interact with chromatin, including components of the transcription apparatus. This implies that it is used to recognize acetylated histones, which means that it is likely to be found in proteins that are involved with gene activation.

Although there is a general correlation in which active chromatin is acetylated while inactive chromatin is methylated on histones, there are some exceptions to the rule. The best characterized is that acetylation of ^{12}Lys of H4 is associated with heterochromatin.

Multiple modifications may occur on the same histone tail, and one modification may influence another. Phosphorylation of a lysine at one position may be necessary for acetylation of a lysine at another position. **Figure 30.19** shows the situation in the tail of H3, which can exist in either of two alternative states. The inactive state has Methyl-^{9}Lys. The active state has Acetyl-^{9}Lys and Phospho-^{10}Ser. These states can be maintained over extended regions of chromatin. The phosphorylation of ^{10}Ser and the methylation of ^{9}Lys are mutually inhibitory, suggesting the order of events shown in the figure. This situation may cause the tail to flip between the active and active states.

30.13 Heterochromatin Depends on Interactions with Histones

Key Concepts

- HP1 is the key protein in forming mammalian heterochromatin, and acts by binding to methylated H3 histone.
- RAP1 initiates formation of heterochromatin in yeast by binding to specific target sequences in DNA.
- The targets of RAP1 include telomeric repeats and silencers at *HML* and *HMR*.
- RAP1 recruits SIR3/SIR4, which interact with the N-terminal tails of H3 and H4.

Histone methylation causes HP1 binding
Histone
deacetylase
SUV39H1 histone
methyltransferase
Ac
H3
H3 Ac
Me
Me
Me
HP1
HP1 Me
Active chromatin
Inactive chromatin

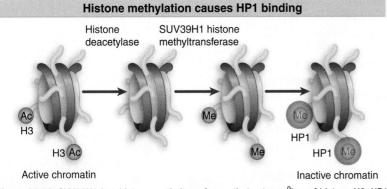

Figure 30.20 SUV39H1 is a histone methyltransferase that acts on ^{9}Lys of histone H3. HP1 binds to the methylated histone.

Inactivation of chromatin occurs by the addition of proteins to the nucleosomal fiber. The inactivation may be due to a variety of effects, including condensation of chromatin to make it inaccessible to the apparatus needed for gene expression, addition of proteins that directly block access to regulatory sites, or proteins that directly inhibit transcription.

Two systems that have been characterized at the molecular level are HP1 in mammals and the SIR complex in yeast. Although there are no detailed similarities between the proteins in each system, the

The chromo domain of HP1 binds to methylated Lys-9 in histone H3

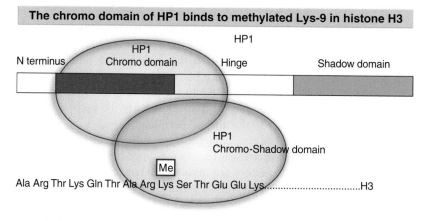

Ala Arg Thr Lys Gln Thr Ala Arg Lys Ser Thr Glu Glu Lys..................................H3

Figure 30.21 Methylation of histone H3 creates a binding site for HP1.

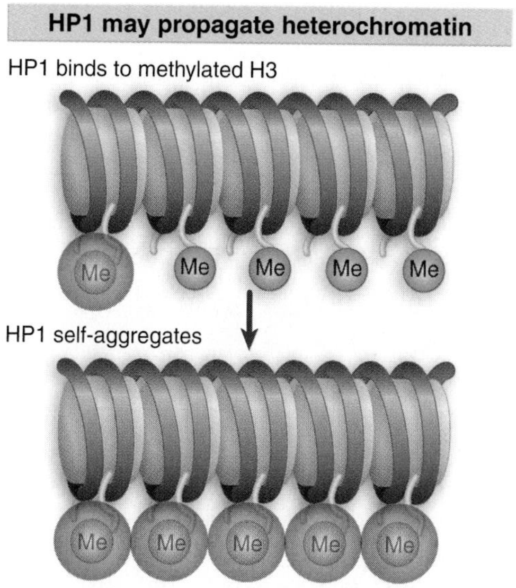

general mechanism of reaction is similar: the points of contact in chromatin are the N-terminal tails of the histones.

One of the most important Su(var) proteins is HP1 (heterochromatin protein 1). It is localized in the heterochromatin of polytene chromosomes. HP1 contains a chromo domain near the N-terminus, and another domain that is related to it, called the chromo-shadow domain, at the C-terminus (see Figure 30.21). The importance of the chromo domain is indicated by the fact that it is the location of many of the mutations in HP1. The original protein identified as HP1 is now called HP1α, since two related proteins, HP1β and HP1γ, have since been found.

HP1 binds to chromatin by using its chromo domain to contact a site in H3 that includes the methylated ^{9}Lys. For ^{9}Lys to be methylated, the acetyl group must be removed from ^{14}Lys (and from ^{9}Lys itself if it is acetylated). This suggests the model for initiating formation of heterochromatin shown in **Figure 30.20**. First the deacetylase acts to remove the modification at ^{14}Lys. Then the SUV39H1 methylase acts on the histone H3 tail to create the methylated ^{9}Lys. **Figure 30.21** expands the reaction to show that the interaction occurs between the chromo domain and the methylated lysine. This is a trigger for forming inactive chromatin. **Figure 30.22** shows that the inactive region may then be extended by the ability of further HP1 molecules to interact with one another.

The existence of a common basis for silencing in yeast is suggested by its reliance on a common set of genetic loci. Mutations in any one of a number of genes cause the silent loci *HML* and *HMR* to become activated, and also relieve the inactivation of genes that have been integrated near telomeric heterochromatin. The products of these loci therefore function to maintain the inactive state of both types of heterochromatin.

Figure 30.23 proposes a model for actions of these proteins. Only one of them is a sequence-specific DNA-binding protein. This is Rap1, which binds to the $C_{1-3}A$ repeats at the telomeres, and also binds to the *cis*-acting silencer elements that are needed for repression of *HML* and *HMR*. Rap1 has the crucial role of identifying the DNA sequences at which heterochromatin forms. It recruits Sir3/sir4, and they interact directly with the histones H3/H4 and also with one another (they may function as a heteromultimer). Sir3/Sir4 interact with the N-terminal tails of the histones H3 and H4. (In fact, the first evidence that histones might be involved directly in formation of heterochromatin was provided by the discovery that mutations abolishing silencing at *HML/HMR* map to genes coding for H3 and H4.)

HP1 may propagate heterochromatin

HP1 binds to methylated H3

HP1 self-aggregates

Figure 30.22 Binding of HP1 to methylated histone H3 forms a trigger for silencing because further molecules of HP1 aggregate on the nucleosome chain.

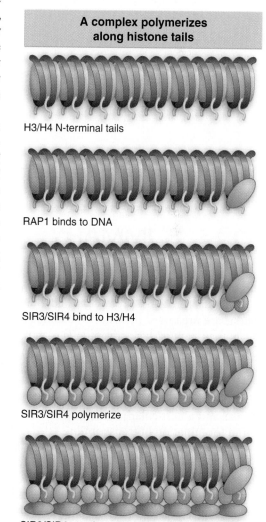

A complex polymerizes along histone tails

H3/H4 N-terminal tails

RAP1 binds to DNA

SIR3/SIR4 bind to H3/H4

SIR3/SIR4 polymerize

SIR3/SIR4 attach to matrix

Figure 30.23 Formation of heterochromatin is initiated when Rap1 binds to DNA. Sir3/4 bind to Rap1 and also to histones H3/H4. The complex polymerizes along chromatin and may connect telomeres to the nuclear matrix.

Once SIR3/SIR4 have bound to histones H3/H4, the complex may polymerize further, and spread along the chromatin fiber. This may inactivate the region, either because coating with SIR3/SIR4 itself has an inhibitory effect, or because binding to histones H3/H4 induces some further change in structure. We do not know what limits the spreading of the complex. The C-terminus of SIR3 has a similarity to nuclear lamin proteins (constituents of the nuclear matrix) and may be responsible for tethering heterochromatin to the nuclear periphery.

A similar series of events forms the silenced regions at *HMR* and *HML* (see also *19.13 Yeast Mating Type Is Changed by Recombination*). Three sequence-specific factors are involved in triggering formation of the complex: RAP1, ABF1 (a transcription factor), and ORC (the origin replication complex). In this case, SIR1 binds to a sequence-specific factor and recruits SIR2,3,4 to form the repressive structure. SIR2 is a histone deacetylase. The deacetylation reaction is necessary to maintain binding of the SIR complex to chromatin.

How does a silencing complex repress chromatin activity? It could condense chromatin so that regulator proteins cannot find their targets. The simplest case would be to suppose that the presence of a silencing complex is mutually incompatible with the presence of transcription factors and RNA polymerase. The cause could be that silencing complexes block remodeling (and thus indirectly prevent factors from binding) or that they directly obscure the binding sites on DNA for the transcription factors. However, the situation may not be this simple, because transcription factors and RNA polymerase can be found at promoters in silenced chromatin. This could mean that the silencing complex prevents the factors from working rather than from binding as such. In fact, there may be competition between gene activators and the repressing effects of chromatin, so that activation of a promoter inhibits spread of the silencing complex.

30.14 SUMMARY

Genes whose control regions are organized in nucleosomes usually are not expressed. In the absence of specific regulatory proteins, promoters and other regulatory regions are organized by histone octamers into a state in which they cannot be activated. This may explain the need for nucleosomes to be precisely positioned in the vicinity of a promoter, so that essential regulatory sites are appropriately exposed. Some transcription factors have the capacity to recognize DNA on the nucleosomal surface, and a particular positioning of DNA may be required for initiation of transcription.

Active chromatin and inactive chromatin are not in equilibrium. Sudden, disruptive events are needed to convert one to the other. Chromatin remodeling complexes have the ability to displace histone octamers by a mechanism that involves hydrolysis of ATP. Remodeling complexes are large and are classified according to the type of the ATPase subunit. Two common types are SWI/SNF and ISW. A typical form of this chromatin remodeling is to displace one or more histone octamers from specific sequences of DNA, creating a boundary that results in the precise or preferential positioning of adjacent nucleosomes. Chromatin remodeling may also change the positions of nucleosomes, sometimes involving sliding of histone octamers along DNA.

N-terminal histone tails are modified by acetylation, methylation, and phosphorylation. Modification is a trigger for controlling chromatin activity. Acetylation is generally associated with gene activation. Histone acetylases are found in activating complexes, and histone deacetylases are found in inactivating complexes. Histone methylation is associated with gene inactivation. Some histone modifications are exclusive or synergistic with others. Histones are usually acetylated in the region of the promoter of an active gene, and usually are deacetylated in regions of heterochromatin, although there are some exceptions.

Chromo domains are common in nonhistone proteins associated with influencing chromatin structure. Bromo domains are used to recognize acetylated sites on histones. The SET domain is part of the active site of histone methyltransferases.

31

Epigenetic Effects Are Inherited

31.1 Introduction

Key Terms

- **Epigenetic** changes influence the phenotype without altering the genotype. They consist of changes in the properties of a cell that are inherited but that do not represent a change in genetic information.

- A **prion** is a proteinaceous infectious agent that behaves as a heritable trait, although it contains no nucleic acid. Examples are PrP^{Sc}, the agent of scrapie in sheep and bovine spongiform encephalopathy, and Psi, which confers an inherited state in yeast.

Key Concept

- Epigenetic effects can result from modification of a nucleic acid after it has been synthesized or by the perpetuation of protein structures.

Epigenetic inheritance describes the ability of different states, which may have different phenotypic consequences, to be inherited without any change in the sequence of DNA. This means that two individuals with the same DNA sequence at the locus that controls the effect may show different phenotypes. The basic cause of this phenomenon is the existence of a self-perpetuating structure in one of the individuals that does not depend on DNA sequence. Several different types of structures have the ability to sustain epigenetic effects:

- A covalent modification of DNA (methylation of a base).
- A proteinaceous structure that assembles on DNA.
- A protein aggregate that controls the conformation of new subunits as they are synthesized.

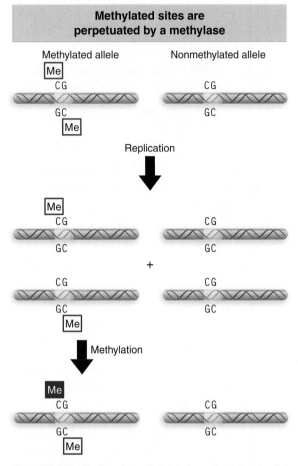

Figure 31.1 Replication of a methylated site produces hemimethylated DNA in which only the parental strand is methylated. A perpetuation methylase recognizes hemimethylated sites and adds a methyl group to the base on the daughter strand. This restores the original situation, in which the site is methylated on both strands. A nonmethylated site remains nonmethylated after replication.

In each case the epigenetic state results from a difference in function (typically inactivation) that is determined by the structure.

In the case of DNA methylation, a methylated DNA sequence may fail to be transcribed, whereas the nonmethylated sequence will be expressed. **Figure 31.1** shows how this situation is inherited. One allele has a sequence that is methylated on both strands of DNA, whereas the other allele has an unmethylated sequence. Replication of the methylated allele creates hemimethylated daughters that are restored to the methylated state by a constitutively active methylase enzyme. Replication does not affect the state of the nonmethylated allele. If the state of methylation affects transcription, the two alleles differ in their state of gene expression, even though their sequences are identical.

Self-perpetuating structures that assemble on DNA usually have a repressive effect by forming heterochromatic regions that prevent the expression of genes within them. Their perpetuation depends on the ability of proteins in a heterochromatic region to remain bound to those regions after replication, and then to recruit more protein subunits to sustain the complex. If individual subunits are distributed at random to each daughter duplex at replication, the two daughters will continue to be marked by the protein, although its density will be reduced to half of the level before replication. **Figure 31.2** shows that the existence of epigenetic effects forces us to the view that a protein responsible for such a situation must have some sort of self-templating or self-assembling capacity to restore the original complex.

It can be the state of protein modification, rather than the presence of the protein *per se*, that is responsible for an epigenetic effect. Usually the tails of histones H3 and H4 are not acetylated in constitutive heterochromatin. However, if centromeric heterochromatin is acetylated, silenced genes may become active. The effect may be perpetuated through mitosis and meiosis, which suggests that an epigenetic effect has been created by changing the state of histone acetylation.

Independent protein aggregates that cause epigenetic effects (called **prions**) work by sequestering the protein in a form in which its normal function cannot be displayed. Once the protein aggregate has formed, it forces newly synthesized protein subunits to join it in the inactive conformation.

31.2 Heterochromatin Propagates from a Nucleation Event

Key Terms

- **Position effect variegation (PEV)** is silencing of gene expression that results from proximity to heterochromatin.
- **Telomeric silencing** describes repression of gene activity in the vicinity of a telomere.

Key Concepts

- Heterochromatin is nucleated at a specific sequence and the inactive structure propagates along the chromatin fiber.

- Genes within regions of heterochromatin are inactivated.

- Because the length of the inactive region varies from cell to cell, inactivation of genes in this vicinity causes position effect variegation.

- Similar spreading effects occur at telomeres and at the silent cassettes in yeast mating type.

An interphase nucleus contains both euchromatin and heterochromatin. The condensation state of heterochromatin is close to that of mitotic chromosomes. Heterochromatin is inert. It remains condensed in interphase, is transcriptionally repressed, replicates late in S phase, and may be localized to the nuclear periphery. Centromeric heterochromatin typically consists of satellite DNAs. However, the formation of heterochromatin is not rigorously defined by sequence. When a gene is transferred, either by a chromosomal translocation or by transfection and integration, into a position adjacent to heterochromatin, it may become inactive as the result of its new location, implying that it has become heterochromatic.

The proteins that form heterochromatin act on chromatin via the histones, and modifications of the histones may be an important feature in the interaction. Once established, such changes in chromatin may persist through cell divisions, creating an epigenetic state in which the properties of a gene are determined by the self-perpetuating structure of chromatin.

A classic example of this type of epigenetic effect is provided by the phenomenon of **position effect variegation (PEV)**, in which genetically identical cells have different phenotypes. **Figure 31.3** shows an example of position effect variegation in the *Drosophila* eye, in which some regions lack color while others are red, because the *white* gene is inactivated by adjacent heterochromatin in some cells, while it remains active in other cells.

The explanation for this effect is shown in **Figure 31.4**. Inactivation spreads from heterochromatin into the adjacent region for a variable distance. In some cells it goes far enough to inactivate a nearby gene, but in others it does not. This happens at a certain point in embryonic development, and after that point the state of the gene is inherited by all the progeny cells. Cells descended from an ancestor in which the gene was inactivated form patches corresponding to the phenotype of loss-of-function (in the case of *white*, absence of color).

The closer a gene lies to heterochromatin, the higher the probability that it will be inactivated. This suggests that the formation of heterochromatin may be a two-stage process: a *nucleation* event occurs at a specific sequence; and then the inactive structure *propagates* along the chromatin fiber. The distance for which the inactive structure extends is not precisely determined, and may be stochastic, being influenced by parameters such as the quantities of limiting protein components. One factor that may affect the spreading process is the activation of promoters in the region; an active promoter may inhibit spreading.

Genes that are closer to heterochromatin are more likely to be inactivated, and will therefore be inactive in a greater proportion of cells. By this model, the boundaries of a heterochromatic region might be terminated by exhausting the supply of one of the proteins that is required.

The effect of **telomeric silencing** in yeast is analogous to position effect variegation in *Drosophila*; genes translocated to a telomeric location

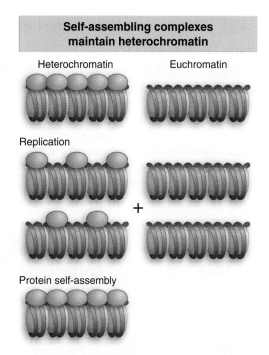

Self-assembling complexes maintain heterochromatin

Heterochromatin   Euchromatin

Replication

+

Protein self-assembly

Figure 31.2 Heterochromatin is created by proteins that associate with histones. Perpetuation through division requires that the proteins associate with each daughter duplex and then recruit new subunits to reassemble the repressive complexes.

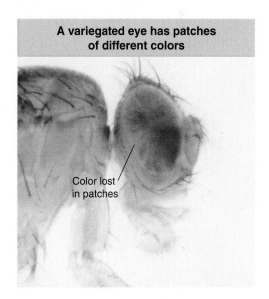

A variegated eye has patches of different colors

Color lost in patches

Figure 31.3 Position effect variegation in eye color results when the *white* gene is integrated near heterochromatin. Cells in which *white* is inactive give patches of white eye, while cells in which *white* is active give red patches. The severity of the effect is determined by the closeness of the integrated gene to heterochromatin. Photograph kindly provided by Steve Henikoff.

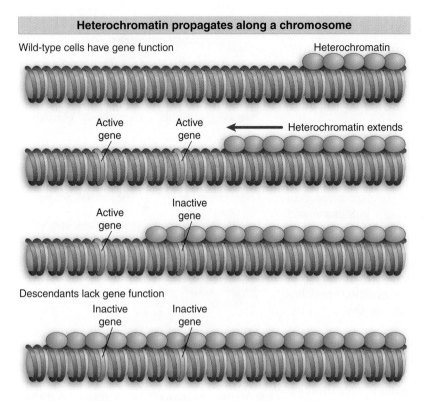

Heterochromatin propagates along a chromosome

Wild-type cells have gene function

Heterochromatin

Active gene   Active gene   ← Heterochromatin extends

Active gene   Inactive gene

Descendants lack gene function

Inactive gene   Inactive gene

Figure 31.4 Extension of heterochromatin inactivates genes. The probability that a gene will be inactivated depends on its distance from the heterochromatin region.

show the same sort of variable loss of activity. This results from a spreading effect that propagates from the telomeres.

The distinction between the recruitment of an inactivating complex to a distinct site, and its linear spread along the chromosome to act on adjacent regions, is common to various types of inactivating event. Another example is provided by the dosage compensation system in *C. elegans*, which ensures that each X chromosome in females is expressed at only half the level of the single X chromosome in males (see *31.4 X Chromosomes Undergo Global Changes*).

A second form of silencing occurs in yeast. Yeast mating type is determined by the activity of a single active locus (*MAT*), but the genome contains two other copies of the mating-type sequences (*HML* and *HMR*), which are maintained in an inactive form. The silent loci *HML* and *HMR* share many properties with heterochromatin, and could be regarded as constituting regions of heterochromatin in miniature (see *19.13 Yeast Mating Type Is Changed by Recombination*).

31.3 Polycomb and Trithorax Are Antagonistic Repressors and Activators

Key Concepts

- Polycomb group proteins (Pc-G) perpetuate a state of repression through cell divisions.
- The PRE is a DNA sequence that is required for the action of Pc-G; it provides a nucleation center from which Pc-G proteins propagate an inactive structure.
- No individual Pc-G protein has yet been found that can bind the PRE.
- Trithorax group proteins antagonize the actions of the Pc-G.

Heterochromatin provides one example of the specific repression of chromatin. Another is provided by the genetics of homeotic genes (which affect the identity of body parts) in *Drosophila*, which have led to the identification of a protein complex that may maintain certain genes in a repressed state. The gene *Pc* is the prototype for a class of loci called the *Pc* group (*Pc-G*); mutations in any of these genes generally have the same result of derepressing homeotic genes.

The Pc-G proteins are not conventional repressors. They are not responsible for determining the initial pattern of expression of the genes on which they act. In the absence of Pc-G proteins, these genes are initially repressed as usual, but later in development the repression is lost without Pc-G group functions. This suggests that *the Pc-G proteins in some way recognize the state of repression when it is established, and they then act to perpetuate it through cell division of the daughter cells.*

Figure 31.5 shows a model in which Pc-G proteins bind in conjunction with a repressor, but the Pc-G proteins remain bound after the repressor is no longer available. This is necessary to maintain repression, so that if Pc-G proteins are absent, the gene becomes activated.

A region of DNA that is sufficient to enable the response to the *Pc-G* genes is called a PRE (*Polycomb* response element). It can be defined operationally by the property that it maintains repression in its vicinity throughout development. The assay for a PRE is to insert it close to a reporter gene that is controlled by an enhancer that is repressed in early development, and then to determine whether the reporter becomes expressed subsequently in the descendants. An effective PRE will prevent such re-expression.

The PRE is a complex structure, ~10 kb. When a locus is repressed by Pc-G proteins, however, the proteins appear to be present over a much larger length of DNA than the PRE itself. Polycomb is found locally over a few kilobases of DNA surrounding a PRE. This suggests that the PRE may provide a nucleation center, from which a structural state depending on Pc-G proteins may propagate. This model is supported by the observation of effects related to position effect variegation (see Figure 31.4); that is, a gene near a locus whose repression is maintained by Pc-G may become heritably inactivated in some cells but not others. Consistent with the pleiotropy of *Pc* mutations, Pc is a nuclear protein that can be visualized at ~80 sites on polytene chromosomes.

The Pc proteins function in large complexes. The PRC1 (polycomb-repressive complex) contains Pc itself, several other Pc-G proteins, and five general transcription factors. The Esc-E(z) complex contains Esc, E(z), other Pc-G proteins, a histone-binding protein, and a histone deacetylase. Pc itself has a "chromo domain" that binds to methylated H3, and E(z) is a methyltransferase that acts on H3. These properties suggest that the Pc-G functions by modifying chromatin.

A working model for Pc-G binding at a PRE is suggested by the properties of the individual proteins. First, two DNA-binding proteins, Pho and Pho1, bind to specific sequences within the PRE. Esc-E(z) is recruited to Pho/Pho1; it then uses its methyltransferase activity to methylate ^{27}Lys of histone H3. This creates the binding site for the PRC, because the chromo domain of Pc binds to the methylated lysine.

The chromo domain was first identified as a region of homology between Pc and the protein HP1 found in heterochromatin (see *30.12 Some Common Motifs Are Found in Proteins that Modify Chromatin*). Binding of the chromo domain of Pc to ^{27}Lys on H3 is analogous to HP1's use of its chromo domain to bind to ^{9}Lys. Since variegation is caused by the spreading of inactivity from constitutive heterochromatin, it is likely that the chromo domain is used by Pc and HP1 in a similar way to induce the formation of heterochromatic or inactive structures. This model implies that similar mechanisms are used to repress individual loci or to create heterochromatin.

Proteins coded by the *trithorax* loci (*trxG*) have the opposite effect to the Pc-G proteins: they act to maintain genes in an active state. There

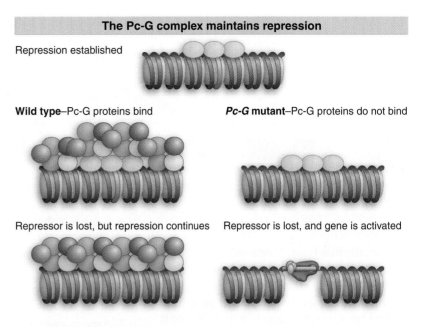

The Pc-G complex maintains repression

Repression established

Wild type—Pc-G proteins bind   ·   *Pc-G* mutant—Pc-G proteins do not bind

Repressor is lost, but repression continues   ·   Repressor is lost, and gene is activated

Figure 31.5 Pc-G proteins do not initiate repression, but are responsible for maintaining it.

may be some similarities in the actions of the two groups: mutations in some loci prevent both Pc-G and trx from functioning, suggesting that they could rely on common components.

31.4 X Chromosomes Undergo Global Changes

Key Concepts

- One of the two X chromosomes is inactivated at random in each cell during embryogenesis of female eutherian mammals.
- In exceptional cases where there are >2 X chromosomes, all but one are inactivated.
- The *Xic* (X inactivation center) is a *cis*-acting region on the X chromosome that is necessary and sufficient to ensure that only one X chromosome remains active.
- *Xic* includes the *Xist* gene, which codes for an RNA that is found only on inactive X chromosomes.
- The mechanism that is responsible for preventing *Xist* RNA from accumulating on the active chromosome is unknown.
- Specific condensins are responsible for condensing inactive X chromosomes in *C. elegans*.

Key Terms

- **Dosage compensation** describes mechanisms employed to compensate for the discrepancy between the presence of two X chromosomes in one sex but only one X chromosome in the other sex.
- **Constitutive heterochromatin** describes the inert state of permanently nonexpressed sequences, usually satellite DNA.
- **Facultative heterochromatin** describes the inert state of sequences that also exist in active copies—for example, one X chromosome in mammalian females.
- The **single X hypothesis** describes the inactivation of one X chromosome in female mammals.
- The **n − 1 rule** states that only one X chromosome is active in female mammalian cells; any others are inactivated.

Sex presents an interesting problem for gene regulation, because of the variation in the number of X chromosomes. If X-linked genes were expressed equally well in each sex, females would have twice as much of each product as males. The importance of avoiding this situation is shown by the existence of **dosage compensation**, which equalizes the level of expression of X-linked genes in the two sexes. Mechanisms used in different species are summarized in **Figure 31.6**:

- In mammals, one of the two female X chromosomes is inactivated completely. The result is that females have only one active X chromosome, which is the same situation found in males. The active X chromosome of females and the single X chromosome of males are expressed at the same level.
- In *Drosophila*, the expression of the single male X chromosome is doubled relative to the expression of each female X chromosome.
- In *C. elegans*, the expression of each female X chromosome is halved relative to the expression of the single male X chromosome.

The common feature in all these mechanisms of dosage compensation is that *the entire chromosome is the target for regulation*. A global change occurs that quantitatively affects all promoters on the chromosome. We know most about the inactivation of the X chromosome in mammalian females, where the entire chromosome becomes heterochromatic.

The twin properties of heterochromatin are its condensed state and associated inactivity. It can be divided into two types:

- **Constitutive heterochromatin** contains specific sequences that have no coding function. Typically these include satellite DNAs and are often found at the centromeres. These regions are invariably heterochromatic because of their intrinsic nature.
- **Facultative heterochromatin** takes the form of entire chromosomes that are inactive in one cell lineage, although they can be expressed in other lineages. The example *par excellence* is the mammalian X chromosome. The inactive X chromosome is perpetuated in a heterochromatic state, while the active X chromosome is euchromatic. So *identical*

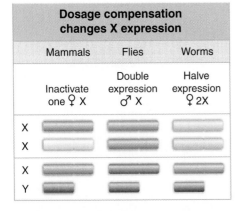

Dosage compensation changes X expression		
Mammals	Flies	Worms
Inactivate one ♀ X	Double expression ♂ X	Halve expression ♀ 2X

Figure 31.6 Different means of dosage compensation are used to equalize X chromosome expression in male and female.

DNA sequences are present in both states. Once the inactive state has been established, it is inherited by descendant cells. This is an example of epigenetic inheritance, because it does not depend on the DNA sequence.

Our basic view of the situation of the female mammalian X chromosomes was formed by the **single X hypothesis** in 1961. Female mice that are heterozygous for X-linked coat color mutations have a variegated phenotype in which some areas of the coat are wild type, but others are mutant. **Figure 31.7** shows that this can be explained *if one of the two X chromosomes is inactivated at random in each cell of a small precursor population.* Cells in which the X chromosome carrying the wild-type gene is inactivated give rise to progeny that express only the mutant allele on the active chromosome. Cells derived from a precursor where the other chromosome was inactivated have an active wild-type gene. In the case of coat color, cells descended from a particular precursor stay together and thus form a patch of the same color, creating the pattern of visible variegation. In other cases, individual cells in a population express one or the other of X-linked alleles; for example, in heterozygotes for the X-linked locus *G6PD*, any particular red blood cell expresses only one of the two allelic forms.

Inactivation of the X chromosome in females is governed by the **n − 1** rule: however many X chromosomes are present, all but one will be inactivated. In normal females there are of course two X chromosomes, but in rare cases where nondisjunction has generated a 3X or greater genotype, only one X chromosome remains active. This suggests a general model in which a specific event is limited to one X chromosome and protects it from an inactivation mechanism that applies to all the others.

A single locus on the X chromosome is sufficient for inactivation. When there is a translocation between the X chromosome and an autosome, this locus is present on only one of the reciprocal products, and only that product can be inactivated. By comparing different translocations, it is possible to map this locus to a 450 kb region, which is called the *Xic* (X-inactivation center). When *Xic* is inserted as a transgene onto an autosome, the autosome becomes subject to inactivation.

Xic is a *cis*-acting locus that contains the information necessary to count X chromosomes and inactivate all copies but one. Inactivation spreads from *Xic* along the entire X chromosome. When *Xic* is present on an X chromosome-autosome translocation, inactivation spreads into the autosomal regions (although the effect is not always complete).

Xic contains a gene, called *Xist*, that is expressed only on the *inactive* X chromosome. The behavior of this gene is effectively the opposite from all other loci on the chromosome, which are turned off. Deletion of *Xist* prevents an X chromosome from being inactivated. However, it does not interfere with the counting mechanism (because other X chromosomes can be inactivated). So we can distinguish two features of *Xic*: an unidentified element(s) required for counting; and the *Xist* gene required for inactivation.

Figure 31.8 illustrates the role of *Xist* RNA in X-inactivation. *Xist* codes for an RNA that lacks open reading frames. The *Xist* RNA "coats" the X chromosome from which it is synthesized, suggesting that it has a structural role. Prior to X-inactivation, it is synthesized by both female X chromosomes. Following inactivation, the RNA is found only on the inactive X chromosome. The transcription rate remains the same before and after inactivation, so the transition depends on posttranscriptional events.

Prior to X-inactivation, *Xist* RNA decays with a half life of ~2 hr. X-inactivation is mediated by stabilizing the *Xist* RNA on the inactive X chromosome. The *Xist* RNA shows a punctate distribution along the

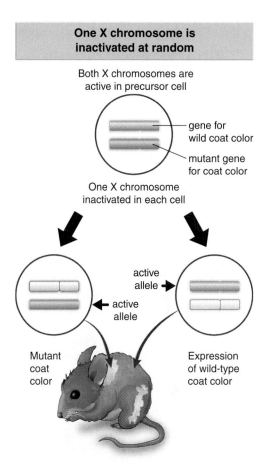

Figure 31.7 X-linked variegation is caused by the random inactivation of one X chromosome in each precursor cell. Cells in which the + allele is on the active chromosome have wild-type phenotype, but cells in which the − allele is on the active chromosome have mutant phenotype.

Xist RNA inactivates one X chromosome

**Both X chromosomes express *XIST*:
RNA is unstable**

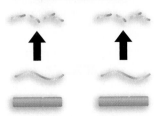

RNA is stabilized and coats one chromosome

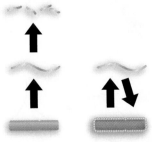

Active X ceases synthesis of XIST RNA

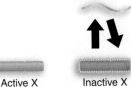

Active X   Inactive X

Figure 31.8 X-inactivation involves stabilization of *Xist* RNA, which coats the inactive chromosome.

X chromosome, suggesting that association with proteins to form particulate structures may be the means of stabilization. We do not know yet what other factors may be involved in this reaction and how the *Xist* RNA is limited to spreading in *cis* along the chromosome. The characteristic features of the inactive X chromosome, which include a lack of acetylation of histone H4, and methylation of CpG sequences (see *24.13 CpG Islands Are Regulatory Targets*), presumably occur later as part of the mechanism of inactivation.

The n − 1 rule suggests that stabilization of *Xist* RNA is the "default," and that some blocking mechanism prevents stabilization at one X chromosome (which becomes the active X). This means that, although *Xic* is necessary and sufficient for a chromosome to be *inactivated*, the products of other loci may be necessary for the establishment of an *active* X chromosome.

Global changes occur in other types of dosage compensation. In *Drosophila*, a complex of proteins is found in males, where it localizes on the X chromosome. In *C. elegans*, a protein complex associates with both X chromosomes in XX embryos, but the protein components remain diffusely distributed in the nuclei of XO embryos. The protein complex contains proteins related to the condensin complexes that cause the condensation of mitotic chromosomes in other species. This suggests that it has a structural role in causing the chromosome to take up a more condensed, inactive state. The complex initially binds to discrete sites on the X chromosome, and then spreads along the chromosome from these sites.

31.5 DNA Methylation Is Perpetuated by a Maintenance Methylase

Key Terms

- A **fully methylated** site is a palindromic sequence that is methylated on both strands of DNA.
- **Hemimethylated** DNA is methylated on one strand of a target sequence that has a cytosine on each strand.
- A **demethylase** is a casual name for an enzyme that removes a methyl group, typically from DNA, RNA, or protein.
- A **methyltransferase (methylase)** is an enzyme that adds a methyl group to a substrate, which can be a small molecule, a protein, or a nucleic acid.
- A **de novo methylase** adds a methyl group to an unmethylated target sequence on DNA.
- A **maintenance methylase** adds a methyl group to a target site that is already hemimethylated.

Key Concepts

- Most methyl groups in DNA are found on cytosine on both strands of the CpG doublet.
- Replication converts a fully methylated site to a hemimethylated site.
- Hemimethylated sites are converted to fully methylated sites by a maintenance methylase.

DNA is methylated at specific sites. In bacteria, methylation is associated with identifying the particular bacterial strain, and also with distinguishing replicated and nonreplicated DNA (see *20.6 Controlling the Direction of Mismatch Repair*). In eukaryotes, its principal known function is connected with the control of transcription; methylation is associated with gene inactivation.

From 2–7% of the cytosines of animal cell DNA are methylated (the value varies with the species). Most of the methyl groups are found in CG "doublets," and, in fact, the majority of the CG sequences are methylated. Usually the C residues on both strands of this short palindromic sequence are methylated, giving the structure

5′ ^{m}CpG   3′
3′   GpC^{m} 5′

Such a site is described as **fully methylated**. But consider the consequences of replicating this site. **Figure 31.9** shows that each daughter duplex has one methylated strand and one unmethylated strand. Such a site is called **hemimethylated**.

The perpetuation of the methylated site now depends on what happens to hemimethylated DNA. If methylation of the unmethylated strand occurs, the site is restored to the fully methylated condition. However, if replication occurs first, the hemimethylated condition will be perpetuated on one daughter duplex, but the site will become unmethylated on the other daughter duplex. **Figure 31.10** shows that the state of methylation of DNA is controlled by **methylases**, which add methyl groups to the 5 position of cytosine, and **demethylases**, which remove the methyl groups. (The more formal name for the enzymes uses **methyltransferase** as the description.)

There are two types of DNA methylase, whose actions are distinguished by the state of the methylated DNA. To modify DNA at a new

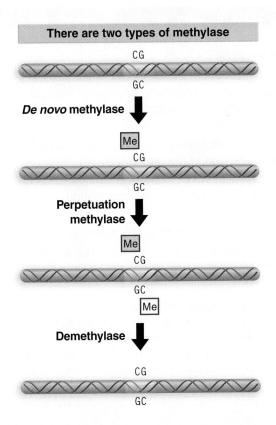

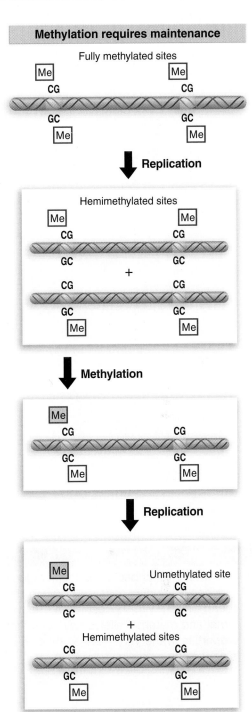

Figure 31.9 The state of methylated sites could be perpetuated by an enzyme that recognizes only hemimethylated sites as substrates.

Figure 31.10 The state of methylation is controlled by three types of enzyme. *De novo* and perpetuation methylases are known, but demethylases have not been identified.

position requires the action of the **de novo methylase**, which recognizes DNA by virtue of a specific sequence. It acts *only* on nonmethylated DNA, to add a methyl group to one strand. There are two *de novo* methylases (Dnmt3A and Dnmt3B) in mouse; they have different target sites, and both are essential for development.

A **maintenance methylase** acts constitutively *only on hemimethylated sites* to convert them to fully methylated sites. Its existence means that any methylated site is perpetuated after replication. There is one maintenance methylase (Dnmt1) in mouse, and it is essential: mouse embryos in which its gene has been disrupted do not survive past early embryogenesis.

Maintenance methylation is virtually 100% efficient, ensuring that the situation shown on the left of Figure 31.9 usually prevails *in vivo*. The result is that, if a *de novo* methylation occurs on one allele but not on the other, this difference will be perpetuated through ensuing cell divisions, maintaining a difference between the alleles that does not depend on their sequences.

Methylation has various types of targets. Gene promoters are the most common target. The promoters are methylated when the gene is inactive, but unmethylated when it is active. The absence of Dnmt1 in mouse causes widespread demethylation at promoters, and we assume this is lethal because of the uncontrolled gene expression. Satellite DNA is another target. Mutations in Dnmt3B prevent methylation of satellite DNA, which causes centromere instability at the cellular level. Mutations in the corresponding human gene cause a disease. The importance of methylation is emphasized by another human disease, which is caused by mutation of the gene for the protein MeCp2 that binds methylated CpG sequences.

31.6 DNA Methylation Is Responsible for Imprinting

Key Term

- **Imprinting** describes a change in a gene that occurs during passage through the sperm or egg with the result that the paternal and maternal alleles have different properties in the very early embryo. This is caused by methylation of DNA.

Key Concepts

- Paternal and maternal alleles may have different patterns of methylation at fertilization.
- Methylation is usually associated with inactivation of the gene.
- When genes are differentially imprinted, survival of the embryo may require that the functional allele is provided by the parent with the unmethylated allele.
- Survival of heterozygotes for imprinted genes is different depending on the direction of the cross.
- Imprinted genes are present in clusters and may depend on a local control site where *de novo* methylation occurs unless specifically prevented.
- Imprinted genes are controlled by methylation of *cis*-acting sites.
- Methylation may be responsible for either inactivating or activating a gene.

All allelic differences are lost when primordial germ cells develop in the embryo; irrespective of sex, the previous patterns of methylation are erased by a genome-wide demethylation, and a typical gene is then unmethylated. Then the pattern specific for each sex is imposed. In males, the pattern develops in two stages. The methylation pattern that is characteristic of mature sperm is established in the spermatocyte. But further changes are made in this pattern after fertilization. In females, the maternal pattern is imposed during oogenesis, when oocytes mature through meiosis after birth. A major question is how the specificity of methylation is determined in each sex.

Systematic changes occur in early embryogenesis. Some sites continue to be methylated, but others are specifically unmethylated in cells in which a gene is expressed. From the pattern of changes, we may infer that individual sequence-specific demethylation events occur during somatic development of the organism as particular genes are activated.

This sex-specific mode of inheritance requires that the pattern of methylation is established specifically during each gametogenesis. The fate of a hypothetical locus in a mouse is illustrated in **Figure 31.11**. In the early embryo, the paternal allele is nonmethylated and expressed, and the maternal allele is methylated and silent. What happens when this mouse itself forms gametes? If it is a male, the allele contributed to the sperm must be nonmethylated, irrespective of whether it was originally methylated or not. So when the maternal allele finds itself in a sperm, it must be demethylated. If the mouse is a female, the allele contributed to the egg must be methylated; so if it was originally the paternal allele, methyl groups must be added.

The specific pattern of methyl groups in germ cells is responsible for the phenomenon of **imprinting**, which describes a difference in behavior between the alleles inherited from each parent. The expression of certain genes in mouse embryos depends upon the sex of the parent from which they were inherited. For example, the allele coding for IGF-II (insulin–like growth factor II) that is inherited from the father is expressed, but the allele that is inherited from the mother is not expressed. The IGF-II gene of oocytes is methylated, but the IGF-II gene of sperm is not methylated, so that the two alleles behave differently in the zygote. This is the most common pattern, but the dependence on sex is reversed for some genes. In fact, the opposite pattern (expression of maternal copy) is shown for IGF-IIR, the receptor for IGF-II.

The consequence of imprinting is that an embryo requires a paternal allele for IGF-II. So in the case of a heterozygous cross where the allele of one parent has an inactivating mutation, the embryo will survive if the wild-type allele comes from the father, but will die if the wild-type allele is from the mother. This type of dependence on the directionality of the cross (in contrast with Mendelian genetics) is an example of epigenetic inheritance, where some factor other than the sequences of the genes themselves influences their effects. Although the paternal and maternal alleles have identical sequences, they display different properties, depending on which parent provided them. These properties are inherited through meiosis and the subsequent somatic mitoses.

Imprinted genes are sometimes clustered. More than half of the 17 known imprinted genes in mouse are contained in two particular regions, each containing both maternally and paternally expressed genes. This suggests the possibility that imprinting mechanisms may function over long distances. Some insights into this possibility come from deletions in the human population that cause the Prader–Willi and Angelman diseases. Most cases are caused by the same 4 Mb deletion, but the syndromes are different, depending on which parent contributed the deletion. The reason is that the deleted region contains at least one gene that is paternally imprinted and at least one that is maternally imprinted.

Imprinting is determined by the state of methylation of a *cis*-acting site near a target gene or genes. These regulatory sites are known as DMDs (differentially methylated domains) or ICRs (imprinting control regions). Deletion of these sites removes imprinting, and the target loci then behave the same in both maternal and paternal genomes.

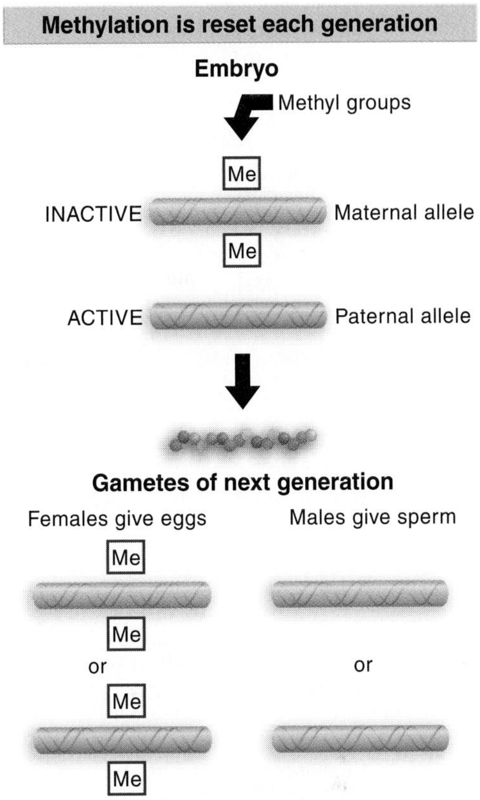

Figure 31.11 The typical pattern for imprinting is that a methylated locus is inactive. If this is the maternal allele, only the paternal allele is active, and is essential for viability. The methylation pattern is reset when gametes are formed, so that all sperm have the paternal type, and all oocytes have the maternal type.

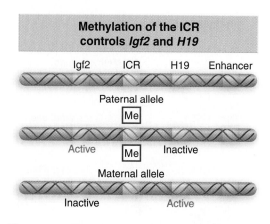

Figure 31.12 ICR is methylated on the paternal allele, where *Igf2* is active and *H19* is inactive. ICR is unmethylated on the maternal allele, where *Igf2* is inactive and *H19* is active.

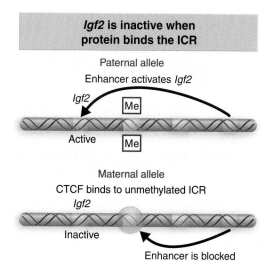

Figure 31.13 The ICR is an insulator that prevents an enhancer from activating Igf2. The insulator functions only when it binds CTCF to unmethylated DNA.

The behavior of a region containing two genes, *Igf2* and *H19*, illustrates the ways in which methylation can control gene activity. **Figure 31.12** shows that these two genes react oppositely to the state of methylation at a site located between them, called the ICR. The ICR is methylated on the paternal allele. *H19* shows the typical response of inactivation. However, *Igf2* is expressed. The reverse situation is found on a maternal allele, where the ICR is not methylated. *H19* now becomes expressed, but *Igf2* is inactivated.

The control of *Igf2* is exercised by an insulator function of the ICR. **Figure 31.13** shows that when the ICR is unmethylated, it binds the protein CTCF. This creates an insulator function that blocks an enhancer from activating the *Igf2* promoter. This is an unusual effect in which methylation indirectly activates a gene by blocking an insulator.

The regulation of *H19* shows the more usual direction of control in which methylation creates an inactive imprinted state. This could reflect a direct effect of methylation on promoter activity.

31.7 Yeast Prions Show Unusual Inheritance

Key Term

- A **prion** is a proteinaceous infectious agent, which behaves as an inheritable trait, although it contains no nucleic acid. Examples are PrP[Sc], the agent of scrapie in sheep and bovine spongiform encephalopathy, and Psi, which confers an inherited state in yeast.

Key Concepts

- The Sup35 protein in its wild-type soluble form is a termination factor for translation.
- It can also exist in an alternative form of oligomeric aggregates, in which it is not active in protein synthesis.
- The presence of the oligomeric form causes newly synthesized protein to acquire the inactive structure.
- Conversion between the two forms is influenced by chaperones.
- The wild-type form has the recessive genetic state *psi⁻* and the mutant form has the dominant genetic state *PSI⁺*.

One of the clearest cases of the dependence of epigenetic inheritance on the condition of a protein is provided by the behavior of **prions**—proteinaceous infectious agents. They have been characterized in two circumstances: by genetic effects in yeast, and as the causative agents of neurological diseases in mammals, including man. A striking epigenetic effect is found in yeast, where two different states can be inherited that map to a single genetic locus, *although the sequence of the gene is the same in both states*. The two different states are [*psi⁻*] and [*PSI⁺*]. A switch in condition occurs at a low frequency as the result of a spontaneous transition between the states.

The *psi* genotype maps to the locus *sup35*, which codes for a translation termination factor. **Figure 31.14** summarizes the effects of the Sup35 protein in yeast. In wild-type cells, which are characterized as [*psi⁻*], the gene is active, and Sup35 protein terminates protein synthesis. In cells of the mutant [*PSI⁺*] type, the factor does not function, causing a failure to terminate protein synthesis properly.

[*PSI⁺*] strains have unusual genetic properties. When a [*psi⁻*] strain is crossed with a [*PSI⁺*] strain, *all of the progeny are* [*PSI⁺*]. This is a pattern of inheritance that would be expected of an extrachromosomal

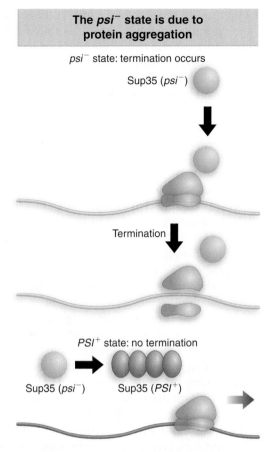

Figure 31.14 The state of the Sup35 protein determines whether termination of translation occurs.

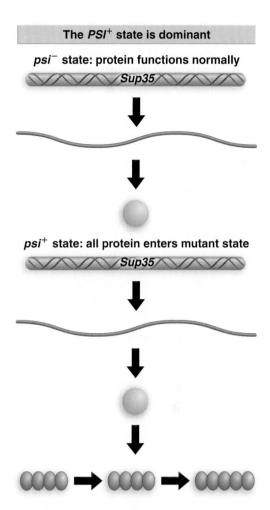

Figure 31.15 Newly synthesized Sup35 protein is converted into the [*PSI*⁺] state by the presence of preexisting [*PSI*⁺] protein.

agent, but the [*PSI*⁺] trait cannot be mapped to any such nucleic acid. The [*PSI*⁺] trait is metastable, which means that, although it is inherited by most progeny, it is lost at a higher rate than is consistent with mutation. Similar behavior is shown also by the locus *URE2,* which codes for a protein required for nitrogen-mediated repression of certain catabolic enzymes. When a yeast strain is converted into an alternative state, called [*URE3*], the Ure2 protein is no longer functional.

The [*PSI*⁺] state is determined by the conformation of the Sup35 protein. In a wild-type [*psi*⁻] cell, the protein displays its normal function. But in a [*PSI*⁺] cell, the protein is present in an alternative conformation in which its normal function has been lost. To explain the unilateral dominance of [*PSI*⁺] over [*psi*⁻] in genetic crosses, we must suppose that *the presence of protein in the [PSI⁺] state causes all the protein in the cell to enter this state.* This requires an interaction between the [*PSI*⁺] protein and newly synthesized protein, probably reflecting the generation of an oligomeric state in which the [*PSI*⁺] protein has a nucleating role, as illustrated in **Figure 31.15**.

A feature common to both the Sup35 and Ure2 proteins is that each consists of two domains that function independently. The C-terminal domain is sufficient for the activity of the protein. The N-terminal domain is sufficient for formation of the structures that make the protein inactive. So yeast in which the N-terminal domain of Sup35 has been deleted cannot acquire the [*PSI*⁺] state; and the presence of a [*PSI*⁺] N-terminal domain is sufficient to maintain Sup35 protein in the [*PSI*⁺] condition. The critical feature of the N-terminal domain is that it is rich in glutamine and asparagine residues.

PSI⁺ protein can convert yeast

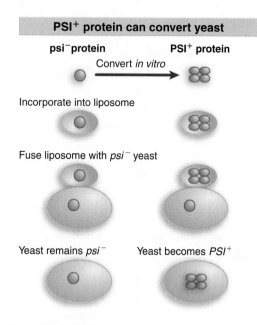

Figure 31.16 Purified protein can convert the [*psi⁻*] state of yeast to [*PSI⁺*].

Loss of function in the [*PSI⁺*] state is due to the sequestration of the protein in an oligomeric complex. Sup35 protein in [*PSI⁺*] cells is clustered in discrete foci, whereas the protein in [*psi⁻*] cells is diffused in the cytosol. Sup35 protein from [*PSI⁺*] cells forms amyloid fibers *in vitro*—these have a characteristic high content of β sheet structures. The involvement of protein conformation (rather than covalent modification) is suggested by the effects of conditions that affect protein structure. Denaturing treatments cause loss of the [*PSI⁺*] state.

Using the ability of Sup35 to form the inactive structure *in vitro*, it is possible to provide biochemical proof for the role of the protein. **Figure 31.16** illustrates a striking experiment in which the protein was converted to the inactive form *in vitro*, put into liposomes (surrounding the protein, in effect, with an artificial membrane), and then introduced directly into cells by fusing the liposomes with [*psi⁻*] yeast. The yeast cells were converted to [*PSI⁺*]. This experiment refutes all of the objections that were raised previously to the conclusion that the protein has the ability to confer the epigenetic state. Experiments in which cells are mated, or in which extracts are taken from one cell to treat another cell, always are susceptible to the possibility that a nucleic acid has been transferred. But when the protein by itself does not convert target cells, but protein converted to the inactive state can do so, the only difference is the treatment of the protein—which must therefore be responsible for the conversion.

The ability of yeast to form the [*PSI⁺*] prion state depends on the genetic background. The yeast must be [*PIN⁺*] in order for the [*PSI⁺*] state to form. The [*PIN⁺*] condition itself is an epigenetic state. It can be created by the formation of prions from any one of several different proteins. These proteins share the characteristic of Sup35 that they have Gln/Asn-rich domains. Overexpression of these domains in yeast stimulates formation of the [*PSI⁺*] state. This suggests that there is a common model for the formation of the prion state that involves aggregation of the Gln/Asn domains.

31.8 Prions Cause Diseases in Mammals

Key Terms

- **Scrapie** is a disease caused by an infective agent made of protein.
- **Kuru** is a human neurological disease caused by prions. It may be caused by eating infected brains.

Key Concepts

- The protein responsible for scrapie exists in two forms, the wild-type noninfectious form PrP^C that is susceptible to proteases and the disease-causing form PrP^Sc that is resistant to proteases.
- The neurological disease can be transmitted to mice by injecting the purified PrP^Sc protein into mice.
- The recipient mouse must have a copy of the *PrP* gene coding for the mouse protein.
- The PrP^Sc protein can perpetuate itself by causing the newly synthesized PrP protein to take up the PrP^Sc form instead of the PrP^C form.
- Multiple strains of PrP^Sc may have different conformations of the protein.

Prion diseases have been found in sheep and man, and, more recently, in cows. The basic phenotype is an ataxia—a neurodegenerative disorder that is manifested by an inability to remain upright. The name of the disease in sheep, **scrapie**, reflects the phenotype: the sheep rub against walls in order to stay upright. Scrapie can be perpetuated by inoculating sheep with tissue extracts from infected animals. The human disease **kuru** was found in New Guinea, where it appeared to be perpetuated by cannibalism, in particular the eating of brains. Related diseases in Western populations with a pattern of genetic transmission include Gerstmann–Straussler syndrome; and the related Creutzfeldt–Jakob disease (CJD) occurs sporadically. Most recently, a disease resembling CJD appears to have been transmitted by consumption of meat from cows suffering from "mad cow" disease.

When tissue from scrapie-infected sheep is inoculated into mice, the disease manifests in the range of 75–150 days. The active component is a protease-resistant protein. The protein is coded by a gene that is normally expressed in brain. The form of the protein in normal brain, called PrP^C, is sensitive to proteases. Its conversion to the resistant form, called Prp^{Sc}, is associated with occurrence of the disease. The infectious preparation has no detectable nucleic acid, is sensitive to UV irradiation at wavelengths that damage protein, and has a low infectivity (1 infectious unit/10^5 PrP^{Sc} proteins). This corresponds to an epigenetic inheritance in which there is no change in genetic information, because normal and diseased cells have the same *PrP* gene sequence, but the PrP^{Sc} form of the protein is the infectious agent, whereas PrP^C is harmless.

The basis for the difference between the PrP^{Sc} and PrP^C forms appears to lie with a change in conformation rather than with any covalent alteration. Both proteins are glycosylated and linked to the membrane by a GPI-linkage. No changes in these modifications have been found. The PrP^{Sc} form has a high content of β sheets, which are absent from the PrP^C form.

The assay for infectivity in mice allows the dependence on protein sequence to be tested. **Figure 31.17** illustrates the results of some critical experiments. In the normal situation, PrP^{Sc} protein extracted from an infected mouse will induce disease (and ultimately kill) when it is injected into a recipient mouse. If the *PrP* gene is "knocked out," a mouse becomes resistant to infection. This experiment demonstrates two things. First, the endogenous protein is necessary for an infection, presumably because it provides the raw material that is converted into the infectious agent. Second, the cause of disease is not the removal of the PrP^C form of the protein, because a mouse with no PrP^C survives normally: the disease is caused by a gain-of-function in PrP^{Sc}.

The existence of species barriers allows hybrid proteins to be constructed to delineate the features required for infectivity. The original preparations of scrapie were perpetuated in several types of animal, but these cannot always be transferred readily. For example, mice are resistant to infection from prions of hamsters. This means that hamster-PrP^{Sc} cannot convert mouse-PrP^C to PrP^{Sc}. However, the situation changes if the mouse *PrP* gene is replaced by a hamster *PrP* gene. (This can be done by introducing the hamster *PrP* gene into the *PrP* knockout mouse.) A mouse with a hamster *PrP* gene is sensitive to infection by hamster PrP^{Sc}. This suggests that the conversion of cellular PrP^C protein into the Sc state requires that the PrP^{Sc} and PrP^C proteins have matched sequences.

There are different "strains" of PrP^{Sc}, which are distinguished by characteristic incubation periods upon inoculation into mice. This implies that the protein is not restricted solely to alternative states of PrP^C and

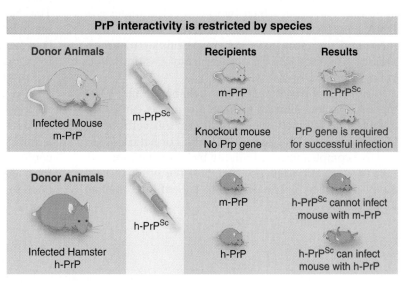

Figure 31.17 A Prp^{Sc} protein can only infect an animal that has the same type of endogenous PrP^C protein.

PrP^{Sc}, but that there may be multiple Sc states. These differences must depend on some self-propagating property of the protein other than its sequence. If conformation is the feature that distinguishes PrP^{Sc} from PrP^C, then there must be multiple conformations, each of which has a self-templating property when it converts PrP^C.

The probability of conversion from PrP^C to PrP^{Sc} is affected by the sequence of PrP. Gerstmann–Straussler syndrome in man is caused by a single amino acid change in PrP. This is inherited as a dominant trait. If the same change is made in the mouse PrP gene, mice develop the disease. This suggests that the mutant protein has an increased probability of spontaneous conversion into the Sc state. Similarly, the sequence of the PrP gene determines the susceptibility of sheep to develop the disease spontaneously; the combination of amino acids at three positions (codons 136, 154, and 171) determines susceptibility.

The prion offers an extreme case of epigenetic inheritance, in which the infectious agent is a protein that can adopt multiple conformations, each of which has a self-templating property. This property is likely to involve the state of aggregation of the protein.

31.9 SUMMARY

Heterochromatin is formed by proteins that bind to specific chromosomal regions (such as telomeres) and that interact with histones. The formation of an inactive structure may propagate along the chromatin thread from an initiation center. Similar events occur in silencing of the inactive yeast mating type loci. Repressive structures that are required to maintain the inactive states of particular genes are formed by the Pc-G protein complex in *Drosophila*. They share with heterochromatin the property of propagating from an initiation center.

Formation of heterochromatin may be initiated at certain sites and then propagated for a distance that is not precisely determined. When a heterochromatic state has been established, it is inherited through subsequent cell divisions. This gives rise to a pattern of epigenetic inheritance, in which two identical sequences of DNA may be associated with different protein structures, and therefore have different abilities to be expressed. This explains the occurrence of position effect variegation in *Drosophila*.

Inactive chromatin at yeast telomeres and silent mating type loci appears to have a common cause, and involves the interaction of certain proteins with the N-terminal tails of histones H3 and H4. Formation of the inactive complex may be initiated by binding of one protein to a specific sequence of DNA; the other components may then polymerize in a cooperative manner along the chromosome.

Inactivation of one X chromosome in female eutherian mammals occurs at random. The *Xic* locus is necessary and sufficient to count the number of X chromosomes. The n − 1 rule ensures that all but one X chromosome are inactivated. *Xic* contains the gene *Xist*, which codes for an RNA that is expressed only on the inactive X chromosome. Stabilization of *Xist* RNA is the mechanism by which the inactive X chromosome is distinguished.

Methylation of DNA is inherited epigenetically. Replication of DNA creates hemimethylated products, and a maintenance methylase restores the fully methylated state. Some methylation events depend on parental origin. Sperm and eggs contain specific and different patterns of methylation, with the result that paternal and maternal alleles are differently expressed in the embryo. This is responsible for imprinting, in which the non-methylated allele inherited from one parent is essential because it is the only active allele; the allele inherited from the other parent is silent. Patterns of methylation are reset during gamete formation in every generation.

Prions are proteinaceous infectious agents that are responsible for the disease of scrapie in sheep and for related diseases in man. The infectious agent is a variant of a normal cellular protein. The PrP^{Sc} form has an altered conformation that is self-templating: the normal PrP^C form does not usually take up this conformation, but does so in the presence of PrP^{Sc}. A similar effect is responsible for inheritance of the *PSI* element in yeast.

32

Genetic Engineering

32.1 Introduction

Key Term

- A **cloning vector** is a plasmid or phage that is used to "carry" inserted foreign DNA for the purposes of producing more material or a protein product.

Genetic engineering was originally used as a term to describe the range of manipulations of DNA that become possible with the ability to clone a gene by placing its DNA into another context, typically a plasmid or a phage, in which it could be propagated. From this beginning, when recombinant DNA was used as a tool to analyze gene structure and expression, we moved to the ability to change the DNA content of bacteria and eukaryotic cells by directly introducing cloned DNA that could become part of the genome. And then by changing the genetic content in conjunction with the ability to develop an animal from an embryonic cell, it became possible to generate mice with deletions or additions of specific genes that are inherited via the germline. We now use genetic engineering to describe a range of activities extending from the manipulation of DNA to the introduction of changes into specific somatic cells within an animal or even to changes in the germline itself.

Figure 32.1 summarizes the steps in cloning a sequence of DNA. The basic principle is that the DNA is inserted into a **cloning vector**, a vehicle that carries it and amplifies it within the target cell. In principle, the reaction is accomplished by cutting the donor DNA to release a fragment containing a gene or some other sequence, cutting a vector DNA (shown in the figure as a circular plasmid), and then joining the ends crosswise to generate a hybrid DNA molecule. The vector is chosen to carry an origin of replication that will let it perpetuate itself in an appropriate host, most often *E. coli* or *S. cerevisiae*. After amplification, hosts carrying the hybrid vector are selected, and the DNA can be isolated.

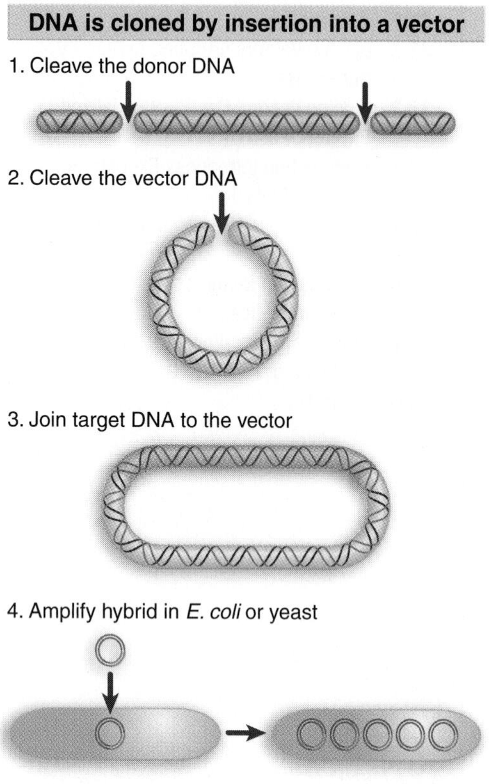

DNA is cloned by insertion into a vector

1. Cleave the donor DNA

2. Cleave the vector DNA

3. Join target DNA to the vector

4. Amplify hybrid in *E. coli* or yeast

Figure 32.1 A donor DNA is generated by cleavage so that its ends can be joined to the ends of a cloning vector. The hybrid molecule is then introduced into a bacterium or yeast in which it can replicate to produce many more copies.

Cloned DNA can enter the genome

Insert cloned DNA into cell

Integrate cloned DNA into chromosome

Recombination

Chromosome

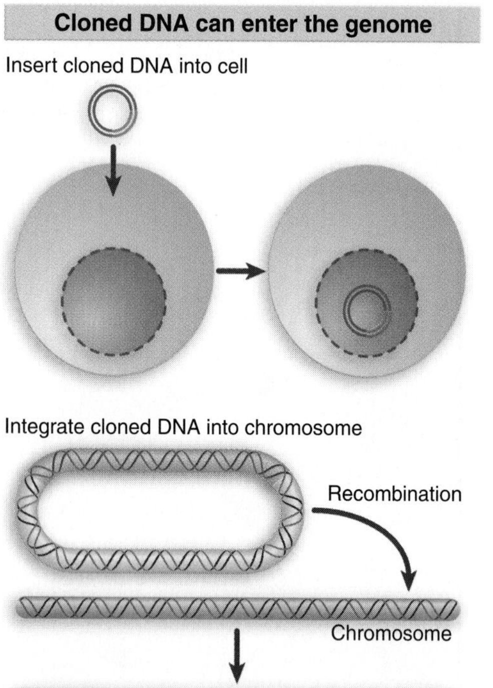

Figure 32.2 Introducing new DNA into the genome requires cloning the donor sequence, delivery of the cloned DNA into the cell, and integration into the genome.

Figure 32.2 gives an overview of the stages used in changing the genome of a target cell. First the DNA that we want to introduce must be cloned as shown in the previous figure. Then we need a delivery method to introduce the cloned DNA into the cell, and it must either be placed directly in the nucleus or be transferred there. Once in the nucleus, the cloned DNA must be integrated into the host chromosome, typically by a recombination event. Depending on the system used, the DNA of the cloning vector may or may not be integrated together with the desired donor DNA sequence.

There is now quite a wide range of methods for introducing DNA into target cells. Some types of cloning vectors use natural methods of infection to pass the DNA into the cell, such as a viral vector that uses the viral infective process to enter the cell. **Figure 32.3** summarizes several other methods.

Liposomes are small spheres made from artificial membranes. The interior can contain DNA or other biological materials. Liposomes can fuse with plasma membranes, the result being to release their contents inside of the cell.

Microinjection uses a very fine needle to puncture the cell membrane. A solution containing DNA can be introduced into the cytoplasm, or directly into the nucleus in the case where the nucleus is large enough to pick out as a target (such as an egg).

The thick cell walls of yeasts and plants are an impediment to many transfer methods, and the "gene gun" was invented as a means for overcoming this obstacle. In principle, it shoots very small particles into the cell by propelling them through the wall at high velocity. The particles can consist of gold or nanospheres coated with DNA. The method now has been adapted for use with mammalian cells.

32.2 Cloning Vectors Are Used to Amplify Donor DNA

Key Terms

- **Cloning** describes propagation of a DNA sequence by incorporating it into a hybrid construct that can be replicated in a host cell.
- A **vector** is a plasmid or phage chromosome that is used to perpetuate a cloned DNA segment.
- A **cosmid** is derived from a bacterial plasmid by incorporating the *cos* sites of phage lambda, which make the plasmid DNA a substrate for the lambda packaging system.

Key Concepts

- Fragments of DNA are generated for cloning by cleavage with restriction enzymes.
- Cloning vectors may be bacterial plasmids, phages, or cosmids, or yeast artificial chromosomes.
- Some cloning vectors have sequences allowing them to be propagated in more than one type of host cell.

The basic principle of gene **cloning** is to insert the DNA of interest into a cloning **vector**, which then replicates to produce many more copies. The key issues are how we identify the donor DNA, what means we use to insert it into the vector, how we select the hybrid molecule that results from the insertion, and the method of amplification.

The ability to identify donor DNA and insert it into a vector has benefited enormously from genome sequencing. Instead of using various indirect techniques, we can work directly on the basis of the sequence. One of the early cloning methods is well adapted to this development. The donor DNA is cleaved with a restriction enzyme that introduces double-strand breaks by making staggered cuts in the two strands of DNA. **Figure 32.4** shows that this generates single-stranded ends on each side of the break. Depending on the pattern of cutting, the protruding ends may be 5′ or 3′. In either case, the single-stranded ends on either side of the break are complementary. Some restriction enzymes cleave at the same point on both strands, and therefore generate blunt ends.

The vector is cleaved with the same restriction enzyme(s) as the donor DNA. **Figure 32.5** shows that the complementary single-stranded ends can then hybridize to join the donor and vector DNAs. By using different enzymes to cleave each end of the donor sequence, we can ensure that the insertion occurs in only one orientation.

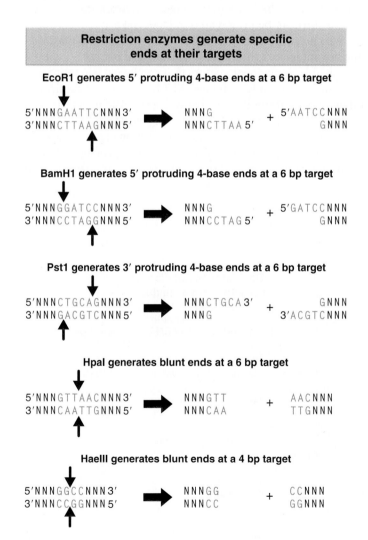

Figure 32.4 Each restriction enzyme cleaves a specific target sequence (usually 4–6 bp long). The sites of cleavage on the two strands may generate 5′ protruding ends, 3′ protruding ends, or blunt ends.

DNA can be introduced into cells in several ways

A viral vector
introduces DNA by infection

Liposomes
may fuse with the membrane

Microinjection
introduces DNA directly into the cytoplasm or nucleus

Nanospheres
can be shot into the cell by a gene gun

Figure 32.3 DNA can be released into target cells by methods that pass it across the membrane naturally, such as a viral vector (in the same way as a viral infection) or by encapsulating it in a liposome (which fuses with the membrane). Or it can be passed manually, by microinjection, or by coating it on the exterior of nanoparticles that are shot into the cell by a gene gun that punctures the membrane at very high velocity.

Restriction enzymes are used to clone DNA

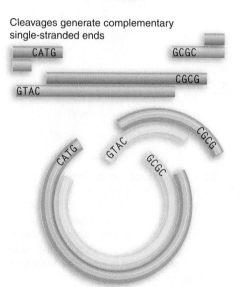

Donor DNA

CATG
GTAC

CGCG
GCGC

Enzyme 1

Enzyme 2

CATG
GTAC

CGCG
GCGC

Cloning vector

Cleavages generate complementary single-stranded ends

CATG

GCGC

GTAC

CGCG

CATG

GTAC

CGCG

GCGC

Join donor and vector DNAs

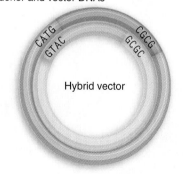

CATG
GTAC

CGCG
GCGC

Hybrid vector

Figure 32.5 Restriction enzymes can generate complementary single-stranded ends in donor and vector DNA by making staggered breaks in their target sequences. Crosswise joining of the fragments generates a hybrid DNA molecule.

The design of the cloning vector is determined by the nature of the experiment. **Figure 32.6** summarizes the properties of the most common classes of cloning vectors. Cloning vectors are usually propagated in bacteria or yeast. Plasmids can be easily engineered to carry long sequences of donor DNA in bacteria, although a disadvantage is that it can be difficult to isolate the DNA. Bacteriophages can be used in bacteria but have the disadvantage that only a limited amount of DNA can be packaged into the viral coat. Another type of vector is the **cosmid**, which propagates like a plasmid but uses the packaging mechanism of phage lambda to isolate the DNA. Lambda DNA ends in *cos* sites that are recognized by the packaging system, which cuts them and inserts the DNA into the phage head. By incorporating *cos* sites into a cloning vector, the lambda system can be used to package and thus isolate the DNA. It is limited to containing 47 kb (the maximum length that can be packaged).

The type of vector most often used for cloning in yeast is the yeast artificial chromosome (YAC). This has an origin to support replication, a centromere to ensure proper segregation, and telomeres to afford stability. In effect it is propagated just like a yeast chromosome. YACs have the largest capacity of any cloning vector, and can propagate with inserts measured in the Mb length range. Somewhat shorter lengths of DNA (~300 kb) can be propagated in bacterial artifical chromosomes (BACs), which are cloning vectors based on the *E. coli* F plasmid; these have the advantage of greater stability than YACs.

It is possible to engineer vectors that can be used in more than one type of host. The example of **Figure 32.7** has origins of replication and selectable markers for both *E. coli* and *S. cerevisiae*. It can replicate as a circular multicopy plasmid in *E. coli*. It has a yeast centromere, and also has yeast telomeres adjacent to BamI restriction cleavage sites, so that cleavage with BamI generates a YAC that can be propagated in yeast. It has a gene interrupted by a restriction site for the enzyme SmaI that can be used to insert donor DNA.

		Several types of cloning vectors are available	
Vector	**Features**	**Isolation of DNA**	**DNA limit**
Plasmid	High copy number	Physical	10 kb
Phage	Infects bacteria	Via phage packaging	20 kb
Cosmid	High copy number	Via phage packaging	48 kb
BAC	Based on F plasmid	Physical	300 kb
YAC	Origin + centromere + telomere	Physical	>1 Mb

Figure 32.6 Cloning vectors may be based on plasmids or phages or may mimic eukaryotic chromosomes.

A vector can be used in both yeast and bacteria

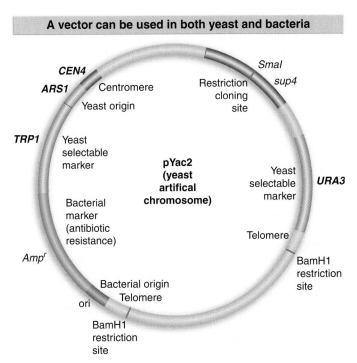

Figure 32.7 pYac2 is a cloning vector with features to allow replication and selection in both bacteria and yeast. Bacterial features include an origin of replication and antibiotic resistance gene. Yeast features include an origin, centromere, two selectable markers, and telomeres.

32.3 Cloning Vectors Can Be Specialized for Different Purposes

Key Concepts

- Amplification of a cloned sequence requires a selective technique to distinguish hybrid vectors from the original vector.
- Promoter activity can be assayed by using reporter genes.
- Exons can be identified by placing inserted DNA into an intron that is flanked by two exons.

Key Term

- A **reporter gene** is a coding unit whose product is easily assayed (such as chloramphenicol transacetylase or beta-galactosidase); it may be connected to any promoter of interest so that expression of the gene can be used to assay promoter function.

Cloning vectors are now customarily designed for particular purposes. Aside from the issue of which type of host cell they use, the main questions are whether they are intended simply to amplify the cloned sequence, to investigate the properties of a promoter (or other regulatory elements), or to identify exons.

When we want simply to amplify the DNA, the main concern is that we should be able to distinguish the hybrid vector, carrying the donor DNA, from the original vector. (Not all copies of the cloning vector will actually get an insertion.) This can be done by selective techniques. **Figure 32.8** shows the approach of using a cloning vector that carries two resistance genes to two different antibiotics. The site where the donor DNA is inserted lies within one of the resistance genes, and the insertion therefore inactivates the gene. The other resistance gene is unaffected by the insertion. The original cloning vector therefore makes its host resistant to both antibiotics, but the hybrid vector is resistant to only

Selective markers are used to distinguish cloning vectors

Select bacteria carrying cloning vector by resistance to ampicillin and neomycin

Select bacteria carrying hybrid cloning vector by sensitivity to ampicillin and resistance to neomycin

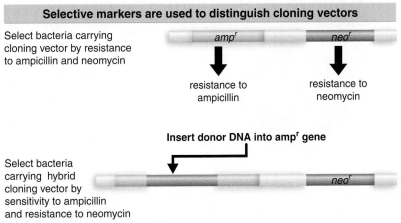

Figure 32.8 Bacteria carrying a cloning vector can be selected by resistance to an antibiotic marker carried by the vector. Bacteria carrying a hybrid cloning vector with an insertion of donor DNA can be distinguished by loss of activity of the vector gene into which the insertion was made.

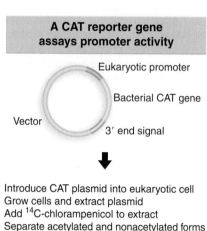

A CAT reporter gene assays promoter activity

Eukaryotic promoter

Bacterial CAT gene

Vector

3′ end signal

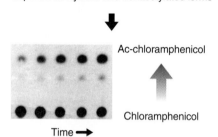

Introduce CAT plasmid into eukaryotic cell
Grow cells and extract plasmid
Add ^{14}C-chloramphenicol to extract
Separate acetylated and nonacetylated forms

Ac-chloramphenicol

Chloramphenicol

Time →

Figure 32.9 The CAT assay can be used to follow the activity of any eukaryotic promoter. The promoter is linked to the gene for chloramphenicol transacetylase in a cloning vector. Extracts are made from target cells carrying the vector. The ability of the extracts to acetylate chloramphenicol is directly proportional to the units of the enzyme, which is determined by the efficiency of the promoter.

A mouse promoter controls tissue-specific expression of lacZ

Figure 32.10 Expression of a *lacZ* gene can be followed in the mouse by staining for β-galactosidase (in blue). In this example, *lacZ* was expressed under the control of a promoter of a mouse gene that is expressed in the nervous system. The corresponding tissues can be visualized by blue staining. Photograph kindly provided by Robert Krumlauf, Stowers Institute for Medical Research.

one. So we can perpetuate the cloning vector itself by selecting bacteria that are resistant to either of the antibiotics (we need to apply selection because otherwise the vector is lost from some bacteria). After insertion of donor DNA, we can identify bacteria carrying the hybrid vector because they are resistant to one of the antibiotics but sensitive to the other.

When our main purpose is to examine the regulation of a gene, usually the crucial issue is how initiation of transcription is controlled at the promoter. In many cases, it would be very difficult to assay the product of the gene. The technique for dealing with this is to connect the promoter to a **reporter gene** whose product can be easily assayed. Usually the function of the promoter is not affected by sequences in the coding region, so from the perspective of regulation, it is quite irrelevant what gene the promoter is connected to. This allows us freedom to design constructs in which the reporter codes for a gene that can be easily assayed in the circumstances in question.

The type of reporter gene that is most appropriate depends on whether we are interested in quantitating the efficiency of the promoter (and, for example, determining the effects of mutations in it or the activities of transcription factors that bind to it) or determining its tissue-specific pattern of expression.

Figure 32.9 summarizes a common system for assaying promoter activity. A cloning vector is created that has a eukaryotic promoter linked to the coding region for the bacterial gene coding for chloramphenicol transacetylase (CAT). Usually a 3′ end termination signal is added to ensure the proper generation of the mRNA. The hybrid vector is introduced into a target cell, the cells are grown (usually for 72 hours), and then an extract containing the enzyme is made. The ability of this extract to convert a substrate of chloramphenicol to acetyl-chloramphenicol is directly proportional to the amount of enzyme that was made, which in turn depends only upon the activity of the promoter.

Some very striking reporters are now available for visualizing gene expression. One of the reactions of β-galactosidase, coded by the *lacZ* gene of the lactose operon, is its ability to convert a substrate called X-gal into a product that is blue. **Figure 32.10** shows what happens when the *lacZ* gene is placed under the control of a promoter that regulates the expression of a gene in the mouse nervous system. The production of β-galactosidase can be followed by supplying the substrate, which causes the nervous system to stain blue. Another reporter that can be used to visualize patterns of gene expression is GFP (green fluorescent protein) obtained from the jellyfish.

When the purpose of cloning is to express a gene in large amounts, the vector is engineered to have a promoter and a translational start sequence adjacent to the site of insertion. Then any open reading frame can be inserted into the vector and expressed without further modification. Specialized vectors may also be used for other purposes, such as identifying exons. **Figure 32.11** shows the use of a vector designed for exon trapping. It contains a strong promoter and has a single intron between two exons. When this vector is transfected into cells, its transcription generates large amounts of an RNA containing the sequences of the two exons. A restriction cloning site lies within the intron and is used to insert genomic fragments from a region of interest. If a fragment does not contain an exon, there is no change in the splicing pattern, and the RNA contains only the same sequences as the parental vector. But if the genomic fragment contains an exon flanked by two partial intron sequences, the splicing sites on either side of this exon are recognized, and the sequence of the exon is inserted into the RNA between the two exons of the vector. This can be detected readily by reverse transcribing the cytoplasmic RNA into cDNA, and using PCR to

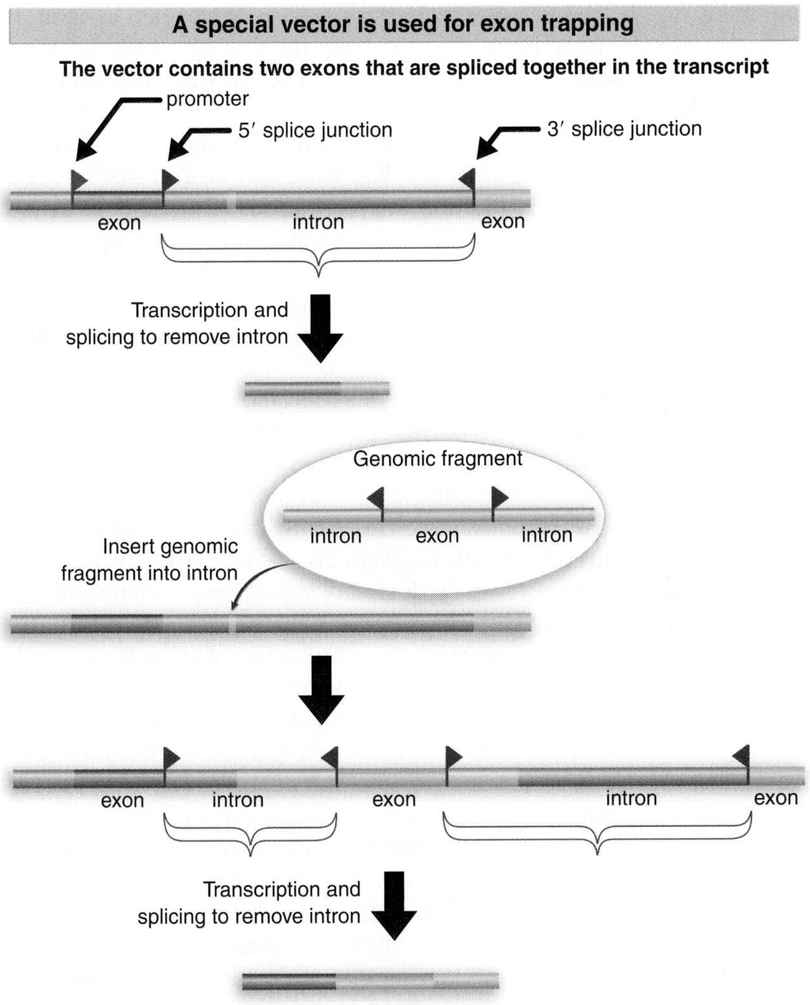

A special vector is used for exon trapping

The vector contains two exons that are spliced together in the transcript

Figure 32.11 A special splicing vector is used for exon trapping. If an exon is present in the genomic fragment, its sequence will be recovered in the cytoplasmic RNA, but if the genomic fragment consists solely of sequences from within the intron, splicing does not occur, and the mRNA is not exported to the cytoplasm.

amplify the sequences between the two exons of the vector. Because introns are usually large and exons are small in animal cells, there is a high probability that a random piece of genomic DNA will contain the required structure of an exon surrounded by partial introns.

32.4 Transfection Introduces Exogenous DNA into Cells

Key Concepts

- DNA that is transfected into a eukaryotic cell forms a large repeating unit of many head to tail tandem repeats.
- The transfected unit is unstable unless it becomes integrated into a host chromosome.
- Genes carried by the transfected DNA can be expressed.

Key Terms

- **Transfection** of eukaryotic cells is the acquisition of new genetic markers by incorporation of added DNA.
- **Unstable transfectants (transient transfectants)** have foreign DNA in an unstable—i.e., extrachromosomal—form.

The problem in delivering DNA to a eukaryotic cell is the lack of any natural uptake system of which we can take advantage. Delivery methods for animal cells fall into two general classes: those based on addition of DNA to cells in a form that relies on some sort of spontaneous uptake mechanism; and those based on mechanical delivery of the DNA, such as microinjection (with a syringe) or electroporation (the use of electrical pulses to create transient holes in the membrane). With plant cells, the thick cell wall poses an obstacle to DNA uptake, and two types of method

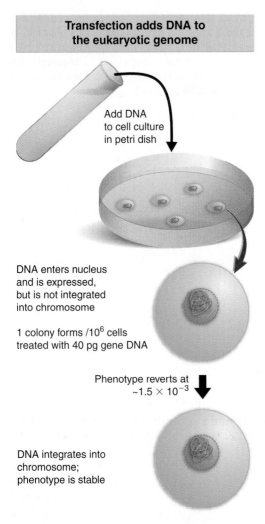

Transfection adds DNA to the eukaryotic genome

Add DNA to cell culture in petri dish

DNA enters nucleus and is expressed, but is not integrated into chromosome

1 colony forms /10^6 cells treated with 40 pg gene DNA

Phenotype reverts at ~1.5×10^{-3}

DNA integrates into chromosome; phenotype is stable

Figure 32.12 DNA added to a cell culture can enter the cell and become expressed in the nucleus, initially as an unstable transfectant. The cell becomes a stable transfectant when the DNA is integrated into a random chromosomal site.

are used: the Ti plasmid of the bacterium *A. tumefaciens*, a natural pathogen of plants, can be used as the basis for cloning vectors; or mechanical methods are employed, often in conjunction with enzymatic removal of the cell wall to form a *protoplast*.

The first procedure that was developed for introducing exogenous donor DNA into recipient cells is called **transfection**. Transfection experiments began with the addition of preparations of metaphase chromosomes to cell suspensions. The chromosomes are taken up rather inefficiently by the cells and give rise to unstable variants at a low frequency. Intact chromosomes rarely survive the procedure; the recipient cell usually gains a fragment of a donor chromosome (which is unstable because it lacks a centromere). Rare cases of stable lines result from integration of donor material into a resident chromosome.

Figure 32.12 shows that similar results are obtained when purified DNA is added to a recipient cell preparation. However, with purified DNA it is possible to add particular sequences instead of relying on random fragmentation of chromosomes. This greatly increases the efficiency of the process. Transfection with DNA yields stable as well as unstable lines, with the former relatively predominant. (These experiments are directly analogous to those performed in bacterial transformation, but are described as transfection because of the historical use of "transformation" to describe changes that allow unrestrained growth of eukaryotic cells.)

Unstable transfectants (sometimes called **transient transfectants**) reflect the survival of the transfected DNA in extrachromosomal form; stable lines result from integration into the genome. The transfected DNA can be expressed in both cases. Because the frequency of transfection is low, the technique is restricted to genes whose transfer produces a selectable phenotype. The classic example was to introduce the thymidine kinase gene into mutant cells that lacked the activity; then when plated on medium on which thymidine kinase is necessary for survival, only the transfectants can survive. An analogous approach allows transfection for the transformed (tumorigenic) phenotype, because the transfected cells can be identified by the change in their morphological properties. This type of protocol has led to the isolation of several cellular *onc* genes, involved in cancer.

Integration to form a stable transfectant occurs by the insertion of the donor DNA into a random site in a host chromosome. The material that is integrated may contain multiple copies of the donor sequence, and if two different types of DNA are "cotransfected" together, usually contains both sequences. For example, when tk^- cells are transfected with a DNA preparation containing both a purified tk^+ gene and the φX174 genome, all the tk^+ transformants have both donor sequences. This is a useful observation, because it allows unselected markers to be introduced routinely by cotransfection with a selected marker.

An individual stable transfectant has only a single site of integration; but the site is different in each case. Probably the selection of a site for integration is a random event; sometimes it is associated with a gross chromosomal rearrangement. Revertants lose the φX174 sequences together with *tk* sequences. So the two types of donor sequence become physically linked during transfection and suffer the same fate thereafter.

The sites at which exogenous material becomes integrated usually do not appear to have any sequence relationship to the transfected DNA. The integration event involves a nonhomologous recombination between the mass of added DNA and a random site in the genome. The recombination event may be provoked by the introduction of a double-strand break into the chromosomal DNA, possibly by the action of DNA repair enzymes that are induced by the free ends of the exogenous DNA.

32.5 Genes Can Be Injected into Animal Eggs

Key Concepts

- DNA that is injected into animal eggs can integrate into the genome.
- Usually a large array of tandem repeats integrates at a single site.
- Expression of the DNA is variable and may be affected by the site of integration and other epigenetic effects.

Key Term

- **Transgenic** animals are created by introducing DNA prepared in test tubes into the germline. The DNA may be inserted into the genome or exist in an extrachromosomal structure.

An exciting development of transfection techniques is their application to introduce genes into animals. An animal that gains new genetic information from the addition of foreign DNA is described as **transgenic**. The approach of directly injecting DNA can be used with mouse eggs, as shown in **Figure 32.13**. Plasmids carrying the gene of interest are injected into the germinal vesicle (nucleus) of the oocyte or into the pronucleus of the fertilized egg. The egg is implanted into a pseudopregnant mouse. After birth, the recipient mouse can be examined to see whether it has gained the foreign DNA, and, if so, whether it is expressed.

The first questions we ask about any transgenic animal are how many copies it has of the foreign material, where these copies are located, and whether they are present in the germline and inherited in a Mendelian manner. The usual result of such experiments is that a reasonable minority (say ~15%) of the injected mice carry the transfected sequence. Usually, multiple copies of the plasmid appear to have been integrated in a tandem array into a single chromosomal site. The number of copies varies from 1–150. They are inherited by the progeny of the injected mouse as expected of a Mendelian locus.

An important issue that can be addressed by experiments with transgenic animals concerns the independence of genes and the effects of the region within which they reside. If we take a gene, including the flanking sequences that contain its known regulatory elements, can it be expressed independently of its location in the genome? In other words, do the regulatory elements function independently, or is gene expression in addition controlled by other effects, such as location in an appropriate chromosomal domain?

Are transfected genes expressed with the proper developmental specificity? The general rule now appears to be that there is a reasonable facsimile of proper control: the transfected genes are generally expressed in appropriate cells and at the usual time. There are exceptions, however, in which a transfected gene is expressed in an inappropriate tissue.

In the progeny of the injected mice, expression of the donor gene is extremely variable; it may be extinguished entirely, reduced somewhat, or even increased. Even in the original parents, the level of gene expression does not correlate with the number of tandemly integrated genes. Probably only some of the genes are active. In addition to the question of how many of the gene copies are capable of being activated, a parameter influencing regulation could

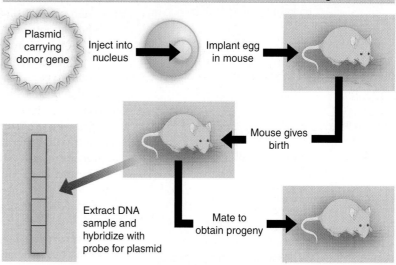

Transfected DNA can be incorporated into the mouse genome

Plasmid carrying donor gene

Inject into nucleus

Implant egg in mouse

Mouse gives birth

Mate to obtain progeny

Extract DNA sample and hybridize with probe for plasmid

Figure 32.13 Transfection can introduce DNA directly into the germline of animals.

Figure 32.14 A transgenic mouse with an active rat growth hormone gene (left) is twice the size of a normal mouse (right). Photograph kindly provided by Dr. Ralph L. Brinster, School of Veterinary Medicine, University of Pennsylvania.

The rat GH gene makes a larger mouse

A transgene can cure a disease

hpg⁻/hpg⁻ mutant is infertile; has deletion in gene

wild-type mouse makes GnRH/GAP; has intact gene

13.5 kb fragment including *hpg⁺* gene microinjected into fertilized eggs

250 eggs

27 mice
2 mice contain transgene (>20 copies/ genome)

Introduce transgene into *hpg⁻* genome by genetic cross

20 out of 48 progeny inherit transgene in heterozygote *hpg⁻/⁺*

Breed 7 *hpg⁻ / hpg⁻* mice with transgene

Endogenous alleles have deletion

Transgenes produce protein

Mice are fertile

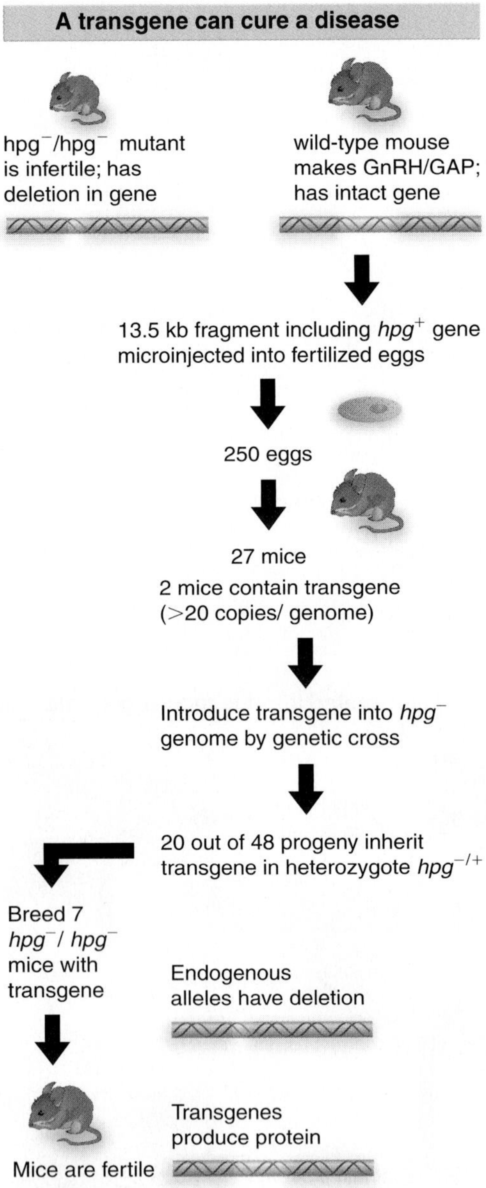

Figure 32.15 Hypogonadism can be averted in the progeny of *hpg* mice by introducing a transgene that has the wild-type sequence.

be the relationship between the gene number and the regulatory proteins: a large number of promoters could dilute out any regulator proteins present in limiting amounts.

What is responsible for the variation in gene expression? One possibility that has often been discussed for transfected genes (and which applies also to integrated retroviral genomes) is that the site of integration is important. Perhaps a gene is expressed if it integrates within an active domain, but not if it integrates in another area of chromatin. Another possibility is the occurrence of epigenetic modification; for example, changes in the pattern of methylation might be responsible for changes in activity. Alternatively, the genes that happened to be active in the parents may have been deleted or amplified in the progeny.

A particularly striking example of the effects of an injected gene is provided by a strain of transgenic mice derived from eggs injected with a fusion consisting of the MT (metallthionein) promoter linked to the rat growth hormone structural gene. Growth hormone levels in some of the transgenic mice were several hundred times greater than normal. The mice grew to nearly twice the size of normal mice, as can be seen from **Figure 32.14**.

Can defective genes be replaced by functional genes in the germline using transgenic techniques? One successful case is represented by a cure of the defect in the hypogonadal mouse. The *hpg* mouse has a deletion that removes the distal part of the gene coding for the polyprotein precursor to GnRH (gonadotropin-releasing hormone) and GnRH-associated peptide (GAP). As a result, the mouse is infertile.

When an intact *hpg* gene is introduced into the mouse by transgenic techniques, it is expressed in the appropriate tissues. **Figure 32.15** summarizes experiments to introduce a transgene into a line of *hpg/hpg* homozygous mutant mice. The resulting progeny are normal. This provides a striking demonstration that expression of a transgene under normal regulatory control can be indistinguishable from the behavior of the normal allele.

Impediments to using such techniques to cure genetic defects at present are that the transgene must be introduced into the germline of the preceding generation, the ability to express a transgene is not predictable, and an adequate level of expression of a transgene may be obtained in only a small minority of the transgenic animals. Also, the large number of transgenes that may be introduced into the germline, and their erratic expression, could pose problems for the animal in cases in which overexpression of the transgene was harmful.

In the *hpg* murine experiments, for example, only 2 out of 250 eggs injected with intact *hpg* genes gave rise to transgenic mice. Each transgenic animal contained >20 copies of the transgene. Only 20 of the 48 offspring of the transgenic mice retained the transgenic trait. When inherited by their offspring, however, the transgene(s) could substitute for the lack of endogenous *hpg* genes. Gene replacement via a transgene is therefore effective only under restricted conditions.

The disadvantage of direct injection of DNA is the introduction of multiple copies, their variable expression, and often difficulty in cloning the insertion site because sequence rearrangements may have been generated in the host DNA. An alternative procedure is to use a retroviral vector carrying the donor gene. A single proviral copy inserts at a chromosomal site, without inducing any rearrangement of the host DNA. It is possible also to treat cells at different stages of development, and thus to target a particular somatic tissue; however, it is difficult to infect germ cells.

32.6 ES Cells Can Be Incorporated into Embryonic Mice

Key Concepts

- ES (embryonic stem) cells that are injected into a mouse blastocyst generate descendant cells that become part of a chimeric adult mouse.
- When the ES cells contribute to the germline, the next generation of mice may be derived from the ES cell.
- Genes can be added to the mouse germline by transfecting them into ES cells before the cells are added to the blastocyst.

A powerful technique for making transgenic mice takes advantage of embryonic stem (ES) cells, which are derived from the mouse blastocyst (an early stage of development, when the embryo consists of 30–150 cells). **Figure 32.16** illustrates the principles of this technique.

ES cells are transfected with DNA in the usual way (most often by microinjection or electroporation). By using a donor that carries an additional sequence such as a drug resistance marker or some particular enzyme, it is possible to select ES cells that have obtained an integrated transgene carrying any particular donor trait. An alternative is to use polymerase chain reaction (PCR) technology to assay the transfected ES cells for successful integration of the donor DNA. By such means, a population of ES cells is obtained in which there is a high proportion carrying the marker.

These ES cells are then injected into a recipient blastocyst. The ability of the ES cells to participate in normal development of the blastocyst forms the basis of the technique. The blastocyst is implanted into a foster mother, and in due course develops into a *chimeric* mouse. Some of the tissues of the chimeric mice are derived from the cells of the recipient blastocyst; other tissues are derived from the injected ES cells. The proportions of tissues in the adult mouse that are derived from cells in the recipient blastocyst

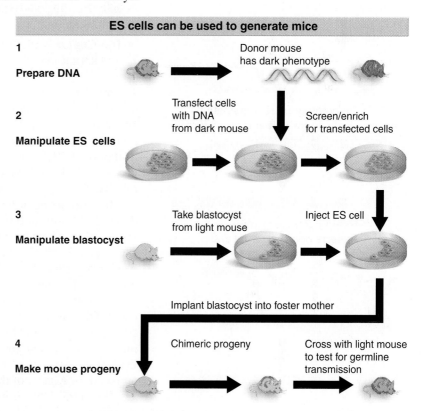

ES cells can be used to generate mice

1
Prepare DNA

Donor mouse has dark phenotype

2
Manipulate ES cells

Transfect cells with DNA from dark mouse

Screen/enrich for transfected cells

3
Manipulate blastocyst

Take blastocyst from light mouse

Inject ES cell

Implant blastocyst into foster mother

4
Make mouse progeny

Chimeric progeny

Cross with light mouse to test for germline transmission

Figure 32.16 ES cells can be used to generate mouse chimeras, which breed true for the transfected DNA when the ES cell contributes to the germline.

and from injected ES cells varies widely in individual progeny; if a visible marker (such as coat color gene) is used, areas of tissue representing each type of cell can be seen.

To determine whether the ES cells contributed to the germline, the chimeric mouse is crossed with a mouse that lacks the donor trait. Any progeny that have the trait must be derived from germ cells that have descended from the injected ES cells. By this means, it is known that an entire mouse has been generated from an original ES cell!

32.7 Gene Targeting Allows Genes to Be Replaced or Knocked Out

Key Terms

- A gene **knockout** is a process in which a gene function is eliminated, usually by replacing most of the coding sequence with a selectable marker *in vitro* and transferring the altered gene to the genome by homologous recombination.

- A gene **knock-in** is a process similar to a knockout, but more subtle mutations are made, such as nucleotide substitutions or the addition of epitope tags.

Key Concepts

- An endogenous gene can be replaced by a transfected gene using homologous recombination.

- The occurrence of a homologous recombination can be detected by using two selectable markers, one of which is incorporated with the integrated gene, the other of which is lost when recombination occurs.

- The Cre/*lox* system is widely used to make inducible knockouts and knock-ins.

The most powerful techniques for changing the genome use gene targeting to delete or replace genes by homologous recombination. In higher eukaryotes, the target is usually the genome of an ES cell, which is then used to generate a mouse with the mutation. When a donor DNA is introduced into the cell, it may insert into the genome by either nonhomologous or homologous recombination. Homologous recombination is relatively rare, probably representing <1% of all recombination events, and thus occurring at a frequency of ~10^{-7}. However, by designing the donor DNA appropriately, we can use selective techniques to identify those cells in which homologous recombination has occurred.

Figure 32.17 illustrates the **knockout** technique that is used to disrupt endogenous genes. The basis for the technique is that nonhomologous and homologous recombination have different effects upon the nature of the donor DNA that is inserted into the target genome. The donor DNA is homologous to a target gene, but has two modifications. First, it is inactivated by interrupting an exon with a marker sequence; most often the *neo* gene that confers resistance to the drug G418 is used. Second, another marker is added on one side of the gene; for example, the *TK* gene of the herpes virus.

When this DNA is introduced into an ES cell, a nonhomologous recombination inserts the whole unit, including the flanking *TK* sequence. These cells are resistant to neomycin, and they also express thymidine kinase, which makes them sensitive to the drug gancyclovir

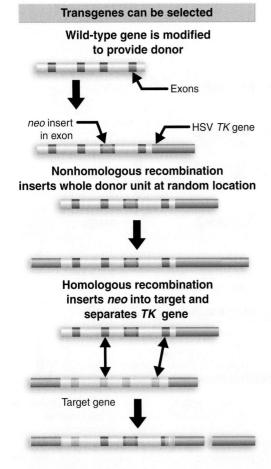

Transgenes can be selected

Wild-type gene is modified to provide donor

Exons

neo insert in exon — HSV *TK* gene

Nonhomologous recombination inserts whole donor unit at random location

Homologous recombination inserts *neo* into target and separates *TK* gene

Target gene

Figure 32.17 A transgene containing *neo* within an exon and *TK* downstream can be selected by resistance to G418 and loss of *TK* activity.

(because they phosphorylate it, making it toxic). But a homologous recombination requires two exchanges within the sequence of the donor gene, as a result of which the flanking *TK* sequence is lost. Cells in which a homologous recombination has occurred therefore gain *neo* resistance in the same way as cells that have nonhomologous recombination, but they do not have thymidine kinase activity, and so are resistant to gangcyclovir. So plating the cells in the presence of neomycin plus gangcyclovir specifically selects those in which homologous recombination has replaced the endogenous gene with the donor gene. If it is not convenient to use a selectable marker such as *TK*, cells can simply be screened by PCR assays for the absence of flanking DNA.

The presence of the *neo* gene in an exon of the donor gene disrupts translation, and thereby creates a null allele. A particular target gene can therefore be "knocked out" by this means; and once a mouse with one null allele has been obtained, it can be bred to generate the homozygote. This is a powerful technique for investigating whether a particular gene is essential, and what functions in the animal are perturbed by its loss.

A major extension of ability to manipulate a target genome has been made possible by using the phage Cre/*lox* system to engineer site-specific recombination in a eukaryotic cell. The Cre enzyme catalyzes a site-specific recombination reaction between two *lox* sites, which are identical 34 bp sequences (see *19.11 Site-Specific Recombination Resembles Topoisomerase Activity*). **Figure 32.18** shows that the consequence of the reaction is to excise the stretch of DNA between the two *lox* sites.

The great utility of the Cre/*lox* system is that it requires no additional components and works when the Cre enzyme is produced in any cell that has a pair of *lox* sites. **Figure 32.19** shows that we can control the reaction to make it work in a particular cell by placing the *cre* gene under the control of a regulated promoter. The procedure starts by making two mice. One mouse has the *cre* gene, typically controlled by a promoter that can be turned on specifically in a certain cell or under certain conditions. The other mouse has a target sequence flanked by *lox* sites. When we cross the two mice, the progeny have both elements of the system; and the system can be turned on by controlling the promoter of the *cre* gene. This allows the sequence between the *lox* sites to be excised in a controlled way.

The Cre/*lox* system can be combined with the knockout technology to give us even more control over the genome. Inducible knockouts can be made by flanking the neomycin gene (or any other gene that is used similarly in a selective procedure) with *lox* sites. After the knockout has been made, the target gene can be reactivated by causing Cre to excise the neomycin gene in some particular circumstance. **Figure 32.20** shows a modification of this procedure that allows a **knock-in** to be created. Basically we use a construct in which some mutant version of the target gene is used to replace the endogenous gene, replying on the usual selective procedures. Then when the inserted gene is reactivated by excising the neomycin sequence, we have in effect replaced the original gene with a different version.

A sophisticated method for introducing new DNA sequences has been developed with *D. melanogaster* by taking advantage of the P element, a naturally occuring transposon in *Drosophila*. The protocol is illustrated in **Figure 32.21**. A defective P element carrying the gene of interest is injected together with an intact P element into preblastoderm embryos. The intact P element provides a transposase that recognizes not only its own ends but also those of the defective element. As a result, either or both elements may be inserted into the genome.

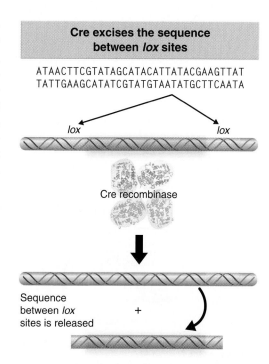

Figure 32.18 The Cre recombinase catalyzes a site-specific recombination between two identical *lox* sites, releasing the DNA between them.

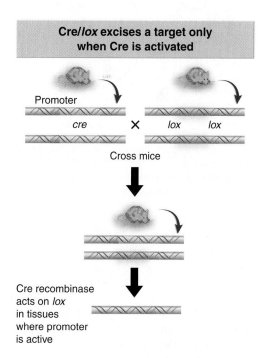

Figure 32.19 By placing the Cre recombinase under the control of a regulated promoter, it is possible to activate the excision system only in specific cells. One mouse is created that has a promoter-*cre* construct, and another that has a target sequence flanked by *lox* sites. The mice are crossed to generate progeny that have both constructs. Then excision of the target sequence can be triggered by activating the promoter.

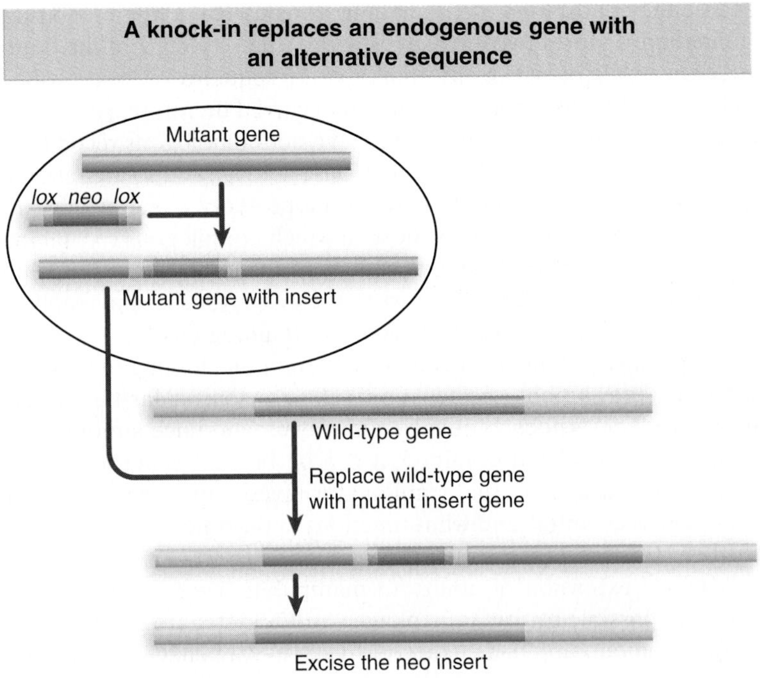

A knock-in replaces an endogenous gene with an alternative sequence

Figure 32.20 An endogenous gene is replaced in the same way as when a knockout is made (see Figure 32.17), but the neomycin gene is flanked by *lox* sites. After the gene replacement has been made using the selective procedure, the neomycin gene can be removed by activating Cre, leaving an active insert.

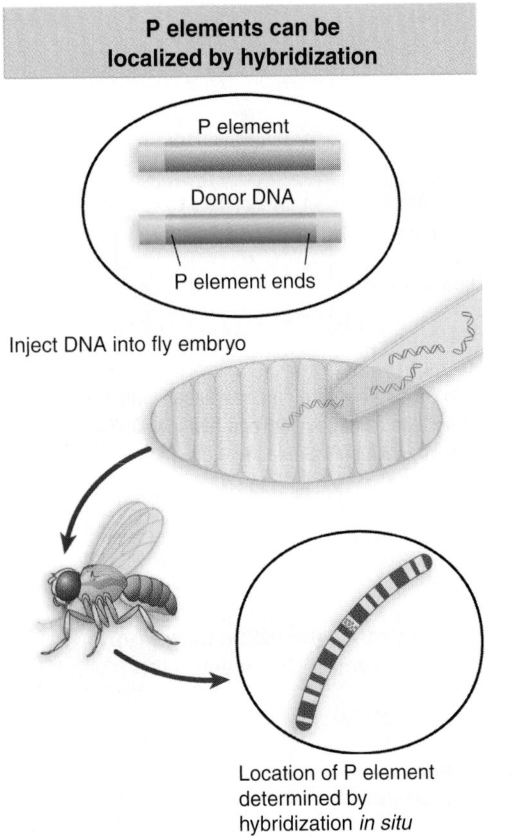

P elements can be localized by hybridization

Figure 32.21 Transgenic flies that have a single, normally expressed copy of a gene can be obtained by injecting *D. melanogaster* embryos with an active P element plus foreign DNA flanked by P element ends.

Only the sequences between the ends of the P DNA are inserted; the sequences on either side are not part of the transposable element. An advantage of this technique is that only a single element is inserted in any one event, so the transgenic flies usually carry only one copy of the foreign gene, a great aid in analyzing its behavior.

With these techniques, we are able to extend from cultured cells to animals the option of examining the regulatory features. The ability to introduce DNA into the genome allows us to make changes in it, to add new genes that have had particular modifications introduced *in vitro*, or to inactivate existing genes. So it becomes possible to delineate the features responsible for tissue-specific gene expression. Ultimately, we may expect routinely to replace defective genes in the genome in a targeted manner.

32.8 SUMMARY

New sequences of DNA may be introduced into a cultured cell by transfection or into an animal egg by microinjection. The foreign sequences may become integrated into the genome, often as large tandem arrays. The array appears to be inherited as a unit in a cultured cell. The sites of integration appear to be random. A transgenic animal arises when the integration event occurs into a genome that enters the germ cell lineage. A transgene or transgenic array is inherited in Mendelian manner, but the copy number and activity of the gene(s) may change in the progeny. Often a transgene responds to tissue and temporal regulation in a manner that resembles the endogenous gene. Under conditions that promote homologous recombination, an inactive sequence can be used to replace a functional gene, thus creating a null locus. Extensions of this technique can be used to make inducible knockouts, where the activity of the gene can be turned on or off, and knock-ins, where a donor gene specifically replaces a target gene. Transgenic mice can be obtained by injecting recipient blastocysts with ES cells that carry transfected DNA.

Glossary

The E pre-splicing complex is converted to the **A complex** by the binding of U2 snRNP to the branch site.

The **A site** of the ribosome is the site that an aminoacyl-tRNA enters to base pair with the codon.

Abortive initiation describes a process in which RNA polymerase starts transcription but terminates before it has left the promoter. It then reinitiates. Several cycles may occur before the elongation stage begins.

The **abundance** of an mRNA is the average number of molecules per cell.

Abundant mRNAs consist of a small number of individual species, each present in a large number of copies per cell.

The **acceptor arm** of tRNA is a short duplex that terminates in the CCA sequence to which an amino acid is linked.

An **acentric fragment** of a chromosome (generated by breakage) lacks a centromere and is lost at cell division.

Acridines are mutagens that act on DNA to cause the insertion or deletion of a single base pair. They were useful in defining the triplet nature of the genetic code.

An **activator** is a protein that stimulates the expression of a gene, typically by acting at a promoter to stimulate RNA polymerase. In eukaryotes, the sequence to which it binds in the promoter is called a response element.

Adenylate cyclase is an enzyme that uses ATP as a substrate to generate cyclic AMP, in which 5′ and 3′ positions of the sugar ring are connected via a phosphate group.

An **alarmone** is a small molecule in bacteria that is produced as a result of stress and which acts to alter the state of gene expression. The unusual nucleotides ppGpp and pppGpp are examples.

An **allele** is one of several alternative forms of a gene occupying a given locus on a chromosome.

Allelic exclusion describes the expression in any particular lymphocyte of only one allele coding for the expressed immunoglobulin. This is caused by feedback from the first immunoglobulin allele to be expressed that prevents activation of a copy on the other chromosome.

Allosteric regulation describes the ability of a protein to change its conformation (and therefore activity) at one site as the result of binding a small molecule to a second site located elsewhere on the protein.

Alternative splicing describes the production of different RNA products from a single product by changes in the usage of splicing junctions.

The **Alu domain** comprises the parts of the 7S RNA of the SRP that are related to Alu RNA.

The **Alu family** is a set of dispersed, related sequences, each ~300-bp long, in the human genome. The individual members have Alu cleavage sites at each end (hence the name).

Amanitin (more fully α-amanitin) is a bicyclic octapeptide derived from the poisonous mushroom *Amanita phalloides*; it inhibits transcription by certain eukaryotic RNA polymerases, especially RNA polymerase II.

The **amber** codon is the triplet UAG, one of the three termination codons that end protein synthesis.

An **aminoacyl-tRNA** is a tRNA linked to an amino acid. The COOH group of the amino acid is linked to the 3′- or 2′-OH group of the terminal base of the tRNA.

Aminoacyl-tRNA synthetases are enzymes responsible for covalently linking amino acids to the 2′- or 3′-OH position of tRNA.

Annealing of DNA describes the renaturation of a duplex structure from single strands that were obtained by denaturing duplex DNA.

Anti-Sm is an autoimmune antiserum that defines the Sm epitope that is common to a group of proteins found in snRNPs that are involved in RNA splicing.

An **antibody** is a protein (immunoglobulin) produced by B lymphocyte cells that recognizes a particular "foreign antigen," and thus triggers the immune response.

The **anticodon** is a trinucleotide sequence in tRNA that is complementary to the codon in mRNA and enables the tRNA to place the appropriate amino acid in response to the codon.

The **anticodon arm** of tRNA is a stem-loop structure that exposes the anticodon triplet at one end.

An **antigen** is any foreign substance whose entry into an organism provokes an immune response by stimulating the synthesis of an antibody (an immunoglobulin protein that can bind to the antigen).

Antiparallel strands of the double helix are organized in opposite orientation, so that the 5′ end of one strand is aligned with the 3′ end of the other strand.

An **antisense gene** codes for an (antisense) RNA that has a complementary sequence to an RNA that is its target.

The **antisense strand (template strand)** of DNA is complementary to the sense strand, and is the one that acts as the template for synthesis of mRNA.

Antitermination is a mechanism of transcriptional control in which termination is prevented at a specific terminator site, allowing RNA polymerase to read into the genes beyond it.

Antitermination proteins allow RNA polymerase to transcribe through certain terminator sites.

Anucleate bacteria lack a nucleoid, but are of similar shape to wild-type bacteria.

An **arm** of tRNA is one of the four (or in some cases five) stem-loop structures that make up the secondary structure.

The **arms** of a lambda phage attachment site are the sequences flanking the core region where the recombination event occurs.

ARS (autonomous replication sequence) is an origin for replication in yeast. The common feature among different *ARS* sequences is a conserved 11 bp sequence called the A-domain.

An **assembly factor** is a protein that is required for formation of a macromolecular structure but is not itself part of that structure.

att sites are the loci on a lambda phage and the bacterial chromosome at which recombination integrates the phage into, or excises it from, the bacterial chromosome.

Attenuation describes the regulation of bacterial operons by controlling termination of transcription at a site located before the first structural gene.

An **attenuator** is a terminator sequence at which attenuation occurs.

Autogenous control describes the action of a gene product that either inhibits (negative autogenous control) or activates (positive autogenous control) expression of the gene coding for it.

An **autonomous controlling element** in maize is an active transposon with the ability to transpose (*compare with* nonautonomous controlling element).

Autosplicing (self-splicing) describes the ability of an intron to excise itself from RNA by a catalytic action that depends only on the sequence of RNA in the intron.

An **axial element** is a proteinaceous structure around which the chromosomes condense at the start of synapsis.

A **B cell** is a lymphocyte that produces antibodies. B cell development occurs primarily in bone marrow.

A **back mutation** reverses the effect of a mutation that had inactivated a gene; thus it restores wild type.

The **background level** of mutation describes the rate at which sequence changes accumulate in the genome of an organism. It reflects the balance between the occurrence of spontaneous mutations and their removal by repair systems, and is characteristic for any species.

Bacteriophages (phages) are viruses that infect bacteria.

Bands of polytene chromosomes are visible as dense regions that contain the majority of DNA. They include active genes.

A **basal factor** is a transcription factor required by RNA polymerase II to form the initiation complex at all promoters. Factors are identified as $TF_{II}X$, where X is a letter.

The level of response from a system in the absence of a stimulus is its **basal level**. (The basal level of transcription of a gene is the level that occurs in the absence of any specific activation.)

Base mispairing is a coupling between two bases that does not conform to the Watson–Crick rule, e.g., adenine with cytosine, thymine with guanine.

Base pairing describes the specific (complementary) interactions of adenine with thymine or of guanine with cytosine in a DNA double helix (thymine is replaced by uracil in double helical RNA).

A **bHLH protein** has a basic DNA-binding region adjacent to the helix-loop-helix motif.

A **bivalent** is the structure containing all four chromatids (two representing each homologue) at the start of meiosis.

A **blocked** reading frame cannot be translated into protein because of the occurrence of termination codons.

Branch migration describes the ability of a DNA strand partially paired with its complement in a duplex to extend its pairing by displacing the resident strand with which it is homologous.

The **branch site** is a short sequence just before the end of an intron at which the lariat intermediate is formed in splicing by joining the 5′ nucleotide of the intron to the 2′ position of an adenosine.

Breakage and reunion describes the mode of genetic recombination, in which two DNA duplex molecules are broken at corresponding points and then rejoined crosswise (involving formation of a length of heteroduplex DNA around the site of joining).

The **breakage-fusion-bridge** cycle is a type of chromosomal behavior in which a broken chromatid fuses to its sister, forming a "bridge." When the centromeres separate at mitosis, the chromosome breaks again (not necessarily at the bridge), thereby restarting the cycle.

The **bromo domain** is found in a variety of proteins that interact with chromatin and is used to recognize acetylated sites on histones.

Buoyant density measures the ability of a substance to float in some standard fluid, for example, CsCl.

A **bZIP** protein has a basic DNA-binding region adjacent to a leucine zipper dimerization motif.

C genes code for the constant regions of immunoglobulin protein chains.

C-bands are generated by staining techniques that react with centromeres. The centromere appears as a darkly staining dot.

The **C-value** is the total amount of DNA in the genome (per haploid set of chromosomes).

The **C-value paradox** describes the lack of relationship between the DNA content (C-value) of an organism and its coding potential.

A **CAAT box** is part of a conserved sequence located upstream of the startpoints of eukaryotic transcription units; it is recognized by a large group of transcription factors.

A **cap** is the structure at the 5′ end of eukaryotic mRNA, introduced after transcription by linking the terminal phosphate of 5′ GTP to the terminal base of the mRNA. The added G (and sometimes some other bases) are methylated, giving a structure of the form $^7MeG5′ppp5′Np\ldots$

A **cap 0** at the 5′ end of mRNA has only a methyl group on 7-guanine.

A **cap 1** at the 5′ end of mRNA has methyl groups on the terminal 7-guanine and the 2′–O position of the next base.

A **cap 2** has three methyl groups (7-guanine, 2′–O position of next base, and N^6 adenine) at the 5′ end of mRNA.

A **capsid** is the external protein coat of a virus particle.

The **carboxy-terminal domain (CTD)** of eukaryotic RNA polymerase II is phosphorylated at initiation and is involved in coordinating several activities with transcription.

A **cascade** is a sequence of events, each of which is stimulated by the previous one. In transcriptional regulation, as seen in sporulation and phage lytic development, it means that regulation is divided into stages, and at each stage, one of the genes that is expressed codes for a regulator needed to express the genes of the next stage.

The **cassette model** for yeast mating type proposes that there is a single active locus (the active cassette) and two inactive copies of the locus (the silent cassettes). Mating type is changed when an active cassette of one type is replaced by a silent cassette of the other type.

The **central dogma** describes the basic nature of genetic information: sequences of nucleic acid can be perpetuated and interconverted by replication, transcription, and reverse transcription, but translation from nucleic acid to protein is unidirectional, because nucleic acid sequences cannot be retrieved from protein sequences.

The **central element** is a structure that lies in the middle of the synaptonemal complex, along which the lateral elements of homologous chromosomes align.

The **centromere** is a constricted region of a chromosome that includes the site of attachment (the kinetochore) to the mitotic or meiotic spindle.

Chemical proofreading describes a proofreading mechanism in which the correction event occurs after addition of an incorrect subunit to a polymeric chain, by reversing the addition reaction.

Chi is an octameric sequence in DNA that provides a hotspot for RecA-mediated genetic recombination in *E. coli*.

A **chiasma** (*pl*. **chiasmata**) is a site at which two homologous chromosomes appear to have exchanged material during meiosis.

Chloroplast DNA (ctDNA) is an independent genome (usually circular) found in a plant chloroplast.

Chromatids are the copies of a chromosome produced by replication. The name is usually used to describe the copies in the period before they separate at the subsequent cell division.

Chromatin describes the state of nuclear DNA and its associated proteins during the interphase (between mitoses) of the eukaryotic cell cycle.

Chromatin remodeling describes the energy-dependent displacement or reorganization of nucleosomes that occurs in conjunction with activation of genes for transcription.

The **chromo domain** is found in nonhistone chromatin proteins and is involved in protein–protein interactions.

The **chromocenter** is an aggregate of heterochromatin from different chromosomes.

Chromomeres are densely staining granules visible in chromosomes under certain conditions, especially early in meiosis, when a chromosome may appear to consist of a series of chromomeres.

A **chromosome** is a discrete unit of the genome carrying many genes. Each chromosome consists of a very long molecule of duplex DNA and an approximately equal mass of proteins. It is visible as a morphological entity only during cell division.

Chromosome pairing is the coupling of the homologous chromosomes at the start of meiosis.

cis configuration describes two sites on the same molecule of DNA.

A *cis*-acting site affects the activity only of sequences on its own molecule of DNA (or RNA); this property usually implies that the site does not code for protein.

A **cistron** is the genetic unit defined by the complementation test; it is equivalent to the gene.

The **clamp loader** is a 5-subunit protein complex that is responsible for loading the β clamp on to DNA at the replication fork.

Class switching describes a change in Ig gene organization in which the C region of the heavy chain is changed but the V region remains the same.

Clonal expansion is the proliferation of mature lymphocytes stimulated by antigen binding. This proliferation stage is necessary for adaptive immune responses because it substantially increases the number of antigen-specific lymphocytes. After proliferation, lymphocytes differentiate into effector cells.

Cloning describes propagation of a DNA sequence by incorporating it into a hybrid construct that can be replicated in a host cell.

A **cloning vector** is a plasmid or phage that is used to "carry" inserted foreign DNA for the purposes of producing more material or a protein product.

The **cloverleaf** describes the structure of tRNA drawn in two dimensions, forming four distinct arm-loops.

Coactivators are factors required for transcription that do not bind DNA but are required for (DNA-binding) activators to interact with the basal transcription factors.

A **coding end** is produced during recombination of immunoglobulin and T-cell receptor genes. Coding ends are at the termini of the cleaved V and (D)J coding regions. The subsequent joining of the coding ends yields a coding joint.

A **coding region** is a part of the gene that represents a protein sequence.

The **coding strand (sense strand)** of DNA has the same sequence as the mRNA and is related by the genetic code to the protein sequence that it represents.

A **codon** is a triplet of nucleotides that represents an amino acid or a termination signal.

Cognate tRNAs (isoaccepting tRNAs) are those recognized by a particular aminoacyl-tRNA synthetase. They all are charged with the same amino acid.

Cohesin proteins form a complex that holds sister chromatids together. They include some SMC proteins.

Coincidental evolution (coevolution) describes a situation in which two genes evolve together as a single unit.

A **cointegrate structure** is produced by fusion of two replicons, one originally possessing a transposon, the other lacking it; the cointegrate has copies of the transposon present at both junctions of the replicons, oriented as direct repeats.

A **colinear** relationship describes the 1:1 representation of a sequence of triplet nucleotides in a sequence of amino acids.

A **compatibility group** of plasmids contains members unable to coexist in the same bacterial cell.

A **complementation group** is a series of mutations unable to complement when tested in pairwise combinations in *trans*; defines a genetic unit (the cistron).

A **complementation test** determines whether two mutations are alleles of the same gene. It is accomplished by crossing two different recessive mutations that have the same phenotype and determining whether the wild-type phenotype can be produced. If so, the mutations are said to complement each other and are probably not mutations in the same gene.

Composite transposons (composite elements) have a central region flanked on each side by insertion sequences, either or both of which may enable the entire element to transpose.

Concerted evolution describes the ability of two related genes to evolve together as though constituting a single locus.

Conjugation is a process in which two cells come in contact and exchange genetic material. In bacteria, DNA is transferred from a donor to a recipient cell. In protozoa, DNA passes from each cell to the other.

A **consensus sequence** is an idealized sequence in which each position represents the base most often found when many actual sequences are compared.

Conserved sequences are identified when many examples of a particular nucleic acid or protein are compared and the same individual bases or amino acids are always found at particular locations.

Constant regions (C regions) of immunoglobulins are coded by C genes and are the parts of the chain that vary least. Those of heavy chains identify the type of immunoglobulin.

A **constitutive** process is one that occurs all the time, unchanged by any form of stimulus or external condition.

Constitutive heterochromatin describes the inert state of permanently nonexpressed sequences, usually satellite DNA.

The **context** of a codon in mRNA refers to the fact that neighboring sequences may change the efficiency with which a codon is recognized by its aminoacyl-tRNA or is used to terminate protein synthesis.

Controlling elements of maize are transposable units originally identified solely by their genetic properties. They may be autonomous (able to transpose independently) or nonautonomous (able to transpose only in the presence of an autonomous element).

Coordinate regulation refers to the common control of a group of genes.

Copy choice is a type of recombination used by RNA viruses, in which the RNA polymerase switches from one template to another during synthesis.

The **copy number** is the number of copies of a plasmid that is maintained in a bacterium (relative to the number of copies of the origin of the bacterial chromosome).

The **core** sequence is the segment of DNA that is common to the attachment sites on both the phage lambda and bacterial genomes. It is the location of the recombination event that allows phage lambda to integrate.

Core DNA is the 146 bp of DNA contained on a core particle that is generated by cleaving the DNA on a nucleosome with microccocal nuclease.

The **core enzyme** is the complex of RNA polymerase subunits needed for elongation. It does not include additional subunits or factors that may be needed for initiation or termination.

A **core histone** is one of the four types (H2A, H2B, H3, H4) found in the core particle derived from the nucleosome (this excludes histone H1).

The **core particle** is a digestion product of the nucleosome that retains the histone octamer and has 146 bp of DNA; its structure is similar to that of the nucleosome itself.

A **core promoter** is the shortest sequence at which an RNA polymerase can initiate transcription (typically at a much lower level than that displayed by a promoter containing additional elements). For RNA polymerase II it is the minimal sequence at which the basal transcription apparatus can assemble, and includes two sequence elements, the InR and TATA box. It is typically ~40 bp long.

A **corepressor** is a small molecule that triggers repression of transcription by binding to a regulator protein.

A **cosmid** is derived from a bacterial plasmid by incorporating the *cos* sites of phage lambda, which make the plasmid DNA substrate for the lambda packaging system.

Cosuppression describes the ability of a transgene (usually in plants) to inhibit expression of the corresponding endogenous gene.

Cotranslational translocation describes the movement of a protein across a membrane as the protein is being synthesized. The term is usually restricted to cases in which the ribosome binds to the channel. This form of translocation may be restricted to the endoplasmic reticulum.

CpG island is a stretch of 1–2 kb in a mammalian genome that is rich in unmethylated CpG doublets.

Crossing-over is a reciprocal exchange of material between chromosomes that occurs during prophase I of meiosis and is responsible for genetic recombination.

Crossover fixation refers to a possible consequence of unequal crossing-over that allows a mutation in one member of a tandem cluster to spread through the whole cluster (or to be eliminated).

Crown gall disease is a tumor that can be induced in many plants by infection with the bacterium *Agrobacterium tumefaciens*.

CRP activator (CAP activator) is a positive regulator protein activated by cyclic AMP. It is needed for RNA polymerase to initiate transcription of many operons of *E. coli*.

A **cryptic satellite** is a satellite DNA sequence not identified as such by a separate peak on a density gradient; that is, it remains present in main-band DNA.

The **cutting periodicity** is the spacing between cleavages on each strand when a duplex DNA immobilized on a flat surface is attacked by a DNAase that makes single-strand cuts.

Cytoplasmic inheritance is a property of genes located in mitochondria or chloroplasts.

Cytotype is a cytoplasmic condition that affects P-element activity. The effect of cytotype is due to the presence or absence of transposition repressors, which are provided by the mother to the egg.

The **D arm** of tRNA has a high content of the base dihydrouridine. The **D segment** is an additional sequence that is found between the V and J regions of an immunoglobulin heavy chain.

A **daughter** strand or duplex of DNA refers to the newly synthesized DNA.

A *de novo* **methylase** adds a methyl group to an unmethylated target sequence on DNA.

A **deacetylase** is an enzyme that removes acetyl groups from proteins.

Deacylated tRNA has no amino acid or polypeptide chain attached because it has completed its role in protein synthesis and is ready to be released from the ribosome.

The **degradosome** is a complex of bacterial enzymes, including RNAase and helicase activities, that may be involved in degrading mRNA.

Delayed early genes in phage lambda are equivalent to the middle genes of other phages. They cannot be transcribed until regulator protein(s) coded by the immediate early genes have been synthesized.

A **deletion** is the removal of a sequence of DNA, the regions on either side being joined together except in the case of a terminal deletion at the end of a chromosome.

A **demethylase** is a casual name for an enzyme that removes a methyl group, typically from DNA, RNA, or protein.

Denaturation of protein describes its conversion from the physiological conformation to some other (inactive) conformation.

A **density gradient** is used to separate macromolecules on the basis of differences in their density. It is prepared from a heavy soluble compound such as CsCl.

A **deoxyribonuclease (DNAase)** is an enzyme that attacks bonds in DNA. It may cut only one strand or both strands.

Deoxyribonucleic acid (DNA) is a nucleic acid molecule consisting of long chains of polymerized (deoxyribo) nucleotides. In double-stranded DNA the two strands are held together by hydrogen bonds between complementary nucleotide base pairs.

The **derepressed** state describes a gene that is turned on because a small molecule corepressor is absent. It has the same effect as the induced state that is produced by a small molecule inducer for a gene that is regulated by induction. In describing the effect of a mutation, *derepressed* and *constitutive* have the same meaning.

A **dicentric chromosome** is the product of fusing two chromosome fragments, each of which has a centromere. It is unstable and may be broken when the two centromeres are pulled to opposite poles in mitosis.

Direct repeats are identical (or closely related) sequences present in two or more copies in the same orientation in the same molecule of DNA; they are not necessarily adjacent.

Divergence is the percent difference in nucleotide sequence between two related DNA sequences or in amino acid sequences between two proteins.

DNA fingerprinting analyzes the differences between individuals of the fragments generated by using restriction enzymes to cleave regions that contain short repeated sequences. Because these are unique to every individual, the presence of a particular subset in any two individuals can be used to define their common inheritance (e.g., a parent–child relationship).

DNA ligase makes a bond between an adjacent 3′–OH and 5′-phosphate end where there is a nick in one strand of duplex DNA.

A **DNA polymerase** is an enzyme that synthesizes a daughter strand(s) of DNA (under direction from a DNA template). Any particular enzyme may be involved in repair or replication (or both).

A **DNA replicase** is a DNA-synthesizing enzyme required specifically for replication.

The **DNA-binding site** of a protein is the region that binds to DNA. Several types of motifs are known for DNA-binding sites. In regulatory proteins, their activities may be controlled by changes in conformation that are triggered by a small molecule binding elsewhere on the protein.

A **domain** of a chromosome may refer *either* to a discrete structural entity defined as a region within which supercoiling is independent of other domains; *or* to an extensive region including an expressed gene that has heightened sensitivity to degradation by the enzyme DNAase I.

A **domain** of a protein is a discrete continuous part of the amino acid sequence that can be equated with a particular function.

A **dominant negative** mutation results in a mutant gene product that prevents the function of the wild-type gene product, causing loss or reduction of gene activity in cells containing both the mutant and wild-type alleles. The effect may result from the titration of another factor that interacts with the gene product or by an inhibiting interaction of the mutant subunit on the multimer.

Dosage compensation describes mechanisms employed to compensate for the discrepancy between the presence of two X chromosomes in one sex but only one X chromosome in the other sex.

A **double-strand break (DSB)** occurs when both strands of a DNA duplex are cleaved at the same site. Genetic recombination is initiated by double-strand breaks. The cell also has repair systems that act on double-strand breaks created at other times.

The **doubling time** is the period (usually measured in minutes) that it takes for a bacterial cell to reproduce.

A **down mutation** in a promoter decreases the rate of transcription.

Downstream identifies sequences proceeding farther in the direction of expression; for example, the coding region is downstream of the initiation codon.

The **E complex** is the first complex to form at a splice site. It consists of U1 snRNP bound at the 5′ splice site together with factor ASF/SF2, U2AF bound at the branch site, and the bridging protein SF1/BBP.

Early genes are transcribed before the replication of phage DNA. They code for regulators and other proteins needed for later stages of infection.

Early infection is the part of the phage lytic cycle between entry and replication of the phage DNA. During this time, the phage synthesizes the enzymes needed to replicate its DNA.

EF-G is an elongation factor needed for the translocation stage of bacterial protein synthesis.

EF-Tu is the elongation factor that binds aminoacyl-tRNA and places it into the A site of a bacterial ribosome.

Elongation is the stage in a macromolecular synthesis reaction (replication, transcription, or translation) when the nucleotide or polypeptide chain is extended by the addition of individual subunits.

Elongation factors (EF in prokaryotes, eEF in eukaryotes) are proteins that associate with ribosomes cyclically, during addition of each amino acid to the polypeptide chain.

Endonucleases cleave bonds within a nucleic acid chain; they may be specific for RNA or for single-stranded or double-stranded DNA.

The **endoplasmic reticulum (ER)** is an organelle involved in the synthesis of lipids, membrane proteins, and secretory proteins. It is a single compartment that extends from the outer layer of the nuclear envelope into the cytoplasm. It has subdomains, such as the rough ER and smooth ER.

An **enhanceosome** is a complex of transcription factors that assembles cooperatively at an enhancer.

An **enhancer** is a *cis*-acting sequence that increases the utilization of (some) eukaryotic promoters, and can function in either orientation and in any location (upstream or downstream) relative to the promoter.

Epigenetic changes influence the phenotype without altering the genotype. They consist of changes in the properties of a cell that are inherited but that do not represent a change in genetic information.

An **episome** is a plasmid able to integrate into bacterial DNA.

Error-prone synthesis occurs when DNA polymerase incorporates noncomplementary bases into the daughter strand.

Euchromatin comprises all of the genome in the interphase nucleus except for the heterochromatin. The euchromatin is less tightly coiled than heterochromatin, and contains the active or potentially active genes.

The **evolutionary clock** is defined by the rate at which mutations accumulate in a given gene.

The **excision** of phage or episome or other sequence describes its release from the host chromosome as an autonomous DNA molecule.

The **excision** step in an excision-repair system consists of removing a single-stranded stretch of DNA by the action of a 5′–3′ exonuclease.

Excision repair describes a type of repair system in which one strand of DNA is directly excised and then replaced by resynthesis using the complementary strand as template.

An **exon** is any segment of an interrupted gene that is represented in the mature RNA product.

Exon definition describes the process when a pair of splicing sites are recognized by interactions involving the 5′ site of the intron and also the 5′ of the next intron downstream.

Exonucleases cleave nucleotides one at a time from the end of a polynucleotide chain; they may be specific for either the 5′ or 3′ end of DNA or RNA.

The **exosome** is a complex of several exonucleases involved in degrading RNA.

Extein sequences remain in the mature protein that is produced by processing a precursor via protein splicing.

The **extra arm** of tRNA lies between the T ψ C and anticodon arms. It is the most variable in length in tRNA, from 3–21 bases. tRNAs are called class 1 if they lack it, and class 2 if they have it.

Extranuclear genes reside outside the nucleus in organelles such as mitochondria and chloroplasts.

The **F plasmid** is an episome that can be free or integrated in *E. coli*, and which in either form can sponsor conjugation.

Facultative heterochromatin describes the inert state of sequences that also exist in active copies, for example, one mammalian X chromosome in females.

Forward mutations inactivate a wild-type gene.

Frameshifts are mutations caused by deletions or insertions that are not a multiple of three base pairs. They change the frame in which triplets are translated into protein.

A **fully methylated** site is a palindromic sequence that is methylated on both strands of DNA.

G-bands are generated on eukaryotic chromosomes by staining techniques and appear as a series of lateral striations. They are used for karyotyping (identifying chromosomal regions by the banding pattern).

A **gain-of-function** mutation usually refers to a mutation that causes an increase in the normal gene activity. It sometimes represents acquisition of certain abnormal properties. It is often, but not always, dominant.

The **GC box** is a common pol II promoter element consisting of the sequence GGGCGG.

A **gene (cistron)** is the segment of DNA specifying a polypeptide chain; it includes regions preceding and following the coding region (leader and trailer) as well as intervening sequences (introns) between individual coding segments (exons).

A **gene cluster** is a group of adjacent genes that are identical or related.

A **gene family** consists of a set of genes within a genome that code for related or identical proteins. The members were derived by duplication of an ancestral gene followed by accumulation of changes in sequence between the copies. Most often the members of a family are related but not identical.

The **genetic code** is the correspondence between triplets in DNA (or RNA) and amino acids in protein.

The **genome** is the complete set of sequences in the genetic material of an organism. It includes the sequence of each chromosome plus any DNA in organelles.

The **glucocorticoid response element (GRE)** is a sequence in a promoter or enhancer that is recognized by the glucocorticoid receptor, which is activated by glucocorticoid steroids.

The **Golgi apparatus** is an organelle that receives newly synthesized proteins from the endoplasmic reticulum and processes them for subsequent delivery to other destinations. It is composed of several flattened membrane disks arranged in a stack.

Gratuitous inducers resemble authentic inducers of transcription but are not substrates for the induced enzymes.

The **GT-AG rule** describes the presence of these constant dinucleotides at the first two and last two positions of introns of nuclear genes.

A **guide RNA** is a small RNA whose sequence is complementary to the sequence of an RNA that has been edited. It is used as a template for changing the sequence of the pre-edited RNA by inserting or deleting nucleotides.

The **haplotype** is the particular combination of alleles in a defined region of some chromosome, in effect the genotype in miniature. Originally used to described combinations of MHC alleles, it now may be used to describe particular combinations of RFLPs, SNPs, or other markers.

Hb anti-Lepore is a fusion gene produced by unequal crossing-over that has the N-terminal part of β globin and the C-terminal part of δ globin.

Hb Lepore is an unusual globin protein that results from unequal crossing-over between the β and δ genes. The genes become fused together to produce a single β-like chain that consists of the N-terminal sequence of δ joined to the C-terminal sequence of β.

HbH disease results from a condition in which there is a disproportionate amount of the abnormal tetramer β_4 relative to the amount of normal hemoglobin ($\alpha_2\beta_2$).

The **headpiece** is the DNA-binding domain of the *lac* repressor.

Heat shock genes are a set of loci that are activated in response to an increase in temperature (and other abuses to the cell). All organisms have heat shock genes. Their products usually include chaperones that act on denatured proteins.

The **heat shock response element (HSE)** is a sequence in a promoter or enhancer that is used to activate a gene by an activator induced by heat shock.

The immunoglobulin **heavy chain** is one of two types of subunits in an antibody tetramer. Each antibody contains two heavy chains. The N-terminus of the heavy chain forms part of the antigen recognition site, whereas the C-terminus determines the subclass (isotype).

A **helicase** is an enzyme that uses energy provided by ATP hydrolysis to separate the strands of a nucleic acid duplex.

The **helix-loop-helix (HLH)** motif is responsible for dimerization of a class of transcription factors called HLH proteins. A bHLH protein has a basic DNA-binding sequence close to the dimerization motif.

The **helix-turn-helix** motif describes an arrangement of two α helices that form a site that binds to DNA, one fitting into the major groove of DNA and other lying across it.

A **helper virus** provides functions absent from a defective virus, enabling the latter to complete the infective cycle during a mixed infection.

Hemimethylated DNA is methylated on one strand of a target sequence that has a cytosine on each strand.

Heterochromatin describes regions of the genome that are highly condensed, are not transcribed, and are late-replicating. Heterochromatin is divided into two types, which are called constitutive and facultative.

Heteroduplex DNA (Hybrid DNA) is generated by base pairing between complementary single strands derived from the different parental duplex molecules; it occurs during genetic recombination.

An **Hfr** cell is a bacterium that has an integrated F plasmid within its chromosome. Hfr stands for high-frequency recombination, referring to the fact that chromosomal genes are transferred from an Hfr cell to an F⁻ cell much more frequently than from an F⁺ cell.

Highly repetitive DNA (simple sequence DNA) is the first component to reassociate after denaturation and is equated with satellite DNA.

Histones are conserved DNA-binding proteins that form the basic subunit of chromatin in eukaryotes. Histones H2A, H2B, H3, H4 form an octameric core around which DNA coils to form a nucleosome. Histone H1 is external to the nucleosome.

Histone acetyltransferase (HAT) enzymes modify histones by the addition of acetyl groups; some transcriptional coactivators have HAT activity.

Histone deacetyltransferase (HDAC) enzymes remove acetyl groups from histones; they may be associated with repressors of transcription.

The **histone fold** is a motif found in all four core histones in which three α-helices are connected by two loops.

An **hnRNP** is the ribonucleoprotein form of hnRNA (heterogeneous nuclear RNA), in which the hnRNA is complexed with proteins. Since pre-mRNAs are not exported until processing is complete, hnRNPs are found only in the nucleus.

A **Holliday** structure is an intermediate structure in homologous recombination, where the two duplexes of DNA are connected by the genetic material exchanged between two of the four strands, one from each duplex. A joint molecule is said to be resolved when nicks in the structure restore two separate DNA duplexes.

The RNA polymerase **holoenzyme (complete enzyme)** is the form that is competent to initiate transcription. It consists of the four subunits of the core enzyme ($\alpha_2\beta\beta'$) and σ factor.

The **homeodomain** is a DNA-binding motif that typifies a class of transcription factors. The DNA sequence that codes for it is called the homeobox.

A **hotspot** is a site in the genome at which the frequency of mutation (or recombination) is very much increased, usually by at least an order of magnitude relative to neighboring sites.

Housekeeping genes (constitutive genes) are those (theoretically) expressed in all cells because they provide basic functions needed for sustenance of all cell types.

Hybrid dysgenesis describes the inability of certain strains of *D. melanogaster* to interbreed, because the hybrids are sterile (although otherwise they may be phenotypically normal).

Hybridization describes the pairing of complementary RNA and DNA strands to give an RNA–DNA hybrid.

Hydrops fetalis is a fatal disease resulting from the absence of the hemoglobin α gene.

Hypermutation describes the introduction of somatic mutations in a rearranged immunoglobulin gene. The mutations can change the sequence of the corresponding antibody, especially in its antigen-binding site.

A **hypersensitive site** is a short region of chromatin detected by its extreme sensitivity to cleavage by DNAase I and other nucleases; it comprises an area from which nucleosomes are excluded.

The **idling reaction** results in the production of pppGpp and ppGpp by ribosomes when an uncharged tRNA is present in the A site; this triggers the stringent response.

IF-1 is a bacterial initiation factor that stabilizes the initiation complex.

IF-2 is a bacterial initiation factor that binds the initiator tRNA to the initiation complex.

IF-3 is a bacterial initiation factor required for 30S subunits to bind to initiation sites in mRNA. It also prevents 30S subunits from binding to 50S subunits.

Immediate early phage genes in phage lambda are equivalent to the early class of other phages. They are transcribed immediately upon infection by the host RNA polymerase.

An **immune response** is an organism's reaction, mediated by components of the immune system, to an antigen.

Immunity in phages refers to the ability of a prophage to prevent another phage of the same type from infecting a cell. It results from the synthesis of phage repressor by the prophage genome.

Immunity in plasmids describes the ability of a plasmid to prevent another of the same type from becoming established in a cell. It results usually from interference with the ability to replicate.

Immunity refers to the ability of certain transposons to prevent others of the same type from transposing to the same DNA molecule.

The **immunity region** is a segment of the phage genome that enables a prophage to inhibit additional phages of the same type from infecting the bacterium. This region has a gene that encodes for the repressor, as well as the sites to which the repressor binds.

An **immunoglobulin (antibody)** is a class of protein that is produced by B cells in response to antigen.

Imprecise excision occurs when the transposon removes itself from the original insertion site, but leaves behind or takes some host DNA of its sequence.

Imprinting describes a change in a gene that occurs during passage through the sperm or egg with the result that the paternal and maternal alleles have different properties in the very early embryo. May be caused by methylation of DNA.

In situ **hybridization (cytological hybridization)** is performed by denaturing the DNA of cells squashed on a microscope slide so that reaction is possible with an added single-stranded RNA or DNA; the added preparation is radioactively labeled and its hybridization is followed by autoradiography.

Incision is a step in a mismatch excision repair system. An endonuclease recognizes the damaged area in the DNA, and isolates it by cutting the DNA strand on both sides of the damage.

Indirect end labeling is a technique for examining the organization of DNA by making a cut at a specific site and isolating all fragments containing the sequence adjacent to one side of the cut; it reveals the distance from the cut to the next break(s) in DNA.

Induced mutations result from the action of a mutagen. The mutagen may act directly on the bases in DNA or it may act indirectly to trigger a pathway that leads to a change in DNA sequence.

An **inducer** is a small molecule that triggers gene transcription by binding to a regulator protein.

The **inducer-binding site** of a repressor or activator is the discrete site on the protein at which the small molecule inducer binds. It affects the structure of the DNA-binding site by an allosteric interaction.

Induction of prophage describes its entry into the lytic (infective) cycle as a result of destruction of the lysogenic repressor, which leads to excision of free phage DNA from the bacterial chromosome.

Induction refers to the ability of bacteria (or yeast) to synthesize certain enzymes only when their substrates are present; applied to gene expression, it refers to switching on transcription as a result of interaction of the inducer with the regulator protein.

Initiation describes the stages of transcription up to synthesis of the first bond in RNA. This includes binding of RNA polymerase to the promoter and melting a short region of DNA into single strands.

The **initiation codon** is a special codon (usually AUG) used to start synthesis of a protein.

An **initiation complex** in bacterial protein synthesis contains a small ribosome subunit, initiation factors, and initiator aminoacyl-tRNA bound to mRNA at an AUG initiation codon.

Initiation factors (IF) (IF in prokaryotes, eIF in eukaryotes) are proteins that associate with the small subunit of the ribosome specifically at the stage of initiation of protein synthesis.

The **Inr** is the sequence of a pol II promoter between −3 and +5 and has the general sequence Py_2CAPy_5. It is the simplest possible pol II promoter.

An **insertion** is the addition of a stretch of base pairs in DNA. Duplications are a special class of insertions.

An **insertion sequence (IS)** is a small bacterial transposon that carries only the genes needed for its own transposition.

An **insulator** is a sequence that prevents an activating or inactivating effect passing from one side to the other.

An **intasome** is a protein–DNA complex between the phage lambda integrase (Int) and the phage lambda attachment site (*attP*).

An **integrase** is an enzyme that is responsible for a site-specific recombination that inserts one molecule of DNA into another.

Integration of viral or another DNA sequence describes its insertion into a host genome as a region covalently linked on either side to the host sequences.

An **intein** is the part that is removed from a protein that is processed by protein splicing.

Interallelic complementation (intragenic complementation) describes the change in the properties of a heteromultimeric protein brought about by the interaction of subunits coded by two different mutant alleles; the mixed protein may be more or less active than the protein consisting of subunits only of one or the other type.

Interbands are the relatively dispersed regions of polytene chromosomes that lie between the bands.

The **intercistronic region** is the distance between the termination codon of one gene and the initiation codon of the next gene.

Interspersed repeats were originally defined as short sequences that are common and widely distributed in the genome. They are now known to consist of transposable elements.

Intrinsic terminators are able to terminate transcription by bacterial RNA polymerase in the absence of any additional factors.

An **intron (intervening sequence)** is a segment of DNA that is transcribed, but later removed from within the transcript by splicing together the sequences (exons) on either side of it.

Intron definition describes the process when a pair of splicing sites are recognized by interactions involving only the 5′ site and the branchpoint/3′ site.

Intron homing describes the ability of certain introns to insert themselves into a target DNA. The reaction is specific for a single target sequence.

Invariant base positions in tRNA have the same nucleotide in virtually all (>95%) of tRNAs.

Inverted terminal repeats are the short related or identical sequences present in reverse orientation at the ends of some transposons.

J segments (joining segments) are coding sequences in the immunoglobulin and T-cell receptor loci. The J segments are between the variable (V) and constant (C) gene segments.

A **joint molecule** is a pair of DNA duplexes that are connected together through a reciprocal exchange of genetic material.

Kinetic proofreading describes a proofreading mechanism that depends on incorrect events proceeding more slowly than correct events, so that incorrect events are reversed before a subunit is added to a polymeric chain.

The **kinetochore** is the structural feature of the chromosome to which microtubules of the mitotic spindle attach. Its location determines the centromeric region.

A gene **knock-in** is a process similar to a knockout, but more subtle mutations are made, such as nucleotide substitutions or the addition of epitope tags.

A gene **knockout** is a process in which a gene function is eliminated, usually by replacing most of the coding sequence with a selectable marker *in vitro* and transferring the altered gene to the genome by homologous recombination.

Kuru is a human neurological disease caused by prions. It may be caused by eating infected brains.

The **lagging strand** of DNA must grow overall in the 3′–5′ direction and is synthesized discontinuously in the form of short fragments (5′–3′) that are later connected covalently.

Lampbrush chromosomes are the extremely extended meiotic bivalents of certain amphibian oocytes.

The **lariat** is an intermediate in RNA splicing in which a circular structure with a tail is created by a 5′–2′ bond.

Late genes are transcribed when phage DNA is being replicated. They code for components of the phage particle.

Late infection is the part of the phage lytic cycle from DNA replication to lysis of the cell. During this time, the DNA is replicated and structural components of the phage particle are synthesized.

A **lateral element** is a structure in the synaptonemal complex. It is an axial element that is aligned with the axial elements of other chromosomes.

The **leader** of a protein is a short N-terminal sequence responsible for initiating passage into or through a membrane.

The **leader (5′ UTR)** of an mRNA is the nontranslated sequence at the 5′ end that precedes the initiation codon.

The **leader peptide** is the product that would result from translation of a short coding sequence used to regulate transcription of the tryptophan operon by controlling ribosome movement.

The **leading strand** of DNA is synthesized continuously in the 5′–3′ direction.

Leaky mutations leave some residual function, for instance when the mutant protein is partially active (in the case of a missense mutation), or when readthrough produces a small amount of wild-type protein (in the case of a nonsense mutation).

The **leucine zipper** is a dimerization motif that is found in a class of transcription factors.

A **licensing factor** is something in the nucleus that is necessary for replication, and is inactivated or destroyed after one round of replication. New licensing factors must be provided for further rounds of replication to occur.

The immunoglobulin **light chain** is one of two types of subunits in an antibody tetramer. Each antibody contains two light chains. The N-terminus of the light chain forms part of the antigen recognition site.

Linkage describes the tendency of genes to be inherited together as a result of their location on the same chromosome; measured by percent recombination between loci.

Linker DNA is all DNA contained on a nucleosome in excess of the 146 bp core DNA. Its length varies from 8–114 bp, and it is cleaved by micrococcal nuclease to leave the core DNA.

The **linking number paradox** describes the discrepancy between the existence of −2 supercoils in the path of DNA on the nucleosome compared with the measurement of −1 supercoil released when histones are removed.

A **locus** is the position on a chromosome at which the gene for a particular trait resides; a locus may be occupied by any one of the alleles for the gene.

The **locus control region (LCR)** that is required for the expression of several genes in a domain.

The **long terminal repeat (LTR)** is the sequence that is repeated at each end of the integrated retroviral genome.

A **loop** is a single-stranded region at the end of a hairpin in RNA (or single-stranded DNA); it corresponds to the sequence between inverted repeats in duplex DNA.

A **loose binding site** is any random sequence of DNA that is bound by the core RNA polymerase when it is not engaged in transcription.

A **loss-of-function** mutation eliminates or reduces the activity of a gene. It is often, but not always, recessive.

Luxury genes are those coding for specialized functions synthesized (usually) in large amounts in particular cell types.

Lysis describes the death of bacteria at the end of a phage infective cycle when they burst open to release the progeny of an infecting phage (because phage enzymes disrupt the bacterium's cytoplasmic membrane or cell wall). The same term also applies to eukaryotic cells; for example, when infected cells are attacked by the immune system.

Lysogeny describes the ability of a phage to survive in a bacterium as a stable prophage component of the bacterial genome.

Lytic infection of a bacterium by a phage ends in the destruction of the bacterium with release of progeny phage.

A **maintenance methylase** adds a methyl group to a target site that is already hemimethylated.

The **major groove** of DNA is 22 Å across.

Maternal inheritance describes the preferential survival in the progeny of genetic markers provided by the female parent.

The **mating type** is a property of haploid yeast that makes it able to fuse to form a diploid only with a cell of the opposite mating type.

A **matrix attachment site (MAR)** is a region of DNA that attaches to the nuclear matrix. It is also known as a scaffold attachment site (SAR).

Mediator is a large protein complex associated with yeast bacterial RNA polymerase II. It contains factors that are necessary for transcription from many or most promoters.

Messenger RNA (mRNA) is the intermediate that represents one strand of a gene coding for protein. Its coding region is related to the protein sequence by the triplet genetic code.

A **methyltransferase (methylase)** is an enzyme that adds a methyl group to a substrate, which can be a small molecule, a protein, or a nucleic acid.

Micrococcal nuclease is an endonuclease that cleaves DNA; in chromatin, DNA is cleaved preferentially between nucleosomes.

MicroRNAs are very short RNAs that may regulate gene expression.

Microsatellite DNAs consist of repetitions of extremely short (typically <10 bp) units.

A **microtubule organizing center (MTOC)** is a region from which microtubules emanate. The major MTOCs in a mitotic cell are the centrosomes.

Middle genes are phage genes that are regulated by the proteins coded by early genes. Some proteins coded by middle genes catalyze replication of the phage DNA; others regulate the expression of a later set of genes.

A **minicell** is an anucleate bacterial (*E. coli*) cell produced by a division that generates a cytoplasm without a nucleus.

The **minichromosome** of SV40 or polyoma is the nucleosomal form of the viral circular DNA.

Minisatellite DNAs consist of ~10 copies of a short repeating sequence. The length of the repeating unit is measured in 10s of base pairs. The number of repeats varies between individual genomes.

The **minor groove** of DNA is 12 Å across.

Minus strand DNA is the single-stranded DNA sequence that is complementary to the viral RNA genome of a plus strand virus.

A **mismatch** describes a site in DNA where the pair of bases does not conform to the usual G · C or A · T pairs. It may be caused by incorporation of the wrong base during replication or by mutation of a base.

Mismatch repair corrects recently inserted bases that do not pair properly. The process preferentially corrects the sequence of the daughter strand by distinguishing the daughter strand and parental strand, sometimes on the basis of their states of methylation.

A **missense suppressor** codes for a tRNA that has been mutated so as to recognize a different codon. By inserting a different amino acid at a mutant codon, the tRNA suppresses the effect of the original mutation.

Mitochondrial DNA (mtDNA) is an independent DNA genome, usually circular, that is located in the mitochondrion.

Moderately repetitive DNA sequences are repeated in the haploid genome, usually in related rather than identical copies.

Modification of DNA or RNA includes all changes made to the nucleotides after their initial incorporation into the polynucleotide chain.

Modified bases are all those except the usual four from which DNA (T, C, A, G) or RNA (U, C, A, G) are synthesized; they result from postsynthetic changes in the nucleic acid.

Monocistronic mRNA codes for one protein.

A plasmid is said to be under **multicopy control** when the control system allows the plasmid to exist in more than one copy per individual bacterial cell.

A **multiforked chromosome** (in a bacterium) has more than one replication fork, because a second initiation has occurred before the first cycle of replication has been completed.

A locus is said to have **multiple alleles** when more than two allelic forms have been found. Each allele may cause a different phenotype.

Mutagens increase the rate of mutation by inducing changes in DNA sequence, directly or indirectly.

A **mutation** is any change in the sequence of genomic DNA.

An **N nucleotide** sequence is a short nontemplated sequence that is added randomly by the enzyme at coding joints during rearrangement of immunoglobulin and T-cell receptor genes. N nucleotides augment the diversity of antigen receptors.

The **n − 1 rule** states that only one X chromosome is active in female mammalian cells; any other(s) are inactivated.

A **nascent protein** has not yet completed its synthesis; the polypeptide chain is still attached to the ribosome via a tRNA.

Nascent RNA is a ribonucleotide chain that is still being synthesized, so that its 3′ end is paired with DNA where RNA polymerase is elongating.

Negative complementation occurs when interallelic complementation allows a mutant subunit to suppress the activity of a wild-type subunit in a multimeric protein.

The default state of genes that are controlled by **negative regulation** is to be expressed. A specific intervention is required to turn them off.

A **neutral** mutation has no significant effect on evolutionary fitness and usually has no effect on the phenotype.

Neutral substitutions in a protein cause changes in amino acids that do not affect activity.

Nonhomologous end-joining (NHEJ) ligates blunt ends. It is common to many repair pathways and to certain recombination pathways (such as immunoglobulin recombination).

Nonallelic genes are two (or more) copies of the same gene that are present at *different* locations in the genome (contrasted with alleles which are copies of the same gene derived from different parents and present at the same location on the homologous chromosomes).

A **nonautonomous controlling element** is a transposon in maize that encodes a nonfunctional transposase; it can transpose only in the presence of a *trans*-acting autonomous member of the same family.

A **nonhistone** is any structural protein found in a chromosome except one of the histones.

Nonreciprocal recombination (unequal crossing-over) results from an error in pairing and crossing-over in which nonequivalent sites are involved in a recombination event. It produces one recombinant with a deletion of material and one with a duplication.

Nonrepetitive DNA shows reassociation kinetics expected of unique sequences.

Nonreplicative transposition describes the movement of a transposon that leaves a donor site (usually generating a double-strand break) and moves to a new site.

A **nonsense suppressor** is a gene coding for a mutant tRNA able to respond to one or more of the termination codons and insert an amino acid at that site.

Nonsense-mediated mRNA decay is a pathway that degrades an mRNA that has a nonsense mutation prior to the last exon.

The **nontranscribed spacer** is the region between transcription units in a tandem gene cluster.

The **nonviral superfamily** of transposons originated independently of retroviruses.

The **nucleation center** of TMV (tobacco mosaic virus) is a duplex hairpin where assembly of coat protein with RNA is initiated.

Nucleic acids are molecules that encode genetic information. They consist of a series of nitrogenous bases connected to ribose molecules that are linked by phosphodiester bonds. DNA is deoxyribonucleic acid, and RNA is ribonucleic acid.

The **nucleoid** is the structure in a prokaryotic cell that contains the genome. The DNA is bound to proteins and is not enclosed by a membrane.

The **nucleolar organizer** is the region of a chromosome carrying genes coding for rRNA.

The **nucleolus (nucleoli)** is a discrete region of the nucleus where ribosomes are produced.

The **nucleosome** is the basic structural subunit of chromatin, consisting of ~200 bp of DNA and an octamer of histone proteins.

Nucleosome positioning describes the placement of nucleosomes at defined sequences of DNA instead of at random locations with regards to sequence.

A **null mutation** completely eliminates the function of a gene.

The **ochre** codon is the triplet UAA, one of the three termination codons that end protein synthesis.

Okazaki fragments are the new polynucleotide stretches of 1000–2000 bases in length that are produced during discontinuous replication; they are later joined into a covalently intact strand.

The **opal** codon is the triplet UGA, one of the three termination codons that end protein synthesis. It has evolved to code for an amino acid in a small number of organisms or organelles.

An **open complex** describes the stage of initiation of transcription when RNA polymerase causes the two strands of DNA to separate to form the "transcription bubble."

An **open reading frame (ORF)** is a sequence of DNA consisting of triplets that can be translated into amino acids starting with an initiation codon and ending with a termination codon.

The **operator** is the site on DNA at which a repressor protein binds to prevent transcription from initiating at the adjacent promoter.

An **operon** is a unit of bacterial gene expression and regulation, including structural genes and control elements in DNA recognized by regulator gene product(s).

An **opine** is a derivative of arginine that is synthesized by plant cells infected with crown gall disease.

Orthologs are corresponding proteins in two species as defined by sequence homologies.

A stretch of **overwound** DNA has more base pairs per turn than the usual average (10 bp = 1 turn). This means that the two strands of DNA are more tightly wound around each other, creating tension.

A **P element** is type of transposon in *D. melanogaster*.

A **P nucleotide** sequence is a short palindromic (inverted repeat) sequence that is generated during rearrangement of immunoglobulin and T-cell receptor V, (D), J gene segments. P nucleotides are generated at coding joints when RAG proteins cleave the hairpin ends generated during rearrangement.

The **P site** of the ribosome is the site that is occupied by peptidyl-tRNA, the tRNA carrying the nascent polypeptide chain, still paired with the codon to which it bound in the A site.

A **palindrome** is a DNA sequence that reads the same on each strand of DNA when the strand is read in the 5′ to 3′ direction. It consists of adjacent inverted repeats.

A **parental** strand or duplex of DNA refers to the DNA that will be replicated.

Patch recombinant DNA results from a Holliday junction being resolved by cutting the exchanged strands. The duplex is largely unchanged, except for a DNA sequence on one strand that came from the homologous chromosome.

Peptidyl transferase is the activity of the ribosomal 50S subunit that synthesizes a peptide bond when an amino acid is added to a growing polypeptide chain. The actual catalytic activity is a property of the rRNA.

Peptidyl-tRNA is the tRNA to which the nascent polypeptide chain has been transferred following peptide bond synthesis during protein synthesis.

The **periplasm** (or periplasmic space) is the region between the inner and outer membranes in the bacterial envelope.

A **periseptal annulus** is an ringlike area where inner and outer membrane appear fused. Formed around the circumference of the bacterium, the periseptal annulus determines the location of the septum.

A **pheromone** is a small molecule secreted by one mating type of an organism in order to interact with a member of the opposite mating type.

Pilin is the subunit that is polymerized into the pilus in bacteria.

A **pilus (pili)** is a surface appendage on a bacterium that allows the bacterium to attach to other bacterial cells. It appears like a short, thin, flexible rod. During conjugation, pili are used to transfer DNA from one bacterium to another.

A **plasmid** is a circular, extrachromosomal DNA. It is autonomous and can replicate itself.

Plus strand DNA is the strand of the duplex sequence representing a retrovirus that has the same sequence as that of the RNA.

A **plus strand virus** has a single-stranded nucleic acid genome whose sequence directly codes for the protein products.

A **point mutation** is a change in the sequence of DNA involving a single base pair.

Poly(A) is a stretch of ~200 bases of adenylic acid that is added to the 3′ end of mRNA following its synthesis.

poly(A)$^+$ mRNA is mRNA that has a 3′ terminal stretch of poly(A).

Poly(A) polymerase is the enzyme that adds the stretch of polyadenylic acid to the 3′ of eukaryotic mRNA. It does not use a template.

Poly(A)-binding protein (PABP) is the protein that binds to the 3′ stretch of poly(A) on a eukaryotic mRNA.

Polycistronic mRNA includes coding regions representing more than one gene.

Polymorphism (more fully genetic polymorphism) refers to the simultaneous occurrence in the population of genomes showing variations at a given position. The original definition applied to alleles producing different phenotypes. Now it is also used to describe changes in DNA affecting the restriction pattern or even the sequence. For practical purposes, to be considered as an example of a polymorphism, an allele should be found at a frequency >1% in the population.

A **polyribosome (polysome)** is an mRNA that is simultaneously being translated by several ribosomes.

Polytene chromosomes are generated by successive replications of a chromosome set without separation of the replicas.

Position effect variegation (PEV) is silencing of gene expression that occurs as the result of proximity to heterochromatin.

Positive control describes a system in which a gene is not expressed unless some action occurs to turn it on.

The **postreplication complex** is a protein–DNA complex in *S. cerevisiae* that consists of the ORC complex bound to the origin.

Posttranslational translocation (posttranslational) is the movement of a protein across a membrane after the synthesis of the protein is completed and it has been released from the ribosome.

ppGpp is guanosine tetraphosphate. Diphosphate groups are attached to both the 5′ and 3′ positions.

pppGpp is a guanosine pentaphosphate, with a triphosphate attached to the 5′ position and a diphosphate attached to the 3′ position.

Precise excision describes the removal of a transposon plus one of the duplicated target sequences from the chromosome. Such an event can restore function at the site where the transposon inserted.

Preinitiation complex in eukaryotic transcription describes the assembly of transcription factors at the promoter before RNA polymerase binds.

A protein to be imported into an organelle or secreted from bacteria is called a "**preprotein**"until its signal sequence has been removed.

The **prereplication complex** is a protein–DNA complex at the origin in *S. cerevisiae* that is required for DNA replication. The complex contains the ORC complex, Cdc6, and the MCM proteins.

A **primary transcript** is the original unmodified RNA product corresponding to a transcription unit.

The **primase** is a type of RNA polymerase that synthesizes short segments of RNA that will be used as primers for DNA replication.

A **primer** is a short sequence (often of RNA) that is paired with one strand of DNA and provides a free 3′–OH end at which a DNA polymerase starts synthesis of a deoxyribonucleotide chain.

The **primosome** describes the complex of proteins involved in the priming action that initiates replication on φX-type origins. It is also involved in restarting stalled replication forks.

A **prion** is a proteinaceous infectious agent, which behaves as an inheritable trait, although it contains no nucleic acid. Examples are PrP^{Sc}, the agent of scrapie in sheep and bovine spongiform encephalopathy, and Psi, which confers an inherited state in yeast.

A **processed pseudogene** is an inactive gene copy that lacks introns, contrasted with the interrupted structure of the active gene. Such genes originate by reverse transcription of mRNA and insertion of a duplex copy into the genome.

Processivity describes the ability of an enzyme to perform multiple catalytic cycles with a single template instead of dissociating after each cycle.

Programmed frameshifting is required for expression of the protein sequences coded beyond a specific site at which a +1 or −1 frameshift occurs at some typical frequency.

A **promoter** is a region of DNA where RNA polymerase binds to initiate transcription.

Proofreading refers to any mechanism for correcting errors in protein or nucleic acid synthesis that involves scrutiny of individual units *after* they have been added to the chain.

Prophage is a phage genome covalently integrated as a linear part of the bacterial chromosome.

Protein splicing is the autocatalytic process by which an intein is removed from a protein and the exteins on either side become connected by a standard peptide bond.

The **proteome** is the complete set of proteins that is expressed by the entire genome. Because some genes code for multiple proteins, the size of the proteome is greater than the number of genes. Sometimes the term is used to describe complement of proteins expressed by a cell at any one time.

Provirus is a duplex sequence of DNA integrated into a eukaryotic genome that represents the sequence of the RNA genome of a retrovirus.

A **puff** is an expansion of a band of a polytene chromosome associated with the synthesis of RNA at some locus in the band.

Puromycin is an antibiotic that terminates protein synthesis by mimicking a tRNA and becoming linked to the nascent protein chain.

A **pyrimidine dimer** is formed when ultraviolet irradiation generates a covalent link directly between two adjacent pyrimidine bases in DNA. It blocks DNA replication.

The **R segments** are the sequences that are repeated at the ends of a retroviral RNA. They are called R-U5 and U3-R.

A **reading frame** is one of the three possible ways of reading a nucleotide sequence. Each reading frame divides the sequence into a series of successive triplets. There are three possible reading frames in any sequence, depending on the starting point. If the first frame starts at position 1, the second frame starts at position 2, and the third frame starts at position 3.

Readthrough at transcription or translation occurs when RNA polymerase or the ribosome, respectively, ignores a termination signal because of a mutation of the template or the behavior of an accessory factor.

rec mutations of *E. coli* cannot undertake general recombination.

Recoding events occur when the meaning of a codon or series of codons is changed from that predicted by the genetic code. It may involve altered interactions between aminoacyl-tRNA and mRNA that are influenced by the ribosome.

The **recognition helix** is the one of the two helices of the helix-turn-helix motif that makes contacts with DNA that are specific for particular bases. This determines the specificity of the DNA sequence that is bound.

A **recombinant joint** is the point at which two recombining molecules of duplex DNA are connected (the edge of the heteroduplex region).

Recombination-repair is a mode of filling a gap in one strand of duplex DNA by retrieving a homologous single strand from another duplex.

Redundancy describes the concept that two or more genes may fulfill the same function, so that no single one of them is essential.

A **regulator gene** codes for a product (typically protein) that controls the expression of other genes (usually at the level of transcription).

A **relaxase** is an enzyme that cuts one strand of DNA, and binds to the free 5′ end.

Relaxed mutants of *E. coli* do not display the stringent response to starvation for amino acids (or other nutritional deprivation).

A **release factor (RF)** is required to terminate protein synthesis to cause release of the completed polypeptide chain and the ribosome from mRNA. Individual factors are numbered. Eukaryotic factors are called eRF.

Renaturation describes the reassociation of denatured complementary single strands of a DNA double helix.

Repair of damaged DNA can take place by repair synthesis, when a strand that has been damaged is excised and replaced by the synthesis of a new stretch. It can also take place by recombination reactions, when the duplex region containing the damaged is replaced by an undamaged region from another copy of the genome.

Replacement sites in a gene are those at which mutations alter the amino acid that is coded.

Replication of duplex DNA takes place by synthesis of two new strands that are complementary to the parental strands. The parental duplex is replaced by two identical daughter duplexes, each of which has one parental strand and one newly synthesized strand. Replication is called semiconservative because the conserved units are the single strands of the parental duplex.

A **replication eye** is a region in which DNA has been replicated within a longer, unreplicated region.

A **replication fork (growing point)** is the point at which strands of parental duplex DNA are separated so that replication can proceed. A complex of proteins including DNA polymerase is found at the fork.

A **replication-defective** virus cannot perpetuate an infective cycle because some of the necessary genes are absent (replaced by host DNA in a transducing virus) or mutated.

Replicative transposition describes the movement of a transposon by a mechanism in which first it is replicated, and then one copy is transferred to a new site.

The **replicon** is a unit of the genome in which DNA is replicated. Each replicon contains an origin for initiation of replication.

The **replisome** is the multiprotein structure that assembles at the bacterial replicating fork to undertake synthesis of DNA. It contains DNA polymerase and other enzymes.

A **reporter gene** is a coding unit whose product is easily assayed (such as chloramphenicol transacetylase or beta-galactosidase); it may be connected to any promoter of interest so that expression of the gene can be used to assay promoter function.

Repression describes the ability of bacteria to prevent synthesis of certain enzymes when their products are present; more generally, refers to inhibition of transcription (or translation) by binding of a repressor protein to a specific site on DNA (or mRNA).

A **repressor** is a protein that inhibits expression of a gene. It may act to prevent transcription by binding to an operator site in DNA, or to prevent translation by binding to RNA.

Resolution occurs by a homologous recombination reaction between the two copies of the transposon in a cointegrate. The reaction generates the donor and target replicons, each with a copy of the transposon.

Resolvase is the enzyme activity involved in site-specific recombination between two transposons present as direct repeats in a cointegrate structure.

A **response element** is a sequence in a eukaryotic promoter or enhancer that is recognized by a specific transcription factor.

Restriction endonucleases recognize specific short sequences of DNA and cleave the duplex (sometimes at target site, sometimes elsewhere, depending on type).

Restriction fragment length polymorphism (RFLP) refers to inherited differences in sites for restriction enzymes (for example, caused by base changes in the target site) that result in differences in the lengths of the fragments produced by cleavage with the relevant restriction enzyme. RFLPs are used for genetic mapping to link the genome directly to a conventional genetic marker.

A **restriction map** is a linear array of sites on DNA cleaved by various restriction enzymes.

A **retroposon (retrotransposon)** is a transposon that mobilizes via an RNA form; the DNA element is transcribed into RNA, and then reverse-transcribed into DNA, which is inserted at a new site in the genome. The difference from retroviruses is that the retroposon does not have an infective (viral) form.

A **retrovirus** is an RNA virus with the ability to convert its sequence into DNA by reverse transcription.

Reverse transcriptase is an enzyme that uses a template of single-stranded RNA to generate a double-stranded DNA copy.

Revertants are derived by reversion of a mutant cell or organism to the wild-type phenotype.

RF1 is the bacterial release factor that recognizes UAA and UAG as signals to terminate protein synthesis.

RF2 is the bacterial release factor that recognizes UAA and UGA as signals to terminate protein synthesis.

RF3 is a protein synthesis termination factor related to the elongation factor EF-G. It functions to release the factors RF1 or RF2 from the ribosome when they act to terminate protein synthesis.

Rho factor is a protein involved in assisting *E. coli* RNA polymerase to terminate transcription at certain terminators (called rho-dependent terminators).

Rho-dependent terminators are sequences that terminate transcription by bacterial RNA polymerase in the presence of the rho factor.

Ribonucleases (RNAase) are enzymes that cleave RNA. They may be specific for single-stranded or for double-stranded RNA, and may be either endonucleases or exonucleases.

Ribosomal DNA (rDNA) is usually a tandemly repeated series of genes coding for a precursor to the two large rRNAs.

Ribosomal RNA (rRNA) is a major component of the ribosome. Each of the two subunits of the ribosome has a major rRNA as well as many proteins.

The **ribosome** is a large assembly of RNA and proteins that synthesizes proteins under direction from an mRNA template. Bacterial ribosomes sediment at 70S, eukaryotic ribosomes at 80S. A ribosome can be dissociated into two subunits.

Ribosome stalling describes the inhibition of movement that occurs when a ribosome reaches a codon for which there is no corresponding charged aminoacyl-tRNA.

A **ribosome-binding site** is a sequence on bacterial mRNA that includes an initiation codon that is bound by a 30S subunit in the initiation phase of protein synthesis.

A **riboswitch** is a catalytic RNA whose activity responds to a small ligand.

A **ribozyme** is an RNA that has catalytic activity.

A helix is said to be **right-handed** if the turns runs clockwise along the helical axis.

RNA editing describes a change of sequence at the level of RNA following transcription.

RNA interference (RNAi) describes the technique in which double-strand RNA is introduced into cells to eliminate or reduce the activity of a target gene. It is caused by using sequences complementary to the double-stranded RNA sequences to trigger degradation of the mRNA of the gene.

An **RNA ligase** is an enzyme that functions in tRNA splicing to make a phosphodiester bond between the two exon sequences that are generated by cleavage of the intron.

RNA polymerases are enzymes that synthesize RNA using a DNA template (formally described as DNA-dependent RNA polymerases).

RNA silencing describes the ability of a dsRNA to suppress expression of the corresponding gene systemically in a plant.

RNA splicing is the process of excising introns from RNA and connecting the exons into a continuous mRNA.

The **rolling circle** is a mode of replication in which a replication fork proceeds around a circular template for an indefinite number of revolutions; the DNA strand newly synthesized in each revolution displaces the strand synthesized in the previous revolution, giving a tail containing a linear series of sequences complementary to the circular template strand.

Rotational positioning describes the location of the histone octamer relative to turns of the double helix, which determines which face of DNA is exposed on the nucleosome surface.

The **S domain** is the sequence of 7S RNA of the SRP that is not related to Alu RNA.

S phase is the restricted part of the eukaryotic cell cycle during which synthesis of DNA occurs.

An **S region** is an intron sequence involved in immunoglobulin class switching. S regions consist of repetitive sequences at the 5′ ends of gene segments encoding the heavy chain constant regions.

Satellite DNA (simple-sequence DNA) consists of many tandem repeats (identical or related) of a short basic repeating unit.

A chromosome **scaffold** is a proteinaceous structure in the shape of a sister chromatid pair, revealed when chromosomes are depleted of histones.

Scarce mRNA (complex mRNA) consists of a large number of individual mRNA species, each present in very few copies per cell. This accounts for most of the sequence complexity in RNA.

Scrapie is an infective agent made of protein.

scRNPs (scyrp) are small cytoplasmic ribonucleoproteins (scRNAs associated with proteins).

Second-site suppression occurs when a second mutation suppresses the effect of a first mutation.

A **sector** is a patch of cells made up of a single altered cell and its progeny.

Selfish DNA describes sequences that do not contribute to the genotype of the organism but have self-perpetuation within the genome as their sole function.

Semiconservative replication is accomplished by separation of the strands of a parental duplex, each then acting as a template for synthesis of a complementary strand.

A **semiconserved (semi-invariant)** position is one where comparison of many individual sequences finds the same type of base (pyrimidine or purine) always present.

The **septal ring (Z-ring)** is a complex of several proteins coded by *fts* genes of *E. coli* that forms at the midpoint of the cell. It gives rise to the septum at cell division. The first of the proteins to be incorporated is FtsZ, which gave rise to the original name of the Z-ring.

A **septum** is the structure that forms in the center of a dividing bacterium, providing the site at which the daughter bacteria will separate. The same term is used to describe the cell wall that forms between plant cells at the end of mitosis.

The **serum response element (SRE)** is a sequence in a promoter or enhancer that is activated by transcription factor(s) induced by treatment with serum. This activates genes that stimulate cell growth.

The **SET domain** is part of the active site of histone methyltransferase enzymes.

The **Shine–Dalgarno** sequence is the polypurine sequence AGGAGG centered about 10 bp before the AUG initiation codon on bacterial mRNA. It is complementary to the sequence at the 3′ end of 16S rRNA.

Sigma factor is the subunit of bacterial RNA polymerase needed for initiation; it is the major influence on selection of promoters.

A **signal end** is produced during recombination of immunoglobulin and T-cell receptor genes. The signal ends are at the termini of the cleaved fragment containing the recombination signal sequences. The subsequent joining of the signal ends yields a signal joint.

Signal peptidase is an enzyme within the membrane of the ER that specifically removes the signal sequences from proteins as they are translocated. Analogous activities are present in bacteria, archaebacteria, and in each organelle in a eukaryotic cell into which proteins are targeted and translocated by means of removable targeting sequences. Signal peptidase is one component of a larger protein complex.

The **signal recognition particle (SRP)** is a ribonucleoprotein complex that recognizes signal sequences during translation and guides the ribosome to the translocation channel. SRPs from different organisms may have different compositions, but all contain related proteins and RNAs.

A **signal sequence** is a short region of a protein that directs it to the endoplasmic reticulum for cotranslational translocation.

Silencing describes the repression of gene expression in a localized region, usually as the result of a structural change in chromatin.

Silent mutations do not change the sequence of a protein because they produce synonymous codons.

A **silent site** in a coding region is one where mutation does not change the sequence of the protein.

Single copy replication describes a control system in which there is only one copy of a replicon per unit bacterium. The bacterial chromosome and some plasmids have this type of regulation.

Single nucleotide polymorphism (SNP) describes a polymorphism (variation in sequence between individuals) caused by a change in a single nucleotide. This is responsible for most of the genetic variation between individuals.

The **single X hypothesis** describes the inactivation of one X chromosome in female mammals.

Single-strand assimilation (single-strand uptake) describes the ability of RecA protein to cause a single strand of DNA to displace its homologous strand in a duplex; that is, the single strand is assimilated into the duplex.

The **single-strand binding protein (SSB)** attaches to single-stranded DNA, thereby preventing the DNA from forming a duplex.

Single-strand exchange is a reaction in which one of the strands of a duplex of DNA leaves its former partner and instead pairs with the complementary strand in another molecule, displacing its homologue in the second duplex.

Site-specific recombination (specialized recombination) occurs between two specific sequences, as in phage integration/excision or resolution of cointegrate structures during transposition.

SL RNA (spliced leader RNA) is a small RNA that donates an exon in the *trans*-splicing reaction of trypanosomes and nematodes.

Small cytoplasmic RNAs (scRNA) are present in the cytoplasm and (sometimes are also found in the nucleus).

A **small nuclear RNA (snRNA)** is one of many small RNA species confined to the nucleus; several of the snRNAs are involved in splicing or other RNA processing reactions.

A **snoRNA** is a small nuclear RNA that is localized in the nucleolus.

snRNPs (snurp) are small nuclear ribonucleoproteins (snRNAs associated with proteins).

Somatic recombination describes the process of joining a V gene to a C gene in a lymphocyte to generate an immunoglobulin or T cell receptor.

A **spacer** is a sequence in a gene cluster that separates the repeated copies of the transcription unit.

The **spindle** guides the movement of chromosomes during cell division. The structure is made up of microtubules.

Splice recombinant DNA results from a Holliday junction being resolved by cutting the nonexchanged strands. Both strands of DNA before the exchange point come from one chromosome; the DNA after the exchange point come from the homologous chromosome.

Splice sites are the sequences immediately surrounding the exon–intron boundaries.

The **spliceosome** is a complex formed by the snRNPs that are required for splicing together with additional protein factors.

A **splicing factor** is a protein component of the spliceosome which is not part of one of the snRNPs.

Spontaneous mutations occur in the absence of any added reagent to increase the mutation rate, as the result of errors in replication (or other events involved in the reproduction of DNA) or by environmental damage.

Sporulation is the generation of a spore by a bacterium (by morphological conversion) or by a yeast (as the product of meiosis).

An **SR protein** has a variable length of an Arg-Ser–rich region and is involved in splicing.

An **sRNA** is a small bacterial RNA that functions as a regulator of gene expression.

Startpoint (startsite) refers to the position on DNA corresponding to the first base incorporated into RNA.

A **stem** is the base-paired segment of a hairpin structure in RNA.

Steroid receptors are transcription factors that are activated by binding of a steroid ligand.

A **stop codon (termination codon)** is one of three triplets (UAG, UAA, UGA) that causes protein synthesis to terminate. They are also known historically as *nonsense codons*. The UAA codon is called ochre, and the UAG codon is called amber, after the names of the nonsense mutations by which they were originally identified.

Strand displacement is a mode of replication of some viruses in which a new DNA strand grows by displacing the previous (homologous) strand of the duplex.

The **stringent factor** is the protein RelA, which is associated with ribosomes. It synthesizes ppGpp and pppGpp when uncharged aminoacyl-tRNA enters the A site.

Stringent response refers to the ability of a bacterium to shut down synthesis of tRNA and ribosomes in a poor-growth medium.

A **structural distortion** is a change in the shape of a molecule.

A **structural gene** codes for any RNA or protein product other than a regulator.

The **structural periodicity** is the number of base pairs per turn of the double helix of DNA.

A **subviral pathogen** is an infectious agent that is smaller than a virus, such as a virusoid.

Super-repressed is a mutant condition in which a repressible operon cannot be de-repressed, so it is always turned off.

Supercoiling describes the coiling of a closed duplex DNA in space so that it crosses over its own axis.

A **superfamily** is a set of genes all related by presumed descent from a common ancestor, but now showing considerable variation.

Suppression occurs when a second event eliminates the effects of a mutation without reversing the original change in DNA.

A **frameshift suppressor** is an insertion or deletion of a base that restores the original reading frame in a gene that has had a base deletion or insertion.

A **suppressor** is a second mutation that compensates for or alters the effects of a primary mutation.

Surveillance systems check nucleic acids for errors. The term is used in several different contexts. One example is the system that degrades mRNAs that have nonsense mutations. Another is the set of systems that react to damage in the double helix. The common feature is that the system recognizes an invalid sequence or structure and triggers a response.

SWI/SNF is a chromatin remodeling complex; it uses hydrolysis of ATP to change the organization of nucleosomes.

Synapsis describes the association of the two pairs of sister chromatids (representing homologous chromosomes) that occurs at the start of meiosis; the resulting structure is called a bivalent.

The **synaptonemal complex** describes the morphological structure of synapsed chromosomes.

Synonym codons have the same meaning in the genetic code. Synonym tRNAs bear the same amino acid and respond to the same codon.

Synteny describes a relationship between chromosomal regions of different species where homologous genes occur in the same order.

Synthetic genetic array analysis (SGA) is an automated technique in budding yeast whereby a mutant is crossed to an array of approximately 5000 deletion mutants to determine if the mutants interact to cause a synthetic lethal phenotype.

Synthetic lethality occurs when two mutations that by themselves are viable, but cause lethality when combined.

T cells are lymphocytes of the T (thymic) lineage; may be subdivided into several functional types. They carry TcR (T-cell receptor) and are involved in the cell-mediated immune response.

The **T-cell receptor (TCR)** is the antigen receptor on T lymphocytes. It is clonally expressed and binds to a complex of MHC class I or class II protein and antigen-derived peptide.

T-DNA is the segment of the Ti plasmid of *Agrobacterium tumefaciens* that is transferred to the plant cell nucleus during infection. It carries genes that transform the plant cell.

TAFs are the subunits of $TF_{II}D$ that assist TBP in binding to DNA. They also provide points of contact for other components of the transcription apparatus.

TATA box is a conserved A·T-rich septamer found about 25 bp before the startpoint of each eukaryotic RNA polymerase II transcription unit; may be involved in positioning the enzyme for correct initiation.

A **TATA-less promoter** does not have a TATA box in the sequence upstream of its startpoint.

Telomerase is the ribonucleoprotein enzyme that creates repeating units of one strand at the telomere, by adding individual bases to the DNA 3′ end, as directed by an RNA sequence in the RNA component of the enzyme.

A **telomere** is the natural end of a chromosome; the DNA sequence consists of a simple repeating unit with a protruding single-stranded end that may fold into a hairpin.

Telomeric silencing describes the repression of gene activity that occurs in the vicinity of a telomere.

A **terminal protein** allows replication of a linear phage genome to start at the very end. The protein attaches to the 5′ end of the genome through a covalent bond, is associated with a DNA polymerase, and contains a cytosine residue that serves as a primer.

A **terminase** enzyme cleaves multimers of a viral genome and then uses hydrolysis of ATP to provide the energy to translocate the DNA into an empty viral capsid starting with the cleaved end.

Termination is a separate reaction that ends a macromolecular synthesis reaction (replication, transcription, or translation), by stopping the addition of subunits, and (typically) causing disassembly of the synthetic apparatus.

A **terminator** is a sequence of DNA that causes RNA polymerase to terminate transcription.

The **ternary complex** in initiation of transcription consists of RNA polymerase and DNA and a dinucleotide that represents the first two bases in the RNA product.

$TF_{II}D$ is the transcription factor that binds to the TATA sequence upstream of the startpoint of promoters for RNA polymerase II. It consists of TBP (TATA binding protein) and the TAF subunits that bind to TBP.

Thalassemia is disease of red blood cells resulting from lack of either α or β globin.

Third base degeneracy describes the lesser effect on codon meaning of the nucleotide present in the third codon position.

The **Ti plasmid** is an episome of the bacterium *Agrobacterium tumefaciens* that carries the genes responsible for the induction of crown gall disease in infected plants.

Tight binding of RNA polymerase to DNA describes the formation of an open complex (when the strands of DNA have separated).

The **TIM complex (TIM)** resides in the inner membrane of mitochondria and is responsible for transporting proteins from the intermembrane space into the interior of the organelle.

Bacterial transposons that contain markers that are not related to their function, e.g., drug resistance, are named as **Tn** followed by a number.

The **TOM complex (TOM)** resides in the outer membrane of the mitochondrion and is responsible for importing proteins from the cytosol into the space between the membranes.

A DNA **topoisomerase** is an enzyme that changes the number of times the two strands in a closed DNA molecule cross each other. It does this by cutting the DNA, passing DNA through the break, and resealing the DNA.

A **trailer (3′ UTR)** is a nontranslated sequence at the 3′ end of an mRNA following the termination codon.

trans configuration of two sites refers to their presence on two different molecules of DNA (chromosomes).

A ***trans*-acting** product can function on any copy of its target DNA. This implies that it is a diffusible protein or RNA.

A **transcript** is the RNA product produced by copying one strand of DNA. It may require processing to generate a mature RNA.

Transcription describes synthesis of RNA on a DNA template.

A **transcription factor** is required for RNA polymerase to initiate transcription at specific promoter(s), but is not itself part of the enzyme.

A **transcription unit** is the distance between sites of initiation and termination by RNA polymerase; may include more than one gene.

The **transcriptome** is the complete set of RNAs present in a cell, tissue, or organism. Its complexity is due mostly to mRNAs, but it also includes noncoding RNAs.

A **transducing virus** carries part of the host genome in place of part of its own sequence. The best-known examples are retroviruses in eukaryotes and DNA phages in *E. coli*.

A **transesterification** reaction breaks and makes chemical bonds in a coordinated transfer so that no energy is required.

Transfection of eukaryotic cells is the acquisition of new genetic markers by incorporation of added DNA.

The **transfer region** is a segment on the F plasmid that is required for bacterial conjugation.

Transfer RNA (tRNA) is the intermediate in protein synthesis that interprets the genetic code. Each tRNA can be linked to an amino acid. The tRNA has an anticodon sequence that complementary to a triplet codon representing the amino acid.

Transformation (oncogenesis) of eukaryotic cells refers to their conversion to a state of unrestrained growth in culture, resembling or identical with the tumorigenic condition.

The **transforming principle** is DNA that is taken up by a bacterium and whose expression then changes the properties of the recipient cell.

Transgenic animals are created by introducing DNA prepared in test tubes into the germline. The DNA may be inserted into the genome or exist in an extrachromsomal structure.

Transient transfectants have foreign DNA in an unstable—i.e., extrachromosomal—form.

A **transition** is a mutation in which one pyrimidine is replaced by the other or in which one purine is replaced by the other.

Translation is synthesis of protein on an mRNA template.

Translational positioning describes the location of a histone octamer at successive turns of the double helix, which determines which sequences are located in linker regions.

A **translocation** is a rearrangement in which part of a chromosome is detached by breakage or aberrant recombination and then becomes attached to some other chromosome.

Protein **translocation** describes the movement of a protein across a membrane. This occurs across the membranes of organelles in eukaryotes, or across the plasma membrane in bacteria. Each membrane across which proteins are translocated has a channel specialized for the purpose.

Translocation describes the stage of nuclear import or export when a protein or RNA substrate moves through the nuclear pore.

Translocation is the movement of the ribosome one codon along mRNA after the addition of each amino acid to the polypeptide chain.

A **translocon** is a discrete structure in a membrane that forms a channel through which (hydrophilic) proteins may pass.

A **transposase** is the enzyme activity involved in insertion of transposon at a new site.

A **transposon (transposable element)** is a DNA sequence able to insert itself (or a copy of itself) at a new location in the genome, without having any sequence relationship with the target locus.

A **transversion** is a mutation in which a purine is replaced by a pyrimidine or vice versa.

$tRNA_f^{Met}$ is the special RNA that is to initiate protein synthesis in bacteria. It mostly uses AUG, but can also respond to GUG and CUG.

$tRNA_i^{Met}$ is the special tRNA used to respond to initiation codons in eukaryotes.

$tRNA_m^{Met}$ inserts methionine at internal AUG codons.

A **true reversion** is a mutation that restores the original sequence of the DNA.

Ty stands for transposon yeast, the first transposable element to be identified in yeast.

A **type I topoisomerase** is an enzyme that changes the topology of DNA by nicking and resealing one strand of DNA.

A **type II topoisomerase** is an enzyme that changes the topology of DNA by nicking and resealing both strands of DNA.

U3 is the repeated sequence at the 3′ end of a retroviral RNA.

U5 is the repeated sequence at the 5′ end of a retroviral RNA.

A stretch of **underwound** DNA has fewer base pairs per turn than the usual average (10 bp = 1 turn). This means that the two strands of DNA are less tightly wound around each other; ultimately this can lead to strand separation.

An **uninducible** mutant is one where the affected gene(s) cannot be expressed.

The **unit cell** describes the state of an *E. coli* bacterium generated by a new division. It is 1.7 μm long and has a single replication origin.

An **up mutation** in a promoter increases the rate of transcription.

Upstream identifies sequences in the opposite direction from expression; for example, the bacterial promoter is upstream of the transcription unit, the initiation codon is upstream of the coding region.

An **upstream activator sequence (UAS)** is the equivalent in yeast of the enhancer in higher eukaryotes.

A **V gene** is sequence coding for the major part of the variable (N-terminal) region of an immunoglobulin chain.

The **variable region (V region)** of an immunoglobulin chain is coded by the V gene and varies extensively when different chains are compared, as the result of multiple (different) genomic copies and changes introduced during construction of an active immunoglobulin.

Variegation of phenotype is produced by a change in genotype during somatic development.

A **vector** is a plasmid or phage chromosome that is used to perpetuate a cloned DNA segment.

The **vegetative phase** describes the period of normal growth and division of a bacterium. For a bacterium that can sporulate, this contrasts with the sporulation phase, when spores are being formed.

The **viral superfamily** comprises transposons that are related to retroviruses. They are defined by sequences that code for reverse transcriptase or integrase.

Virion is the physical virus particle (irrespective of its ability to infect cells and reproduce).

A **viroid** is a small infectious nucleic acid that does not have a protein coat.

Virulent phage mutants are unable to establish lysogeny.

A **virusoid (satellite RNA)** is a small infectious nucleic acid that is encapsidated by a plant virus together with its own genome.

VNTR (variable number tandem repeat) regions describe very short repeated sequences, including microsatellites and minisatellites.

The **wobble hypothesis** accounts for the ability of a tRNA to recognize more than one codon by unusual (non-G · C, non-A · T) pairing with the third base of a codon.

The **zinc finger** is a DNA-binding motif that typifies a class of transcription factor.

A **zoo blot** describes the use of Southern blotting to test the ability of a DNA probe from one species to hybridize with the DNA from the genomes of a variety of other species.

Index